## Units of the Fundamental Dimensions in Three Unit Systems

| Fundamental Dimension | Système International (SI) Unit | Centimeter-Gram-Second (cgs) Unit | English Unit |
|---|---|---|---|
| Mass | Kilogram (kg) | Gram (g) | Slug |
| Force | Newton (N) | Dyne (dyn) | Pound (lb) |
| Length | Meter (m) | Centimeter (cm) | Foot (ft) |
| Time | Second (s) | Second (s) | Second (s) |
| Temperature | Kelvin (K) | Degree Celsius (° C) | Degree Fahrenheit (° F) |

## Conversion Factors for Non-SI Units

| Abbreviation | Unit | Conversion Factor | SI unit |
|---|---|---|---|
| ac | acre | $\times 4047$ | $= m^2$ |
| Btu | British thermal unit | $\times 1055$ | $= J$ |
| cal | calorie | $\times 4.187$ | $= J$ |
| cm | centimeter | $\times 1 \times 10^{-2}$ | $= m$ |
| cm $H_2O$ | centimeter of water | $\times 98.07$ | $= Pa$ |
| cP | centipoise | $\times 1 \times 10^{-3}$ | $= Pa \cdot s$ |
| day | day | $\times 86,400$ | $= s$ |
| ° | degree of angle | $\times 0.01745$ | $= rad$ |
| ° C | degree Celsius | $+ 273.16$ | $= K$ |
| ° F | degree Fahrenheit | $(\quad -32) \times 5/9$ $+ 237.16$ | $= K$ |
| ° R | degree Rankine | $\times 5/9$ | $= K$ |
| dyn | dyne | $\times 1 \times 10^{-5}$ | $= N$ |
| ft | foot | $\times 0.3048$ | $= m$ |
| gal | gallon | $\times 0.003785$ | $= m^3$ |
| g | gram | $\times 1 \times 10^{-3}$ | $= kg$ |
| ha | hectare | $\times 1 \times 10^4$ | $= m^2$ |
| hp | horsepower | $\times 746$ | $= W$ |
| hr | hour | $\times 3600$ | $= s$ |
| in. | inch | $\times 0.0254$ | $= m$ |
| in. Hg | inch of mercury | $\times 3386$ | $= Pa$ |
| L | liter | $\times 1 \times 10^{-3}$ | $= m^3$ |
| mi | mile | $\times 1609$ | $= m$ |
| mb | millibar | $\times 100$ | $= Pa$ |
| min | minute | $\times 60$ | $= s$ |
| lb | pound | $\times 4.448$ | $= N$ |
| Pa | Pascal | $\times 1$ | $= N\,m^{-2}$ |
| rev | revolution | $\times 6.282$ | $= rad$ |
| yr | year | $\times 31,557,600$ | $= s$ |

# Physical Hydrology

## second edition

### S. LAWRENCE DINGMAN
*University of New Hampshire*

**WAVELAND**
**PRESS, INC.**
Long Grove, Illinois

> *To Jane, again,*
> *and to Francis R. Hall*

For information about this book, contact:
> Waveland Press, Inc.
> 4180 IL Route 83, Suite 101
> Long Grove, IL  60047-9580
> (847) 634-0081
> info@waveland.com
> www.waveland.com

**Cover Photo Credits:** Center: Delta River, Alaska, photo by author. Top: Earth surface composite, courtesy NASA. Bottom left: meteorological station, photo by author. Bottom right: watershed water balance drawing, courtesy of R. S. Pierce, U.S. Forest Service.

# Contents

# Preface

The goal of the first edition of *Physical Hydrology* was to provide a comprehensive text for upper-level undergraduates and graduate students that treats hydrology as a distinct geoscience. It attempted to develop an understanding of the conceptual basis of hydrology and an introduction to the quantitative relations that implement that understanding in answering scientific and water-resources-management questions. The text seemed to fulfill a need, and I have been pleased with its reception by my colleagues and students.

My primary goals in revising *Physical Hydrology* have been to incorporate significant advances in the rapidly developing field of hydrologic science, to provide a more explicit connection of that science to hydrologic modeling, and to make more complete and useful the treatment of the relation between scientific hydrology and water-resources management. The major changes that have resulted are the following:

Chapter 2 (Basic Hydrologic Concepts) now concludes with an introduction to hydrologic modeling, including discussions of model use, modeling terminology, and the process of model development. It also introduces the BROOK90 model, a physically based, lumped-parameter model that can be readily accessed on the World-Wide Web for student use. Discussions of the ways in which BROOK90 incorporates the physical relations discussed in the text are included as boxes in many of the subsequent chapters.

Chapter 3 (Global Climate, Hydrologic Cycle, Soils, and Vegetation) now includes a tabulation of documented trends in global change of climatic and hydrologic quantities.

In Chapter 4 (Precipitation), I have added a more extensive discussion of precipitation recycling and a new section on methods for handling missing data—an almost universal problem in hydrologic analysis. Also, the discussion of methods for estimating areal precipitation has been streamlined somewhat (one of the few places in which I was able to cut!).

In Chapter 5 (Snow and Snowmelt), I have updated the discussion of ways of estimating energy-balance components and added a discussion of hybrid snowmelt models that combine energy-balance and temperature-index approaches.

Chapter 6 (Water in Soils) now introduces the concepts of soil-moisture diffusivity and sorptivity, adds a discussion of equilibrium soil-moisture profiles, and expands the discussion of moisture redistribution.

Chapter 7 (Evapotranspiration) now contains a brief discussion of soil evaporation as well as updates of the treatments of lake evaporation and energy-budget estimation.

In Chapter 8 (Ground Water), the discussion of ground-water—surface-water relations has been expanded to include hyporheic flow and the Dupuit approximation for unconfined aquifers draining to streams.

Chapter 9 (Stream Response to Water Input) has been reorganized so that the discussion of the mechanisms of stream response to water-input events now precedes the sections on rainfall-runoff modeling. The treatments of both mechanisms and modeling have been substantially revised and updated, and much of the detailed discussion of open-channel flow has been moved to Appendix B.

Chapter 10 (Hydrology and Water Resources) has been entirely rewritten and expanded. It now includes a more complete and modern treatment of water-resource management goals and processes; a more detailed discussion of water supply and demand, including the concept of "safe yield" in various ground-water and surface-water settings and an expanded discussion of the estimation and application of flow-duration curves; a more complete discussion of water-quality issues; an expanded section on floods, including flood-frequency analysis; a completely new section on drought and low-flow analysis; and a concluding section on current and projected United States and global water use.

Appendix A (Hydrologic Quantities) has been reorganized and largely rewritten to provide a more logical presentation of dimensions, units, and significant figures.

Appendix B (Water as a Substance) now contains the detailed treatment of open-channel flow that was formerly in Chapter 9.

Appendix C (Statistics) now includes discussions of approaches to fitting probability distributions to data and estimating parameters of distributions, which incorporate the application of L-moments, and a section on statistical criteria for calibrating and validating models.

In Appendix D (Water and Energy in the Atmosphere), the treatment of turbulent transfer has been substantially revised.

In Appendix F (Stream Gaging), the section on slope-area estimation of discharge has been completely revised.

Appendix G is new. It contains links to hydrologic information on the World-Wide Web and is found on the CD accompanying the text.

In keeping with my goal of treating hydrology as a science and of providing an entrée to the literature of the field, this edition continues the practice of supporting its discussion with extensive reference citations, in the style of a journal article rather than that of most textbooks. In the revision, over 200 reference citations have been added, and they now total over 800.

In carrying out the primary goals of the revision, I also saw opportunities to improve several other aspects of the text:

- Cross referencing is facilitated by use of a decimal numbering system for headings.
- Many of the exercises have been revised so that they provide more opportunity for student exploration of topics rather than simple "plug-and-chug" work.
- The spreadsheets on the CD that accompanies the book are now in EXCEL, and their formats have been improved and regularized.
- More modern units are used (kPa instead of mb; J and W instead of cal).
- The notation, especially for statistical quantities, is generally more conventional (and, I hope, less cumbersome).
- The "·" symbol for multiplication is used throughout; this allows the use of multi-letter symbols without ambiguity.

Of course, I have also corrected any errors that I've detected or that have been pointed out to me. How-

ever, I remain responsible for those that remain and any new ones I've introduced!

I thank again those who provided comments that guided me in writing the first edition: J. M. Harbor (*Kent State University*) and present and/or former University of New Hampshire colleagues R. L. A. Adams, W. A. Bothner, C. V. Evans, S. E. Frolking, F. R. Hall, R. C. Harriss, S. L. Hartley, E. Linder, D. S. L. Lawrence, M. A. Person, and G. A. Zielinski.

Many people have assisted me in this revision. I am especially grateful for the complete reviews of the first edition and suggestions for changes provided by Marc Parlange, *Johns Hopkins University*; David Huntley, *San Diego State University;* Benjamin S. Levy, *University of Maryland—College Park*; Guido D. Salvucci, *Boston University*; Kaye L. Brubaker, *University of Maryland—College Park*; Michael E. Campana, *University of New Mexico;* David L. Brown, *California State University—Chico;* Richard Kattleman, *Colorado State University*; Richard H. Hawkins, *University of Arizona*; and James Buttle, *Trent University*.

J. Matthew Davis, *University of New Hampshire,* David Tarboton, *Utah State University,* Richard Vogel, *Tufts University,* and W. Breck Bowden, *Landcare, New Zealand,* provided reviews and comments on individual chapters or substantial portions of the text. Richard H. Cuenca, *Oregon State University,* and James R. Wallis, *Yale University,* gave helpful criticisms of many of the exercises and spreadsheet programs, and Barry Keim, *University of New Hampshire,* provided information on sources of climatic data. Mark Person, *Indiana University,* and Gerry Lang, *Natural Resources Conservation Service,* made special efforts to provide original illustrations. Kristin Green tracked down the hydrologic web sites listed in Appendix G.

Special thanks go to C. Anthony Federer, *U.S. Forest Service,* retired, for reviewing the sections discussing the BROOK90 model and for his help in facilitating the links between the model and the text.

As with the first edition, I could not have completed this work without the love, support, and hours of detailed editorial assistance of my wife, Dr. Jane Van Zandt.

S. Lawrence Dingman

# 1

# Introduction to Hydrologic Science

## 1.1  DEFINITION AND SCOPE OF HYDROLOGY

Hydrology is broadly defined as the geoscience that describes and predicts the occurrence, circulation, and distribution of the water of the earth and its atmosphere. It has two principal focuses.

**The global hydrologic cycle:** the distribution and spatial and temporal variations of water substance in the terrestrial, oceanic, and atmospheric compartments of the global water system.

**The land phase of the hydrologic cycle:** the movement of water substance on and under the earth's land surfaces, the physical and chemical interactions with earth materials accompanying that movement, and the biological processes that conduct or affect that movement.[1]

Figure 1-1 shows the components of the global hydrologic cycle, and Figure 1-2 shows the storages and flows of energy and water that constitute the land phase of the cycle.

Horton (1931, p. 192) characterized the range of scales on which hydrologic processes operate:

> Any natural exposed surface may be considered as a unit area on which the hydrologic cycle operates. This includes, for example, an isolated tree, even a single leaf or twig of a growing plant, the roof of a building, the drainage basin of a river-system or any of its tributaries, an undrained glacial depression, a

swamp, a glacier, a polar ice-cap, a group of sand dunes, a desert playa, a lake, an ocean, or the earth as a whole.

Figure 1-3 gives a quantitative sense of the range of time and space scales in the domain of hydrologic science.

Figure 1-4 shows the position of hydrologic science in the spectrum from basic sciences to water-resource management. Hydrology is built upon the basic sciences of mathematics, physics, chemistry, and biology, and it uses them as tools. It is an interdisciplinary geoscience, built also upon its sister geosciences, but differing from them in the subject matter on which it focuses. Much of the motivation for answering hydrologic questions has come, and will continue to come, from the practical need to manage water resources and water-related hazards. Thus, hydrologic science is (or should be) the basis for engineering hydrology and, along with economics and related social sciences, for water-resources management.

## 1.2  DEVELOPMENT OF SCIENTIFIC HYDROLOGY

Humans have been concerned with managing water as a necessity of life and as a potential hazard at least since the first civilizations developed along the banks of rivers. Hydraulic engineers built functioning canals, levees, dams, subterranean water con-

---

[1] This definition is consistent with that given by the U.S. Committee on Opportunities in the Hydrologic Sciences (Eagleson *et al.*, 1991, p. 57–58).

**FIGURE 1-1**
Principal storages (boxes) and pathways (arrows) of water in the global hydrologic cycle.

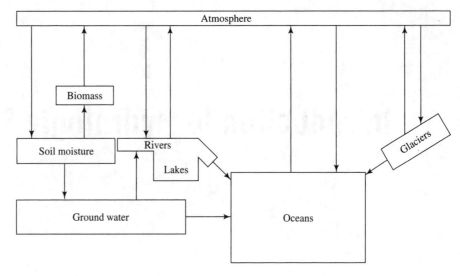

**FIGURE 1-2**
Principal storages (boxes) and pathways (arrows) of water in the land phase of the hydrologic cycle.

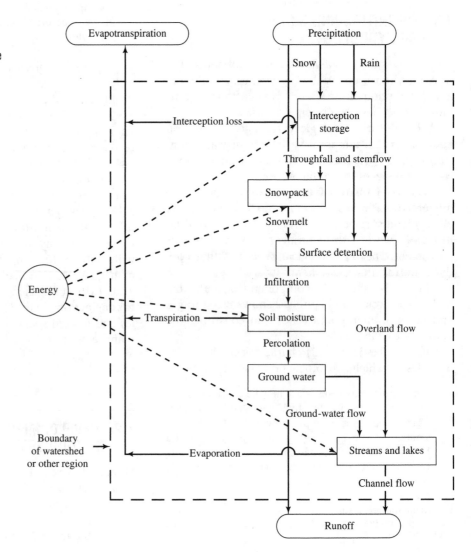

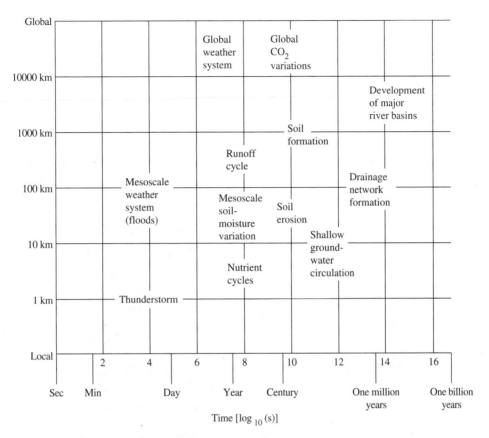

**FIGURE 1-3**
Range of space and time scales of hydrologic processes (After Eagleson et al (1991).

duits, and wells along the Indus in Pakistan, the Tigris and Euphrates in Mesopotamia, the Hwang Ho in China, and the Nile in Egypt as early as 5,000–6,000 years ago (B.P.). Hydroclimatologic information became vital to these civilizations; monitoring of river flows was begun by the Egyptians about 3,800 B.P., and the first known rainfall measurements are by Kautilya of India by 2,400 B.P. (Eagleson et al., 1991).

The concept of a global hydrologic cycle dates from at least 3,000 B.P. (Nace 1974), when Solomon wrote in Ecclesiastes 1:7 that

> All the rivers run into the sea; yet the sea is not full; unto the place from whence the rivers come, thither they return again.

Early Greek philosophers such as Thales, Anaxagoras, Herodotus, Hippocrates, Plato, and

Aristotle also embraced the basic idea of the hydrologic cycle. However, while some of them had reasonable understandings of certain hydrologic processes, they postulated various fanciful underground mechanisms by which water returned from sea to land and entered rivers. The Romans had extensive practical knowledge of hydrology and, especially, hydraulics and developed extensive aqueduct systems; their "scientific" ideas, however, were based very closely on those of the Greeks.

The theories of the Greek philosophers continued to dominate western thought until the Renaissance, when Leonardo da Vinci (ca. 1500) in Italy and, most notably, Bernard Palissy (ca. 1550) in France asserted, on the basis of field observations, that the water in rivers comes from precipitation (Adams 1938; Biswas 1970). The modern scientific approach to the hydrologic cycle which they initiated was taken up in the 17th century by

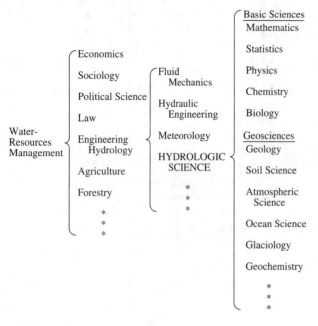

**FIGURE 1-4**
Hydrologic science in the hierarchy from basic sciences to water-resources management (After Eagleson et al (1991).

the Frenchmen Pierre Perrault and Edmé Mariotte: In the 1670s and 1680s, they published measurements and calculations that quantitatively verified the rainfall origin of streamflow. Shortly after (ca. 1700), the English scientist Edmund Halley extended the quantification of the hydrologic cycle by estimating the amounts of water involved in the ocean–atmosphere–rivers–ocean cycle of the Mediterranean Sea and its surrounding lands.

The 18th century saw considerable advances in applications of mathematics to fluid mechanics and hydraulics by Pitot, Bernoulli, Euler, Chézy, and others in Europe. Use of the term "hydrology" in approximately its current meaning began about 1750. By about 1800, the work of the English physicist and chemist John Dalton had firmly established the nature of evaporation and the present concepts of the global hydrologic cycle (Nace 1974), and Lyell, Hutton, and Playfair had published scientific work on the fluvial erosion of valleys. Routine network measurements of precipitation began before 1800 in Europe and the United States and were well established there and in India by 1820.

One of the barriers to understanding the hydrologic cycle was ignorance of the process of ground-water flow. This ignorance lasted until 1856, when the French engineer Henry Darcy established the basic phenomenological law of flow through porous material. The 1800s also saw many advances in fluid mechanics, hydraulics, and sediment transport by Poiseuille, DuPuit, DuBoys, Stokes, Manning, Reynolds, and others whose names have become associated with particular laws or principles.

Treatises on various aspects of hydrology, beginning with the Englishman Nathaniel Beardmore's *Manual of Hydrology* in 1862, appeared with increasing frequency in the last half of the 19th century. Many of these works examined relations between rainfall amounts and streamflow rates, answering the need to estimate flood flows for the design of bridges and other structures. This was the beginning of a close association between hydrology and civil engineering; the first English-language texts in hydrology were those of Daniel Mead in 1904 and Adolf Meyer in 1919, which were written for civil engineers (Eagleson et al. 1991). This association has in some respects enhanced, but in other respects has possibly inhibited, the development of hydrology as a science (Klemeš 1986).

The first half of the twentieth century saw great progress in many aspects of hydrology and, with the formation of the Section of Scientific Hydrology in the International Union of Geodesy and Geophysics (1922) and the Hydrology Section of the American Geophysical Union (1930), the first formal recognition of the scientific status of hydrology. Among those contributing notably to advances in particular areas in the early and middle decades of the century were the following: A. Hazen, E.J. Gumbel, H.E. Hurst, and W.B. Langbein in the application of statistics to hydrologic data; O.E. Meinzer, C.V. Theis, C.S. Slichter, and M.K. Hubbert in the development of the theoretical and practical aspects of ground-water hydraulics; L. Prandtl, T. Von Kármán, H. Rouse, V.T. Chow, G.K. Gilbert, and H.A. Einstein in stream hydraulics and sediment transport; R.E. Horton and L.B. Leopold in the understanding of runoff processes and quantitative geomorphology; W. Thornthwaite and H.E. Penman in the understanding of climatic aspects of hydrology and in the modeling of evapotranspiration; and A. Wolman and R.S. Garrels in the understanding and modeling of water quality.

Advances in all these areas, and others, are currently accelerating and new scientific questions are emerging. There are, in fact, great opportunities for progress in physical hydrology in many areas, including the determination of regional evapotranspiration rates, the movement of ground water in rock fractures, the relation between ground water and surface water, the relations between hydrologic behavior at different scales, the relation of hydrologic regimes to past and future climates, and the interaction of hydrologic processes and landform development (Eagleson et al. 1991).

As the twenty-first century begins, research is rapidly intensifying in both components of the scope of hydrology:

1. The ability to understand and model hydrologic processes at continental and global scales is becoming increasingly important because of the need to predict the effects of large-scale changes in land use and in climate. Thus exploration of approaches for transferring knowledge of physical hydrologic processes at a "point" to larger areas is becoming a major focus of hydrologic research, using newly available remote-sensing platforms and geographic-information systems.

2. In the land phase of the hydrologic cycle, it is interesting that detailed field studies to understand the mechanisms by which water enters streams began to proliferate only in the 1960s, pioneered by T. Dunne and others. Because of the temporal and spatial variability of natural conditions, this understanding is still far from complete. Research into these mechanisms is accelerating, motivated in part by concerns about the effects of land use on water quality and quantity and spurred by advances in technology that allow the use of a suite of chemical and isotopic tracers.

## 1.3 APPROACH AND SCOPE OF THIS BOOK

This text has four principal themes. First, the basic concepts underlying the science of hydrology are

introduced in Chapter 2. Chapter 3 then provides an overview of global climate, the global hydrologic cycle, and the relation of hydrology to soils and vegetation. The major part of the text—Chapters 4 through 9—deals with the second component of the scope of the science: the land phase of the hydrologic cycle. These chapters proceed more or less sequentially through the processes shown in Figure 1-2. Finally, Chapter 10 provides an overview of water-resource-management principles and introduces some of the ways in which hydrologic analysis is applied in that context.

The treatment in Chapters 4–9 draws on your knowledge of basic science (mostly physics, but also chemistry, geology, and biology) and of mathematics to develop a sound intuitive and quantitative sense of the way in which water moves through the land phase of the hydrologic cycle. In this process, we focus on quantitative, conceptually sound, but relatively simple representations of physical hydrologic processes and on approaches to the field measurement of the quantities of water and rates of flow involved in those processes.

The material covered in Chapters 4–9 constitutes the foundation of hydrologic science, and the advances in the science that will come in the next decades—in understanding watershed response to rain and snowmelt, in forecasting the hydrologic effects of land use and climatic change over a range of spatial scales, in understanding and predicting water chemistry, and in other areas—will be built upon this foundation. However, we must be aware that an understanding of the basic physics of such processes as evaporation, snowmelt, and infiltration as they occur instantaneously at a given "point" (i.e., a small, relatively homogeneous region of the earth's surface) does not always extrapolate easily to an understanding of the hydrology of a finite area, such as a drainage basin, over a finite time. An important reason for this problem of scale is that hydrologic quantities and the factors that control them vary greatly in both space and time, and it is difficult and expensive to obtain data to characterize this variability.

Another reason for the difficulty in extrapolating from knowledge of processes at small space and time scales to larger scales is that, in general, rates of water movement are *nonlinear* functions of the controlling quantities. If $q$ represents an instantaneous water-movement rate and $x$ the instantaneous value(s) of the variable(s) that control that rate, we can often use physical principles to derive the functional relation between them:

$$q = f(x). \tag{1-1}$$

However, we are usually interested in the rate $q$ averaged over a region (e.g., a watershed) or a period of time (e.g., a day) or both. If relation (1-1) is linear, we can measure spatial or temporal averages of $x$ (denoted as $\bar{x}$) and compute the average flow rate ($\bar{q}$) as

$$\bar{q} = f(\bar{x}). \tag{1-2}$$

If, however, the relation is nonlinear,[2]

$$\bar{q} \neq f(\bar{x}). \tag{1-3}$$

This means that, even if we have information about $\bar{x}$ (and the use of satellite-based and other remotely-sensed information is an increasingly valuable source of such information) and good knowledge of $f(x)$, we might not be able to make good estimates of $\bar{q}$.

Thus, although the basic physical principles described in this text are powerful tools, the degree of knowledge that can be obtained with them is bounded, almost always by the limited spatial and temporal availability (and often the quality) of the data that characterize the field situation, and sometimes by the inherent problem of scaling as represented by Equation (1-3). Hydrologists must be as aware of these limitations as they are of the tools themselves. Thus, I have tried to point out the assumptions behind each conceptual approach and the difficulties in applying it, because

> It ain't so much the things we don't know that gets us in trouble. It's the things we know that ain't so.[3]

---

[2] You can easily demonstrate that relation (1-3) is true by some calculations with a simple nonlinear relation such as $q = x^2$ or $q = \ln(x)$.

[3] I have seen this quote attributed to three American humorists: Artemus Ward (pseudonym of Charles Farrar Browne), Mark Twain, and Will Rogers. Take your pick.

# 2

# Basic Hydrologic Concepts

The concepts discussed in this chapter are so frequently applied in hydrology that they can be considered basic hydrologic concepts. Although hydrology is not a fundamental science in the sense that physics and chemistry are, its basic concepts are for the most part extensions of basic physical laws such as the conservation of mass.

## 2.1 PHYSICAL QUANTITIES AND LAWS

Hydrology is a quantitative geophysical science, and hydrologic relationships are usually expressed most usefully and concisely as relations between or among the numerical values of hydrologic quantities. In principle, these numerical values are determined by either

1. counting, in which case the quantity takes on a value that is a positive integer or zero; or

2. measuring, in which case the quantity takes on a value corresponding to a point on the real number scale that is the ratio of the magnitude of the quantity to the magnitude of a standard unit of measurement.[1]

Quantities determined by counting are **dimensionless** (dimensional quality expressed as [1]); measurable quantities have a dimensional quality

expressed in terms of the fundamental physical dimensions force [F] (or mass [M]), length [L], time [T], and temperature [Θ]. Appendix A is a review of the rules for the treatment of dimensions and units in equations, for converting between different systems of units, and for handling significant figures.

The basic relations of physical hydrology are derived from fundamental laws of classical physics, particularly those listed in Table 2-1. Derivations begin with a statement of the appropriate fundamental law(s) in a mathematical form and with boundary and (if required) initial conditions appropriate to the situation under study and are carried out by using mathematical operations (algebra and calculus). This is the approach that we will usually follow in the discussions of hydrologic processes in this text.

The properties of water dictate how it responds to the forces that drive the hydrologic cycle; these properties are summarized in Appendix B.

## 2.2 HYDROLOGIC SYSTEMS

Several basic hydrologic concepts are related to the simple model of a **system** shown in Figure 2-1. For present purposes, a system is any conceptually defined region of space that is capable of receiving a sequence of **inputs** of a **conservative** quantity, storing some amount of that quantity, and discharg-

---

[1]Common temperature scales are *interval*, rather than ratio, scales and hence require designation of an arbitrary zero point as well as of a unit of measurement.

**TABLE 2-1**

Summary of Basic Laws of Classical Physics Most Often Applied in Hydrologic Analyses

#### Conservation of Mass

Mass is neither created nor destroyed.

#### Newton's Laws of Motion

1. The momentum of a body remains constant unless the body is acted upon by a net force (conservation of momentum).
2. The rate of change of momentum of a body is proportional to the net force acting on the body and is in the same direction as the net force. (Force equals mass times acceleration.)
3. For every net force acting on a body, there is a corresponding force of the same magnitude exerted by the body in the opposite direction.

#### Laws of Thermodynamics

1. Energy is neither created nor destroyed (conservation of energy).
2. No process is possible in which the sole result is the absorption of heat and its complete conversion into work.

#### Fick's First Law of Diffusion

A diffusing substance moves from where its concentration is larger to where its concentration is smaller at a rate that is proportional to the spatial gradient of concentration.

ing **outputs** of that quantity. The region is sometimes called the **control volume.** Note that a control volume can be defined to include regions that are not physically contiguous (e.g., the world's glaciers).

A conservative quantity is one that cannot be created or destroyed within the system. In the branch of physics known as mechanics, there are three conservative quantities: (1) mass ([M] or $[F\,L^{-1}T^2]$); (2) momentum ($[M\,L\,T^{-1}]$ or $[F\,T]$); and (3) energy ($[M\,L^2T^{-2}]$ or $[F\,L]$). In many hydrologic analyses, it is reasonable to assume that the mass density (mass per unit volume, $[M\,L^{-3}]$) of water is effectively constant; in these cases, volume $[L^3]$ (i.e., $[M]\,/\,[M\,L^{-3}]$) is treated as a conservative quantity. However, mass density is a function of temperature (Section B.2.1), so this assumption might not always be warranted.

The storages and flows in Figures 1-1 and 1-2 are linked systems. The outer dashed line in Figure 1-2 indicates that any group of linked systems can be aggregated into a larger system; the smaller systems could then be called **subsystems.**

## 2.3   THE CONSERVATION EQUATIONS

The basic conservation equation can be stated in words as follows:

> The amount of a conservative quantity entering a control volume during a defined time period, minus the amount of the quantity leaving the volume during the time period, equals the change in the amount of the quantity stored in the volume during the time period.

Thus, the basic conservation equation is a generalization of the conservation of mass, Newton's first law of motion (when applied to momentum), and the first law of thermodynamics (when applied to energy). (See Table 2-1.)

In condensed form, we can state the conservation equation as

$$\text{Amount In} - \text{Amount Out} = \text{Change In Storage,} \qquad \textbf{(2–1)}$$

but we must remember that the equation is true *only* (1) for conservative substances, (2) for a defined control volume, and (3) for a defined time period.

If we designate the amount of a conservative quantity entering a region in time period $\Delta t$ by $I$, the amount leaving during that period by $Q$, and the change in storage over that period as $\Delta S$, we can write Equation (2–1) as

$$I - Q = \Delta S. \qquad \textbf{(2–2)}$$

Another useful form of the basic conservation equation can then be derived by dividing each of the terms in Equation (2–2) by $\Delta t$:

$$\frac{I}{\Delta t} - \frac{Q}{\Delta t} = \frac{\Delta S}{\Delta t}. \qquad \textbf{(2–3)}$$

If we now define the average rates of inflow, $m_I$, and outflow, $m_Q$, for the period $\Delta t$ as follows:

$$m_I \equiv \frac{I}{\Delta t}, \qquad \textbf{(2–4)}$$

$$m_Q \equiv \frac{Q}{\Delta t}, \qquad \textbf{(2–5)}$$

we can write Equation (2–2) as

**FIGURE 2-1**
Conceptual diagram of a system. *i* is input rate, *q* is output rate, and *S* is storage.

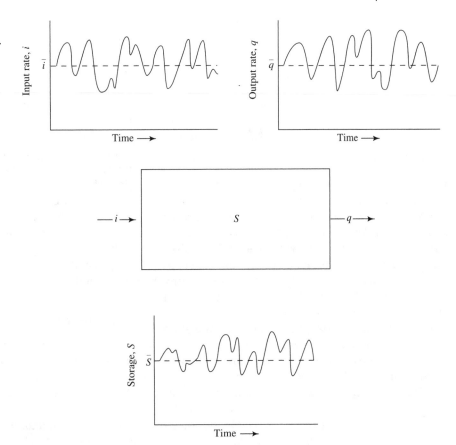

$$m_I - m_Q = \frac{\Delta S}{\Delta t}. \qquad (2\text{–}6)$$

Equation (2–6) states that the average rate of inflow, minus the average rate of outflow, equals the average rate of change of storage.

Another version of the conservation equation can be developed by defining the instantaneous rates of inflow, *i*, and outflow, *q*, as

$$i \equiv \lim_{\Delta t \to 0} \frac{I}{\Delta t} \qquad (2\text{–}7)$$

and

$$q \equiv \lim_{\Delta t \to 0} \frac{Q}{\Delta t}, \qquad (2\text{–}8)$$

respectively. Substituting these into Equation (2–3) allows us to write

$$i - q = \frac{dS}{dt}, \qquad (2\text{–}9)$$

which states that the instantaneous rate of input, minus the instantaneous rate of output, equals the instantaneous rate of change of storage.

All three forms of the conservation equation, Equations (2–2), (2–6), and (2–9), are applied in various contexts throughout this text. They are called **water-balance equations** when applied to the mass of water moving through various portions of the hydrologic cycle; control volumes in these applications range in size from infinitesimal to global, and time intervals range from infinitesimal to annual or longer (Figure 1-3). A special application of these equations, the regional water balance, is discussed later in this chapter. As indicated in Figure 1-2, energy fluxes are directly involved in evaporation and snowmelt, and the application of the conservation equation in the form of **energy-balance equations** is essential to the understanding of those processes developed in Chapters 5 and 7. Consideration of the conservation of momentum is important in the analysis of fluid flow, and this principle is applied in the discussion of turbulent exchange of heat and water vapor with the atmosphere (Section D.6).

## 2.4   THE WATERSHED (DRAINAGE BASIN)

### 2.4.1   Definition

Hydrologists commonly apply the conservation equation in the form of a water-balance equation to a geographical region in order to establish the basic hydrologic characteristics of the region. Most commonly, the region is a **watershed** (also called **drainage basin, river basin,** or **catchment**), defined as the area that appears on the basis of topography to contribute all the water that passes through a given cross section of a stream (Figure 2-2). The surface trace of the boundary that delimits a watershed is called a **divide.** The horizontal projection of the area of a watershed is called the **drainage area** of the stream at (or above) the cross section.

The watershed concept is of fundamental importance because the water passing through the stream cross section at the watershed outlet originates as precipitation on the watershed[2] and because the characteristics of the watershed control the paths and rates of movement of water as it moves to the stream network. Hence, watershed geology, topography, and land cover determine the quality of ground water and of surface water as well as the magnitude and timing of streamflow and of ground-water outflow. As William Morris Davis stated in 1899,

> [O]ne may fairly extend the "river" all over its [watershed] and up to its very divides. Ordinarily treated, the river is like the veins of a leaf; broadly viewed it is like the entire leaf.

Thus the watershed can be viewed as a natural landscape unit, integrated by water flowing through the land phase of the hydrologic cycle and, although political boundaries do not generally follow watershed boundaries, water-resource and land-use planning agencies recognize that effective management of water quality and quantity requires a watershed perspective.

The location of the stream cross section that defines the watershed is determined by the purpose of the analysis. Hydrologists are most often interested in delineating watersheds above stream-gaging stations (where streamflow is measured; see Appendix F), or above points at which some water-resource activity takes place (e.g., a hydroelectric plant, a reservoir, a waste-discharge site, or a location where flood damages are of concern).

There are an infinite number of points (cross sections) along a stream, so an infinite number of watersheds can be drawn for any stream. As indicated in Figure 2-2, upstream watersheds are nested within, and are part of, downstream watersheds.

### 2.4.2   Delineation

The conventional manual method of watershed delineation requires a topographic map (or stereoscopically viewed aerial photographs). To trace the divide, start at the location of the chosen stream cross section, then draw a line away from the left or right bank, *maintaining it always at right angles to the contour lines*. Continue the line until it is generally above the headwaters of the stream network and its trend is generally opposite to the direction in which it began. Finally, return to the starting point and trace the divide from the other bank, eventually connecting it with the first line.

Frequent visual inspection of the contour pattern is required as the divide is traced out to assure that an imaginary drop of water falling streamward of the divide would, if the ground surface were imagined to be impermeable, flow downslope and eventually enter the stream network upstream of the starting point. A divide can never cross a stream, though there are rare cases where a divide cuts through a wetland (or, even more rarely, a lake) that has two outlets draining into separate stream systems. The lowest point in a drainage basin is always the basin outlet, i.e. (the starting point for the delineation). The highest point is usually, but not necessarily, on the divide.

Increasingly, topographic information is becoming available in the form of **digital elevation models** (DEMs). These are computer data files that give land-surface elevations at grid points. Although it is not an entirely straightforward exercise,

---

[2]As discussed in Section 2.5.2, there are situations in which ground-water inflow from adjacent watersheds contributes a portion (usually minor) of the flow of a stream.

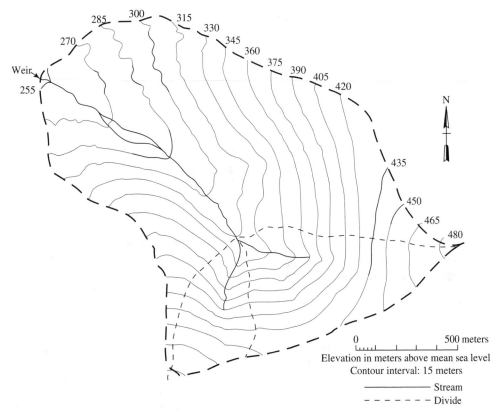

**FIGURE 2-2**
Watersheds delineated on a topographic map. Large area is the watershed of Glenn Creek in Fox, AK, above a streamflow-measurement weir; watersheds of tributaries to Glenn Creek are also shown.

it is possible to develop computer programs that can trace out stream networks and drainage divides by analyzing DEMs (e.g., Fairfield and Leymarie 1991; Martz and Garbrecht 1992; Tarboton 1997; McKay and Band 1998). This automated approach to watershed delineation allows the concomitant rapid extraction of much hydrologically useful information on watershed characteristics (such as the distribution of elevation and slope) that previously could be obtained only by very tedious manual methods.

## 2.5 THE REGIONAL WATER BALANCE

The regional water balance is the application of the water-balance equation to a watershed (or to any land area, such as a state or continent). Thus, the watershed area delimited by the divide (or other surface area) is the upper surface of the control volume; the sides of the volume extend vertically downward from the divide some indefinite distance that is assumed to reach below the level of significant ground-water movement.

In virtually all regional hydrologic analyses, it is reasonable to assume a constant density of water and to treat volume $[L^3]$ as a conservative quantity. For many such analyses, it is convenient to divide the volumes of water by the surface area of the region, so that the quantities have the dimension $[L]( = [L^3]/[L^2])$; often, it is convenient also to divide by the duration of the measurement period, to obtain $[L\ T^{-1}]$. [Compare Equations (2-2) and (2-6).]

In this section, we first develop a conceptual regional water balance from which we can define some useful terms and show the importance of climate in determining regional water resources; we

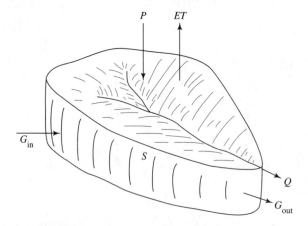

**FIGURE 2-3**
Schematic diagram of a watershed, showing the components of the regional water balance: *P* = precipitation, *ET* = evapotranspiration, *Q* = stream outflow, $G_{in}$ = ground-water inflow, $G_{out}$ = ground-water outflow.

then show how actual water-balance measurements are used to estimate regional evapotranspiration (defined in the next section).

### 2.5.1  The Water-Balance Equation

Consider the watershed shown in Figure 2-3. For any time period of length $\Delta t$, we can write the water-balance equation as

$$P + G_{in} - (Q + ET + G_{out}) = \Delta S, \quad \text{(2–10)}$$

where $P$ is precipitation (liquid and solid), $G_{in}$ is ground-water inflow (liquid), $Q$ is stream outflow (liquid), $ET$ is evapotranspiration[3] (vapor), $G_{out}$ is ground-water outflow (liquid), and $\Delta S$ is the change in all forms of storage (liquid and solid) over the time period. The dimensions of these quantities are $[L^3]$ (or, if divided by drainage area, $[L]$). If we average these quantities over a reasonably long time period (say, many years) in which there are no significant climatic trends or geological changes and no anthropogenic inputs, outputs, or

storage modifications, we can usually assume that net changes in storage will be effectively zero and write the water balance as

$$\mu_P + \mu_{Gin} - [\mu_Q + \mu_{ET} + \mu_{Gout}] = 0, \quad \text{(2–11)}$$

where $\mu$ denotes the time average of the subscript quantity and the dimensions are now $[L^3 T^{-1}]$ or $[L T^{-1}]$.

The total amount of liquid water leaving the region is called the **runoff**,[4] *RO*, for the region. Therefore,

$$\mu_{RO} \equiv \mu_Q + \mu_{Gout}. \quad \text{(2–12)}$$

The amount of liquid water actually "produced" in the region is called the **hydrologic production**, $\Pi$:

$$\mu_\Pi \equiv \mu_Q + \mu_{Gout} - \mu_{Gin} = \mu_P - \mu_{ET}. \quad \text{(2–13)}$$

From Equations (2-11) and (2-12),

$$\mu_{RO} = \mu_P + \mu_{Gin} - \mu_{ET}. \quad \text{(2–14)}$$

Because watersheds are defined topographically and ground-water flow is driven by gravity,[5] we can usually assume that $G_{in}$ is negligible and write the water-balance equation as

$$\mu_{RO} \equiv \mu_Q + \mu_{Gout} = \mu_P - \mu_{ET}. \quad \text{(2–15)}$$

Evaluation of the terms in the water-balance equation provides the most basic information about a region's hydrology. The runoff represents the water potentially available for human use and management and hence, the quantity of water resource available from a given region. However, as we will explore later in this chapter, the temporal variability of runoff must be evaluated in assessing actual water-resource availability.

As we will see in Chapter 7, evapotranspiration is determined largely (but not solely) by meteorologic variables (solar radiation, temperature, humidity, and wind speed), so both precipitation and evapotranspiration can be considered to be externally imposed climatic 'boundary conditions.' Thus,

---

[3]**Evapotranspiration** is the total of all water that leaves a region via direct evaporation from surface-water bodies, snow, and ice, plus that which is evaporated after passing through the vascular systems of plants (**transpiration**; the process is described in Chapter 7).

---

[4]Note that hydrologists also use the term "runoff" to denote overland flow, which is discussed in Chapter 9.
[5]The geometry of regional ground-water flows is extensively described in Chapter 8.

from Equation (2-15), runoff is a residual or difference between two climatically-determined quantities.

### 2.5.2 Estimation of Regional Evapotranspiration

#### Basic Approach

Perhaps the most common form of hydrologic analysis is the estimation of the long-term average value of regional evapotranspiration via the water-balance equation. This application arises because, although there are techniques for determining the precipitation over an area (discussed in Section 4.3) and the streamflow from an area (Appendix F), it is virtually impossible to measure areal evapotranspiration directly. (This is discussed in detail in Section 7.8.) In such analyses, it is usually assumed that ground-water flows either are negligible or cancel out and that $\Delta S$ is negligible, so that Equation (2-15) becomes

$$m_{ET} = m_P - m_Q, \qquad (2\text{–}16)$$

where $m$ indicates the average of the subscript quantity for the period of measurement [rather than the true long-term averages as in Equation (2-15)].

Equation (2-16) is straightforward, but there are two types of errors that potentially affect the accuracy of $m_{ET}$ estimates made with it: **model error**, which refers to the omission of potentially significant terms from the equation, and **measurement error** in the quantities $m_P$ and $m_Q$, which is unavoidable. Together, these errors introduce uncertainty that propagates into the estimate of evapotranspiration and hence is critical for assessing the validity of that estimate. Thus, it is worthwhile to examine further these sources of error.

#### Model Error

**Ground-Water Flows** Ground-water flows are usually considered negligible in water-balance computations of regional evapotranspiration. However, the consideration of regional ground-water flows in Chapter 8 makes it clear that it is often unwise to assume that ground-water outflow is negligible. The higher the relief of a given watershed and the more hydraulically conductive its geologic composition (which is often not well known), the more likely it is to lose water by subsurface flow. Streams draining larger watersheds tend to receive the subsurface outflows of their smaller constituent watersheds, so the importance of ground-water outflow generally decreases as one considers larger and larger watersheds. However, such generalities can be obviated by particular geologic situations.

**Storage Changes** The net change in storage over a period of measurement is the difference between the amount of water in storage in the watershed (as ground water and as water in rivers, lakes, soil, vegetation, and snow and ice) at the end of the period and the amount in storage at the beginning of the period. No change-in-storage term appears in Equation (2-16), and this quantity is almost always assumed to be negligible.

Measurements of watershed storage are usually lacking, so the storage residual cannot be directly evaluated. Instead, hydrologists attempt to minimize its value by (1) using long measurement periods and (2) selecting the time of beginning and end of the measurement period such that storage values are likely to be nearly equal. (See Box 2-1.)

To minimize the storage residual in annual water-balance computations in the United States, the U.S. Geological Survey begins the **water year** on 1 October, on the assumption that by this time transpiration by plants will have largely ceased and soil-moisture and ground-water storage will have been recharged to near their maximum levels. However, as is suggested by the analysis in Box 2-1, other water-year spans may be more appropriate for specific regions—for example, the time of disappearance of the annual snowpack in northern areas.

#### Measurement Error

Uncertainty due to measurement error is always present in hydrologic computations, and sources of such error are reviewed in this text where various measurement techniques are discussed. Here we briefly characterize the uncertainty in estimating areal precipitation and streamflow and show quantitatively how this uncertainty is propagated into estimates of regional evapotranspiration when Equation (2-16) is assumed to be correct (i.e., when there is no model error).

**Accuracy of Regional Precipitation Values** In the application of Equation (2-16), it is assumed that $m_P$ can

---

## BOX 2-1
· · · · · · ·
### Evaluation of Changes in Watershed Storage

If $S_i$ represents the watershed storage at the end of year $i$ and $\Delta S_i$ the change in storage over year $i$, then the average change in storage over an $N$-year period, $m_{\Delta S}$, is

$$m_{\Delta S} = \frac{S_N - S_0}{N}. \qquad \text{(2B1-1)}$$

Thus. $m_{\Delta S}$ will be small if $S_N - S_0$ is small or $N$ is large. As noted in the text, $S_N - S_0$ can be minimized by choosing the beginning of the water year to be at a time when the year-to-year variability of storage is minimal.

Using the BROOK watershed model (Section 2.9.5) to simulate monthly evapotranspiration and soil-moisture storage over a 50-year period in New Hampshire, Hartley (1990) explored the consequences of using water years beginning on the first of each month. As shown in the table, she found a considerable variation in the sta-

tistics of annual storage changes depending on the month chosen as the beginning of the water year: choosing April gave the least year-to-year variability; choosing September gave the most. However, it turned out that the error in estimating average evapotranspiration introduced by assuming $\Delta S = 0$ was less than 5% after no more than two years regardless of the month selected.

Thus this study suggested that, although choosing different beginning times for water years leads to very different $\Delta S$ values, assuming $\Delta S = 0$ does not introduce significant error into water-balance estimates of evapotranspiration in this region if the averaging period exceeds a few years. However, ground-water and lake or wetland storage were not included in the simulation, and the conclusion might be different where these are important.

| | **Water Year Beginning 1st of** | | | | | | | | | | | |
| --- | --- | --- | --- | --- | --- | --- | --- | --- | --- | --- | --- | --- |
| | **Jan** | **Feb** | **Mar** | **Apr** | **May** | **Jun** | **Jul** | **Aug** | **Sep** | **Oct** | **Nov** | **Dec** |
| $m_{\lvert \Delta S \rvert}$ (mm) | 29.3 | 30.5 | 13.6 | 9.6 | 14.4 | 25.0 | 26.3 | 27.4 | 36.3 | 24.7 | 15.5 | 20.7 |
| $s_{\Delta S}$ (mm)[a] | 37.4 | 41.6 | 19.1 | 12.7 | 17.5 | 30.0 | 33.5 | 38.2 | 43.7 | 32.4 | 19.6 | 27.5 |
| Maximum $\lvert \Delta S \rvert$ (mm) | 93.3 | 117.2 | 62.8 | 32.9 | 42.2 | 62.3 | 96.3 | 106.8 | 77.9 | 78.0 | 52.4 | 81.9 |
| Years for $< 5\%$ | | | | | | | | | | | | |
| Error in $m_{ET}$ | 2 | 2 | 1 | 1 | 1 | 2 | 2 | 2 | 2 | 2 | 1 | 1 |

[a]Standard deviation of $\Delta S$.

---

be calculated as the spatial average of temporally averaged precipitation-gage measurements made for many years in or near the region.

As discussed in Section 4.2.2, measurements of precipitation at individual gages are subject to error from several causes, and additional error is introduced in the process of computing areal averages. Thus, estimates of long-term (annual or longer) total (or average) areal precipitation typically have relative uncertainties on the order of 10% (i.e., the true value is taken to be within 10% of the estimated value) (Winter 1981). In regions of high relief or with few or poorly distributed gages, or for shorter measurement periods, the uncertainty can be considerably larger.

**Accuracy of Streamflow Values**   In the application of Equation (2-16), it is assumed that $m_Q$ can be calculated as the temporal average of streamflow measurements made at the watershed outlet for many years.

Winter (1981) estimated that the measurement uncertainty for long-term average values of streamflow at a gaging station is on the order of $\pm 5\%$. (The accuracy of such measurements is discussed further in Section F.2.4.) Where $\mu_Q$ is estimated for locations other than carefully maintained gaging stations, the uncertainty can be much greater.

**Propagation of Measurement Errors**   In general, both model and measurement error are present and are propagated into estimates of $\mu_{ET}$. To simplify the discussion here, we show how uncertainty in the estimate of $\mu_{ET}$ via Equation (2-16) can be assessed quantitatively, given information about the uncertainties in the measurements of $m_P$ and $m_Q$. To do this, we make the assumptions (1) that model errors in Equation (2-16) are negligible (i.e., that $m_{Gin}$, $m_{Gout}$, and the storage residual are negligible) and (2) that the terms in the equation refer to water-balance quantities for the period in which both $P$ and $Q$ were measured—that is, that we are not treating the data as samples from an indefinitely long time period.

Potential measurement errors are usually assumed to be distributed symmetrically about the true value (equal chance of under- or over-estimation) and to follow the bell-shaped **normal distribution** described in Appendix C: the further a measured value is from the true value (i.e., the larger is the error), the smaller is the probability that it will occur (Figure 2-4). The spread, or variation, of the potential measured values about the true value is expressed as the **standard deviation** of the potential errors.

We can apply to Equation (2-16) the general rule that, if a quantity ($m_{ET}$) is the sum or difference of two measured quantities ($m_P$ and $m_Q$), and the errors in the two terms are normally distributed, then the errors in $m_{ET}$ are also normally distributed. Furthermore, if the measurement errors of $P$ and $Q$ are not related (a reasonable assumption), then the standard deviations of the errors due to measurement of the quantities are related as

$$\sigma_{ET}^2 = \sigma_P^2 + \sigma_Q^2, \qquad \textbf{(2–17)}$$

where $\sigma$ indicates the standard deviation of the measurement errors of the subscripted quantity.[6] Thus, to evaluate the error in $ET$ due to measurement errors in $P$ and $Q$ via Equation (2-17), we must estimate $\sigma_P$ and $\sigma_Q$.

Statements of measurement uncertainty are usually expressed probabilistically—for example, as

"I am $100 \cdot p\%$ sure that the true value of precipitation is within $u_P \cdot m_P$ of the measured value."

$$\textbf{(2–18a)}$$

---

[6]Equation (2-17) can be derived directly from the definition of the standard deviation [Equation (C-19)].

Here, $m_P$ is the estimate of average precipitation and $u_P$ is the **relative uncertainty** in the estimate (e.g., if the measurement uncertainty is stated to be 10%, $u_P = 0.1$). The **absolute uncertainty** in $m_P$ is $u_P \cdot m_P$. In conventional probability notation, Equation (2-18a) is written as

$$\Pr\{(m_P - u_P \cdot m_P) \leq \text{true precipitation}$$
$$\leq (m_P + u_P \cdot m_P)\} = p,$$
$$\textbf{(2–18b)}$$

where $\Pr\{\ \}$ indicates the probability of the statement in braces.

The probability $p$ should be close to 1, but it is not realistic to assume that $p = 1$; that would be equivalent to stating, "I am absolutely certain that the true value is within $\pm u_P \cdot m_P \ldots$". Typically, statements of measurement error are given with $p = 0.95$, so the statement would be "I am 95% sure. . . ."

Given that potential measurement errors follow the normal distribution, we can find from the properties of that distribution, summarized in Table C-5, that there is a 95% probability that an observation will be within $\pm 1.96$ standard deviations of the central (true) value. Thus we can write Equation (2-18b) equivalently as

$$\Pr\{(m_P - 1.96 \cdot s_P) \leq \text{true precipitation}$$
$$\leq (m_P + 1.96 \cdot s_P)\} = 0.95,$$
$$\textbf{(2–19)}$$

where $s_P$ is the estimated error standard deviation for precipitation.

From comparison of Equations (2-18) and (2-19), we see that when $p = 0.95$,

$$u_P \cdot m_P = 1.96 \cdot s_P, \qquad \textbf{(2–20)}$$

so that

$$s_P = \frac{u_P \cdot m_P}{1.96}. \qquad \textbf{(2–21a)}$$

Because $u_P$ must be an estimate and because hydrologic measurements can seldom be made with three-significant-figure precision (see Section A.3), it makes sense to approximate Equation (2-21a) as

$$s_P = \frac{u_P \cdot m_P}{2}. \qquad \textbf{(2–21b)}$$

The statements of Equations (2-18)–(2-21) can also be made for streamflow and its measurement errors, so we can also conclude that

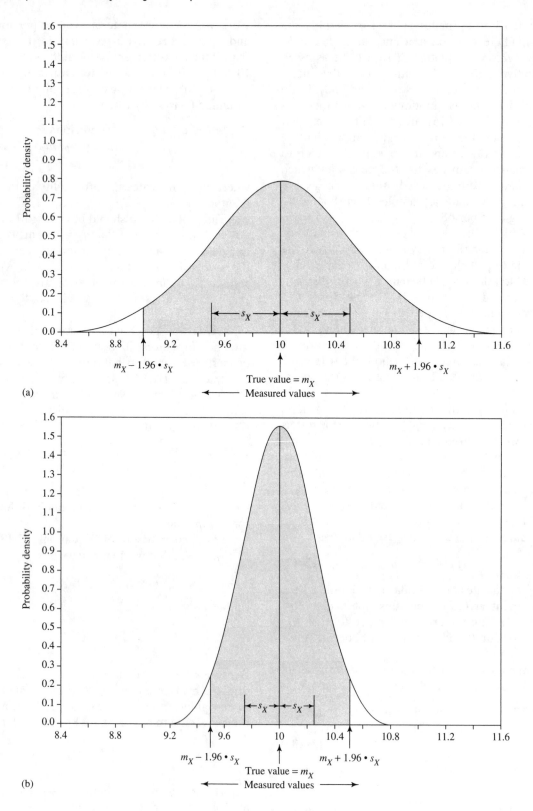

(a)

(b)

$$s_Q = \frac{u_Q \cdot m_Q}{2}, \qquad \textbf{(2–22)}$$

where $u_Q$ is the relative error in streamflow measurement (and $p = 0.95$).

Thus, if reasonable estimates of $u_P$ and $u_Q$ can be obtained, $s_{ET}$ is readily calculated via Equation (2-17). By analogy with Equations (2-21) and (2-22), the relative error in the evapotranspiration estimate, $u_{ET}$, is then

$$u_{ET} = \frac{2 \cdot s_{ET}}{m_{ET}} = \frac{(u_P^2 \cdot m_P^2 + u_Q^2 \cdot m_Q^2)^{1/2}}{m_P - m_Q}. \qquad \textbf{(2–23)}$$

---

### EXAMPLE 2-1

For the period 1961–1985, average annual precipitation for the Oyster River drainage basin, NH, was 1066 mm yr$^{-1}$, and average annual streamflow was 551 mm yr$^{-1}$. Assume the relative measurement errors for precipitation and streamflow are $u_P = 0.1$ and $u_Q = 0.05$. Estimate (a) the average annual evapotranspiration for that period and (b) the relative and absolute uncertainties in that estimate.

**Solution:** (a) With $m_P = 1066$ mm yr$^{-1}$ and $m_Q = 551$ mm yr$^{-1}$, and assuming that $m_{Gin}$, $m_{Gout}$, and the storage residual are all negligible, Equation (2-16) yields

$$m_{ET} = 1066 \text{ mm yr}^{-1} - 551 \text{ mm yr}^{-1} = 515 \text{ mm yr}^{-1}.$$

(b) From Equations (2-21b) and (2-22),

$$s_P = \frac{0.1 \times 1066 \text{ mm yr}^{-1}}{2} = 53.3 \text{ mm yr}^{-1};$$

$$s_Q = \frac{0.05 \times 551 \text{ mm yr}^{-1}}{2} = 13.8 \text{ mm yr}^{-1}.$$

Using these values in Equation (2-17) gives

$$s_{ET}^2 = (53.3 \text{ mm yr}^{-1})^2 + (13.8 \text{ mm yr}^{-1})^2$$
$$= 3031.33 \text{ mm}^2 \text{ yr}^{-2};$$
$$s_{ET} = 55.1 \text{ mm yr}^{-1}.$$

Thus, from Equation (2-23),

$$u_{ET} = \frac{2 \times 55.1 \text{ mm yr}^{-1}}{515 \text{ mm yr}^{-1}} = 0.214.$$

By analogy with Equation (2-19),

$$\Pr\{(515 \text{ mm yr}^{-1} - 0.214 \times 515 \text{ mm yr}^{-1}) \le \mu_{ET}$$
$$\le (515 \text{ mm yr}^{-1} + 0.214 \times 515 \text{ mm yr}^{-1})\} = 0.95;$$

$$\Pr\{405 \text{ mm yr}^{-1} \le \mu_{ET} \le 625 \text{ mm yr}^{-1}\} = 0.95.$$

---

The result of Example 2-1 is quite general: the relative uncertainty in estimates of $\mu_{ET}$ found via regional water balances is usually considerably greater than the uncertainties in the measured quantities, even when there are no unmeasured water-balance components and when storage residuals are negligible. More generally, the uncertainty in *any* quantity found as the difference of measured quantities is larger than the uncertainties in the measured quantities.

The studies by Lesack (1993) and Cook et al. (1998) are among the few published attempts to assess uncertainty in regional water balances.

---

## 2.6   SPATIAL VARIABILITY

Rates of input and output—and many other hydrologically relevant properties—vary spatially over the geographic regions that constitute control volumes for many types of hydrologic analyses (e.g., watersheds). Thus it is essential that hydrologists become familiar with methods for describing and comparing spatial as well as temporal variability.

Descriptions of spatial variability of some variables—notably precipitation—are based on measurements made over time at discrete points (precipitation gages). These measurements, expressed as average rates or total amounts, constitute spatial as well as temporal samples, and the

---

**FIGURE 2-4**

Probability distribution of potential measurement errors of a quantity *X*, shown as having a true value $m_X = 10$. Such errors are usually assumed to be symmetrical about the true value and to follow the normal distribution (see Appendix C). (a) shows the case where one is 95% sure that the true value is within 10% of the measured value, so that $s_X = 0.1 \cdot m_X/1.96 \approx 0.5$. (b) Shows the case where one is 95 % sure the true value is within 5 % of the measured value, so that $s_X = 0.05 \cdot m_X/1.96 \approx 0.25$. In both cases the shaded area = 0.95 (the probability *p*).

values can be contoured to produce a model of the quantity's continuous spatial variation.

Traditional statistical methods, such as those described in Appendix C, can also be used to compute spatial averages and measures of spatial variability from the point values. However, precipitation gages are usually unevenly distributed over any given region, and the point values are therefore an unrepresentative sample of the true precipitation field. Because of this, and because of the importance of accurately quantifying variables such as precipitation, basic statistical concepts have been incorporated into special techniques for characterizing and accounting for spatial variability. These techniques are introduced in Section 4.3; however, they apply not only to rainfall, but to the spatial variability of infiltration, of ground-water levels, and of other spatially distributed quantities as well.

## 2.7   TEMPORAL VARIABILITY

The inputs, storages, and outputs in Figures 1-1, 1-2, and 2-1 are all **time-distributed variables**—quantities that can vary with time. Thus the concept of time variability is inherent to the concept of the system, and we have seen how time averaging is applied to develop alternative forms of the conservation equations.

In particular, the streamflow rate at a given location is highly variable in time. Even in humid regions, it typically varies annually over three or more orders of magnitude as a result of seasonal fluctuations of rainfall and evapotranspiration; in arid regions, the annual fluctuations are even greater. Year-to-year weather variations cause further temporal variability, reflected in fluctuations in annual mean flows and in the occurrences of floods and droughts.

From the human viewpoint, the long-term average streamflow rate, $\mu_Q$, is highly significant: it represents the maximum rate at which water is potentially available for human use and management, and is therefore a measure of the ultimate water resources of a watershed or region. However, because of the large time variability of streamflow, we generally cannot rely on the mean flow to be available most of the time. The rate at which water

is *actually* available for use is best measured as the streamflow rate that is available a large percentage—say 95%—of the time. This value is designated $q_{.95}$. Where streamflow records are available, $q_{.95}$ is readily determined by constructing a flow-duration curve, as described in Sections 2.7.2 and 10.2.5.

Streamflow variability is directly related to the seasonal and interannual variability of runoff (and hence of the climate of precipitation and evapotranspiration) and inversely to the amount of storage in the watershed. Humans can increase water availability by building storage reservoirs, as discussed in Sections 2.8 and 10.2.5. Humans can also attempt to increase $\mu_P$ through "rain-making" (Section 4.4.5) and to decrease $\mu_{ET}$ by modifying vegetation (Sections 7.6.4 and 10.2.5). However, such interferences in the natural hydrologic cycle usually have serious environmental, social, economic, and legal consequences. Some of the consequences involved in exploiting ground water are considered in Sections 8.6 and 10.2.4.

Because streamflow is the difference between two climatically determined quantities [Equation (2-15)], it is clear that climate change, whether natural or anthropogenic, will affect runoff and hence water resources. The BROOK90 model introduced in Section 2.9.5 can be used for detailed study of climatic and land-use effects; some simpler approaches to evaluating these effects are given in Chapter 3 (Boxes 3-4 and 3-5).

In the next section, we introduce (1) the basic approach for constructing and analyzing samples of time-distributed variables and (2) the duration curve, a widely applicable approach to characterizing time variability.

### 2.7.1   Time Series

Clearly, it will be useful to be able to describe and compare time-distributed variables in terms of their average value, variability, and perhaps other characteristics. Such descriptions and comparisons usually are made by applying the statistical methods described in Appendix C. However, these methods are applicable only to discrete sequences of values obtained from the time trace of the variable of interest, each value of which is associated with a particular time in a sequence of times. Such a sequence is called a **time series.**

Some time-series variables are obtained by counting—for example, the number of days with more than 1 mm rain in each year at a particular location. Such variables are inherently **discrete**, and it is a straightforward matter to construct a time series for such variables if the relevant measurements are available.

However, many hydrologic variables—including the inputs, outputs, and storages in Figure 2-1—are **continuous** time traces: they take on values at every instant in time. Thus there are infinitely many values of the variable associated with each time interval. In order to construct a time series for a continuous variable, one must convert it to discrete form. To do this, first select a time interval, $\Delta t$, and divide the total period of interest into increments of length $\Delta t$. (The value of $\Delta t$ is determined by the purpose of the analysis; in hydrologic studies it is often 1 day, 1 month, or 1 year.) Then, the single value of the variable of interest associated with each interval is determined either as (1) the average, (2) the largest, or (3) the smallest value of the variable that occurred during the interval (for inputs and outputs) or as (4) the value of the variable

at the beginning or end of each $\Delta t$ (for storages). (See Figure 2-5.) Example 2-2 and Figure 2-6 show examples of three time series developed from the continuous measurements at a streamflow gaging station.

---

### EXAMPLE 2-2

Table 2-2 lists, and Figure 2-6 plots, three time series developed from the continuous streamflow record obtained at the **stream-gaging station** operated by the U.S. Geological Survey on the Oyster River in Durham, NH. In all three plots, $\Delta t = 1$ yr, and the ordinate is a streamflow rate, or **discharge**, $[L^3 T^{-1}]$. However, the discretization of the continuous record was done differently for each series: Series *a* is the average streamflow for the year, Series *b* is the highest instantaneous flow rate for the year, and Series *c* is the lowest of the flow rates found by averaging over all the seven-consecutive-day periods within each year. (These data are also in the spreadsheet file `Table2-2.xls` on the disk accompanying this text.)

Note that the lines connecting the time-series values in each graph do not represent a time trace; they serve only to

**FIGURE 2-5**
Schemes for converting a continuous time trace into a discrete time series. For each $\Delta t$, one may select the average, $E(Q)$, maximum, $M(Q)$, or minimum, $m(Q)$.

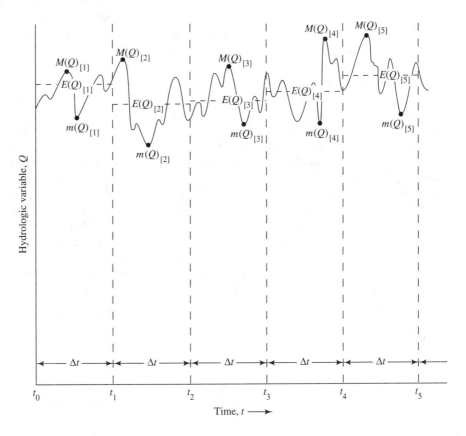

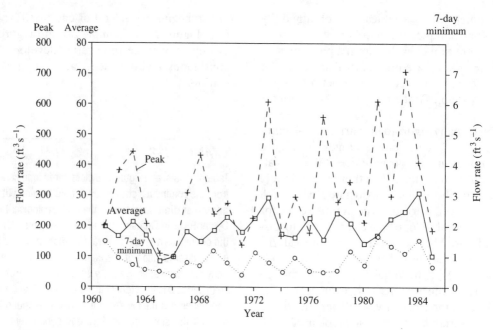

**FIGURE 2-6**
Plots of the time series in Table 2-2.

connect the point values to provide a visual impression of the nature of the series.

Time series are usually treated as more or less representative **samples** of the long-term behavior of the variable and are described and compared on the basis of their statistical attributes. For example, the temporal variability of a time series can be characterized in absolute terms by its **interquartile range** or by its **standard deviation**, and in relative terms by the ratio of its interquartile range to its **median** or by its **coefficient of variation** (Sections C.2.4 and C.2.5).

It is important to note that time series developed from a single continuous time trace by choosing different discretizing schemes (as in Example 2-2) or different $\Delta t$ values will in general have very different statistical characteristics. Box 2-2 explains how one can obtain time series of peak and daily average streamflows measured at U.S. Geological Survey gaging stations.

### 2.7.2 Duration Curves

One conceptually simple but highly informative way to summarize the variability of a time series is by means of a **duration curve**—a cumulative-fre-quency curve that shows the fraction (percent) of time that the magnitude of a given variable exceeds a specified value over a period of observation that is long enough to include a wide range of seasonal and inter-annual variability. Duration curves are most commonly used to depict the temporal variability of streamflow; such curves are then called **flow-duration curves** (FDCs) (Figure 2-7).

Searcy (1959) and Vogel and Fennessey (1994; 1995) have provided comprehensive reviews of FDCs. Here we examine the general characteristics and interpretation of FDCs; their construction for stream locations with and without long-term streamflow records is discussed in Section 10.2.5. FDCs can be developed for flows averaged over periods ($\Delta t$) of any length—days, months, or years. However, we will focus exclusively on FDCs of daily average flows ($\Delta t = 1$ day), which are by far the most commonly used.

### Statistical Interpretation

In statistical terms, the FDC is a graph plotting the magnitudes, $q$, of the variable $Q$ (average daily flow, $y$-axis) vs. the fraction of time, $EP_Q(q)$, that $Q$ exceeds any specified value $Q = q$ ($x$-axis). $EP_Q(q)$ is called the **exceedence probability** (or **exceedence**

**TABLE 2-2**

Three Time Series Developed by Discretization of the Continuous Streamflow Record for the Oyster River for the Period 1961–1985[a]

| Year | Annual Average | Peak | Annual Seven-Day Minimum |
|------|------|------|------|
| 1961 | 20.5 | 213 | 1.56 |
| 1962 | 17.5 | 386 | 0.97 |
| 1963 | 22.0 | 450 | 0.76 |
| 1964 | 17.5 | 213 | 0.61 |
| 1965 | 9.1 | 110 | 0.55 |
| 1966 | 10.2 | 106 | 0.43 |
| 1967 | 18.8 | 309 | 0.86 |
| 1968 | 15.3 | 440 | 0.74 |
| 1969 | 18.9 | 240 | 1.26 |
| 1970 | 23.3 | 280 | 0.79 |
| 1971 | 18.0 | 140 | 0.43 |
| 1972 | 22.5 | 233 | 1.17 |
| 1973 | 29.7 | 610 | 0.83 |
| 1974 | 17.8 | 169 | 0.52 |
| 1975 | 16.8 | 299 | 1.01 |
| 1976 | 23.2 | 179 | 0.55 |
| 1977 | 15.6 | 567 | 0.52 |
| 1978 | 24.9 | 278 | 0.56 |
| 1979 | 21.2 | 348 | 1.19 |
| 1980 | 14.1 | 210 | 0.73 |
| 1981 | 17.3 | 615 | 1.64 |
| 1982 | 22.5 | 287 | 1.34 |
| 1983 | 25.0 | 709 | 1.13 |
| 1984 | 31.5 | 410 | 1.57 |
| 1985 | 10.3 | 190 | 0.68 |

[a] All streamflows are in ft³s⁻¹. See Example 2-2 for explanation.

**frequency**), and it is related to the $p$th quantile of streamflow, $q_p$, as

$$EP_Q(q_p) = 1 - F_Q(q_p) = \Pr\{Q > q_p\}, \quad \textbf{(2–24)}$$

where $F_Q$ designates the cumulative distribution function (Section C.2) of $Q$ and $\Pr\{\ \}$ denotes the probability of the condition within the braces.

For FDCs, exceedence frequency refers to the frequency or probability of exceedence in a "suitably long" period rather than the probability of exceedence on any specific day. Seasonal effects and hydrological persistence (Section C.9) cause exceedence probabilities of daily flows to vary as a function of time of year and antecedent conditions; the FDC does not account for those dependencies. By contrast, exceedence probabilities for flood flows and low flows are usually calculated on an an-

nual basis ($\Delta t = 1$ yr) and do not vary from year to year.

It can be shown that the integral of the FDC is equal to the average flow for the period plotted; if that period is the period of measurement, the integral equals $m_Q$ in the regional water balance [Equations (2-11)–(2-16)][7]. The flow exceeded on 50% of the days, $q_{.50}$, is the median flow. The distribution of daily streamflows is almost always highly skewed, so the mean flow is much larger than the median flow—in many humid regions, the mean flow is exceeded on only 20 to 30% of the days. The steepness of the FDC is proportional to the variability of the daily flows.

### Comparison and Controlling Factors

For comparing flow characteristics among streams, it is often useful to plot $q/A$, (where $A$ is drainage area) or $q/m_Q$ on the vertical axis of the FDC. Using either $q/A$ or $q/m_Q$ eliminates the effect of drainage-basin size; using $q/m_Q$ also eliminates the effect of differing per-unit-area precipitation and/or evapotranspiration rates. Either approach can reveal similarities in FDCs for streams in a region that might be useful in inferring the FDC for streams lacking long-term flow records (Box 10-3).

For streams unaffected by diversion, regulation, or land-use modification, the slope of the upper end of the FDC is determined principally by the regional climate, and the slope of the lower end by the geology and topography. The slope of the upper end of the FDC is usually relatively flat where snowmelt is a principal cause of floods and for large streams where floods are caused by storms that last several days. "Flashy" streams, where floods are typically generated by intense storms of short duration, have steep upper-end slopes. At the lower end of the FDC, a flat slope usually indicates that flows come from significant storage in ground-water aquifers (Section 8.5.3) or in large lakes or wetlands; a steep slope indicates an absence of significant storage. However, Dingman (1978b, 1981a) found that more frequent inputs of precipitation can also produce relatively flat slopes in the low ends of FDCs. The effect of reser-

---

[7]Graphically, the area under the FDC is proportional to $m_Q$ if both axes have arithmetic scales.

---

## BOX 2-2
. . . . . . .
### Obtaining U.S. Geological Survey Streamflow Data in Spreadsheet Format

The Water Resources Division (WRD) of the U.S. Geological Survey (USGS) maintains over 5000 stream gages throughout the United States and publishes data on peak flows (floods), daily average flows, and water-quality parameters measured at these gages. The data are published annually in books entitled "Water Resources of [state(s)]" that are available from district offices of the WRD and electronically via the Internet. Spreadsheet files of these data can readily be obtained by downloading the data by using the following steps:

1. Create a directory, or insert a formatted floppy disk in your computer to receive the data.
2. Use your Internet browser to access the address

   http://h2o.usgs.gov/nwis-w/US

   to go to the national data base, from which you can select the state of interest; or replace the "US" with a state abbreviation to go directly to data files for gages in that state.
3. Click your mouse on the appropriate area to access historical streamflow data.
4. Select the gaging station of interest from a map, a statewide list, or lists organized by county or river basin.
5. Select data on peak flows or daily average flows. For peak flows, you can choose between (a) all **peaks**

**above a base value** that the USGS has determined for the gage (there could be more than one peak for each water year, or none); and (b) **annual peaks** (that is, the highest peak flow in each water year).

6. The entire available period of record is displayed; you can select it or change the beginning or ending dates of interest.
7. For spreadsheet downloading, select "tab-delimited" rather than "punchcard image" format.
8. Select a date format.
9. Select "Retrieve Data". An image of the header information and data will appear on your screen.
10. Select "Save As" from the "File" menu of your Program Manager menu bar, designate the directory and file destination, and click "OK". This will save the information in plaintext format.
11. Exit from your browser, open your spreadsheet, and import the file.
12. To work with the data, you will have to convert the plaintext dates to date format and numerical values to numbers. If you have EXCEL 5.0 or higher, the "Text Wizard" "General" format command does all the conversions appropriately.
13. Save the converted file in appropriate spreadsheet format (`*.xls for EXCEL`).

---

voir regulation on FDCs is discussed in Section 10.2.5.

### *Significance*

The streamflow rate is itself highly significant as an indicator of water availability for instream and withdrawal uses, and many other quantities (e.g., water-quality determinants such as dissolved-oxygen, dissolved-solids, and suspended-sediment concentrations; stream-habitat determinants such as velocity and depth; and stream erosive power) are at least partly dependent on flow rate. Thus FDCs are extremely versatile and useful tools for analysis of water-resource problems, as described in detail in Section 10.2.5.

## 2.8 STORAGE, STORAGE EFFECTS, AND RESIDENCE TIME

### 2.8.1 Storage

The control volumes in Figures 1-1 and 1-2 represent storage in the hydrologic cycle, and the entire watershed or region within the dashed boundaries can also be thought of as a storage reservoir. The term "storage" often connotes a static situation, but, in reality, water is always moving through each control volume. In fact, it can be said that *water in the hydrologic cycle is always in motion AND always in storage.*

In many hydrologic reservoirs, such as lakes, segments of rivers, ground-water bodies, and water-

**FIGURE 2-7**

Flow-duration curve of mean daily flows for the Oyster River near Durham, NH, plotted on logarithmic-probability paper.

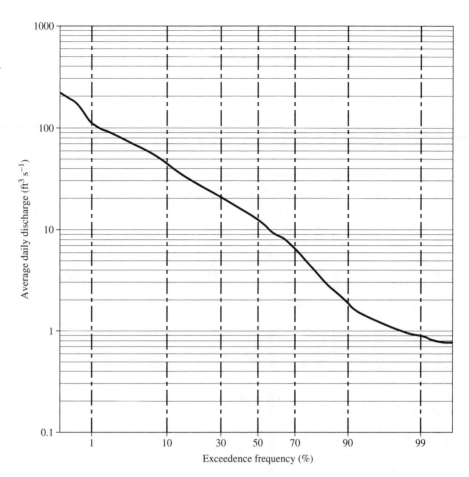

sheds, the outflow rate increases as the amount of storage increases.[8] For these situations, we can model the relation between outflow rate, $q$, and storage volume, $S$, as

$$q = f(S). \qquad (2\text{--}25)$$

In some cases, the nature of the function $f(S)$ can be developed from the basic physics of the situation; in others, such as natural watersheds, Equation (2-25) is merely a conceptual model. The simplest version of Equation (2-25) describes a **linear reservoir**:

$$q = k \cdot S, \qquad (2\text{--}26)$$

where $k$ is a positive constant. Although no natural reservoir is strictly linear, Equation (2-26) is often a

useful approximation of hydrologic reservoirs. It is used as the basis of the "convex watershed model" in Section 9.1.5.

### 2.8.2 Storage Effects

Where Equation (2-25) applies, storage has two effects on outflow time series:

**1.** It decreases the **relative variability** of the outflows relative to the inflows. Standard statistical measures such as the coefficient of variation (ratio of standard deviation to mean; Section C.2.5) or simple ratios determined from FDCs (e.g., $q_{.50}/q_{.95}$) can be used to characterize relative variability quantitatively.

**2.** It increases the **persistence**—the tendency for high values of a time-distributed variable to be followed by high values, and low values by low values—of the outflow time series relative to the inflow time series. As is explained in Section C.9, persistence can be characterized by the **autocorrelation coefficient** of a time series.

---

[8]Note that there are many hydrologic reservoirs for which this relation does *not* hold, e.g., a melting snowpack, the global atmosphere, the global ocean.

### 2.8.3 Residence Time

**Residence time**, or **average transit time**, is a universal relative measure of the storage effect of a reservoir. It is equal to the average length of time that a "parcel" of water spends in the reservoir. If outflow rate and storage are related as in Equation (2-25), the relative effect of the storage on the variability and persistence of the outflows increases as residence time increases. Figure 2-8 illustrates these effects for linear reservoirs [Equation (2-26)].

Residence time can be calculated by dividing the average mass (or volume) of the substance of interest in the reservoir, $S$, by the average rate of outflow, $m_Q$, or inflow, $m_I$, of that substance:

$$T_R = \frac{S}{m_Q} = \frac{S}{m_I}, \qquad (2\text{--}27)$$

where $T_R$ is residence time. Note that Equation (2-27) holds only when $m_Q = m_I$ —that is, when the average rate of change of storage is 0; as noted above, this assumption can usually be made for natural reservoirs over the long term. Note also that Equation (2-27) is dimensionally correct. Residence time is also called **turn-over time**, because it is a measure of the time it takes to completely replace the substance in the reservoir.

For many hydrologic reservoirs, such as lakes, values of $S$ and of $m_Q$ or $m_I$ can be determined readily, and computation of residence time is

**FIGURE 2-8**
(a) Ratio of relative variability of outflows to relative variability of inflows as a function of residence time for a linear reservoir [Equation (2-26)]; and (b) Persistence of outflows (expressed as the autocorrelation coefficient) as a function of residence time for a linear reservoir when inflows have no persistence.

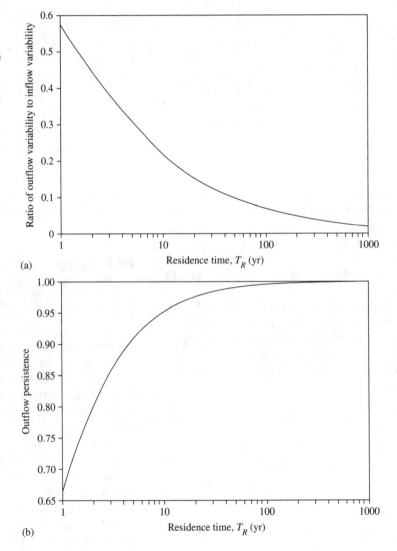

straightforward. For others, such as watersheds and ground-water bodies, it may be difficult to determine the value of $S$ with precision; in these cases, we can usually speak of residence times in relative terms—for example, under natural conditions, outputs from most ground-water reservoirs enter streams and provide the bulk of the streamflow, at least between rain and snowmelt events (Sections 8.5.3 and 9.1). Thus, under similar climatic regimes, streams receiving water from ground-water reservoirs with large residence times (i.e., with large volumes of storage per unit watershed area) will tend to have less variable and more persistent streamflow than those receiving water from reservoirs with shorter residence times.

## 2.9 HYDROLOGIC MODELING

Much current research in hydrology is directed at improving our ability to predict or forecast the effects of land-use and climate changes on the water balance, ground-water levels, streamflow variability, and water quality of regions ranging from hillslopes or landfills to river basins to entire continents. These, and most other applications of hydrology to practical problems of design and forecasting, require the use of hydrologic models. A principal motivation for understanding the physics of hydrologic processes as developed in this text is to provide a sound basis for development and application of such models.

This text explicitly discusses the modeling of snowmelt (Section 5.5), infiltration (Section 6.6), interception (Section 7.6.3), evapotranspiration (Section 7.8), ground-water flow (Section 8.1.6), open-channel flow (Section 9.3), and runoff to streams (Section 9.5). Although our focus will be on the science underlying such modeling, the centrality of modeling to the study and application of hydrology requires that we explore approaches to and issues in modeling more broadly, and that exploration is the goal of this section.

We begin with a consideration of what a hydrologic model is; then we consider how and why models are used in hydrologic science and engineering, the various types of models, and the modeling process. The discussion concludes with an introduc-

tion to the BROOK90 model, which links the land-surface hydrologic processes that are the focus of this text into an integrated watershed model that can be used for exploring the impacts of climate and land use on regional hydrology.

### 2.9.1 What Is a Model?

In this text, the term "model" will refer to **simulation models**. Such a model is a representation of a portion of the natural or human-constructed world "which is simpler than the prototype system and which can reproduce some but not all of the characteristics thereof" (Dooge 1986). The essential feature of a simulation model is that it produces an output or a series of outputs in response to an input or series of inputs. The characteristics and use of the three major classes of simulation models that have contributed to scientific and applied hydrology are described in the following paragraphs.

A **physical model** is a tangible constructed representation of a portion of the natural world. If it is constructed at a larger or smaller scale than the natural system, formal rules of scaling based on dimensional analysis (see Appendix A; King et al. 1960) are used to relate observations on the model to the real world. Physical models have been important means to understanding problems of hydraulics and fluid mechanics, and they are often used to help design complex engineering structures, particularly those involving open-channel flow. Ground-water hydrologists have used physical models to simulate two-dimensional ground-water flow under various boundary conditions. One-to-one-scale physical models in the form of sprinkler-plot studies have been used to understand the process of infiltration (e.g., Nassif and Wilson 1975; see Figure 6-19), and small-scale physical models of watersheds have been used to elucidate some basic characteristics of watershed response to rainfall (Amorocho and Hart 1965; Grace and Eagleson 1966; Dickinson et al. 1967; Chery 1967, 1968; Black 1970, 1975).

**Analog models** use observations of one process to simulate a physically analogous natural process. For example, the flow of electricity as given by Ohm's Law is exactly analogous to Darcy's Law of ground-water flow, so the distribution of electrical potentials (voltage) on specially designed conductive paper can be used to determine the patterns of ground-water potentials (and hence of ground-water flow) under various boundary conditions.

A **mathematical model** is an explicit sequential set of equations and numerical and logical steps (or rules or "recipes") that converts numerical inputs representing flow rates or states of storages to numerical outputs representing other flow rates or storage states. The "guts" of a mathematical model are the equations whose forms represent the qualitative behavior of the flows and storages and the **parameters**—numerical constants—in these equations that dictate the quantitative behavior.

As the availability of more powerful digital computers, modeling techniques, and software has rapidly increased, the use of both physical and analog models in hydrology has been largely replaced by that of computer-implemented mathematical models, which are usually cheaper and much more flexible. Such models are usually mathematical representations of 'box-and-arrow' diagrams similar to those shown in Figures 1-1 and 1-2. Unless otherwise stated, our subsequent discussion of models will focus on mathematical models.

Perhaps the best metaphor for hydrologic models of all types is a map: A model is to hydrologic reality as a map is to the actual landscape. A mental comparison of a map of a region you're familiar with to the actual region gives a good sense of how a model approximates reality. The map metaphor also makes clear two essential characteristics of models:

**1.** *A model, like a map, is designed for a specific purpose.* A model emphasizes features appropriate to its purpose, while omitting other features: a road map shows road types, route numbers, and locations of cities and towns, but usually does not show topography, land cover, or other features that might be extremely important for purposes other than finding your way by car from point A to point B. (Actually, topography might be very important for such a purpose, yet still be omitted from your map!)

**2.** *A model, like a map, is constructed at a particular scale.* Neither represents features that are not visible at that scale—and some of the omitted features could be important in many contexts.

The issue of scale in modeling is an active area of research and discussion in the hydrological liter-

ature (e.g. Giorgi and Avissar 1997; Bergstrom and Graham 1998).

### 2.9.2 Purposes of Models

Hydrologic simulation models are developed either (1) to guide the formulation of water-resource-management strategies (including the design of structures) or (2) as tools of scientific inquiry.

Virtually all applications of hydrology to practical water-resource problems involve the use of models. These applications require either predictions or forecasts:

**Predictions** are estimates of the magnitude of some hydrologic quantity (e.g., the peak flow) that is either (1) associated with a particular exceedence probability or statistic of the quantity or (2) produced by a hypothetical rainfall or snowmelt event (often called the **design storm**).[9] Predictions are the basis for the design of civil engineering works such as reservoirs and reservoir spillways and of land-use plans (e.g., floodplain zoning) and for the assessment of the hydrologic impacts of land-use and climate changes.

**Forecasts** are estimates of the response to an actual anticipated event; e.g., the peak flow rate that will result from the rain that is expected in the next 24 hr on a given watershed. Forecasts are used to guide the operation of reservoir systems and to provide flood warnings to floodplain occupants.[10]

In the scientific context, models are used along with data to test hypotheses about the processes operating in some portion of the hydrologic cycle (Beven 1989). Models are developed to give explicit form to concepts of hydrologic function, and comparison of modeled hydrologic response with observed responses might confirm, or suggest revision

---

[9]Design storms are often specified as the event with a particular exceedence frequency (i.e., by a quantile of the input event such as the 10-yr, 1-hr rainfall on a given watershed). Approaches to estimating exceedence probabilities of rainfalls of specified durations are discussed in Section 4.4.4.

[10]Forecasting models are also commonly used for **hindcasts** (or **backcasts**), in which the objective is to estimate an unmeasured hydrologic response to a past event.

of, hypotheses about the importance of various processes or the ways in which those processes are related. For example, you might assume, from a priori knowledge of the topography and geology of a watershed, that ground-water outflow (deep seepage) is not significant, develop a model of runoff processes that omitted deep seepage, and find that measured streamflow was consistently less than that predicted by the model. This might lead you to conduct further field studies to identify the reason for the discrepancy and perhaps ultimately to revise your concept of watershed processes by including ground-water outflow. Wigmosta and Burges (1997) reported an excellent example of the interactive use of models and field measurement in understanding runoff processes.

Development of hydrologic models has been a natural consequence of the widespread need for predictions and forecasts, along with (1) the complexity and spatial and temporal variability of hydrologic processes and (2) the limited availability of spatially and temporally distributed hydrologic, climatologic, geologic, pedologic, and land-use data.

### 2.9.3  Types of Models

Any given model can usually be described by one or more terms from each of the six categories in Table 2-3. The terms denoting physical domain and process should be reasonably self-explanatory (and will become more so as you progress through this text); those in the other categories are defined in Box 2-3.

### 2.9.4  The Modeling Process

The modeling process is diagrammed in Figure 2-9; its major elements are (1) conceptualization of the problem, (2) selection or development of the appropriate model, (3) parameter estimation ("calibration"), and (4) acceptance testing ("validation").[11]

---

[11]Matalas and Maddock (1976) suggested that the terms "parameter estimation" and "acceptance testing" are preferable to "calibration" and "validation," respectively, because they "remind us that a model is an abstraction of the physical process and not the physical process per se".

#### Conceptualization of the Problem

The most important element in the modeling process is the determination of the overall form and essential components of the model. These decisions must be based on a clear idea of the scientific or engineering purpose of the model, and this idea must be translated into an explicit formulation of the nature and form of the model output that is required—specifically,

- the type of information required (e.g., peak flows, flow volumes, ground-water heads, soil-water contents, evapotranspiration rates);
- the required accuracy and precision of the output;
- the locations for which the output is required;
- the time intervals for which the output is required.

The model conceptualization is dictated also by the nature and form of the information that is available about the system being modeled and the nature and form of the available input data—and, we must not forget, by the resources and time available to collect needed information that is not already available.

#### Model Selection or Development

The hydrologic literature abounds with descriptions of models; once the basic model requirements are established, one can usually develop the requisite software by implementing approaches developed by others with modifications to apply to the situation of interest. Many models are readily available as computer software designed to be easily modified to apply to a particular situation; descriptions of many of these were given by DeVries and Hromadka (1992) for streamflow and watershed models and by Anderson et al. (1992) for subsurface-flow models.

#### Parameter Estimation and Acceptance Testing

Both parameter estimation and acceptance testing require measured values of input and output quantities for the prototype system of interest and involve numerical and/or graphical comparison of measured outputs to modeled outputs. This requires splitting the measured data into a "parame-

**TABLE 2-3**
Terms Used to Characterize Hydrologic Models. See Box 2-3 for definitions of terms.

| Physical Domain | Process | Simulation Basis |
| --- | --- | --- |
| Vegetative canopy | Interception | Physically based |
| Snowpack | Snowmelt | Conceptual |
| Unsaturated zone | Infiltration | Empirical/regression |
| Aquifer | Overland flow | Stochastic—time series |
| Hillslope | Unsaturated flow | |
| Stream reach | Transpiration | |
| Stream network | Ground-water flow/head | |
| Lake or reservoir | Evaporation | |
| Watershed | Open-channel flow | |
| Region/continent | Stream hydrograph | |
| | Integrated watershed/region | |

| Spatial Representation | Temporal Representation | Method of Solution |
| --- | --- | --- |
| Lumped | Steady state | 0-dimensional |
| Distributed | Steady state—seasonal | Formal-analytical |
| Coordinate system | Single event | Formal-numerical |
|   1-dimensional | Continuous |   Finite difference |
|   2-dimensional | |   Finite element |
|   3-dimensional | |   Other |
| | | Hybrid |

ter-estimation set" and an "acceptance-testing set." There are no firm rules for the proportion of the total data allocated to each set, but usually no more than half the data is allocated for acceptance testing. For some types of data the allocation can be done randomly, but for data representing a time sequence it is usually necessary to select a continuous period for parameter estimation and a prior or subsequent period for acceptance testing.

In both parameter estimation and acceptance testing, it is important to focus on the purposes of the model. For example, the following aspects of model output might be more or less important in different contexts:

- the ability to reproduce the long-term or spatial mean value of a quantity;
- the ability to reproduce overall variability (e.g., the standard deviation or range of the quantity);
- the ability to minimize *actual errors* (errors as measured in the units in which the quantity is measured; e.g., streamflows in $m^3s^{-1}$);
- the ability to minimize *relative errors* (errors expressed as a percentage of the mean);
- the ability to reproduce high values of a quantity (e.g., peak streamflows);
- the ability to reproduce low values of a quantity (e.g., drought streamflows);

- the ability to reproduce patterns of seasonal or spatial variability.

It is also important to remember that measured values of model inputs and outputs are themselves more or less subject to errors due to instrumental limitations and the inherent inability of observational networks like rain gages or wells to capture the spatial variability of input or output quantities. To the extent that such errors exist, parameter selection and evaluation of model performance will be subject to error.

**Parameter Estimation**    The objective of parameter estimation is to determine appropriate values for model parameters whose values are not known a priori—for example, hydraulic conductivity in Darcy's Law of ground-water flow [Equation (8-1)] or the proportionality constant in the linear-reservoir equation [Equation (2-26)]. The input data of the parameter-estimation set are entered into the model and the values of parameters are systematically adjusted to determine which values give the "best" fit between the modeled and the measured outputs according to predetermined criteria. Fit can be judged qualitatively by visual comparison of measured and simulated hydrographs or flow-duration curves or of scatter plots of simulated vs. actual output quantities. Numerical measures of best fit are reviewed by Martinec and Rango (1989) and discussed in Section C.10.

## BOX 2-3

· · · · · · ·

### Definitions of Modeling Terms in Table 2-3

**Simulation Basis**

***Physically Based***  Uses equations derived from basic physics [e.g., conservation of mass, energy, or momentum; force balance; diffusion (see Table 2-1)] to simulate flows and storages.

***Conceptual***  Uses "reasonable" a priori relationships to simulate flows and storages. Example: Outflow from storage modeled as proportional to the amount of water in storage [Equation (2-26)].

***Empirical/Regression***  Uses approximate relationships developed from observations to simulate flows and storages. Statistical regression techniques are often used to develop these relationships. Example: Snowmelt rates modeled as proportional to air temperature [Equation (5-57)].

***Stochastic Time Series***  Uses formal time-series-analysis techniques to characterize the behavior of flows and storages. Example: Estimation of ground-water recharge via impulse-response analysis [Equation (8-39); see Salas (1992) for a review of these techniques.]

**Spatial Representation**

***Lumped***  Region or watershed represented as a point. Spatially varying inputs, soils, vegetation, topography, and so on are each characterized by a single parameter that is an average or representative value. Also called '0-dimensional' models.

***Distributed***  Provides some representation of the spatial variability in a region. Can range from dividing the region into two subregions (e.g., upland and lowland) to more elaborate representations of spatial variability (e.g., subdividing the region by a grid system, with model parameters that vary from cell to cell). Sometimes accomplished by linking lumped models.

***Coordinate System***  Formal coordinate system of 1, 2, or 3 dimensions is used to represent space. Used where formal mathematical relations are basis for model. Coordinate system is usually orthogonal (Cartesian), but radial coordinates are used for ground-water models involving flows to or from wells.

**Temporal Representation**

***Steady State***  Outputs are one or more values that represent a long-term average or the ultimate magnitude of a quantity. Example: Global average temperature, simulated as response to changes in solar output or changes in land or cloud cover (Box 3-1).

***Steady-State Seasonal***  Outputs are long-term seasonal averages of a quantity. Example: The Thornthwaite water-balance model (Box 7-3) predicts average monthly values of evapotranspiration in response to average monthly values of precipitation and temperature.

***Single-Event***  Simulates time-varying response of a system to an isolated input. Example: SCS model (Section 9.6.2) simulates streamflow response to a single design storm.

***Continuous***  Outputs are a sequence of responses to a sequence of inputs over a specific period. The time step of the sequence may be days, months, years, or other periods. Example: The BROOK90 model (Section 2.9.5) simulates one or more years of daily hydrologic responses to specified daily weather data.

**Method of Solution**

***0-Dimensional***  Computations not based on formal coordinate system, usually employed in lumped models. Examples: Thornthwaite model of evapotranspiration (Box 7-3); SCS flood-hydrograph model (Section 9.6.2).

***Formal–Analytical***  Basic differential equations in coordinate system solved *analytically*. Example: Philip solution of Richards Equation of infiltration [Equation (6-31)].

***Formal–Numerical***  Basic differential equations in coordinate system solved by *finite-difference* or *finite-element* discretization schemes. (See Wang and Anderson 1982; Bear and Verruijt 1987.) Example: Regional ground-water flow models (Section 8.2).

***Hybrid***  0-dimensional and formal solutions used for different processes within model. Example: BROOK90 model (Section 2.9.5) uses formal–numerical solutions for soil-water movement (Box 6-3) and 0-dimensional methods for other processes (Box 4-1, 5-1, 7-2, 7-4).

**FIGURE 2-9**
Flow chart for the modeling
process.

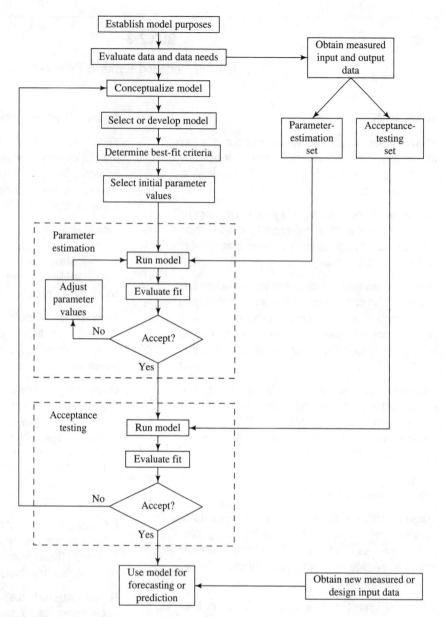

Although conceptually straightforward, the parameter-estimation process is often fraught with
difficulty and ambiguity, especially in multiparame-
ter models:

- Very different sets of parameter values may
  give nearly equivalent fits.
- Model outputs may be insensitive to the values of one or more parameters.
- One or more best-fit values may differ greatly
  from what seems intuitively reasonable.
- Best-fit values may differ in different time
  periods.

When these situations occur, confidence in a
model's ability to simulate the situation of interest
is diminished.

**Acceptance Testing**  Once the parameters are selected, performance testing leading to acceptance
or rejection of the model for a particular application should be evaluated by graphical and/or numerical comparison of modeled and measured
outputs *for situations not used in parameter estimation*. As an example, Figures 2-10 and 2-11 compare
streamflows simulated by the BROOK90 model
(described in Section 2.9.5) with measured values

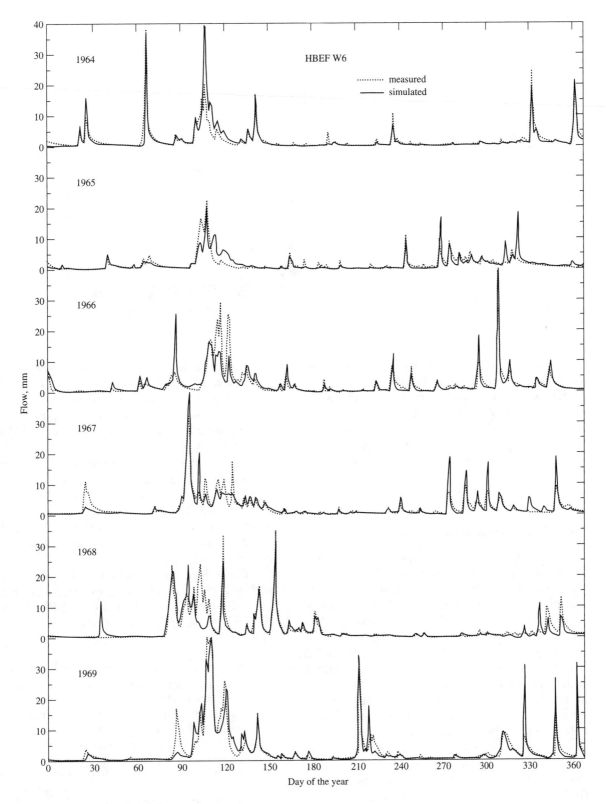

**FIGURE 2-10**
Comparison of measured daily flows at Hubbard Brook Experimental Forest, West Thornton, NH, (dotted lines) with flows simulated by BROOK90 (solid lines) for 6 years. From Federer (1995).

**FIGURE 2-11**

Scatter plot of monthly flows simulated by BROOK90 vs. measured flows for 1964-1969 at Hubbard Brook Experimental Forest, West Thornton, NH. Line shows 1:1 ratio. From Federer (1995).

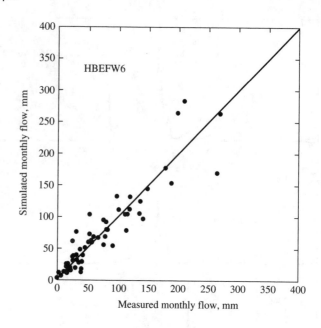

in a research watershed. If a model does not satisfactorily simulate the measured values, a new model, perhaps based on a revised conceptualization of the situation, should be developed.

Klemeš (1986b) provided an excellent discussion of the philosophy and process of testing simulation models, and studies comparing models are published by the World Meteorological Organization (1986b, 1992) and Perrin (2001).

## 2.9.5 The BROOK90 Model

This book emphasizes the scientific understanding of individual processes (precipitation, snowmelt, infiltration, evapotranspiration, and others) in the land phase of the hydrologic cycle. In order to provide a means of synthesizing this understanding into a form that can be used to answer hydrologic questions, the text also describes a model called BROOK90. This model incorporates our scientific understanding into an integrated model that connects these processes to simulate the behavior of a watershed. We also give information about how BROOK90 can be accessed on the Internet (in Exercise 2-7) and how it can be used to predict watershed responses to climatic inputs and land-use changes. Here we introduce the basic features of BROOK90; a more detailed discussion of how the model simulates each specific process is given in the chapter that treats that process.

As can be seen in Table 2-3, BROOK90 is an integrated model that provides detailed simulations of the processes of interception, snowmelt, infiltration, unsaturated soil-water flow, transpiration, and soil evaporation and more approximately simulates overland flow and ground-water flow. Its structure is shown in Figure 2-12—note the similarity to Figure 1-2. It is largely a physically-based, spatially-lumped, continuous-time (daily time step) model that uses 0-dimensional computational approaches for processes other than soil-water movement; the latter is simulated by using a formal finite-difference numerical approach.

BROOK90 was developed by Dr. C.A. Federer, formerly of the U.S. Forest Service, who actively updates the model and maintains the BROOK Web site. The following overview is taken from Federer (1995):

The hydrologic model BROOK90 simulates the water budget on a unit land area at a daily time step. Given precipitation at daily or shorter intervals and daily weather variables, the model estimates interception and transpiration from a single-layer plant canopy, soil and snow evaporation, snow accumulation and melt, soil-water movement through multiple soil layers, stormflow by source area or pipe-flow mechanisms, and delayed flow from soil drainage and a linear ground-water storage.

BROOK90 simulates the land phase of the precipitation-evaporation-streamflow part of the hydro-

**FIGURE 2-12**

Flow chart for the BROOK90 model. Terms are defined below. Boxes are accumulated storages. All quantities expressed in mm of water. From Federer (1995).

PREC = Precipitation.
EVAP = Evapotranspiration.
SEEP = Ground-water outflow.
FLOW = Stream outflow.
INTS = Snow on vegetative canopy
INTR = Rain on vegetative canopy.
SNOW = Snowpack on ground.
SWATI = Soil-water storage (1 to N layers.)
GWAT = Ground-water storage.

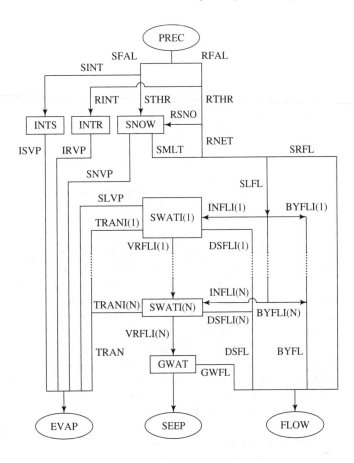

logic cycle for a point or for a small, uniform (lumped parameter) watershed. There is no provision for spatial distribution of parameters in the horizontal. There is no provision for lateral transfer of water to adjacent downslope areas. Instead, BROOK90 concentrates on detailed simulation of evaporation processes, on vertical water flow, and of local generation of stormflow. Below ground, the model includes one to many soil layers, which may have differing physical properties.

BROOK90 has been designed to be applicable to any land surface. The model has numerous parameters, but all parameters are provided externally, are physically meaningful, and have default values. Parameter fitting is not necessary to obtain reasonable results. However, a procedure for modifying important parameters to improve the fit of simulated to measured streamflow is described.

BROOK90 is designed to fill a wide range of needs: as a research tool to study the water budget and water movement on small plots, as a teaching tool for evaporation and soil-water processes, as a water-budget model for land managers and for predicting climate-change effects, and as a fairly complex water-budget model against which simpler models can be tested.

Some published studies that have used current or earlier versions of the BROOK model include Federer and Lash (1978), Devillez and Laudelout (1986), Hornbeck et al. (1986), Hornbeck et al. (1987), Forster and Keller (1988), Yanai (1991), and Lawrence et al. (1995).

### 2.9.6 Final Words of Caution

Developing and working with computer simulation models is challenging and fun, and it can be done in a comfortable room with a coffee cup at hand. Collection of data in the field is also challenging, but often uncomfortable, tedious, frustrating, and expensive. Thus, although computer models have greatly facilitated the science of hydrology and its application and will play an increasing role in the future, we must continually remind ourselves that the goal is to understand

and predict nature, not to demonstrate our cleverness. As it was nicely phrased by Dooge (1986, p. 49S),

> Many . . . modelers seem to follow . . . the example of Pygmalion, the sculptor of Cyprus, who carved a statue so beautiful that he fell deeply in love with his own creation. It is to be feared that a number of hydrologists fall in love with the models they create. In hydrology, . . . the proliferation of models has not been matched by the development of criteria for the evaluation of their effectiveness in reproducing the relevant properties of the prototype.[12]

Models are essential tools for almost all practical applications of hydrology and can be powerful aids in scientific analysis. However, anyone seeking to use a model to provide predictions or forecasts that will be used for critical design or operational applications or scientific decisions should first review the discussions by Matalas and Maddock (1976), Klemeš (1986b), Dooge (1986), Beven (1993), Oreskes et al. (1994), and Perrin et al. (2001). The collective wisdom of these discussions can be summarized as follows:

> Although acceptable parameter values can be determined for almost any model, in most cases the parameters are not unique. In addition, because of the inevitable errors in measured data and the impossibility of representing the space-time continuum of nature as a finite array of space-time points, no model can be validated as a true simulation of nature.

## EXERCISES

Exercises marked with ** have been programmed in EXCEL on the CD that accompanies this text. Exercises marked with * can advantageously be executed on a spreadsheet, but you will have to construct your own worksheets to do so.

**\*2-1**  Using Equation (2-16) and assuming no model error, compute (a) the estimated evapotranspiration and

[12]Note that quantitative criteria for evaluating models are discussed in Section C. 10.

(b) the absolute and relative uncertainties in the estimate for the following situations:

| Location | a Connecticut River, USA | b Yukon River, Canada | c Euphrates River, Iraq | d Mekong River, Thailand |
|---|---|---|---|---|
| Watershed Area $(km^2)^*$ | 20,370 | 932,400 | 261,100 | 663,000 |
| Precipitation, $m_P (mm\ yr^{-1})^*$ | 1,100 | 570 | 300 | 1,460 |
| Relative Error in $P$, $u_P^+$ | 0.1 | 0.2 | 0.1 | 0.15 |
| Streamflow, $m_Q (m^3 s^{-1})^*$ | 386 | 5,100 | 911 | 13,200 |
| Relative Error in $Q$, $u_Q^+$ | 0.05 | 0.1 | 0.1 | 0.05 |

\* Published values. +Assumed - actual uncertainty unknown.

**\*\*2-2**  Using the methods in Box C-1, compute the sample median and 0.25- and 0.75-quantile values of the three time series in Table 2-2. The data for this table are in file `Table 2-2.xls` on the CD accompanying this text.

**\*\*2-3**  Using the methods in Box C-2, compute the mean, standard deviation, coefficient of variation, and skewness of the three time series in Table 2-2. The data for this table are in file `Table 2-2.xls` on the CD accompanying this text. Which time series is the most variable, relatively speaking? Which is the most skewed?

**\*\*2-4**  Using the methods of Box C-5, compute the sample autocorrelation coefficients of the three time series in Table 2-2. The data for this table are in file `Table 2-2.xls` on the CD accompanying this text. Which time series shows the most persistence? (Hint: If there are $N$ items in the time series, you can calculate the autocorrelation coefficient by using the spreadsheet's correlation function: specify the first data range as values 1 through $N - 1$, the second data range as values 2 through $N$.)

**\*2-5**  Using the instructions in Box 2-2, obtain a time series of annual peak flows for a stream of interest. Using the methods of Box C-2, estimate the mean, standard deviation, and coefficient of variation of the time series.

**2-6**  (a) Obtain appropriate topographic maps and trace out the watershed of a stream of interest. (b) Measure the drainage area by using a digitizer, planimeter, or grid overlay. (c) What geologic information is available that could help you assess whether ground-water outflow is significant?

**2-7**  Access the "Compass Brook" home page on the World Wide Web at

http://users.rcn.com/compassbrook/compassb.htm

and click on "BROOK90". Read the information about the BROOK90 model and how to download it and obtain the documentation. Check with your instructor to see whether your institution has a site license and a copy of the documentation.[13] Read the first chapter of the documentation (Chapter B90). Follow the instructions to download Version 3.2*x* into a directory B90V3_2 that you have created on your hard drive. (*x* designates the current version of the program; *x* = 5 at the time of writing.) In your File Manager, double click on B90V3_2x.EXE to execute the model. In the "Run time" box, input 2192 days. This will allow a continuous run for six years (1964–1969) for Watershed 6 (W6) at Hubbard Brook Experimental Forest (HBEF) in West Thornton, NH. Click on "Output" and select Annual, Monthly, and Daily values of *PREC, MESFL, FLOW, EVAP,* and *SNOW* (note all values are in mm). Click on "Run" –"New run" and watch the hydrologic years 1964–1969 unfold. Identify what each color graph is showing and note the annual patterns of the quantities.

(a) Compare the curves representing simulated streamflow (*FLOW*) and measured streamflow (*MESFL*). Note the relative values of *FLOW* and *MESFL*. Are there consistent patterns in these values? What might cause the patterns observed?

(b) Make a scatterplot of monthly estimated streamflow (*FLOW*) vs. measured streamflow (*MESFL*). Assess the goodness of fit qualitatively by reference to the graphical printout. Supplement this qualitative analysis by computing the Nash–Sutcliffe criterion (Section C.10).

(c) Make separate scatterplots of *FLOW* vs. *MESFL* for each of the four seasons: Winter (Dec, Jan, Feb); Spring (Mar, Apr, May); Summer (Jun, Jul, Aug); Fall (Sep, Oct, Nov), and make the same kinds of qualitative and quantitative analyses as for the aggregated data. Which season gives the best results, and the worst? What do these results suggest about the model's shortcomings?

(d) Compute the water balance for each year (*PPTN, EVAP, MESFL*) and the change in storage at the end of each year. Note any differences between the balances for the wettest and driest years. Is the storage change at the end of each year significant?

(e) Qualitatively and quantitatively compare *FLOW* and *MESFL* for the wettest and driest years with the overall results of part (b).

(f) Compute *PPTN − MESFL* for all months and compare with monthly evapotranspiration (*EVAP*). Generate a time-series graph of both values and analyze the data for temporal patterns. Try to assess what accounts for these patterns.

(g) Describe and discuss the causes of the temporal patterns of soil moisture (*SWAT*) on the graphical printout.

---

[13]You do not need a license for one-time use in an exercise, but *you should obtain a license if you plan to use the model regularly for educational or scientific purposes.* See the web site for instructions.

# 3

# Climate, the Hydrologic Cycle, Soils, and Vegetation: A Global Overview

## 3.1 BASIC ASPECTS OF GLOBAL CLIMATE

### 3.1.1 The Energy Budget of the Earth

The sun radiates energy approximately as a black-body with a temperature of 6000 K (Figure 3-1, curve $a$); its radiation spectrum extends from the ultraviolet to the infrared, with a maximum in the visible range. (See Section D.1 for a review of the physics of radiation.) However, gases in the earth's atmosphere are strong absorbers of energy at specific wavelengths in this range, so that the radiation striking the earth's surface is depleted in portions of the spectrum (Figure 3-1, curve $b$). In particular, normal oxygen ($O_2$) and ozone ($O_3$) in the lower stratosphere shield terrestrial biota from much of the energy in the ultraviolet range, which is damaging to most forms of life. Water vapor also absorbs some of the sun's energy in the "near infrared" range. Thus virtually all the sun's energy arriving at the surface is at wavelengths less than 4 μm; this energy is referred to as **solar radiation** or **shortwave radiation**.

The sun's energy arrives at the outer edge of the atmosphere at an average rate, $S$, of $1.74 \times 10^{17}$ W ($4.16 \times 10^{16}$ cal s$^{-1}$). This quantity, divided by the area of the planar projection of the earth, $1.28 \times 10^{14}$ m$^2$, is called the **solar constant**, $I_{sc}$.[1] Thus $I_{sc} = 1367$ W m$^{-2}$ (2821 cal cm$^{-2}$ day$^{-1}$). To simplify our discussion, we take $S = 100$ units of radiant energy input and trace out the fate of this energy in the earth–atmosphere system prior to its ultimate reflection or reradiation back to space (Figure 3-2). All these energy values are estimates of globally and seasonally integrated averages of values that are highly variable in time and space.[2]

Of the 100 units of incident energy, 26 are reflected from the atmosphere (20 by clouds) back to space. Clouds absorb 4, and atmospheric gases about 16, of the remaining units, so 54 units are incident upon the earth's surface. The surface reflects 4 of these, so 50 units are absorbed at the surface to cause warming, evaporation of water, and melting of snow and ice.[3]

To a high degree of approximation, the rate of energy output from the earth–atmosphere system equals the rate of input; thus the 70 units of solar energy absorbed by the earth and atmosphere are eventually reradiated to outer space.[4] The overall temperature of the earth–atmosphere system (the **planetary temperature**) is about 253 K (−20 °C) (Miller et al. 1983), so this radiation is at a much

---

[1]Recent measurements and analyses suggest that the energy output of the sun varies about 1 percent over an 80- to 90-year cycle (Reid 1987).

[2]A detailed model for estimating the daily clear-sky solar radiation incident on a sloping portion of the earth's surface is developed in Appendix E.
[3]Recent research (Cess et al. 1995) strongly suggests that clouds absorb 15 units of radiant energy, rather than 4. This would reduce the amount absorbed at the surface to 39 units and commensurately reduce the transfer from the surface to the atmosphere.
[4]There is reason to believe that the earth–atmosphere system is warming. Although this warming could be climatologically and hydrologically significant (see Section 3.2.9), it represents only a negligibly small fraction of the total energy budget and thus can be ignored for purposes of energy-budget analysis.

**FIGURE 3-1**
Spectra of energy (a) emitted by a blackbody at 6000 K, (b) received at the earth's surface (global average), (c) emitted by a blackbody at 290 K, (d) emitted to space by the earth–atmosphere system (global average). Upper graph shows absorption spectrum for principal absorbing gases in the atmosphere. Modified from Barry and Chorley (1982) and Miller *et al.* (1983).

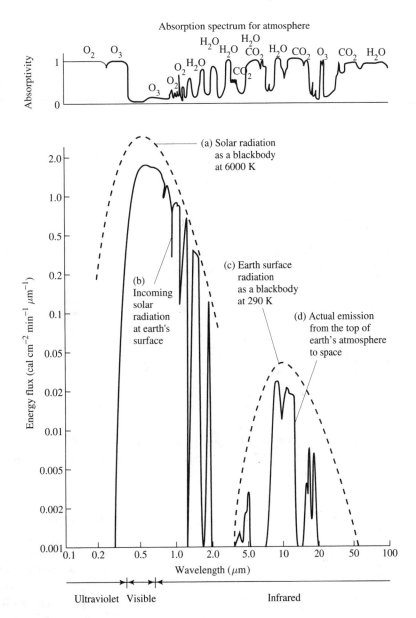

smaller rate and at much longer wavelengths than is the solar radiation [Equations (D-1) and (D-3)]. The equality of the total incoming solar energy and outgoing terrestrial radiation is reflected in the equality of the areas under the two spectral curves when plotted on arithmetic scales as in Figure 3-3.

The steps by which solar radiation is transformed into earth radiation are the critical determinants of the earth's climate. As noted above, 50 units of energy are absorbed to heat the surface and provide latent heat for evaporation and melting. Because the earth's surface, like the system as a whole, is in essential balance, this energy must be transferred away. This transfer is accomplished via three modes: (1) radiation (20 units); (2) latent-heat transfer (24 units); and (3) conduction/convection (6 units). The basic physics of these energy-transfer modes is described in Appendix D.

The average temperature of the earth's surface is about 290 K (17 °C), so the surface radiates approximately as a blackbody at this temperature and emits 20 units of energy in the infrared range between wavelengths of 4 and 50 μm (Figure 3-1, curve *c*). Energy in this wavelength band is referred

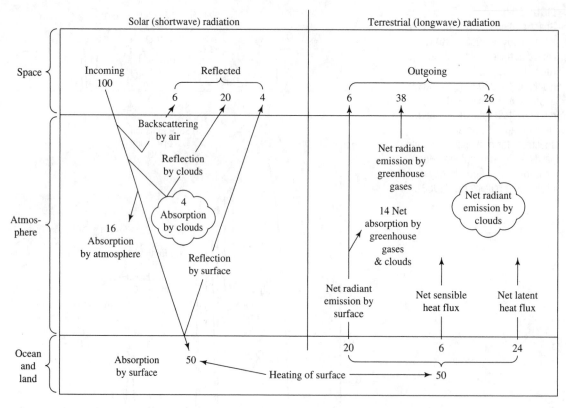

**FIGURE 3-2**
Average global energy balance of the earth–atmosphere system. Numbers indicate relative energy fluxes; 100 units equals the solar constant, 1367 W m⁻². Modified from Shuttleworth (1991); data from Peixoto and Oort (1992).

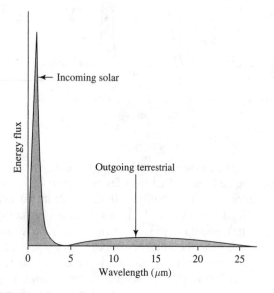

**FIGURE 3-3**
Spectra of incoming solar and outgoing terrestrial radiation plotted on arithmetic scales. Compare Figure 3-1. After Barry and Chorley (1982).

to as **terrestrial** or **longwave radiation**. As shown in the upper portion of Figure 3-1, many naturally occurring and human-introduced gases strongly absorb longwave radiation, so that 14 of the 20 units radiated by the surface are absorbed to heat the atmosphere. The absorption of this energy is called the **greenhouse effect**; the most important "greenhouse gases" are water vapor (which accounts for 65% of the absorption), carbon dioxide (33%), and methane, nitrous oxide, ozone, and chlorinated fluorocarbons (2% combined).

The components of the atmosphere also radiate energy in all directions. The net effect of the exchange of radiant energy between the surface and the atmosphere is the upward transfer of the 20 units of energy, of which 6 are radiated directly to outer space.

The transfer of latent heat via evaporation (mostly from the oceans) adds another 24 units of energy to the atmosphere; this is the largest source of atmospheric energy. Sensible heat transfer (con-

---

## BOX 3-1

· · · · · · ·

### Energy-Balance Model of Global Radiational Temperature

We can apply some of the basic physics of energy transfer discussed in Section D.1 to develop a simple "zero-dimensional" energy-balance model of the temperature of the earth–atmosphere system. The long-term average rate of energy input to this system, $i$ [E T$^{-1}$], equals the solar flux, $S$ [E T$^{-1}$], minus the fraction, $a_p$, of this arriving energy that is reflected by the system:

$$i = S \cdot (1 - a_p). \qquad \textbf{(3B1-1)}$$

To maintain equilibrium, the average rate at which the system radiates energy to outer space, $q$, must equal $i$. From the Stefan-Boltzmann Law (Equation D-1) this rate is

$$q = \sigma \cdot T_p^4 \cdot A, \qquad \textbf{(3B1-2)}$$

where $\sigma$ is the Stefan-Boltzmann constant, $T_p$ is the effective radiating temperature of the system (assuming an emissivity of 1), and $A$ is the surface area of the system.

Equating $i$ and $q$ and solving for $T_p$ yields

$$T_p = \left[ \frac{S \cdot (1 - a_p)}{\sigma \cdot A} \right]^{1/4}. \qquad \textbf{(3B1-3)}$$

Equation (3B1-3) shows how the radiating temperature of the earth–atmosphere system (called the **planetary temperature**) depends on the solar constant and the reflectivity of the system (called the **planetary albedo**).

$S = 1.74 \times 10^{17}$ W, $\sigma = 5.78 \times 10^{-8}$ W m$^{-2}$ K$^{-4}$, and $A$ is $5.10 \times 10^{14}$ m$^2$. From Figure 3-2, we see that the value of $a_p$ is 0.3. Inserting these values in Equation (3B1-3) gives $T_p = 253.6$ K, which is quite close to the value of 253 K quoted in the text. (The difference is due to rounding errors in the numbers in Figure 3-2 and to uncertainty in the true value of $a_p$.)

Exercise 3-1 gives you an opportunity to explore this question by using this simple model. A simple model for calculating the earth's surface temperature is described in Box 3-2.

---

duction/convection) contributes another 6 units because the surface is, on average, warmer than the overlying air.

If we focus on the energy balance of the atmosphere (including clouds) in Figure 3-2, we see that 69% of the input, 44 units, comes from the earth's surface, while only 31% (20 units) is absorbed directly from solar radiation. The 22% due to absorption of longwave radiation via the greenhouse gases is a critical determinant of the earth's climate: As noted by Ramanathan (1988), without the greenhouse effect the earth's surface would have a temperature of −18 °C and be covered with ice. The potential climatic and hydrologic effects of the increases in concentrations of greenhouse gases due to industrial activity and the clearing of forests are reviewed in Section 3.2.9.

The emission spectrum of the earth–atmosphere system as viewed from space is shown in Figure 3-1 (curve *d*); this curve represents the emission spectrum of the surface depleted by absorption by greenhouse gases, and it is this radiation to outer

space that completes the overall energy balance of the earth–atmosphere system.

Box 3-1 develops a simple energy-balance model that shows how the planetary temperature of the earth is related to the solar constant and the reflectivity of the system, and Box 3-2 describes a model for estimating the earth's surface temperature based on the energy balance for a two-layer atmosphere. These models can be used to explore the sensitivity of these temperatures to changes in the solar constant, the albedo, and other factors (Exercises 3-1 and 3-3).

### 3.1.2   Latitudinal Energy Transfer

Figures 3-4 and 3-5 summarize the geometrical relations of the earth's orbit that cause seasonal and latitudinal variations in the receipt of solar energy. Figure 3-4 shows how a given energy flux is spread out over larger areas at high latitudes because the earth is a sphere. This strictly latitudinal effect is modified seasonally because the earth's axis of ro-

## BOX 3-2
. . . . . . . .
### Energy-Balance Model of Earth-Surface Temperature

This model is a modification of the one described by Harte (1985). Harte (1985) shows that it is appropriate, given the physics of radiation, to divide the earth's atmosphere into a lower layer (extending to an altitude of 1.8 km and containing 20% of the air and 50% of the water vapor), which is the major absorber of the terrestrial radiation and an upper layer (containing 80% of the air and 50% of the water). The model is developed by formulating three energy-balance equations: (1) one for the earth–atmosphere system as a whole; (2) one for the upper layer of the atmosphere; and (3) one for the lower layer of the atmosphere. All the energy terms are expressed as long-term average fluxes $[E\ T^{-1}]$, all temperatures are absolute, all emissivities are equal to 1, and it is assumed that each of these systems is in equilibrium.

Energy enters the earth–atmosphere system from above at the rate $S$ and from below at the rate $W$ (which represents the heat generated from nuclear and fossil fuels). Energy leaves the system by three routes: (1) reflected solar radiation; (2) thermal radiation from the top of the atmosphere; and (3) the portion of thermal radiation from the surface that is not absorbed in the atmosphere, $(1 - f) \cdot \sigma \cdot T_s^4 \cdot A$, where $f$ is the fraction of surface radiation absorbed in the atmosphere, $\sigma$ is the Stefan-Boltzmann constant, $A$ is the area of the earth,

and $T_s$ is the surface temperature (absolute). Thus the energy balance for the system is

$$S + W = a_p \cdot S + \sigma \cdot T_u^4 \cdot A$$
$$+ (1 - f) \cdot \sigma \cdot T_s^4 \cdot A, \qquad \text{(3B2-1)}$$

where $a_p$ is the planetary albedo and $T_u$ is the absolute temperature of the upper atmospheric layer.

The upper atmospheric layer absorbs a fraction $k_u$ of the solar radiation that strikes it, and it also receives (1) energy radiated upward from the lower layer and (2) one-half the latent heat that accompanies evaporation from the surface (because this layer holds one-half the atmospheric water vapor). The upper layer loses energy by thermal radiation upward to outer space and downward to the lower layer. Thus the energy balance for the upper layer is

$$k_u \cdot S + \sigma \cdot T_l^4 \cdot A + 0.5 \cdot Q_e = 2 \cdot \sigma \cdot T_u^4 \cdot A, \qquad \text{(3B2-2)}$$

where $T_l$ is the absolute temperature of the lower layer and $Q_e$ is the latent-heat flux from the surface.

Energy enters the lower atmospheric layer from above by the absorption of a fraction $k_l$ of the solar radiation that enters it and by thermal radiation from the

tation is tilted at an angle of 23.5° to the orbital plane (Figure 3-5); the seasons are in fact caused by the contrasts in solar radiation receipt as the northern and southern hemispheres are alternately tilted toward (summer) and away from (winter) the sun.[5] Figure 3-6 quantifies the seasonal and latitudinal variations of solar radiation incident at the top of the atmosphere.

The top curve of Figure 3-7 shows the difference between solar radiation received and the terrestrial radiation emitted at each latitude. The net radiation balance is positive for latitudes below about 35° and negative poleward of that. Because total energy inputs and outputs must be in balance at all latitudes, there is a net poleward transfer of energy from the regions of surplus to those of deficit; the magnitude of this transfer is indicated by the total flux curve in the lower part of Figure 3-7.

This poleward, or meridional, energy transfer is accomplished by air and ocean currents. Roughly two-thirds of the total transfer occurs as sensible- and latent-heat transfer in the atmosphere and one-third as sensible-heat transfer in the oceans; the latitudinal importance of these modes is also indicated in the lower part of Figure 3-7.

---

[5]The orbital tilt is known to vary between 22.1° and 24.3°, with a periodicity of about 40,000 yr. This variability and other periodic fluctuations in the geometry of the earth's orbit affect the amount of solar radiation received seasonally in the two hemispheres over time. It is now widely accepted that these orbital variations controlled the timing of the glacial and interglacial periods of at least the last ice age (Hays et al. 1976; Lamb 1982).

upper layer. From below, energy enters from (1) the absorbed portion of thermal radiation from the surface, $f \cdot \sigma \cdot T_s^4 \cdot A$, (2) one-half the latent-heat flux from the surface, $Q_e$, (3) the sensible-heat flux from the surface, $Q_h$, and (4) the anthropogenic heat flux. Energy is lost from this layer by upward and downward radiation. Thus the energy balance for the lower layer is

$$k_l \cdot S + \sigma \cdot T_u^4 \cdot A + f \cdot \sigma \cdot T_s^4 \cdot A$$
$$+ \, 0.5 \cdot Q_e + Q_h + W = 2 \cdot \sigma \cdot T_l^4 \cdot A. \quad \textbf{(3B2-3)}$$

Equations (3B2-1) to (3B2-3) are a system of three equations in three unknowns, the temperatures $T_u$, $T_h$, and $T_s$; the other quantities are parameters whose values must be given. The values of the temperatures can be found via the following steps:

1. Solve Equation (3B2-1) for $T_u$ in terms of $T_s$ and parameters.
2. Solve Equation (3B2-2) for $T_l$ in terms of $T_u$ and parameters.
3. Substitute the results of Step 1 into the results of Step 2 to give an equation for $T_l$ in terms of $T_s$ and parameters.

4. Solve Equation (3B2-3) for $T_s$ in terms of $T_u$, $T_h$, and parameters.
5. Put the results of Step 1 and the results of Step 3 into the results of Step 4 and simplify to give the equation for $T_s$ as a function of parameters.

The resulting expression is

$$T_s = \left[ \frac{(3 - 3 \cdot a_p - 2 \cdot k_u - k_l)}{\cdot S - 1.5 \cdot Q_e - Q_h + 2 \cdot W} \middle/ (3 - 2 \cdot f) \cdot \sigma \cdot A \right]^{1/4}. \quad \textbf{(3B2-4)}$$

The values given by Harte (1985), which are generally consistent with those in Figure 3-2, are:

| | |
|---|---|
| $S = 1.74 \times 10^{17}$ W | $a_p = 0.3$ |
| $Q_e = 4.08 \times 10^{16}$ W | $k_u = 0.18$ |
| $Q_h = 8.67 \times 10^{15}$ W | $k_l = 0.075$ |
| $W = 1.07 \times 10^{13}$ W | $f = 0.950.$ |

With these values, Equation (3B2-4) gives $T_s = 290.4$ K, close to the actual value of 290 K.

Exercise 3-2 asks you to derive Equation (3B2-4) by following the above steps. Exercise 3-3 gives you an opportunity to use the model to explore the greenhouse effect.

### 3.1.3 The General Circulation and the Distribution of Pressure and Temperature[6]

The unequal latitudinal distribution of radiation and the requirement for the conservation of angular momentum on the rotating earth give rise to a system of three circulation cells in the latitude bands 0°–30°, 30°–60°, and 60°–90° in each hemisphere, along with the jet streams and characteristic prevailing surface-wind directions (Figure 3-8). This system is called the **general circulation** of the at-

mosphere. The cell nearest the equator is responsible for most of the poleward energy transfer between latitudes 0° and 30°, but mechanisms other than the general circulation dominate the atmospheric transfer at higher latitudes. As indicated in Figure 3-9, winds circulating in large-scale horizontal eddies—both the quasi-stationary zones of high and low pressure discussed later and the moving cyclonic storms that dominate weather systems in the mid-latitudes—are the major agents of transport above latitude 30° (Barry and Chorley 1982).

Figure 3-8 shows that the general circulation results in regions of rising air near the equator and near latitude 60°, and descending air near latitude 30° and the poles. We would expect the zones of ascent to be characterized by relatively low atmos-

---

[6]Much of the discussion in this section is based on Miller et al. (1983).

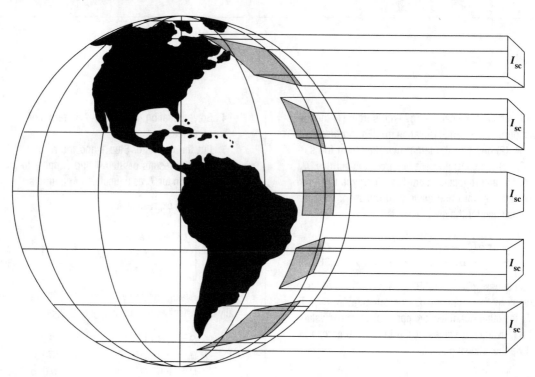

**FIGURE 3-4**
Variation of solar radiation intensity ($[E\ L^{-2}\ T^{-1}]$) with angle of incidence. At higher angles (higher latitudes), a given energy flux is spread over a larger area. From Day and Sternes (1970), used with permission.

pheric pressures at the surface, and those of descent by high pressures. Maps of average sea-level pressures (Figure 3-10) generally confirm these expectations, though the zones of high and low pressure actually occur as cells rather than continuous belts.

Horizontal pressure gradients are the basic driving force for winds, and the resultant of these pressure forces with forces produced by the motion itself (centrifugal forces, the Coriolis effect due to the earth's rotation, and friction) produces surface winds that move approximately parallel to the isobars, but with a tendency to spiral inward toward low-pressure centers and outward from high-pressure centers. In the northern hemisphere, the sense of circulation is clockwise around highs (**anticyclonic circulation**) and counterclockwise around lows (**cyclonic circulation**) (Figure 3-11); the circulations are in the opposite senses in the southern hemisphere.

The subtropical high-pressure zone exists as cells over the Pacific and Atlantic Oceans; these cells are especially well defined in the summer, and

occur farther to the north in the summer than in the winter. Winds are moving clockwise around these highs, so the coastal areas of southwestern North America and Europe are subject to dry, cool northerly winds in the summer. Conversely, the southeastern United States, Hawaii, the Philippines, and southeast Asia are subject to warm, moist winds from the tropics and have warm, humid summers with frequent rain.

The subpolar low-pressure zone occurs as cells over the northern Pacific and Atlantic, which are especially evident in winter. These cells, called the Aleutian low (Pacific) and Icelandic low (Atlantic), are "centers of action", where major mid-latitude cyclonic storms develop their greatest intensities.

Figure 3-12 shows the global distribution of mean temperature in January and July. Clearly, this distribution is strongly related to latitude and hence to the average receipt of solar radiation, but it is modified by the distribution of the continents and oceans. Because of water's very high heat capacity (Section B.2.4), the annual temperature range of

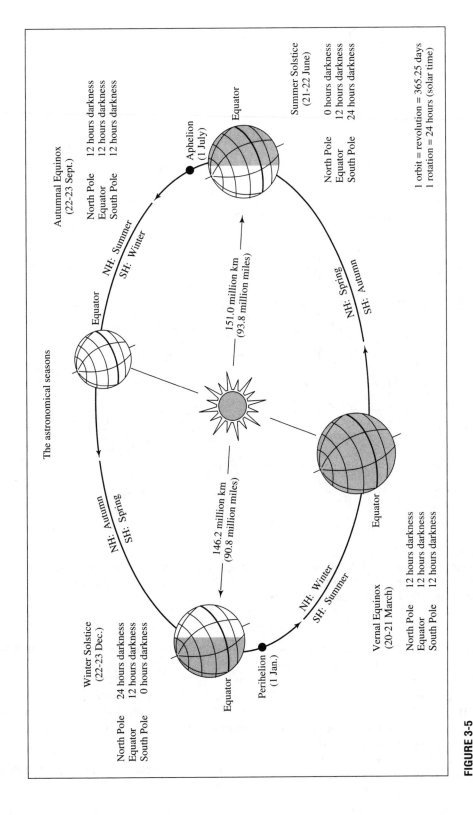

**FIGURE 3-5**

Revolution of the earth around the sun, showing that the occurrence of summer and winter in the northern and southern hemispheres is determined by the 23.5° tilt of the rotational axis toward or away from the sun. Reprinted with permission of Macmillan Publishing Co. from *Climatology* by Oliver and Hidore, © 1984 by Bell & Howell Co.

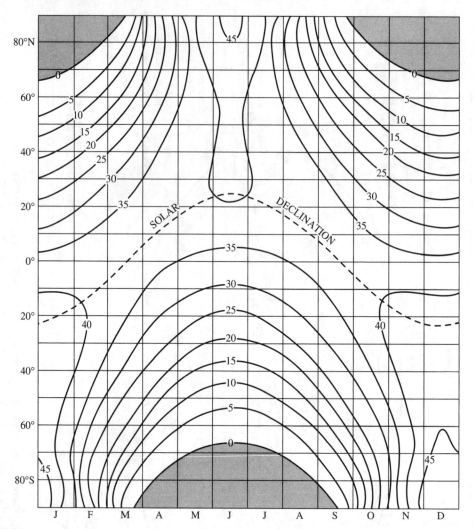

**FIGURE 3-6**
Daily total receipt of solar radiation (MJ m$^{-2}$) at the top of the atmosphere as a function of latitude and time of year. From *The Physics of Climate,* by J.P. Peixoto and A.H. Oort © 1992, used with permission of the American Institute of Physics.

the oceans is much less than that of the continents. This is reflected in the equatorward dip of the isotherms over the ocean in summer (oceans cooler than land), and the poleward dip in winter (oceans warmer than land).

The very cold winter temperatures in the centers of the North American and Asiatic land masses are due to radiational cooling and distance from the relatively warm oceans; these low temperatures produce cells of high density and high pressure. [See Equation (D-5).] The situation is reversed in summer, when extensive radiational heating occurs,

and these continents are then sites of generally low pressure. Note particularly the summertime trough of low pressure over southern Asia; winds associated with this trough carry the monsoon rains on which the agricultural economy of this vast and populous region depends.

### 3.1.4 Teleconnections: El Niño and the Southern Oscillation

A **teleconnection** is a climatic anomaly that is a distant consequence of another climatic anomaly. In

**FIGURE 3-7**

(a) The long-term average radiation balance of the earth–atmosphere system as a function of latitude. (b) The total poleward energy flux crossing each latitude and the portions of total flux effected by sensible and latent heat in the atmosphere and sensible heat in the oceans. Reprinted with permission from *Physical Climatology* by W.D. Sellers © 1965, University of Chicago Press.

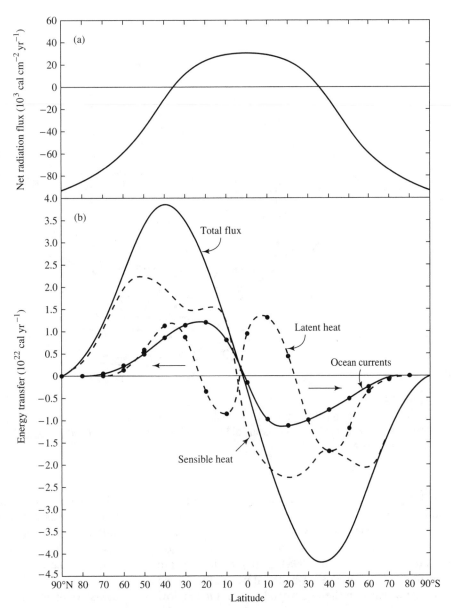

the late 1960s, oceanographic and atmospheric measurements, global observations via satellite, and careful study of historical records began to establish some definite, though spatially and temporally variable, relations among certain atmospheric and oceanic climatic anomalies.

The best known system of teleconnections is called the "El Niño–Southern Oscillation" (ENSO), a quasi-cyclic phenomenon that occurs every three to seven years and has persisted for at least the last 450 years (Rasmussen 1985; Enfield 1989). This phenomenon consists of an oscillation between (1) a warm phase ("El Niño"), during which abnormally high sea-surface temperatures (SSTs) occur off the coast of Peru[7] accompanied by low atmospheric pressure over the eastern Pacific and high pressure in the western Pacific and (2) a cold phase ("La Niña" or "El Viejo") with low SSTs

---

[7]The term "El Nino" refers to the Christ Child and was given by Peruvian fishermen (whose catches were adversely affected by the phenomenon) because the abnormal warming usually becomes pronounced around Christmas.

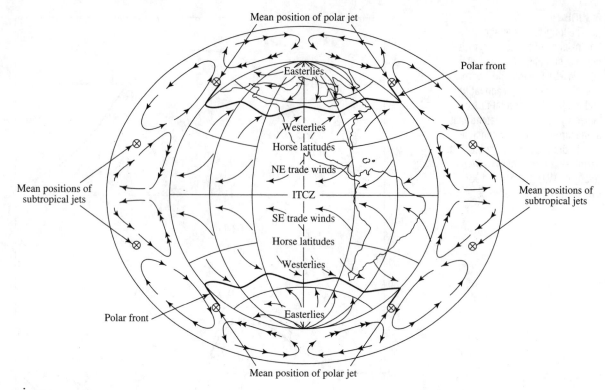

**FIGURE 3-8**
The general circulation of the atmosphere. Double-headed arrows in cross section indicate that the wind has a component from the east. Reprinted with permission of Prentice-Hall from *Elements of Meteorology*, 4th ed., by Miller et al. © 1984 by Bell & Howell Co.

in the eastern Pacific and the opposite pressure anomalies.

The typical ENSO warm episode evolves and declines over an approximately two-year period (Harrison and Larkin 1998). It begins in the late spring to fall of year 1 with abnormally strong westerly winds in the equatorial Indian Ocean, low pressures in eastern Australia, and warming SSTs in the South Pacific. As winter progresses, a tongue of abnormally warm water forms off the coast of Peru; this intensifies and builds westward along the equator during the spring and summer of year 2. The peak of the cycle usually occurs between July and December of year 2, with abnormally high SSTs extending westward to the International Date Line. These are accompanied by abnormal westerly winds and strong convergence along the equator, high pressures and low-ered sea levels in the western Pacific and

Indonesia, and low pressures and elevated sea levels in the eastern Pacific. A pool of abnormally cold water and enhanced westerly winds also occurs near latitude 45° in the North Pacific. The declining phase typically begins in January to April of year 3, when the equatorial pool of high SSTs begins to shrink, and most of the SST, wind, and pressure anomalies dissipate by the end of the summer of year 3.

There is considerable evidence that ENSO is an inherently oscillatory phenomenon that requires no outside forcing. The end of an ENSO episode begins when the eastward waves of warm water are reflected off South America and, in a complicated process that involves poleward circulation of the reflected westward-moving surface water and atmospheric processes, the SST returns to its original levels and the easterly trade-wind flow is re-established (Enfield 1989). Continued cooling of SSTs

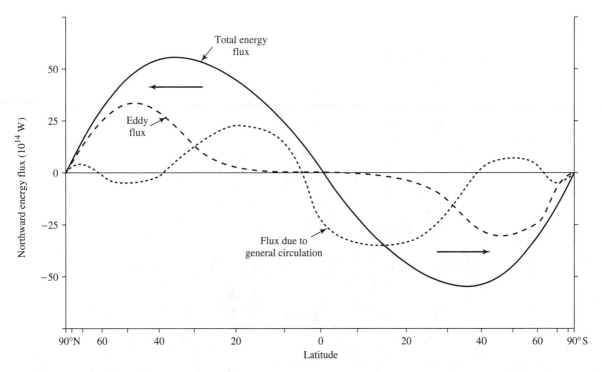

**FIGURE 3-9**
Total poleward energy flux and components of total flux carried as latent and sensible heat in the general circulation of the atmosphere and in winds associated with horizontal eddies. After Barry and Chorley (1982).

in the eastern Pacific leads to the cold phase of ENSO.

Although ENSO is essentially the product of large-scale, long-period waves in the surface of the tropical Pacific Ocean, it shifts the jet stream in the eastern North Pacific and North America to the south (warm phase) or north (cold phase). These shifts can steer unusual weather systems into low- and mid-latitude regions around the world. The result is unusually warm or cold winters in particular regions, drought in normally productive agricultural areas, and torrential rains in normally arid regions (Rasmussen 1985). Some of the teleconnections associated with ENSO episodes are indicated in Figures 3-13 and 3-14; the most consistent are the severe droughts in Australia and northern South America and heavy rainfall in Ecuador and northern Peru. In other places the effects can vary from episode to episode depending on the state of the atmosphere. For example, the 1976–1977 event was associated with drought along the west coast of the United States; that of 1982–1983 produced

increased storminess (Enfield 1989). The severe drought in the north-central United States in the summer of 1988 (Figure 10-34c) was a consequence of the 1986–1987 ENSO event (Trenberth et al. 1988). The strong 1997–1998 event produced warm and dry conditions from India to northern Australia (leading to extensive forest fires in Indonesia); dry conditions in the eastern Amazon region; a wet winter with considerable flooding along the West Coast, Gulf Coast, and south Atlantic Coast of the United States; and a warm winter in the northeastern United States. A source for current information on ENSO is given in Appendix G.

A number of studies have found relationships between streamflows and ENSO cycles (e.g., Dracup and Kahya 1994; Eltahir 1996; Piechota et al. 1997; Amarasekera et al. 1997). ENSO anomaly patterns persist for several months, so useful long-range hydrological forecasts can be made for the regions shown in Figures 3-13 and 3-14 (Halpert and Ropelewski 1992).

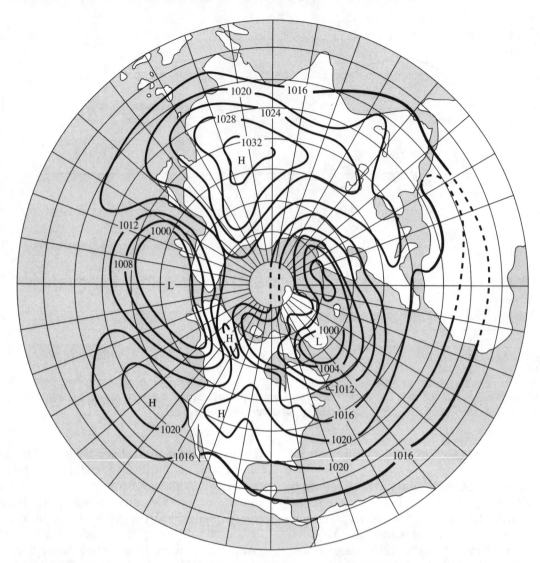

**FIGURE 3-10**
Normal sea-level pressures (mb) in the northern hemisphere in (a) January and (b) July. Reprinted with permission of Prentice-Hall from *Elements of Meteorology*, 4th ed., by Miller *et al.* © 1984 by Bell & Howell Co.

## 3.2  THE GLOBAL HYDROLOGIC CYCLE

### 3.2.1  Stocks and Fluxes in the Global Cycle

Figure 3-15 is a snapshot of the global hydrologic cycle in action. This "cycle" is actually a complex web of continual flows, or **fluxes,** of water among the major "reservoirs", or **stocks** of water (Figure 3-16). The sun provides the energy that causes evaporation and mixes water vapor in the atmosphere and thereby drives the cycle against the pull of gravity.

As is shown in Tables 3-1 and 3-2 and Figure 3-17, 96.5% of the water on earth is in the oceans. Of the fresh water, 69% is in solid form in glaciers[8]

[8]The proportion of the earth's water in glaciers was, of course, considerably larger as recently as 18,000 years ago, when the last glaciation was at its peak and the total volume of glacier ice was about three times its present value; at other periods of earth history there were no glaciers.

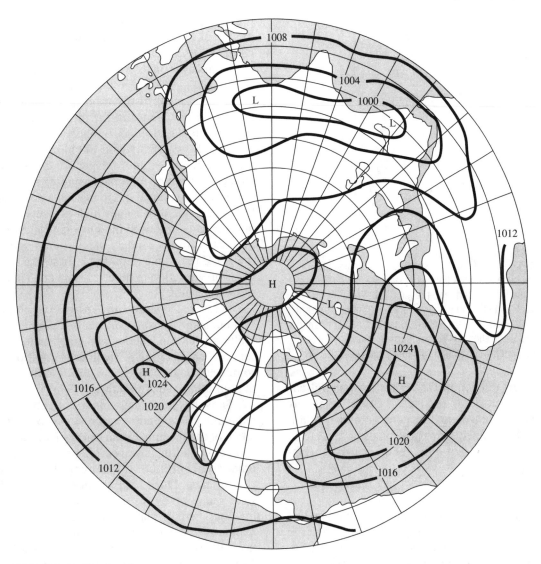

**FIGURE 3-10** (Continued)

and 30% is ground water; only 1% is in surface-water bodies.

The major features of the global cycle are: (1) the oceans lose more water by evaporation than they gain by precipitation; (2) the land surfaces receive more water as precipitation than they lose by evapotranspiration; and (3) the excess of water on the land returns to the oceans as runoff, balancing the deficit in the ocean–atmosphere exchange. The oceanic fluxes dominate the cycle: oceans receive 79% of the global precipitation and contribute 88% of the global evapotranspiration.

As was noted in Section 2.8.1, the flow of water from one reservoir to another implies that the water *within* each of the reservoirs is also continually in motion. The average residence time (average length of time a molecule of water is in a given reservoir) can be calculated from Equation (2-27); Exercise 3-8 asks you to calculate the residence times for the various stocks in Figure 3-16.

### 3.2.2   Distribution of Precipitation

Regions characterized by rising air tend to have relatively high average precipitation, and those characterized by descending air tend to have low precipitation. (See the discussion of precipitation

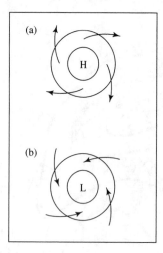

**FIGURE 3-11**
Arrows show general directions of surface winds in (a) anticyclonic circulation (around high-pressure cells) and (b) cyclonic circulation (around low-pressure cells) in the northern hemisphere. Solid lines are **isobars:** lines of equal atmospheric pressure. Circulations in the southern hemisphere are reversed (i.e., circulation around cyclones is clockwise).

mechanisms in Section D.5.) Thus the general circulation (Figure 3-8) produces belts of relatively high precipitation near the equator and 60° latitude, and relatively low precipitation near 30°, where most of the world's great deserts occur (Figure 3-18).

The equatorial belt of high precipitation is especially pronounced because warm easterly winds from both hemispheres carrying large amounts of moisture evaporated from tropical oceans converge in this zone; this phenomenon is called the **intertropical convergence zone** (ITCZ). The peaks of precipitation coincident with the mid-latitude zone of rising air are produced mainly by extra-tropical cyclonic storms that tend to develop along the polar front.

Because precipitation rates are influenced by topography, air temperatures, frontal activity, and wind directions in relation to moisture sources, global precipitation patterns (Figure 3-19) show significant deviations from the general latitudinal distribution depicted in Figure 3-18. The major causes of these deviations are mountain ranges, such as the Rocky Mountain–Andean chain, the Alps, and the Himalayas. These ranges induce high rates of precipitation in their immediate vicinity and, typically, produce "rain-shadow" zones of reduced precipitation over large areas leeward of the prevailing winds. Note, for example, the dry zone in the Great Plains of North America extending from latitude 20° to above latitude 60°, and the effects of the Himalayas in blocking moisture-laden winds from reaching the interior of Asia.

The seasonal distribution of precipitation, including its occurrence in the form of rain or snow, has important hydrologic implications and, as is discussed in Section 3.3, significant impacts on soil formation and vegetation. Figure 3-20 shows the global distribution of seven general precipitation regimes. The reversal of circulation associated with the development of winter high pressure and summer low pressure over the huge land mass of Asia interacts with the topography and the migration of the ITCZ to produce a particularly strong seasonality of precipitation in much of Asia and Africa; this is the **monsoon.**

Figure 3-21 shows the global distribution of perennial and seasonal snow and ice. Note that virtually all the land above 40° north latitude has a seasonal snow cover of significant duration; in the southern hemisphere, snow occurs only in mountainous areas and Antarctica.

Snow has important climatic effects, helping to maintain colder temperatures (1) by reflecting much of the incoming solar energy (see Table D-2), and (2) in melting, by absorbing energy that would otherwise contribute to warming the near-surface environment. During the ice ages, the reflection of solar radiation by ice and snow was an important feed-back effect that contributed to creating and maintaining a colder climate. Under present conditions, the surface cooling induced by snow has profound effects on surface and air temperatures, global circulation patterns and storm tracks, and precipitation (Berry 1981; Walsh 1984; Barnett et al. 1988; Leathers and Robinson 1993; Groisman et al. 1994). A number of studies (Dey and Kumar 1983; Dickson 1984) have shown an inverse relation between the extent of winter snow cover in Eurasia and the amount of rainfall in India during the ensuing summer monsoon (Figure 3-22).

Snow also acts as an insulating blanket that helps to retain heat in the soil, which is important hydrologically as well as biologically: if soil is prevented from freezing, its ability to accept infiltrating water is generally enhanced (Dingman 1975). However, the principal hydrologic effect of snow is to delay the input of precipitated water into the land phase of the hydrologic cycle and thus to af-

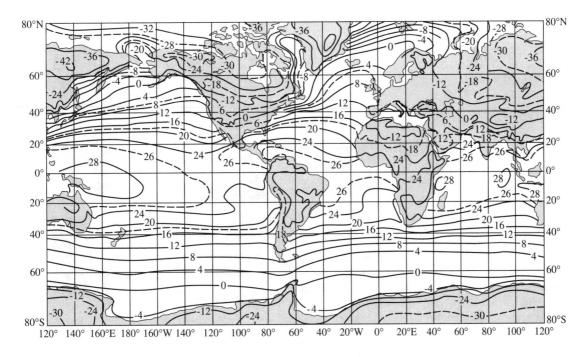

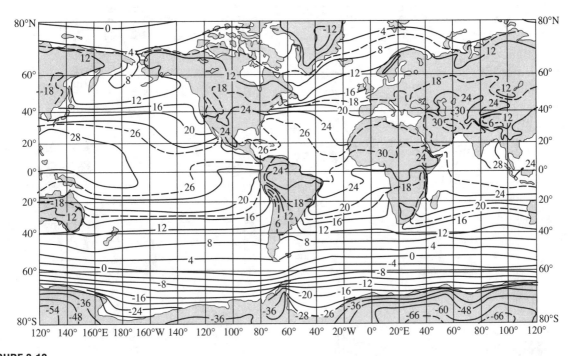

**FIGURE 3-12**

Distribution of mean temperature (°C) in (a) January; and (b) July. From *The Physics of Climate,* by J.P. Peixoto and by A.H. Oort © 1992, used with permission of American Institute of Physics.

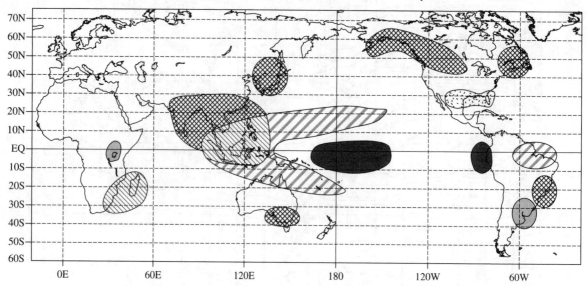

Warm Episode Relationships   December–February

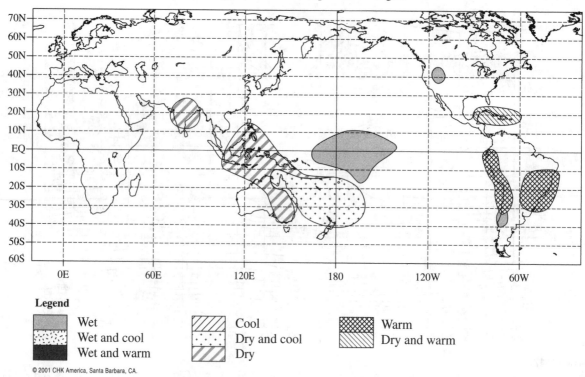

Warm Episode Relationships   June–August

**Legend**

| | | | | | |
|---|---|---|---|---|---|
| Wet | | Cool | | Warm | |
| Wet and cool | | Dry and cool | | Dry and warm | |
| Wet and warm | | Dry | | | |

© 2001 CHK America, Santa Barbara, CA.

**FIGURE 3-13**
Typical climatic anomalies associated with the warm (El Niño) phase of ENSO. (a) winter (December–February); (b) summer (June–August). From U.S. National Atmospheric and Oceanographic Administration Climate Prediction Center website http://nic.fb4.noaa.gov/products/analysis_mon (1998).

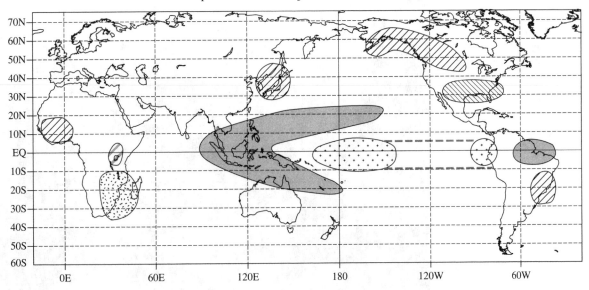

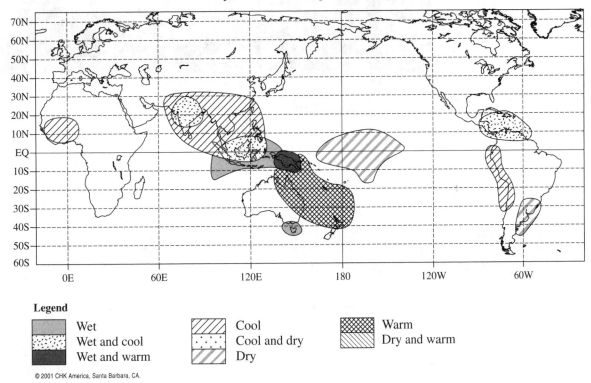

**Legend**

| | | | | | |
|---|---|---|---|---|---|
| Wet | | Cool | | Warm | |
| Wet and cool | | Cool and dry | | Dry and warm | |
| Wet and warm | | Dry | | | |

© 2001 CHK America, Santa Barbara, CA.

**FIGURE 3-14**

Typical climatic anomalies associated with the cold (La Niña) phase of ENSO. (a) winter (December–February); (b) summer (June–August). From U.S. National Atmospheric and Oceanographic Administration Climate Prediction Center website http://nic.fb4.noaa.gov/products/analysis_mon (1998).

**FIGURE 3-15**
The global hydrologic cycle in action. Photo courtesy of U.S. National Aeronautics and Space Agency. *From NASA/Science Source/Photo Researchers. Inc.*

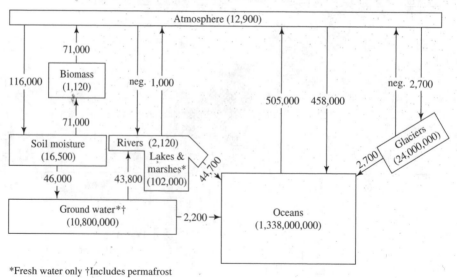

*Fresh water only †Includes permafrost

**FIGURE 3-16**
Schematic diagram of stocks and annual fluxes in the global hydrologic cycle. Based on data of Shiklomanov and Sokolov (1983) (Table 3-1). Inflows and outflows may not balance for all compartments due to rounding.

**TABLE 3-1**
Stocks in the Global Hydrologic Cycle.[a]

| | | | Share of World Reserves (%) | |
| Form of Water | Area Covered (km²) | Volume (km³) | Of Total Water Reserves | Of Fresh-Water Reserves |
|---|---|---|---|---|
| World oceans | 361,300,000 | 1,338,000,000 | 96.5 | — |
| Ground waters | 134,800,000 | 23,400,000 | 1.7 | — |
|   Fresh ground water | | 10,530,000 | 0.76 | 30.1 |
| Soil moisture | 82,000,000 | 16,500 | 0.001 | 0.05 |
| Glaciers and permanent | | | | |
|   snowpack: | 16,227,500 | 24,064,100 | 1.74 | 68.7 |
|     Antarctica | 13,980,000 | 21,600,000 | 1.56 | 61.7 |
|     Greenland | 1,802,400 | 2,340,000 | 0.17 | 6.68 |
|     Arctic islands | 226,100 | 83,500 | 0.006 | 0.24 |
|     Mountain areas | 224,000 | 40,600 | 0.003 | 0.12 |
| Ground ice in zone of | | | | |
|   permafrost strata | 21,000,000 | 300,000 | 0.022 | 0.86 |
| Water reserves in lakes: | 2,058,700 | 176,400 | 0.013 | — |
|   Fresh-water lakes | 1,236,400 | 91,000 | 0.007 | 0.26 |
|   Saltwater lakes | 822,300 | 85,400 | 0.006 | — |
| Marsh water | 2,682,600 | 11,470 | 0.0008 | 0.03 |
| Water in rivers | 148,800,000 | 2,120 | 0.0002 | 0.006 |
| Biologic water | 510,000,000 | 1,120 | 0.0001 | 0.003 |
| Atmospheric water | 510,000,000 | 12,900 | 0.001 | 0.04 |
| Total water reserves | 510,000,000 | 1,385,984,610 | 100 | — |
| Fresh water | 148,800,000 | 35,029,210 | 2.53 | 100 |

[a]Illustrated in Figure 3-16, page 54.
Data from Shiklomanov and Sokolov (1983).

fect the seasonal distribution of runoff (see Section 3.2.4 and 10.2.5).

### 3.2.3   Distribution of Evapotranspiration

**Evapotranspiration** includes all processes involving the phase change from liquid (or solid) to water vapor. Globally, its principal components are evapo-ration from the oceans and transpiration by land vegetation. The latitudinally averaged evapotranspiration (Figure 3-23) has a maximum near the equator and near-zero values at the poles. The general pattern is similar to that of the radiation balance and temperature, reflecting the importance of the availability of energy to supply the latent heat that accompanies the phase change (Sections D.6.6 and 7.1.1).

**TABLE 3-2**
Stocks and Annual Fluxes for Major Compartments of the Global Hydrologic Cycle. Data from Shiklomanov and Sokolov (1983).

| Stock | Volume[a] | Percentage of All Water | Sources | Input Flux[b] | Sinks | Output Flux[b] |
|---|---|---|---|---|---|---|
| Oceans | 1338 | 96.5 | Pptn: | 458 | Evap: | 505 |
| | | | Runoff: | 47 | | |
| Atmosphere | 0.013 | 0.001 | Land evap: | 72 | Pptn: | 577 |
| | | | Ocean evap: | 505 | | |
| Land | 48 | 3.46 | Pptn: | 119 | Evap: | 72 |
| | | | | | Runoff: | 47 |
| Total | 1386 | 100 | | | | |

[a] Stocks in $10^6$ km³.

[b] Fluxes in $10^3$ km³ yr⁻¹.

Data from Shiklomanov and Sokolov (1983).

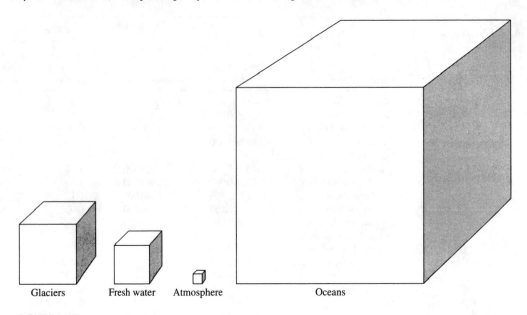

**FIGURE 3-17**
Relative volumes of water in oceans, glaciers, fresh water, and atmosphere.

**FIGURE 3-18**
Latitudinal distribution of average precipitation rate. From Barry and Chorley (1982).

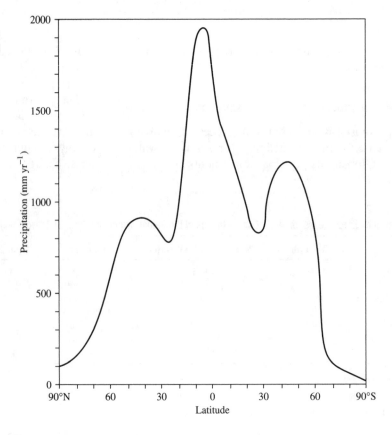

**FIGURE 3-19**

Average precipitation rate over the globe. Reprinted with permission of Prentice-Hall from *Climatology* by Oliver and Hidore, © 1984 by Bell & Howell Co.

Legend:
<25 cm yr$^{-1}$
25 – 50
50 – 100
100 – 150
150 – 200
>200

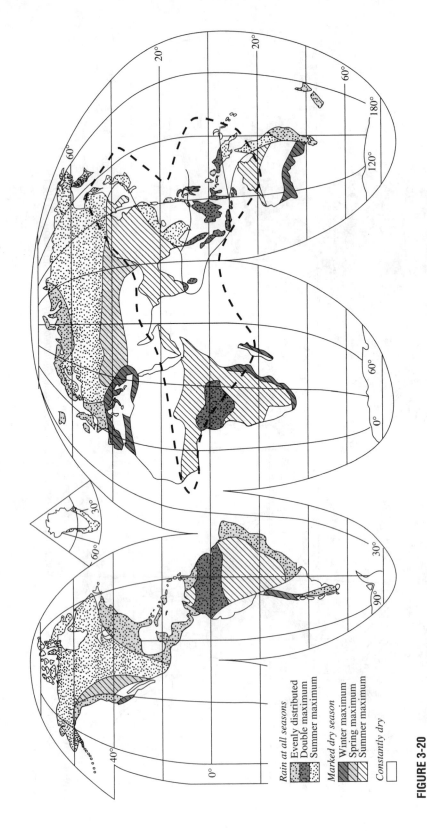

**FIGURE 3-20**
Generalized map of seasonal precipitation regimes. Heavy dashed line encloses the monsoon region of Asia and Africa. Reprinted with permission of Prentice-Hall from *Climatology* by Oliver and Hidore, © 1984 by Bell & Howell Co.

*Rain at all seasons*
▨ Evenly distributed
▦ Double maximum
▒ Summer maximum

*Marked dry season*
▧ Winter maximum
▨ Spring maximum
▨ Summer maximum

*Constantly dry*
☐

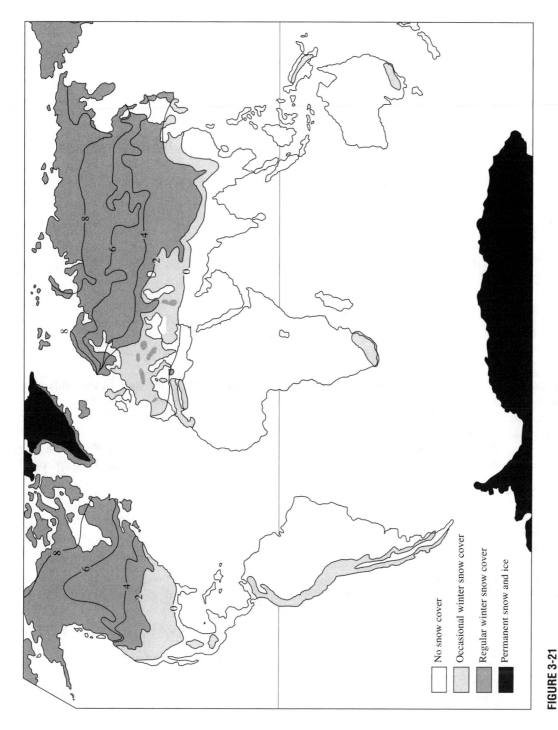

**FIGURE 3-21**
Global distribution of snow cover. Numbered lines indicate normal duration of seasonal snow cover in months. From Walsh (1984), used with permission of Sigma Xi, the Scientific Research Society.

No snow cover

Occasional winter snow cover

Regular winter snow cover

Permanent snow and ice

**FIGURE 3-22**
Summer monsoon rainfall in India and preceding winter snow-cover in the Himalayas for 1971–1980. From Walsh (1984) used with permission of Sigma Xi, the Scientific Research Society.

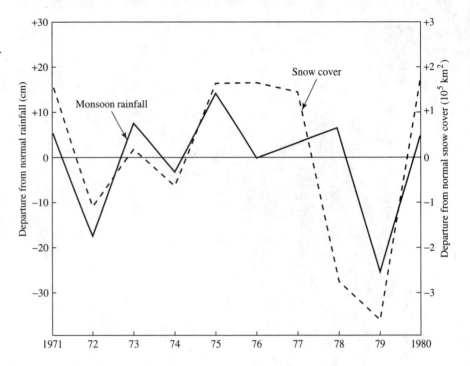

The latitudinal distribution of evapotranspiration is also influenced by the latitudinal distribution of land and oceans and of land precipitation. In lower to middle latitudes more water evaporates from the oceans, where precipitation does not limit the availability of water, than from the continents.

The slight oceanic minimum at the equator is due to the generally lower winds and high humidity in this zone. For the continents, the equatorial peak is due to large heat inputs and high water availability. The minor mid-latitude peaks reflect strong prevailing westerly winds and elevated water availability

**FIGURE 3-23**
Latitudinal distribution of average annual evaporation from the oceans, evapotranspiration from the continents, and globally averaged evapotranspiration. From Barry and Chorley (1982).

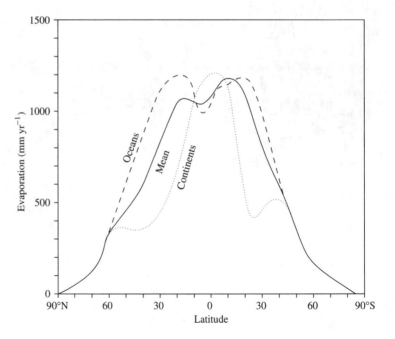

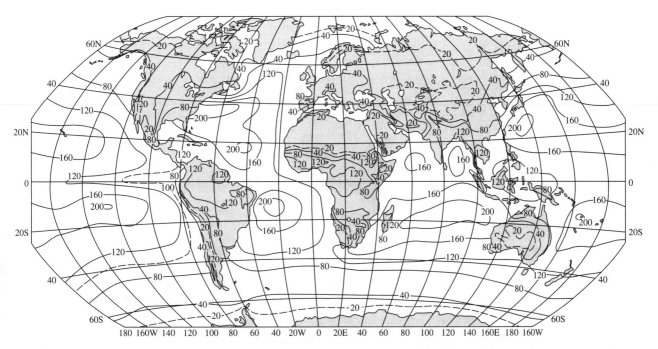

**FIGURE 3-24**

Global distribution of oceanic evaporation and continental evapotranspiration (cm yr$^{-1}$). From *The Physics of Climate*, by J.P. Peixoto and A.H. Oort © 1992, used with permission of American Institute of Physics.

(Figure 3-18) (Barry and Chorley 1982); the minor minima near 30° are due to the scarcity of water in the extensive deserts in that latitude band.

Figure 3-24 shows the global distribution of oceanic evaporation and continental evapotranspiration. As expected, there is a general correlation between mean annual temperature and mean annual evapotranspiration (compare Figures 3-12 and 3-24.) In the oceans, the basic latitudinal patterns are distorted largely by the effects of surface currents (for example, note the effect of the warm Gulf Stream off northern Europe and the equatorward-flowing cold currents off western South America and North America). The highest continental values are in the tropical rain forests of South America, Africa, and southeast Asia; the lowest are in the Sahara Desert, Antarctica, arctic North America, and central Asia.

### 3.2.4 Distribution of Runoff

Figure 3-25 shows the global distribution of annual runoff (i.e., the difference between precipitation and evapotranspiration) for the continents. Not surprisingly, comparison of Figures 3-19 and 3-25 shows a close correspondence between average

runoff and average precipitation: Virtually all the zones with the highest runoff also have the highest precipitation, and regions with low precipitation have low runoff. The highest average runoff rates, near 3000 mm yr$^{-1}$, occur on the east coast of the Bay of Bengal; the Amazon Basin of northern South America contains the largest region with runoff exceeding 1000 mm yr$^{-1}$.

The seasonal pattern of runoff is commonly quite different from that of precipitation due to the seasonality of evapotranspiration and the storage of precipitation as snow. In the northeastern United States, for example, precipitation is equally distributed through the year, but 25% of the annual runoff typically occurs in one spring month and only 10% in the three months of summer. The effect of snowmelt in concentrating the period of runoff becomes more pronounced the longer the annual snowcover persists; in northern Alaska, one-half of the annual runoff occurs in a three- to ten-day period (Dingman et al. 1980).

Figure 3-26 shows the types of runoff regimes classified by L'vovich (1974). In this classification, regimes are identified by (1) the season in which the most runoff occurs (spring, summer, winter, fall) and (2) the degree to which runoff is concentrated

**FIGURE 3-25**
Global distribution of average runoff rate. From Gregory and Walling (1973).

mm yr$^{-1}$
1000
500
50

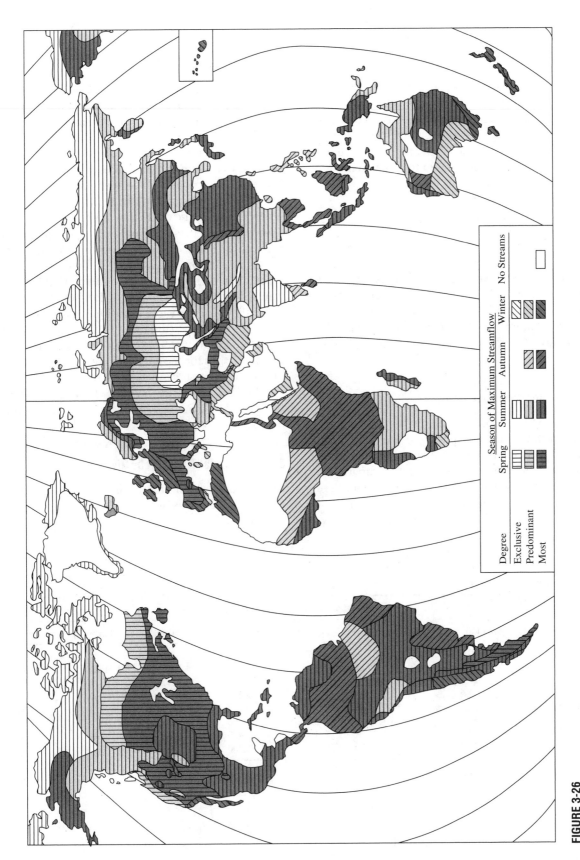

**FIGURE 3-26**

Seasonal runoff regimes. Modified from L'vovich (1974); used with permission of the American Geophysical Union.

Season of Maximum Streamflow

| Degree | Spring | Summer | Autumn | Winter | No Streams |
|---|---|---|---|---|---|
| Exclusive | | | | | |
| Predominant | | | | | |
| Most | | | | | |

in that season (more than 80%, 50 to 80%, or less than 50%). In most areas that have a seasonal snow cover or are glacierized, the maximum runoff occurs in the melt season: summer in arctic, subarctic, and alpine regions, and spring at lower latitudes. A summer maximum also occurs in regions with monsoonal climates, such as India and southeast Asia, and other areas with summer precipitation maximums (Figure 3-20). Fall and winter runoff maxima are also directly related to concurrent rainfall maxima.

### 3.2.5 Continental Water Balances

It is clear from Figures 3-19, 3-24, and 3-25 that the components of the hydrologic cycle vary considerably in magnitude over the continents. As shown in Table 3-3, South America is by far the wettest continent in terms of both precipitation and runoff per unit area; Antarctica is the driest in terms of precipitation, and Australia has by far the lowest runoff per unit area.

### 3.2.6 Major Rivers and Lakes

Rivers are the major routes by which "surplus" water on the continents returns to the oceans; the rate of direct runoff of ground water to the oceans is not well established, but is small compared to river flows (Table 3-1). Table 3-4 shows the average discharges and drainage areas of the 16 largest rivers (ranked by discharge); Figure 3-27 shows

their locations. Together, these rivers drain 22.9% of the world's land area and contribute 32.8% of the total runoff to the oceans; the Amazon River alone delivers 13% of the total runoff. Note that only rivers draining directly to the ocean are included in Table 3-4; there are tributaries of the Amazon that have larger discharges than many of the rivers listed.

From the point of view of the global hydrologic cycle, lakes are simply wide places in rivers. The main hydrologic functions of natural and man-made lakes are (1) to provide storage that reduces the time variability of flow in the rivers that drain them (Figure 2-8) and (2) to increase evaporation by providing large evaporating surfaces. However, on a global scale, the evaporation from lakes and wetlands is small, amounting to only about 3% of the total land evapotranspiration (L'vovich 1974). It should also be noted that lakes play important roles with respect to sediment transport (discussed in the following section), temporarily storing particulate sediment and providing sites for the chemical precipitation and biological uptake of dissolved materials.

Table 3-5 lists the world's 25 largest natural lakes (ranked by area), and Figure 3-27 shows their locations.

### 3.2.7 Material Transport by Rivers

In addition to their role in the global water cycle, rivers are the means by which the products of continental weathering are carried to the oceans. Thus

**TABLE 3-3**

Water Balances of the Continents.

| Continent | Area ($10^6$ km²) | Precipitation (km³ yr⁻¹) | Precipitation (mm yr⁻¹) | Evapotranspiration (km³ yr⁻¹) | Evapotranspiration (mm yr⁻¹) | Runoff (km³ yr⁻¹) | Runoff (mm yr⁻¹) |
|---|---|---|---|---|---|---|---|
| Europe | 10.0 | 6,600 | 657 | 3,800 | 375 | 2,800 | 282 |
| Asia | 44.1 | 30,700 | 696 | 18,500 | 420 | 12,200 | 276 |
| Africa | 29.8 | 20,700 | 695 | 17,300 | 582 | 3,400 | 114 |
| Australia[a] | 7.6 | 3,400 | 447 | 3,200 | 420 | 200 | 27 |
| North America | 24.1 | 15,600 | 645 | 9,700 | 403 | 5,900 | 242 |
| South America | 17.9 | 28,000 | 1,564 | 16,900 | 946 | 11,100 | 618 |
| Antarctica | 14.1 | 2,400 | 169 | 400 | 28 | 2,000 | 141 |
| Total land[b] | 148.9 | 111,100[c] | 746 | 71,400 | 480 | 39,700[c] | 266 |

[a]Not including New Zealand and adjacent islands.

[b]Including New Zealand and adjacent islands.

[c]Estimate differs from that of Table 3-2.

Data from Baumgartner and Reichel (1975)

**TABLE 3-4**

The World's 16 Largest Rivers in Terms of Average Discharge. (See Figure 3-27 for locations.)

| River | Drainage Area | | Discharge | | | | Runoff Ratio[c] |
|---|---|---|---|---|---|---|---|
| | $(10^3 \text{ km}^2)$ | %[a] | $(\text{m}^3 \text{ s}^{-1})$ | $(\text{km}^3 \text{ yr}^{-1})$ | $(\text{mm yr}^{-1})$ | %[b] | |
| 1. Amazon | 7,180 | 4.8 | 190,000 | 6,000 | 835 | 13.0 | 0.47 |
| 2. Congo | 3,822 | 2.6 | 42,000 | 1,330 | 340 | 2.9 | 0.25 |
| 3. Yangtzekiang | 1,970 | 1.3 | 35,000 | 1,100 | 560 | 2.4 | 0.50 |
| 4. Orinoco | 1,086 | 0.7 | 29,000 | 915 | 845 | 2.0 | 0.46 |
| 5. Brahmaputra | 589 | 0.4 | 20,000 | 630 | 1,070 | 1.4 | 0.65 |
| 6. La Plata | 2,650 | 1.8 | 19,500 | 615 | 235 | 1.3 | 0.20 |
| 7. Yenesei | 2,599 | 1.7 | 17,800 | 565 | 215 | 1.2 | 0.42 |
| 8. Mississippi | 3,224 | 2.2 | 17,700 | 560 | 175 | 1.2 | 0.21 |
| 9. Lena | 2,430 | 1.6 | 16,300 | 515 | 210 | 1.1 | 0.46 |
| 10. Mekong | 795 | 0.8 | 15,900 | 500 | 630 | 1.1 | 0.43 |
| 11. Ganges | 1,073 | 0.7 | 15,500 | 490 | 455 | 1.1 | 0.42 |
| 12. Irrawaddy | 431 | 0.3 | 14,000 | 440 | 1,020 | 1.0 | 0.60 |
| 13. Ob | 2,950 | 2.0 | 12,500 | 395 | 135 | 0.9 | 0.24 |
| 14. Sikiang | 435 | 0.3 | 11,500 | 365 | 840 | 0.8 | — |
| 15. Amur | 1,843 | 1.2 | 11,000 | 350 | 190 | 0.8 | 0.32 |
| 16. Saint Lawrence | 1,030 | 0.7 | 10,400 | 330 | 310 | 0.7 | 0.33 |
| Totals | 34,107 | 22.9 | 478,100 | 15,100 | | 32.8 | |

[a]Percent of total earth land area $(148.9 \times 10^6 \text{ km}^2)$.

[b]Percent of total runoff to oceans $(46 \times 10^3 \text{ km}^3 \text{ yr}^{-1})$.

[c]Ratio of long-term average discharge to long-term average precipitation ($w$ in the model in Box 3-4).

Data from Baumgartner and Reichel (1975), Wigley and Jones (1985), and L'vovich (1974).

they are a crucial link in the "tectonic cycle" in which rock material is formed deep in the earth's crust, raised to the surface by tectonic processes, eroded and transported to the oceans, and ultimately subducted to become resorbed into the lower crust and upper mantle (Howell and Murray 1986). The materials transported by rivers are also parts of the biogeochemical cycles involving carbon, oxygen, nitrogen, hydrogen, phosphorus, sulphur, and many other elements [see, for example, Deevey (1970)] that are essential for the maintenance of the earth's ecosystems.

Rivers transport material as individual ions or molecules in solution (**dissolved load**) or as solid particles (**particulate load**). The particulate load can be further classified as **suspended load,** which is carried above the channel bottom by turbulent eddies, or **bed load,** which moves in contact with the bottom.

In discussing material transport, it is important to distinguish between **concentration,** which is usually expressed as the mass (or weight) of material constituent per unit volume of water ($[M \text{ L}^{-3}]$ or $[F \text{ L}^{-3}]$), and **load,** which is the rate of discharge of

the material constituent ($[M \text{ T}^{-1}]$ or $[F \text{ T}^{-1}]$). The relation between the two quantities is

$$L_x = C_x \cdot Q, \qquad (3\text{-}1)$$

where $L_x$ is the load of constituent $x$, $C_x$ is the concentration of constituent $x$, and $Q$ is the rate of discharge of water ($[L^3 \text{ T}^{-1}]$). As with water discharge, it is often useful to compare loads in different rivers on a per-unit-drainage-area basis ($[M \text{ T}^{-1} \text{ L}^{-2}]$ or $[F \text{ T}^{-1} \text{ L}^{-2}]$); this quantity is called **sediment yield.**

In this section, we review (1) some of the global relations between material transport and climate, geology, topography, and vegetation and (2) current estimates of the global rates of material transport to the oceans. Walling and Webb (1987) have recently reviewed much of the literature on these topics, and our overview relies heavily on their work.

### Dissolved Material

Table 3-6 lists the estimated mean composition of the river waters of the world. However, composi-

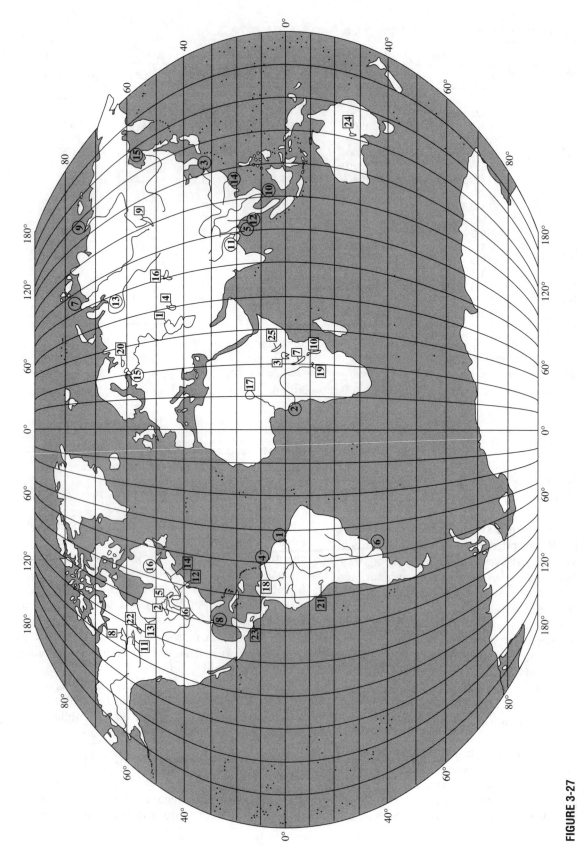

**FIGURE 3-27**
Locations of the world's 16 largest rivers shown by numbers in circles; refer to Table 3-4. Locations of the world's 25 largest lakes shown by numbers in squares; refer to Table 3-5.

**TABLE 3-5**

The World's 25 Largest Natural Lakes in Terms of Surface Area. (See Figure 3-27 for locations.)

| Lake | Area (km²) | Maximum Depth (m) | Elevation (m) |
|---|---|---|---|
| 1. Caspian Sea | 371,800 | 995 | −28 |
| 2. Superior | 82,400 | 406 | 183 |
| 3. Victoria | 69,500 | 81 | 1,134 |
| 4. Aral Sea[a] | 65,500 | 68 | 53 |
| 5. Huron | 59,600 | 229 | 177 |
| 6. Michigan | 58,000 | 281 | 177 |
| 7. Tanganyika | 32,900 | 1,436 | 773 |
| 8. Great Bear | 31,800 | 413 | 156 |
| 9. Baikal | 30,500 | 1,620 | 455 |
| 10. Nyasa | 29,600 | 679 | 473 |
| 11. Great Slave | 28,400 | 614 | 156 |
| 12. Erie | 25,700 | 64 | 174 |
| 13. Winnipeg | 24,500 | 18 | 217 |
| 14. Ontario | 19,700 | 245 | 75 |
| 15. Ladoga | 17,700 | 225 | 4 |
| 16. Balkhash[b] | 17,400 | 26 | 340 |
| 17. Chad[b] | 16,300 | 7 | 240 |
| 18. Maracaibo | 13,300 | 35 | 0 |
| 19. Bangwelu | 9,800 | ? | 1,150 |
| 20. Onega | 9,600 | 110 | 33 |
| 21. Titicaca | 8,300 | 281 | 3,813 |
| 22. Athabasca | 8,100 | 124 | 213 |
| 23. Nicaragua | 8,000 | 70 | 32 |
| 24. Eyre[b] | 7,700 | 1 | −16 |
| 25. Rudolf | 6,400 | 61 | 375 |

[a] The Aral Sea has been much reduced in size by disastrous water-resource developments (Micklin 1988).

[b] Area varies significantly in response to seasonal and longer-term precipitation fluctuations. Data mostly from Todd (1970).

tion varies widely in different regions of the world, largely because of variations in rock type and climate. Gibbs (1970) showed that rivers draining areas with high annual precipitation and runoff tend to have low total dissolved concentrations and compositions similar to that of the precipitation (i.e., they are relatively rich in sodium [Na] and chlorine [Cl]) and largely independent of rock type (Figure 3-28). In climates with moderate precipitation and runoff, concentrations are at moderate levels and composition is dominated by rock type and tends to be high in calcium (Ca) and bicarbonate ($HCO_3$). As one moves toward drier climates, water chemistry becomes increasingly controlled by fractional crystallization due to evaporation: Concentrations increase, and the composition shifts from Ca–$HCO_3$ toward Na–Cl. The ultimate end-member in this progression is sea water.

Walling and Webb (1987) examined data for some 500 rivers worldwide and found average dissolved-sediment yields ranging from less than 1 T $km^{-2}$ $yr^{-1}$ to 750 T $km^{-2}$ $yr^{-1}$; the average was about 40 T $km^{-2}$ $yr^{-1}$. Figure 3-29 shows the global variation of total dissolved load for major river basins. (Data are sparse for other areas.) The high loads in southern Asia reflect the high discharges in those regions (Figure 3-26); those in central Europe are

**TABLE 3-6**

Mean Composition of River Water of the World.

| Constituent | Concentration (mg/L) |
|---|---|
| Silica ($SiO_2$) | 10.4 |
| Calcium (Ca) | 13.4 |
| Magnesium (Mg) | 3.35 |
| Sodium (Na) | 5.15 |
| Potassium (K) | 1.3 |
| Bicarbonate ($HCO_3$) | 52 |
| Sulfate ($SO_4$) | 8.25 |
| Chloride (Cl) | 5.75 |
| Total dissolved solids | 73.2 |

Data from Hem (1985).

**FIGURE 3-28**
Processes controlling the chemistry of world rivers, according to Gibbs (1970).

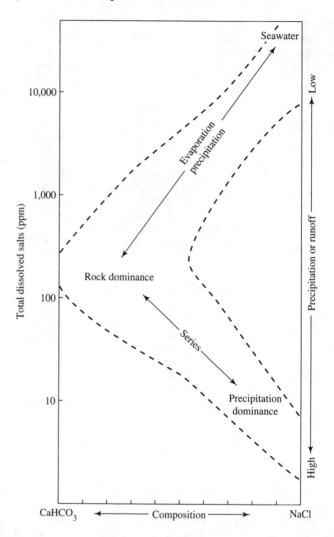

due to widespread soluble rocks, especially limestones. The very high loads of the Irrawaddy River in Southeast Asia and those on New Guinea are produced by a combination of readily weathered rocks, high rates of weathering due to high temperatures and precipitation, and high discharges. The presence of crystalline rocks with low solubilities in much of Africa and Australia gives rise to generally low dissolved loads on those continents.

### Particulate Material

The natural rate of erosion of particulate material is determined by climate, rock type, topography, tectonic activity, and vegetation. However, it appears that human activity has doubled global sediment transport in historic times (Milliman and

Syvitski 1992). Thus all these factors, plus the effects of lakes and reservoirs acting as sediment traps (also greatly increased by humans; see Vörösmarty et al. 1993), affect the global patterns of particulate-sediment yield and make it difficult to generalize about these patterns. However, Dedkov and Mozzherin (1984) have related sediment yields to vegetation, after first classifying rivers with respect to topography and size (Figure 3-30). These vegetative zones can be generally related to climatic factors, as discussed in Section 3.3.2. Note that, outside of glacierized regions, the highest yields tend to occur in "Mediterranean" climates, where vegetation is sparse because annual rainfall is low, but the rain is concentrated in a few months of the year. In another global survey that included smaller streams, Milliman and Syvitski (1992)

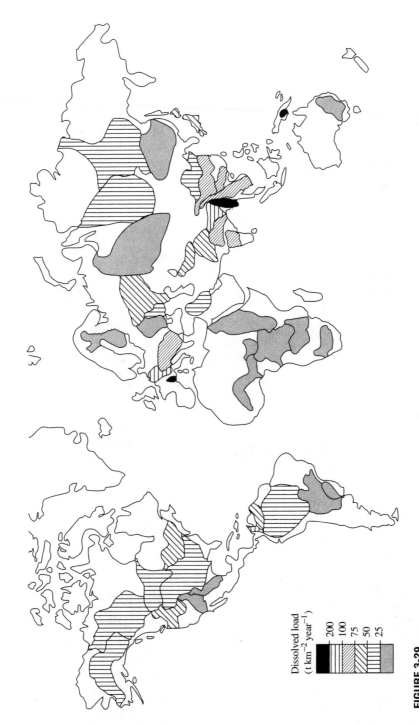

**FIGURE 3-29**
Dissolved-sediment yields in major river basins of the world. From Walling and Webb (1987).

Dissolved load
(t km$^{-2}$ year$^{-1}$)

200
100
75
50
25

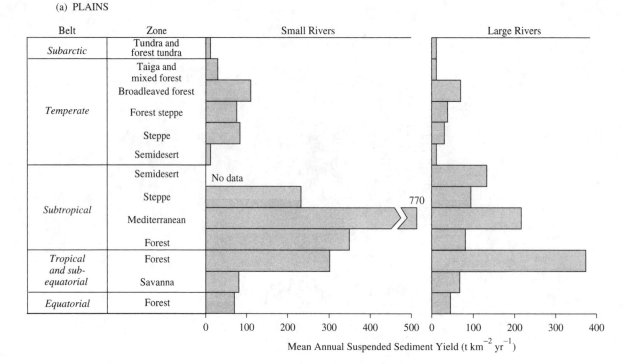

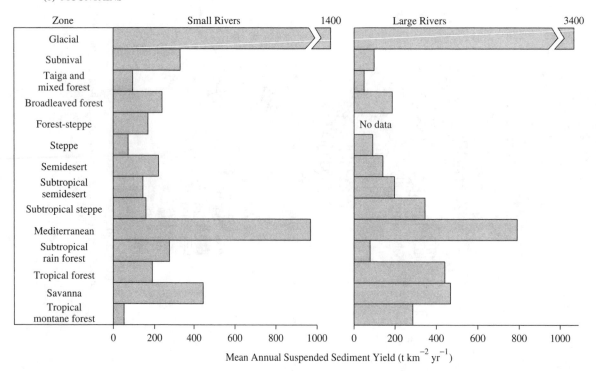

**FIGURE 3-30**
Relation between particulate-sediment yields and vegetation, as proposed by Dedkov and Mozzherin (1984). The boundary between "small" and "large" rivers is at a drainage area of 5000 km$^2$. From Walling and Webb (1987).

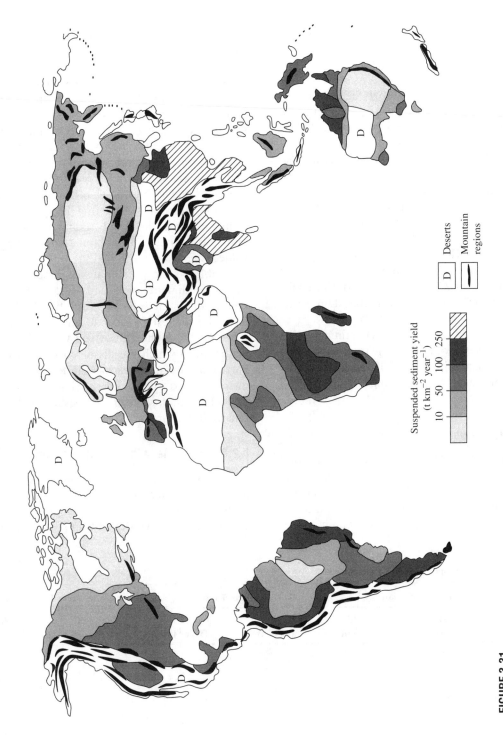

**FIGURE 3-31**

Global distribution of particulate-sediment yields, as mapped by Dedkov and Mozzherin (1984). From Walling and Webb (1987).

Suspended sediment yield
(t km$^{-2}$ year$^{-1}$)
10  50  100  250

D  Deserts

Mountain regions

found that sediment yields ranged from 1.2 to 36,000 T km$^{-2}$ yr$^{-1}$ and were positively related to drainage area, maximum drainage-basin elevation, and runoff.

Figure 3-31 shows the global distribution of particulate-sediment yields. The highest yields are in areas with seasonal-rainfall climates (Figure 3-20) coupled with active mountain building (India) or highly erodible soils (China); high yields are also associated with mountain belts in Alaska, the Andes, and the western Mediterranean region. Milliman and Syvitski (1992) found the highest yields in Taiwan, the Philippines, and New Zealand, where human activity also plays a significant role in sediment production. The areas of lowest yields (outside of deserts) are in northern North America and Eurasia, equatorial Africa, and eastern Australia, where low relief is coupled with resistant rocks and/or extensive vegetative cover.

Dedkov and Mozzherin (1984) estimated that bed load averages about 8% of the total particulate load in large plains rivers and 23% in large mountain rivers, but these values are highly variable and uncertain.

### Total Material Transport to the Oceans

Walling and Webb (1987) estimated that the total load of dissolved plus particulate material to the oceans is 17.2 × 10$^9$ T yr$^{-1}$. When allowance is made for the amount of sediment being trapped in reservoirs, the total rate of sediment movement is between 19.0 × 10$^9$ T yr$^{-1}$ and 20.0 × 10$^9$ T yr$^{-1}$, of which about 80% is particulate and 20% is dissolved. Under the assumption of an average rock density of 2500 kg m$^{-3}$, this total sediment yield represents the removal of 7.8 × 10$^9$ m$^3$ yr$^{-1}$ (about 0.05 mm yr$^{-1}$ worldwide). However, it is not clear how well this value reflects natural erosion rates. Trimble (1975) cautioned that agriculture and other human activities may have greatly accelerated erosion, and that sediment yields could be as little as 5% of erosion rates due to the storage of particulates as colluvial and alluvial sediment. Glasby (1988) cited estimates that the total sediment yield before human intervention was about one-half the present value.

Table 3-7 shows estimates of dissolved and particulate sediment loads and yields for the continents. Interestingly, Oceania and the Pacific Islands have the highest particulate and total yields. Europe has the highest dissolved yield; it is the only continent in which the dissolved load exceeds the particulate load. Africa has the lowest particulate, dissolved, and total sediment yields due to its generally low relief, widespread resistant rocks, and extensive desert areas.

### 3.2.8   Your Role in the Global Hydrologic Cycle

Western culture tends to view human beings as separate from nature. While we may recognize that we are connected in an ecological sense to the rest of the world—that we depend upon it to supply the

**TABLE 3-7**

Estimated Sediment Loads and Yields by Continent.

| Continent | Particulate | | Dissolved | | Total | |
|---|---|---|---|---|---|---|
| | Load (10$^6$ T yr$^{-1}$) | Yield (T yr$^{-1}$ km$^{-2}$) | Load (10$^6$ T yr$^{-1}$) | Yield (T yr$^{-1}$ km$^{-2}$) | Load (10$^6$ T yr$^{-1}$) | Yield (T yr$^{-1}$ km$^{-2}$) |
| Africa | 530 | 35 | 201 | 13 | 731 | 48 |
| Asia | 6,433 | 229 | 1,592 | 57 | 8,052 | 286 |
| Europe | 230 | 50 | 425 | 92 | 655 | 142 |
| North and Central America | 1,462 | 84 | 758 | 43 | 2,220 | 127 |
| Oceania and Pacific Islands[a] | 3,062 | 589 | 293 | 56 | 3,355 | 645 |
| South America | 1,788 | 100 | 603 | 34 | 2,391 | 134 |

[a] Includes Australia and the large Pacific Islands.

Data from Walling and Webb (1987).

**TABLE 3-8**

Water content and water intake of humans.

|  | **Man** | **Woman** |
|---|---|---|
| Percentage of weight as water | 60 | 50 |
| Weight of water in body (kg) | 42 | 25 |
| Average intake (kg day$^{-1}$) |  |  |
| in milk | 0.30 | 0.20 |
| in tap water | 0.15 | 0.10 |
| in other fluids | 1.50 | 1.10 |
| as free water in food | 0.70 | 0.45 |
| from oxidation of food | 0.35 | 0.25 |
|  | 3.00 | 2.10 |

Data from Harte (1985).

food, clothing, and shelter that are essential for our existence—we tend to think of the environment as being "out there." In fact, each of us is part of the great biogeochemical cycles that have moved matter and energy through the global ecosystem for billions of years. None of the atoms that currently constitute your body was part of you at birth; each atom has a finite residence time within you before it leaves to continue its ceaseless cycling.

Table 3-8 shows the amounts of water in typical humans and the rates of intake of water from various sources. For adults, the average rate of output (via breathing, perspiration, urine, and feces) is essentially equal to the average rate of input. As for other reservoirs, we can calculate the average residence time of water in the typical human male and female by using Equation (2-26). If you perform this computation with the data in Table 3-8, you will see that the residence time of water in both sexes is about 14 days.

Thus, on the average, the water in your body is completely replaced every two weeks. The hydrologic cycle is flowing through you, as well as through the rivers, aquifers, glaciers, oceans, and atmosphere of the world.

### 3.2.9  Climate Change and the Hydrologic Cycle

As is discussed in Section D.2.1, measurements show that atmospheric concentrations of carbon dioxide and other greenhouse gases have been increasing throughout this century. The concentration of carbon dioxide ($CO_2$) has increased from 280 parts per million (ppm) in 1850 to 353 ppm in 1990 and is projected to reach 500 ppm or higher by 2050—a level that is higher than any the earth has experienced in the last million or so years. The concentrations of other greenhouse gases, especially methane and chlorofluorocarbons, are increasing at even faster relative rates (Ramanathan 1988), and they are many times more effective (per molecule) than $CO_2$ at absorbing longwave radiation. It is virtually certain that most, if not all, of these increases are caused by human activities—particularly the burning of fossil fuels and the clearing of forests—and these activities show every sign of continuing, and perhaps accelerating, for at least the next several decades.

Elaborate models of the earth's climate system indicate that the combined effects of these increases in greenhouse gases will be a rise in the average earth-surface temperature of about 1 to 2 C° by 2050 and by as much as 5 C° by 2100.[9] If this warming occurs—and there is mounting evidence that it is under way (Mitchell et al. 1995; Santer et al. 1996; Tett et al. 1996; Harris and Chapman 1997; Kaufmann and Stern 1997)—the earth will become warmer than it has been at any time in human history. This warming will certainly be accompanied by significant changes in the hydrologic cycle, in vegetative patterns, and in sea level, changes that will force major adjustments in the magnitudes, timing, and locations of water demands, supplies, quality, and hazards. Complete examination of the likely hydrologic and water-resources impacts of these climate changes would require a separate book-length treatment, but we provide here a brief introduction and overview of the major considerations.

### Historical Variability of Streamflow

As was indicated in Section 2.5.1, streamflow is climatically determined and is inherently highly variable. Therefore it is instructive to examine historical data on streamflow variability to provide a context for detecting and predicting the effects of climate change on the hydrologic cycle.

As is suggested by Figure 3-32, relative variability of streamflow is much higher in arid regions (Upper and Lower Plains, Southwest) than in humid regions (Northeast, Northwest). Interestingly, time

---

[9]These increases can be simulated via the model described in Box 3-2 by increasing $f$, the fraction of longwave energy emitted by the surface that is absorbed in the atmosphere. (See Exercise 3-3.)

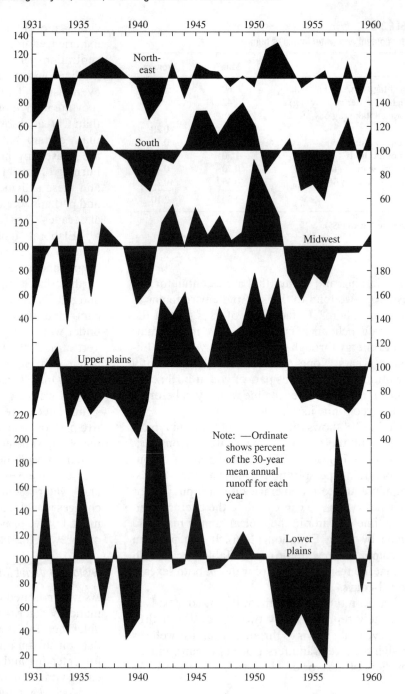

**FIGURE 3-32**
Standardized regional variations in streamflow for the United States for 1931–1960. From Busby (1963).

series of river discharge show considerable synchronism over the United States (Figure 3-32) and at larger scales (Figures 3-33 and 3-34), reflecting large-scale and fairly persistent climatic patterns.

Climate-related persistence is evident in the records of large rivers with long records. The longest streamflow record in the world is that of the Nile River, for which information is available from 622 to 1520 and from 1700 to the present. Riehl and Meitin (1979) found three contrasting patterns of variability in this record: (1) from 622 to about 950, periods of high flow alternated with

**FIGURE 3-32**
(Continued)

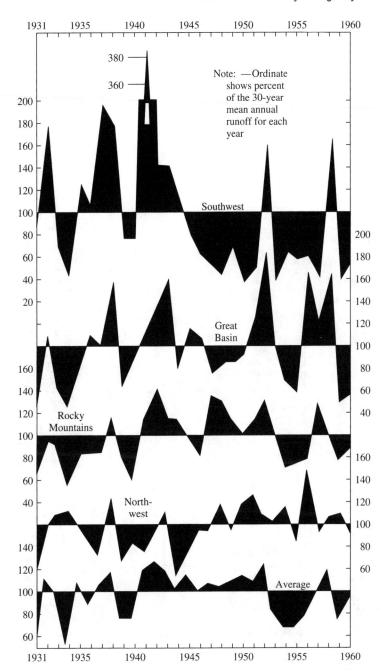

Note: —Ordinate shows percent of the 30-year mean annual runoff for each year

periods of low flow, with each cycle lasting 50 to 90 yr and having a moderate amplitude; (2) from 950 to 1225 there were no major trends or cycles; (3) for the remainder of the record, there were again alternating periods of high and low flow, but having cycles of from 100 to 180 yr and of much higher amplitude than in the first pattern. These very pro-

nounced changes in the pattern of variability appear to be related to global climatic fluctuations; for example, 950–1225 corresponds to the "little climatic optimum," a period of reduced storminess. Subsequent studies have found that Nile flows are influenced by the ENSO cycle (Eltahir 1996).

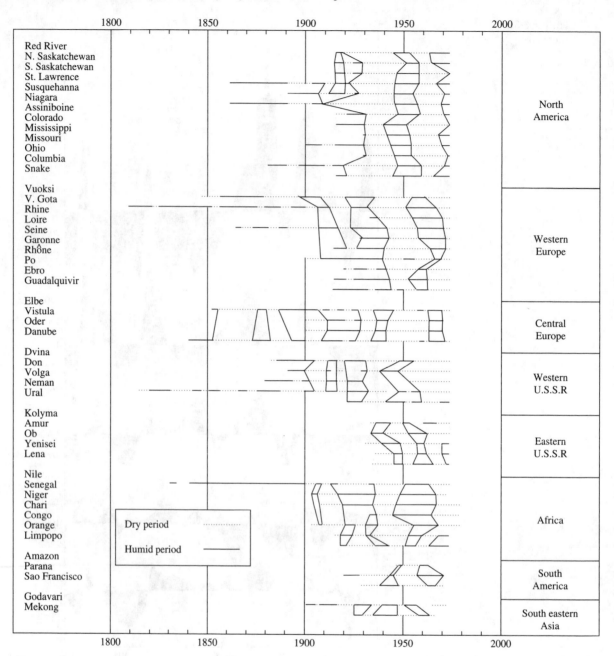

**FIGURE 3-33**
Wet and dry periods in historical streamflow records for 50 major rivers of the world. From Probst and Tardy (1987), used with permission.

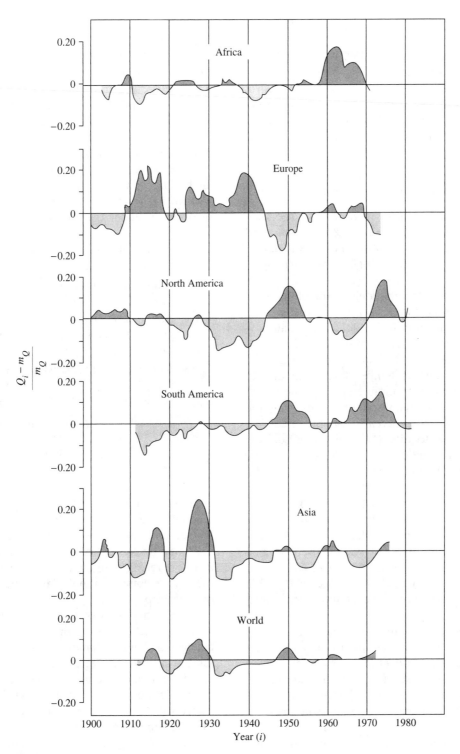

**FIGURE 3-34**
Standardized fluctuations in total runoff for the continents and the world. From Probst and
Tardy (1987), used with permission.

Richey et al. (1989) examined the discharge record of the Amazon which, as we have seen (Table 3-4), contributes some 13% of the total global runoff. They could find no indication of climate or land-use change over the period of record, 1903–1985. However, they did find a two- to three-year period of declining flow following the warm phase of the ENSO cycle; periods of high flow were coincident with the ENSO cold phase.

### Future Climate Change and Water Resources

Box 3-3 summarizes documented recent large-scale and global changes in the hydrologic cycle, and Loaiciga et al. (1996) review the findings and limitations of predictions of hydrologic responses to global warming.

Recent results from models of the global general circulation indicate that projected increases in the $CO_2$ concentration will produce a globally averaged temperature increase of from 1 to 5 C°, with the largest increases at high latitudes (Figure 3-35). These warmer surface temperatures will tend to increase evapotranspiration and the amount of water vapor in the atmosphere, intensifying the hydrologic cycle and increasing global precipitation by from 3% to 15% (Mitchell 1989; Loaiciga et al. 1996). The models predict generally lower precipitation at latitudes below 30° and higher precipitation in the mid-latitudes (Figure 3-36); interestingly, the patterns reported in Box 3-3 are generally consistent with those predicted by the general-circulation models. One can also infer that the higher temperatures will reduce the extent of snow cover, affecting the seasonality of streamflow at latitudes above 40° N. As is noted in Box 3-3, studies have documented recent snowpack reductions.

Experiments indicate that higher $CO_2$ concentrations tend to reduce water use by plants (Lemon 1983), and this could offset increases in evapotranspiration from land surfaces that are due to the temperature effect. Thus one plausible scenario is that evaporation from the oceans will increase, while land evapotranspiration will change little or perhaps even decrease. Interestingly, evaporation from measurement pans in the United States and former Soviet Union has been declining since about 1950 (Box 3-3).

Changes in long-term average runoff can be estimated from changes in precipitation and evapo-transpiration via the water-balance equation (Equation 2-15). Using historical data, Karl and Riebsame (1989) examined the sensitivity of streamflow in the United States to changes in temperature and precipitation. They concluded that 1- to 2-C° temperature changes typically have little effect on streamflow, whereas a given relative change in precipitation is amplified to a one- to six-fold change in relative streamflow.

These conclusions are generally consistent with the findings of Wigley and Jones (1985), who, on the basis of studies with the simple model described in Box 3-4, concluded the following:

1. Changes in runoff are everywhere more sensitive to changes in precipitation than to changes in evapotranspiration (i.e., $\partial q/\partial p > \partial q/\partial e$).[10]

2. The relative change in runoff is always greater than the relative change in precipitation (i.e., $q > p$).

3. Runoff is most sensitive to climatic changes in arid and semi-arid regions, where the runoff ratio, $w$, is small (Table 3-5).

4. The relative change in runoff exceeds the relative change in evapotranspiration (i.e., $q > e$) only in regions where $w < 0.5$.

One set of results from Wigley and Jones (1985) is shown in Figure 3-37; other results can be generated by using the model in Box 3-4 (Exercise 3-13). Wigley and Jones (1985) concluded that, overall, one might expect "very large" increases in average runoff in response to the predicted warming, unless there is a compensatingly large increase in land evapotranspiration.

Box 3-5 describes another simple water-balance approach for estimating the effects of climate changes or land-use changes on the global hydrologic cycle. This model can be used to explore the effects of an increase in ocean evaporation that accompanies no change (or a decrease) in land evapotranspiration. Completion of Exercise 3-12 suggests that a given percentage increase in ocean evaporation (e.g., 6%) gives rise to a smaller relative increase in land precipitation (3.4%) and a larger relative increase in runoff (8%).

---

[10]Symbols are defined in Box 3-4.

## BOX 3-3

· · · · · · ·

## Recent Large-Scale and Global Changes in the Hydrologic Cycle

*Cloud cover*  Cloud cover increased over wide areas of the globe since 1900 (Henderson-Sellers 1992; Karl et al. 1993; Dai et al. 1997).

*Precipitation*  Precipitation increased in mid-latitudes and decreased in low latitudes over the last 30 to 40 yr (Bradley et al. 1987).

Precipitation increased in many areas since 1900 (Karl et al. 1993; Wilmott and Legates 1991; Dai et al. 1997).

Precipitation increased in southern Canada by 13% and in the United States by 4% during the last 100 years; the greatest increases were in eastern Canada and adjacent regions of the United States (Groisman and Easterling 1994).

Precipitation increased by up to 20% in Canada north of latitude 55° (Groisman and Easterling 1994).

Decadal to multidecadal variability of global precipitation increased since 1900 (Tsonis 1996).

Proportion of precipitation occurring in extreme one-day events increased in the United States in the last 30 to 80 yr (Karl et al. 1995).

Fall precipitation increased in the central United States between 1948 and 1988 (Lettenmaier et al. 1994).

*Snow*  Areal snow cover in the northern hemisphere declined 10% in the past 20 yr (Groisman et al. 1994).

Areal snow cover in North America declined 8% in the past 19 yr (Karl *et al.* 1993).

*Glaciers*  Most arctic glaciers experienced a net loss of water since 1940, contributing 0.13 mm yr$^{-1}$ to sea-level rise (Dowdeswell et al. 1998).

*Evapotranspiration*  Pan evaporation in the United States and former Soviet Union declined since 1950 (Peterson et al. 1995).

Plant growth in northern high latitudes increased from 1981 to 1991 (Myneni et al. 1997).

26% of global evapotranspiration was directly used by humans (Postel et al. 1996).

*Streamflow*  Streamflow increased in the European part of the former Soviet Union (Georgievsky et al. 1995).

Winter–spring streamflow strongly increased at over 50% of United States gaging stations from 1948 to 1988, with the strongest trends in the north-central region (Lettenmaier et al. 1994).

Streamflow increased, especially in fall and winter, during past 50 yr in most of the conterminous United States (Lins and Michaels 1994).

54% of geographically and temporally accessible streamflow was directly used by humans (Postel et al. 1996).

77% of the flow of the 139 largest river systems in the United States, Canada, Europe, and the former Soviet Union was moderately to strongly affected by reservoir regulation, diversion, and irrigation (Dynesius and Nilsson 1994).

Volume of water in global river systems was increased 700% due to dams (Vörösmarty et al. 1997a).

Average residence time of water in global river systems was tripled due to dams, which caused changes in flow regimes and water quality (Vörösmarty et al. 1997a).

*Sediment Transport*  Global sediment transport to oceans doubled in the last 2500 yr (Milliman and Syvitski 1992); however, 16% of the current sediment being transported is trapped in reservoirs (Vörösmarty et al. 1997b).

*Sea Level*  Sea level increased at a rate of 2.4 mm yr$^{-1}$ throughout the 20th century (Peltier and Tushingham 1991).

At least 0.54 mm yr$^{-1}$ of net sea-level rise (20 to 30% of total) was caused by human intervention in the hydrologic cycle (ground-water mining, reduction in volume of Aral and Caspian seas, desertification, deforestation, and wetland drainage, minus reservoir storage; Sahagian et al. 1994).

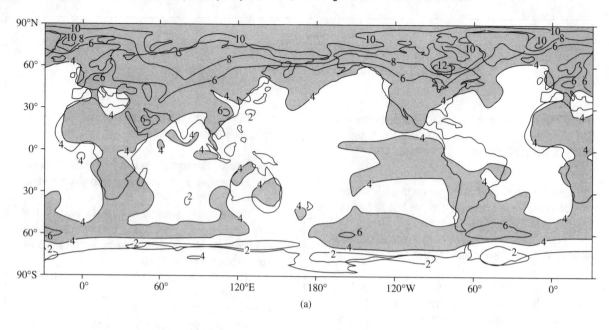

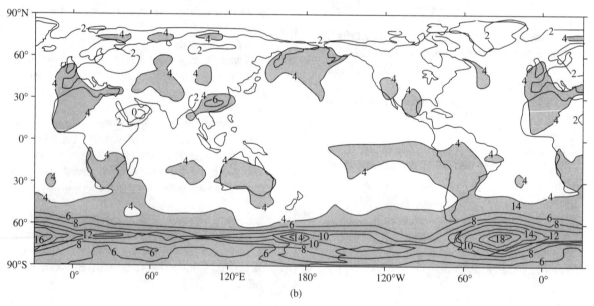

**FIGURE 3-35**
Global distribution of changes in surface temperature due to a doubling of $CO_2$ as projected by one model. (a) December–February average; (b) June–August average. Contours every 2 C°; increases > 4 C° are stippled. From Mitchell (1989), used with permission of the American Geophysical Union.

Even though the models in Boxes 3-4 and 3-5 are extremely simple, they give results that are in general agreement with those of very complex general-circulation models: the higher surface temperatures will increase global precipitation [in the range of from 3% to 11% (Wigley and Jones 1985)]; this increase will probably lead to even greater relative increases in runoff. More detailed studies in the mid-latitudes predicted that global warming will lead to shorter winters, reduced snowpacks and snowmelt runoff, larger winter floods, drier summers, and increased temporal variability. Interest-

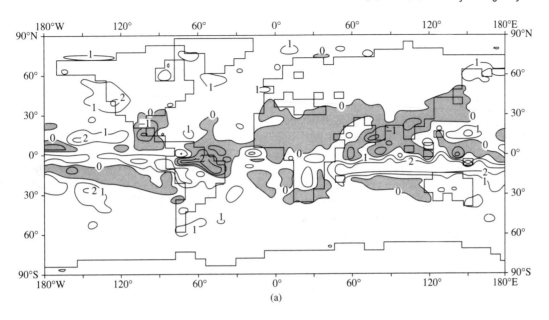

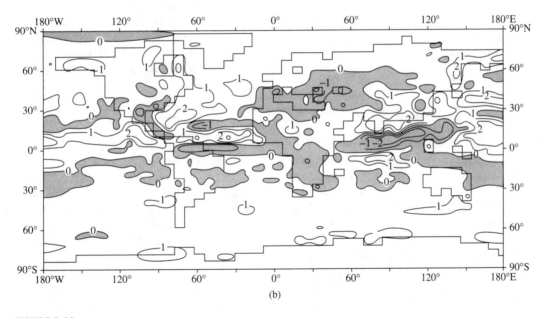

**FIGURE 3-36**

Global distribution of changes in precipitation due to a doubling of $CO_2$ as projected by one model. (a) December–February average; (b) June–August average. Contours are 0, $\pm 1$, and $\pm 2$ mm day$^{-1}$; areas of decrease are stippled. From Mitchell (1989), used with permission of the American Geophysical Union.

ingly, many observed changes, including increasing streamflow trends, are consistent, at least in direction, with model predictions (Box 3-3).

However, we must keep in mind that there is considerable uncertainty in the predictions of even the most elaborate general-circulation models, because they operate on very large grid scales (8° lati-

tude by 10° longitude) and contain only crude representations of important hydrologic processes, especially cloud formation and evapotranspiration. The proportion of the increased atmospheric water vapor that becomes clouds, and the nature of those clouds, cannot be predicted with certainty, but they have pronounced effects on the earth's heat bal-

*[handwritten note at top:]* $w$ = long term average discharge / long term average precipitation

---

## BOX 3-4
· · · · · · ·
### Model for Estimating Effects of Climatic Change on Runoff

Following Wigley and Jones (1985), we begin with the water-balance equation for a drainage basin [Equation (2-16)]:

$$Q = P - ET, \qquad \text{(3B4-1)}$$

where $Q$, $P$, and $ET$ are long-term average values of runoff, precipitation, and evapotranspiration, respectively. Designating the **runoff ratio,** defined in Table 3-4, as $w$, we have

$$Q = w \cdot P \qquad \text{(3B4-2)}$$

and

*[handwritten: $ET = $ ...]*

$$ET = (1 - w) \cdot P. \qquad \text{(3B4-3)}$$

Now suppose a change in climate causes both precipitation and evapotranspiration to change by the relative amounts $p$ and $e$, respectively, so that

$$P_1 = p \cdot P_0 \qquad \text{(3B4-4)}$$

and

$$ET_1 = e \cdot ET_0, \qquad \text{(3B4-5)}$$

where subscript 0 indicates present values and subscript 1 indicates the new values. Combining Equations (3B4-1)–(3B4-5), we can write

$$
\begin{aligned}
q &\equiv \frac{Q_1}{Q_0} = \frac{P_1 - ET_1}{P_0 - ET_0} \\
&= \frac{p \cdot P_0 - (1 - w) \cdot e \cdot P_0}{P_0 - (1 - w) \cdot P_0} \\
&= \frac{p - (1 - w) \cdot e}{w}.
\end{aligned}
\qquad \text{(3B4-6)}
$$

Equation (3B4-6) gives the relative change in runoff as a function of the present runoff ratio and the relative changes in precipitation and evapotranspiration. Table 3-4 gives values of $w$ for the world's largest rivers, and Exercise 3-13 gives you an opportunity to experiment with Equation (3B4-6).

---

ance (Figure 3-2). The magnitude, and even the direction, of the change in evapotranspiration is difficult to assess because many factors are involved besides the direct responses to higher temperatures and increased $CO_2$, including changes in length of growing season, in area of plant cover, in plant species, in wind speed, and in cloudiness. Uncertainty is increased because there is considerable feedback: Modeling studies show that land evapotranspiration can strongly influence global temperature and precipitation (Shukla and Mintz 1982; Loaiciga et al. 1996).

Some of the intricate complexities of the global hydrologic cycle that further confound predictions can be appreciated in the modeling studies described by Eagleson (1986). In these studies, a general-circulation model of the earth was used to "trace" the water vapor introduced into the atmosphere in a one-day pulse of evapotranspiration from selected regions (rectangles of 8° latitude by 10° longitude), to see where it fell as precipitation over the subsequent two months. The results for three cases are shown in Figure 3-38. These cases

suggest that deforestation of the Amazon will have its greatest effect on precipitation in South America (Figure 3-38a), that deforestation in Southeast Asia could affect precipitation over much of the northern hemisphere (Figure 3-38b), and that drainage of huge wetlands in the Sudan could affect precipitation over much of Africa and Europe (Figure 3-38c).

Thus we see that the large-scale land-use changes currently taking place could interact with changes due to global warming, perhaps reinforcing them in some areas and weakening them in others. Clearly there is much to learn about the global hydrologic cycle and its complex feedbacks with human activities, and there are many potentially fruitful avenues of study. Eagleson's (1986) summary comments are an apt conclusion for this brief overview of our current understanding:

> Because of humanity's sheer numbers and its increasing capacity to affect large regions, the hydrologic cycle is being altered on a global scale with consequences for the human life support system

**FIGURE 3-37**
Runoff changes due to various percentage changes in precipitation as a function of runoff ratio, *w*, assuming that future evapotranspiration is 70 % of the present value ($e = 0.7$). For a region with $w = 0.2$, a 20% increase in precipitation yields a more than 200-% increase in runoff; for $w = 0.6$, the increase would be about 60%. Reprinted with permission from *Nature* (Wigley and Jones 1985), © 1985, Macmillan Magazines Ltd.

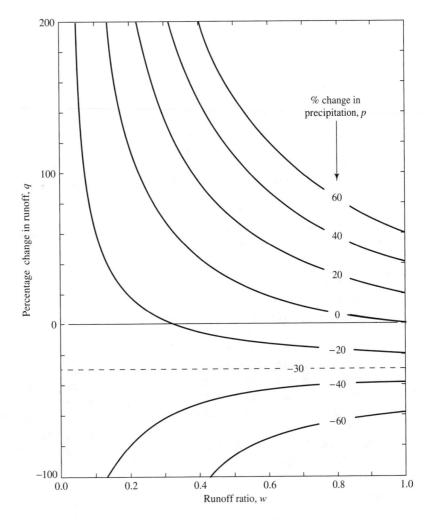

that are often counterintuitive. There is a growing need to assess comprehensively our agricultural, urban, and industrial activities, and to generate a body of knowledge on which to base plans for the future. It seems safe to say that these actions must come ultimately from global-scale numerical models of the interactive physical, chemical, and biological systems of the earth. Of central importance among these systems is the global hydrologic cycle, and its representation in these models presents many analytical and observational challenges for hydrologists.

As was noted in Section 1.3, a major challenge for hydrologists is to establish the linkage between local-scale and global-scale processes, and this relationship is the subject of much current research. The subsequent chapters of this book develop the basic aspects of the processes that control the land phase of the hydrologic cycle and provide the foundation necessary to establish those links.

## 3.3   CLIMATE, SOILS, AND VEGETATION

### 3.3.1   Climate and Soils[11]

Soils are formed by the physical and chemical breakdown of rock, and the types and rates of these processes depend largely on temperature and the availability of water. Thus climate, along with the type of geologic parent material, the actions of biota (which are largely determined by climate), the

---

[11]Much of the discussion in this section is based on Donahue et al. (1983).

---

## BOX 3-5

· · · · · · ·

### Global Water-Balance Model

Following Harte (1985), we formulate a simplified model based on the global water balance depicted in Figure 3-16 and Table 3-1. This model can be used to evaluate the effects of changes in evapotranspiration from land and evaporation from the oceans on other flows and stocks. These changes might be due to alterations of climate (as discussed in the text) or of land use (deforestation would reduce evapotranspiration from the land.)

To develop this model, we define the following terms:

$P =$ global precipitation rate;

$P_L =$ rate of precipitation on land;

$P_S =$ rate of precipitation on sea;

$Q =$ rate of runoff from land to sea;

$E =$ global evapotranspiration rate;

$E_L =$ rate of evapotranspiration from land;

$E_{LS} =$ rate of evapotranspiration from land of water that falls as precipitation on the sea;

$E_{LL} =$ rate of evapotranspiration from land of water that falls as precipitation on the land;

$E_S =$ rate of evapotranspiration from sea;

$E_{SS} =$ rate of evapotranspiration from sea of water that falls as precipitation on the sea;

$E_{SL} =$ rate of evapotranspiration from sea of water that falls as precipitation on the land.

All the above quantities have dimensions $[L^3 \, T^{-1}]$ and represent long-term average flux rates.

We can write the following water-balance equations. For the sea,

$$P_S = E_{SS} + E_{SL} - Q; \qquad \text{(3B5-1)}$$

for the land,

$$P_L = E_{LS} + E_{LL} + Q. \qquad \text{(3B5-2)}$$

The flux of water from land to sea must balance that from sea to land, so

$$Q + E_{LS} = E_{SL}, \qquad \text{(3B5-3)}$$

and it also must be true from the above definitions that

$$P_L = E_{LL} + E_{SL} \qquad \text{(3B5-4)}$$

and

$$P_S = E_{LS} + E_{SS}. \qquad \text{(3B5-5)}$$

Taking values of $P_S$, $P_L$, and $Q$ as given in Figure 3-16, we can use the above relations and one additional equation to compute the values of the remaining water-balance components under present conditions. The additional equation required expresses the ratio of land evapotranspiration ($E_L$) that subsequently falls as precipitation on land ($E_{LL}$) to that which falls on the sea ($E_{LS}$)—that is, the value of $k$ in

$$E_{LL} = k \cdot E_{LS}. \qquad \text{(3B5-6)}$$

The value of $k$ is not known very precisely (but note the discussion in the text of climatic models that attempt to trace the destination of water evaporated from portions of the continents). Following Harte (1985), we will initially assume $k = 3$; Exercise 3-12 allows you to investigate the consequences of assuming other values.

---

effects of topography, and time, is one of the principal factors determining the nature of the soil at any location.

The classification of soils is a complex topic, one that we can explore only briefly in this text. A widely accepted taxonomic scheme defines 10 **soil orders** covering all the soils of the world; classification into these orders is based largely on the degree of development of characteristic **horizons** that result from the operation of soil-forming processes

over time. A very general description of these typical horizons is given in Figure 3-39, and the major features characterizing the soils in each order are given in Table 3-9.

The influence of climate on soil type increases with the passage of time, reducing the influences of parent material and topography. Thus we would expect a reasonably strong relation between climate and soil type on a global scale. This is confirmed by Table 3-9 and by Figure 3-40, which shows the

The computations of the present-day water balance can now proceed by the following steps, which are incorporated in Exercise 3-12:

**P1.** Find $P_L$, $P_S$, and $Q$ from Figure 3-16.
**P2.** Substitute Equation (3B5-6) in Equation (3B5-2) and solve for $E_{LS}$.
**P3.** Use this value of $E_{LS}$ to find $E_{SS}$ from Equation (3B5-5).
**P4.** Compute $E_{LL}$ from Equation (3B5-6).
**P5.** Use this value of $E_{LL}$ to compute $E_{SL}$ from Equation (3B5-4).

The present-day residence times for air, sea, and land are computed via Equation (2-27) by using the values for the stocks given in Table 3-1.

Again following Harte (1985), we can use the above equations as a model to calculate water-balance quantities under conditions in which change in land or sea evaporation is due to changes in climate or land use. For example, we might postulate that (1) if the climate warms, both $E_L$ and $E_S$ will increase or (2) large-scale deforestation might reduce $E_L$. The following steps, which are also incorporated in Exercise 3-12, can be used to compute future values of components [which are indicated by primes (')]:

**F1.** Assume future values of evaporation are related to present values as

$$E_{SS}' = K_S \cdot E_{SS},$$
$$E_{SL}' = K_S \cdot E_{SL},$$
$$E_{LS}' = K_L \cdot E_{LS},$$

and

$$E_{LL}' = K_L \cdot E_{LL}.$$

**F2.** Specify $K_S$ and $K_L$ and use the relations in Step F1 to compute future evaporation quantities.
**F3.** Use Equation (3B5-4) to compute $P_L'$.
**F4.** Use Equation (3B5-5) to compute $P_S'$.
**F5.** Use Equation (3B5-2) to compute $Q'$.

As a first approximation, the model assumes that the residence times of water in the atmosphere and on land do not change under the new conditions; this allows computation of the new volumes of water in those stocks from Equation (2-27) as

$$V_A' = P' \cdot T_{RA} \qquad \text{(3B5-7)}$$

and

$$V_L' = P_L' \cdot T_{RL}, \qquad \text{(3B5-8)}$$

where the $V'$ are the new volumes, the $T_R$ are the unchanged residence times, and the subscripts $A$ and $L$ refer to the atmosphere and land, respectively. The new volume of water in the oceans is computed under the assumption that the total volume of water on earth does not change. Note that these assumptions do not account for the melting of ice caps and glaciers or thermal expansion of water due to any temperature increase or for any other climatic feedback effects.

world-wide distribution of soil orders; this map can be compared with Figures 3-12 and 3-19.

The occurrence of Entisols and some Inceptisols is determined primarily by recent geologic history and topography rather than climate, so soils of these orders are found in many regions. Note, however, that Inceptisols are widespread in the Arctic and Subarctic, where soil-forming processes proceed only slowly. Inceptisols are also found on recent alluvial and colluvial deposits like those of the Mississippi and Amazon valleys and the Himalayas.

The development of soils of the remaining orders is determined mostly by climatic factors, particularly annual temperature, annual precipitation, and seasonal distribution of precipitation. Brief descriptions of the distributions of these soils and their relation to climate follow.

**FIGURE 3-38**
Shaded regions are places where water evaporated in the black rectangles during one day ultimately fell as precipitation over the following two-month period, according to a general-circulation model. (a) Amazon basin in March; (b) Southeast Asia in March; (c) Sudan in January. From Eagleson (1986), used with permission of the American Geophysical Union.

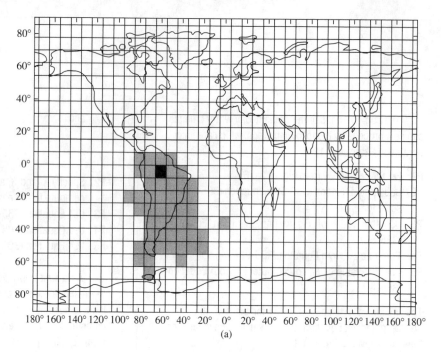

(a)

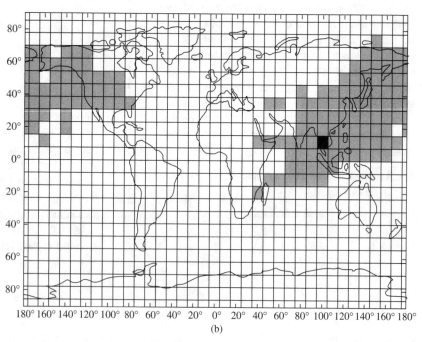

(b)

**Histosols** are concentrated where more than 80% of the growing season (defined as months with average temperature > 10° C) has > 40 mm of precipitation (Lottes and Ziegler 1994). The largest zones of Histosols are north of latitude 50° N (Canada, British Isles).

**Aridisols** occur in desert regions, which are concentrated near 30° latitude. In South America, however, the zone of Aridisols extends southward from 30° in the rain shadow of the Andes, and in Asia these soils are found near 40° in the shadow of the Himalayas.

**FIGURE 3-38**
(Continued)

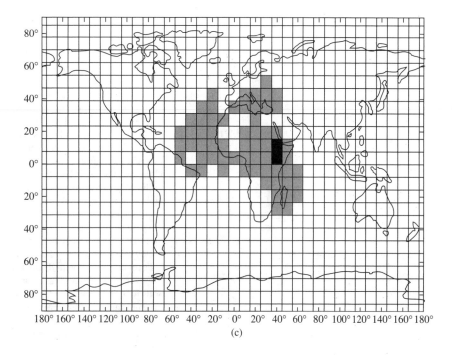

(c)

**FIGURE 3-39**
General features of typical horizons resulting from soil-forming processes. These horizons vary in thickness in various soils (and may be absent in some). Transition zones between horizons can often be identified. [See, for example, Donahue et al. (1983).]

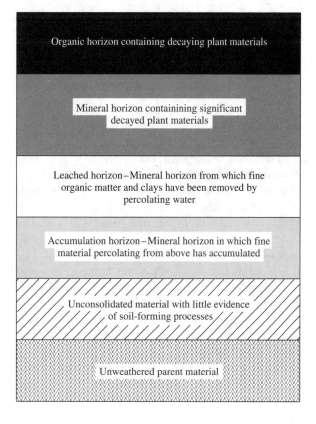

Organic horizon containing decaying plant materials

Mineral horizon containining significant decayed plant materials

Leached horizon–Mineral horizon from which fine organic matter and clays have been removed by percolating water

Accumulation horizon–Mineral horizon in which fine material percolating from above has accumulated

Unconsolidated material with little evidence of soil-forming processes

Unweathered parent material

**TABLE 3-9**
Brief characterization of the 10 world soil orders.

| Order | General Features |
|---|---|
| Entisols | Unconsolidated deposits with virtually no soil development (e.g., recent alluvium, volcanic ash, desert sands). Cover 8.3% of world land surface. |
| Inceptisols | Usually moist soils with weak to moderate development of horizons due to cold climate, waterlogging, and/or lack of time. Cover 8.9% of world land surface. |
| Histosols | Organic soils (peat and muck) consisting largely of plant remains in bogs, marshes, and swamps. Cover 0.9% of world land surface. |
| Aridisols | Usually dry soils with little organic matter; form in dry climates. Cover 18.8% of world land surface. |
| Mollisols | Soils with deep organic horizon; usually associated with grasslands and some broadleaf forests. Cover 8.6% of world land surface. |
| Vertisols | Soils with deep organic horizon and high concentrations of clay minerals that swell when wet and shrink when dried. Form in climates with distinct wet and dry seasons. Cover 1.8% of world land surface. |
| Alfisols | Soils with well-developed accumulation horizon and sometimes a leached horizon. Form where precipitation averages 500 to 1300 mm $yr^{-1}$, usually under forests. Cover 13.2% of world land surface. |
| Spodosols | Soils with well-developed organic, leached, and accumulation horizons. Usually form in cool, wet climates under forests. Cover 4.3% of world land surface. |
| Ultisols | Usually moist, extensively weathered soils with well-developed leached and accumulation layers. Form in humid tropical or subtropical climates under forest or savanna. Cover 5.6% of world land surface. |
| Oxisols | Usually moist, excessively weathered soils consisting mostly of clay minerals containing few mineral nutrients. Form in humid tropical or subtropical areas, usually under hardwood forests. Cover 8.5% of world land surface. |

Largely from Donahue et al. (1983).

**Mollisols**, which include some of the naturally most productive and hence most widely cultivated soils, occur in climates ranging from temperate to cool and semiarid to humid. They are concentrated in the grassland belts north of the Aridisol belts of the northern hemisphere, and are also found near 30° S in central South America.

The development of **Vertisols** depends on the presence of clay minerals that swell when wet and shrink when dry; hence, it is determined in part by the nature of the parent material. However, these soils are most commonly associated with climates that experience a pronounced alternation of wet and dry seasons.

**Alfisols** are naturally fertile soils that occur in large regions to the north of the Mollisols in the northern hemisphere, as well as in several regions between about 35° N and S. These areas have subhumid to humid climates and typically support grassland, savanna, or hardwood forests.

**Spodosols** develop in well-drained sites in cool, wet climates under hardwood and conifer forests. They are widespread in the northeastern United States and southeastern Canada and in a large belt north of 60° latitude in Scandinavia and the former Soviet Union.

Most **Ultisols** are confined to within 20° of the equator, where climates are humid and subtropical or tropical and soil-forming processes are intense. There are also large areas of these soils in the southeastern United States and in southeastern China.

The extensively-weathered **Oxisols** are confined to the tropical and subtropical rain forests on ei-

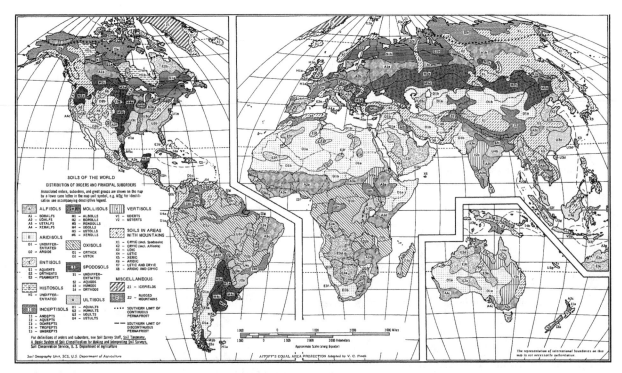

**FIGURE 3-40**
Global distribution of soil orders. Map prepared by U.S. Department of Agriculture.

ther side of the equator, where intense leaching has been occurring for long periods of geologic history.

The seasonal or continuous occurrence of soil temperatures below 0 °C is a climatic factor with important hydrologic implications, because water in the solid state is essentially immobile. Hydrologically significant seasonal freezing of soil occurs in many winters over much of the northern hemisphere land areas above 40° latitude (Figure 3-40). However, the depth and extent of seasonal freezing are highly dependent on local surface conditions, especially vegetative cover and snow depth, and on the severity of winter temperatures. Impermeable frozen ground can significantly accelerate runoff from rain or snowmelt, and hence exacerbate flooding (Dingman 1975).

**Permafrost** is the condition in which soils and/or their underlying parent materials remain at temperatures below 0 °C throughout the year, with only a thin surface layer thawing in the summer. Figure 3-40 delineates areas in which this condition is spatially continuous and those in which it is dis-

continuous; in the latter areas, permafrost is typically present under north-facing slopes and absent under south-facing slopes (in the northern hemisphere). Permafrost bottom depths range from 60–90 m at the southern edge of the continuous-permafrost zone to up to 1000 m in northern Alaska and arctic Canada (Brown and Péwé 1973). Permafrost is almost always a barrier to the movement of water (Williams and van Everdingen 1973), so its presence controls the percolation of infiltrated water and the movement of ground water and thereby exerts a major influence on the hydrologic cycle (Dingman 1973).

### 3.3.2   Climate and Vegetation

Whittaker (1975) identified six major structural types of land vegetation: forest; woodland (dominated by small trees, generally widely spaced and with well developed undergrowth); shrubland (dominated by shrubs, with total plant coverage exceeding 50% of the land area); grassland; scrubland (dominated by shrubs, with plant coverage between 10 and 50%); and desert (plant coverage below

**TABLE 3-10**

Biome types[a] identified by Whittaker (1975). Numbers correspond to Figure 3-42.

| Structural Types | | | | | |
|---|---|---|---|---|---|
| **Forest** | **Woodland** | **Shrubland** | **Grassland** | **Scrubland** | **Desert** |
| Tropical rain forests (1) | Elfin woods (7) | Temperate shrublands (11) | Savanna (12) | Warm semidesert scrublands (17) | True deserts (20) |
| Tropical seasonal forests (2) | Tropical broadleaf woodlands (8) | Alpine shrublands (14) | Temperate grasslands (13) | Cool semideserts (18) | Arctic-alpine deserts (21) |
| Temperate rain forests (3) | Thornwoods (9) | Tundra (16) | Alpine grasslands (15) | Arctic-alpine semideserts (19) | |
| Temperate deciduous forests (4) | Temperate woodlands (10) | | | | |
| Temperate evergreen forests (5) | | | | | |
| Taiga (6) | | | | | |

[a]Biomes are realizations of biome types on particular continents.

10%). The occurrence of these structural types in various climatic zones produces 21 major terrestrial biological communities, called **biome-types** (Table 3-10). The global distribution of these biome-types is summarized in Figure 3-41.

Climate is the dominant control on the geographical distribution of plants (Woodward 1987), and each biome-type is associated with a particular range of mean annual temperature and mean annual precipitation (Figure 3-42). The exact mechanism by which climate affects vegetation type is the object of current research. Eagleson (1982) has developed a theory in which climate, soil, and vegetative type evolve synergistically: In drier climates, where the availability of water is limiting, the character of the vegetative cover adjusts to maximize soil moisture; in moist climates, where available radiant energy is limiting, there is an ecological pressure toward maximization of biomass productivity. Recent studies using hydrometeorologic models suggest that vegetation type may be determined by the balance between precipitation and evapotranspiration, along with thermal controls on growth (Woodward 1987). In North America, Currie and Paquin (1987) found a high correlation between the numbers of tree species and average annual evapotranspiration, and Wilf et al. (1998) documented a close relation between average leaf area and mean annual precipitation across a range of climates.

## EXERCISES

Exercises marked with ** have been programmed in EXCEL on the CD that accompanies this text. Exercises marked with * can advantageously be executed on a spreadsheet, but you will have to construct your own.

*3-1.  Use the model described in Box 3-1 to explore and compare the sensitivity of planetary temperature ($T_p$) to changes in (a) planetary albedo ($a_p$) and (b) the solar flux ($S$). For comparisons, it is most meaningful to express sensitivity in relative terms, as the fractional change in $T_p$ in response to a given fractional change in $a_p$ and $S$.

3-2.  Following the steps described in Box 3-2, derive Equation (5B2-4); then derive the expressions for $T_l$ and $T_u$ in terms of $T_s$ and parameters.

**3-3.  In the model described in Box 3-2, the greenhouse effect can be modeled by increasing the fraction, $f$, of longwave radiation from the surface that is absorbed in the atmosphere. Use the EXCEL program SURFTEMP.XLS to explore the sensitivity of $T_s$ to increases in $f$. Graph $T_s$ as a function of $f$ ($f < 1.00$).

3-4.  For the region in which you live, obtain information from the U.S. Geological Survey, the U.S. Weather Service, or other appropriate federal or state agencies (see Appendix G for Internet information sources) to establish (a) the long-term average precipitation, runoff, and

**FIGURE 3-41**

Global distribution of biome types identified by Whittaker (1975). (See Table 3-10; some biome types are combined for simplification.) Reprinted with permission of Macmillan Publishing Co. from *Communities and Ecosystems*, 2nd ed., by Whittaker, © 1975 by R.H. Whittaker

Polar ice cap

Tundra and Alpine

Boreal forest (Taiga)

Temperate forest

Mediterranean Sclerophyll types

Woodland

Temperate grassland

Tropical savanna

Desert and semidesert

Tropical thornwood

Tropical seasonal forest

Tropical rainforest

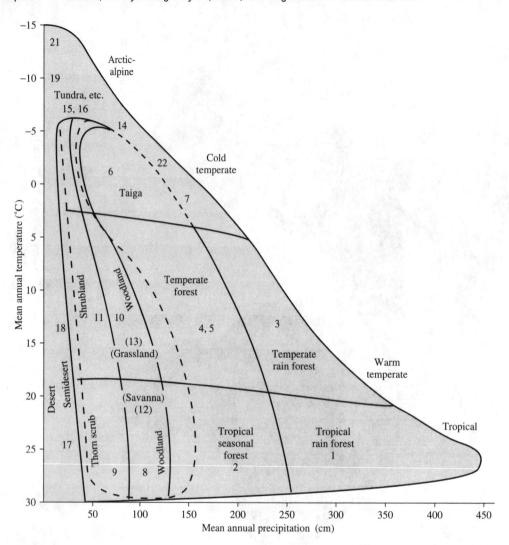

**FIGURE 3-42**

Relation of world biome types to mean annual temperature and mean annual precipitation. (Numbers as in Table 3-10.) For climates within the dot-and-dash line, maritime vs. continental climates, soil types, and fire history can shift the balance between woodland, shrubland, and grassland. Reprinted with permission of Macmillan Publishing Co. from *Communities and Ecosystems*, 2nd ed., by Whittaker, © 1975 by R.H. Whittaker.

evapotranspiration, (b) the seasonal distribution of precipitation and runoff, and (c) the runoff ratio (*w* in Box 3-4). Compare these values with those shown in Figures 3-19, 3-20, 3-21, 3-24, 3-25, 3-26, and Table 3-2.

**3-5.**   For the region in which you live, obtain information from the U.S. Natural Resources Conservation Service (formerly Soil Conservation Service) or from other appropriate federal or state agencies to determine the dominant types of soils (see Appendix G for Internet information sources). Which of the 10 soil orders in Table

3-9 do they belong to? Are they consistent with the distributions shown in Figure 3-40?

**3-6.**   What type of natural vegetation dominates the region in which you live? (See Table 3-10.) Is this consistent with the average precipitation and temperature ranges shown in Figure 3-42?

**3-7.**   Why don't the values in the columns of Tables 3-1 and 3-2 add up to exactly the totals that are given?

**3-8.**   Calculate the residence times for all the global reservoirs in Figure 3-16.

**3-9.** For the region in which you live, obtain information on the typical concentration and composition of dissolved solids in river waters from reports of the U.S. Geological Survey or other appropriate federal or state agency. (See Appendix G for Internet information sources.) Compare these values with the information in Tables 3-5 and 3-6 and Figures 3-28 and 3-29.

**3-10.** For the region in which you live, obtain information on the typical concentration of particulate matter in river waters from reports of the U.S. Geological Survey or other appropriate federal or state agency. (See Appendix G for Internet information sources.) Compare these values with the information in Table 3-7 and Figures 3-30 and 3-31.

**3-11.** Use data from Tables 3-1 and 3-7 to determine (a) what fraction of the global annual runoff passes through your body in a year; (b) what fraction of the earth's fresh water passes through your body in a typical lifetime of 70 yr. Assuming you use 100 gal day$^{-1}$ of water for various purposes, (c) What fraction of the global annual runoff do you use in a year? (d) What fraction of the earth's fresh water do you use in a lifetime?

**\*\*3-12.** The model described in Box 3-5 can be used to simulate the effects of changes in land evapotranspiration and/or ocean evaporation on the global water balance. (a) Use the EXCEL program WATBALEX.XLS to estimate the response of land precipitation and runoff to increases of up to 12% in ocean evaporation (which might be induced by global warming); show results on a graph. (b) Use the same program to explore how increases and decreases in land evapotranspiration would enhance or weaken the effects of increased ocean evaporation. (c) Repeat parts (a) and (b), using different values of $k$, to determine the sensitivity of the results to that parameter.

**\*\*3-13.** The model described in Box 3-4 can be used to simulate the effects of changing precipitation or evapotranspiration (or both) on runoff from a particular drainage basin or region. (a) Refer to Figure 3-37 (which is for $e = 0.7$) and estimate the changes in runoff due to a range of changes in precipitation for the region in which you live. (Use the value of $w$ determined in Exercise 3-4.) Use the EXCEL program DRODPDE.XLS to create graphs similar to Figure 3-37, but for (b) $e = 1$; (c) $e = 1.1$.

# 4
# Precipitation

All water enters the land phase of the hydrologic cycle as precipitation. Thus in order to assess, predict, and forecast hydrologic responses, hydrologists need to understand how the amount, rate, duration, and quality of precipitation are distributed in space and time.

Global patterns of precipitation were outlined in Section 3.2.2, and the physics of precipitation formation is described in Section D.5. This chapter begins with a survey of the meteorology of precipitation. Following that, we focus on methods of measuring precipitation at a point and of estimating precipitation over a region. Estimates of regional precipitation are critical inputs to water-balance and other types of models used in water-resource management.

Sound interpretation of the predictions of such models requires an assessment of the uncertainty associated with their output, which in turn depends in large measure on the uncertainty of the input values, as discussed in Section 2.5.2. The uncertainty associated with a value of regional precipitation consists, in turn, of two parts: (1) that due to errors in point measurements and (2) that due to uncertainty in converting point-measurement data into estimates of regional precipitation. Thus, a central goal of this chapter is to develop an understanding of these errors and uncertainties.

The remainder of the chapter discusses the climatology of precipitation (including methods of characterizing seasonality and extreme values), gives a brief assessment of inadvertent and intentional human influences on precipitation and their

implications for hydrologic analyses, and concludes with an overview of precipitation quality and the phenomenon called "acid rain."

## 4.1  METEOROLOGY OF PRECIPITATION

The formation of precipitation requires a four-step process: (1) cooling of air to approximately the dew-point temperature; (2) condensation on nuclei to form cloud droplets or ice crystals; (3) growth of droplets or crystals into raindrops, snowflakes, or hailstones; and (4) importation of water vapor to sustain the process (Section D.5.4). Radiation, mixing, conduction, and horizontal movement from high to low pressure regions can produce cooling, but not at rates sufficient to produce hydrologically significant precipitation. Such rates occur only when cooling is due to vertical uplift: When a parcel of air rises it cools **adiabatically,** i.e., without the loss of heat. (See Section D.2.3.) Rising dry air cools at a fixed rate of 1 C°/100 m (the dry adiabatic **lapse rate**); moist air cools at a variable but lower lapse rate (about 0.5 C°/100 m) due to the liberation of latent heat as condensation occurs.

Here, we discuss the three meteorological situations in which significant rates of adiabatic cooling by vertical uplift occur and the characteristic spatial and temporal precipitation patterns associated with each.

### 4.1.1 Uplift Due to Convergence

As discussed in Section 3.1.3, horizontal air flow is induced by pressure gradients and movement is generally toward regions of low pressure. Thus low-pressure areas are loci of **convergence;** the air converging from several directions is forced to rise, and the rising produces adiabatic cooling. **Frontal convergence** (or **cyclonic convergence**) is characteristic of the mid-latitudes and occurs at the boundaries between **air masses** of contrasting temperature and/or humidity. **Nonfrontal convergence** is largely a tropical phenomenon that occurs within a mass of warm, moist air.

#### *Extra-Tropical Cyclonic and Frontal Convergence*

Outside the tropics, between approximately 30° and 60° N and S latitudes, much of the precipitation is associated with **extra-tropical cyclones.** As noted in Section 3.1.2, these cyclonic storms are the principal agents of poleward energy transfer in the midlatitudes.

Extra-tropical cyclones are dynamic wave-like features that form at the boundaries between air masses with contrasting temperatures, moisture contents, and densities. Air circulating for several days over a surface with roughly uniform temperature and moisture conditions takes on the characteristics of that surface and can be identified as a particular type of air mass. Table 4-1 gives the general characteristics of the major air-mass types and their regions of origin. These air masses are of subcontinental scale, covering hundreds of thousands of square kilometers. In general, the centers of air masses are areas of high pressure (**anticyclones**), so they are also the centers of clockwise circulation in the Northern Hemisphere.[1]

Figure 4-1 depicts the typical development of an extra-tropical cyclone, a process called **cyclogenesis.** Few actual cyclones follow this sequence exactly, but it provides a framework for understanding the major precipitation-producing processes in the mid-latitudes. Cyclogenesis takes place at an air-mass boundary, or **front,** which is a region about 100 to 200 km wide in which temperature and pressure gradients are relatively large. Atmospheric pressure decreases away from the air-mass centers, and the **isobars** (lines of equal atmospheric pressure) are subparallel. Thus the front develops in a low-pressure trough with relatively cold, denser air to the north and warm, less dense air to the south (Figure 4-1a). If the front is not migrating, it is a **stationary front.** Because the circulations of the air masses are clockwise, wind blows in opposite directions on either side of the front and so creates a zone of **wind shear.** Often, the eddies generated in this zone trigger instabilities in the boundary that appear as waves when viewed in plan.

Cyclonic storms form when the frontal wave develops further, such that the counterclockwise circulation around the wave apex intensifies (Figure 4-1b). To the west of the apex, cold air moves over the surface displacing the warm air, which rises along the frontal surface. This portion of the boundary is now a **cold front.** To the east of the apex, the warm air is displacing cold air at the surface to create a **warm front,** and the warm air rises along the frontal surface here also.

Cloud formation, often leading to precipitation, commonly occurs along one or both frontal surfaces (Figure 4-2a) and at the apex, where the air converging at the center of the counterclockwise circulation must also rise. Typically, cold fronts have relatively steep slopes, about 1 in from 30 to 40, whereas warm fronts have slopes of 1 in from 60 to 120; thus precipitation is usually more intense and areally more concentrated at cold fronts than at warm fronts. Note that the isobars now form a quasi-circular pattern around a low-pressure center at the apex; typically the diameter of a fully developed cyclonic circulation is on the order of 1500 km.

Cold fronts usually move faster than warm fronts, so the evolution of an extra-tropical cyclone typically follows a sequence something like the one shown in Figures 4-1c, d. Where the cold front overtakes the warm front, so that there is colder air everywhere at the surface with warm air above, the front is said to be **occluded.** Although the air-mass contrast at the surface is now weaker, it is still present aloft (Figure 4-2b), and the rising of the warm air over the cold may continue to generate precipitation—in fact, the maximum rainfall intensity usually occurs during the early stages of occlusion (Miller et al. 1983).

---

[1] In the remainder of this section, cyclonic development will be described for the northern hemisphere. For the southern hemisphere, the directions north and south are to be reversed, and the circulations described are in the opposite sense: cyclones circulate clockwise, anticyclones counter-clockwise.

**TABLE 4-1**

Characteristics and Origins of Major Air-Mass Types in the Northern Hemisphere.

| Air Mass | Characteristics | Source Regions | |
|---|---|---|---|
| | | Winter | Summer |
| Continental polar (cP) | Cold, dry | Arctic Ocean | Northern Canada |
| | | Canada—northern United States | Northern Asia |
| | | Eurasia | |
| Continental tropical (cT) | Warm, dry | California-Arizona-Mexico | Nevada-Arizona-northern Mexico |
| | | Northern Africa-Arabia-northern India | Northern Africa-Mediterranean-Arabia-central Asia |
| Maritime polar (mP) | Cold, moist | Northwest Atlantic Ocean | Northernmost Atlantic Ocean-Arctic Ocean |
| | | Northwest Pacific Ocean | Northernmost Pacific Ocean |
| Maritime tropical (mT) | Warm, moist | Central Atlantic Ocean | Central Atlantic Ocean |
| | | Central Pacific Ocean | Central Pacific Ocean |
| | | Arabian Sea-Bay of Bengal | |

Data from Barry and Chorley (1982).

Ultimately, several processes combine to bring the rain production to a halt: (1) the temperature difference between the air masses is decreased by adiabatic cooling of the rising warm air and warming of descending cold air; (2) the pressure contrast that initiated the convergence is reduced; and (3) the inflow of moisture is reduced.

Evolution to the stage shown in Figure 4-1c usually takes 12 to 24 hr, and the process typically is completed in another two or three days (Miller et al. 1983). Because the development takes place in the zone of westerly winds, the low-pressure center at the apex of the cyclonic circulation moves generally eastward, at a speed of about 1000 km day$^{-1}$, as the cyclone evolves.

Figure 4-3 is a weather map showing a typical extra-tropical cyclonic storm over North America, generating precipitation over an area of some 500,000 km$^2$. Precipitation associated with extra-tropical cyclones thus covers wide areas, and persists for from tens of hours to days at a given location. Because rates of uplift are relatively low, precipitation intensities are generally low to moderate. As noted above, precipitation at cold fronts usually covers a smaller area and is more intense than warm-front precipitation; on occasion, rates of uplift at cold fronts are similar to those produced by thermal convection (see below) and produce bands of intense thunderstorms.

On a given winter day there are usually about 10 cyclonic storms in various stages of development through the mid-latitudes. The net result of these storms, in addition to the production of precipitation, is the climatically critical equatorward transfer of colder air and the poleward transfer of warmer air and latent heat of condensation (Figure 3-7).

### Non-frontal Convergence

Non-frontal convergence is associated with (1) the quasi-permanent intertropical convergence zone (ITCZ; see Section 3.1.3) and (2) temporary storms that form seasonally over the tropical oceans.

**Convergence at the ITCZ** The low-latitude Hadley cells (Figure 3-8) create a zone of convergence that circles the globe in tropical regions. The ITCZ migrates seasonally and is spatially discontinuous and intermittent (Barry and Chorley 1982), but it is persistent enough to create the equatorial band of heavy precipitation that is apparent in Figures 3-18 and 3-19.

**Tropical Cyclones (Hurricanes)** Tropical cyclones are cyclonic storms, not associated with fronts, that form over the oceans between 5° and 20° latitude in both hemispheres. They have the potential to develop into extremely intense storms, which are called **hurricanes** (North America), **typhoons** (eastern Asia), **cyclones** (Indian Ocean), and **baguios** (China Sea) (Miller et al. 1983).

Tropical-cyclone formation begins with a small low-pressure disturbance in a maritime tropical air mass. Sea-surface temperatures of at least 27 °C (Miller et al. 1983) are required to induce high rates of evaporation into the converging and rising air;

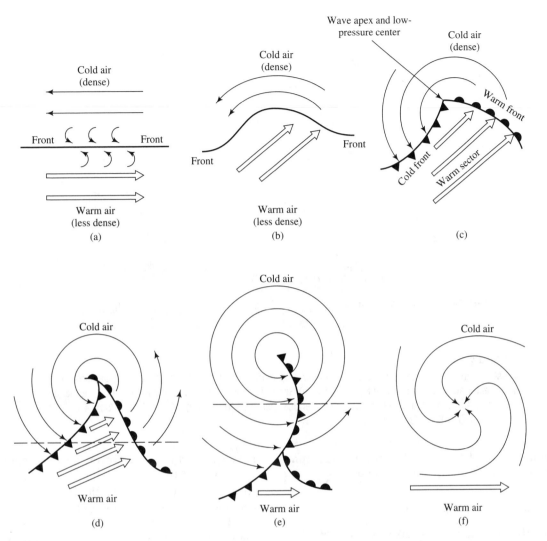

**FIGURE 4-1**

Typical sequence of development of extra-tropical cyclones (map view). Arrows indicate wind flow. (a) Stationary front between two air masses, showing eddies due to wind shear. (b) Initial wave development from eddy. (c) Intermediate stage with distinct cold and warm fronts and warm sector between. (d) Late stage with occlusion beginning. Dashed line is location of cross section in Figure 4-2a. (e) Front largely occluded. Dashed line is location of cross section in Figure 4-2b. (f) Final stage, with fronts dissipated. Used with permission of Prentice Hall, Inc., modified from *Elements of Meteorology*, 4th ed., by Miller et al., © 1983 by Bell & Howell Co.

the cooling of this air then triggers condensation, and the accompanying release of latent heat further fuels the uplift. Thus the cyclone is fed by evaporation and driven by condensation; if conditions are right, the circulation will intensify until winds near the center reach speeds as high as 65 m s$^{-1}$ (140 mi hr$^{-1}$). A fully developed hurricane (by convention, a tropical storm becomes a hurricane when its winds exceed 33 m s$^{-1}$) has the structure shown in Figures 4-4 and 4-5.

Because they form in the belt of easterly winds, the initial movement of hurricanes is usually westward; however, they often move poleward into the zone of westerly winds and may be swept well into the mid-latitudes (Figure 4-6). Because of the reduction in evaporation, their intensity lessens when they move over colder water or, especially, land. However, they can persist for thousands of kilometers over land, moving at speeds of about 5 to 7 m s$^{-1}$ (10 to 15 mi hr$^{-1}$) and delivering very high rates

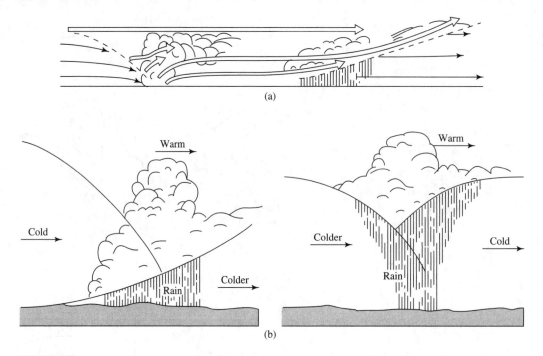

**FIGURE 4-2**
Cross-sectional view of extra-tropical cyclone development. Arrows indicate wind flow. (a) Section through warm sector along dashed line of Figure 4-1d. (b) Sections for two types of occluded fronts along dashed line in Figure 4-1e. Reprinted with permission of Prentice Hall, Inc., from *Elements of Meteorology*, 4th ed., by Miller et al., © 1983 by Bell & Howell Co.

of rainfall, as well as destructive winds, over hundreds of thousands of square kilometers.

### 4.1.2 Uplift Due to Convection

Convective precipitation occurs as a result of adiabatic cooling associated with "parcels" of air that rise because they are less dense than the air that surrounds them. **Thermal convection** is illustrated in Figure 4-7. Line E is the actual temperature in the atmosphere (its slope is the **environmental lapse rate**); the temperature at the surface is initially $T_0$. During the day, intense solar radiation heats the ground surface, and locally the air in contact with the surface is warmed by conduction to a temperature $T_1$. This cell of heated air is now warmer and therefore less dense than the surrounding air, so it rises. As it rises, it cools initially at the dry adiabatic rate (curve D). If the air is relatively dry, it will continue to rise and cool at the dry adiabatic rate until its temperature reaches the ambient temperature (point S). At this point, the atmosphere is stable: further rising would make the parcel colder and denser than the surrounding air, so convection

ceases at this level. However, unstable conditions will persist if the parcel contains enough moisture so that its dew point is reached during uplift, and cooling occurs at the moist adiabatic rate (curve M). In this case, the parcel's temperature remains higher than ambient, and uplift and condensation continue, often leading to significant precipitation. Figure 4-8 shows cumulus clouds formed due to thermal convection over Cuba.

Thermal-convection cells usually cover areas of a few square kilometers. Rates of uplift due to thermal convection can be very high, reaching from 10 to 30 m s$^{-1}$ (Barry and Chorley 1982), in contrast to about 1 m s$^{-1}$ along a typical frontal surface. However, the frictional drag caused by the falling rain acts as negative feedback and reduces the uplift velocity. Thus, these cells produce very intense rain, often accompanied by lightning, thunder, and hail, covering small areas and lasting less than an hour.

### 4.1.3 Uplift Due to Orography

In most regions of the world, long-term mean precipitation increases with elevation; this relationship

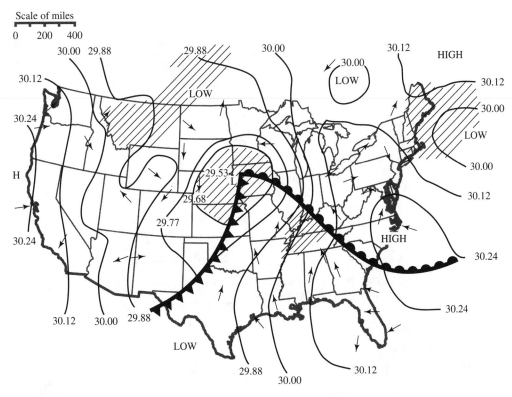

**FIGURE 4-3**
Weather map showing a cyclonic storm over the United States on 22 April 1970. Solid lines are isobars
(pressure in in. Hg). Shaded area is region of precipitation. Arrows show wind direction.

is referred to as the **orographic effect.** Although this effect is widespread, the rates of increase of precipitation with elevation vary widely from region to region (Figure 4-9) and can even reverse at the highest elevations, especially in tropical regions (Barry and Chorley 1982). Barros and Lettenmaier (1994) reported orographic enhancement of precipitation by factors of from 150 to 185% for hills of up to a few hundred meters relief and of more than 300% when relief exceeds 1–2 km.

In the simplest situations, the orographic effect occurs because air moving horizontally encounters a topographic barrier and acquires a vertical component of motion as it flows over that barrier. The rate and degree of cooling are determined by the horizontal wind speed, the wind direction relative to the barrier, the steepness and height of the barrier, and the temperature and humidity conditions that control convective stability.

Clouds and precipitation form on the windward slope of the mountains, and peak precipitation often occurs windward of the topographic crest. The effect commonly persists across the crest for at least a short distance above the leeward slope, but the downward air movement on the lee side causes adiabatic warming. This warming tends to dissipate the clouds and thus turn off the precipitation-producing process and so produces a **rain shadow** (Figure 4-10). Within a region of orographic effect, the precipitation–elevation relation varies as a function of the aspect and inclination of individual slope facets and their relation to local topographic barriers.

The classic orographic situation exists on the central west coast of North America, where several successive mountain ranges are all nearly perpendicular to persistent moist westerly winds from the Pacific Ocean. Figure 4-11 shows the close relation between long-term average precipitation and topography over a distance of 850 km in this region; the reduced precipitation east of the Fraser Valley is presumably due to the fact that the air has lost most of its moisture by this point.

One of the most pronounced orographic effects occurs in conjunction with the monsoon over the

**FIGURE 4-4**
Vertical section of a fully developed hurricane showing patterns of wind, pressure, and rain. Reprinted with permission of Prentice Hall, Inc., from *Elements of Meteorology*, 4th ed., by Miller et al., © 1983 by Bell & Howell Co.

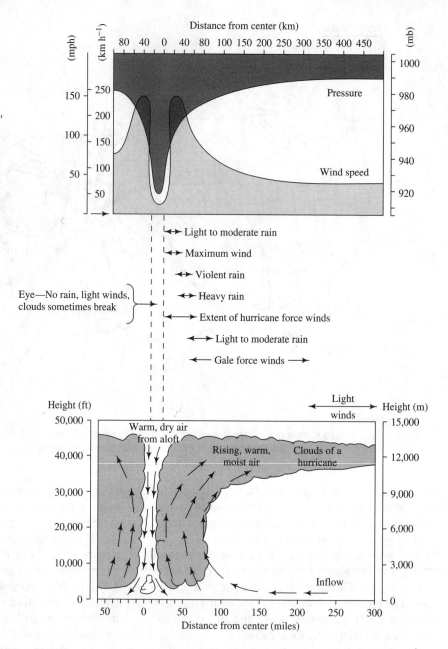

Indian subcontinent. In summer, a northward shift in the position of the intertropical convergence zone (see Section 3.1.3) and accompanying changes in circulation patterns induce moist southerly winds from the Bay of Bengal. These winds converge and rise over the Khasi Hills in Assam, India (north of Bangladesh), producing persistent heavy rains and giving the city of Shillong (elevation 1598 m) one of the highest average annual rainfalls in the world, 2200 mm, more than half of which occurs in June, July, and August.

In many, if not most, situations, orographic effects are the result of convective, frontal, or cyclonic mechanisms interacting with topography rather than a separate precipitation-generating mechanism. Even in the classic situation shown in Figure 4-11, much of the precipitation comes from extra-tropical cyclones; the extra uplift provided by the mountains increases the cyclonic precipitation on windward slopes and the downdrafts reduce it on leeward slopes (Bruce and Clark 1966).

**FIGURE 4-5**
Satellite images of Hurricane Andrew moving across the southeastern United States, 24–26 August 1992. *From NASA Headquarters.*

Under some circumstances, the frictional resistance offered by topographic rises of only a few meters can cause sufficient uplift to produce precipitation (Barry and Chorley 1982); Figure 4-12 shows cumulus clouds induced by relatively small hills in the southern Great Plains.

An increase in precipitation with elevation can be due to (1) an increase in the fraction of time precipitation falls at higher elevations, which could be called the pure orographic effect, and/or (2) higher intensities at higher elevations when precipitation is occurring at all elevations, which would represent enhancement of precipitation triggered by convective or cyclonic disturbances. The first effect is shown in Figure 4-13 for an intensively gaged New England region; it was also found by Hendrick et al. (1978), who reported a three-fold increase in hours of winter precipitation between 400 m and 1200 m elevation in another area of northern Vermont. However, Hendrick et al. (1978) attributed about

75% of the difference in precipitation at the two elevations to higher intensities at the higher elevation when precipitation was falling at both elevations, so that enhancement accounted for most of the orographic effect in that region.

In regions with significant relief and otherwise similar climate, the orographic effect commonly accounts for much of the spatial variation of precipitation. For example, Dingman et al. (1988) found that elevation accounted for 78% of the variation in long-term mean precipitation in New Hampshire and Vermont. Thus one can often exploit the effect in estimating the average precipitation over a region, as in the hypsometric method discussed in Section 4.3.2. The orographic effect can also carry over into many hydrological relations: Dingman (1981) showed that average streamflows, dry-season streamflows, and floods were all significantly related to elevation in central New England.

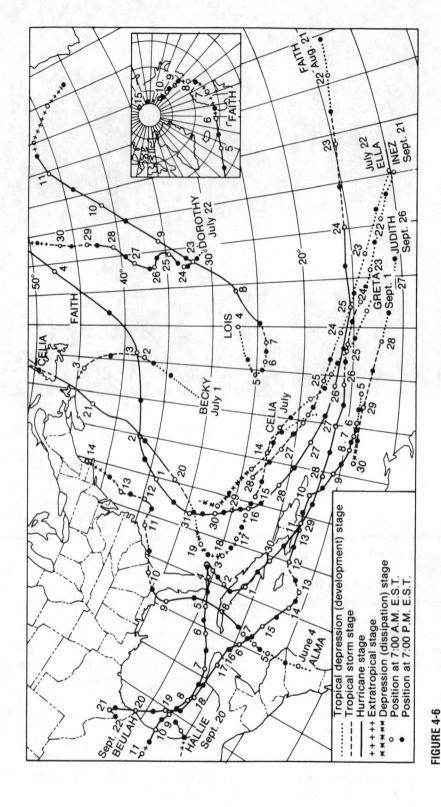

**FIGURE 4-6**

Tracks of 1966 tropical depressions, storms, and hurricanes and of Hurricane Beulah (1967) in the North Atlantic. Positions shown at 12-hr intervals. Reprinted with permission of Prentice Hall, Inc., from *Elements of Meteorology*, 4th ed., by Miller et al., © 1983 by Bell & Howell Co.

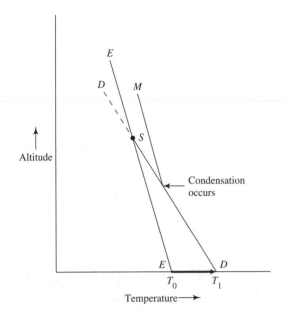

**FIGURE 4-7**
Convection due to surface heating. Line E is the environmental lapse rate (the actual atmospheric temperature as a function of elevation); line D is the dry adiabatic lapse rate; and line M is the moist adiabatic lapse rate. Air at the surface is heated from temperature $T_0$ to $T_1$ by contact with the ground. See text for full explanation.

**FIGURE 4-8**
Fair-weather cumulus clouds formed by thermal convection over Cuba. Note that the clouds are present only over the land—the heat capacity of the sea prevents it from heating sufficiently to induce convection. Photo by author.

### 4.1.4   Critical Temperature for Rain-Snow Transition

Much of the precipitation that falls outside the tropics originally forms as snow or hail in super-cooled clouds (Section D.5.3). The form of precipitation reaching the surface is determined largely by the height of the 0 °C-temperature surface: Rain occurs if that surface is high enough to allow complete melting; otherwise the snow reaches the ground.

Input data for many watershed models consist of the amounts of precipitation falling in a given time period, along with other surface meteorological information for the period, such as temperature, humidity, wind speed, and solar radiation. Information about the form of the precipitation typically is not readily available; therefore one of the most critical decisions the model must make is whether precipitation is rain or snow. A mistake in this decision will result in an erroneous snowpack water balance and will cause the model either to predict a groundwater recharge and/or streamflow event when none occurred or to fail to predict an event that did occur.

The most natural criterion for this decision is the air temperature at the site of interest. We know that air temperature usually decreases with altitude (Section D.2.2) but, almost always, we have information only about surface temperature. Thus it is not immediately obvious what the temperature criterion should be, and we must turn to empirical information for guidance. Auer (1974) examined about 1000 weather observations in which the surface temperature was recorded and the solid or liquid nature of the precipitation was clearly indicated. (Occurrences of freezing rain were omitted.) He calculated the frequency of rain vs. snow reports at each temperature and constructed the graph shown in Figure 4-14, which indicates that rain is virtually never recorded when temperature is less than 0 °C, snow is never observed when temperature exceeds 6 °C, and the probabilities of rain and snow are equal when temperature equals 2.5 °C.

It should be noted that the geographical applicability of Auer's (1974) analysis is not clear, and it may not apply in all regions. Thus, if at all possible, one should attempt to determine the relationship appropriate for a region of interest by examining local data. The method used in the BROOK90 model to separate rain and snow is described in Box 4-1.

**FIGURE 4-9**
Relations between mean annual precipitation and elevation in several regions of North America. Data from Barry and Chorley (1982) and Dingman et al. (1971; 1981).

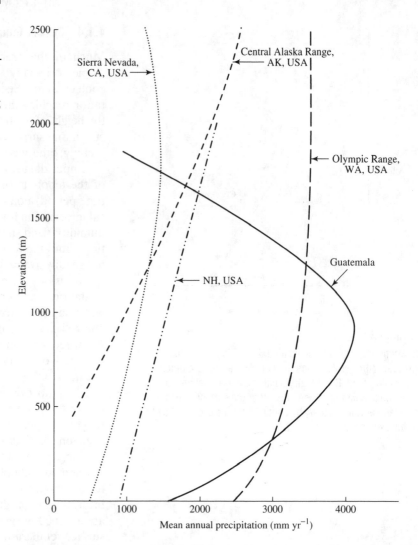

### 4.1.5   Moisture Sources and Precipitation Recycling

For a given region, the fraction of precipitation that entered the atmosphere as evapotranspiration from that region is called the **recycling ratio.** Determining the value of this ratio is important in assessing the potential for human modification of climate. Prior to the 1940s, it was widely thought that most continental precipitation was recycled. This belief underlay several proposals for altering land-use practices in order to increase precipitation and thus prevent recurrences of conditions like the disastrous Dust Bowl of the early 1930s in the south-central U.S. However, a study by Holzman (1937) showed (1) that most of the precipitation falling on the United States comes from maritime air masses that obtain their moisture from the oceans and (2)

most of the evapotranspiration from the land enters continental air masses which move off the continent without yielding much precipitation; thus the recycling ratio is relatively low.

As is shown in Table 4-2, Holzman's (1937) conclusions have been supported in more detailed studies of the Mississippi River basin and Eurasia. However, two large environmentally sensitive regions have been found to have greater degrees of recycling: the Amazon Basin, where it is feared that the destruction of the rain forest will significantly affect at least the regional hydrologic cycle (Lean and Warrilow 1989), and the Sahel region of sub-Saharan Africa, where there are concerns that recycling produces positive feedbacks that will amplify the desertification processes initiated by over-intensive land use (Savenije 1995).

**FIGURE 4-10**
View westward toward the Rocky Mountains in Colorado, showing orographically-induced clouds persisting for many tens of kilometers leeward of the mountains before dissipating. Photo by author.

Precipitation recycling also appears to be at least seasonally important in other regions. Evaporation from large lakes, such as the Laurentian Great Lakes, gives rise to localized regions of high precipitation on their downwind shores, especially in late fall and early winter. Moisture evapotranspired from extensive wetlands during summer days may contribute a significant proportion of that season's precipitation in central Alaska.

The variations in estimates for the same regions and the limited geographical scope of the studies cited in Table 4-2 suggest that more research on recycling is needed to refine our knowledge of the likely impacts of land-use changes on climate, especially in arid regions where the recycling ratio may be high (Eltahir and Bras 1996). Modeling studies like that illustrated in Figure 3-38 are one approach to the problem; other approaches are used in the studies listed in Table 4-2.

## 4.2 MEASUREMENT AT A POINT

Because precipitation is the input to the land phase of the hydrologic cycle, its accurate measurement is the essential foundation for quantitative hydrologic analyses. Unfortunately, there are often many reasons for concern about the accuracy of precipitation data, and these reasons must be understood and accounted for in both scientific and applied hydrologic analyses.

Almost always, hydrologists are concerned about the amount or rate of precipitation over an area such as a drainage basin, so there are two parts to the problem of accuracy: (1) How accurate are point measurements of precipitation? and (2) How accurately can point measurements be converted to measurements over an area? This section addresses the first question.

### 4.2.1 Types of Precipitation Gages

Conceptually, the measurement of precipitation at a point is straightforward: Simply place a vessel open to the air (a **rain gage** or **precipitation gage**) at the point of observation, and periodically measure or continuously record the quantity of water it collects. The volume collected is divided by the area of the opening and recorded as the depth of precipitation.[2]

Nonrecording **storage gages** may be simple straight-sided cylinders or more elaborate devices with funnels and collecting vessels. A **weighing-recording gage** works by introducing the collected water to a vessel on a scale and recording the accumulated weight. A second type of recording gage, known as a **tipping-bucket gage,** introduces the water to one of a pair of vessels with a known small capacity that are balanced on a fulcrum; when one vessel is filled, it tips and empties and records the time of this event, and the other vessel is brought into position for filling. Details of the construction and operation of several types of storage and recording gages were given by Gilman (1964), Brakensiek et al. (1979), and Shaw (1988).

The measurement of snow in precipitation gages is fraught with difficulties. The hydrologist is interested in the amount of water substance that falls, so the catch of frozen precipitation must be melted before measurement; this is often done by placing a charge of antifreeze in storage gages or by activating heating elements in tipping-bucket gages. Less tractable problems arise when snow piles up at the gage orifice and subsequently blows off and, as

---

[2] There is evidence that this approach has been in use at least since 400 B.C.E. in India, where the rainfall data were used in managing agriculture (Neff 1977).

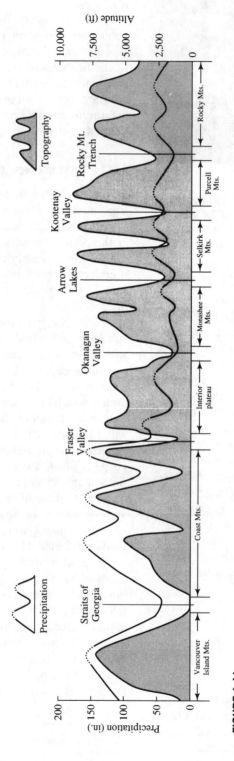

**FIGURE 4-11**

Relation between mean annual precipitation and topography across southern British Columbia. After Bruce and Clark (1966).

**FIGURE 4-12**
Cumulus clouds induced by uplift over small hills in the southern Great Plains of the United States. Photo by author.

will be discussed more fully below, when gage-induced wind eddies prevent significant amounts of snow from entering the gage. For these reasons, it is usually preferable to measure the water content of snow by means other than standard precipitation gages, as discussed in Section 5.2.

A promising new technology is the **optical precipitation gage,** which measures precipitation rate as proportional to the disturbance to a beam between an infrared light-emitting diode and a sensor. These gages can measure rain or snow; they avoid the problems of wind, splash, evaporation, and mechanical failure; and they can record directly to a data-storage device that can be accessed by computer. Although relatively expensive, they appear to be well suited to accurate measurement at remote locations under a wide range of conditions.

**FIGURE 4-13**
Number of days and hours with precipitation as a function of elevation in the Sleepers River Watershed, Vermont. From Engman and Hershfield (1969).

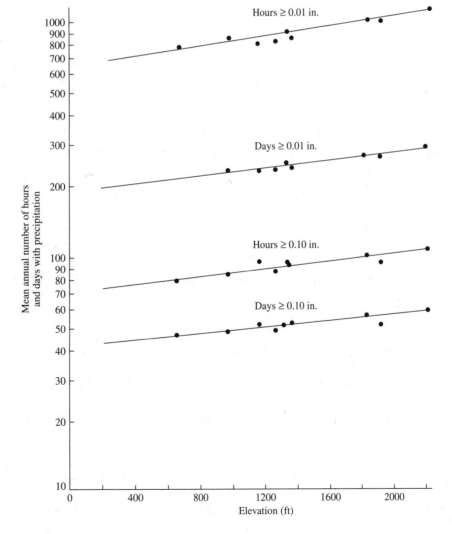

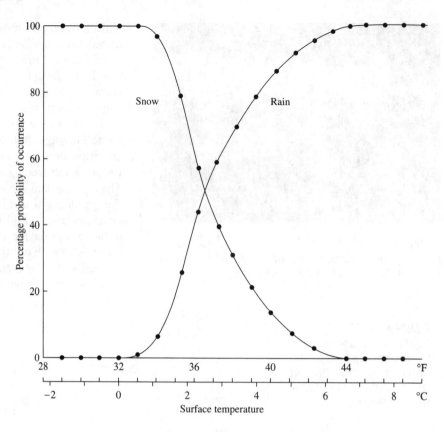

**FIGURE 4-14**
Probability of occurrence of rain or snow as a function of surface (2-m) air temperature. From Auer (1974).

### 4.2.2 Factors Affecting Measurement Accuracy

For conventional gages, serious concerns arise as to the fraction of the true precipitation that enters the gage (the **gage catch**), and we need to address the following questions:

- What size orifice should be used?
- How should the plane of the orifice be oriented?
- How much should the gage protrude above the ground surface?
- Should the gage be installed with a device to reduce wind effects (a **wind shield**)?
- How far should the gage be from other projections (trees, buildings, etc.)?
- How can we prevent water from splashing in or out?
- How can we prevent evaporation of the collected water?

Perhaps surprisingly, there are no universally accepted answers to these questions—virtually every country has its own standards for rain-gage installations and observations (see Table 4-3), and these standards may even differ from agency to agency in a given country. However, the member nations of the World Meteorological Organization (WMO) have agreed to install at least one "interim reference precipitation gage" (IRPG) which can be used as a standard to which measurements in other types of gages can be related (Shaw 1988); the characteristics of this gage are also given in Table 4-3.

In the following sections, we summarize the results of several studies that provide guidance in addressing the questions itemized above.

#### Orifice Size

Studies by Huff (1955) indicate that orifice diameter should not be less than about 30 mm (area about 700 mm$^2$); above this limit, the size of the opening itself has little effect on gage catch of rain in most circumstances. Brakensiek et al. (1979) cited several studies that support this conclusion. However, there are several caveats that should be kept in mind when using small-diameter gages: (1) they are not suitable for snow; (2) the true size of the opening of a small-diameter gage can be significantly affected

---

**BOX 4-1**

. . . . . . .

**Treatment of Precipitation Inputs in the BROOK90 Model**

BROOK90 accepts precipitation input (*PREC*, mm) at daily or more frequent time steps.

The fraction of precipitation that is snow for a given day (*SNOFRC*) is calculated from the daily maximum temperature (*TMAX*, °C) and daily minimum temperature (*TMIN*, °C) as

$$SNOFRC = \frac{RSTEMP - TMIN}{TMAX - TMIN} ; \qquad \textbf{(4B1-1a)}$$

$$SNOFRC = 1, TMAX \leq RSTEMP ; \qquad \textbf{(4B1-1b)}$$

and

$$SNOFRC = 0, TMIN \geq RSTEMP. \qquad \textbf{(4B1-1c)}$$

Here *RSTEMP* (°C) is a base-temperature parameter with a default value of *RSTEMP* = −0.5 °C, which Federer (1995) found to work best at the Hubbard Brook Experimental Forest in central New Hampshire.

---

*Orifice Orientation*

For virtually all situations, the gage orifice should be level because the depth of precipitation falling on a horizontal surface represents the hydrological input. As described by McKay (1970), there may be rare conditions (rains commonly accompanying very strong upslope winds) when such an orientation will result in seriously deficient gage catch; in these cases, it may be appropriate to orient the orifice parallel to the ground surface.

*Orifice Height and Wind Shielding*

Rain gages that project above the ground surface cause wind eddies that tend to reduce the catch of the smaller raindrops and snowflakes (Figure 4-15). These effects are the most common and serious causes of precipitation-measurement errors and have been the subject of several studies (Larson and Peck 1974; Legates and DeLiberty 1993; Yang et al. 1998).

Figure 4-16 shows gage-catch deficiencies for the standard U.S. 8-in. gage for rain, snow, and mixed precipitation, as determined in comparative studies by the World Meteorological Organization. The curves indicate that deficiencies of 10% for rain and well over 50% for snow are common in unshielded gages. To correct for this, daily measured values (after corrections for evaporation, wetting losses, and other factors have been applied, as is discussed subsequently) should be multiplied by the following factors, $K$ (with subscripts indicating rain ($r$), snow ($s$), or mixed precipitation ($m$); and unshielded ($u$) or Alter-shielded ($s$) gages):

**Rain in unshielded gage**

$$K_{ru} = 100 \cdot \exp(-4.605 + 0.062 \cdot v_a^{0.58}) ; \qquad \textbf{(4-1a)}$$

**Rain in gage with Alter shield**

$$K_{rs} = 100 \cdot \exp(-4.606 + 0.041 \cdot v_a^{0.69}) ; \qquad \textbf{(4-1b)}$$

**Mixed precipitation in unshielded gage**

$$K_{mu} = \frac{1}{1.008 - 0.0834 \cdot v_a} ; \qquad \textbf{(4-2a)}$$

by small variations in manufacture, so such gages should be calibrated volumetrically or by weighing; (3) water adhering to the walls of the gage ("wetting loss"), which is increased in plastic gages, can be a significant percentage of the total catch; and (4) evaporation has a greater relative effect on catch, so such gages should be read and emptied frequently.

**TABLE 4-2**

Estimates of Annual Precipitation-Recycling Ratios in Various Regions.

| Region | Recycling Ratio (%) | Source |
|---|---|---|
| Amazon | 25 | Brubaker et al. (1993) |
| Amazon | 25 | Eltahir and Bras (1994) |
| Amazon | 35 | Eltahir and Bras (1994) |
| Mississippi basin | 10 | Benton et al. (1950) |
| Mississippi basin | 24 | Brubaker et al. (1993) |
| Eurasia | 11 | Budyko (1974) |
| Eurasia | 13 | Brubaker et al. (1993) |
| Sahel | 35 | Brubaker et al. (1993) |
| Sahel | >90 | Savenije (1995) |
| West Africa | 44 | Cong and Eltahir (1996) |

**TABLE 4-3**

Specifications for Standard Rain Gages Used by Meteorological Services in Four Countries and by the World Meteorological Organization (WMO).

| Country | Orifice Diameter (mm) | Gage Height (mm) | Height of Orifice above Ground (mm) | Wind Shield? |
|---|---|---|---|---|
| Canada | 91 | 305 | 305 | No |
| United States | 203 | 660 | 787 | No |
| United Kingdom | 127 | 489 | 300 | No |
| | 138 | | 300 | No |
| | 309 | | 300 | No |
| WMO IRPG[a] | 127 | 457 | 1,000 | Alter[b] |

[a]Interim reference precipitation gage (see text).

[b]See text and Figure 4-15.

Information from Shaw (1988) and other sources.

**FIGURE 4-15**

Wind effects of projecting rain gages. (a) Without wind shielding, upward-moving air in eddies prevents many snowflakes from entering the gage. Rigid Nipher-type shields (b) or hinged Alter-type shields (c) reduce this effect. After Bruce and Clark (1966).

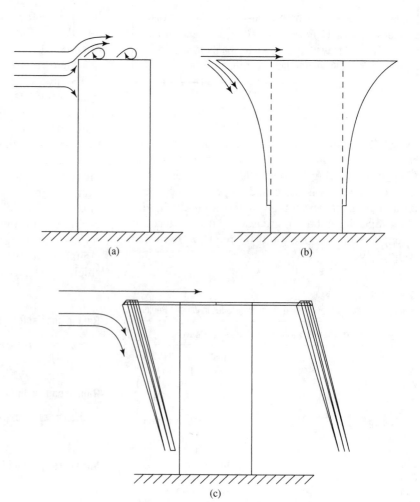

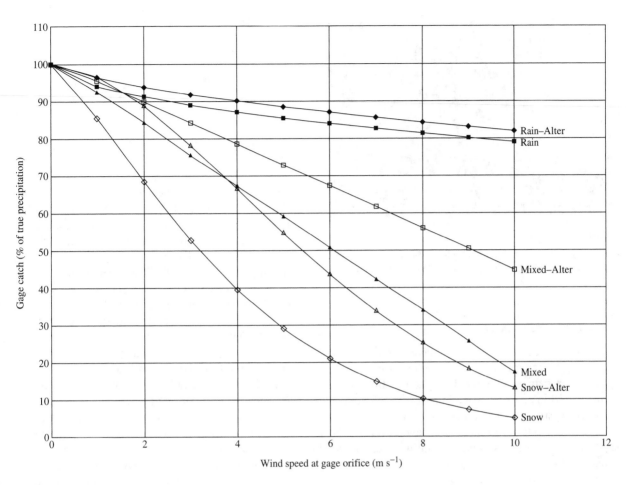

**FIGURE 4-16**
Gage-catch deficiencies for rain, mixed precipitation, and snow as a function of wind speed for U.S. standard 8-in. gages with and without Alter wind shields [Equations (4-1)–(4-3)].

**Mixed precipitation in gage with Alter shield**

$$K_{ms} = \frac{1}{1.010 - 0.0562 \cdot v_a}; \quad \textbf{(4-2b)}$$

**Snow in unshielded gage**

$$K_{su} = 100 \cdot \exp(-4.606 + 0.157 \cdot v_a^{1.28}); \quad \textbf{(4-3a)}$$

**Snow in gage with Alter shield**

$$K_{ss} = 100 \cdot \exp(-4.606 + 0.036 \cdot v_a^{1.75}). \quad \textbf{(4-3b)}$$

In the preceding equations, $v_a$ is wind speed at the gage orifice in m s$^{-1}$ (Yang et al. 1998).

Even though precipitation measured in gages that project above the ground surface is usually significantly less than true precipitation, there are practical considerations that militate against the conversion to ground-level gages such as shown in Figure 4-17. One of these considerations is, of course, cost: It is considerably more expensive to install a ground-level gage than it is to set up a standard projecting gage, and it would be very costly to convert existing gage networks completely. A second consideration is that ground-level gages are useless when a snowpack accumulates. Finally, a conversion to a new gage type would complicate the analysis of historical rainfall data.

**FIGURE 4-17**
Rain gage with orifice at ground level and surrounded by plastic "egg-crate" structure to eliminate in-splashing. Photo by author.

### Distance to Obstructions

Although few studies seem to have been done to determine the optimal siting configuration, some guidelines can be stated. The best location for a gage is within an open space in a fairly uniform enclosure of trees, shrubs, fences, or other objects, so that wind effects are reduced. However, none of the surrounding objects should extend into the conical space defined by a 45° angle centered on the gage (Figure 4-18) (Brakensiek et al. 1979).

Individual trees, buildings, fences, or other isolated objects can produce wind eddies that can significantly affect (usually reduce) gage catch, especially if they are appreciably taller than the gage. As a general rule, an isolated obstruction should not be closer to the gage than twice (preferably four times) its height above the gage (Brakensiek et al. 1979).

### Splash, Evaporation, and Wetting Losses

If the surface of the water captured in a rain gage is too near the orifice, in-falling drops can cause water to splash out. This loss can largely be prevented by the use of deep gages or gages with walls that are vertical or slope outward below the orifice, by conducting the collected water to a covered vessel, and/or by emptying the gage frequently. (See Shaw 1988.)

For gages whose orifices are at the ground surface, one must also prevent in-splashing from drops falling near the gage. This is done either (1) by placing the gage in a small excavation so that its orifice is coincident with the general surface, but well above the immediately adjacent ground; (2) by placing the gage in a shallow excavation in which the near-surface soil has been replaced by an "egg-crate" structure (Figure 4-17), or (3) by covering the soil surface within several meters of the gage with a mat of coarse fiber that prevents splashing [e.g., Neff (1977); Helvey and Patric (1983)].

For small-orifice gages and all non-recording gages that are read only at intervals of several days or more, one must prevent catch deficiencies due to evaporation from the water surface and the walls of the gage (wetting losses). Again, use of a gage in which the water is conducted to a closed vessel is one remedy. Another commonly used technique is to introduce a nonvolatile immiscible oil that will prevent evaporation by floating on the collected water.

Errors due to splashing and evaporation in U.S. standard gages usually are small and can be neglected; however, evaporative losses can be significant where low-intensity precipitation is common (Yang et al. 1998). Correction for wetting losses in U.S. standard gages can be made by adding 0.03 mm

**FIGURE 4-18**
A space defined by a 45° cone centered on the rain gage should contain no trees or other obstructions.

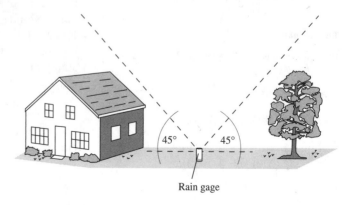

Rain gage

for each rainfall event and 0.15 mm for each snow-fall event (Groisman and Legates 1994).

### Instrument Errors

Systematic errors are often associated with recording-rain-gage measurements simply because of the mechanics of their operation. Tipping-bucket gages often under-record during heavy rains, and the use of heating elements to record snowfall greatly increases evaporative losses (Groisman and Legates 1994). Weighing gages tend to have reduced sensitivity as the weight of collected water increases (Winter 1981). Temperature variations can also affect gage responses and, of course, there is always the potential for mechanical or electrical failure. Thus it is normal practice to install a nonrecording gage adjacent to each recording gage to assure that at least the total precipitation can be determined. Overall, Winter (1981) estimated that instrument errors are typically 1–5% of total catch.

### Observer Errors

Although difficult to quantify and often undetected, errors in measurement and in the recording and publishing of precipitation observations are fairly common. Such errors can often be detected by periodically comparing the records from a number of gages in a region and noting observations that do not conform to the overall areal pattern of precipitation. If a particular observation seems anomalously high, one can check weather maps to see if, for example, small-scale thunderstorm activity was occurring during the period of interest. It may also be possible to examine streamflow records for the region to see whether high flows occurred. If a probable error is detected, one might be able to correct the record objectively by using techniques such as are described in Section 4.2.3. Usually, however, some subjectivity is involved in identifying an observer error and in deciding what the "true" value is.

### Errors Due to Differences in Observation Time

Most precipitation measurements, in the United States and worldwide, are made in nonrecording gages that are observed daily. The total gage catch at the time of observation is recorded as the precipitation associated with that calendar day. In analyzing such data, and in checking for errors, one must be aware that the times of daily observations can vary widely from gage to gage within a region, as in the example shown in Figure 4-19. Thus the precipitation from a storm on a given day might be recorded as occurring on that day at some gages, and as occurring on the next day at other gages. If a large storm occurred on the last day of a month, a discrepancy in observation times could lead to significantly different reported monthly precipitation.

### Errors Due to Occult Precipitation

**Occult precipitation** is precipitation that is induced when clouds encounter trees or other vegetation; it is, therefore, not captured by precipitation gages in clearings. It occurs in both liquid (**fog drip**) and solid (**rime**) forms. Fog drip occurs when clouds move through forests, cloud droplets are deposited on vegetative surfaces, and the water drips to the ground. Rime is formed when supercooled clouds encounter exposed objects, such as trees, that provide nucleation sites for ice-crystal formation and the build-up of ice, much of which eventually falls to the ground in solid or liquid form.

Several studies have shown that occult precipitation is hydrologically important in many high-elevation areas and in certain other environments. For example, fog drip amounts to about 450 mm $yr^{-1}$ (20% of annual precipitation) at elevations above 1200 m in northern New England (Lovett et al. 1982); it constitutes about 880 mm $yr^{-1}$ (about 30% of the annual precipitation) in a Douglas fir forest in Oregon (Harr 1982); and it is the sole source of water in "cloud forests" on the rainless coast of Peru (Lull 1964). Berndt and Fowler (1969) made observations suggesting that rime contributes from 38 to 50 mm $yr^{-1}$ to water inputs in the Cascade Mountains in Washington, and Gary (1972) estimated that from 20 to 30 mm $yr^{-1}$ of precipitation occurs in this form in the mountains of New Mexico.

Measurement of fog drip and rime requires special techniques, which are described in the studies just cited. Failure to include these inputs can result in another form of gage-catch deficiency in environments where they are important.

### Errors Due to Low-Intensity Precipitation

A final source of error can occur in environments where a significant fraction of precipitation comes as very low-intensity rain and snow. In these situations, the observer will often note that some precipitation has entered the gage since the last ob-

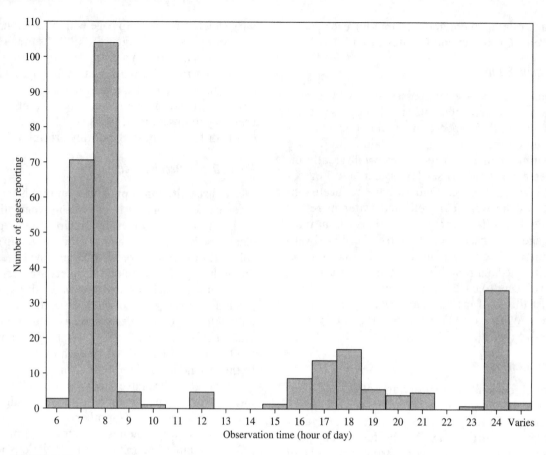

**FIGURE 4-19**
Distribution of observation times for the 281 daily-precipitation stations in New England as of January 1983.

servation, but not enough to measure accurately. For example, in the United States, observers measure rainfall in standard nonrecording gages to the nearest 0.01 in. (0.25 mm), and an observation of less than 0.005 in. (0.13 mm) is called a "trace". Traces are counted as zeros in totaling rainfall, so if there are many traces recorded, reported rainfall totals could be significantly less than the actual input.

The North Slope of Alaska is one location where this form of observation error can be important: Brown et al. (1968) found that traces accounted for 10% of the summer precipitation on average, and could be as much as one-third of the total in some years. Thus corrections for this effect have to be made in assessing the region's water balance (Dingman et al. 1980; Yang et al. 1998). These corrections can be made by multiplying the number of trace observations by one-half the maximum amount designated as a "trace" (i.e., by 0.0025 in. or

0.065 mm in the United States) and adding the result to the reported total precipitation.

### Summary

Although conceptually simple to make, point measurements of precipitation are subject to significant errors of several types. The most common and largest errors are those due to wind effects for gages that project above the ground surface, both shielded and unshielded. These errors are especially large for measurements of snow in standard cylindrical gages.[3]

From a thorough review of the problem, Winter (1981) concluded that errors in point measurements can be in the range of from 5 to 15% for long-term data and as high as 75% for individual

---

[3] Other approaches for measuring water inputs due to snow are discussed in Section 5.2.

storms. Rodda (1985) stated similarly discouraging findings: Catch deficiencies for rain collected in standard gages were found to vary by from 3 to 30% for annual totals and over a wider range for individual storms, with far larger errors for snow. In the United States, the under-measurement of annual-average precipitation at gages averages 9%; it ranges from less than 6% where snow is absent to over 14% where snow is important (Groisman and Legates 1994). However, the underestimation varies seasonally and from year to year and so should be corrected on a monthly or daily basis.

Thus, although it is often tempting to accept reported precipitation data uncritically, it is usually unwise to do so. The implications of errors in precipitation data for water-balance computations are discussed in detail in Section 2.5.2. Clearly, using erroneous precipitation values as inputs to hydrologic models can lead to faulty calibration and validation decisions and faulty predictions and forecasts.

### 4.2.3   Estimating Missing Data

When undertaking an analysis of precipitation data, especially from gages where only daily observations are made, you will virtually always find days when no observations are recorded at one or more gages. These missing days may be isolated occurrences or may extend over long periods. In order to compute precipitation totals and averages, one must estimate the missing values.

Here, we describe several approaches to developing such estimates based on observations at other gages in the region. In the following, the estimate for a particular day at the gage with missing data is designated $\hat{p}_0$, and we consider that we have observed values, designated $p_g$, for the corresponding day at $g = 1, 2, ..., G$ "nearby" gages. The definition of "nearby" must be based on meteorological judgment.

#### Station-Average Method

In this approach, we compute $\hat{p}_0$ as the simple average of the values at the nearby gages:

$$\hat{p}_0 = \frac{1}{G} \cdot \sum_{g=1}^{G} p_g. \qquad \textbf{(4-4)}$$

McCuen (1998) recommends using this method only when the annual precipitation value at each of

the $G$ gages differs by less than 10% from that for the gage with missing data.

#### Normal-Ratio Method

Where the annual precipitation at the gages in the region differs by more than 10%, one can estimate the values at the gage with missing data by weighting the observations at the $G$ gages by their respective annual-average values, $P_g$:

$$\hat{p}_0 = \frac{1}{G} \cdot \sum_{g=1}^{G} \frac{P_0}{P_g} \cdot p_g. \qquad \textbf{(4-5)}$$

Here, $P_0$ is the annual-average precipitation at the gage with missing values.

If there is insufficient information for computing $P_0$, it might be estimated by constructing an isohyetal map of mean annual precipitation (as described in Section 4.3.2) or from regional relations between annual precipitation and elevation or other factors.

#### Inverse-Distance Weighting

This approach weights the $p_g$ values only by their distances, $d_g$, from the gage with the missing data and so does not require information about average annual precipitation at the gages. First, we decide whether the weights should be inversely proportional to distance ($b = 1$) or to distance squared ($b = 2$) and compute

$$D = \sum_{g=1}^{G} d_g^{-b}. \qquad \textbf{(4-6)}$$

Then, we estimate the missing values as

$$\hat{p}_0 = \frac{1}{D} \cdot \sum_{g=1}^{G} d_g^{-b} \cdot p_g. \qquad \textbf{(4-7)}$$

As was pointed out by McCuen (1998), including all regional gages in computing $\hat{p}_0$ could fail to take into account redundant information if some of the $G$ gages are clustered together, and so could bias the estimate of $\hat{p}_0$. To reduce this redundancy, he recommended constructing a coordinate system with the gage that is missing data as origin and selecting in each quadrant only the gage that is closest to the origin for computing weights in the above equations. (This approach could also be used to reduce redundancy in the station-average and normal-ratio methods.) If one quadrant lacks a nearby gage, the estimate can be done with three gages.

## EXAMPLE 4-1

Figure 4-20 shows the locations of 11 precipitation gages in a research watershed. Measurements are missing at gage F for a rain storm during which the amounts shown in the following table were measured at the other gages. Estimate the value at gage F by using (a) the station-average, (b) the normal-ratio, and (c) the inverse-distance methods.

| Gage | Distance from Gage F (km) | Average Annual Precip. (mm) | Measured Storm Precip. (mm) |
|------|---------------------------|------------------------------|------------------------------|
| A | 5.13 | 1373 | 14.4 |
| B | 4.43 | 1452 | 12.2 |
| C | 2.93 | 1404 | 11.6 |
| D | 3.36 | 1433 | 14.8 |
| E | 6.63 | 1665 | 13.3 |
| F | 0.00 | 1137 | — |
| G | 2.13 | 1235 | 12.3 |
| H | 6.10 | 1114 | 11.5 |
| I | 4.95 | 1101 | 11.6 |
| J | 4.40 | 1086 | 11.2 |
| K | 5.93 | 1010 | 9.7 |

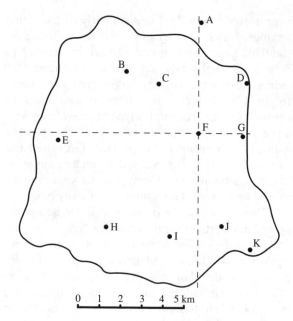

**FIGURE 4-20**
Map showing gage locations for Example 4-1.

**Solution**  First, we construct a coordinate system centered at gage F. We orient the system N-S and E-W, but other orientations could be chosen to give quasi-equal numbers of gages in each quadrant. The gages nearest gage F in each quadrant are D, G, I, and C.

**(a) Station-average method**  Using all gages in Equation (4-4), we compute the average storm rainfall as $\hat{p}_0 = 12.3$ mm. If we select only the nearest gages in each quadrant to average, we find $\hat{p}_0 = 12.6$ mm.

**(b) Normal-ratio method**  Using all gages in Equation (4-5), we compute the following values of $P_0/P_g$:

| Gage: | A | B | C | D | E |
|-------|---|---|---|---|---|
| $P_0/P_g$: | 0.828 | 0.783 | 0.810 | 0.793 | 0.683 |

| Gage: | G | H | I | J | K |
|-------|---|---|---|---|---|
| $P_0/P_g$: | 0.921 | 1.021 | 1.033 | 1.047 | 1.126 |

Substituting these values and the other appropriate quantities into Equation (4-5) gives $\hat{p}_0 = 10.9$ mm. If only the nearest gages in each quadrant are used, we find $\hat{p}_0 = 11.1$ mm.

**(c) Inverse-distance method**  We select $b = 2$ and compute the inverse of the squared distances as follows:

| Gage: | A | B | C | D | E |
|-------|---|---|---|---|---|
| $d_g^{-2}$: | 0.038 | 0.051 | 0.117 | 0.089 | 0.023 |

| Gage: | G | H | I | J | K |
|-------|---|---|---|---|---|
| $d_g^{-2}$: | 0.221 | 0.027 | 0.041 | 0.052 | 0.028 |

With the use of all gages, the sum of these values is $D = 0.686$ km$^{-2}$. Substituting the appropriate values in Equation (4-7) then yields $\hat{p}_0 = 12.4$ mm. If only the nearest gage in each quadrant is used, $D = 0.467$ km$^{-2}$ and $\hat{p}_0 = 12.5$ mm.

## Regression

If relatively few values are missing at the gage of interest, it may be possible to estimate $\hat{p}_0$ for those days as

$$\hat{p}_0 = b_0 + b_1 \cdot p_1 + b_2 \cdot p_2 + \ldots + b_G \cdot p_G, \quad \text{(4-8)}$$

where the coefficients $b_0, \ldots, b_G$ are calculated by conventional least-squares regression methods, using data for a large number of days when observations are available at all gages.

Programs to compute the coefficients in regression equations are widely available in spreadsheets and other software, but one must be cautious about using regression relations without a sound understanding of the principles underlying them. [See, for example, Draper and Smith (1981) or Helsel and Hirsch (1992).] For example, in order to be statistically representative, the data should be chosen to represent separate storms in the season of the missing observations. Also, if daily precipitation at one or more pairs of the gages used to estimate the missing data are highly correlated (as they might be, particularly if some gages are clustered), the computed $\hat{p}_0$ values could be significantly in error.

### 4.2.4 Checking the Consistency of Point Measurements

Changes in the type, exact location, or environment of the gage associated with a weather station with a particular name designation are quite common. For example, many United States first-order weather stations were relocated from metropolitan centers to suburban airports between 1930 and 1960, many of the gages have been shifted between ground-level and roof-top sites, and many have had wind shields added or removed (Groisman and Legates 1994). At cooperative stations in the U.S. network, the exact gage location may change when a new observer is appointed, and trees may grow up or buildings are constructed around a gage. The preceding discussion makes it clear that such changes can significantly affect the gage catch. Thus if a review of station history suggests that changes in the type, location, or immediate environment of a gage have occurred, it is prudent for the hydrologist to determine whether the precipitation record is affected and, if so, to adjust the data to make a more consistent record.

The most common technique for detecting and correcting for inconsistent precipitation data is the use of a **double-mass curve.** Such a curve is a plot, on arithmetic graph paper, of cumulative annual[4] precipitation collected at a gage where measurement conditions may have changed significantly against the average of the cumulative annual precipitation for the same period of years collected at several gages in the same region. A change in the proportionality between the measurements at the suspect station and those in the region is reflected in a change in the slope of the trend of the plotted points. Note that slope breaks in double-mass curves can occur because of climatic shifts; thus adjustments should only be made if there is reason to believe the break is due to a change in the measurement conditions.

If a double-mass curve reveals a change in slope that is significant (see the following discussion) and is due to changed measurement conditions at a particular station, the annual values of the earlier portion of the record should be adjusted to be consistent with the later portion before computing regional averages by the methods discussed in Section 4.3. This adjustment is accomplished simply by multiplying the data for the period before the slope change by the factor $K$, where

$$K \equiv \frac{\text{slope for period AFTER slope change}}{\text{slope for period BEFORE slope change}}. \quad \text{(4-9)}$$

Most analysts recommend that a break in slope be considered significant only if it persists for five or more years, and then only (1) if it is clearly associated with a change in measurement conditions and (2) if it is determined to be statistically significant (Searcy and Hardison 1960). Statistical significance of slope changes in double-mass curves can be detected by techniques called analyses of variance and covariance. These techniques are described in most introductory statistical texts, and their application to double-mass curves was given by Searcy and Hardison (1960).[5] Chang and Lee (1974) described

---

[4] The double-mass curve analysis is virtually always applied only to annual precipitation data. Data for shorter periods are naturally highly variable and tend to mask the kinds of measurement inconsistencies one is trying to identify.

[5] However, the usual statistical tests of significance for differences in slope for regressions are not directly applicable for double-mass curves, because the process of accumulation introduces a high degree of correlation between successive values. Because

a computerized objective approach to double-mass-curve analysis that incorporates analyses of variance and covariance.

Example 4-2 illustrates the basic procedure for double-mass-curve analysis.

---

### EXAMPLE 4-2

Table 4-4 shows the annual precipitation values for from 1926 to 1942 at five gages in a region. Gage E was moved in 1931. Determine how the annual values for 1926–1930 should be adjusted to be consistent with the subsequent period of record.

**Solution** The cumulative annual values for Station E are plotted against the cumulative values of the mean of the other four stations as a double-mass curve in Figure 4-21. The slope of a straight line through the points prior to 1931 has a slope of 0.77, compared to a slope of 1.05 for a line through the points for succeeding years. Here we assume that this difference is statistically significant and compute

$$K = \frac{1.05}{0.77} = 1.36 .$$

Thus the annual precipitation values at Station E before 1931 are multiplied by 1.36 to produce a consistent record for the entire period of measurement. The appropriately adjusted values are given in the last column of the upper portion of Table 4-4.

---

In a thorough study of meteorological data in Iowa, Carlson et al. (1994) found many inhomogeneities due to changes in measurement conditions; they emphasized the importance of examining data quality before undertaking climatic analyses. Unfortunately, it appears that inconsistencies in precipitation data may be even more common than a review of station histories would reveal: In an examination of precipitation records in West Virginia, Chang and Lee (1974) found that

> In some instances, inconsistencies in station records [detected via double-mass-curve analysis] corresponded with changes in station locations or other circumstances revealed by the station histories. But frequently, such correspondence was lacking be-

cause station changes of assumed minor importance were not recorded. On-site visits and the interrogation of observers were sometimes helpful. In one instance, for example, the only evidence of a site change was . . . a large . . . tree stump within 8 m of the gage.

---

## 4.3   AREAL ESTIMATION

Hydrologists are almost always interested in precipitation over a region, such as a drainage basin, rather than at a point. Several approaches have been devised for estimating regional precipitation from point values (totals for a storm or for a specific time period, or averages over some period of time). Some of these approaches provide estimates only of the spatially averaged precipitation; others provide an estimate of the spatial distribution of precipitation within the region of interest as well as of the spatial average.

To discuss these approaches in a consistent framework, consider the region depicted in Figure 4-22. Conceptually, the average precipitation over this region, $P$, is represented as

$$P = \frac{1}{A} \cdot \iint\limits_{A} p(x,y) \cdot dx \cdot dy , \qquad \textbf{(4-10)}$$

where $A$ is the area of the region, $x$ and $y$ represent the rectangular coordinates of points in the region, and $p(x,y)$ is the precipitation at all such points. Measurements of precipitation for the period of interest have been made at $G$ gages, designated $g = 1$, 2, ..., $G$. Approaches to computing an estimate of $P$, $\hat{P}$, can be divided into two classes: (1) those that compute $\hat{P}$ directly as a weighted average of the measured values; and (2) surface-fitting methods that first use the measured values to estimate the precipitation at points in $A$, and then use some scheme to approximate the integral in Equation (4-10). Table 4-5 lists the various methods that have been developed for estimating areal precipitation. Here we present the general principles used in these methods, describe some of the simplest of them in detail, and introduce the concept of spatial correlation, which underlies some of the most widely used of the numerically intensive approaches. The references given in Table 4-5 should be consulted for detailed descriptions of other methods.

---

of this correlation, the conventional tests will often fail to detect statistically significant slope changes. (See Matalas and Benson 1961.)

**TABLE 4-4**

Example of Double-Mass–Curve Analysis: Measured Annual and Cumulative Precipitation at Stations A–E (in inches).

| Year | A | B | C | D | E | Adjusted[a] E |
|------|------|------|------|------|------|------|
| 1926 | 39.75 | 45.70 | 30.69 | 37.36 | 32.85 | 44.68 |
| 1927 | 39.57 | 38.52 | 40.99 | 30.87 | 28.08 | 38.19 |
| 1928 | 42.01 | 48.26 | 40.44 | 42.00 | 33.51 | 45.57 |
| 1929 | 41.39 | 34.64 | 32.49 | 39.92 | 29.58 | 40.23 |
| 1930 | 31.55 | 45.13 | 36.72 | 36.32 | 23.76 | 32.31 |
| 1931 | 55.54 | 53.28 | 62.35 | 36.61 | 58.39 | |
| 1932 | 48.11 | 40.08 | 47.85 | 38.61 | 46.24 | |
| 1933 | 39.85 | 29.57 | 32.74 | 26.89 | 30.34 | |
| 1934 | 45.40 | 41.68 | 36.13 | 32.44 | 46.78 | |
| 1935 | 44.89 | 48.13 | 30.73 | 41.56 | 38.06 | |
| 1936 | 32.64 | 39.48 | 35.40 | 31.32 | 42.82 | |
| 1937 | 45.87 | 44.11 | 39.16 | 44.14 | 37.93 | |
| 1938 | 46.05 | 38.94 | 43.27 | 50.62 | 50.67 | |
| 1939 | 49.76 | 41.58 | 49.85 | 41.09 | 46.85 | |
| 1940 | 47.26 | 49.66 | 47.86 | 39.01 | 50.52 | |
| 1941 | 37.07 | 31.92 | 32.15 | 34.45 | 34.38 | |
| 1942 | 45.89 | 38.16 | 52.39 | 47.32 | 47.60 | |

| Year | Average A–D | Cumulative A–D | Cumulative E |
|------|------|------|------|
| 1926 | 38.38 | 38.38 | 32.85 |
| 1927 | 37.49 | 75.87 | 60.93 |
| 1928 | 43.18 | 119.05 | 94.44 |
| 1929 | 37.11 | 156.16 | 124.02 |
| 1930 | 37.43 | 193.59 | 147.78 |
| 1931 | 51.95 | 245.53 | 206.17 |
| 1932 | 43.66 | 289.19 | 252.41 |
| 1933 | 32.26 | 321.46 | 282.75 |
| 1934 | 38.91 | 360.37 | 329.53 |
| 1935 | 41.33 | 401.70 | 367.59 |
| 1936 | 34.71 | 436.41 | 410.41 |
| 1937 | 43.32 | 479.73 | 448.34 |
| 1938 | 44.72 | 524.45 | 499.01 |
| 1939 | 45.57 | 570.02 | 545.86 |
| 1940 | 45.95 | 615.96 | 596.38 |
| 1941 | 33.90 | 649.86 | 630.76 |
| 1942 | 45.94 | 695.80 | 678.36 |

[a]Measured value times 1.36 (see text).

Data from Searcy and Hardison (1960).

## 4.3.1  Direct Weighted Averages

Methods of this type use the general equation

$$\hat{P} = \sum_{g=1}^{G} w_g \cdot p_g , \qquad \text{(4-11)}$$

where the $w_g$ are weights, which satisfy

$$\sum_{g=1}^{G} w_g = 1, \qquad 0 \leq w_g \leq 1 , \qquad \text{(4-12)}$$

and the $p_g$ are the values of precipitation measured at each gage. We will examine three relatively simple methods of this type, which differ only in the way in which the weights $w_g$ are estimated. The comparative study by Singh and Chowdhury (1986) can be consulted for references that provide full descriptions of methods not discussed here.

### Arithmetic Average

In this method, the weights in Equation (4-11) are assigned a value $(1/G)$ for all gages in the region. Thus the regional average becomes the arithmetic mean of the values measured at all the gages in the region:

**FIGURE 4-21 CORRECTED**

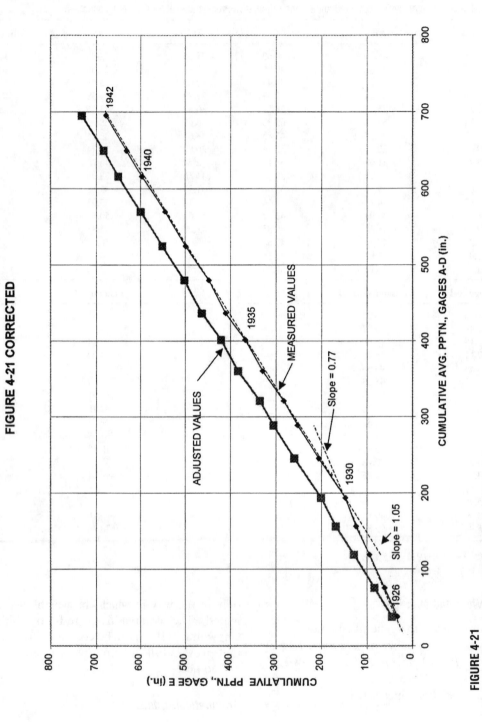

**FIGURE 4-21**
Double-mass curve for Station E. Data in Table 4-4.

120

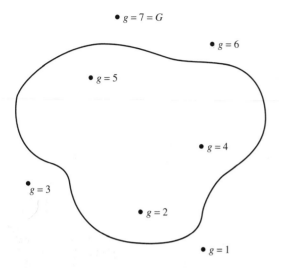

**FIGURE 4-22**
The general problem of areal estimation from point measurements: We wish to estimate the average precipitation over the enclosed region from values measured at points $g = 1, 2, ...,$ $G$. Here, $G = 7$.

$$\hat{P} = \frac{1}{G} \cdot \sum_{g=1}^{G} p_g . \qquad \text{(4-13)}$$

Stations adjacent to the region of interest can be included, if appropriate.

### Thiessen Polygons

In this approach, the region is divided into $G$ subregions, approximately centered on each of the $G$ rain gages. The subregions are defined such that all points in each subregion are closer to their central gages than they are to any other gage. Once these $G$ subregions are identified and their areas, $a_g$, measured, the weights are determined as $w_g = a_g/A$ and the spatial average is computed as

$$\hat{P} = \frac{1}{A} \cdot \sum_{g=1}^{G} a_g \cdot p_g . \qquad \text{(4-14)}$$

Equation (4-12) is satisfied because

$$\sum_{g=1}^{G} a_g = A .$$

The method for delineating the boundaries of the subregions that are closest to each of the $G$ rain gages was developed by Thiessen (1911) (Figure 4-23). First, straight lines are drawn between the locations of adjacent gages to form a network of triangles; then perpendicular bisectors of each line are

constructed and extended until they intersect to form irregular polygons. When ambiguities arise in constructing polygon boundaries, they can be resolved by simply comparing the distances to the gages in question.

Diskin (1970) and Croley and Hartmann (1985) have developed computer algorithms for calculating Thiessen subregions. Subregion sizes can be measured by counting evenly spaced points on an overlaid grid, by planimeter, or, most conveniently and accurately, by using a computer-attached digitizing tablet.

### Two-Axis Method

The two-axis method, developed by Bethlahmy (1976), derives $w_g$ values that reflect the nearness of each gage to the center of the region, rather than the relative area associated with each gage (Figure 4-24). To derive the weights, one first draws the longest straight line that can be drawn on a map of the region. The perpendicular bisector of this longest line is then drawn; this is called the **minor axis**. The **major axis** is then drawn as the perpendicular bisector of the minor axis. One next draws two lines from each of the gages, one line to the farther end of the major axis, the other to the farther end of the minor axis. The angle between these two lines, $\alpha_g$, is measured (it is always < 90°). The sum of all the angles, $\mathring{A}$, is then computed:

$$\mathring{A} = \sum_{g=1}^{G} \alpha_g . \qquad \text{(4-15)}$$

The weights are then defined as $w_g = \alpha_g/\mathring{A}$, and the average precipitation is calculated as

$$\hat{P} = \frac{1}{\mathring{A}} \cdot \sum_{g=1}^{G} \alpha_g \cdot p_g . \qquad \text{(4-16)}$$

### 4.3.2 Surface-Fitting Methods

In surface-fitting methods, the measured values are used to identify a surface intended to represent the precipitation at all points of the region of interest. This surface is thus a model of the spatial variability of the particular value of precipitation (average or total for some period or total for a storm), which can be depicted on a map. The various approaches differ in the methods used in constructing the surface. In some methods, the surface does not pass exactly through the points (Figure 4-25a); these are called **smoothing methods.** In **interpolation**

**TABLE 4-5**

Classification of Methods for Estimating Areal Precipitation from Point Measurements.

| Method | Classification | Computational Complexity | Source for Full Description |
|--------|----------------|--------------------------|----------------------------|
| Arithmetic average | Direct | Very low | This text |
| Thiessen polygons | Direct | Low | This text |
| Bethlahmy's two-axis | Direct | Low | This text |
| Eyeball isohyetal | Surface-fitting Deterministic Smoothing | Low to moderate | This text |
| Polynomial surface | Surface-fitting Deterministic Smoothing | Moderate | Tabios & Salas (1985) |
| Lagrange polynomial surface | Surface-fitting Deterministic Smoothing | Moderate to high | Tabios & Salas (1985) |
| Spline surface | Surface-fitting Deterministic Interpolation | Moderate to high | Creutin & Obled (1982) Lebel et al. (1987) |
| Inverse-distance interpolation | Surface-fitting Deterministic Interpolation | Moderate to high | Tabios & Salas (1985) |
| Multiquadric interpolation | Surface-fitting Deterministic Interpolation | Moderate to high | Shaw (1988) Creutin & Obled (1982) |
| Optimal interpolation/kriging | Surface-fitting Statistical Interpolation | High | Creutin & Obled (1982) Tabios & Salas (1985) Lebel et al. (1986) |
| Empirical orthogonal functions | Surface-fitting Statistical Interpolation | High | Creutin & Obled (1982) |
| Hypsometric | Surface-fitting Deterministic Smoothing | Low to moderate | This text |

**methods,** the precipitation mapped at each point representing a gage location exactly equals the value measured at that gage (Figure 4-25b).

A second classification of surface-fitting approaches distinguishes between **statistical** (or **stochastic**) methods and **deterministic** methods. Statistical approaches are based on the principle of minimizing estimation variances (also called interpolation errors) at ungaged points; deterministic approaches construct surfaces based on other mathematical criteria (Creutin and Obled 1982). One advantage of the statistical approaches is that they can provide an estimate of the interpolation error at all points in the region.

### General Features

Surfaces representing precipitation values over an area are usually depicted in the form of maps showing contours of equal precipitation (isohyets) (Figure 4-26). Regardless of the method used to

develop such maps, they can be used to estimate the areal average precipitation, $\hat{P}$, by considering that the isohyets serve as boundaries of $I$ subregions within the region, with all points in a subregion assigned a precipitation value equal to the average of the values associated with its boundary isohyets:

$$\hat{p}_i = 0.5 \cdot (p_{i-} + p_{i+}) . \qquad \textbf{(4-17)}$$

In this equation, $\hat{p}_i$ is the precipitation at all points in the $i$th subregion and $p_{i-}$ and $p_{i+}$ are the values of the isohyets that bound the $i$th subregion. The regional average is then estimated as

$$\hat{P} = \frac{1}{A} \cdot \sum_{i=1}^{I} a_i \cdot \hat{p}_i , \qquad \textbf{(4-18)}$$

where $a_i$ is the area between the two contours within the region.

In the past, contours were often sketched by hand, incorporating the subjective judgment of the analyst (the "eyeball" method discussed below).

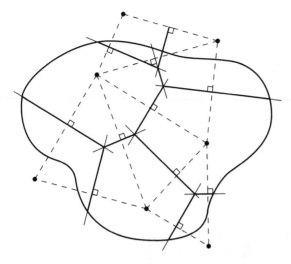

**FIGURE 4-23**
Construction of Thiessen polygons. The short-dashed lines connect adjacent gages, the long-dashed lines are perpendicular bisectors of those lines, and the solid lines are the portions of the bisectors that constitute polygon boundaries. Points in each polygon are closer to the gage near the polygon center than to any other gage.

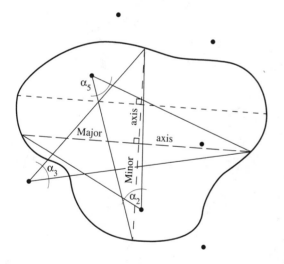

**FIGURE 4-24**
Construction of angles for the two-axis method. The short-dashed line is the longest line that can be drawn in the region of interest; the shorter-dashed line is its perpendicular bisector (the minor axis), and the perpendicular bisector of the minor axis is the major axis. The solid lines are drawn from each gage to the farther end of the two axes to define the angles $\alpha_g$. Angles are shown only for gages 2, 3, and 5, for clarity.

Such a practice may still be adequate for some applications, but the wide availability of programs for personal as well as mainframe computers that will construct contours when given a set of $p_g$ values and their $(x,y)$ coordinates is making "objectively" constructed isohyetal maps a potentially more attractive choice. These automated contouring routines use one of the surface-fitting methods described below to estimate precipitation at each of $J$ grid nodes covering the region of interest and use some algorithm for constructing the isohyets to reflect the grid-point values.

Surface-fitting and contouring algorithms follow these steps:

**1.** A grid covering the region of interest is established. The grid spacing is at the discretion of the analyst, but is usually on the order of one-tenth the average distance between rain gages. We designate the total number of grid nodes in the region as $J$.

**2.** Values of precipitation at each of the grid points are estimated by using the measured values; that is,

$$\hat{p}_j = \sum_{g=1}^{G} w_{jg} \cdot p_g, \qquad \textbf{(4-19)}$$

where $\hat{p}_j$ is the estimated precipitation at the $j$th point, $p_g$ is the precipitation measured at station $g$, and $w_{jg}$ is the weight assigned to station $g$ for point $j$. To assure an unbiased estimate of $P$, most methods constrain the sum of the weights for each point as

$$\sum_{g=1}^{G} w_{jg} = 1 . \qquad \textbf{(4-20)}$$

Many contouring routines use only the $p_g$ within a certain radius around a point $j$ to estimate $p_j$, with $w_{jg} = 0$ for all measured values outside that radius. The essential differences among the various surface-fitting methods lie in the way in which the weights $w_{jg}$ are established.

**3.** Once the $\hat{p}_j$ values are estimated, another algorithm examines the difference between values at adjacent nodes and determines exactly how the contour lines are to be drawn.

Thus, although objective in one sense, all automated contouring routines involve an arbitrary choice of method for choosing the weights in Equation (4-19) and for contouring, and different algo-

**FIGURE 4-25**
Smoothing (a) vs. interpolation (b) methods of surface fitting. In (a) and (b), lower grids show ($x, y$) coordinates; dots show location of gages; heights of solid vertical lines are proportional to measured precipitation; upper surface represents the model of spatial distribution of precipitation, $p(x,y)$; and crosses show the precipitation assigned to gage locations by the method. (a) Surfaces estimated by smoothing methods do not fit measured values exactly. (b) Surfaces estimated by interpolation methods do fit measured values exactly.

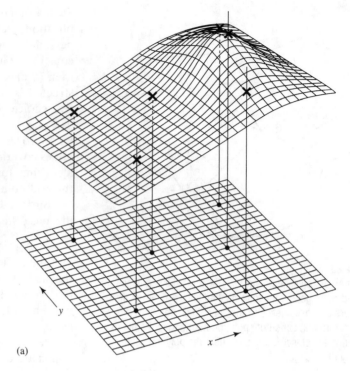

(a)

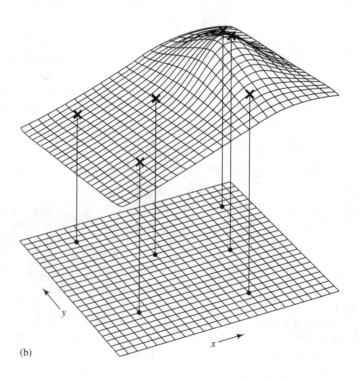

(b)

**FIGURE 4-26**
The isohyetal method of integration. Dots are precipitation-gage locations; thin lines are isohyets; shaded zone is the area $a_i$ between the $p_{i-}$ and $p_{i+}$ isohyets.

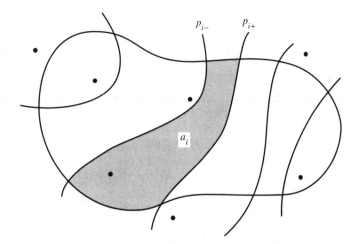

rithms produce different isohyetal patterns—so some subjective judgment is involved in any case.

In using the surface-fitting methods described below, one should also keep the following points firmly in mind:

**1.** The point values used in estimating the surface may contain significant errors, as discussed in Section 4.2.2. Therefore,

**a.** determine the potential for each of the sources of error, making corrections as appropriate; and

**b.** don't use a method that is more elaborate than is warranted by the quality of the data.

**2.** Some "objective" methods are based on mathematical or statistical criteria that have little relation to the actual spatial variability of precipitation in a given area. It will, for example, often be unwise to draw isohyets based only on the $p_g$ values: If precipitation in a region is dependent on factors such as elevation or relation to a topographic barrier, those factors should be taken into account when constructing the isohyetal map. Thus one must use physical (meteorologic) judgment in examining the surface produced by a given method, and apply the results only if they appear physically realistic.

Once a satisfactory map is constructed, average precipitation can be determined via Equation (4-17), with areas measured by the same techniques described for the Thiessen method. Alternatively, if one has access to sophisticated digitizing software, such as a geographic information system (GIS), it

may be possible to compute $\hat{P}$ directly from the $\hat{p}_j$ values generated at each grid node as

$$\hat{P} = \frac{1}{J} \cdot \sum_{j=1}^{J} \hat{p}_j, \qquad \textbf{(4-21)}$$

where $J$ is the number of grid nodes in the region.

The most widely used approaches for deriving the weights in Equation (4-19) are listed in Table 4-5. We describe three of these approaches here; details on the others can be found in the references listed in Table 4-5.

### "Eyeball" Isohyetal Method

This term refers to the free-hand drawing of isohyets, guided by the measured values plotted at the gage locations and the analyst's judgment. In most hydrology texts, this method is called simply the "isohyetal method". As noted above, there are circumstances in which this approach is appropriate because of the nature of the data or because it allows incorporation of information other than the measured values.

### Optimal Interpolation/Kriging

Optimal interpolation is a general term for a group of statistical methods in which the weights in Equation (4-19) are derived by minimizing the variance of the interpolation error, $S^2(e_j)$, where

$$S^2(e_j) = S^2(\hat{p}_j - p_j), \qquad \textbf{(4-22)}$$

where $\hat{p}_j$ is given by Equation (4-19) and $p_j$ are values measured at gages. [Variance is defined in Equation (C-19); Box C-2 describes how sample variance is calculated.] The term **kriging** (pronounced "kreegh-ing" or "kreej-ing") is also used to refer to this class of methods.

This variance function is minimized under the assumptions that there is no spatial trend to precipitation, that a particular pattern of spatial correlation relates values in the region (see next paragraph, Figure 4-27, and Box 4-2) and that the sum of the weights equals unity [Equation (4-20)]. This leads to a system of simultaneous equations that can be solved for the weights and another set of equations that expresses the magnitude of the interpolation error variance at each grid point.

The differences between precipitation values measured at nearby points are usually smaller on average than those between widely separated points. One way of characterizing this spatial relation is in the form of the spatial correlation coefficient,[6] $R_s(d)$, between precipitation values at various separation distances, $d$. The solutions to the simultaneous equations that minimize $S^2(e_j)$ in Equation (4-22) depend on the exact way in which $R_s(d)$ decreases with $d$, and the computations are numerically complex. Kitanidis (1992) provides a useful introduction to the mathematics of kriging and sources of computer software for its implementation.

### Conventional Hypsometric Method

Conceptually, the hypsometric method is a deterministic, smoothing, surface-fitting method. However, it differs from other surface-fitting methods in that the spatial average precipitation is not computed from isohyets [Equations (4-17) and (4-18)]. The method is appropriate only for regions in which orographic effects are important, so that the precipitation for the period of interest is a strong function of elevation. A contour map of topography or a digital elevation model (DEM) is required to apply the method.

The first step in applying this approach is to plot the measured $p_g$ values against station elevation, $z_g$, and to establish a relationship between precipitation ($p$) and elevation ($z$). This functional relation, the "orographic equation," can be sketched by eye or determined by statistical techniques, such as regression analysis. Commonly, it is approximated as the simple linear relation

$$p(z) = a + b \cdot z \qquad \textbf{(4-23)}$$

(e.g., Dingman et al. 1988). In some cases, the relation between $p(z)$ and $z$ varies systematically in the region—the windward side of a mountain range can have a more rapid elevational increase of precipitation than the leeward side [e.g., Troxell (1942)]—and in these cases the hypsometric method should be applied separately for each identified subregion.

Once the precipitation–elevation relation is established, a graph relating elevation, $z$, to the area above a given elevation, $A(z)$, called a **hypsometric curve,** is constructed. These curves can be developed from a topographic map by laying a grid over the region, reading the elevation at grid nodes, and constructing a cumulative-frequency curve giving the percentage of points above a given elevation. Typically, one would determine elevations at at least 100 points in the region; Wallis and Bowden (1962), Bonnor (1975), and Omernik and Kinney (1983) discussed accuracies associated with grid-point sampling. If one has access to a DEM for the region of interest and appropriate software, the hypsometric curve can be readily calculated without the tedium of tabulating individual point elevations from a topographic map.

Average precipitation for the region is then computed via the following steps:

1. Select an elevation interval, $\Delta z$, and divide the total elevation range into $H$ increments of $\Delta z$.
2. From the hypsometric curve, determine the fraction of total area within each increment of elevation; designate these fractions as $a_h$, $h = 1, 2, ..., H$.
3. Use the precipitation–elevation relation [Equation (4-23)] to estimate the precipitation, $\hat{p}(z_h)$, at each elevation $z_h$, where $z_h$ is the central elevation of each elevation increment.
4. Compute the estimated areal average precipitation as

$$\hat{P} = \sum_{h=1}^{H} \hat{p}(z_h) \cdot a_h . \qquad \textbf{(4-24)}$$

---

[6] The correlation coefficient is defined in Equations (C-33) and (C-34), and the method for estimating its value from measured data is given in Box C-5.

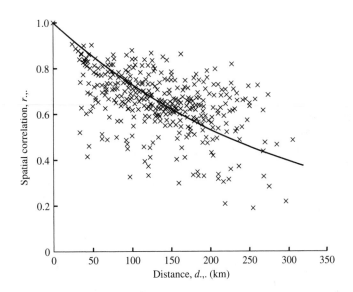

**FIGURE 4-27**
Spatial correlation of annual precipitation for the 29 stations studied by Tabios and Salas (1985). Curve is Equation (4B2-5c) with $c = 0.0031$ km$^{-1}$. Reprinted with permission of the American Water Resources Association.

The following example illustrates the procedure.

---

## EXAMPLE 4-3

Estimate the mean annual precipitation for the Delta River drainage basin in central Alaska, where precipitation gages supplemented by glaciological studies indicate a strong orographic effect (Figure 4-28a). Topographic maps of the basin are available and a hypsometric curve has been constructed (Figure 4-28b).

**Solution** The lowest elevation in the basin is very close to 1000 ft, and the highest is close to 9000 ft; for convenience, we select $H = 8$ and $\Delta z = 1000$ ft. Using Figures 4-28a and 4-28b, we then construct the following table:

| $h$ | Elevation Range (ft) | $z_h$ (ft) | $a_h$ | $\hat{p}(z_h)$ (in. yr$^{-1}$) | $a_h \cdot \hat{p}(z_h)$ (in. yr$^{-1}$) |
|---|---|---|---|---|---|
| 1 | 1000-2000 | 1500 | 0.185 | 11.0 | 2.0 |
| 2 | 2000-3000 | 2500 | 0.135 | 25.0 | 3.4 |
| 3 | 3000-4000 | 3500 | 0.290 | 42.0 | 12.2 |
| 4 | 4000-5000 | 4500 | 0.160 | 57.0 | 9.1 |
| 5 | 5000-6000 | 5500 | 0.110 | 71.0 | 7.8 |
| 6 | 6000-7000 | 6500 | 0.085 | 83.0 | 7.1 |
| 7 | 7000-8000 | 7500 | 0.025 | 93.0 | 2.3 |
| 8 | 8000-9000 | 8500 | 0.010 | 100.0 | 1.0 |
| | | | | Total | 44.9 |

Thus our estimate of the basin mean annual precipitation is 44.9 in. yr$^{-1}$.

---

Even if the precipitation–elevation relation is determined by a statistical analysis such as regression, this method is not statistical in the sense described above because the statistical criterion does not involve minimization of estimation errors at the ungaged points. And, because the precipitation represented by Equation (4-23) does not exactly fit the measured values, the hypsometric approach is a deterministic smoothing method.

### Algorithmic Hypsometric Methods

Computer algorithms have recently been developed that use gridded elevation data (DEMs) to estimate precipitation in mountainous regions. One of these, called PRISM (Precipitation–elevation REgressions on Independent Slopes Model; Taylor et al. 1993), first generates a smoothed version of the actual topography (the "orographic elevations"). It then estimates precipitation for each grid cell as a function of its orographic elevation, with the functions varying depending on location and orientation. This is also a deterministic smoothing method, but it does have a procedure for estimating the uncertainty of its estimates. Another algorithm, called ANUSPLIN, uses a spline interpolator along with gridded latitude, longitude, and elevation data to estimate precipitation (Custer et al. 1996).

### 4.3.3 Comparison of Methods and Summary

Several studies have attempted to assess the strengths and weaknesses of the various methods

## BOX 4-2

. . . . . . .

### Spatial Correlation

Consider a region in which we have $G$ gages at which precipitation has been measured for $T$ years. At each gage, we can measure the total precipitation for each year, $p_{gt}$, and compute the average annual precipitation, $m_{pg}$, and the standard deviation of the annual precipitation, $s_{pg}$, as

$$m_{pg} = \frac{1}{T} \cdot \sum_{t=1}^{T} p_{gt} \qquad \text{(4B2-1)}$$

and

$$s_{pg} = \left[ \frac{1}{T-1} \cdot \sum_{t=1}^{T} (p_{gt} - m_{pg})^2 \right]^{1/2}. \qquad \text{(4B2-2)}$$

(See Box C-2.)

There are a total of $G \cdot (G - 1)/2$ pairs of gages. For each pair, say the two gages $g = a$ and $g = b$, the correlation between annual values $r_{a,b}$ is

$$r_{a,b} = \frac{\sum_{t=1}^{T} (p_{at} - m_{pa}) \cdot (p_{bt} - m_{pb})}{(T-1) \cdot s_{pa} \cdot s_{pb}}. \qquad \text{(4B2-3)}$$

(See Box C-5.)

If $|r_{a,b}|$ exceeds a critical value $r^*$ that depends on $T$ [Equation (CB5-2)], we conclude that there is a linear relation between annual values of precipitation at points $a$ and $b$: positive values of $r_{a,b}$ indicate a tendency for a given year to have relatively high or relatively low values of precipitation at both gages, and negative values indicate a tendency for precipitation at the two gages to vary in opposite ways—a low year at one gage tends to be a high year at the other, and vice versa.

When $r_{a,b}$ has been calculated for all pairs of gages, the values accepted as significant can be plotted against the distance between each pair, $d_{a,b}$, which is given by

$$d_{a,b} = [(x_a - x_b)^2 + (y_a - y_b)^2]^{1/2}, \qquad \text{(4B2-4)}$$

where $(x_a, y_a)$ and $(x_b, y_b)$ are the map coordinates of gages $a$ and $b$, respectively.

Figure 4-27 shows an $r_{.,.}$ vs. $d_{.,.}$ plot for 29 precipitation gages (406 gage pairs) in Nebraska and Kansas. As is common for such plots, it shows considerable scatter. However, in order for us to use the optimal interpolation methods described in the text, the relation between $r_{.,.}$ and $d_{.,.}$ has to be described as a simple function. Tabios and Salas (1985) suggested the alternative functional forms

$$r_{.,.} = [1 + c \cdot d_{.,.}]^{-1}, \qquad \text{(4B2-5a)}$$
$$r_{.,.} = [1 + c \cdot d_{.,.}]^{-1/2}, \qquad \text{(4B2-5b)}$$

and

$$r_{.,.} = \exp[-c \cdot d_{.,.}], \qquad \text{(4B2-5c)}$$

where $c$ is a parameter to be determined from the data. Figure 4-27 shows Equation (4B2-5c) with $c = 0.0031$ km$^{-1}$ fitted to the data of Tabios and Salas (1985). (See Example 4-4.)

for estimating areal precipitation from point values, some by simply comparing estimates among themselves and others by using statistical criteria to make more nearly absolute comparisons.

Shaw and Lynn (1972) compared estimates of annual, monthly, and storm precipitation for three drainage basins via the arithmetic-average, Thiessen-polygon, computer-based-isohyetal (using inverse-distance weighting), polynomial-surface, and multiquadric methods. There was little difference among the estimates from Thiessen, isohyetal, and multiquadric values for the annual data, or between the Thiessen and multiquadric methods for

the monthly data. They concluded that the multiquadric method is a practicable and efficient approach to areal precipitation estimation.

Court and Bare (1984) compared annual precipitation estimates for five years in eight drainage basins using the arithmetic-average, Thiessen, eyeball-isohyetal, hypsometric, and two-axis methods. They found that the hypsometric method tended to give high values, the isohyetal method low values, the arithmetic average varied from low to high, and the Thiessen and two-axis methods tended to give the more central estimates. Thus they concluded that the two-axis method is as acceptable as any

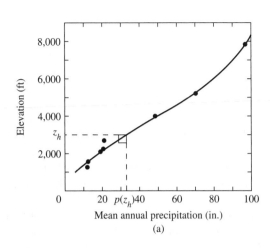

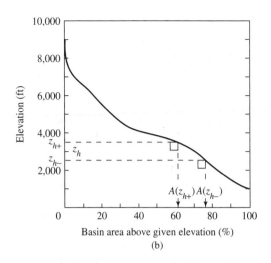

**FIGURE 4-28**

Application of the hypsometric method to the Delta River watershed, central Alaska (Dingman et al. 1971). (a) The estimated relation between mean annual precipitation and elevation. (b) The hypsometric curve. $z_{h-}$ and $z_{h+}$ are boundaries of elevation increment, $A(z_{h-})$ and $A(z_{h+})$ are areas above given elevations, and $a_h = [A(z_{h-}) - A(z_{h+})]/A$. Other symbols as defined in text.

of the other methods and, because it is easy to use and flexible (i.e., it is easy to adjust the weights when a gage is added or deleted), that it could often be the method of choice for estimating mean basin precipitation.

Creutin and Obled (1982) applied the Thiessen, arithmetic-mean, spline-function, empirical-orthogonal-functions (EOF) approaches, and two forms of optimal interpolation to data for 81 storm events in France. A total of 99 gages were available, and comparisons were made using 73 of these to estimate values at the remaining 26 via the various methods, and then comparing the estimates with the values measured at the 26. Their analysis showed that the more sophisticated statistical methods—optimal interpolation and EOF—were superior; the arithmetic average and Thiessen methods performed significantly less well by many of their test criteria.

Tabios and Salas (1985) used long-term average precipitation data at 29 gages to compare Thiessen polygons, two types of least-squares surfaces, two types of inverse-distance interpolation, multiquadric interpolation, and several forms of optimal interpolation. Their tests consisted of using the data from 24 of the gages to estimate the values at the remaining five via each of the methods. Their conclusions were similar to those of Creutin and Obled (1982): The optimal interpolation techniques are superior. However, they found that the multiquadric method performed almost as well. They also concluded that the inverse-distance and Thiessen approaches were acceptable, but that the least-squares polynomial surface was not.

Singh and Chowdhury (1986) compared estimates of mean areal daily, monthly, and annual rainfall at two locations in New Mexico and one in Great Britain, using 10 direct weighted-average methods and three surface-fitting methods. They found that, with few exceptions, the values given by all methods were within 10% of each other and concluded that there was no evidence that any particular method was superior, although there might be reasons for selecting a particular method in a particular situation.

Lebel et al. (1987) compared the Thiessen, spline, and optimal interpolation/kriging approaches for estimating areal precipitation for rainfalls of durations from one to 24 hr. The comparisons were made in a region of France with a high density of gages and utilized varying numbers of gages to investigate the effects of a range of gage densities. The optimal interpolation/kriging method was judged by far the most accurate at all gage densities and for all durations.

Taylor et al. (1993) found that the PRISM algorithm had lower bias and better precision than several variations of kriging in Oregon; they also found that it gave "excellent results" when applied to the entire coterminous United States.

Custer et al. (1996) compared isohyetal maps of average annual precipitation in Montana constructed by "eyeball" and by a spline-based contouring algorithm that accounted for elevation and found that the two maps differed by more than 10% over almost half the area.

In summary, the choice of a method for computing areal precipitation depends on several factors, including (1) the objective of the analysis, (2) the nature of the region, and (3) the time and computing resources available. If the objective is to develop only an estimate of spatial average precipitation, any of the techniques discussed can be used. When time and resources are limited, or only reconnaissance-level estimates are required, one would usually apply one of the direct weighted-average methods or the hypsometric or eyeball isohyetal method. The arithmetic-average, Thiessen, or two-axis methods are not well suited to regions where topographic factors strongly influence precipitation unless gages are well distributed, but simple weighting approaches accounting for topography can be developed. (See Singh and Chowdhury 1986.) Otherwise, the hypsometric or surface-fitting methods can be used. The widespread availability of contouring programs (which use inverse-distance interpolation) for personal computers makes "objective" construction of isohyetal maps the preferred choice over the eyeball method in many situations.

For many applications, regional precipitation values are used as input to hydrologic models (e.g., in the calculation of regional evapotranspiration in Section 2.5.2 or the BROOK90 model discussed throughout this text). As noted earlier, sound application of model predictions or forecasts requires an assessment of the uncertainty associated with outputs. Because this uncertainty depends heavily on the uncertainty of inputs, it is as important to have an estimate of the uncertainty of regional precipitation estimates as it is to have the estimates themselves. Where this is true, one of the statistical interpolation methods is indicated, because these provide explicit estimates of the interpolation error. (Note, however, that neither these methods nor any others account for any errors in point measurements, so these must always be evaluated separately.)

Most of the comparative studies reviewed above have concluded that optimal-interpolation/kriging methods provide the best estimates of regional precipitation in a variety of situations. Presumably, this is so because these methods are based on the spatial correlation structure of precipitation in the region of application rather than on essentially arbitrary spatial structures. The optimal-interpolation methods have the additional advantages of providing a model of the spatial variability of precipitation and providing estimates of uncertainty, so they emerge as the methods of choice when resources are available. A number of inexpensive optimal-interpolation/kriging programs for personal computers are now available (e.g., Grundy 1988; Kitanidis 1992), and so the feasibility of using these approaches is much enhanced.

Recent experiences with DEM-based models like PRISM suggest that these may be the best approaches for generating climatologic maps of precipitation (and other climatic parameters) and inputs for hydrologic models. These models reflect the spatial structure of precipitation by means of empirical relations between precipitation and topography, rather than via statistical relations, but they also evaluate uncertainty.

### 4.3.4 Precipitation-Gage Networks

#### Existing Networks

In general, precipitation for a given region will have higher spatial variability (see Box 4-2) for hourly or daily values than for monthly or annual values, and regions of convective rainfall and varying topography will tend to have relatively high variability. On the basis of these general considerations, the World Meteorological Organization (1974) recommended the minimum gage densities shown in Table 4-6 for general hydrometeorological purposes in various climatic regions. In fact, as can be seen from Figure 4-29, there are many regions of the world—including much of the United States—where these criteria are not met.

The National Weather Service surface-monitoring network in the contiguous United States includes 278 primary stations (mainly at airports; Figure 4-30a), which are staffed on a 24-hr basis by paid technicians, and over 8000 cooperative stations (Figure 4-30b), which are operated mostly by volunteers. Hourly precipitation is recorded at 241

**TABLE 4-6**

Minimum Precipitation-Gage Densities Recommended by the World Meteorological Organization (1981) for Various Climatic Situations. (Numbers in parentheses are "provisional norms" acceptable under difficult conditions.)

| Geographic Region | km²/gage | gage/km² |
|---|---|---|
| Small mountainous islands with irregular precipitation | 140-300 | 0.00333-0.00714 |
| Temperate, Mediterranean, and tropical mountainous regions | 300-1000 (1000-10,000) | 0.00100-0.00333 (0.000100-0.00100) |
| Flat areas in temperate, Mediterranean, and tropical regions | 1000-2500 (3000-10,000) | 0.000400-0.00100 (0.000100-0.000333) |
| Arid and polar regions | 5000-20,000 | 0.0000500-0.000200 |

first-order stations and about 2600 cooperative stations; the rest report daily precipitation (Groisman and Legates 1994). Chang (1981) reported that there were 3036 recording precipitation gages in the conterminous 48 states in 1975 and that, for individual states, gage density is highly correlated with population.

Precipitation data are archived by the National Climatic Data Center (NCDC) in Asheville, NC, and can be obtained from the NCDC and from regional data centers. (See Appendix G.) Table 4-7 describes the categories in which these data are summarized.

Figure 4-31 shows the distribution of gage records of various lengths for the United States; the longest gage records as of 1975 were at Charleston, SC (184 yr, nonrecording gage) and New Bedford, MA (162 yr, recording gage). As of 1975, there were 5690 nonrecording gages with records of 25 yr or more, and Chang (1981) concluded that this network was probably sufficient for the study of spatial variation in all parts of the country except the Rocky Mountains and the Pacific Northwest. However, as discussed in the next section, there are significant biases in the data arising largely from gage-catch deficiencies.

### Information Analysis of Gage Networks

The information provided by a set of precipitation measurements can be evaluated in terms of **bias** and **precision**.

Bias, $B(\hat{P})$, is defined as the difference between the estimate of mean areal precipitation made from measured values, $\hat{P}$, and the true mean, $P$:

$$B(\hat{P}) \equiv \hat{P} - P. \qquad (4\text{-}25)$$

As was discussed in Section 4.2.2, wind and other factors cause widespread undermeasurement of precipitation in gages of the type used in the United States, so measured values are typically on the order of 10% low for rain and more than 50% low for snow (Figure 4-16). As is shown in Figure 4-32, this leads to underestimates of mean annual precipitation of more than 10% over much of the country, with the largest effect where snow is most important (Legates and DeLiberty 1993). Furthermore, unless special efforts are made to account for them, orographic effects usually cause underestimation bias in areal estimates in topographically diverse regions because most gages are located at lower elevations. In general, neither source of underestimation is accounted for in precipitation data published in U.S. National Weather Service summaries.

Precision refers to the random uncertainty in estimates of areal precipitation due to the sparseness of the gage network. This effect is illustrated in Figure 4-33, which compares 1985 precipitation in North Carolina based on the first-order network (Figure 4-33a) with that from the cooperative network (Figure 4-33b) and illustrates that the density of a precipitation-gage network can make a large difference in estimates of areal precipitation: The differences exceed 762 mm in the mountainous regions and are over 254 mm even in the relatively flat piedmont portions of the state (Figure 4-33c). As was discussed in Section 2.5.2, precision is expressed as a confidence interval about the estimate, which is computed from the standard deviation of the estimate (standard error), $S(\hat{P})$. Precision can be improved by increasing the density of the gage network and by extending the period of measurement. The trade-off between the increase in information and gage density or time takes the form shown in Figure 4-34.

We can use the concepts of autocorrelation (persistence)[7] and spatial correlation (Box 4-2) to explore the precision of precipitation-gage networks quantitatively. Specifically, we can see how

---

[7]See Equation (C-64) and Box C-6.

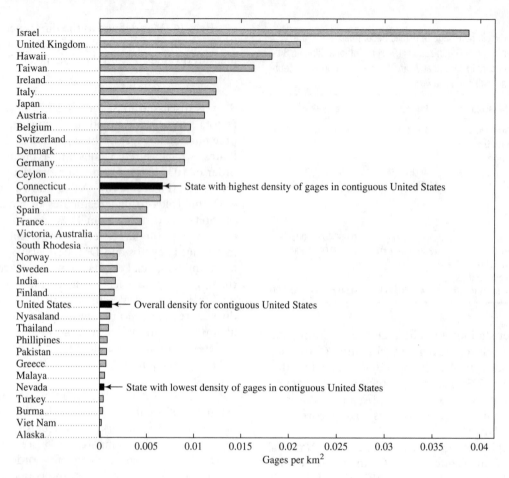

**FIGURE 4-29**
Precipitation-gage densities for various countries and states. The data are for the mid-1960s, but they are believed to approximate the current situation. From Gilman (1964).

the estimation variance $\hat{S}^2(\hat{P})$ is reduced by (1) making measurements for longer time periods or (2) adding more stations to a network. Following Bras and Rodriguez-Iturbe (1984), we write

$$\hat{S}^2(\hat{P}) = F_1(T) \cdot F_2(N) \cdot \hat{S}^2(p) , \quad \textbf{(4-26)}$$

where $\hat{S}^2(p)$ is the aggregate variance of the annual precipitation measurements at all gages in a region over the period of record; $F_1(T)$ is a variance-reduction function that depends on the number of years of observation, $T$, and on the average value of the autocorrelation coefficient for all gages; and $F_2(N)$ is a variance-reduction function that depends on the number of stations, $N$, the spatial correlation of precipitation, and the spatial distribution of the gages. [We assume a random

distribution; Morrissey et al. (1995) explore the effects of other network configurations.]

Figure 4-35 shows how $F_1(T)$ decreases with $T$ for various values of autocorrelation, $\rho_1 p$. These curves are calculated from Equation (C-44) for the standard error of the mean, accounting for the effect of autocorrelation on the effective record length [Equation (C-66)], so that the variance reduction is less as $\rho_1 p$ increases for a given $T$.

In order to evaluate $F_2(N)$, we first need to determine the spatial correlation as described in Box 4-2. To be consistent with Bras and Rodriguez-Iturbe (1984), we use Equation (4B2-5c) to express the relation between correlation coefficient and distance; then we use the quantity $A \cdot c^2$, where $A$ is the area of the region of interest, as a dimensionless

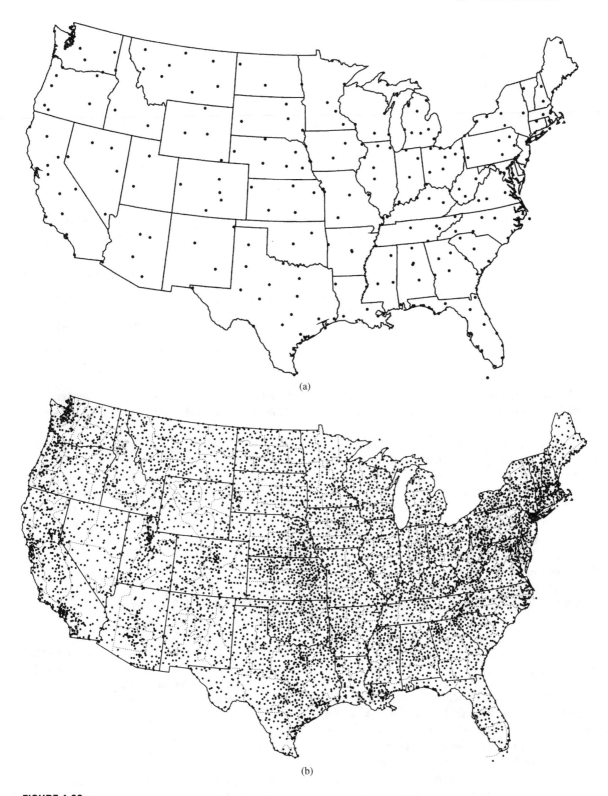

**FIGURE 4-30**
Distribution of National Weather Service stations in the contiguous United States. (a) Primary surface-monitoring network.
(b) Cooperative network. From Karl and Quayle (1988).

**TABLE 4-7**

Precipitation Data Summaries Maintained by the U.S. National Climatic Data Center (NCDC), National Oceanographic and Atmospheric Administration (NOAA). Information from Groisman and Legates (1994).

| Summary | Data for | Description |
|---------|----------|-------------|
| Local Climatological Data (LCD) | 272 1st-order stations | Published monthly since 1948. Most stations relocated prior to 1960. 20-40% have Alter wind shields. |
| Climatological Data (CD) | ~8000 cooperative stations | Daily values. Published monthly by state. |
| Hourly Precipitation Data (HPD) | 241 1st-order stations ~2600 cooperative stations | Hourly values. Published since 1950. Most stations relocated prior to 1960. 20–40% have Alter wind shields. |
| Historical Climatology Network (HCN) | 1221 1st-order and cooperative stations | Monthly values of long-record stations evenly distributed over U.S. with relatively undisturbed environments and minimal relocation. |
| Climate Division Data Base (CDDB) | 344 Climatological Divisions | Monthly values for divisions averaged from > 6000 stations (which may change over time). |

variable. Now we can refer to Figure 4-36 and see how $F_2(N)$ decreases (i.e., more variance reduction) as $N$ and $A \cdot c^2$ increase.

The important implication of Equation (4-26) and Figures 4-35 and 4-36 is that we can increase the precision of areal precipitation measurements either (1) by increasing the period of observation, $T$; or (2) by increasing the number of stations, $N$. An example will show how the accuracy gains from each strategy can be compared and how we can calculate the degree to which we can substitute gage density for length of record in attempting to improve the precision of areal precipitation estimates.

## EXAMPLE 4-4

For the region of eastern Nebraska and northern Kansas ($A = 52,000$ km$^2$) studied by Tabios and Salas (1985), mean annual precipitation has been measured at 29 stations in the region, each with 30 yr of record. The overall mean value (mean of the means at all stations) $\hat{P} = 619$ mm yr$^{-1}$ and the overall variance $\hat{S}^2(\hat{P}) = 25,445$ mm$^2$ yr$^{-2}$. Assume that $r_1(p) = 0$. (a) Compute the uncertainty in the estimate of mean annual precipitation, $\hat{S}^2(\hat{P})$ and the 95% confidence interval for $P$; and (b) compare

the value, in terms of variance reduction, of additional years of record with the value of additional gages in this situation.

**Solution** We read values of $F_1(T)$ directly from Figure 4-35; these are plotted in Figure 4-37. To use Figure 4-36, we first note that $c = 0.0031$ km$^{-1}$ (Box 4-2) and calculate $A \cdot c^2 = 0.50$. Entering Figure 4-36 at this value, we read the values of $F_2(N)$ for $N = 1, 2, 3, 5, \ldots$ and plot these on Figure 4-37.
(a) From Figure 4-37, $F_1(30) = 0.032$ and $F_2(29) = 0.67$. Thus, from Equation (4-26),

$$\hat{S}^2(\hat{P}) = (25,445 \text{ mm}^2 \text{ yr}^{-2})$$
$$\times 0.032 \times 0.67 = 546 \text{ mm}^2 \text{ yr}^{-2},$$

and the standard deviation $\hat{S}(\hat{P})$ is

$$\hat{S}(\hat{P}) = [\hat{S}^2(\hat{P})]^{1/2} = 23.4 \text{ mm yr}^{-1}.$$

Assuming the normal distribution of measurement error and using reasoning discussed in Section 2.5.2, we can state that we are 95% sure that the true value of the mean $P$ is within $1.96 \cdot \hat{S}(\hat{P})$ of the sample mean, so we are 95% sure that

$$(619 - 1.96 \times 23.4) \text{ mm yr}^{-1} < P$$
$$< (619 + 1.96 \times 23.4) \text{ mm yr}^{-1};$$
$$573 \text{ mm yr}^{-1} < P < 665 \text{ mm yr}^{-1}.$$

(b) Comparing the two curves in Figure 4-37, we see that an additional year of record is generally more valuable than an additional gage. In fact, for estimating mean annual precipitation,

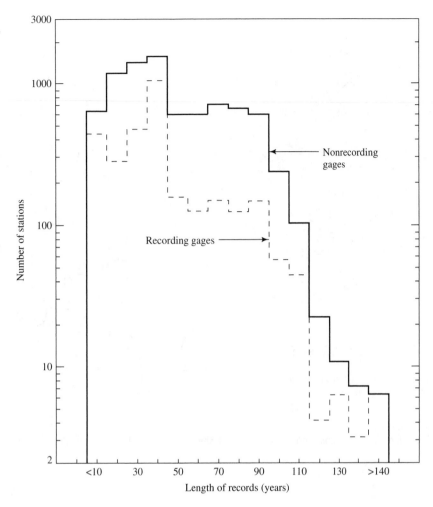

**FIGURE 4-31**
Number of precipitation gages for every 10-yr category of record length in the coterminous United States. Note the logarithmic scale. From Chang (1981), used with permission of the American Water Resources Association.

there is little to be gained by having more than about 20 gages in the region, and additional years of record beyond the present 30 add variance reduction only very slowly.

Of course, we would not conclude from Example 4-4 (b) that there is no need for any more data collection in this region. That analysis applies only to mean annual precipitation, and we are interested in values for individual storms, days, months, and so on. Also, the analysis applies only for a situation in which climate is not changing, and we are interested in maintaining measurements to detect such change. However, the example illustrates an important conclusion:

We can trade time versus space in hydrologic data collection when we do not reduce the time interval too much, but no miracles can be expected in short

times even from the most dense of all possible networks [Bras and Rodriguez-Iturbe (1984) p. 349].

### 4.3.5 Radar and Satellite Observation

#### *Radar Observation*

Ground-based radar can be used to estimate the areal distribution of instantaneous precipitation rates in clouds, and these rates can be electronically integrated to provide estimates of total precipitation for any time period. Energy in the wavelength band between 1 and 10 cm (microwaves; see Figures D-1, D-7) is reflected by raindrops and snowflakes (collectively called **hydrometeors**). The strength of the signal reflected from a given distance depends on the size distribution, number per unit volume, state (liquid or

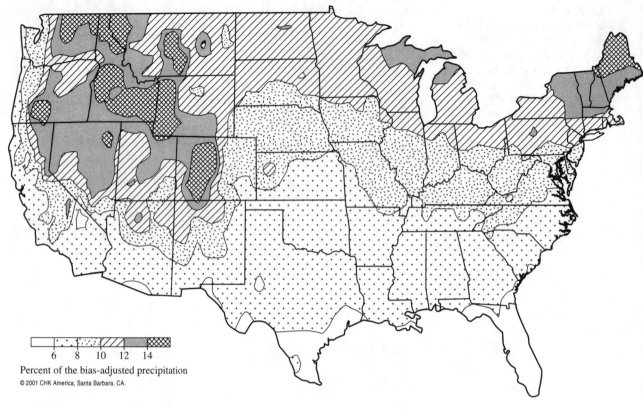

Percent of the bias-adjusted precipitation

© 2001 CHK America, Santa Barbara, CA.

**FIGURE 4-32**
Estimated bias of precipitation measurements in the coterminous United States. From Groisman and Legates (1994).

solid), and shape of the hydrometeors, along with characteristics of the radar system, and is related to precipitation intensity by empirical relations. Doppler radar systems also measure the horizontal velocity of the hydrometeors and hence the movement of the storm.

Because of the various factors that affect the radar image of precipitation, plus the evaporation of falling rain and the distortion of the precipitation field by below-cloud winds, the image cannot provide a precise estimate of precipitation amounts. Linsley et al. (1982) stated that radar measurements of rainfall within 110 km of the radar are between one-half and twice the gage measurements (which, as we have seen, can themselves be in substantial error). However, radar gives a picture of the areal extent of precipitation that can be very useful when used in conjunction with gage data to improve the accuracy of areal estimates.

In the United States, the NEXRAD system of WSR-88D weather-radar stations provides information on precipitation distribution at over 130 sites, each with a range of 230 km. This system yields real-time images with qualitative indications of precipitation intensity and summary quantitative estimates of daily or storm total precipitation with a spatial resolution of about $2 \times 2$ km. The locations of these systems and the information they produce can be accessed at Internet sites maintained by the U.S. National Weather Service and commercial weather enterprises. (See Appendix G.)

### Satellite Observation

Satellites obtain information about the distribution and amounts of precipitation by both direct and indirect means (Arkin and Ardanuy 1989). Direct observation is by passive sensing of microwave energy

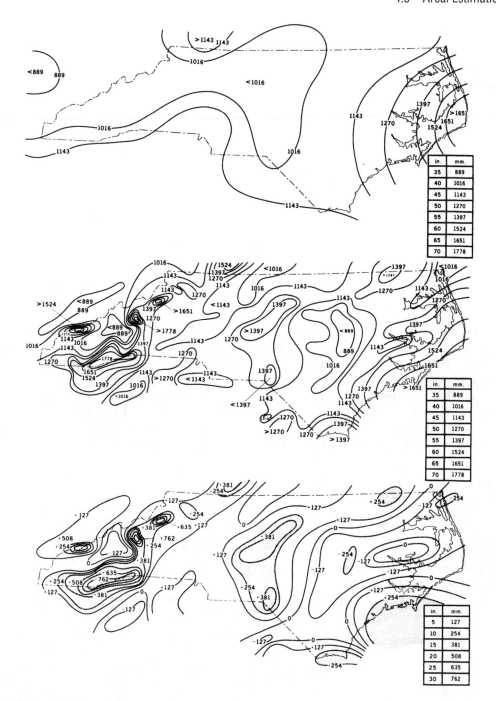

**FIGURE 4-33**
1985 annual precipitation in North Carolina, (a) from the primary network, (b) from the cooperative network. (c) Differences between the two estimates. From Karl and Quayle (1988).

**FIGURE 4-34**
General relation between accuracy and cost of a gage network. After Bras and Rodriguez-Iturbe (1976).

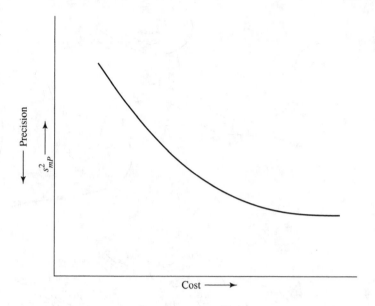

absorbed and scattered by hydrometeors and conversion of these observations into estimates of rainfall rates by accounting for the background radiation from the earth's surface and making assumptions about the size distribution of the hydrometeors. Indirect observation is by sensing infra-red radiation emitted by clouds, converting the radiation flux to cloud-top temperature via the

Stefan-Boltzmann Law [Equation (D-1)], and making use of empirical correlations of the spatial and temporal coverage of clouds with temperatures below a threshold value and rainfall.

Satellite estimates of precipitation are made at "climatic scales"—that is, they are averaged over at least $10^4$ km$^2$ (about 1° latitude by 1° longitude) and over time periods of 5 days or more (Arkin and Ar-

**FIGURE 4-35**
Variance-reduction function $F_1(T)$ as a function of number of years of record, $T$, and average regional autocorrelation, $r_{1p}$. From Rodriguez-Iturbe and Mejia (1974), used with permission of the American Geophysical Union.

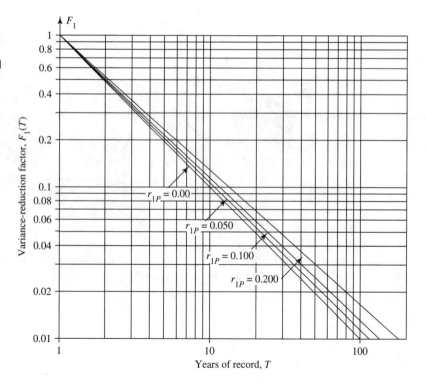

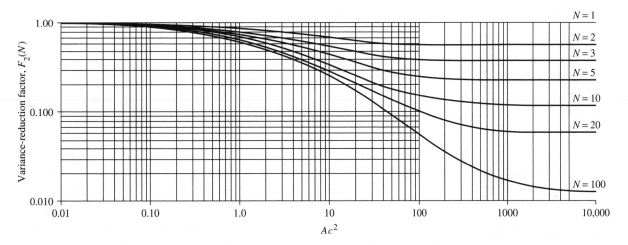

**FIGURE 4-36**

Variance-reduction function $F_2(N)$ as a function of number of stations, $N$; area, $A$; and spatial-correlation factor [$c$ in Equation (4B2-5c)]. From Rodriguez-Iturbe and Mejia (1974), used with permission of the American Geophysical Union.

dunuy 1989). And because of the calibrations and assumptions involved in both direct and indirect approaches, the estimates have large uncertainties. In spite of this, satellites have an extremely important and growing role in assessing global precipitation. The earth is 70% ocean, and much of the land surface is virtually uninhabited; as a result, the percentage of the earth's surface on which precipitation-gage measurements are available is very small, and gage densities are generally very sparse. Thus it is only with the advent of satellites that scientists have been able to obtain a global perspective on

precipitation and other aspects of the hydrologic cycle.

Recently, major programs have been instituted to refine satellite estimates of precipitation, including the Global Precipitation Climatology Project (GPCP), which collects and synthesizes information from polar-orbiting and geostationary satellites and surface stations; a new series of U.S. (NOAA) polar-orbiting satellites with an Advanced Microwave Sounding Unit (AMSU); and the U.S. NASA Earth Observing System (EOS). NASA's recent Tropical Rainfall Measuring Mission (TRMM) is a low-

**FIGURE 4-37**

Variance-reduction functions $F_1(T)$ and $F_2(N)$ for Example 4-4.

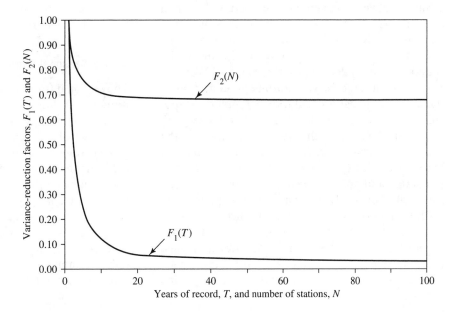

orbiting satellite that includes precipitation radar as well as passive sensing of visible, infra-red, and microwave radiation. Thus satellite-based information will play an increasing role in providing precipitation data for hydrologic science and applications.

## 4.4 PRECIPITATION AND RAINFALL CLIMATOLOGY

As was discussed in Sections 2.5.1 and 3.3, long-term and seasonal average precipitation are fundamental determinants of regional hydrology, soils, and vegetation. In addition, a large class of hydrologic models requires estimates of regional **extreme values** of precipitation as inputs for predicting magnitudes of flood flows, as a basis for designing such structures as culverts, bridges, storm sewers, and dams for flood control and other purposes and for delineating areas subject to flooding. This section describes how these climatological averages and extreme values are developed and how these values are related to general meteorological conditions.

### 4.4.1 Long-Term Average Precipitation Rates

**Average annual precipitation** at a gage site is calculated by totaling the precipitation catch for a number of complete years of record (corrected for changes in gage exposure as described earlier) and dividing the sum by the number of years used; the result is usually expressed in mm yr$^{-1}$ or in. yr$^{-1}$. The years used in the computation should always be stated. In the United States, the National Weather Service defines the **30-year normal precipitation** as the average annual precipitation computed for specific 30-yr periods which are updated every 10 years; thus we have the 1951–1980 normal, the 1961–1990 normal, etc. These values are published in a series called *Climatography of the United States* (by state) and are also available through the U.S. National Climatic Data Center. (See Appendix G for Internet access to this information.)

Average annual or 30-yr normal precipitation is often used as an estimate of the long-term average precipitation rate at a station; areal values can be computed from values for individual gages by using the methods described in the previous section. The **inter-annual variability** of precipitation can be expressed as the standard deviation (see Box C-2) of the annual values.

The global distribution of average precipitation and its relation to large-scale climatic and topographic features were described in Section 3.2.2 (Figure 3-19). More detailed relationships between average precipitation rate and the time and space characteristics of the precipitation types discussed above are reflected in the map of the average annual precipitation for the 48 contiguous United States (Figure 4-38). Note the strong orographic effects in western Washington and Oregon and the Sierra Nevada of California, along the crest of the Appalachians from northern Georgia to Pennsylvania, and in the Adirondacks of New York.

The importance of the Gulf of Mexico as a source of moisture for the cyclonic systems that move across the country is clear in the roughly concentric pattern of isohyets extending northward from the Gulf Coast and westward to about longitude 105° W. The role of the Atlantic Ocean as a moisture source is also apparent, but, because the prevailing storm movements are from the west, its effect is limited to the immediate coastal area.

The irregular isohyetal patterns between longitude 105° W and the mountains of California and western Washington and Oregon reflect the influence of orography in the Rockies and other mountains and the importance of local convective storms; the low average values in that region are due to the rain-shadow effects of the West Coast mountains and to distance from the Gulf of Mexico.

### 4.4.2 Seasonal Variability of Precipitation

The seasonal (or **intra-annual**) variability of precipitation is an important aspect of hydroclimatology because it largely determines the seasonality of other hydrologic quantities, such as streamflow and ground-water recharge. As was discussed in Section 3.2.2, the seasonal patterns of relative heating of the continents and the migration of large-scale circulation features largely control the seasonality of precipitation on a global scale. Monsoon regions in particular have pronounced seasonal variability of precipitation: Figure 4-39 shows the percentage of normal annual precipitation that occurs in June through September in two of these regions.

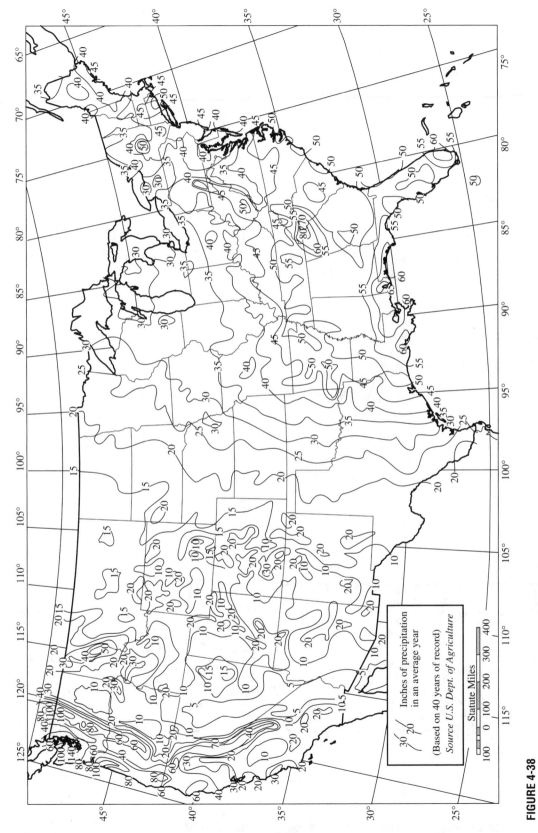

**FIGURE 4-38**
Long-term average annual precipitation for the contiguous United States. Source: U.S. Department of Agriculture.

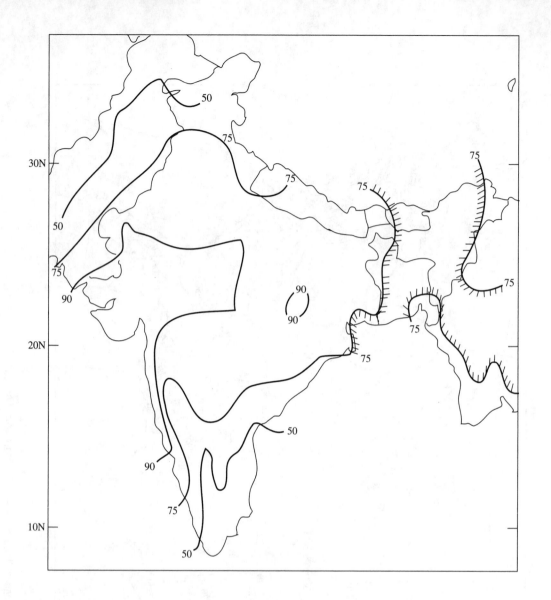

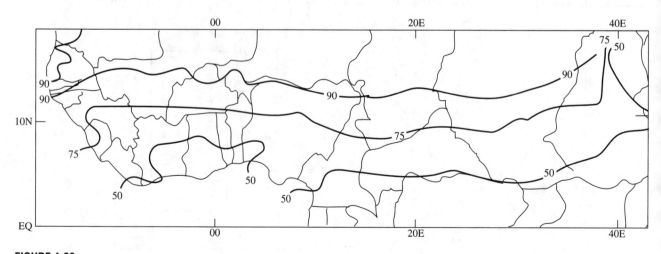

**FIGURE 4-39**
Percentage of normal annual precipitation that occurs in June through September in (a) India and Pakistan and (b) the Sahel region of sub-Saharan Africa. From U.S. National Oceanographic and Atmospheric Administration Climate Analysis Center (1989).

## BOX 4-3

· · · · · · ·

## Circular Statistics for Quantifying Seasonality of Precipitation and Other Hydrologic Variables

Circular, or directional, statistics is useful for quantifying the time of occurrence of events when time is measured on a circle—like a clock or a year. The quantification involves calculating two values: one expresses the **average time of occurrence;** the other expresses the degree to which the events tend to be concentrated in time, called the **seasonality index.** The first quantity is analogous to the arithmetic mean; the second is analogous to the standard deviation for noncircular variables. (See Section C.2.5.). Following Markham (1970), we apply circular statistics to the quantification of precipitation seasonality.

To do this, time through the year is represented on a circle and each month is assigned the angle, measured clockwise from 1 January, of its mid-month date (Table 4-8). Average monthly precipitation at a station is then considered to be a vector quantity, with a length proportional to the amount and a direction given by Table 4-8. The 12 monthly vectors can be added to give a resultant vector with a direction $\phi_R$ and a magnitude $P_R$, where

$$\phi_R' = \tan^{-1}\left(\frac{S}{C}\right) \qquad \text{(4B3-1)}$$

$$P_R = (S^2 + C^2)^{1/2}, \qquad \text{(4B3-2)}$$

$$S \equiv \sum_{m=1}^{12} P_m \cdot \sin(\phi_m), \qquad \text{(4B3-3)}$$

$$C \equiv \sum_{m=1}^{12} P_m \cdot \cos(\phi_m), \qquad \text{(4B3-4)}$$

the $P_m$ are the 12 monthly precipitation amounts, and the $\phi_m$ are the 12 monthly time angles given in Table 4-8.

The average time of occurrence, $\phi_R$, is given by the value of $\phi_R'$ and the signs of $S$ and $C$, which determine which quadrant of the circle the resultant vector is in:

$$\phi_R = \phi_R' \qquad \text{if } S > 0 \text{ and } C > 0; \quad \text{(4B3-5a)}$$
$$\phi_R = \phi_R' + 180° \qquad \text{if } C < 0; \quad \text{(4B3-5b)}$$
$$\phi_R = \phi_R' + 360° \qquad \text{if } S < 0 \text{ and } C > 0. \quad \text{(4B3-5c)}$$

$\phi_R$ can then be converted to the month in which it occurs by reference to the last column of Table 4-8.

The seasonality index is the ratio of the resultant vector $P_R$ to the average annual precipitation (the arithmetic total of the 12 monthly values). This ratio ranges from 0 (when the precipitation is exactly equal in all months) to 1 (when all the precipitation occurs in a single month).

One way of quantitatively describing seasonality is by means of **circular statistics** (Box 4-3). Markham (1970) applied this approach to monthly precipitation data in the United States to characterize seasonality in terms of (1) the **average time of occurrence** and (2) the **seasonality index** (Figure 4-40). Example 4-5 shows how these values are calculated for this station.

---

### EXAMPLE 4-5

The table that follows lists the 1931–1960 average monthly precipitation (in inches) for San Francisco, CA. We want to compute the average time of occurrence and the seasonality index of precipitation for this station.

| Jan | Feb | Mar | Apr | May | Jun | Jul |
|------|------|------|------|------|------|------|
| 4.01 | 3.48 | 2.69 | 1.30 | 0.48 | 0.11 | 0.01 |

| Aug | Sep | Oct | Nov | Dec | Year |
|------|------|------|------|------|-------|
| 0.02 | 0.19 | 0.74 | 1.57 | 4.09 | 18.69 |

**Solution:** Referring to Box 4-3, we first calculate $S$ by multiplying the values in the preceding table by the sine value for the corresponding month given in Table 4-8 and summing [Equation (4B3-3)]:

$$S = 4.01 \times 0.272 + 3.48 \times 0.705 + \ldots$$
$$+ 4.09 \times (-0.256) = 4.69 \text{ in.}$$

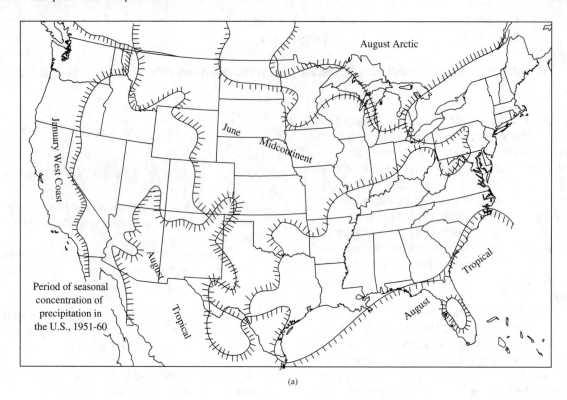

(a)

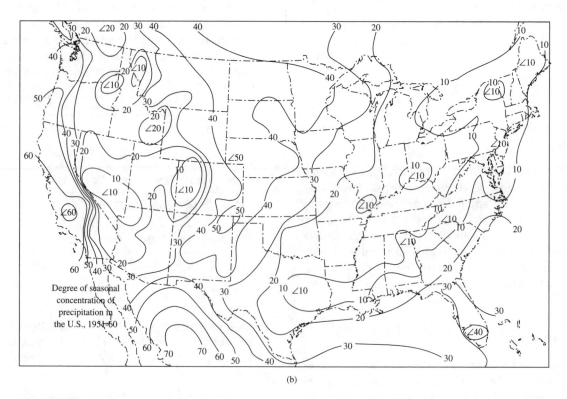

(b)

**FIGURE 4-40**

(a) Average month of occurrence and (b) seasonality index of annual precipitation calculated form monthly precipitation data by using methods of circular statistics (Box 4-3). From Markham (1970).

**TABLE 4-8**

Angles and Angle Functions for Computing Seasonality Index and Average Time of Occurrence for Monthly Data.

| Month | Mid-Month | | | | First of Month | |
| | Day of Year | Angle, $\phi_m$ (degrees) | Sine | Cosine | Day of Year | Angle (degrees) |
| --- | --- | --- | --- | --- | --- | --- |
| Jan | 16.0 | 15.8 | 0.272 | 0.962 | 1 | 1.0 |
| Feb | 45.5 | 44.9 | 0.705 | 0.709 | 32 | 31.6 |
| Mar | 75.0 | 74.0 | 0.961 | 0.276 | 60 | 59.2 |
| Apr | 105.5 | 104.1 | 0.970 | −0.243 | 91 | 89.8 |
| May | 136.0 | 134.1 | 0.718 | −0.696 | 121 | 119.3 |
| Jun | 166.5 | 164.2 | 0.272 | −0.962 | 152 | 149.9 |
| Jul | 197.0 | 194.3 | −0.246 | −0.969 | 182 | 179.5 |
| Aug | 228.0 | 224.9 | −0.705 | −0.709 | 213 | 210.1 |
| Sep | 258.5 | 255.0 | −0.966 | −0.260 | 244 | 240.7 |
| Oct | 289.0 | 285.0 | −0.966 | 0.259 | 274 | 270.2 |
| Nov | 319.5 | 315.1 | −0.706 | 0.708 | 305 | 300.8 |
| Dec | 350.0 | 345.2 | −0.256 | 0.967 | 335 | 330.4 |

$C$ is then computed in the same way, using the cosine values [Equation (4B3-4)]:

$$C = 4.01 \times 0.962 + 3.48 \times 0.709 + \ldots + 4.09 \times 0.967 = 11.50 \text{ in.}$$

Then from Equation (4B3-2) we have

$$P_R = (4.69^2 + 11.50^2)^{1/2} = 12.41 \text{ in.}$$

Dividing $P_R$ by the total annual precipitation gives the seasonality index:

$$\text{Seasonality Index} = 12.41/18.69 = 0.66.$$

The average time of occurrence is found from Equation (4B3-1) as

$$\phi_R' = \tan^{-1}(4.69/11.50) = 22.2°.$$

$S$ and $C$ are both $> 0$, so $\phi_R = \phi_R'$ [Equation (4B3-5a)]. By referring to the last column of Table 4-8, we can see that this time angle falls in January. (Only monthly data were used in this computation, so more-than-monthly precision is not warranted.)

Figure 4-40a shows a January maximum of precipitation on the West Coast. This is associated with the position and intensity of the North Pacific high-pressure cell and the Aleutian low-pressure cell. As is shown in Figure 3-10a, there is an eastward wind-flow parallel to the isobars in winter, bringing moisture-laden air; in the summer (Figure 3-10b), the high-pressure cell has expanded and causes subsidence and a northerly wind flow parallel to the coast.

The June maximum of precipitation in the mid-continent is produced by the springtime northward extension of moist air from the Gulf of Mexico into the zone of intense cyclogenesis in the westerly winds. By July this zone has migrated farther north with the Arctic front (Barry and Chorley 1982).

The August maximum in the tropics is due to the northward migration of moist air from the Gulf of Mexico and Gulf of California, increased thermal-convective thunderstorm activity, and, in the Southeast, rain from tropical cyclones. The mild August Arctic maximum results from the northward position of cyclogenesis associated with the Arctic front and the increased amount of moisture that can be held in the warmer air of late summer.

### 4.4.3   Storm Climatology

#### Thunderstorms

Figure 4-41 shows the average number of days with thunderstorms for the 48 contiguous United States; most of these storms occur in summer and are due to thermal convection.

#### Tropical Storms

Figures 4-42 and 4-43 show the global frequency and northern-hemisphere seasonality of tropical cyclones. The concentration of storms in the summer and early fall reflects the necessity of high sea-surface temperatures for triggering hurricane development.

Figure 4-44 shows the number of hurricanes in the North Atlantic between 1886 and 1992. Most

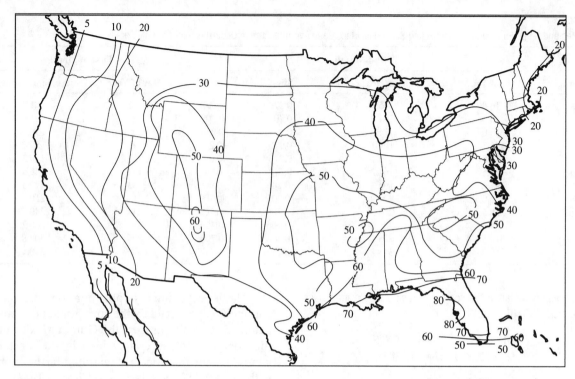

**FIGURE 4-41**
Average annual number of days with thunderstorms in the contiguous United States. Reprinted with permission of Prentice Hall, Inc., from *Elements of Meteorology*, 4th ed., by Miller et al. © 1983 by Bell & Howell Co.

of the highest floods of medium- to large-size drainage basins along the eastern seaboard of the United States have been caused by hurricanes. Two notable examples are the infamous " '38 Hurricane" (21 September 1938), which dumped between 3 and 6 inches of rain in 24 hr on most of New England (Brooks 1940) and caused massive flooding; and Hurricane Agnes, which spent almost a week over the Middle Atlantic states in June 1972 and produced exceptionally serious flooding in the Carolinas, near-record floods in Virginia, and record-breaking floods in central Pennsylvania and western New York (Hopkins 1973; Figure 4-45).

Interestingly, the '38 Hurricane had other lasting hydrologic effects: So many trees were destroyed by its winds that evapotranspiration was reduced in central New England, causing a significant increase in the annual flow of rivers draining the region in the following three years (Patric 1974).

### 4.4.4  Extreme Rainfall Amounts

Extreme values of rainfall are of prime interest as inputs to simulation models used to estimate design floods. The design flood is then used as a basis for designing drainage systems, culverts, and flood-control structures or floodplain-management and land-use plans. We begin by examining the record amounts of rainfall that have been recorded for various durations, then introduce the deterministic concept of the "probable maximum precipitation," and conclude with the statistical treatment of extreme rainfall values.

#### Record Rainfalls

The largest amounts of rainfall recorded for various durations are listed in Table 4-9 and plotted in Figure 4-46. The envelope curve shown in Figure 4-46, which appears to describe an upper bound for the values, is given by

$$R = 425 \cdot D^{0.47}, \tag{4-27}$$

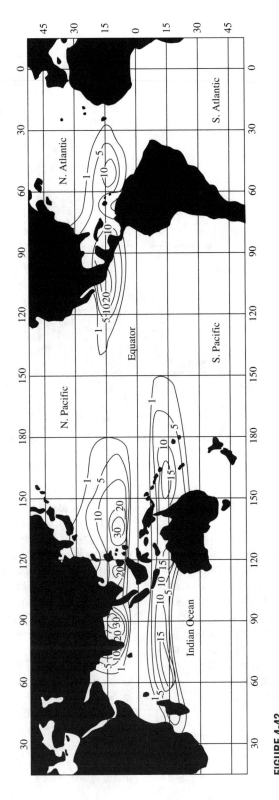

**FIGURE 4-42**
Numbers of tropical cyclones generated during 1958–1978. From Simpson and Riehl (1981).

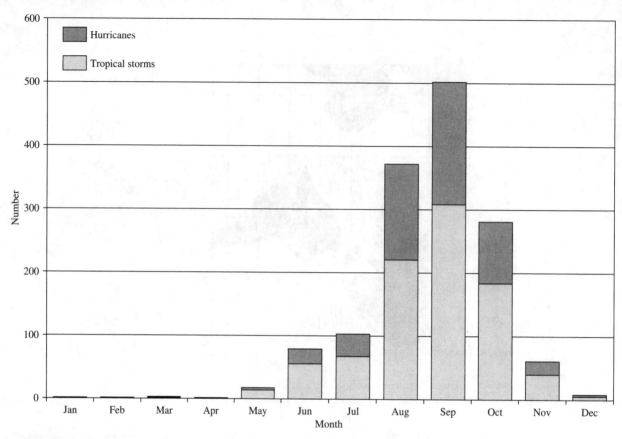

**FIGURE 4-43**
Numbers of North Atlantic tropical storms (lower bars) and hurricanes (upper bars) observed in each month, 1886-1992. Data from Neumann et al. (1993).

where $R$ is the maximum recorded rainfall depth in mm and $D$ is duration in hours.

The record rainfalls for various durations can be related directly to the characteristics of the meteorological processes that produced them. For durations up to about 4.5 hr, the record amounts were produced by convective storms; for those between 9 hr and 8 days, they were from tropical cyclones (enhanced by orographic effects; see Paulhus 1965); for the longest durations, they are the product of the unusually intense orographic enhancement described earlier for Cherrapunji, India.

### Probable Maximum Precipitation

The **probable maximum precipitation** (abbreviated PMP) is "theoretically the greatest depth of precipitation for a given duration that is physically possi-

ble over a given size of storm area at a particular geographical location at a certain time of year" (Hansen 1987). This quantity is used as input to hydrological models to provide an indication of the largest flood that could occur from a given drainage basin—the **probable maximum flood** (PMF). Such floods are used in determining the required capacity of the emergency spillway of a dam whose failure would cause massive economic damage and loss of lives.

PMP is usually estimated either (1) by examining rainfall data for the largest flood-producing storms in and near the region of interest and using meteorological reasoning to estimate the combination of conditions that could have produced the largest possible rainfall rates from those storms on the drainage basin of interest or (2) by using generalized PMP maps based on meteorological analysis

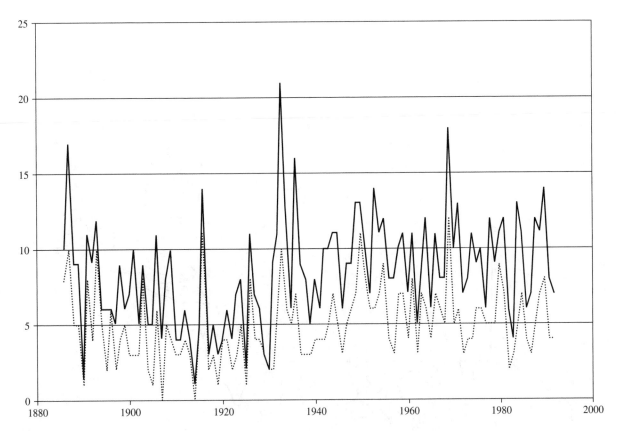

**FIGURE 4-44**
Annual numbers of North Atlantic tropical storms (upper line) and hurricanes (lower line), 1886-1992. Data from Neumann et al. (1993).

of the largest storms in a large region. Details of the first approach were described in World Meteorological Organization (1973); here we will briefly examine the generalized approach developed by the U.S. National Weather Service.

Schreiner and Reidel (1978) developed generalized PMP maps for portions of the U.S. east of the 105th meridian, where orographic effects are negligible. Maps were given for storms of 6- 12-, 24-, 48-, and 72-hr duration and for areas of 10, 200, 1,000, 5,000, 10,000, and 20,000 mi². Figure 4-47 shows three of these maps. The report of the National Academy of Sciences (1983) identified sources for estimating PMP for regions west of the 105th meridian.

The storm size and duration appropriate for a given situation are determined by the size of the drainage basin of interest: In general, the critical storm size is approximately equal to the basin size, and the critical duration is approximately equal to the "time of concentration" of the basin (the time it takes storm runoff to travel from the most distant part of the drainage basin to the outlet; see Section 9.1.5.) As a general guide, Figure 4-48 gives approximate relationships between time of concentration and drainage area. Schreiner and Reidel (1978) gave methods for developing PMP estimates for durations and areas other than those mapped.

Hydrologic modeling requires information on the time and areal distribution of rainfall inputs, not just the total rainfall amount at a point. To provide this information, Hansen et al. (1982) gave detailed instructions for developing the **probable maximum storm;** the approach was summarized by Hansen (1987).

The concepts of PMP and PMF are controversial: Can we really specify an upper bound to the amount of rain that can fall in a given time? The envelope curve in Figure 4-46 suggests that it is reasonable to do so, but we must recognize that the

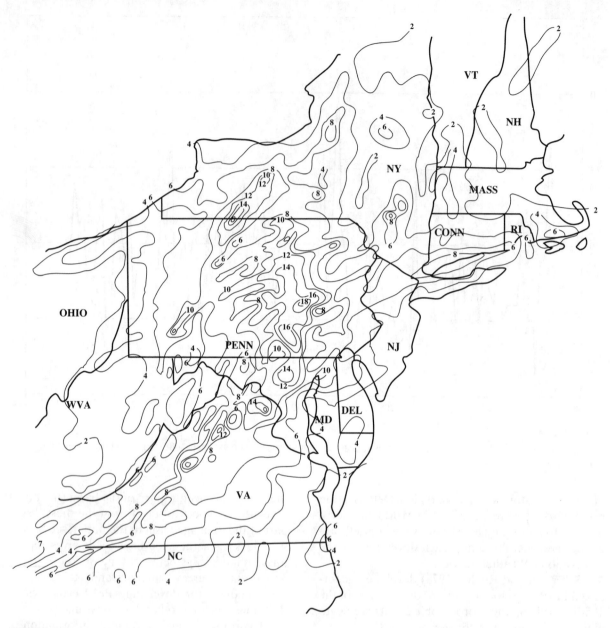

**FIGURE 4-45**
Total rainfall (in.) from Hurricane Agnes, 18 to 25 June 1972, in the Middle Atlantic region. From Hopkins (1973).

plotted values are only those that have been observed in this century at the infinitesimal fraction of the earth covered by rain gages, and higher amounts must have fallen at ungaged locations at other times and places—and, conceptually, we can always imagine that a few more molecules of water could fall beyond any limit that is specified.

Nevertheless, specific numerical values are required for design purposes, and the concepts of PMP and PMF are useful in many practical situations. The reasonableness of current approaches to estimating PMP can be verified by direct comparisons of PMP with observed rainfalls. Reidel and Schreiner (1980) gave a summary of all storms in the United States with total rainfalls of at least 50% of estimated PMP, and a portion of their results is summarized in Table 4-10.

**TABLE 4-9**

Locations, Dates, and Amounts of Record Rainfalls for Various Durations. Data are Plotted in Figure 4-46.

| Duration | Depth | | Location | Date |
|----------|-------|-------|----------|------|
| | in. | mm | | |
| 1 min | 1.50 | 38 | Barot, Guadeloupe | Nov. 26, 1970 |
| 8 min | 4.96 | 126 | Füssen, Bavaria | May 25, 1920 |
| 15 min | 7.80 | 198 | Plumb Point, Jamaica | May 12, 1916 |
| 20 min | 8.10 | 206 | Curtea-de-Arges, Roumania | July 7, 1889 |
| 42 min | 12.00 | 305 | Holt, MO | June 22, 1947 |
| 2 hr 10 min | 19.00 | 483 | Rockport, WV | July 18, 1889 |
| 2 hr 45 min | 22.00 | 559 | D'Hanis, TX (17 mi NNW) | May 31, 1935 |
| 4 hr 30 min | 30.8 | 782 | Smethport, PA | July 18, 1942 |
| 9 hr | 42.79 | 1,087 | Belouve, Réunion | Feb. 28, 1964 |
| 12 hr | 52.76 | 1,340 | Belouve, Réunion | Feb. 28–29, 1964 |
| 18 hr 30 min | 66.49 | 1,689 | Belouve, Réunion | Feb. 28–29, 1964 |
| 24 hr | 73.62 | 1,870 | Cilaos, Réunion | Mar. 15–16, 1952 |
| 2 days | 98.42 | 2,500 | Cilaos, Réunion | Mar. 15–17, 1952 |
| 3 days | 127.56 | 3,240 | Cilaos, Réunion | Mar. 15–18, 1952 |
| 4 days | 146.50 | 3,721 | Cherrapunji, India | Sept. 12–15, 1974 |
| 5 days | 151.73 | 3,854 | Cilaos, Réunion | Mar. 13–18, 1952 |
| 6 days | 159.65 | 4,055 | Cilaos, Réunion | Mar. 13–19, 1952 |
| 7 days | 161.81 | 4,110 | Cilaos, Réunion | Mar. 12–19, 1952 |
| 8 days | 162.59 | 4,130 | Cilaos, Réunion | Mar. 11–19, 1952 |
| 15 days | 188.88 | 4,798 | Cherrapunji, India | June 24–July 8, 1931 |
| 31 days | 366.14 | 9,300 | Cherrapunji, India | July 1861 |
| 2 mo | 502.63 | 12,767 | Cherrapunji, India | June–July 1861 |
| 3 mo | 644.44 | 16,369 | Cherrapunji, India | May–July 1861 |
| 4 mo | 737.70 | 18,738 | Cherrapunji, India | Apr.–July 1861 |
| 5 mo | 803.62 | 20,412 | Cherrapunji, India | Apr.–Aug. 1861 |
| 6 mo | 884.03 | 22,454 | Cherrapunji, India | Apr.–Sept. 1861 |
| 11 mo | 905.12 | 22,990 | Cherrapunji, India | Jan.–Nov. 1861 |
| 1 year | 1,041.78 | 26,461 | Cherrapunji, India | Aug. 1860–July 1861 |
| 2 years | 1,605.05 | 40,768 | Cherrapunji, India | 1860–1861 |

Data are plotted in Figure 4-46. From *Hydrology for Engineers* (3rd ed.), by R. K. Linsley, Jr., M. A. Kohler, and J. L. H. Paulhus. Copyright © 1982 by McGraw-Hill Book Company. Reprinted with permission of McGraw-Hill Book Company.

## Depth–Duration–Frequency Analysis

Most engineering and land-use planning situations do not involve the risk of catastrophic economic damage or loss of life that warrants design to PMF levels. For these less extreme circumstances, design is based on the flood that is estimated to have a specified **exceedence probability** (or frequency), the inverse of which is the **return period** [defined in Equations (C-31) and (C-32)]. One way to estimate the flood with a given return period is by means of a simulation model, using as input the rainfall with the appropriate return period and duration.

Estimation of the rainfall depths with a given return period for various storm durations at a given rain-gage location is called **depth–duration–frequency** (DDF) **analysis.** An equivalent procedure differs only in using rainfall intensities (depth divided by duration) rather than depth and is called **intensity–duration–frequency** (IDF) **analysis.**

Standard engineering practice dictates the return period appropriate for a given situation. For example, the flood with a return period of 25 yr (exceedence probability of 0.04 in any year) is commonly used for designing culverts; the 100-yr flood (annual exceedence probability of 0.01) is used to delineate floodplains for land-use planning. The appropriate duration is that which produces the largest peak flood from the area of interest. In general, the critical duration increases with drainage area; Figure 4-48 indicates a range of typical values. DDF/IDF analyses are usually carried out for several return periods and durations and the appropri-

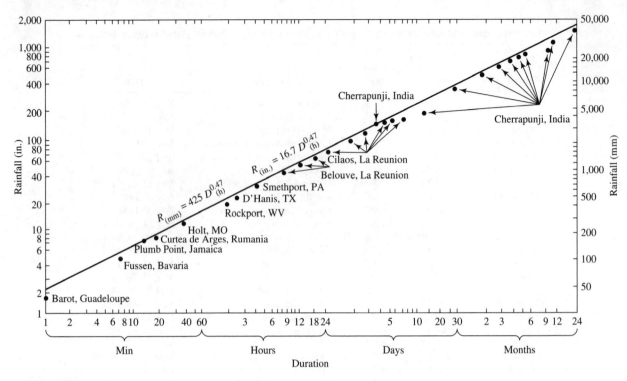

**FIGURE 4-46**
Maximum amounts of recorded rainfall as a function of duration. Data are listed in Table 4-9. From *Hydrology for Engineers*, 3rd ed., by R.K. Linsley, Jr., M.A. Kohler, and J.L.H. Paulhus. Copyright © 1982 by McGraw-Hill Book Co. Reprinted with permission of McGraw-Hill Book Co.

ate duration determined by comparing the floods predicted by the simulation model.

In most cases, the hydrologist requires rainfall frequency analysis for a finite area, such as a watershed. Thus there are two parts to DDF/IDF procedure: (1) determining relations for one or more representative points (rain-gage locations) in the region of interest; and (2) adjusting the point values to give estimates for the area of interest.

**DDF/IDF Analysis at Points**   The first step in the analysis is the development of a time series of annual maximum values of rainfall for selected durations at selected gages in the region of interest. The analysis procedure consists of the estimation of quantiles for this time series, following the method of Box C-1, and is described via Example 4-6. (This example is done in terms of depth; Exercise 4-9 asks you to repeat the analysis in terms of intensity.)

## EXAMPLE 4-6

We wish to estimate the depths of rainfalls of durations of 1 to 24 hr with return periods of 2, 5, 10, and 25 yr for the Chicago Airport weather station.

***Solution***   The analysis should begin with a review of the history of the weather station to assure that measurement conditions have not changed significantly during the period of record. Assuming that conditions have been stable, we examine the rainfall records to determine the annual maximum rainfalls for each duration of interest for the period of record. (This is by far the most time-consuming step in the process.) Table 4-11 lists these values for the Chicago Airport station for 1949–1972. (In practice, one would use the complete record.)

The next step is to rank the values for each duration from highest to lowest and to compute the estimated quantile, *q*, for each value via Equation (CB1-3) (Table 4-12). Interpolation to determine the depths associated with the return periods of in-

**FIGURE 4-47**

Generalized all-season PMP (in.)
for the eastern United States for a
storm of 200-mi$^2$ area and durations of (a) 6 hr, (b) 12 hr,
(c) 24 hr. Shaded areas require
correction for orographic effects.
From Schreiner and Reidel
(1978).

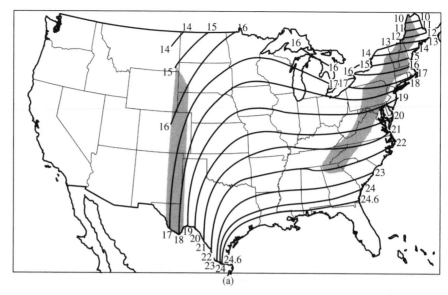

(a)

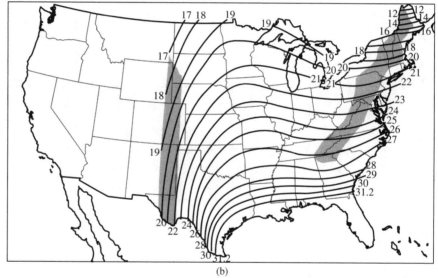

(b)

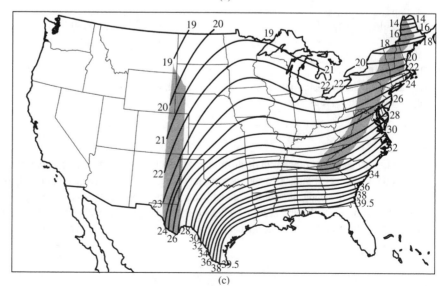

(c)

**TABLE 4-10**
Numbers of Storms Exceeding Various Percentages of PMP for Two Regions of the United States.

| Region | Percentage of PMP | | | | | Total Number of Storms |
|---|---|---|---|---|---|---|
| | **50** | **60** | **70** | **80** | **90** | |
| East of 105th meridian | 75 | 35 | 18 | 6 | 3 | 75 |
| West of Continental Divide | 67 | 33 | 10 | 4 | 0 | 67 |

Data from Hansen (1987).

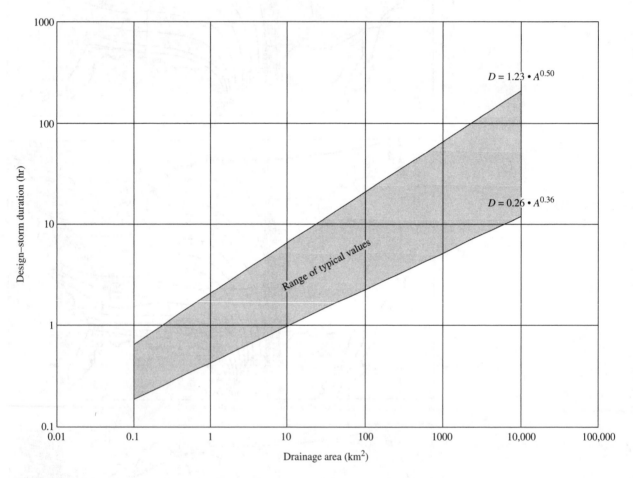

**FIGURE 4-48**
Typical values of design-storm duration as a function of drainage area. Envelope curves based on relations in McCuen (1998) and Pilgrim and Cordery (1992).

terest is usually done on a graph with depth or intensity plotted on a logarithmic or arithmetic scale (whichever gives a smoother and more nearly straight-line pattern) and exceedence probability on a probability scale. (See Box 10-8.) For the present case, the logarithmic-probability plots appear to fit the data best, and smooth curves are fit by eye through the points for each duration (Figure 4-49). The 2-, 5-, 10-, and 25-yr rainfalls are then read from the 50-, 20-, 10-, and 4-% exceedence probabilities, respectively [Equation (C-32)]. These values (in.) are tabulated here:

**TABLE 4-11**
Annual Maximum 1-, 6-, and 24-hr Rainfalls for Chicago Airport Weather Station for 1949–1972.

| Year | 1-hr (in.) | 6-hr (in.) | 24-hr (in.) |
|------|-----------|-----------|------------|
| 1949 | 0.79 | 1.58 | 2.73 |
| 1950 | 1.69 | 2.50 | 3.51 |
| 1951 | 1.50 | 2.15 | 2.92 |
| 1952 | 1.19 | 1.60 | 1.60 |
| 1953 | 1.02 | 2.42 | 2.42 |
| 1954 | 1.41 | 3.93 | 5.55 |
| 1955 | 1.13 | 1.69 | 3.09 |
| 1956 | 1.07 | 1.43 | 1.57 |
| 1957 | 2.08 | 5.23 | 6.24 |
| 1958 | 1.04 | 2.03 | 2.25 |
| 1959 | 2.06 | 4.58 | 4.58 |
| 1960 | 1.82 | 2.05 | 2.06 |
| 1961 | 1.38 | 2.52 | 2.63 |
| 1962 | 1.53 | 1.82 | 1.82 |
| 1963 | 0.91 | 2.19 | 2.57 |
| 1964 | 0.65 | 1.56 | 2.09 |
| 1965 | 0.86 | 1.64 | 1.80 |
| 1966 | 1.78 | 3.24 | 5.39 |
| 1967 | 1.55 | 2.54 | 2.95 |
| 1968 | 1.61 | 3.73 | 3.83 |
| 1969 | 1.55 | 2.84 | 3.29 |
| 1970 | 1.75 | 2.48 | 2.97 |
| 1971 | 1.16 | 1.84 | 1.91 |
| 1972 | 2.32 | 3.65 | 3.68 |

**TABLE 4-12**
Ranks and Exceedence Probabilities for the Annual Maximum Rainfalls in Table 4-11.

| Rank | $\widehat{EP}(x)$ | 1-hr (in.) | 6-hr (in.) | 24-hr (in.) |
|------|------|-----------|-----------|------------|
| 1 | 0.04 | 2.32 | 5.23 | 6.24 |
| 2 | 0.08 | 2.08 | 4.58 | 5.55 |
| 3 | 0.12 | 2.06 | 3.93 | 5.39 |
| 4 | 0.16 | 1.82 | 3.73 | 4.58 |
| 5 | 0.20 | 1.78 | 3.65 | 3.83 |
| 6 | 0.24 | 1.75 | 3.24 | 3.68 |
| 7 | 0.28 | 1.69 | 2.84 | 3.51 |
| 8 | 0.32 | 1.61 | 2.54 | 3.29 |
| 9 | 0.36 | 1.55 | 2.52 | 3.09 |
| 10 | 0.40 | 1.55 | 2.50 | 2.97 |
| 11 | 0.44 | 1.53 | 2.48 | 2.95 |
| 12 | 0.48 | 1.50 | 2.42 | 2.92 |
| 13 | 0.52 | 1.41 | 2.19 | 2.73 |
| 14 | 0.56 | 1.38 | 2.15 | 2.63 |
| 15 | 0.60 | 1.19 | 2.05 | 2.57 |
| 16 | 0.64 | 1.16 | 2.03 | 2.42 |
| 17 | 0.68 | 1.13 | 1.84 | 2.25 |
| 18 | 0.72 | 1.07 | 1.82 | 2.09 |
| 19 | 0.76 | 1.04 | 1.69 | 2.06 |
| 20 | 0.80 | 1.02 | 1.64 | 1.91 |
| 21 | 0.84 | 0.91 | 1.60 | 1.82 |
| 22 | 0.88 | 0.86 | 1.58 | 1.80 |
| 23 | 0.92 | 0.79 | 1.56 | 1.60 |
| 24 | 0.96 | 0.65 | 1.43 | 1.57 |

| Return Period (yr) | Duration (hr) 1 | 6 | 24 |
|------|------|------|------|
| 2 | 1.44 | 2.25 | 2.70 |
| 5 | 1.80 | 3.27 | 4.00 |
| 10 | 2.05 | 4.10 | 5.00 |
| 25 | 2.32 | 5.20 | 6.15 |

In the parlance of the profession, we would say that the "2-yr, 1-hr rainfall" for Chicago is 1.44 in., the "25-yr, 24-hr rainfall" is 6.15 in., and so on.

To allow interpolation to any duration between 1 and 24 hr, the data from the above table are plotted, usually on logarithmic graph paper (Figure 4-50). From this figure we would, for example, estimate the 10-yr, 2-hr rainfall as 3.1 in.

Example 4-6 describes the essential aspects of DDF analysis, but such analyses are usually done for more than three durations, and many situations require consideration of durations as small as 5 min. Also, as noted earlier, the design return period

may be as high as 100 yr. The plotting-position approach is not appropriate for estimating return periods greater than the length of record, so in these cases more sophisticated statistical analysis may be required to arrive at a frequency distribution for the rainfall data (Section C.5).

It has been found that relations between intensity, duration, and frequency in most locations can be represented by equations of the form

$$I = \frac{A}{(D + C)^B} \tag{4-28a}$$

or

$$I = \frac{A}{D^B + C}, \tag{4-28b}$$

where $I$ is intensity, $D$ is duration, $A$ is a constant for a given return period, and $B$ and $C$ are constants that do not depend on return period (Wenzel 1982). These equations have no theoretical basis; they are purely empirical devices that are sometimes useful for expressing relations such as those shown in Figure 4-49 (with intensity substituted for depth). As

**FIGURE 4-49**
Logarithmic-probability plots of depths of 1-, 6-, and 24-hr rainfall data from Tables 4-11 and 4-12.

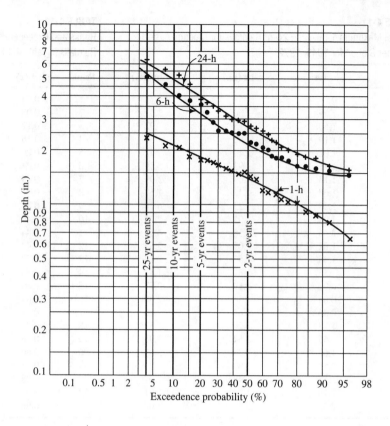

indicated in Table 4-13, the constants in these equations have a strong geographic variation and must be determined by analysis of data for the location of interest.

In the United States, isohyetal maps showing depths of rainfall for various return periods have been prepared by the National Weather Service and its predecessors. These maps are generalizations of analyses like that of Example 4-6 done in many locations; they are intended for use in design situations where local data are unavailable or where a complete DDF analysis like that of Example 4-6 is not economically justified. Hershfield (1961) covered the entire country for durations of 0.5 to 24 hr and return periods up to 100 yr; Miller et al. (1973) updated Hershfield's maps for the 11 western

**FIGURE 4-50**
Depth–duration–frequency plot for Chicago Airport for durations from 1 to 24 hr and return periods from 2 to 25 yr (Example 4-6).

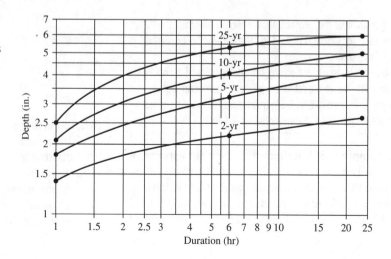

**TABLE 4-13**

Values of Empirical Constants *A*, *B*, and *C* in Equations (4-28a) and (4-28b) for 10-yr-Return-Period Storms at Various U.S. Locations.[a]

| Location | Equation (4-28a) | | | Equation (4-28b) | | |
|---|---|---|---|---|---|---|
| | **A** | **B** | **C** | **A** | **B** | **C** |
| Chicago | 60.9 | 0.81 | 9.56 | 94.9 | 0.88 | 9.04 |
| Denver | 50.8 | 0.84 | 10.50 | 96.6 | 0.97 | 13.90 |
| Houston | 98.3 | 0.80 | 9.30 | 97.4 | 0.77 | 4.80 |
| Los Angeles | 10.9 | 0.51 | 1.15 | 20.3 | 0.63 | 2.06 |
| Miami | 79.9 | 0.73 | 7.24 | 124.2 | 0.81 | 6.19 |
| New York | 51.4 | 0.75 | 7.85 | 78.1 | 0.82 | 6.57 |
| Olympia | 6.3 | 0.40 | 0.60 | 13.2 | 0.64 | 2.22 |
| Atlanta | 64.1 | 0.76 | 8.16 | 97.5 | 0.83 | 6.88 |
| Helena | 30.8 | 0.81 | 9.56 | 36.8 | 0.83 | 6.46 |
| St. Louis | 61.0 | 0.78 | 8.96 | 104.7 | 0.89 | 9.44 |
| Cleveland | 47.6 | 0.79 | 8.86 | 73.7 | 0.86 | 8.25 |
| Santa Fe | 32.2 | 0.76 | 8.54 | 62.5 | 0.89 | 9.10 |

[a]These constants are appropriate when intensity is in inches per hour and duration is in minutes.

From Wenzel (1982).

states; and Frederick et al. (1977a) gave maps for the central and eastern states for durations of from 5 min to 1 hr. Examples of these maps are given in Figure 4-51. One must use caution in accepting these generalized map values (1) in areas where precipitation is highly variable spatially because of variations in relief and exposure, and (2) in large metropolitan areas where the urban "heat-island" effects described in Section 4.4.5 increase the frequency of heavy rainfalls.

**Adjustment of Point Estimates to Areal Estimates**   The DDF analysis described in Example 4-6 and the generalized maps developed from such analyses are for precipitation at a point. For use as input to simulation models, the point values of depth for a given duration and return period must be multiplied by an **areal reduction factor,** $K(D, A)$, to give the corresponding average depth over the drainage area being modeled. As the notation suggests, this factor is a function of the duration ($D$) and area ($A$).

Studies to determine appropriate values of $K(D, A)$ have been of limited geographical extent, and hydrologists usually make use of the curves developed by the U.S. National Weather Service and thought to be generally applicable to the United States (Figure 4-52). Eagleson (1972) presented an empirical relation that can be used to approximate these curves:

$$K(D, A) = 1 - \exp(-1.1 \cdot D^{0.25})$$

$$+ \exp(-1.1 \cdot D^{0.25} - 0.01 \cdot A), \quad \textbf{(4-29)}$$

where $D$ is duration in hours and $A$ is area in mi$^2$. Example 4-7 shows how these curves are used.

---

### EXAMPLE 4-7

Estimate the 25-yr, 3-hr rainfall for input to a simulation model used to design a drainage system for an area of 50 mi$^2$ near the Chicago Airport.

*Solution*   From Figure 4-50, the 25-yr, 3-hr point rainfall in this region is 4.50 in. From Figure 4-52, the reduction factor for a 3-hr storm on an area of 50 mi$^2$ is 0.90. Thus the design rainfall depth is

$$4.50 \text{ in.} \times 0.90 = 4.05 \text{ in.}$$

Using Equation (4-29), we find that

$$K(D, A) = 1 - \exp(-1.1 \times 3^{0.25})$$
$$+ \exp(-1.1 \times 3^{0.25} - 0.01 \times 50) = 0.91.$$

---

Note that the relationships shown in Figure 4-52 reflect the meteorological characteristics of various types of storms discussed earlier: The heaviest short-duration rainfalls come from convective storms that are almost always of small areal extent, and the heaviest longer-duration storms are usually generated by cyclonic storms that cover thousands of square miles.

Rodriguez-Iturbe and Mejia (1974b) developed a theory relating the areal reduction factor to the spatial correlation of precipitation, and Omolayo (1993) compared factors for Australia, the

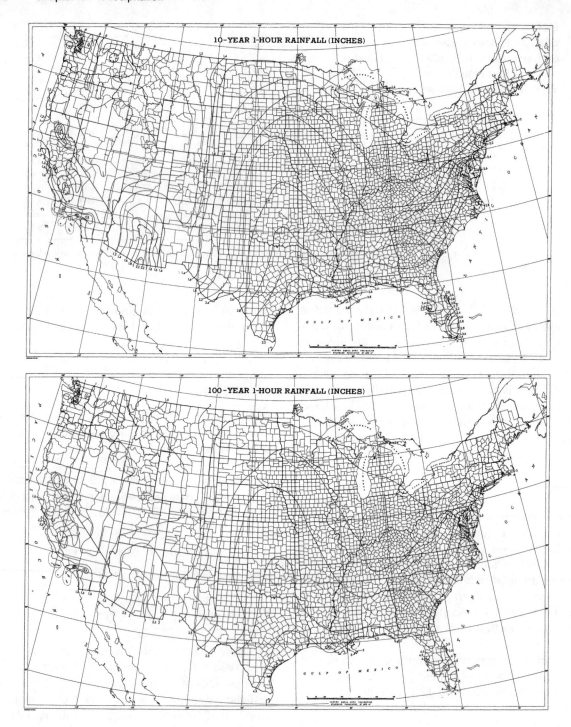

**FIGURE 4-51**
Examples of generalized depth–duration–frequency maps for the United States: (a) 10-yr, 1-hr rainfall; (b) 100-yr, 1-hr rainfall; (c) 10-yr, 24-hr rainfall; (d) 100-yr, 24-hr rainfall. From Hershfield (1961).

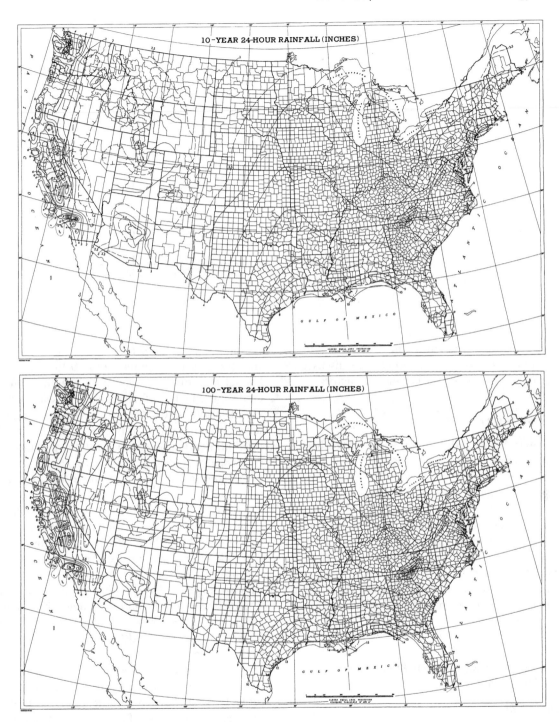

**FIGURE 4-51**
Continued

**FIGURE 4-52**
Reduction of point rainfalls for application to drainage areas up to 400 mi$^2$, as recommended by the National Weather Service. From Miller et al. (1973).

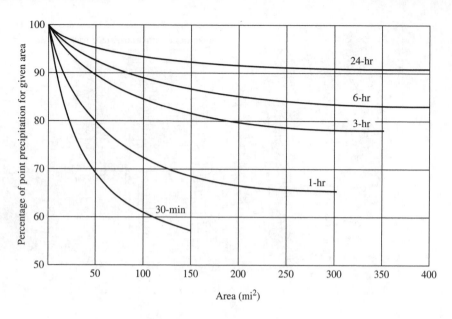

### Time Distribution of Rainfall

Some hydrologic models require the time distribution of rainfall inputs, not just the total storm rainfall. The U.S. Soil Conservation Service (now Natural Resources Conservation Service) has developed typical patterns of time distribution of rainfall for design storms in the United States; these are described in many engineering hydrology texts (e.g., McCuen 1998). Various approaches to developing appropriate time distributions for design storms, and other aspects of developing design storms for hydrologic analyses, were described by Wenzel (1982).

### 4.4.5 Anthropogenic Effects on Precipitation Climatology

### Inadvertent Effects

In its policy statement on planned and inadvertent weather modification, the American Meteorological Society (AMS 1992) found that agricultural practices, urbanization, industrial activity, and condensation trails ("contrails") from jet aircraft modify local and regional weather.

United Kingdom, and the United States. Huff (1995) found that the urbanization effects discussed in the next section alter areal reduction factors.

As we have seen (Table 4-2), regional evapotranspiration contributes from 10 to over 40% of annual precipitation over large regions. Thus precipitation can be affected by large-scale land-use modifications associated with agriculture and urban development. Irrigation increases the potential for precipitation by increasing humidity and modifying surface albedo and temperature; overgrazing and deforestation have the opposite effect by reducing humidity.

Urban effects on precipitation have been studied in detail in the United States. Huff and Changnon (1973) and Changnon (1981) found increases of from 9 to 17% in total precipitation and an increase in the number of heavy (> 25-mm) rainstorms associated with six of eight large United States urban areas. These effects were greater in summer than in winter and were greatest over and immediately downwind of the urban centers. Although these heavy storms were associated largely with cold fronts, the major cause of the increase appears to be the enhanced convective uplift induced by higher surface temperatures. These higher temperatures are due in turn to the greater absorption of solar radiation by buildings and pavements and to heat emissions from industries and motor vehicles. The increased production of precipitation nuclei by these heat sources also contributes to the phenomenon.

This urban "heat-island" effect on precipitation was found to be of definite hydrologic significance:

It was reflected in increased summertime stream-flow, increased ground-water recharge, increased flow in sewers, and increased flooding in the affected areas (Huff 1977).

### Precipitation Augmentation as a Water-Resource Management Tool

Human attempts to modify weather—particularly to increase rainfall when it is needed for agriculture—are as old as civilization. However, only since the 1940s, when scientific understanding of the precipitation process accelerated, has it been possible to consider precipitation augmentation a realistic management alternative.

The main goals of precipitation-augmentation activities in the United States have been to increase (1) the snowfall from winter cyclonic/orographic storms and (2) the rainfall from summertime convective clouds. Both cases involve **cloud seeding**—the introduction of artificial nuclei in the form of crystals of silver iodide (AgI) (which serve as templates for the formation of ice crystals) or of common salt (NaCl) (which induce droplet growth by hygroscopic action). (See Section D.5.2.)

In the first case, it has been found that snowfall from orographic clouds with temperatures from $-10\,°C$ to $-23\,°C$ can be increased by providing the additional nuclei; however, seeding colder clouds in otherwise similar conditions can cause **overseeding,** which reduces precipitation because the available cloud water is spread over the more abundant nuclei to the extent that few of the ice crystals grow large enough to fall. In the case of supercooled convective clouds, the additional nuclei trigger the formation of ice crystals; the resulting production of latent heat accelerates uplift within the cloud and triggers the process of precipitation formation.

One of the requirements for the formation of significant precipitation is the continuing importation of water vapor into the precipitating cloud (Section D.5.4). This requirement imposes limitations on the usefulness of cloud seeding for precipitation enhancement—where this importation does not occur, the introduction of nuclei may simply cause most of the cloud moisture to precipitate, resulting in minor precipitation and the disappearance of the cloud. In fact, cloud seeding is sometimes done under these conditions in order to dissipate fogs and stratiform clouds. Attempts also

have been made to exploit the overseeding phenomenon to prevent or reduce hail formation and to reduce the intensity of hurricanes. (See Mather 1984; AMS 1992.)

Recent reviews of the status of weather-modification (AMS 1992; Orville 1995) found the following:

- Several experiments using AgI seeding have produced significant precipitation increases.
- Seasonal precipitation increases of 10%–30% have been produced by seeding super-cooled orographic clouds.
- Seeding may increase or decrease precipitation from convective clouds, and the conditions contributing to positive or negative results are not well understood.
- Rainfall can be increased by seeding warm-based ($> 10\,°C$) convective clouds under certain conditions; especially consistent increases have been reported in South Africa.
- More consistent results would probably be achieved by improving seeding methodology.
- Precipitation increases or decreases may be induced beyond the target areas.
- The economic viability of precipitation augmentation has not been demonstrated.
- Scientific credibility of precipitation-augmentation programs is limited by incomplete understanding of the processes involved.
- Long-term scientific studies of weather modification have declined, but short-term studies are active in many countries, including Russia, Israel, South Africa, and Italy.

The AMS (1992) review further concluded that the ecological, hydrological, socioeconomic, and legal impacts of weather modification are potentially far-reaching and that assessment of these impacts should be included in field studies. Certainly, the potential for conflicts of interest in deciding whether, how, and where to increase precipitation and for issues of liability resulting from increasing (or failing to increase) precipitation in a given instance would appear to be limitless. As Mather (1984) noted, lawyers might well reap the only real benefits of cloud seeding.

Interestingly, these reviews of weather modification have not discussed the potentially most significant impacts of the operational use of weather modification to enhance water resources: undermining the value of hydrologic statistics as a scientific basis for water-resource management (Kazmann 1988). Most water-management decisions rely in significant part on statistical quantities such as averages, return periods, and probable-maximum values of precipitation and streamflow. These statistics are based on past records and are useful only insofar as they are indicative of future behavior. If future precipitation is to be governed in part by administrative actions, court decisions, or private economic interests, the uncertainty in these statistics could grow to the point of making them meaningless.

Without meaningful statistics, we have no basis for designing water-supply systems, bridges, wastewater treatment plants, or flood-damage reduction measures. The economic costs of losing this basis would appear both as money wasted in overdesigning and as money spent in repairing measures that are too frequently damaged due to their underdesign. These costs, and the potential legal entanglements, should be included in assessing the economic

viability and overall wisdom of weather-modification schemes.

## 4.5 PRECIPITATION QUALITY

Every raindrop and snowflake is formed around a solid nucleus; the most common nuclei are sea salt, clay- and silt-sized windblown dust, volcanic debris, smoke particles, meteoric dust, and various pollutants (Section D.5.2). The relative abundances and solubilities of these materials are the first determinants of the quality of precipitation. Subsequently, the chemical composition of droplets changes as they dissolve gases from the surrounding air at rates proportional to the solubility and concentration of each gas. Further changes occur when droplets grow to raindrops and scavenge natural and anthropogenic aerosols as they fall.

Figure 4-53 shows typical concentrations of the major ions in continental and coastal/marine rain;

**FIGURE 4-53**

Typical concentration ranges for major ions in continental (straight lines) and marine (wavy lines) rain. The dashed lines indicate the ranges of concentrations found in areas of high air pollution. Data from Berner and Berner (1987).

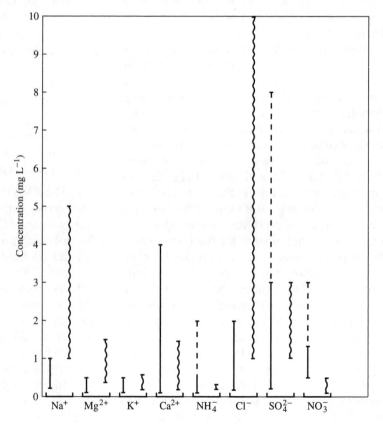

**TABLE 4-14**
Sources of Dissolved Constituents in Precipitation.

| Ion | Origin | | |
|---|---|---|---|
| | **Marine Inputs** | **Terrestrial Inputs** | **Pollution Inputs** |
| $Na^+$ | Sea salt | Soil dust | Burning vegetation |
| $Mg^{2+}$ | Sea salt | Soil dust | Burning vegetation |
| $K^+$ | Sea salt | Biogenic aerosols | Burning vegetation |
| | | Soil dust | Fertilizer |
| $Ca^{2+}$ | Sea salt | Soil dust | Cement manufacture |
| | | | Fuel burning |
| | | | Burning vegetation |
| $H^+$ | Gas reaction | Gas reaction | Fuel burning to form gases |
| $Cl^-$ | Sea salt | | Industrial HCl |
| | Gas release from sea salt | | |
| $SO_4^{2-}$ | Sea salt | $H_2S$ from biological decay | Burning of fossil fuels to $SO_2$ |
| | Marine gases | Volcanoes | Forest burning |
| | | Soil dust | |
| | | Biogenic aerosols | |
| $NO_3^-$ | $N_2$ plus lightning | $NO_2$ from biological decay | Gaseous auto emissions |
| | | $N_2$ plus lightning | Combustion of fossil fuels |
| | | | Forest burning |
| | | | Nitrogen fertilizers |
| $NH_4^+$ | | $NH_3$ from bacterial decay | Ammonia fertilizers |
| | | | Decomposition of human and animal wastes |
| | | | Combustion |
| $PO_4^{3-}$ | | Soil dust | Burning vegetation |
| | | Biogenic aerosols | Fertilizer |
| | | Absorbed on sea salt | |
| $HCO_3^-$ | $CO_2$ in air | $CO_2$ in air | |
| | | Soil dust | |
| $SiO_2$, Al, Fe | | Soil dust | Land clearing |

From Elizabeth Kay Berner and Robert A. Berner, *The Global Water Cycle: Geochemistry and Environment,* © 1987, p. 71. Reprinted by permission of Prentice Hall, Inc., Upper Saddle River, NJ.

Table 4-14 gives the origins of these ions. Where not strongly affected by pollution, rainfall is a very dilute solution dominated by the cations sodium ($Na^+$), magnesium ($Mg^{+2}$), and calcium ($Ca^{+2}$) and the anions chloride ($Cl^-$), sulfate ($SO_4^{-2}$), and nitrate ($NO_3^-$), with a total concentration typically between 5 and 20 mg $L^{-1}$. The pH of pure water is 7 (Section B.1.4), but, because of the dissolution of carbon dioxide, the pH of natural rainfall is somewhat acid, typically about pH 5.7. At some times and places, natural rain pH may be somewhat higher due to dissolved calcium or ammonia or lower due to naturally occurring acids (Berner and Berner 1987).

The concentrations of virtually all the constituents of precipitation are increased by human activity, but it is the anthropogenic production of the gases sulfur dioxide ($SO_2$) and nitrogen oxides ($NO_x$) that is chiefly responsible for the phenomenon of **acid rain,** which is defined as rain with a pH less than 5.7. Table 4-15 summarizes estimates of the inputs of these gases from the earth's surface and shows that over 50% of the gaseous sulfur and 70% of the nitrate–nitrogen entering the atmosphere are anthropogenic. These gases dissolve in cloud droplets and result in precipitation with pH between 4 and 5 over much of the earth's land surfaces—especially in eastern North America, western Europe, and Japan.

Berner and Berner (1987) provided an excellent review of the chemistry of acid rain and its relation to the global atmospheric cycles of sulfur and nitrogen. An entire issue of the journal *Ambio* (v. 18, no. 3, 1989) is devoted to a broad global and historical perspective of the problem, its impact on ecosystems, and the actions required to ameliorate

**TABLE 4-15**
Sources of Atmospheric Sulfur and
Nitrate Nitrogen.[a]

| | Sulfur | | |
|---|---|---|---|
| **Source** | $10^{12}$ g S yr$^{-1}$ | | **Percentage of Total** |
| Fossil-fuel burning[*] | 65–70 | | 40 |
| Forest burning[*] | 7 | | 4 |
| Volcanoes | 2 | | 1 |
| Dust | 1 | | 0.6 |
| Biogenic (from land) | 20 | | 12 |
| Biogenic (from oceans) | 28 | | 17 |
| Sea spray | 44 | | 24 |

| | Nitrate-nitrogen | | |
|---|---|---|---|
| **Source** | $10^{12}$ g N yr$^{-1}$ | | **Percentage of Total** |
| Lightning | 4 | | 9 |
| Conversion of $N_2O$ in stratosphere | 0.3 | | 1 |
| Oxidation of $NH_4^+$ | <1 | | <1 |
| Soil microbes | 5 | | 12 |
| Fossil-fuel burning[*] | 21 | | 49 |
| Forest burning[*] | 12 | | 28 |

[a]Asterisks indicate largely or entirely anthropogenic sources.

Data from Berner and Berner (1987).

it. One of the articles in that issue (Galloway 1989) concludes with the following overview:

> Current emissions of $SO_2$ and $NO_x$ have caused regional scale acidification of the atmosphere and sensitive ecosystems in North America and Europe. Over the next several decades, Europe and North America will not experience major increases in the emissions of $SO_2$ and $NO_x$. However, environmental effects in Europe and North America will still continue, and in some cases, will get worse due to delayed acidification of aquatic and terrestrial ecosystems.
>
> Over the next several decades, Asia, Africa and South America will probably experience major increases in $SO_2$ and $NO_x$ emissions ... Given the reality of population growth in these areas and the potential for industrial expansion, future emissions of $SO_2$ and $NO_x$ could greatly exceed current emissions to the global atmosphere. Some of the consequences to the environment of the increased emissions are well known from our current knowledge. However, given the magnitude of the potential change in emissions, other effects are possible.

One of these effects, recently detected in northern New England, has been the decline of forest growth, probably because calcium required for growth was depleted from the soil by acid precipitation in previous decades (Likens et al. 1996). Thus calcium and other basic cations are no longer available to buffer acidic anions—so even though emissions of acid pollutants have declined, the problems of acidity in stream and lake water could persist for some time.

# EXERCISES

Exercises marked with ** have been programmed in EXCEL on the CD that accompanies this text. Exercises marked with * can advantageously be executed on a spreadsheet, but you will have to construct your own worksheets to do so.

**4-1.** Identify and locate the nearest precipitation-gaging station. If possible, visit the station to note the type of gage, type of wind shield (if any), exposure with respect to surrounding obstructions, and frequency and timing of observation. Find out from the observer, state climatologist, or weather-agency official the history of the station.

**4-2.** Obtain from the state climatologist, weather-agency official, or other source a map showing the locations of precipitation gages in your region. How does the gage density compare with those shown in Table 4-6 and Figure 4-29? Does the network appear to sample the region adequately? (Consider topographic effects, distance from coast, storm directions, and so on.)

**4-3.** Select a portion of the precipitation-gage map you found for Exercise 4-2 containing at least six gages for which average annual (or monthly) precipitation data are available. Compare the regional average precipitation using (a) the arithmetic average, (b) Thiessen polygons, (c) the two-axis method, and (d) the eyeball isohyetal method.

**4-4.** If you have access to a computer-based surface-generating and contouring program, enter the data used in Exercise 4-3 and compare the contour patterns with those of Exercise 4-3d. Compare the regional average calculated from the computer-generated isohyets with the regional averages computed in Exercise 4-3.

**\*4-5.** The tables that follow give area-elevation data for a drainage basin and the average annual precipitation measured at six gages in the basin. The basin has an area of 269 km$^2$, a minimum elevation of 311 m, and a maximum elevation of 1600 m. (a) Compute the average annual precipitation for the basin using the hypsometric method. (b) Compare the value computed in (a) with the arithmetic average.

Area-Elevation Data

| Elevation Range (m) | Fraction of Area Within Range |
|---|---|
| 311–400 | 0.028 |
| 400–600 | 0.159 |
| 600–800 | 0.341 |
| 800–1000 | 0.271 |
| 1000–1200 | 0.151 |
| 1200–1400 | 0.042 |
| 1400–1600 | 0.008 |

Precipitation Data

| Gage | Elevation (m) | Average Annual Precipitation (mm) |
|---|---|---|
| 1 | 442 | 1392 |
| 2 | 548 | 1246 |
| 3 | 736 | 1495 |
| 4 | 770 | 1698 |
| 5 | 852 | 1717 |
| 6 | 1031 | 1752 |

**\*\*4-6.** The table that follows gives annual precipitation measured over a 17-yr period at five gages in a region. Gage C was moved at the end of 1974. Carry out a double-mass-curve analysis to check for consistency in that gage's record, and make appropriate adjustments to correct for any inconsistencies discovered. The data are in text-disk file 2MassCurve.xls.

Annual Precipitation (mm) at Station

| Year | A | B | C | D | E |
|---|---|---|---|---|---|
| 1970 | 1010 | 1161 | 780 | 949 | 1135 |
| 1971 | 1005 | 978 | 1041 | 784 | 970 |
| 1972 | 1067 | 1226 | 1027 | 1067 | 1158 |
| 1973 | 1051 | 880 | 825 | 1014 | 1022 |
| 1974 | 801 | 1146 | 933 | 923 | 821 |
| 1975 | 1411 | 1353 | 1584 | 930 | 1483 |

Annual Precipitation (mm) at Station

| Year | A | B | C | D | E |
|---|---|---|---|---|---|
| 1976 | 1222 | 1018 | 1215 | 981 | 1174 |
| 1977 | 1012 | 751 | 832 | 683 | 771 |
| 1978 | 1153 | 1059 | 918 | 824 | 1188 |
| 1979 | 1140 | 1223 | 781 | 1056 | 967 |
| 1980 | 829 | 1003 | 782 | 796 | 1088 |
| 1981 | 1165 | 1120 | 865 | 1121 | 963 |
| 1982 | 1170 | 989 | 956 | 1286 | 1287 |
| 1983 | 1264 | 1056 | 1102 | 1044 | 1190 |
| 1984 | 1200 | 1261 | 1058 | 991 | 1283 |
| 1985 | 942 | 811 | 710 | 875 | 873 |
| 1986 | 1166 | 969 | 1158 | 1202 | 1209 |

**4-7.** Consider a drainage basin of area = 1611 km$^2$ in which three precipitation gages have been operating for 20 yr. The average annual precipitation, $\hat{P}$, at the three gages is 1198 mm yr$^{-1}$ and the overall variance of annual precipitation, $\hat{S}^2(p)$, is 47,698 mm$^2$ yr$^{-2}$. An analysis of the spatial correlation structure of annual precipitation in the region gives a value of $c = 0.0386$ km$^{-1}$ [Equation (4B2-5c)]. Compute the variance reduction achieved by this measurement program and compare the variance reduction that would have been achieved by having four, rather than three, gages in the region during the 20-yr period.

**\*\*4-8.** Compute and compare the seasonality index and average time of occurrence for monthly precipitation at the following stations (the data are in text-disk file Pptn Seas.xls):

Average Monthly Precipitation (mm)

| Station | Jan | Feb | Mar | Apr | May | Jun |
|---|---|---|---|---|---|---|
| New Orleans, LA, USA | 97 | 102 | 135 | 117 | 112 | 112 |
| Seattle, WA, USA | 145 | 107 | 97 | 61 | 43 | 41 |
| Fairbanks, AK, USA | 23 | 13 | 10 | 8 | 18 | 36 |
| Mexico City, Mexico | 5 | 8 | 13 | 18 | 48 | 104 |
| Belem, Brazil | 318 | 358 | 358 | 320 | 259 | 170 |
| Leningrad, USSR | 25 | 23 | 23 | 25 | 41 | 51 |
| Addis Ababa, Ethiopia | 13 | 38 | 66 | 86 | 86 | 137 |
| Shillong, India | 14 | 24 | 53 | 126 | 277 | 464 |

Average Monthly Precipitation (mm)

| Station | Jul | Aug | Sep | Oct | Nov | Dec |
|---|---|---|---|---|---|---|
| New Orleans, LA, USA | 170 | 135 | 127 | 71 | 84 | 104 |
| Seattle, WA, USA | 20 | 25 | 53 | 102 | 137 | 160 |
| Fairbanks, AK, USA | 46 | 56 | 28 | 23 | 15 | 13 |
| Mexico City, Mexico | 114 | 109 | 104 | 41 | 13 | 8 |
| Belem, Brazil | 150 | 112 | 89 | 84 | 66 | 155 |
| Leningrad, USSR | 64 | 71 | 53 | 46 | 36 | 30 |
| Addis Ababa, Ethiopia | 279 | 300 | 191 | 20 | 15 | 5 |
| Shillong, India | 358 | 332 | 332 | 175 | 33 | 6 |

**\*\*4-9.** Redo the depth–duration–frequency analysis of Example 4-6 as an intensity–duration–frequency problem. The data are in text-disk file ChiAnn Max.xls.

# 5

# Snow and Snowmelt

Over much of the world's land areas, a significant portion of precipitation falls as snow and is stored on the surface for periods of time ranging from hours to months before melting and continuing through the land phase of the hydrologic cycle. (See Figure 3-21.) In many of these areas, snowmelt is the main source of surface-water supply (Figure 3-26) and ground-water recharge and the main cause of flooding.

The goal of this chapter is to develop an understanding of the hydrologically important aspects of snow and snowmelt. Following an examination of the nature of snow as a material and of techniques for the measurement of snow and snowmelt, we survey the distribution of snow at various scales and the role of snow in the hydrologic cycle. We then discuss the processes that control the melting of snow (relying extensively on the physical principles discussed in Appendix D) and the ways in which meltwater moves through the snowpack to the ground surface. An understanding of these processes is the basis for models used to predict and forecast snowmelt runoff and, next, we review approaches to and results of snowmelt modeling. The chapter concludes with a brief discussion of water-quality aspects of snowmelt.

## 5.1 MATERIAL CHARACTERISTICS OF SNOW

### 5.1.1 Snow Properties

Snow is a granular porous medium consisting of ice and pore spaces. When the snow is **cold** (i.e., its temperature is below the melting point of ice, 0 °C), the pore spaces contain only air (including water vapor). At the melting point, the pore spaces can contain liquid water as well as air, and snow becomes a three-phase system.

Consider a representative portion of a snowpack of height (depth) $h_s$ and area $A$ (Figure 5-1). Using the symbols $M$ to designate mass, $V$ for volume, $h$ for height, and $\rho$ for mass density; and the subscripts $s$ for snow, $i$ for ice, $w$ for liquid water, $m$ for water substance (liquid water plus ice), and $a$ for air; we can define quantities that are useful in characterizing a snowpack. The snow volume is

$$V_s = V_i + V_w + V_a = h_s \cdot A. \qquad \text{(5-1)}$$

The **porosity,** $\phi$, is defined as the ratio of pore volume to total volume:

$$\phi \equiv \frac{V_a + V_w}{V_s}; \qquad \text{(5-2)}$$

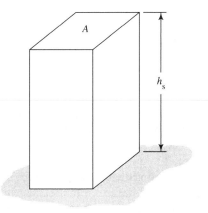

**FIGURE 5-1**
Dimensions of a representative portion of a snowpack, used in defining snowpack properties. $A$ is the area of the upper surface; $h_s$ is the snow depth.

therefore,

$$V_i = (1 - \phi) \cdot V_s. \qquad \text{(5-3)}$$

The **liquid-water content,** $\theta$, is defined as the ratio of the volume of liquid water in the snowpack to the total volume of snow:

$$\theta \equiv \frac{V_w}{V_s}. \qquad \text{(5-4)}$$

**Snow density,** $\rho_s$, is defined as the mass per unit volume of snow, so

$$\rho_s = \frac{M_i + M_w}{V_s} = \frac{\rho_i \cdot V_i + \rho_w \cdot V_w}{V_s}. \qquad \text{(5-5)}$$

Combining Equations (5-2)–(5-5) allows us to relate density, liquid-water content, and porosity as

$$\rho_s = (1 - \phi) \cdot \rho_i + \theta \cdot \rho_w, \qquad \text{(5-6)}$$

where $\rho_i = 917$ kg m$^{-3}$ and $\rho_w = 1000$ kg m$^{-3}$.

For the hydrologist, the most important property of a snowpack is the amount of water substance it contains, because this is the amount of water that will ultimately enter the land phase of the hydrologic cycle. This amount is called the **water equivalent** of the snowpack, and it is expressed as the depth of water that would result from the complete melting of the snow in place, $h_m$. Thus

$$h_m \equiv \frac{V_m}{A}, \qquad \text{(5-7)}$$

where $V_m$ is the volume of water resulting from the complete melting of the snow.

To relate the water equivalent to previously defined quantities, we note that

$$V_m = V_w + V_i \cdot \frac{\rho_i}{\rho_w}. \qquad \text{(5-8)}$$

Substituting Equations (5-3) and (5-4) into Equation (5-8) yields

$$V_m = \theta \cdot V_s + (1 - \phi) \cdot V_s \cdot \frac{\rho_i}{\rho_w}. \qquad \text{(5-9)}$$

We can now use Equations (5-1) and (5-7) to rewrite Equation (5-9) as

$$h_m = \theta \cdot h_s + (1 - \phi) \cdot h_s \cdot \frac{\rho_i}{\rho_w}, \qquad \text{(5-10)}$$

$$= \left[ \theta + (1 - \phi) \cdot \frac{\rho_i}{\rho_w} \right] \cdot h_s. \qquad \text{(5-11)}$$

Finally, we see from Equation (5-6) that Equation (5-11) can be written as

$$h_m = \frac{\rho_s}{\rho_w} \cdot h_s. \qquad \text{(5-12)}$$

In words, Equation (5-12) is often expressed as "water equivalent equals density times depth," where density is understood to mean relative density (i.e., specific gravity). This relation is used in snow-course measurements. (See Example 5-1.)

### 5.1.2   Snowpack Metamorphism

The density of new-fallen snow is determined by the configuration of the snowflakes, which is largely a function of air temperature, the degree of supersaturation in the precipitating cloud (Mellor 1964), and the wind speed at the surface of deposition. Higher wind speeds tend to break snowflakes that formed in stellar or needle-like shapes and to pack them together into denser layers. Observed relative densities ($\rho_s/\rho_w$) of freshly fallen snow range from 0.004 to 0.34 (McKay 1970), with the lower values occurring under calm, very cold conditions and higher values accompanying higher winds and higher temperatures. However, the usual range is 0.07 to 0.15 (Garstka 1964). Because of the difficulty of measuring the density of new snow, an average relative density of 0.1 is often assumed to apply when converting snowfall observations to water equivalents in the United States. The user of water-equivalent data should be aware of the potential for significant errors in estimates based on an assumed density (Goodison et al. 1981).

As soon as snow accumulates on the surface, it begins a process of metamorphism that continues until melting is complete. Four mechanisms are largely responsible for this process: (1) gravitational settling, (2) destructive metamorphism, (3) constructive metamorphism, and (4) melt metamorphism.

Gravitational settling in a given snow layer takes place at rates that increase with the weight of the overlying snow and the temperature of the layer and decrease with the density of the layer. According to relations given by Anderson (1976), one could expect gravitational settling to increase density at rates on the order of 2 to 50 kg m$^{-3}$ day$^{-1}$ in shallow snowpacks. On glaciers, the pressure of thick layers of accumulating snow is the principal factor leading to the formation of solid ice.

**Destructive metamorphism** occurs because vapor pressures are higher over convex surfaces with smaller radii of curvature, so the points and projections of snowflakes tend to evaporate, and the vapor deposits on nearby less convex surfaces. These events lead to the formation of larger, more spherical snow grains with time. This process occurs most rapidly in snowflakes that have recently fallen, causing the density of a new-snow layer to increase at about 1% per hour. The process ceases to be important when densities reach about 250 kg m$^{-3}$ (Anderson 1976).

**Constructive metamorphism** is the most important pre-melt densification process in seasonal snowpacks. Over short distances, this process occurs by **sintering**, in which water molecules are deposited in concavities where two snow grains touch, gradually building a "neck" between adjacent grains. Over longer distances, constructive metamorphism can occur as a result of vapor transfer within a snowpack due to temperature gradients; sublimation occurs in warmer portions of the snowpack, and the water vapor moves toward colder portions where condensation occurs. Very cold air overlying a relatively shallow snowpack often produces a strong upward-decreasing temperature gradient within the snow, with a concomitant upward-decreasing vapor-pressure gradient. Under these conditions, snow near the base of the pack evaporates at a high rate, often resulting in a basal layer of characteristic large planar crystals with very low density and strength called **depth hoar.**

**Melt metamorphism** occurs via two processes. In the first, liquid water formed by melting at the surface, or introduced as rain, freezes within the cold snowpack. This process results in densification and may produce layers of essentially solid ice that extend over long distances. The freezing at depth also liberates latent heat, which contributes to the warming of the snowpack and the acceleration of vapor transfer. The second metamorphic process accompanying melt is the rapid disappearance of smaller snow grains and growth of larger grains that occurs in the presence of liquid water. Because of this phenomenon, an actively melting snowpack is typically an aggregation of rounded grains the size of coarse sand (1 to 2 mm in diameter) (Colbeck 1978).

Except for the temporary formation of depth hoar, all the processes of metamorphism lead to a progressive increase in density through the snow-accumulation season (Figure 5-2). Figures 5-3 and 5-4 depict the regional variation in average seasonal snowpack density in North America and the former Soviet Union. It should be noted, however, that there is much year-to-year variability in snowpack characteristics and that both snowfall and the processes causing metamorphism occur at highly variable rates over short distances, due largely to differences in slope, aspect, and vegetative cover.

At the beginning of the melt season, the snowpack is typically vertically heterogeneous as well, with perhaps several layers of markedly contrasting densities. During melt, density continues to increase and the vertical inhomogeneities tend to disappear. During this period, density can fluctuate on an hourly or daily time scale due to the formation and drainage of melt water. Snowpacks that are at 0 °C and well drained tend to have relative densities near 0.35 (McKay 1970).

## 5.2 MEASUREMENT OF SNOW AND SNOWMELT

Discussion of measurement of snow and snowmelt requires definition of several terms:

**Precipitation** is the depth of rainfall plus the water equivalent of snow, sleet, and hail falling during a given storm or measurement period.

**Snowfall** is the incremental depth of snow and other forms of solid precipitation that accumu-

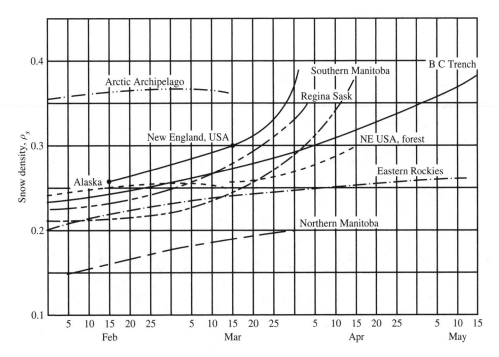

**FIGURE 5-2**
Seasonal variation in snowpack relative densities in various regions of North America. Modified from McKay (1970).

lates on the surface during a given storm event or measurement period.

**Snowpack** refers to the accumulated snow on the ground at a time of measurement. The snowpack water equivalent and areal extent are of particular hydrologic importance; information on depth and density is also useful.

**Snowmelt** is the amount of liquid water produced by melting that leaves the snowpack during a given time period; it is usually expressed as a depth.

**Ablation** is the total loss of water substance (snowmelt plus evaporation/sublimation) from the snowpack during a given time period; it is usually expressed as a depth.

**Water output** is the total amount (depth) of liquid water (rain plus snowmelt) leaving the snowpack during a given time period.

Snow drifting during storms can produce large variations in snow depth and density over short distances, and variations in subsequent snow metamorphism, melting, and evaporation due to local wind, temperature, radiation, and other microclimatic conditions can further modify the distribution of these properties. Thus snow characteristics will be highly variable in space, mainly because of variations in vegetative cover, slope, and aspect (slope orientation); consequently, obtaining a representative picture of the distribution of snow and snow properties is important and usually difficult. Remotely-sensed information that simply shows areas with and without snowcover within the region of interest can be extremely valuable in assessing the amount of water present in the form of snow. Peck (1997) has emphasized that particular care must be taken in obtaining reliable hydrometeorological measurements in cold regions and has stressed that a smaller number of high-quality records may be more valuable than a larger number of records of questionable quality.

Approaches to measurement of each of the quantities listed previously are discussed next and summarized in Table 5-1.

### 5.2.1 Precipitation

#### Standard Gages

As discussed in Section 4.2.1, precipitation is measured by collecting gages in which the snow and ice have melted or that weigh the total in-falling water

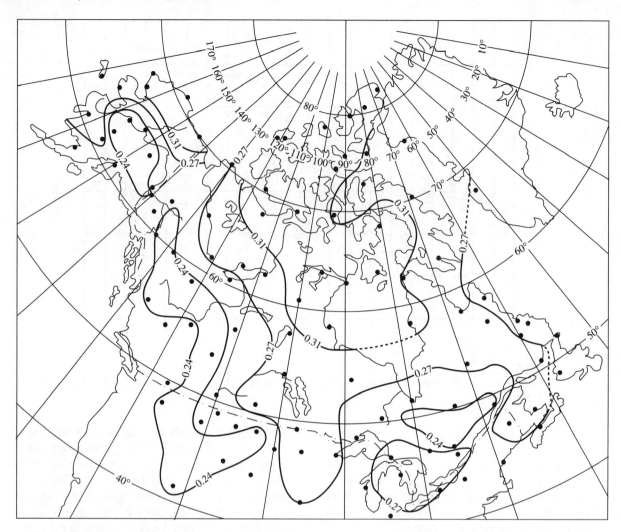

**FIGURE 5-3**
Average seasonal snowpack relative densities ($\rho_s/\rho_w$) for North America. From Bilello (1984).

substance. In most cases, the form of the precipitation is not recorded.

We have seen that measurements in such gages are subject to several sources of error, the most important of which is due to wind. Most network precipitation gages in the United States lack windshields, and even those equipped with Alter-type shields suffer from generally significant catch deficiencies for snow [Figure 4-16; Equations (4-3)]. Studies in Canada indicate that a specially designed gage with a Nipher windshield (Figure 5-5) maintains an adequate gage catch in winds up to about 5.5 m s$^{-1}$, but as of 1980 only 14% of Canadian gages were equipped with such shields (Goodison

et al. 1981). Gages used in the former Soviet Union and most other northern countries also suffer from wind-induced catch deficiencies, so that standard network measurements must be assessed with caution for periods in which snow is an important component of precipitation (Legates and DeLiberty 1993; Yang et al. 1998).

### Universal Gage

Cox (1971) developed a "universal surface precipitation gage" that measures snowfall and snowpack water equivalent by weight and collects and measures water output, so that all of the quantities

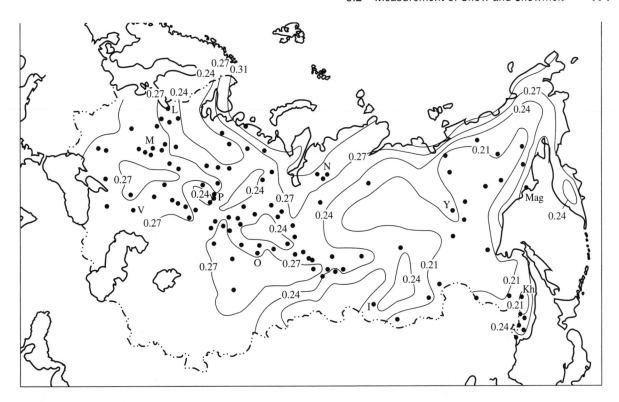

**FIGURE 5-4**
Average seasonal snowpack relative densities $(\rho_s/\rho_w)$ for the former Soviet Union. From Bilello (1984).

**TABLE 5-1**

Methods of Measuring Depth, Water Equivalent, and Areal Extent of Precipitation, Snowfall, Snowpack and Snowmelt-Water Output. Letters in parentheses indicate ground based (G), aircraft based (A), satellite based (S).

| Parameter | Depth | Water Equivalent | Areal Extent |
|---|---|---|---|
| Precipitation | — | Standard storage gages (G) Universal gage (G) | Gage networks (G) Radar (G) |
| Snowfall | Ruler, board (G) | Melt snow on board (G) Use estimated density (G) Universal gage (G) Snow pillow (G) | Observation networks (G) Radar (G) Visible/Infrared (S) |
| Snowpack | Snow stake (G,A) Snow tube (G) Acoustic gage (G) | Universal gage (G) Snow tube (G) Snow pillow (G) Artificial radioisotope gage (G) Natural gamma radiation (G,A) Microwave/radar (A,S) | Snow surveys (G) Visible/Infrared (S) Microwave/radar (A,S) Snow-pillow network (G) |
| Snowmelt and Water Output | | Snow pillow (G) Lysimeter (G) Universal gage (G) | |

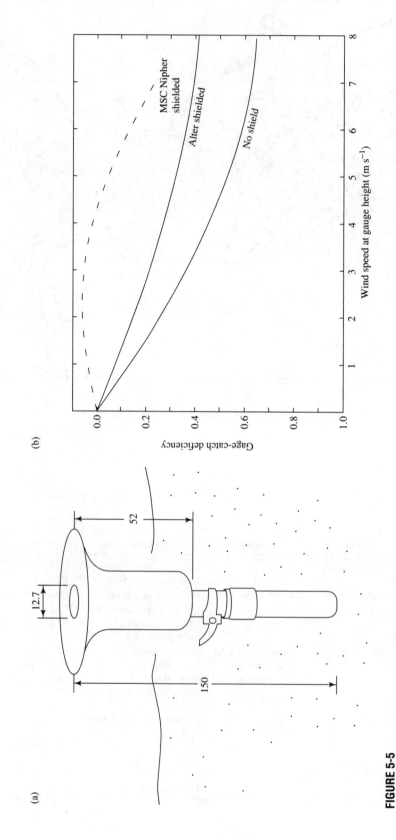

**FIGURE 5-5**

Canadian precipitation gage with Nipher windshield for areas with significant snowfall. (a) Dimensions of gage (cm). The gage is adjusted so that its orifice is maintained at 150 cm above the snow surface. (b) Comparison of gage-catch deficiency for snow as a function of wind speed for Canadian Nipher shield, Alter shield, and no shield. Data from Goodison et al. (1981) and Larson and Peck (1974).

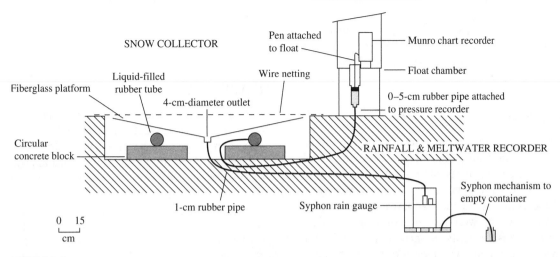

**FIGURE 5-6**
Diagram of the universal gage, which can measure precipitation, snowfall, water equivalent, and water output. From Waring and Jones (1980).

listed in Table 5-1 are simultaneously determined and recorded. Waring and Jones (1980) modified Cox's design to make it more suitable for shallow snowpacks (Figure 5-6). Clearly, these gages are considerably more elaborate and expensive to install than are standard gages, and they are therefore not widely used in observation networks.

### Radar

As described in Section 4.3.5, radar observations can be used to determine the areal extent and type of precipitation and to provide general information on precipitation rates.

### 5.2.2   Snowfall

#### Standard Methods

Snowfall depth is usually measured simply by placing a ruler vertically on a board that had been set on the surface of the previous snowpack. The water equivalent of snowfall can be determined by melting the snow collected on the board and measuring its volume, although it is more common simply to estimate this value by assuming a density, usually $100 \text{ kg m}^{-3}$.

### Universal Gage

The universal gage (Figure 5-6) measures snowfall water equivalent by recording an increase in weight on the collector in the absence of near-simultaneous production of water output. Similarly, an increase in the weight recorded by a snow pillow (see later) would usually indicate a snowfall event, although rainfall that stayed in the snowpack would also cause a weight increase.

### Radar

Snowflakes reflect radar waves more strongly than do raindrops, so radar images can distinguish between the two forms of precipitation and can be used to delineate the areal extent of snowfall for a given storm. However, radar measurement of snowfall rates is even more difficult than of rainfall rates, because the various forms of snowflakes have greatly differing radar reflectivities (Goodison et al. 1981).

### 5.2.3   Snowpack

#### Snow Stakes

The depth of snow can be observed simply by inserting a ruler or similar device through the snow to the ground surface, or by observing the height of the snow surface against a fixed ruler, called a **snow stake,** with its zero point at the ground surface. In some remote areas, permanent snow stakes are designed with large markings so that readings can be made from aircraft.

### Snow Surveys

The most important snow information for the hydrologist is the water equivalent of the snowpack. Network measurements of this quantity are most commonly obtained via periodic **snow surveys** at fixed locations, called **snow courses.** A snow course is a path between two fixed end points over which a series of measurements of snow depth and water equivalent are made. The length of the path is typically 150 to 250 m, with measurements made at about six points (more if snow conditions are highly variable) spaced at a fixed interval of at least 30 m. At each point, a coring tube equipped with a toothed cutting rim, called a **snow tube** (Figure 5-7), is inserted vertically to the surface. After the snow depth is read against markings on the outside of the tube, the tube is pushed a few centimeters into the soil and twisted to secure a small plug of soil that retains the snow in the tube. The tube is then extracted and weighed on a specially calibrated scale that is pre-tared and reads directly in centimeters or inches of water equivalent. Density at each point can be calculated via Equation (5-12), and water equivalent for the course is the average of the values at the measurement points.

---

### EXAMPLE 5-1

Snow surveyors measured the snowpack depths and water equivalents given in the accompanying table using a snow tube at five points along a snow course in the Kootenai River Basin, Montana, on 27 February 1958. Compute the average depth, water equivalent, and density for this location.

| Point | 1 | 2 | 3 | 4 | 5 |
|---|---|---|---|---|---|
| Depth (cm) | 98 | 108 | 102 | 109 | 105 |
| Water Equiv. (cm) | 27 | 30 | 27 | 30 | 29 |

**Solution** The average depth and the water equivalent are found simply as the arithmetic averages of the five observations: 104 cm and 29 cm, respectively. The density at each point is found via Equation (5-12), with $\rho_w = 1000$ kg m$^{-3}$:

| Point | 1 | 2 | 3 | 4 | 5 |
|---|---|---|---|---|---|
| Density (kg m$^{-3}$) | 280 | 280 | 260 | 280 | 280 |

Thus, the average density is 280 kg m$^{-3}$.

---

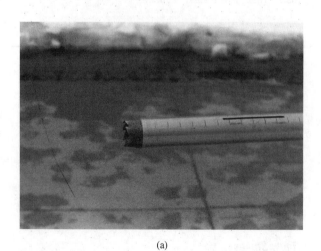

(a)

(b)

**FIGURE 5-7**
(a) Snow tube showing cutter. (b) Determining water equivalent with a snow tube. Photo by author.

Several different designs of snow tubes are available; the tubes may be made of aluminum or fiberglass and range in diameter from 3.8 to 7.6 cm. Comparisons of measurements taken by snow tubes with measurements taken by carefully excavating and weighing snow have shown that most snow tubes tend to overestimate water equivalent by up to 10% (Work et al. 1965; Goodison et al. 1981).

In shallow (i.e., less than about 1 m) snowpacks, depth and density have been found to be essentially independent, and there is typically less temporal and spatial variability in density than in depth (Goodison et al. 1981); this is clearly the case in Example 5-1. Under these conditions, little precision is lost and considerable time may be gained by making more depth measurements than water-equivalent measurements. Rikhter (1954) suggested a rough field guide to variations in density: Snow with a relative density of 0.32 to 0.35 will support an adult without skis; at 0.35 to 0.38, an adult's foot leaves only a slight impression; and above 0.4, the foot leaves no mark on the surface.

A snow-course network, like a precipitation-gage network, should be designed to provide a representative picture of the snowpack in the region of interest. However, since measurements are labor intensive, snow courses are usually considerably more widely spaced than gages and are usually read at longer time intervals—e.g., every two weeks during the snow season. Because snowpack conditions are largely determined by local conditions, it is usually sound strategy to design the network to sample representative ranges of land use (vegetative cover), slope, aspect, and elevation. Areal averages may then be estimated by extrapolating from these measurements on the basis of the distribution of the various conditions in the region of interest. However, for operational purposes such as forecasting runoff, measurement agencies commonly rely on only a few snow-course sites that have been "calibrated" over a period of years to provide an index, rather than a sample estimate, of the watershed snowcover.

### Snow Pillows

The water equivalent of snowpack can also be measured with **snow pillows,** which are circular or octagonal membranes made of rubber or flexible

**FIGURE 5-8**

A snow pillow (foreground) and a water-output lysimeter (background) installed at the U.S. National Weather snow-research site at Danville, VT. Buried lines transmit the fluid pressure from the pillow to a sensor, and the water released by the snowpack to a measuring device, both of which are in an instrument shelter. The metal ring is electrically heated and is melted down through the snowpack to isolate the cylinder of snow above the lysimeter. Photo by author.

metal that contain a liquid with a low freezing point (Figure 5-8). The weight of the snow on the pillow controls the pressure of the liquid, which is recorded or monitored via a manometer or pressure transducer.

The diameter of snow pillows ranges from 1 to 4 m, with larger diameters recommended for deeper snowpacks (Barton 1974). Several factors influence the accuracy and continuity of readings, including (1) leaks; (2) temperature variations that affect the density of the liquid; (3) the formation of ice layers within the snowpack, which can support a portion of the snow and lead to under-measurement of water equivalent (called "bridging"); (4) disruption of the contact between the snow and the ground, which can distort the snowpack energy and water balances; and, in remote installations, (5) instability of power supply to sensors and recorders. Detailed considerations for installation and maintenance of snow pillows were given by Davis (1973) and Cox et al. (1978), and measurement problems were discussed by McGurk and Azuma (1992) and McGurk et al. (1993).

Snow pillows are well suited for remote installation, with pressure readings transmitted by telemetry. In Alaska and the 12 western states of the U.S. mainland, the U.S. Natural Resources Conservation Service (NRCS) operates the SNOTEL network of 560 snow pillows, which provides telemetered data on water equivalent with an accuracy of about 2.5 mm (Schaefer and Werner 1966). If read frequently enough, snow pillows can be used to measure the water equivalent of individual snowfalls.

The universal gage (Figure 5-6) measures snowpack water equivalent using the same basic principle as that of the snow pillow.

### Acoustic Gages

Chow (1992) described a system that measures snow depth ultrasonically. The system is commercially available at relatively low cost and is well suited for installation as part of an automatic weather station.

### Radioactive Gages

Several types of instruments exploiting the attenuation of gamma rays or neutrons by water substance can be used for nondestructive measurement of the water equivalent of the snowpack. One version involves an artificial gamma-ray source ($^{60}$Co or $^{137}$Cs) and a detector, one of which is at the ground surface and the other suspended above the ground; the readings from the detector are typically transmitted by telemetry from a remote location to the observer. Bland et al. (1997) reported a method by which a portable gamma-ray source is inserted into permanent structures in the field at the time of measurement, and a hand-held detector is used to make a nondestructive determination of water equivalent with a precision of 3 mm.

For snowpacks with water equivalents less than about 40 cm, it is also possible to measure the attenuation by snow of natural gamma radiation emitted from the soil surface using a detector either fixed a few meters above the surface (Bissell and Peck 1973) or mounted on an aircraft (Loijens and Grasty 1973). Use of an airborne detector requires low-altitude (< 150 m) flights along a route over which the snow-free gamma emission has been previously determined; corrections must then be made to account for soil moisture and radioactive emissions from the air (Goodison et al. 1981; Foster et al. 1987). However, work by Grasty (1979) suggested that a simpler, single-flight technique can give results of high accuracy. Carroll and Voss

(1984) found good correlation between water equivalents determined from airborne gamma-radiation sensors and snow tubes in forested regions of the northern United States and of Canada (Figure 5-9a), as did Bergstrom and Brandt (1985) in Sweden. The U.S. National Weather Service now routinely uses low-flying aircraft to estimate water equivalents from natural gamma radiation (Carroll and Carroll 1989).

### Airborne Microwave and Radar

Microwave radiation (wavelengths of 0.1 to 50 cm), including radar, can be used to remotely measure the water equivalent, areal extent, and other properties of the snowpack. Airborne systems exploiting these wavelengths have the advantage of being able to "see through" clouds; however, many variables affect the observations, and methods for interpreting data are still being worked out.

The flux of microwave radiation emitted by a snowpack depends on its temperature, grain size, and the underlying soil conditions, as well as on its density. Thus considerable information about ground conditions is required for translating "passive" microwave emissions to estimates of water equivalent, and research is currently underway to provide reliable algorithms for interpretation (Foster et al. 1987).

Radar observation involves directing a beam of microwave radiation at the snowpack and measuring the strength of the reflected energy. This radiation can penetrate into the pack, so it can be used to provide information about snowpack stratigraphy and liquid-water content, as well as water equivalent, if sufficient information about surface cover and topography is available.

### Satellites

Satellite imagery using visible, infrared, and microwave wavelengths provides information on the areal extent of snow cover for large areas. Although careful interpretation is required to distinguish snow from clouds and to identify snow in areas of forest and highly reflective land surfaces, the most accurate maps of areal snow cover to date have been produced from visible-wavelength images. These images have been the basis for weekly maps that have been produced since 1978 for northern-hemisphere land, with a resolution of about 1.2 km$^2$ (Robinson et al. 1993).

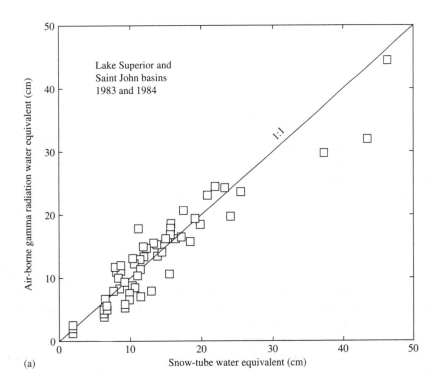

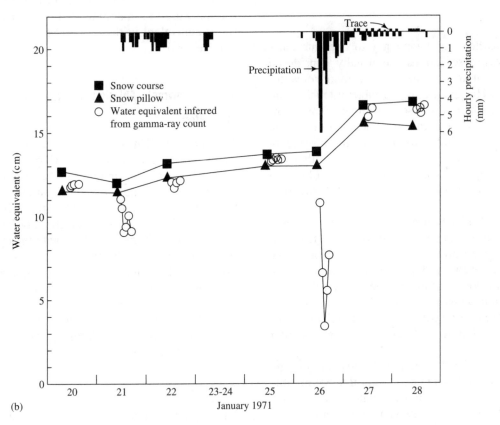

**FIGURE 5-9**

(a) Comparison of water equivalent determined by airborne gamma-radiation sensors and snow-tube measurements in the Lake Superior and St. John River Basins, United States and Canada. From Carroll and Voss (1984). (b) Water equivalent of snowpack at Danville, VT, measured over nine days by snow tube, snow pillow, and attenuation of natural gamma radiation. The false low readings of the radiation detector were due to radioactivity deposited during snowstorms and could easily be corrected for. From Bissell and Peck (1973); used with permission of the American Geophysical Union.

More recently, algorithms have been developed for mapping the distribution of snow cover over wide areas by automated analysis of satellite-borne radiometer data. These algorithms use reflected and emitted energy in several visible and infrared wavelengths to differentiate between clouds and snow and to correctly interpret variations produced by forest cover and shading. Rosenthal and Dozier (1996) described an algorithm for mapping mountain snow-cover extent from the Landsat Thematic Mapper satellite, which has a resolution of about 800 m². A new methodology for automated analysis of visible and infrared radiation collected by the AVHRR (Advanced Very High Resolution Radiometer) satellite can separate clouds from snow and clear land under most conditions, and the results can be used to produce daily continental-scale maps of the areal extent of snow cover with a resolution of 1.1 km² and 97% accuracy (Simpson et al. 1998).

### Overview

In spite of their slight tendency to over measure, snow-survey observations are usually considered the most accurate "routine" measurements of water equivalent. However, they are labor intensive and impractical for routine use in remote areas. Snow pillows are accurate and are widely used in the western United States for remote monitoring of mountain snowpacks; however, they are subject to bridging, temperature effects, and failures of instrument components (Goodison et al. 1981; McGurk et al. 1993).

Figure 5-9b compares water equivalents measured via snow tube, snow pillow, and a fixed radioisotope gage over several days at one location. Natural radioactivity deposited with falling snow caused false low readings by the gamma detector, but this radioactivity was found to decay rapidly and could readily be corrected for (Bissell and Peck 1973). Thus, all three methods appear to give similar results.

Goodison (1981) compared snow-survey and snowfall data in Canada and found that compatible estimates of regional water equivalents were possible only if (1) snow-survey data are weighted to account for the variability of water equivalent as a function of land use and (2) precipitation-gage measurements of snowfall are corrected for gage-catch deficiencies due to wind.

Remotely-sensed observations via aircraft or satellite using active or passive microwave, infrared, or visible wavelengths are becoming increasingly relied on for information on the areal extent of snow cover. Maps of snow-cover extent in the northern hemisphere derived from satellite observations, beginning in 1978, are available from the U.S. National Oceanic and Atmospheric Administration (NOAA) as the NOAA-NESDIS Weekly Northern Hemisphere Snow Charts, and similar maps based on newer satellite sensors and interpretation algorithms are becoming available. Such observations of the areal extent of snow cover, along with water-equivalent information developed from telemetered remote snow pillows and airborne detection of gamma radiation, are widely used for water-resource management decisions, especially in the western United States.

Techniques involving spatial correlation, similar to those described for mapping of areal precipitation (Box 4-2), have been developed and are routinely used for mapping the areal distribution of water equivalent in the United States (Carroll and Cressie 1996).

### 5.2.4 Snowmelt, Ablation, and Water Output

#### Lysimeters

The most straightforward method for measuring water output is via a **lysimeter** (Figure 5-8), which collects the water draining from the overlying snow and directs it to a device that measures and records the flow. This instrument may be fitted with a circular metal ring that can be electrically heated and lowered through the snow to isolate the cylinder of snow above the collecting surface; this method avoids gaining or losing water that might be moving horizontally along ice layers in the snowpack.

Note that, as with snow pillows, snow conditions above a lysimeter may differ from those in the natural snowpack due to interruption of the snow–ground connection.

#### Snow Pillows

Snow pillows detect ablation as a decrease in weight (assuming the water runs off the pillow); in many cases, evaporation can be considered negligible and the weight change can be attributed to water output.

## Universal Gage

As noted previously, universal gages collect and measure water output. Water output occurring at the same time as a corresponding weight decrease indicates snowmelt; water output in the absence of a weight decrease indicates rainfall; and a weight decrease in the absence of water output indicates evaporation.

## Pans

Specific measurement of evaporation and sublimation of the snowpack can be made using pans that are periodically weighed. Slaughter (1966) reviewed studies that employed various types of pans and concluded that good measurements can be obtained using pans that are made of plastic or metal, as long as the edge of the pan is flush with the snow surface and the surface roughness of the snow in the pan is the same as that of the surrounding snowpack. The pan should be at least 10 cm deep to avoid absorption of radiation by the bottom of the pan and, if significant melt is occurring, should be designed to allow meltwater to drain into a collector for separate measurement.

---

## 5.3  HYDROLOGIC IMPORTANCE AND DISTRIBUTION OF SNOW

### 5.3.1  Water Input

Precipitation in the form of snow is usually stored for a significant period on the ground surface and does not become an input to the land phase of the hydrologic cycle until it melts. Thus it is hydrologically useful to define **water input** as the sum of rainfall plus snowmelt for a given time period. When a snowpack is present, water input equals the water output from the snowpack.

In areas with a seasonal snowpack, the amount and timing of **surplus water input** (water input minus evapotranspiration), rather than of **surplus precipitation** (precipitation minus evapotranspiration), largely determines the amount and timing of streamflow and ground-water recharge (Figure 5-10). However, in spite of the strong relationship between water-input climate and hydrologic cli-

mate, there have been few studies characterizing water-input climatology. Hendrick and DeAngelis (1976) showed that water-input climatology in New England is determined by latitude and elevation; they developed a method for determining climatic and synoptic patterns of water input from standard network observations of temperature and precipitation. Further research along these lines would clearly be useful for many hydrologic purposes, including predicting the effects of climate change on streamflow.

### 5.3.2  Distribution of Snow

#### Global

Virtually all the land above 40° N latitude has a seasonal snow cover of significant duration (Figure 3-21); this amounts to about 42% of the land in the northern hemisphere. Figure 5-11 shows the steep latitudinal increase in the fraction of average annual precipitation falling as snow in North America; this fraction reaches about 0.65 on the north coast of Alaska (Dingman et al. 1980). Location relative to oceans and elevation also influences the portion of precipitation occurring as snow.

Average dates of formation and disappearance of snow cover are shown in Figure 5-12 for North America and in Figure 5-13 for the former Soviet Union. Figure 5-14 gives average seasonal maximum snow depths in the northern hemisphere. As noted earlier, maps and statistics of weekly northern hemisphere snow-cover extent since 1978 are available and show that snow cover has been declining over the last two decades (Robinson et al. 1993; Box 3-3).

Because of the lack of detailed knowledge of runoff processes (Chapter 9), it is difficult to quantify the fraction of runoff derived from snowmelt. However, a smaller proportion of snowfall than of rainfall is evaporated and transpired, so it is clear that snowfall contributes proportionally more to runoff. L'vovich (1974) estimated that more than half the annual runoff is derived from snowmelt in much of the northern hemisphere (Figure 5-15).

#### Elevational

Because of the general decrease of air temperature with altitude (Figure D-2), the fraction of precipitation falling as snow, and the water equivalent of

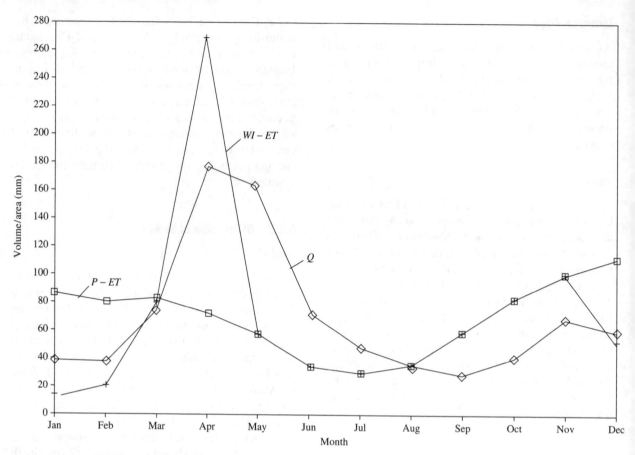

**FIGURE 5-10**

Average monthly precipitation minus evapotranspiration ($P - ET$) and water input minus evapotranspiration ($WI - ET$) compared with streamflow ($Q$) in the Pemigewasset River basin, NH. The timing of streamflow is more closely related to $WI - ET$ than to $P - ET$.

snow, are usually strong functions of elevation in a region (Figure 5-16). For mountain regions, as much as 85% of the annual runoff may come from snowmelt (Shafer and Dezman 1982).

Rates of increase of water equivalent with elevation vary regionally and with local factors such as aspect (i.e., north- or south-facing slopes) and vary from year to year at a given location. Meiman (1968) reviewed a number of studies on the elevational distribution of snow in North America and reported rates of increase of water equivalent ranging from 5.8 to 220 mm per 100 m elevation. Caine (1975) found that the year-to-year variability of water equivalent decreased with elevation in the southern Rocky Mountains, USA.

### Local

As noted in Section 5.2, snow properties are highly dependent on local site factors such as aspect (slope orientation) and vegetation cover. In general, local variability will be greatest in regions where periods of melting occur during the winter, where there are pronounced spatial changes in land cover and topography, and where much of the heat input to the snow is from solar radiation.

The main effect of aspect is on energy inputs from solar radiation, resulting in faster densification and melting on south-facing slopes. Aspect may also affect the wind microclimate, which in turn affects snow deposition and densification and energy exchanges of sensible and latent heats. These en-

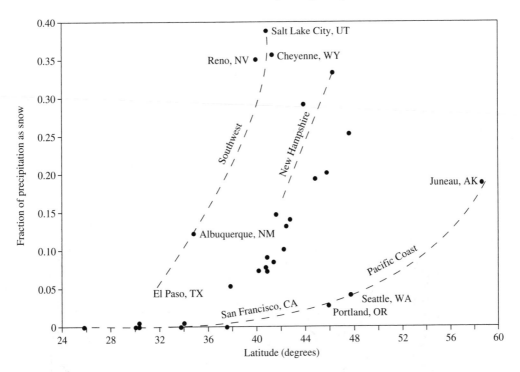

**FIGURE 5-11**
Fraction of precipitation occurring as snow as a function of latitude in the United States. City data from Todd (1970); unidentified points are for cities east of the Rocky Mountains.

ergy exchanges are discussed quantitatively in Section 5.4.

The accumulation of precipitation on the leaves and branches of vegetation is called **canopy interception.** Some intercepted snow eventually falls to the ground either before or after melting and is added to the snowpack; the rest evaporates and is called **canopy interception loss.**[1] Clearly, deciduous trees intercept less snow than do conifers, and various conifer species differ in their capture of snow. Schmidt and Gluns (1991) found that (1) individual conifer branches intercepted 11 to 80% of snow (water equivalent) in 22 storms, (2) the fraction intercepted was inversely related to snow density ($\rho_s/\rho_w$) and to total storm precipitation, and (3) the maximum intercepted water equivalent was about 7 mm. Thus in forests a large proportion of snowfall is intercepted, but most studies have found that it is of minor hydrologic importance (Hoover

1971; Tennyson et al. 1974), because most intercepted snow falls to the ground or melts rather than evaporates.

Forest clearings and thinnings disrupt the typical upward-increasing wind velocities above a canopy (see Section D.6.3) and generally lead to increased snow deposition. Many studies have shown that snow accumulation is greater in small (i.e., diameters less than $20 \cdot z_f$, where $z_f$ is the height of surrounding trees) forest clearings than in the surrounding forest. The relative importance of wind redistribution and interception in producing these differences changes from region to region. However, for clearings larger than $20 \cdot z_f$, the accumulation pattern tends to be reversed, because the wind speed tends to be higher than in the forest; this redistributes snow into the surrounding forest and causes higher evaporation in the clearing (Golding and Swanson 1986).

Watershed-scale experiments have shown that both selective and clear-cutting tree harvesting tend to increase snowpack water equivalent and snowmelt runoff (Schmidt and Troendle 1989). This

---

[1] Rainfall interception and interception loss are discussed in detail in Section 7.6.

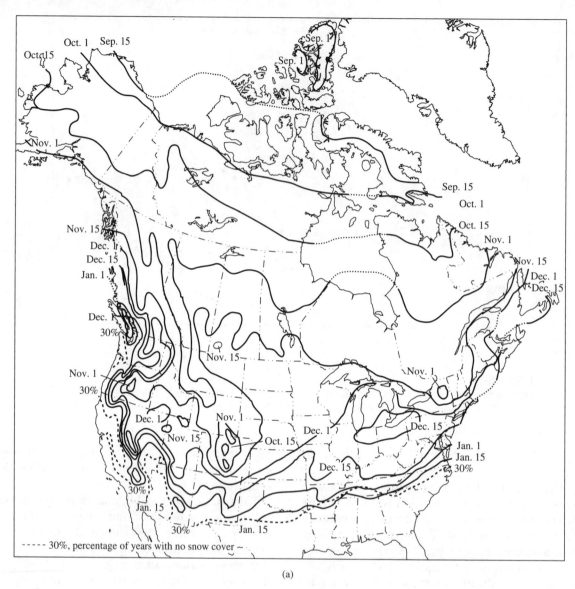

**FIGURE 5-12**
Average date of snow cover (a) formation and (b) disappearance in North America. From McKay and Gray (1981).

increase is attributed to a reduction in the evaporation of intercepted snow and increased snow deposition into clearings and thinned forests, partially offset by increased evaporation from the ground snowpack.

Figure 5-17 shows the variability of seasonal peak depth, density, and water equivalent on a range of vegetation types in Ontario, Canada. Overall, these observations are consistent with those just described for forest clearings: The highest depths and water equivalents were in an open forest with shrub understory (Vegetation Zone B), and the lowest values were in areas without forest cover, including grass (Zone A) and marsh (Adams 1976). Density tended to vary little with land-cover type. Note also that there was considerable year-to-year variability in the relative values.

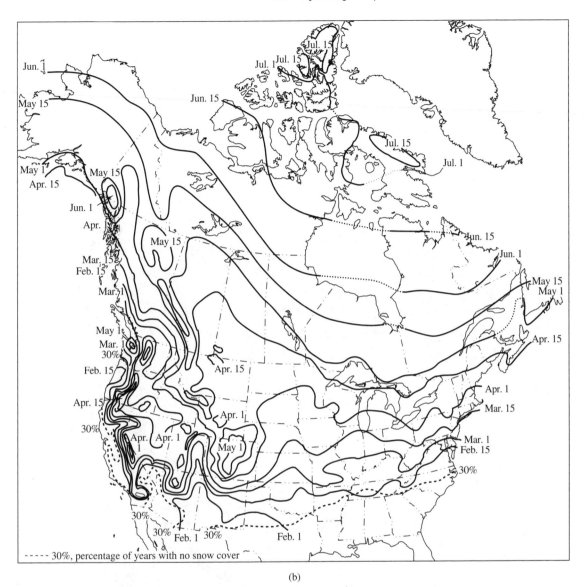

(b)

**FIGURE 5-12**
Continued

On an even more local scale, Woo and Steer (1986) presented data on variations of snow depth around individual trees in a subarctic spruce forest in northern Ontario. The data were used, along with information on tree spacing, to compute the average snow depth for the forest. As shown in Figure 5-18, depth increases nonlinearly away from the trunk and reaches the clearing value at a distance of 2 to 4 m from the tree. Presumably,

this pattern is produced by snow interception and by added heat inputs due to longwave radiation from the tree trunk, which can accelerate the processes that increase snow density and produce melt.

Donald et al. (1995) developed relations between land-cover types and water equivalent and other snow properties in Ontario that are useful for snowmelt modeling.

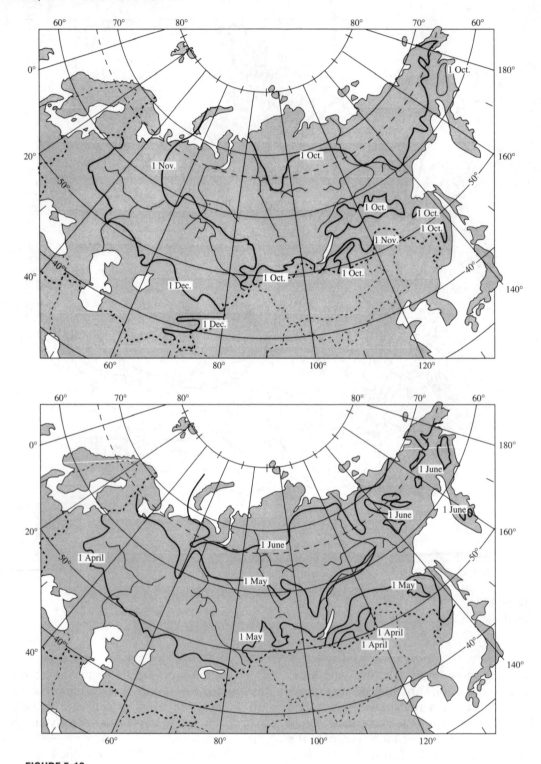

**FIGURE 5-13**
Average date of snow cover (a) formation and (b) disappearance in the former Soviet Union. From Wilson (1969).

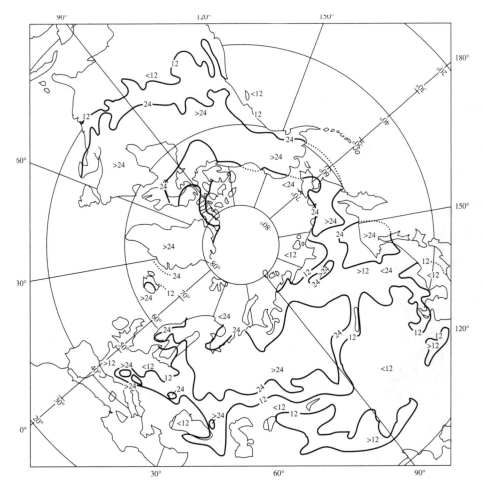

**FIGURE 5-14**
Average seasonal maximum snow depth (in.) for the northern hemisphere. From Bates and Bilello (1966).

## 5.4 SNOWMELT PROCESSES[2]

### 5.4.1 Phases of Snowmelt

The period of general increase of snowpack water equivalent prior to the melt period is called the **accumulation period.** During this period, the net inputs of energy are generally negative and the average snowpack temperature is decreasing; water

equivalent typically has an increasing trend during this period. The **melt period** of a seasonal snowpack begins when the net input of energy to it becomes more or less continually positive, and it can usually be separated into three phases:[3]

the **warming phase,** during which the average snowpack temperature increases more or less steadily until the snowpack is isothermal at 0 °C;

---

[2] Much of our current understanding of snowmelt processes and the forecasting of snowmelt runoff is based on an intensive research program conducted by the U.S. Army Corps of Engineers in the western United States in the early 1950s. The results of this research are extensively summarized in *Snow Hydrology* (U.S. Army Corps of Engineers 1956).

---

[3] In some regions and during some years, snow covers may form and melt more than once during the winter; in these cases, each successive snowpack will go through an accumulation period and the three phases of the melt period.

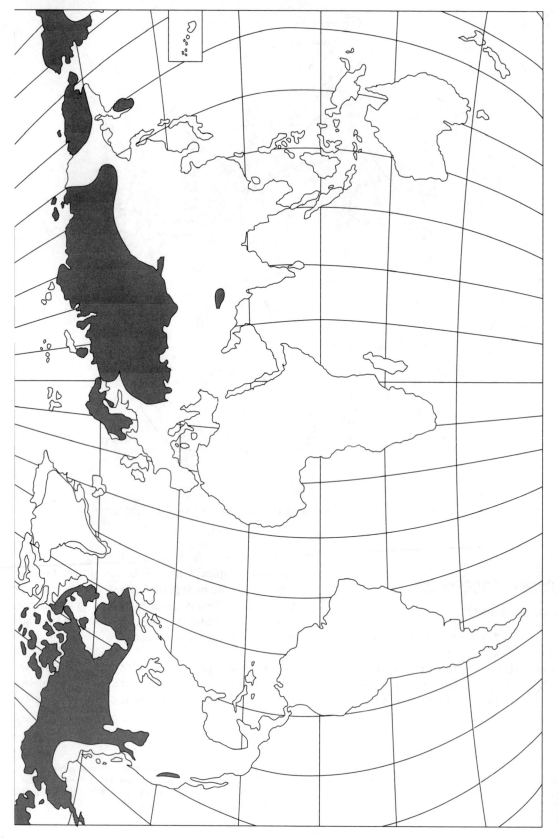

**FIGURE 5-15**

Regions in which more than half of the annual runoff is derived from snowmelt. From L'vovich (1974); used with permission of the American Geophysical Union.

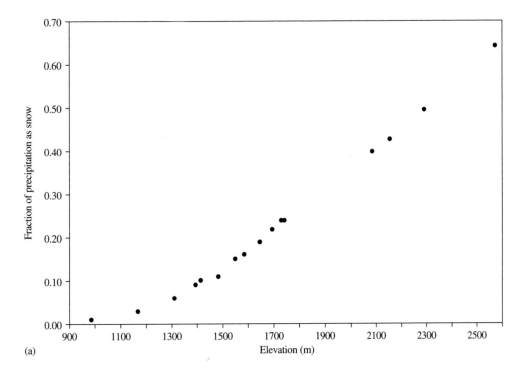

(a)

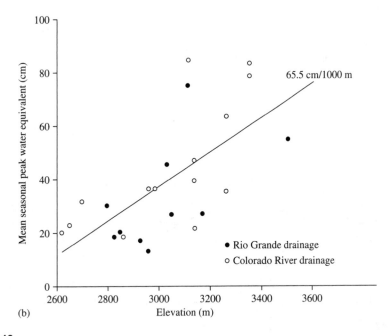

(b)

**FIGURE 5-16**
(a) Fraction of precipitation occurring as snow as a function of elevation in the San Bernardino Mountains of southern California. Data from Minnich (1986). (b) Annual peak water equivalent as a function of elevation, San Juan Mountains, CO, USA. From Caine (1975); used with permission of the American Water Resources Association.

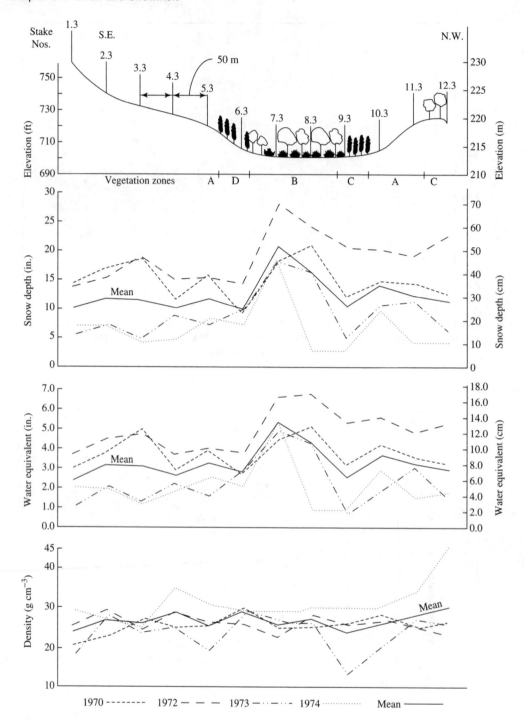

**FIGURE 5-17**
Peak seasonal snow properties along a transect near Peterborough, Ontario, for four winters. Vegetation Zones: A = grass; B = open deciduous forest with shrubs understory; C = moderately dense deciduous and coniferous forest; D = dense cedar forest. From Adams (1976); used with permission of the American Geophysical Union.

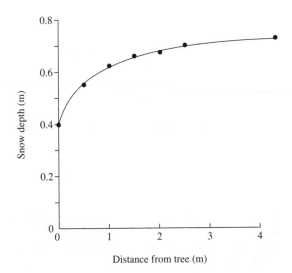

**FIGURE 5-18**
Snow depth as a function of distance from a tree in a spruce forest in northern Ontario. From data of Woo and Steer (1986).

the **ripening phase,** during which melting occurs, but the meltwater is retained in the snowpack. At the end of this phase, the snowpack is isothermal at 0 °C, cannot retain any more liquid water, and is said to be **ripe;** and

the **output phase,** during which further inputs of energy produce water output.

In many situations, the snowpack does not progress steadily through this sequence. Some melting usually occurs at the surface of a snowpack from time to time prior to the ripening phase, when the air temperature rises above 0 °C for periods of hours or days. The meltwater thus produced percolates into the cold snow at depth and refreezes, releasing latent heat that raises the snow temperature. Similarly, snow-surface temperatures may fall below freezing during the melt period, and the surface layer must warm again before melting can continue. Even where daytime temperatures are continuously above freezing, temperatures commonly fall below 0 °C at night, and it may take several hours for the snowpack to warm and resume melting each day (Bengtsson 1982; Tseng et al. 1994). Nevertheless, the sequence of three phases provides a useful context for understanding the melt process. As described next, the amounts of net energy inputs required for each of the melt phases can be readily computed.

### Warming Phase

The **cold content,** $Q_{cc}$, of a snowpack is the amount of energy required to raise its average temperature to the melting point, so

$$Q_{cc} = -c_i \cdot \rho_w \cdot h_m \cdot (T_s - T_m), \qquad (5\text{-}13)$$

where $c_i$ is the heat capacity of ice (2102 J kg$^{-1}$ K$^{-1}$),[4] $T_s$ is the average temperature of the snowpack, $T_m$ is the melting-point temperature (0 °C), and the other symbols are as previously defined. The cold content can be computed at any time prior to the ripening phase, and the net energy input required to complete phase 1, $Q_{m1}$, equals the cold content at the beginning of the melt period. Note that $Q_{cc}$ has dimensions [E L$^{-2}$].

---

### EXAMPLE 5-2

Given a snowpack with the average conditions computed in Example 5-1 and an average temperature of −9 °C, what is its cold content?

**Solution** The constants in Equation (5-13) have the following values: $c_i$ = 2102 J kg$^{-1}$ K$^{-1}$, $\rho_w$ = 1000 kg m$^{-3}$, and $T_m$ = 0 °C. From Example 5-1, $h_m$ = 29 cm = 0.29 m and $T_s$ = −9 °C. Substituting these values in Equation (5-13) yields

$$\begin{aligned} Q_{cc} = &-2102 \text{ J kg}^{-1}\text{ K}^{-1} \\ &\times 1000 \text{ kg m}^{-3} \times 0.29 \text{ m} \\ &\times (-9°\text{C} - 0°\text{C}) = 5.49 \text{ MJ m}^{-2} \end{aligned}$$

as the net input of energy required to raise this snowpack to the melting point.

---

### Ripening Phase

Phase 2 of melting begins when the snowpack becomes isothermal at 0 °C. After this time, further net inputs of energy produce meltwater, which is initially retained in the pore spaces by surface-tension forces (Figure 5-19). The liquid-water-retaining capacity, $h_{wret}$, of a snowpack is given by

$$h_{wret} = \theta_{ret} \cdot h_s, \qquad (5\text{-}14)$$

---

[4] Since the Kelvin and Celsius temperature scales have identical degree intervals and differ only in the location of the zero point, units involving temperature intervals have identical values in both scales; that is, 2102 J kg$^{-1}$ K$^{-1}$ = 2102 J kg$^{-1}$ °C$^{-1}$. (See Example A-7.)

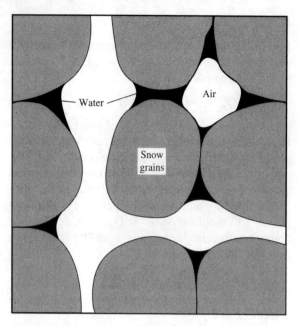

**FIGURE 5-19**
An idealized thin section of snow showing snow grains, water retained by surface tension, and continuous pores filled with air. After Colbeck (1971).

where $\theta_{ret}$ is the maximum volumetric water content that the snow can retain against gravity.[5] $\theta_{ret}$ for ripe snow can be estimated from empirical studies summarized by Eagleson (1970) as

$$\theta_{ret} = -0.0735 \cdot \left(\frac{\rho_s}{\rho_w}\right) + (2.67 \times 10^{-4}) \cdot \left(\frac{\rho_s^2}{\rho_w}\right),$$

$$(5\text{-}15)$$

where the densities are in kg m$^{-3}$. This relation is graphed in Figure 5-20.

We can use previously derived expressions to compute the proportion of pore spaces that are filled with water when $\theta = \theta_{ret}$. Consider a typical ripe snowpack with $\rho_s = 500$ kg m$^{-3}$. Equation (5-15) gives $\theta_{ret} = 0.03$. Substituting these quantities into Equation (5-6) allows us to compute the corresponding porosity $\phi = 0.49$. The ratio $\theta_{ret}/\phi$ is the proportion of pore spaces filled with water at the end of phase 2 [see Equations (5-2) and (5-4)], which for this case is 0.03/0.49 = 0.061. Thus it is clear that only about 6% of the pore spaces in a ripe

---

[5] $\theta_{ret}$ is directly analogous to the "field capacity" of a soil, defined in Section 6.4.1.

snowpack contain water, and such snowpacks are far from being saturated.

The net energy input required to complete the ripening phase, $Q_{m2}$, can be computed as

$$Q_{m2} = h_{wret} \cdot \rho_w \cdot \lambda_f = \theta_{ret} \cdot h_s \cdot \rho_w \cdot \lambda_f, \quad (5\text{-}16)$$

where $\lambda_f$ is the latent heat of fusion (0.334 MJ kg$^{-1}$).

---

### EXAMPLE 5-3

Assume that the snowpack considered in Examples 5-1 and 5-2 has become isothermal at the melting point and its density has increased to 400 kg m$^{-3}$. How much energy is required to bring it to a ripe condition?

**Solution** Assuming the water equivalent has not changed, the current snow depth is 72.5 cm [Equation (5-12)]. From Equation (5-15), $\theta_{ret} = 0.013$. With $\lambda_f = 0.334$ MJ kg$^{-1}$, we compute

$$Q_{m2} = 0.013 \times 0.725 \text{ m} \times 1000 \text{ kg m}^{-3}$$
$$\times 0.334 \text{ MJ kg}^{-1} = 3.15 \text{ MJ m}^{-2}$$

via Equation (5-16).

---

### Output Phase

Once the snowpack is ripe, further net energy inputs produce meltwater that cannot be held by surface-tension forces against the pull of gravity, and water begins to percolate downward, ultimately to become water output. This flow is more fully described in Section 5.4.3.

The net energy input required to complete the output phase, $Q_{m3}$, is the amount of energy needed to melt the snow remaining at the end of the ripening phase:

$$Q_{m3} = (h_m - h_{wret}) \cdot \rho_w \cdot \lambda_f. \quad (5\text{-}17)$$

### 5.4.2 The Energy Balance

Consider again the representative element of a snowpack shown in Figure 5-1. The energy balance for this element is

$$\Delta t \cdot S = \Delta Q, \quad (5\text{-}18)$$

where $S$ [E L$^{-2}$ T$^{-1}$] is the net rate of energy exchanges into this element by all processes over a

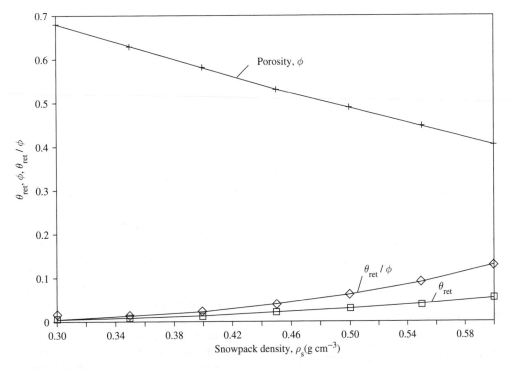

**FIGURE 5-20**
Empirical relations between porosity, $\phi$, and relative snow density, $\rho_s/\rho_w$ [Equation (5-6)]; liquid-water-holding capacity, $\theta_{ret}$, and $\rho_s$ [Equation (5-15)]; and $\theta_{ret}/\phi$ and $\rho_s/\rho_w$. From data plotted in Eagleson (1970).

time period $\Delta t$ [T], and $\Delta Q$ [E L$^{-2}$] is the change in heat energy absorbed by the snowpack during $\Delta t$.

The equations developed in this section provide a means of modeling the progress of snowmelt based on information about the net rate of energy input, $S$, during successive time periods. This approach to modeling is explicitly explored in Section 5.5.1. The several components of $S$ and the ways in which their magnitudes can be estimated are discussed later in this section.

During the warming phase, $\Delta Q$ is reflected in an increase in average temperature of the snowpack, $\Delta T_s$:

$$\Delta Q = c_i \cdot \rho_w \cdot h_m \cdot \Delta T_s. \qquad (5\text{-}19)$$

[Compare Equation (5-19) with Equation (5-13).] Substitution of Equation (5-19) into Equation (5-18) relates the temperature change to the net rate of energy inputs:

$$\Delta T_s = \frac{\Delta t \cdot S}{c_i \cdot \rho_w \cdot h_m}. \qquad (5\text{-}20)$$

In the second and third phases, $\Delta Q$ is reflected in melting. During the ripening phase, this melting causes an increase in the liquid water held in the snowpack, so

$$\Delta Q = \rho_w \cdot \lambda_f \cdot \Delta h_w \qquad (5\text{-}21)$$

and

$$\Delta h_w = \frac{\Delta t \cdot S}{\rho_w \cdot \lambda_f}. \qquad (5\text{-}22)$$

During the output phase,

$$\Delta Q = -\rho_w \cdot \lambda_f \cdot \Delta h_m, \qquad (5\text{-}23)$$

so

$$\Delta h_m = -\frac{\Delta t \cdot S}{\rho_w \cdot \lambda_f} \qquad (5\text{-}24)$$

and

$$\Delta w = -\Delta h_m, \qquad (5\text{-}25)$$

where $\Delta w$ is the increment of water output from the snowpack.

---

### EXAMPLE 5-4

Given a ripe snowpack, compute the amount of water output produced on a day when the total net energy input, $S$, is 5.11 MJ m$^{-2}$.

**Solution** According to Equations (5-24) and (5-25),

$$\Delta w = \frac{1 \text{ day} \times 5.11 \text{ MJ m}^{-2} \text{ day}^{-1}}{1000 \text{ kg m}^{-3} \times 0.334 \text{ MJ kg}^{-1}}$$
$$= 0.0153 \text{ m} = 15.3 \text{ mm}.$$

---

### Energy-Exchange Processes

The energy exchanges that determine the progress of snowmelt occur via the following processes (the symbols will be used to represent the net rates of energy input, or **fluxes**, [E L$^{-2}$ T$^{-1}$], by each process):

shortwave (solar) radiation input, $K$;

longwave radiation exchange, $L$;

turbulent exchange of sensible heat with the atmosphere, $H$;

turbulent exchange of latent heat with the atmosphere, $LE$;

heat input by rain, $R$;

conductive exchange of sensible heat with the ground, $G$.

Thus,

$$S = K + L + H + LE + R + G. \quad \textbf{(5-26)}$$

The basic physics of each of these components of the energy budget, and the approaches to determining their magnitudes, are discussed in the following sections.

**Shortwave Radiation Input**  As indicated in Figure 3-1, the sun's energy is electromagnetic radiation with wavelengths less than 4 μm; most of this energy is concentrated between 0.4 and 0.7 μm wavelength.

$K$ is the net flux of solar energy entering the snowpack, so

$$K = K_{in} - K_{out} = K_{in} \cdot (1 - a), \quad \textbf{(5-27)}$$

where $K_{in}$ is the flux of solar energy incident on the snowpack surface, $K_{out}$ is the reflected flux, and $a$ is the shortwave reflectance, or albedo (see Table D-2).[6]

$K$ includes both the energy in the direct solar beam and the diffuse solar radiation due to scattering by atmospheric gases and aerosols (Section E.1). The diffuse component makes up about 10% of total incident shortwave radiation under clear-sky conditions, increasing to 100% when there is complete overcast (Hay 1976). However, the diffuse component is difficult to estimate, and since it seldom accounts for more than a small fraction of the total energy balance, we will not treat it separately in our development.

$K$ can be measured with instruments called **pyranometers** [described in Iqbal (1983)], one facing upward to measure $K_{in}$ and one downward to measure $K_{out}$ in a representative location. However, pyranometers and the associated data loggers are installed at only a few permanent locations (see Figures E-1 and E-2) and research stations, so that it is common to estimate $K$. To do this, we note that $K_{in}$ represents the clear-sky shortwave radiation flux, $K_{cs}$, adjusted for the slope and aspect (i.e., the azimuth of a line running downslope) of the land surface and reduced by cloud cover and shading by vegetation:

$$K_{in} = K_{cs}(\Lambda, J) \cdot f_1(\Lambda, \beta, \alpha) \cdot f_2(C) \cdot f_3(F).$$
$$\textbf{(5-28)}$$

Here, $\Lambda$ is the latitude, $J$ is the day of year (the number of days since the calendar year began), $\beta$ is the slope inclination angle, $\alpha$ is the slope azimuth angle (measured clockwise from north), $C$ is the fraction of sky covered with clouds, $F$ is the fraction of sky obscured by forest canopy, and $f_1, f_2$, and $f_3$ are functions expressing the effects of the quantities in parentheses.

Combining Equations (5-27) and (5-28), we have

---

[6] The albedo equals the integrated reflectance for shortwave radiation, for which the symbol $\rho(\lambda)$ is used in Section D.1 [Equation (D-4)]. The symbol $a$ is used for albedo throughout the text to avoid confusion with the symbol used for densities.

$$K = K_{cs}(\Lambda, J) \cdot f_1(\Lambda, \beta, \alpha)$$
$$\cdot f_2(C) \cdot f_3(F) \cdot (1 - a). \qquad \textbf{(5-29)}$$

Methods for estimating the quantities in Equation (5-29) are outlined next.

***Clear-Sky Shortwave Radiation***   The complete set of equations and the algorithm for computing clear-sky shortwave radiation on horizontal and sloping surfaces without vegetative cover are given in Appendix E; only the general approach is described here.

The total daily flux of clear-sky radiation falling on a horizontal surface is a function of the latitude ($\Lambda$) and the declination angle ($\delta$) of the sun, which is a function of the time of year ($J$). At a given latitude, this quantity varies approximately sinusoidally through the year (Figure 5-21).

As anyone living in a region with seasonal snow cover knows, there is a great difference between the amounts of solar radiation received on north- and south-facing slopes. This difference is illustrated by the data in Table 5-2, which compares the clear-sky shortwave radiation inputs on north- and south-facing slopes at Danville, VT, during the snowmelt season. The function $f_1(\Lambda, \beta, \alpha)$ is derived using the concept of **equivalent latitude** and is given in Section E.2.

***Effect of Cloud Cover***   The effect of cloud cover can be estimated using empirical relations such as

$$f_2(C) = 0.355 + 0.68 \cdot (1 - C). \qquad \textbf{(5-30)}$$

This expression was used successfully by Croley (1989) in predicting evaporation from lakes; the U.S. Army Corps of Engineers (1956) suggested an alternative relation involving cloud height:

$$f_2(C) = 1 - (1 - k_s) \cdot C. \qquad \textbf{(5-31)}$$

In this equation,

$$k_s = 0.18 + (7.9 \times 10^{-5}) \cdot Z_C, \qquad \textbf{(5-32)}$$

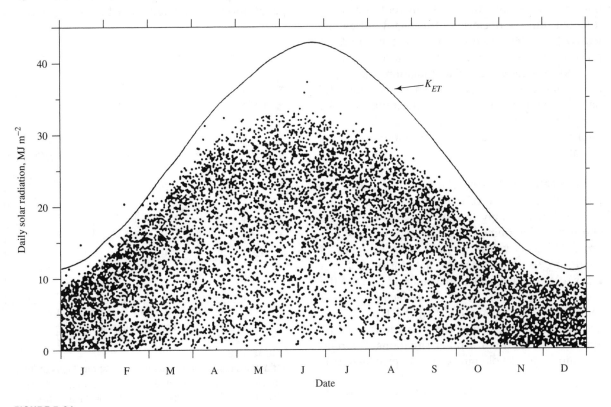

**FIGURE 5-21**
Extraterrestrial (potential), $K_{ET}$, and incident, $K_{in}$, solar radiation on a horizontal surface at West Thornton, NH, for 1960 through 1987. Each dot represents measured $K_{in}$ for a day. From Federer et al. (1990).

**TABLE 5-2**

Total Daily Incident Clear-Sky Solar Radiation Received on North- and South-Facing Slopes of Varying Inclinations at Danville, VT (latitude 44.5° N) on 17 March (MJ m⁻²).

| Slope Inclination Angle, $\beta$ (°) | North-facing slope ($\alpha = 0°$) | South-facing slope ($\alpha = 180°$) |
|:---:|:---:|:---:|
| 0 | 20.6 | 20.6 |
| 5 | 18.7 | 22.4 |
| 10 | 16.6 | 24.1 |
| 15 | 14.3 | 25.5 |
| 20 | 12.0 | 26.8 |
| 25 | 9.5 | 27.9 |
| 30 | 7.0 | 28.7 |
| 35 | 4.5 | 29.3 |
| 40 | 1.9 | 29.7 |

where $Z_C$ is height of the cloud base in meters.

A slightly different approach to estimating the effect of cloud cover is based on the ratio of the "percentage of possible sunshine," which is the ratio of the duration of bright sunshine during a day to the length of time between sunrise and sunset. A nomograph for this approach was developed by Hamon et al. (1954).

However, manual observations of $C$ are no longer available in the United States because human weather observers are being replaced by automated stations. Because of this fact, Lindsey and Farnsworth (1997) recommended that, where $K_{in}$ is not directly measured, it is best determined from gridded estimates of solar radiation developed from Geostationary Operational Environmental Satellite (GOES) imagery. Useful estimates of $K_{in}$ can also be developed using daily maximum and minimum temperature and precipitation data (Hunt et al. 1998).

***Effect of Forest Canopy*** The value of $F$ can be determined as the ratio of the horizontally projected area of forest canopy to the total area of interest and is best determined from air photographs. Figure 5-22 shows values of $f_3(F)$ for four types of conifer forest. No comparable data for other forest types seem to have been published, and because the effect of vegetative cover in reducing incident shortwave radiation depends on the type, height, and spacing of the plants, these relations must be applied with caution for other situations. The relationship for lodgepole pine can be approximated as

$$f_3(F) = \exp(-3.91 \cdot F). \quad (5\text{-}33)$$

***Albedo*** As shown in Table D-2, snowpack albedo ranges from about 0.45 for old snow to about 0.85 for a fresh snow surface; thus this factor is an important determinant of the net input of energy to the snowpack. Figure 5-23 shows the empirical relation between albedo and the age of snow surface for the melt and accumulation seasons as measured by the U.S. Army Corps of Engineers (1956); these curves are commonly used in energy-balance models of snow melt.

Because of the porous nature of snow, solar radiation is reflected not at the surface plane, but over a finite depth. Studies have shown that there is little penetration of solar radiation below about 10 cm,

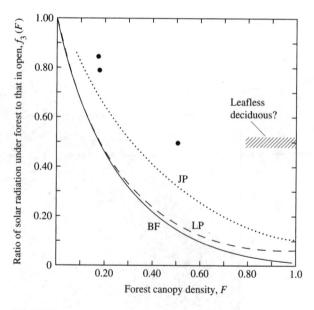

**FIGURE 5-22**

Ratio of incident solar radiation under various types of forest canopies to that received in the open for four forest types: BF = balsam fir, JP = jack pine, LP = lodgepole pine, circles = open boreal spruce forest. From *Water in Environmental Planning*, by Thomas Dunne and Luna B. Leopold. Copyright © 1978 by W.H. Freeman and Company. Reprinted with permission.

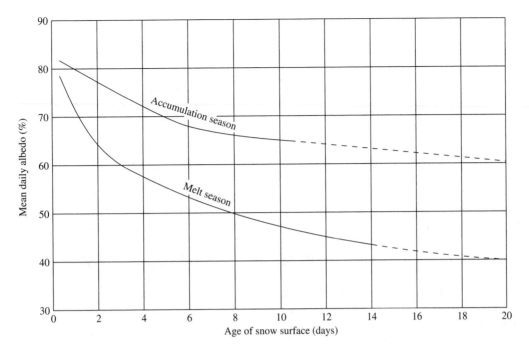

**FIGURE 5-23**
Albedo as a function of age of snow surface (i.e., time since last snowfall). From U.S. Army Corps of Engineers (1956).

so Figure 5-23 applies only when $h_s > 10$ cm. For shallower snowpacks, significant amounts of radiation are absorbed by the ground and may heat the snow from below. Leonard and Eschner (1968) found that albedo measured for snow intercepted on a conifer forest was considerably lower ($a \approx 0.2$) than for a ground snowpack, resulting in more rapid melting and greater evaporation for intercepted snow.

**Summary** The following example shows how net solar radiation flux to a snowpack is typically estimated by use of the preceding relationships.

## EXAMPLE 5-5

Compute the total daily net input of solar radiation, $K$, to a snowpack in a conifer forest with a canopy closure of $F = 0.8$ on a north-facing slope with an inclination of 20° on 17 March at the latitude of Danville, VT. The cloud cover is 0.5, and the last snow storm occurred on 12 March.

**Solution** The product of the first two terms in Equation (5-29) can be found in Table 5-2 as 14.3 MJ m⁻² day⁻¹. (For other days, locations, slopes, and aspects, we would use the relations given in Appendix E.) We compute $f_2(C)$ using Equation (5-30) and $f_3(F)$ using Equation (5-33):

$$f_2(C) = 0.355 + 0.68 \times (1 - 0.5) = 0.695;$$
$$f_3(F) = \exp(-3.91 \times 0.8) = 0.044.$$

Since the snowpack age is 5 days, the albedo estimated from Figure 5-23 is $a = 0.55$. Substituting all these values into Equation (5-29) gives

$$K = 14.3 \text{ MJ m}^{-2} \times 0.695$$
$$\times 0.044 \times (1 - 0.55) = 0.20 \text{ MJ m}^{-2} \text{ day}^{-1}.$$

**Longwave Radiation Exchange** Longwave, or terrestrial, radiation is electromagnetic radiation with wavelengths of 4 to 20 μm emitted by materials at near-earth-surface temperatures. (See Figure 3-1 and Section D.1.) The net input of longwave energy, $L$, is the difference between the incident flux, $L_{in}$, emitted by the atmosphere, clouds, and overlying forest canopy and the outgoing radiation from the snowpack, $L_{out}$:

$$L = L_{in} - L_{out}. \tag{5-34}$$

Longwave radiation can be measured directly by means of **pyrgeometers**; alternatively, it can be found as the difference between all-wave radiation measured by a **radiometer** and shortwave radiation measured by a pyranometer. However, such instruments are seldom installed except at sites of intensive research, and routine measurements of longwave radiation are available at only a few locations. Thus, as with shortwave radiation, one is usually forced to estimate the longwave component of the energy balance from more readily available meteorological information; this estimation is based on the considerations described next.

The flux of electromagnetic radiation emitted by a surface is given by the Stefan-Boltzmann equation [Equation (D-1)]. Hence we can write

$$L_{in} = \varepsilon_{at} \cdot \sigma \cdot (T_{at} + 273.2)^4, \qquad \textbf{(5-35)}$$

where $\varepsilon_{at}$ is the integrated effective emissivity of the atmosphere and canopy, $\sigma$ is the Stefan-Boltzmann constant ($\sigma = 4.90 \times 10^{-9}$ MJ m$^{-2}$ day$^{-1}$ K$^{-4}$), and $T_{at}$ is the effective radiating temperature of the atmosphere and canopy (°C). The outgoing flux is the sum of the radiation emitted by the snow surface and the portion of the incident radiation reflected by the surface. Since the longwave reflectivity of a surface equals one minus its longwave emissivity, we have

$$L_{out} = \varepsilon_{ss} \cdot \sigma \cdot (T_{ss} + 273.2)^4 + (1 - \varepsilon_{ss}) \cdot L_{in}, \qquad \textbf{(5-36)}$$

where the subscript $ss$ designates the values of emissivity and temperature for the snow surface.

Combining Equations (5-34), (5-35), and (5-36) and expanding and simplifying gives

$$\begin{aligned} L = \varepsilon_{ss} \cdot \varepsilon_{at} \cdot \sigma \cdot (T_{at} + 273.2)^4 \\ - \varepsilon_{ss} \cdot \sigma \cdot (T_{ss} + 273.2)^4; \end{aligned} \qquad \textbf{(5-37a)}$$

however, since the emissivity of snow is very close to 1 (Table D-1), Equation (5-37a) can be simplified to

$$L = \varepsilon_{at} \cdot \sigma \cdot (T_{at} + 273.2)^4 - \sigma \cdot (T_{ss} + 273.2)^4. \qquad \textbf{(5-37b)}$$

The major problem in applying Equation (5-37) is to find expressions for $\varepsilon_{at}$ and $T_{at}$ or, equivalently, to estimate the value of $L_{in}$ under various conditions of cloudiness and forest cover. This problem is addressed in the following sections.

**Clear-Sky Longwave Radiation** Expressions for estimating $L_{in}$ are usually developed by noting that the most important absorbers and emitters of longwave radiation in the atmosphere are carbon dioxide and water vapor (Figure 3-1). Since the concentration of carbon dioxide is quite constant over time periods of interest to snowmelt modeling, variations in the downward flux of longwave radiation under clear skies and no forest canopy are due largely to fluctuations in humidity. Several empirical functions expressing this relation have been developed; we select one that was derived from considering the vertical distribution of water vapor in the atmosphere (Brutsaert 1975):

*clear sky, no forest canopy*

$$\varepsilon_{at} = 1.72 \cdot \left( \frac{e_a}{T_a + 273.2} \right)^{1/7}, \qquad \textbf{(5-38)}$$

where $e_a$ is near-surface (2-m height) atmospheric vapor pressure (kPa) and $T_a$ is 2-m air temperature (°C).

**Effect of Cloud Cover** Clouds are black bodies emitting longwave radiation at a rate determined by the temperature of the cloud base, and their presence greatly increases the effective emissivity of the atmosphere. Thus under cloudy conditions $\varepsilon_{at}$ will be determined by the degree of cloud cover, $C$; an approximate empirical equation expressing this relation was given by Kustas et al. (1994):

*cloudy sky, no forest canopy*

$$\varepsilon_{at} = 1.72 \cdot \left( \frac{e_a}{T_a + 273.2} \right)^{1/7} \cdot (1 + 0.22 \cdot C^2). \; \textbf{(5-39)}$$

**Effect of Forest Canopy** As is the case with clouds, trees are very nearly blackbodies with respect to longwave radiation (Table D-1), and they can be considered to be emitting radiant energy at a rate determined by their temperature. Since their temperature is close to the near-surface air temperature, their effect on the total integrated atmospheric emissivity can be modeled as in Equation (5-40):

*cloudy sky, forest canopy (general equation)*

$$\begin{aligned} \varepsilon_{at} = (1 - F) \cdot 1.72 \cdot \left( \frac{e_a}{T_a + 273.2} \right)^{1/7} \\ \cdot (1 + 0.22 \cdot C^2) + F. \end{aligned} \qquad \textbf{(5-40)}$$

Here, $F$ is as defined for shortwave radiation. Note that Equation (5-40) is general; it reduces to Equation (5-39) for cloudy conditions with no forest canopy and to Equation (5-38) for clear-sky conditions with no canopy.

***Longwave Radiation Emitted by Snow Surface*** The second term on the right side of Equation (5-37) is the radiation flux emitted by the snow surface. Brubaker et al. (1996) showed that the average daily snow-surface temperature is well approximated by

$$T_{ss} = \min[(T_a - 2.5),\ 0], \qquad \textbf{(5-41)}$$

where temperatures are in °C. During phases 2 and 3 of melting, the snow surface is at the freezing point: $T_{ss} = 0$ °C.

***Summary*** The foregoing discussion indicates that, in practice, $T_{at} = T_a$, and, when $T_a > 0$ °C, $T_{ss} = 0$ °C and Equation (5-37b) becomes

$$L \approx \varepsilon_{at} \cdot \sigma \cdot (T_a + 273.2)^4 - 27.3, \qquad \textbf{(5-42)}$$

where the energy fluxes are in MJ m$^{-2}$ day$^{-1}$.

In most situations, $\varepsilon_{at} < 1$ and $L$ is negative. However, $\varepsilon_{at}$ can be > 1 and $L$ can be positive if the air temperature and cloud cover or forest cover have large values. Example 5-6 depicts a case in which the latter situation occurs.

---

### EXAMPLE 5-6

Compute the net longwave radiation exchange, $L$, for the situation described in Example 5-5, assuming a ripe snowpack, an air temperature, $T_a$, of 4 °C, and a relative humidity, $W_a$, of 0.80.

**Solution** We use Equation (5-40) to compute the atmospheric emissivity $\varepsilon_{at}$. To do this, we must first compute the vapor pressure $e_a$. From Equation (D-7), we find that the saturation vapor pressure, $e_a^*$, is

$$e_a^* = 0.611\ \text{kPa} \times \exp\left(\frac{17.3 \times 4\,°\text{C}}{4\,°\text{C} + 237.3\,°\text{C}}\right)$$
$$= 0.814\ \text{kPa},$$

and from Equation (D-10), we find that the actual vapor pressure is

$$e_a = 0.80 \times 0.814\ \text{kPa} = 0.651\ \text{kPa}.$$

Then

$$\varepsilon_{at} = (1 - 0.80) \times 1.72 \times \left(\frac{0.651}{4 + 273.2}\right)^{1/7}$$
$$\times (1 + 0.22 \times 0.5^2) + 0.80 = 0.953.$$

We now use this value in Equation (5-42):

$$L = 0.953 \times 4.90 \times 10^{-9}\ \text{MJ m}^{-2}\ \text{day}^{-1}\ \text{K}^{-4}$$
$$\times [(4 + 273.2)\ \text{K}]^4 - 27.3\ \text{MJ m}^{-2}\ \text{day}^{-1}$$
$$= 0.27\ \text{MJ m}^{-2}\ \text{day}^{-1}.$$

---

**Turbulent Exchange of Sensible Heat with the Atmosphere** The physics of sensible-heat exchange near the earth's surface is developed in Section D.6, leading to Equation (D-49) for neutral atmospheric conditions:

$$H = -\rho_a \cdot c_a \cdot \frac{k^2}{\left[\ln\left(\dfrac{z_a - z_d}{z_0}\right)\right]^2} \cdot v_a \cdot (T_a - T_s).$$

$$\textbf{(D-49)}$$

Here, $\rho_a$ is the density of air, $c_a$ is the heat capacity of air, $k = 0.4$, $z_d$ is the zero-plane displacement height, $z_0$ is the surface-roughness height, $v_a$ is the windspeed, $T_a$ is the air temperature, $z_a$ is the height above the snow surface at which $v_a$ and $T_a$ are measured, and $T_s$ is the surface temperature.

To modify this equation for snow, we first drop the minus sign, since we consider energy inputs as positive. Second, we can assume that $z_d$ is negligibly small. The roughness height, $z_0$, depends on the irregularity of the snow surface and thus can be highly variable from place to place and with time at a given location. For example, Anderson (1976) measured values between 0.0001 and 0.038 m for his research site in Vermont. However, his data show a strong decrease in $z_0$ as the season progressed, and values during the melt season did not exceed 0.005 m. In the absence of other information, a value between 0.0005 and 0.005 m may be selected; however, it should be noted that for special situations, such as vegetation projecting above the snow surface or patchy snow, $z_0$ could be considerably higher.

With the preceding modifications, and assuming $z_a = 2.00$ m, $z_0 = 0.0015$ m, $c_a = 0.00101$ MJ kg$^{-1}$ K$^{-1}$, and $\rho_a = 1.29$ kg m$^{-3}$, Equation (D-49) becomes

$$H = (4.03 \times 10^{-6}) \cdot v_a \cdot (T_a - T_{ss}) \text{ MJ m}^{-2} \text{ s}^{-1},$$
$$\text{(5-43a)}$$

or

$$H = 0.348 \cdot v_a \cdot (T_a - T_{ss}) \text{ MJ m}^{-2} \text{ day}^{-1}, \quad \text{(5-43b)}$$

where $v_a$ is in m s$^{-1}$ and temperatures are in °C.

There are two additional considerations in the application of Equation (5-43). First, those equations apply only to conditions of neutral atmospheric stability—i.e., when the actual temperature gradient in the air near the surface equals the adiabatic lapse rate (Section D.6.8). If the gradient is steeper than adiabatic, turbulent exchange will be enhanced by buoyant forces (unstable conditions); if it is less steep than adiabatic, turbulent exchange will be suppressed (stable conditions). (See Figure D-12.) Stable conditions are common over snowpacks, with temperature not uncommonly *increasing* with elevation (a "temperature inversion"). For these conditions, Equation (5-43) is divided by **stability-correction factors** $\Phi_M$ and $\Phi_H$, as in Equation (D-54). These factors are computed via Equation (D-55) and Table D-5.

The second consideration in applying Equation (5-43) is that wind speeds are virtually always measured in fields or clearings, and such measurements must be adjusted for calculating turbulent exchange in forested areas. Few data are available on which to base an adjustment factor; Dunne and Leopold (1978) suggested the following:

$$v_{aF} = (1 - 0.8 \cdot F) \cdot v_{aO}. \quad \text{(5-44)}$$

Here, $F$ is the fractional forest cover, and the subscripts $F$ and $O$ indicate wind speed inside and outside of the forest, respectively.

---

### EXAMPLE 5-7

Calculate the average daily turbulent sensible-heat exchange for the conditions given in Examples 5-5 and 5-6. The average wind speed measured in the open, $v_{aO}$, is 6.00 m s$^{-1}$; the measurement height, $z_a$, is 2.00 m.

**Solution** In the absence of other information, we assume $z_0 = 0.0015$ m; thus Equation (5-43b) can be used directly. Since the site of interest is forested with $F = 0.80$, we apply Equation (5-44) to adjust the wind speed:

$$v_{aF} = (1 - 0.8 \times 0.80) \times 6.00 \text{ m s}^{-1}$$
$$= 2.16 \text{ m s}^{-1}.$$

Direct use of Equation (5-43b) yields

$$H = 0.348 \times 2.16 \text{ m s}^{-1} \times (4°\text{C} - 0°\text{C})$$
$$= 3.01 \text{ MJ m}^{-2} \text{ day}^{-1}.$$

To compute the stability-correction factors, we first calculate the Richardson number, $Ri$, via Equation (D-55):

$$Ri = \frac{2 \times 9.81 \times (2 - 0.0015) \times (4 - 0)}{(277.2 + 273.2) \times (2.16 - 0)^2} = 0.061.$$

As expected, $Ri > 0$, reflecting a stable condition. From Table D-5, we calculate the stability-correction factors:

$$\Phi_H = \Phi_M = (1 - 5.2 \times 0.061)^{-1} = 1.47.$$

Dividing the original value of $H$ by $\Phi_H$ and $\Phi_M$ then gives the stability-adjusted value:

$$H = \frac{3.01 \text{ MJ m}^{-2} \text{ day}^{-1}}{1.47 \times 1.47} = 1.40 \text{ MJ m}^{-2} \text{ day}^{-1}.$$

Thus the stable atmospheric conditions significantly suppress turbulent transfer.

---

**Turbulent Exchange of Latent Heat with the Atmosphere**
Latent-heat exchange with the atmosphere is governed by the same turbulent process that produces sensible-heat exchange; this process is described in Section D.6 and leads to Equation (D-42):

$$LE = -\lambda_v \cdot \frac{0.622 \cdot \rho_a}{P} \cdot \frac{k^2}{\left[ \ln \left( \dfrac{z_a - z_d}{z_0} \right) \right]^2}$$
$$\cdot v_a \cdot (e_a - e_s). \quad \text{(D-42)}$$

In this equation, $\lambda_v$ is the latent heat of vaporization, $P$ is the atmospheric pressure, $e_a$ is the air vapor pressure, $e_s$ is the surface vapor pressure, and the other symbols are as given previously for Equation (D-42).

If the vapor-pressure gradient is directed upward ($e_s > e_a$), water substance will move from the snow to the air (evaporation); if it is directed downward ($e_s < e_a$), water substance will move from the air to the snow (condensation). In applying this equation to snow, however, we note that *two* phase changes may be involved. For cold snowpacks ($T_{ss} < 0\ °C$), evaporation and condensation involve the solid–vapor or vapor–solid phase change (sublimation); then the latent heat involved is the sum of the latent heats of vaporization, $\lambda_v$, and fusion, $\lambda_f$. For melting snowpacks ($T_{ss} = 0\ °C$), no solid–liquid or liquid–solid phase change occurs and only $\lambda_v$ is involved (Anderson 1968).

To put Equation (D-42) into a form directly applicable to the snow-surface energy balance, we again drop the minus sign, so that energy inputs to the snowpack are positive, and make assumptions analogous to those used in developing Equation (5-43): $\lambda_v = 2.47$ MJ kg$^{-1}$, $\lambda_f = 0.334$ MJ kg$^{-1}$, $k = 0.4$, $z_a = 2.00$ m, $z_0 = 0.0015$ m, $P = 101.1$ kPa, and $\rho_a = 1.29$ kg m$^{-3}$. With these assumptions, we arrive at the following equations:

*cold snow ($T_{ss} < 0\ °C$)*

$$LE = (6.86 \times 10^{-5}) \cdot v_a \cdot (e_a - e_{ss})\ \text{MJ m}^{-2}\,\text{s}^{-1},$$
**(5-45a)**

$$LE = 5.93 \cdot v_a \cdot (e_a - e_{ss})\ \text{MJ m}^{-2}\,\text{day}^{-1};$$ **(5-45b)**

*melting snow ($T_{ss} = 0\ °C$)*

$$LE = (6.05 \times 10^{-5}) \cdot v_a \cdot (e_a - e_{ss})\ \text{MJ m}^{-2}\,\text{s}^{-1},$$
**(5-46a)**

$$LE = 5.22 \cdot v_a \cdot (e_a - e_{ss})\ \text{MJ m}^{-2}\,\text{day}^{-1}.$$ **(5-46b)**

In these equations, $v_a$ is in m s$^{-1}$ and vapor pressures are in kPa.

As with the turbulent exchange of sensible heat, Equation (5-44) may be appropriate for estimating wind speeds in forests, and stability effects may be accounted for by using the approach described in Section D.6.8 and Example 5-7.

---

### EXAMPLE 5-8

Compute the turbulent exchange of latent heat for the conditions given in Example 5-7.

**Solution** We first decide whether evaporation or condensation is occurring by comparing $e_a$ and $e_{ss}$. The snow is at 0 °C, so $e_{ss}$ = 0.611 kPa. In Example 5-6, we found that $e_a$ = 0.651 kPa; since $e_a > e_{ss}$, we have condensation on melting snow and Equation (5-46b) is applicable. The appropriate value of wind speed is as found in Example 5-7: $v_a$ = 2.16 m s$^{-1}$. Thus

$LE = 5.22 \times 2.16$ m s$^{-1}$
  $\times$ (0.651 kPa − 0.611 kPa) = 0.451 MJ m$^{-2}$ day$^{-1}$.

As in Example 5-7, this value must be adjusted by the stability factors, giving

$$LE = \frac{0.451\ \text{MJ m}^{-2}\,\text{day}^{-1}}{1.47 \times 1.47} = 0.210\ \text{MJ m}^{-2}\,\text{day}^{-1}.$$

---

Various studies have found significantly different values for the "constants" in the turbulent-exchange equations. [See, for example, Male and Gray (1981).] This variability may be due to differences in (1) the periods over which measurements are averaged, (2) measurement heights, (3) roughness heights, (4) the ways in which stability effects are accounted for, or (5) the degree to which the assumed wind profiles coincide with actual conditions. Because of these factors, one must recognize that energy-budget values calculated with these equations are approximate.

In forests, snow evaporation occurs from the snow intercepted on the canopy, and evaporation from the ground becomes important only after the intercepted snow has disappeared by ablation or falling or blowing off (Lundberg and Halldin 1994). There is evidence that evaporation of intercepted snow may occur at rates of up to 3.3 mm day$^{-1}$ and may be an important component of the snow ablation in forests—perhaps amounting to 200 mm or more per winter (Lundberg et al. 1997).

**Heat Input by Rain** When rain falls on a snowpack that is at the freezing point ($T_s = T_m$), the rainwater is cooled to the snow temperature and the heat given up by the water is used in melting. Thus for this situation, we can calculate the heat contributed by rain, $R$, as

$$R = \rho_w \cdot c_w \cdot r \cdot (T_r - T_m),$$ **(5-47a)**

where $c_w$ is heat capacity of water ($4.19 \times 10^{-3}$ MJ kg$^{-1}$ K$^{-1}$), $r$ is the rainfall rate [L T$^{-1}$], and $T_r$ is the temperature of the rain.

When rain falls on snow that is below freezing, it will first cool to the freezing point, giving up sensible heat according to Equation (5-47a), and then freeze, liberating latent heat. In this case we have

$$R = \rho_w \cdot c_w \cdot r \cdot (T_r - T_m) + \rho_w \cdot \lambda_f \cdot r.$$ **(5-47b)**

If humidity information as well as the air temperature is available, $T_r$ can be estimated as the dew-point temperature, which is given by Equation

(D-11). However, since relative humidity usually is close to 1 during rain, and since $R$ is rarely a large contributor to the energy balance, the usual practice is to assume $T_r = T_a$.

## EXAMPLE 5-9

Compute the heat contributed by 25 mm of rain falling on the snowpack considered in the previous examples.

**Solution** Since the snow is at 0 °C, Equation (5-47a) is applicable. We assume that $T_r = T_a$ and that interception loss on the forest canopy is negligible. (See Section 7.6.) Thus

$$R = 1000\ \text{kg m}^{-3} \times (4.19 \times 10^{-3}\ \text{MJ kg}^{-1}\ \text{K}^{-1})$$
$$\times\ 0.025\ \text{m day}^{-1} \times (4\ ^\circ\text{C} - 0\ ^\circ\text{C})$$
$$=\ 0.419\ \text{MJ m}^{-2}\ \text{day}^{-1}.$$

Note that $R$ in this case is relatively large compared with the other energy-balance components computed in Examples 5-5–5-8. This unusual situation arises because the rainfall rate was high and because $K$, $H$, and $LE$ are severely reduced on a steeply sloping north-facing slope with a dense forest cover.

**Conductive Exchange of Sensible Heat with the Ground** Temperatures in the soil under snowpacks usually increase downward due to thermal energy stored during the summer and geothermal heat. Thus, in these circumstances, heat is conducted upward to the base of the snowpack at a rate $G$ given by

$$G = k_G \cdot \frac{dT_G}{dz}, \qquad \text{(5-48)}$$

where $k_G$ is the thermal conductivity of the soil [E $\text{L}^{-1}\ \text{T}^{-1}\ \theta^{-1}$], $T_G$ is the soil temperature, and $z$ is the distance below the ground surface.

Thermal conductivities of soils depend on soil texture, soil density, and moisture content and vary widely spatially and temporally. For example, the U.S. Army Corps of Engineers (1956) reported a more than tenfold increase in $k_G$ during the melt season in the soil they studied, from $8.37 \times 10^{-3}$ to 0.100 MJ m$^{-1}$ K$^{-1}$ day$^{-1}$. This variability, along with the general lack of information about thermal conductivity and ground-temperature gradients, usually precludes accurate computation of $G$. This uncertainty has little practical effect on energy-balance estimates during the snowmelt season, however, because $G$ is usually negligible compared with other terms.

In spite of its generally negligible contribution during the melt season, the energy conducted to the snowpack from the ground during the accumulation season can be hydrologically significant. Studies from various localities (Federer and Lash 1978; Male and Gray 1981) indicate that this heat produces continual melting at the base of the snowpack, called **groundmelt,** at rates up to 2 mm day$^{-1}$. Groundmelt can add significantly to moisture in the soil, increasing the percentage of snowmelt that will run off during the snowmelt season and, in regions like New England, groundmelt may be the principal source of flow in upland streams during the winter.

### Case Study of Snowmelt at Danville, VT, USA

The accumulation and melt of a snowpack during a typical winter (1972–1973) at the U.S. National Weather Service–U.S. Agricultural Research Service Snow Research Station in Danville, VT, are traced in Figure 5-24. Although the average snowpack temperature was not measured, the beginnings and ends of the accumulation period and the phases of the melt period can be approximated from the traces of depth, average snowpack density, water equivalent, and average air temperature.

The air temperature was above 0 °C only occasionally between mid-November and late February, and the water equivalent of the pack increased more or less continually during this period due to snowfalls and minor rain. The maximum snow depth of 72 cm was reached in late February. The density was initially about 100 to 150 kg m$^{-3}$, jumped to about 250 kg m$^{-3}$ in early December, and increased gradually thereafter to about 300 kg m$^{-3}$ when the melt season began. Density increases were due largely to constructive metamorphism and occasionally to refreezing at depth of rain and surface melt.

The accumulation period ended when air temperature began a final rise on 26 February; temperature was above freezing from 4 to 17 March, and water equivalent began to decline on 4 March. Significant water output was measured in a snowmelt lysimeter from 9 through 18 March. Density climbed from about 300 kg m$^{-3}$ to 400 kg m$^{-3}$ during the first two phases of melt; it then fluctuated between 350 and 450 kg m$^{-3}$ and eventually reached 520 kg m$^{-3}$ just before melting was complete.

Table 5-3 gives the amounts of energy involved in each of the components of the energy budget during the accumulation and melt periods for six

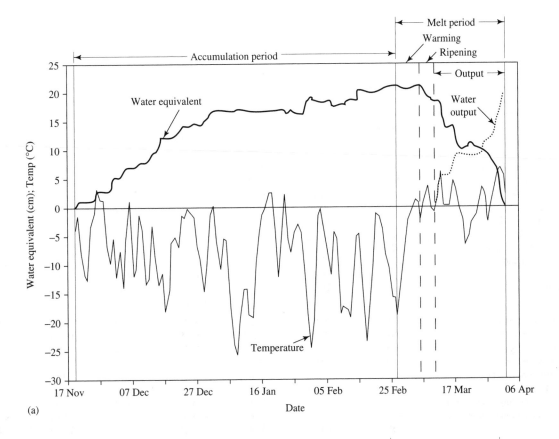

(a)

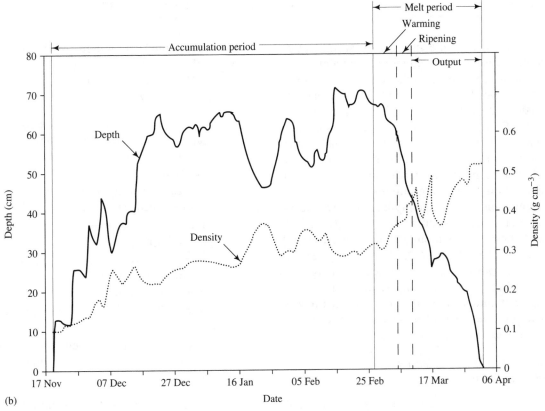

(b)

**FIGURE 5-24**

(a) Snowpack water equivalent, average air temperature, and cumulative water output; and (b) snowpack depth and density at the Sleepers River Research Watershed in Danville, VT, USA, for 1972–1973. The accumulation period and three phases of the melt period are shown; the boundary between the warming and ripening phases is uncertain. Data from Anderson et al. (1977).

**TABLE 5-3**

Energy-Balance Components (MJ m$^{-2}$) for Six Seasons at the U.S. National Weather Service–U.S. Agricultural Research Service Snow Research Station, Danville, VT. Data from Anderson (1976).

| | 68–69 | 69–70 | 70–71 | 71–72 | 72–73 | 73–74 | Average |
|---|---|---|---|---|---|---|---|
| | | | | **Accumulation Season** | | | |
| Net shortwave radiation, $K$ | 167.94 | 171.50 | 195.28 | 169.70 | 115.14 | 191.76 | 168.55 |
| Net longwave radiation, $L$ | −264.49 | −259.64 | −282.87 | −238.95 | −174.39 | −225.85 | −241.03 |
| Net radiation, $K + L$ | −96.55 | −88.14 | −87.59 | −69.25 | −59.25 | −34.08 | −72.48 |
| Heat from rain, $R$ | 0.08 | 0.84 | 0.25 | 0.38 | 0.71 | 1.00 | 0.54 |
| Heat from ground, $G$ | 30.77 | 53.89 | 32.24 | 25.00 | 26.29 | 29.73 | 32.99 |
| $K + L + R + G$ | −65.69 | −33.41 | −55.10 | −43.88 | −32.24 | −3.35 | −38.95 |
| Turbulent exchange, sensible, $H$ | 96.76 | 84.37 | 112.92 | 92.49 | 65.40 | 81.86 | 88.97 |
| Turbulent exchange, latent, $LE$ | −11.97 | −25.88 | −27.13 | −24.45 | −16.08 | −20.39 | −20.98 |
| $H + LE$ | 84.79 | 58.49 | 85.79 | 68.04 | 49.32 | 61.47 | 67.98 |
| Net heat input, $S$ | 19.09 | 25.08 | 30.69 | 24.16 | 17.08 | 58.12 | 29.04 |
| | | | | **Melt Season** | | | |
| Net shortwave radiation, $K$ | 191.97 | 129.00 | 162.83 | 150.73 | 149.69 | 168.78 | 158.83 |
| Net longwave radiation, $L$ | −101.12 | −67.49 | −77.29 | −74.78 | −92.45 | −109.99 | −87.19 |
| Net radiation, $K + L$ | 90.86 | 61.51 | 85.54 | 75.95 | 57.24 | 58.79 | 71.65 |
| Heat from rain, $R$ | 0.46 | 0.63 | 0.71 | 1.26 | 0.67 | 1.17 | 0.82 |
| Heat from ground, $G$ | 5.23 | 3.56 | 5.07 | 4.27 | 9.55 | 9.30 | 6.16 |
| $K + L + R + G$ | 96.55 | 65.69 | 91.32 | 81.48 | 67.45 | 69.25 | 78.62 |
| Turbulent exchange, sensible, $H$ | 66.57 | 43.13 | 67.12 | 65.86 | 56.32 | 60.80 | 59.96 |
| Turbulent exchange, latent, $LE$ | −28.35 | −17.88 | −23.66 | −28.85 | −35.00 | −31.90 | −27.61 |
| $H + LE$ | 38.23 | 25.25 | 43.46 | 37.01 | 21.31 | 28.89 | 32.36 |
| Net heat input, $S$ | 134.78 | 90.94 | 134.78 | 118.49 | 88.76 | 98.14 | 110.98 |

seasons at the Danville site. For all accumulation seasons, $L$ was negative and more than balanced $K$, resulting in negative net radiation. $H$ was positive and $LE$ negative in all accumulation seasons, but the magnitude of $H$ was several times greater than that of $LE$, so there was a net input of heat from turbulent exchange. There was a very small contribution from rain. The overall positive net input is largely due to ground heat, which was sufficient to produce about 87 mm of groundmelt.

In all melt seasons, the input from $K$ was about twice the loss via $L$, resulting in a strongly positive net radiation. Energy input was augmented slightly by $G$ and negligibly by $R$. Again, sensible-heat exchange was positive, and latent-heat exchange was negative but of considerably smaller magnitude, giving a net input from turbulent-exchange processes. The positive net heat input for melt seasons is the energy used in melting.

### Comparison of Energy Balances in Different Environments

Kuusisto (1986) reviewed over 20 studies of melt-period energy balances; his summary is given in Table 5-4. The comparable information for Ander-

son's (1976) study (computed for the melt season from Table 5-3) has also been added to Table 5-4. Kuusisto (1986) made the following generalizations based on his survey:

- Net radiation and turbulent exchange play a major role in the energy balance, and heat from rain and from the ground are small or negligible.
- Net radiation and sensible-heat exchange are positive during snowmelt in most locations.
- Latent-heat exchange is positive in some places, but negative in others.
- The net radiation is the most important component in forests, probably due to the reduced wind speeds.
- On cloudy or rainy days, turbulent heat exchange dominates.
- Very high areal snowmelt rates are usually caused by intense positive turbulent heat exchange.

The last generalization has implications for forecasting conditions that can cause snowmelt flooding: very warm, humid air and high winds

**TABLE 5-4**

Relative Contributions of Energy-Balance Components to Snowmelt in Different Environments[a]

| Study | Site[b] | Period[c] | K + L | H | LE | R | G |
|---|---|---|---|---|---|---|---|
| | | | Percent Contribution from | | | | |
| 1 | Open field, CA, 37°N | | 72 | 28 | −18 | | |
| 2 | Open field, Canada, 45°N (100 m) | Mar 59 (0.7) | 75 | 25 | −74 | | |
| 3 | Open field, AK, 67°N | Mar–Apr 66 | 86 | 14 | 24 | | |
| 4 | Open field in mts., CA, 37°N | Apr–May 47–51 (daytime) | 73 | 23 | 4 | | |
| 5 | Open field, MI, 46°N | 23 Jan 69 (1.5) | 17 | 47 | 36 | | |
| 6 | Forest opening, CO, 39°N (3260 m) | Jun 68 (5.0) | 56 | 44 | −3 | 0 | |
| 7 | Open field in mts., France, 46°N (3550 m) | Jul 68 (1.6) | 85 | 15 | −15 | | |
| 8 | Open field in mts., Spain, 41°N (1860 m) | Apr 70 (1.0) | 100 | −11 | −42 | | |
| 9 | Open field, AK, 71°N (10 m) | Jun 71 | 100 | −19 | −10 | | −22 |
| 10 | Open field in prairie, Canada, 51°N | Melt season 74 (0.8) | 59 | 41 | −10 | | −6 |
| 11 | Open field in prairie, Canada, 51°N | Melt season 75 (0.5) | 95 | 5 | −29 | | −1 |
| 12 | Open field in prairie, Canada, 51°N | Melt season 76 (0.3) | 67 | 33 | −14 | | −4 |
| 13 | Deciduous forest, Ontario, 46°N | Apr 78 (1.0) | 100 | 0 | 0 | | |
| 14 | Open field, Finland (60 m) | Melt seasons 68–73 (0.7) | 46 | 53 | −4 | 1 | |
| 15 | Open field in mts., Norway, 60°N | Apr–May 79–80 (1.2) | 35 | 65 | 0 | | |
| | | cloudy days (2.3) | 20 | 54 | 26 | | |
| | | clear days (0.7) | 37 | 63 | −24 | | |
| 16 | Small basin, 23% forest, Switzerland, 47°N (800 m) | Days with intense snowmelt, 77–80 (2.3) | 8 | 65 | 20 | 7 | |
| 17 | Open field, AK, 65°N | Apr 80 (0.3) | 67 | 33 | −68 | | 0 |
| 18 | Open field, Finland, 61°N | Days with intense snowmelt, 59–78 (1.4) | 48 | 47 | 2 | 3 | |
| 19 | Open field, Finland, 67°N | Days with intense snowmelt, 59–78 (1.5) | 58 | 40 | −23 | 2 | |
| 20 | Open tundra, N.W.T., Canada, 79°N (200 m) | Melt seasons 69–70 | 100 | −90 | −77 | | −45 |
| 21 | Open field in mts., New Zealand, 43°S (1500 m) | Oct–Nov 76–80 | 30 | 57 | 13 | <1 | |
| | | Rainy days | 17 | 55 | 25 | 3 | |
| | | Days with greatest heat input | 16 | 60 | 23 | 1 | |
| 22 | Open field in mts., New Zealand, 43°S (1450 m) | Melt season 82 (3.1) | 16 | 57 | 25 | 2 | |
| 23 | Open field, WI, 3 sites, 43–45°N | Melt seasons 53–64 | −31 | 100 | −12 | | |
| 24 | Small basin, 82% forest, Finland 64°N (120 m) | Melt seasons 71–81 (0.5) | 86 | 14 | −13 | <1 | |
| 25 | Open field, hilly area, VT, 45°N (550 m) | Melt seasons 69–74 (1.3) | 52 | 43 | −20 | <1 | 4 |

[a]100% is the sum of all positive components. Data from Kuusisto (1986): see that paper for references to individual studies. Data from Anderson (1976) added as study 25.

[b]Number in parentheses is site elevation.

[c]Number in parentheses is average melt in cm day$^{-1}$.

above a ripe snowpack. The potential for flooding under these conditions is increased because this situation would typically exist generally over a watershed, whereas rapid melting due to solar radiation is restricted largely to south-facing nonforested slopes. When rain accompanies warm winds, the flooding potential may be further exacerbated by the rain to produce a "rain-on-snow event," such as the one that generated the record floods of March 1936 in central New England. Note that the heat introduced by the rain plays only a small role in generating melt; in some cases the rain itself, with little melt, can produce floods (Singh et al. 1997).

### 5.4.3  Movement of Water through Snow

#### Percolation and Water-Output Production

We saw earlier that a ripe snowpack typically retains only a small fraction of its water equivalent as liquid water, which is present as thin films held by surface tension occupying less than 10% of the pore space (Figures 5-19 and 5-20). As additional water is produced during the last phase of melting, the water can no longer be held against gravity and downward percolation begins. Natural snowpacks are seldom homogeneous and usually contain discontinuous layers of varying density that temporarily store and horizontally divert the percolating water (Marsh and Woo 1985; Conway and Benedict 1994). However, ice layers disappear rapidly as melt progresses, and melting snowpacks tend to become fairly uniform assemblages of semi-spherical grains of from 1 to 3 mm in diameter. Thus to gain a basic understanding of the physical processes involved, it is appropriate to treat melting snowpacks as **homogeneous porous media,** physically identical to coarse-grained soils.

As noted, the pores of snowpacks are only partially filled with liquid water.[7] Thus the vertical percolation of water in the snowpack is a form of **unsaturated porous-media flow,** which is discussed in detail in Section 6.3 and Box 6-1. Colbeck (1978) gave a detailed description and experimental validation of the application of these relations to snowmelt percolation. He showed that capillary forces in snow are usually negligible in relation to

gravitational forces, so the equation of motion [Darcy's Law, Equation (6B1-6)] can be simplified to

$$V_{z'} = K_h(\theta_w), \qquad (5\text{-}49)$$

where $V_{z'}$ is the vertical flux rate (volume of water per unit horizontal area) and $K_h$ is the hydraulic conductivity of the snow. As for soils, $K_h$ is a function of the liquid–water content, $\theta_w$, which can be represented as

$$K_h(\theta_w) = K_h^* \cdot \left( \frac{\theta_w - \theta_{ret}}{\phi - \theta_{ret}} \right)^c, \qquad (5\text{-}50)$$

where $K_h^*$ is the hydraulic conductivity for completely saturated snow and $c$ is an empirical exponent with a value $c \approx 3$.[8] $K_h^*$, in turn, is a function of snow density:

$$K_h^* = 0.0602 \cdot \exp(-0.00957 \cdot \rho_s), \qquad (5\text{-}51)$$

where $K_h^*$ is in m s$^{-1}$ and $\rho_s$ is in kg m$^{-3}$ (Sommerfield and Rocchio 1993).

Energy inputs to the surface of a ripe snowpack vary diurnally, usually approximating a sine curve with a peak input in the early afternoon. This variation generates a daily wave of meltwater originating near the surface and percolating downward (Figure 5-25). As discussed by Dunne et al. (1976) and Colbeck (1978), the rate of travel, $U_z$, of a wave is related to the snow properties and flux rate, $V_{z'}$, as

$$U_z = \frac{c}{\phi - \theta_{ret}} \cdot K_h^{*1/c} \cdot V_{z'}^{(c-1)/c}. \qquad (5\text{-}52)$$

Because the speed of these melt waves increases as the flux (melt) rate increases, water produced during the period of peak melting near midday overtakes water produced earlier in the day. Thus the waves tend to accumulate water and develop a sharp wave front, as in Figures 5-25 and 5-26.

Example 5-10 shows how Equation (5-52) can be used to estimate the time of travel of a melt wave through a snowpack and how Equations (5-50) and (5-51) can be used to estimate the liquid-water content of a melting snowpack.

---

[7] This discussion applies to snowpacks above any basal saturated zone that may form where meltwater accumulates at an ice layer or at the ground surface.

[8] Equation (5-50) is analogous to Equation (6-13b).

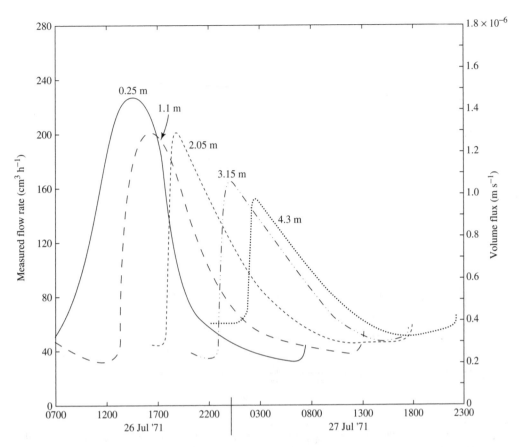

**FIGURE 5-25**
Volume rates of flow at different depths in a homogeneous snowpack produced by a day of intense fair-weather melting. From Colbeck (1978).

## EXAMPLE 5-10

Compute (a) the time lag of water output and (b) the liquid-water content for the snowpack considered in Examples 5-5–5-9.

**Solution** (a) If groundmelt is negligible, the total energy input for the day considered, $S$, is

$$S = K + L + H + LE + R = 0.19 + 0.27 + 1.40 + 0.21 + 0.42 = 2.49 \text{ MJ m}^{-2} \text{ day}^{-1}.$$

The melt rate is found via Equation (5-24) as

$$h_m = \frac{2.49 \text{ MJ m}^{-2} \text{ day}^{-1}}{1000 \text{ kg m}^{-3} \times 0.334 \text{ MJ kg}^{-1}}$$
$$= 0.00746 \text{ m day}^{-1} = 7.5 \text{ mm day}^{-1}.$$

Adding to this result the 25-mm rain, the total water output is 32.5 mm day$^{-1}$ = $3.76 \times 10^{-7}$ m s$^{-1}$. This value is the flux rate

$V_{z'}$, in Equations (5-49) and (5-52). Using the density $\rho_s = 400$ kg m$^{-3}$, we find from Figure 5-20 that $\phi = 0.58$ and $\theta_{ret} = 0.013$ and compute

$$K_h^* = 0.0602 \cdot \exp(-0.00957 \cdot 400)$$
$$= 1.31 \times 10^{-3} \text{ m s}^{-1}$$

from Equation (5-51). Inserting these values into Equation (5-52) and taking $c = 3$ yields

$$U_z = \frac{3}{0.58 - 0.013} \times (1.31 \times 10^{-3} \text{ m s}^{-1})^{1/3}$$
$$\times (3.76 \times 10^{-7} \text{ m s}^{-1})^{2/3} = 3.02 \times 10^{-5} \text{ m s}^{-1}$$
$$= 2.61 \text{ m day}^{-1} = 0.11 \text{ m hr}^{-1}.$$

If the snow depth is 0.725 m, it will take 6.7 hr for the meltwater wave to travel through the snowpack.

(b) To find the water content associated with this melt wave, note that Equations (5-49) and (5-50) can be combined and solved for $\theta$:

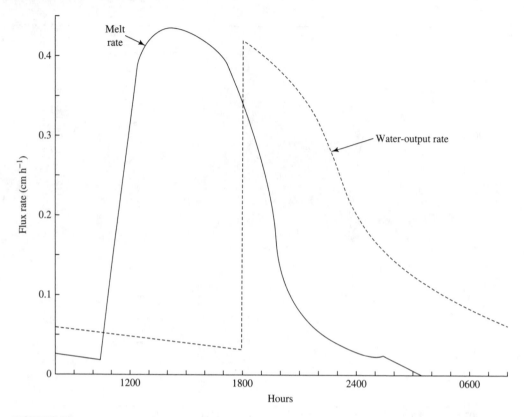

**FIGURE 5-26**
Comparison of timing of the rate of melting at the surface (the "input") and the rate of vertical unsaturated flow (water output) at the base of a 101-cm-deep tundra snowpack. From Dunne et al. (1976); used with permission of the American Geophysical Union.

$$\theta = (\phi - \theta_{ret}) \cdot \left( \frac{V_{z'}}{K_h^*} \right)^{1/c} + \theta_{ret}.$$

Thus

$$\theta = (0.58 - 0.013) \times \left( \frac{3.76 \times 10^{-7} \text{ m s}^{-1}}{1.31 \times 10^{-3} \text{ m s}^{-1}} \right)^{1/3}$$
$$+ 0.013 = 0.05.$$

One can calculate that the peak of the wave in Figure 5-25 moves at a velocity of 0.2 to 0.3 m hr$^{-1}$, which is in the general range found in Example 5-10 and in other studies (Figure 5-27). Thus for most seasonal snowpacks, the peak water output will occur within a few hours of the peak melt rate.

Anderson (1973) developed an empirical expression for the **time lag** of snow melt, i.e., the elapsed time between the beginning of daily melt and the beginning of water output. This expression was designed for use with a model using a 6-hr time step and is given by

$$t_{lag} = 5.33 \cdot \left[ 1.00 - \exp\left( -\frac{0.03 \cdot h_m}{\Delta h_{w6}} \right) \right], \quad \textbf{(5-53)}$$

where $t_{lag}$ is the time lag (hr), $h_m$ is the water equivalent of the snowpack when melt begins (m), and $\Delta h_{w6}$ is the amount of melt generated in a 6-hr period (m).

### Snowmelt-Runoff Generation

Water arriving at the bottom of the snowpack infiltrates into the soil and/or accumulates to form a saturated zone at the base of the snowpack, moving toward a surface-water body by one of the paths illustrated in Figure 5-28. In case (a) of that figure, the water table is at depth, the ground surface is unsaturated, and all of the water output infiltrates and moves streamward as subsurface flow. In case (b), the surface cannot accept all the water output, and a basal saturated zone develops within the snowpack through which water flows toward the stream.

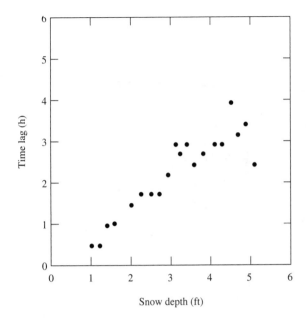

**FIGURE 5-27**
Approximate time lag between the time of peak surface melting and the time of peak flow from the bottom of a snowpack as a function of snow depth. From Anderson (1968); used with permission of the American Geophysical Union.

In case (c), the ground conditions are similar to those of case (a), but the water table has risen above the ground surface on the lower part of the slope, so that water moves streamward by both surface and subsurface routes.

The process of infiltration is discussed in detail in Chapter 6, and runoff generation by processes similar to those shown in Figure 5-28 is examined in Section 9.2. For now, we explore aspects of snowmelt-runoff generation by the process shown in Figure 5-28b. Assuming a constant snow depth, the daily wave of water output arrives at the base of the snowpack at the same time all along the slope. This input produces daily waves that travel downslope in the basal saturated zone at a velocity, $U_s$, where

$$U_s = \frac{K_h^*}{\phi - \theta_{ret}} \cdot S_s. \qquad (5\text{-}54)$$

In this equation, $K_h^*$ is the saturated hydraulic conductivity, $\phi$ is the porosity of the basal snow layer, and $S_s$ is the tangent of the slope angle.

As for percolation, $K_h^*$ can be estimated from the snow density via Equation (5-51). However, because of more rapid metamorphism in the saturated

zone, the size of snow grains can be significantly larger there than in the unsaturated zone. For example, for a subarctic snowpack, Dunne et al. (1976) found $d = 6$ mm in the basal saturated zone and $d = 2$ mm in the unsaturated zone. Thus, Equation (5-51) may significantly underestimate $K_h^*$ in the basal saturated zone.

Using the wave velocity $U_s$, one can calculate the average time of travel of water in the basal saturated zone, $t_s$, as

$$t_s = \frac{L}{2 \cdot U_s}, \qquad (5\text{-}55)$$

where $L$ is the slope length. $t_s$ is typically on the order of 1% of the travel time through the unsaturated zone (Male and Gray 1981). Thus, where a basal saturated zone forms, the lag time between peak melt rates and peak inputs to small upland streams is determined largely by the travel time associated with the vertical percolation through the snowpack and is typically on the order of several hours. In contrast, infiltrating snowmelt that percolates through the soil to the ground water (Figure 5-28a) may not appear in streamflow for months (Bengtsson 1988).

Singh et al. (1997) developed a model for the flow of water in a basal saturated zone; we will explore the phenomenon further in Section 9.2.3.

## 5.5 SNOWMELT MODELING

### 5.5.1 Snowmelt at a Point

#### Energy-Balance Approach

The energy-balance relations given by Equations (5-18)–(5-48) form the basis for one class of models of snowmelt at a point. These models are described as "physically based," because the energy balance is a fundamental physical principle (conservation of energy) and because they use equations that describe the physics of processes in each component of the energy balance. The general logic of a typical energy-balance model is diagrammed in Figure 5-29.

Much of the discussion in this section is based on the work of Anderson (1976), as this work appears to be the most thorough and extensive study

**FIGURE 5-28**
Three modes of snowmelt-runoff generation. (a) The top of the saturated zone in the soil (water table) is at depth; percolating meltwater infiltrates and percolates to the saturated zone to raise the water table and thereby induce increased ground-water flow to the stream. (b) The water table is at the soil surface, or the soil surface is impermeable (perhaps due to solid soil frost); percolating meltwater accumulates to form a basal saturated zone through which water drains to the stream. (c) The lower portion of the water table has risen above the ground surface into the snowpack; water in the upper part of the slope moves as in (a), and water in the lower part moves as in (b). Modified after Dunne and Leopold (1978).

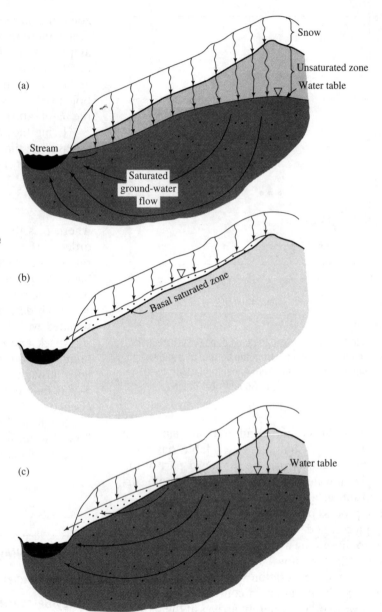

and testing of point snowmelt-modeling techniques. Anderson's studies were conducted in a large clearing at an elevation of 550 m in Danville, VT, over a period of six years. Energy-balance modeling of snowmelt under a forest cover has not been carefully tested.

**Time Step**   In order to develop a computer model based on Equation (5-18), one must first decide on the time step, $\Delta t$, to be used. Anderson (1976)

compared results using $\Delta t = 1, 3$, and 6 hr and found only minor differences.

Time periods of up to 24 hr will probably give reasonably satisfactory results, because the energy-balance components are linear or nearly linear functions of the meteorologic and other independent variables [Equations (5-27)–(5-48)]. Linearity guarantees that

$$E_t\{f[x(t)]\} = f[E_t\{x(t)\}], \qquad \textbf{(5-56)}$$

**FIGURE 5-29**
Flowchart showing how the energy balance is used in modeling snowmelt. The subscript $i$ is the counter for successive time periods of length $\Delta t$.

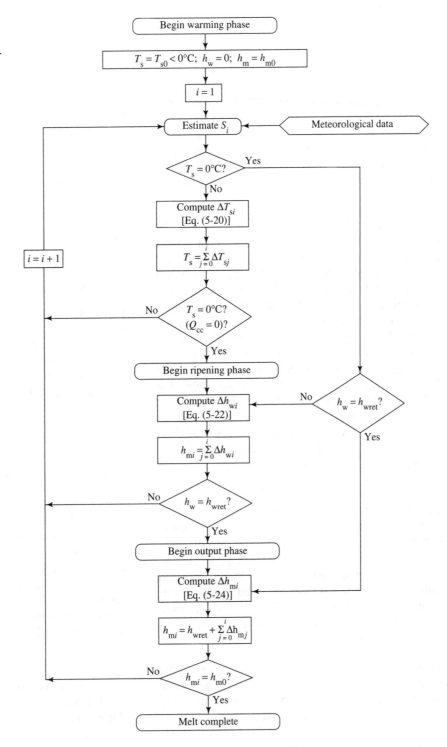

where $E_t\{\ \}$ indicates a time average, $f[\ ]$ is a linear function, and $x(t)$ represents one or more independent variables that are functions of time.

Potential exceptions to this near linearity are in the turbulent-exchange equations, where wind speed is multiplied by another variable; distortions due to time averaging for these equations will occur in proportion to the degree of independent variation of the variables involved. Even though Equation (5-37) appears highly nonlinear, the relation is nearly linear over the range of temperatures encountered.

**Layer Thickness**  Water and energy move vertically in a snowpack, so development of a snowmelt model also requires a decision about the thickness of the layers within the snowpack to be represented in the model. Anderson (1976) compared results using thicknesses of 1, 2.5, 5, and 10 cm, and found that the predicted water equivalent at a given time increased as the thickness increased because the thicker layers introduced distortions in the diurnal warming–cooling cycles. Although there is little further evidence on which to base definitive guidelines, one could probably expect significant distortions to appear in energy-balance models using thicknesses exceeding 50 to 100 cm. However, many models, including the BROOK90 model introduced in Section 2.9.5, treat the snowpack as a single layer of varying depth.

**Data Requirements**  Clearly, use of the energy-balance approach requires measurement of a large number of variables, including at least air temperature, humidity, wind speed, cloud cover, precipitation, snow-surface temperature, and, ideally, incident and outgoing short- and longwave radiation. Measurements of these variables must be made at representative locations and with sufficient frequency to calculate at-least-daily averages.

**Validation**  The soundness of any model must be established by "validation"—that is, by comparing its forecasts with actual events that were not used in developing the model. (See Section 2.9.4.) Anderson (1976) compared snowmelt and water equivalent estimated by the energy-balance model with those measured at the intensively instrumented Danville site. These comparisons are presented following discussion of the temperature-

index approach to snowmelt modeling in the next section.

### Temperature-Index Approach

Because of the difficulty and expense of fulfilling the data requirements of the energy-balance approach—and, in fact, the virtual impossibility of collecting such data at enough locations to be spatially representative of even a moderate-sized watershed—the empirical **temperature-index approach** is incorporated in most models used to predict or forecast snowmelt runoff. This approach estimates snowmelt, $\Delta w$, for a daily or longer time period as a linear function of average air temperature:

$$\Delta w = M \cdot (T_a - T_m), \qquad T_a \geq T_m;$$
$$\Delta w = 0, \qquad T_a < T_m. \qquad \text{(5-57)}$$

Here, $M$ is called a **melt coefficient, melt factor,** or **degree-day factor.**

The logic of this approach is clear when one recognizes that, during melting, the snow-surface temperature is at or near 0 °C, that energy inputs from longwave radiation and turbulent exchange are approximately linear functions of air temperature, and that there is a general correlation between solar radiation and air temperature. Figure 5-30 shows that a relation like Equation (5-57) reasonably approximates daily melt at the Danville snow-research site—although there is considerable scatter. The value of $M$ that best fits these data is $M = 3.6$ mm day$^{-1}$ °C$^{-1}$, and the intercept value, $-1.4$ mm day$^{-1}$, is not significantly different from 0.

However, the value of $M$ varies with latitude, elevation, slope inclination and aspect, forest cover, and time of year and ideally should be empirically estimated for a given watershed. Male and Gray (1981) cited a study suggesting that, in the absence of site-specific data, $M$ can be estimated as

$$M = 4.0 \cdot (1 - a) \cdot \exp(-4 \cdot F) \cdot f_{sl}, \qquad \text{(5-58)}$$

where $M$ is in mm day$^{-1}$ °C$^{-1}$, $a$ is albedo (see Figure 5-23), $F$ is the fraction of forest cover, and $f_{sl}$ is the **slope factor,** i.e., the ratio of solar radiation received on the site of interest to that on a horizontal surface [Equation (E-27), a function of latitude, day of year, slope inclination, and slope aspect]. Federer and Lash (1978) expressed the melt factor for forests in the eastern United States as

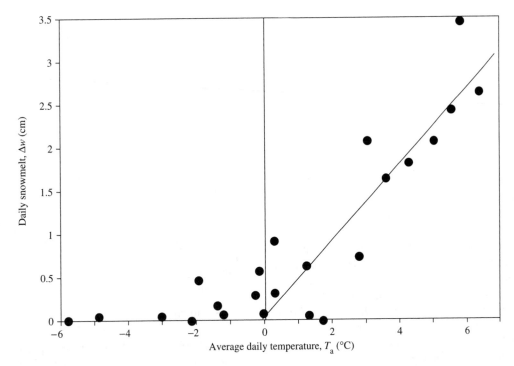

**FIGURE 5-30**
Daily snowmelt, $\Delta w$, as a function of daily average air temperature, $T_a$, at Danville, VT, in March 1973. The line is the best-fit linear relation between the two variables for days when $T_a > 0$ °C. Data from Anderson et al. (1977).

$$M = f_F \cdot (0.7 + 0.0088 \cdot J) \cdot f_{sl}, \quad J < 183, \quad \text{(5-59)}$$

where $f_F$ is a vegetative-cover factor equal to 30.0 for open areas, 17.5 for hardwood forests, and 10.0 for conifer forests, and the other symbols are as in Equation (5-58).

Kuusisto (1980) recommended relating $M$ to density, which generally increases during the melt season. For the forest,

$$M = 10.4 \cdot \frac{\rho_s}{\rho_w} - 0.7; \quad \text{(5-60a)}$$

in the open,

$$M = 19.6 \cdot \frac{\rho_s}{\rho_w} - 2.39. \quad \text{(5-60b)}$$

In both equations, $M$ is in mm day$^{-1}$ °C$^{-1}$.

Further adjustments in $M$ must be made to account for differences in the value of air temperature at the measurement site and the value applicable to the area being modeled.

## Hybrid Approach

The variability in melt factor as a function of time of year, land cover, and slope factor largely reflects the importance of solar radiation on snowmelt. To reduce this variability and improve prediction accuracy while retaining practical data requirements, Kustas et al. (1994) and Brubaker et al. (1996) evaluated an approach that computes daily snowmelt, $\Delta w$, as

$$\Delta w = \frac{K + L}{\rho_w \cdot \lambda_f} + M_r \cdot T_a, \quad \text{(5-61)}$$

where $M_r$ is a "restricted" melt factor with a constant value of 2.0 mm day$^{-1}$ °C$^{-1}$. In this method, the radiation terms are measured or evaluated via Equations (5-29)–(5-42), and $M_r$ accounts for the turbulent-exchange processes. Brubaker et al. (1996) showed that the value of $M_r \approx 2.0$ mm day$^{-1}$ °C$^{-1}$ can be derived from the basic equations for those processes [Equations (D-46) and (D-47)].

### Evaluation of Point Snowmelt Models

The report by Anderson (1976) covering six snow seasons at Danville, VT, is the most thorough comparison of point energy-balance and temperature-index modeling with actual data. Anderson's complete model included simulation of snowpack settling and compaction as well as the snowpack energy balance and was used for both accumulation and melt seasons. The temperature-index model, calibrated for the Danville site, is applicable only for the output phase of the melt seasons.

Figure 5-31 compares snow-course measurements of snowpack water equivalent and density with values simulated by Anderson's complete energy-balance model for the accumulation season and the warming and ripening phases of the melt season in 1972–1973. Figure 5-32 shows the same comparison for the output phase and includes the predictions of the temperature-index model. Computed daily water output is compared with values measured in a lysimeter in Figure 5-33.

Clearly, both models perform very well in simulating density and water equivalent for the

1972–1973 snow season; this is also the case for 1971–1972. In three of the other melt seasons examined, the energy-balance model gave somewhat better results, while the temperature-index model was slightly better during one season. Except for two of the highest output days, predictions of water output were also good.

Figure 5-34 compares lysimeter-measured snowmelt with melt estimated via a complete energy-balance approach, the temperature-index approach with seasonally increasing melt coefficient, and the hybrid approach with radiation terms estimated as in Equations (5-29)–(5-42). The accuracy of the hybrid approach was almost identical to that of the complete energy-balance approach, and both methods performed considerably better than the temperature-index approach.

These results lead to the following conclusions:

- Our understanding of snowpack processes at a point as reflected in Equations (5-26)–(5-48) is sound, and these equations can be used to simulate point snowmelt to a high degree of precision, given careful measurements

**FIGURE 5-31**

Comparison of observed and simulated (energy-balance model) snowpack water equivalent and density for the 1972–1973 accumulation season and warming and ripening phases of the melt season at Danville, VT. Observed values are snow-course observations. From Anderson (1976).

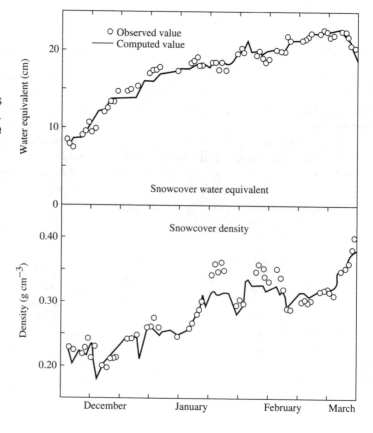

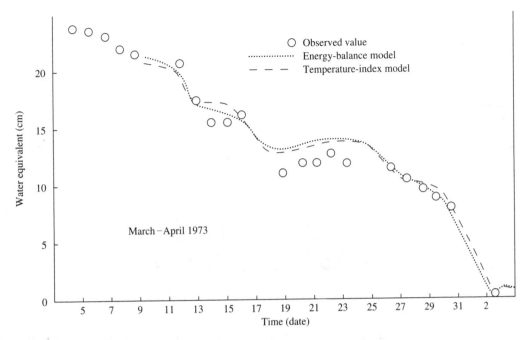

**FIGURE 5-32**
Comparison of observed and simulated (energy-balance and temperature-index models) snowpack water equivalent and density for the output phase of the 1973 melt season at Danville, VT. Observed values are snow-course observations. From Anderson (1976).

**FIGURE 5-33**
Observed vs. simulated (energy-balance model) daily water output for March and April at Danville, VT. From Anderson (1976).

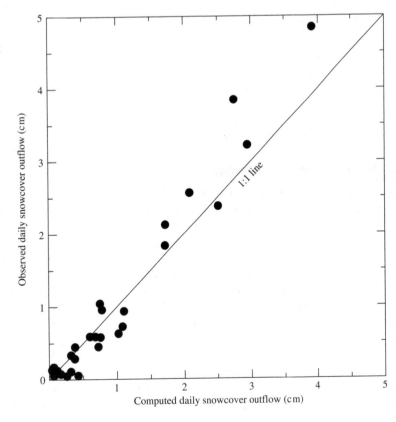

**FIGURE 5-34**
Daily snowmelt measured by lysimeter and simulated by the temperature-index approach, the hybrid approach, and the energy-balance approach at Weiss-fluhjoch, Switzerland, for the 1985 melt season. From Kustas et al. (1994); used with permission of the American Geophysical Union.

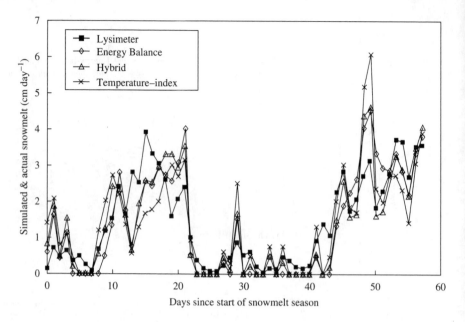

of the meteorological inputs at the point of interest.

• The temperature-index approach can provide useful estimates of daily snowmelt, but the value of the melt coefficient should reflect local conditions and seasonal changes.

• The hybrid approach appears to predict daily snowmelt with precision equivalent to that of the complete energy-balance approach and, since good estimates of the radiation components of the energy balance can usually be made with commonly available data via Equations (5-29)–(5-42), this approach will often be attractive for modeling.

### 5.5.2  Watershed Snowmelt Modeling

As discussed in Section 5.3.2, snowmelt is a significant contributor to runoff over much of the northern hemisphere, and changes in snowmelt runoff will be one of the most pronounced hydrologic responses to global warming. Many agencies in the countries of that region have developed models for forecasting the water supply and/or floods generated by snowmelt. The models are specifically designed for the regions of interest and the purposes of the forecasts, and hence differ in many respects. However, they all attempt to simulate the melt processes discussed in this chapter on the basis of meteorological data; to account for the typically wide range of topography, land use, and weather

over watersheds; and to integrate the processes of water transmission over hillslopes and in stream channels.

Because of the wide variability of conditions, the usual sparseness of meteorological data, and the need to provide operational forecasts at relatively short time intervals (typically from 6 to 24 hr), these models make extensive use of empirical and semi-empirical relations rather than rigorous physically based models of melt and water movement. In particular, most watershed-scale models use some form of temperature-index model or hybrid approach to predict melt.

Important considerations in modeling snowmelt over large areas include the following:

1. Successfully accounting for the form of precipitation. As noted in Section 4.1.4, weather stations typically measure and report the water equivalent of precipitation without specifying whether the precipitation is in the form of rain or snow. (See Figure 4-14.) Incorrect specification of the type of precipitation will cause major errors to be made in predicting runoff and the areal extent of snow cover. This problem is exacerbated in watersheds that have a wide range of elevation.

2. Successfully accounting for the typically large range of variations in topography, elevation, and land use. This is usually accom-

plished by making adjustments in the melt factor. [See the discussion of Equations (5-59) and (5-60).]

3. Successfully accounting for spatial and temporal variations in the areal extent of snow cover. As melt proceeds, the area with snow cover decreases; over- or under-estimation of water output will be proportional to over- or under-estimation of this area. Increasingly, satellite imagery is used to update information on areal snow cover in snowmelt models.

4. Successfully accounting for the movement of water output to the watershed outlet. Various empirical and semi-empirical techniques, some of which are discussed in Section 9.6, are used for this accounting.

Dunne and Leopold (1978) summarized the equations developed by the U.S. Army Corps of Engineers for forecasting watershed snowmelt, and Anderson (1973) and Morris (1986) also examined this subject in detail. Anderson (1976) concluded that in heavily forested watersheds, a temperature-index model that includes a way to account for decreases in areal snow cover during melt should give results similar to those of an energy-balance model. However, an energy-balance model should perform better than a temperature-index model when applied to a relatively open (unforested) watershed where there is considerable variability in meteorological conditions, and in watersheds with considerable physiographic and climatic variability. Anderson found that the minimal data requirements for use of an energy-balance model are an accurate and representative estimate of incoming solar radiation, plus measurements of air temperature, vapor pressure, and wind speed.

Figure 5-35 compares streamflow from a small (8.42-km$^2$) watershed at the Sleepers River, VT, research site with flow predicted with a model called the Snowmelt Runoff Model (SRM) (Brubaker et al. 1996) for six seasons. Two versions of the SRM are compared, one with melt predicted by the seasonally varying temperature-index approach [Equation (5-61)] and the other with the hybrid approach [Equation (5-62)]. The hybrid approach predicted daily runoff better in only two of the six years, but predicted total runoff volume significantly better in all six years.

The most comprehensive survey of operational snowmelt-runoff models was made by the World Meteorological Organization (1986), which compared 11 temperature-index models using common data sets for six different watersheds. Rango and Martinec (1994) further compared the predictions of seven of these models for a large Canadian watershed. Blöschl et al. (1991a; 1991b) compared the performance of spatially-distributed and lumped models (see Box 2-5) in predicting snowmelt runoff from a small Austrian watershed.

Other recent studies that develop promising approaches to snowmelt that would be especially useful in predicting the effects of global warming on snow and snowmelt runoff include (1) a model of the areal depletion of snow cover in a forested catchment (Buttle and McDonnell 1987), (2) a model that simulates the growth and disappearance of the seasonal snow cover from daily air-temperature and precipitation data (Motoyama 1990), and (3) a model that uses areally averaged versions of the basic energy-balance relations to predict areal snow cover and snowmelt runoff (Horne and Kavvas 1997).

Box 5-1 summarizes the treatment of snow and snowmelt in the BROOK90 model introduced in Section 2.9.5.

## 5.6  WATER-QUALITY ASPECTS[9]

As described in Section 4.5, precipitation quality is determined first by the material that forms the nucleus of each raindrop or snowflake—usually sea salt, terrestrial or meteoric dust, volcanic debris, smoke particles, or various pollutants. The solubility of gases in ice is much lower than in liquid water, so gases have little effect on the quality of snowfall. However, snowflakes have large surface-to-volume ratios, and their surfaces can form hydrogen bonds with compounds in the atmosphere and are thus efficient scavengers of natural and anthropogenic atmospheric constituents as they fall.

Snowpacks in remote mountainous regions typically have pH > 5.2 and low concentrations of dis-

[9]Material in this section is based largely on the review by Hornbeck (1986); other papers on the subject can be found in Morris (1986).

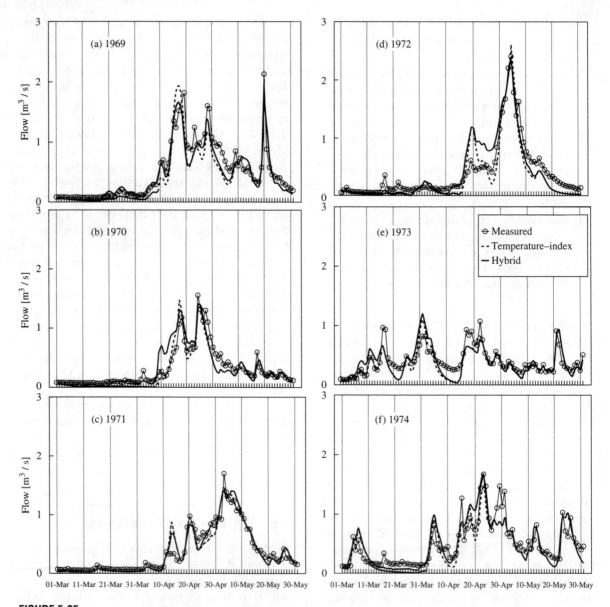

**FIGURE 5-35**
1969–1974 snowmelt runoff hydrographs at Sleepers River Research Watershed, VT. Open circles = measured; dashed line = temperature-index approach; solid line = hybrid approach. From Brubaker et al. (1996); used with permission.

solved ions, while those in areas affected by anthropogenic emissions have pH as low as 3.8 and ionic strengths 10 to 25 times greater. Regional variations in the relative concentration of various constituents in snow depend on storm patterns and distance from sources of contamination; on a local scale, composition can vary depending on whether the snow passed through a forest canopy or fell in the open.

After snow falls, its chemistry is further altered by natural and anthropogenic particulates that accumulate on the surface (a process called **dry deposition**) and are buried by successive snowfalls, and by rain. The percolation of rain and meltwater and their refreezing at depth further alter the distribution of chemicals in the snowpack.

At the microscopic level, the processes of freeze–thaw and vapor migration during metamor-

## BOX 5-1.

. . . . . . .

### Snow and Snowmelt in the BROOK90 Model

All snow quantities are expressed as water equivalents (mm), and we assume a daily time step, with the day counter indicated by $i$ where necessary. The snow water equivalent on the ground at the end of day $i$ is $SNOW(i)$, and the overall computations are

$$\underset{Snowpack(i)}{SNOW(i)} = \underset{=Snowpack(i-1)}{SNOW(i-1)} + \left(\underset{Snow\ throughfall}{STHR(i)} - \underset{-Groundmelt}{GRDMLT(i)}\right)$$

$$\underset{-Snow\ evaporation}{-SNVP(i)} - \underset{-Snowmelt}{SMLT(i)} + \underset{Rain\ on\ snow}{RSNO(i)}\right) \cdot 1\ day. \quad \text{(5B1-1)}$$

These quantities are determined as follows:

1. Daily precipitation ($PREC$) is input to the model. Snowfall ($SFAL$) is computed as

$$SFAL = SNOFRC \cdot PREC, \quad \text{(5B1-2)}$$

where the algorithm for determining the fraction of precipitation that is snow ($SNOFRC$) is described in Box 4-1.

2. Snow throughfall ($STHR$) is found as

$$STHR = SFAL - SINT. \quad \text{(5B1-3)}$$

The snowfall interception rate, $SINT$, is determined by two input parameters that describe the canopy density: the leaf-area index, $LAI$, (one-sided surface area of leaves divided by ground area) and the stem-area index, $SAI$, (projected cylindrical surface area of stems and branches divided by ground area). (See Section 7.5.2.) The relation is

$$SINT = (FSINTL \cdot LAI + FSINTS \cdot SAI)$$
$$\cdot SFAL, \quad \text{(5B1-4)}$$

with default values $FSINTL = 0.04$ and $FSINTS = 0.04$. Interception continues at this rate until the canopy-storage capacity, $INTSMX$, is reached where

$$INTSMX = CINTSL \cdot LAI$$
$$+ CINTSS \cdot SAI. \quad \text{(5B1-5)}$$

Default values are $CINTSL = 0.6$ and $CINTSS = 0.6$.

3. The groundmelt rate $GRDMLT$ is a parameter with a default value of 0.35 mm day$^{-1}$.

4. Snow evaporation/condensation ($SNVP$) is computed from the latent-heat exchange relation [Equations (5-46) and (D-42)], with roughness and zero-plane displacement heights determined by the vegetation height and density.

5. Snowmelt is computed by a temperature-index approach that accounts for slope, aspect, and vegetative density. If air temperature $TA > 0\ °C$, the energy available for melting, $SNOEN$ (MJ m$^2$ day$^{-1}$), is

$$SNOEN = MELFAC \cdot 2 \cdot DAYLEN$$
$$\cdot SLFDAY \cdot TA \cdot \exp(LAIMLT \cdot LAI$$
$$+ SAIMLT \cdot SAI), \quad \text{(5B1-6)}$$

where $MELFAC$ is a melt factor for a 12-hr day on a horizontal surface with no plant cover, with a default value of 1.5 MJ m$^{-2}$ day$^{-1}$ K$^{-1}$. $DAYLEN$ is the fraction of the day the sun is above the horizon, which is calculated from relations given in Section E.2.2, and $SLFDAY$ is the ratio of extraterrestrial solar radiation on the sloping watershed to that on a horizontal plane (the "slope factor," $f_{sh}$ discussed in Section E.2.4). The last two terms are empirical adjustments for vegetation density, where $SAIMLT = -0.5$ and $LAIMLT = -0.2$ are the default values. Partitioning of $SNOEN$ is discussed in Step 8.

6. If some of the precipitation is rain, the rain throughfall $RTHR$ is added to the snowpack. If $TA < 0\ °C$, some of the rain is refrozen and warms the snowpack. Any remaining liquid water is used to increase the liquid-water content $SNOWLQ$ up to a maximum value $MAXLQ$ ($= \theta_{ret}$), which has a default value of 0.05; this quantity $= RSNO$. The remaining water infiltrates into the soil.

7. After $STHR$ is added to the snowpack, if $TA < 0\ °C$, the cold content, $CC$, of the new snowpack is computed as

$$CC(i) = CC(i-1)$$
$$- TA \cdot CVICE \cdot STHR \cdot 1\ day, \quad \text{(5B1-7)}$$

where $CVICE$ is the volumetric heat capacity of ice.

8. After $GRDMLT$ and $SNVP$ are removed, the energy available to produce a temperature change or melt, $EQEN$, is computed as

$$EQEN = (SNOEN + RTHR$$
$$\cdot RMAX(TA,\ 0) \cdot CVLQ)/LF,$$
$$\text{(5B1-8)}$$

where $RMAX(TA,0)$ is the rain temperature (the maximum of $TA$ and 0 °C) and $CVLQ$ is the heat capacity and $LF$ the latent heat of freezing of water. If $EQEN < 0$, the energy is used to refreeze any liquid water and then increase $CC$. If $EQEN > 0$, the energy is used first to reduce $CC$ and then to produce melt, $SMLT$.

phosis result in the expulsion of ions from the interiors of snow grains and a marked accumulation at grain boundaries. Because of this segregation, the ionic concentration of the first pulse of water output is usually substantially higher than the average concentration in the snowpack: Typically the first 20 to 30% of water output removes from 50 to 80% of the dissolved constituents originally in the snowpack. The degree of differential leaching is greatest in locations and years with a single concentrated melt, and least when there are several periods of melting; it also varies with ionic species for a given location and time.

Differential leaching is clearly shown for hydrogen ions ($H^+$) in Figure 5-36. The "acid shock" that is produced by the high $H^+$ concentrations in the early meltwater contributions to streams in many locations can reduce the reproductive success of fish and amphibians and has caused massive fish kills. Sulfate ($SO_4^{-2}$), chloride ($Cl^-$), nitrate ($NO_3^-$), trace metals, and organic compounds are also differentially leached and can contribute to severe impacts on aquatic biota during the early stages of melt.

## EXERCISES

Exercises marked with ** have been programmed in EXCEL on the CD that accompanies this text. Exercises marked with * can be advantageously executed on a spreadsheet, but you will have to construct your own worksheets to do so.

**FIGURE 5-36**
Hydrogen-ion concentration vs. time during snowmelt in Quebec, Canada, showing the differential leaching that can produce acid shock to aquatic ecosystems. After Stein et al. (1986).

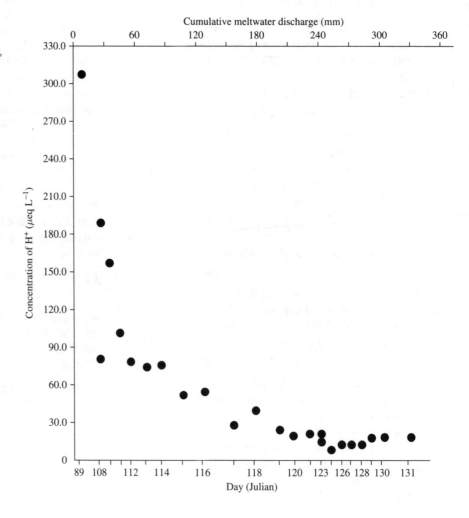

**\*5-1.**  Snow surveyors using a snow tube and thermometer recorded the following data from a snow course on two different dates (the temperature is taken at the midpoint of the snow depth to represent the average snowpack temperature):

| Date | | Point | | | | |
|------|------|---|---|---|---|---|
| | | **1** | **2** | **3** | **4** | **5** |
| 2 March | Depth (cm) | 92 | 94 | 105 | 93 | 96 |
| | Water Equiv. (cm) | 29 | 30 | 33 | 29 | 32 |
| | Temperature (°C) | −6 | −5 | −6 | −6 | −6 |
| 17 March | Depth (cm) | 88 | 89 | 102 | 88 | 91 |
| | Water Equiv. (cm) | 35 | 36 | 40 | 35 | 37 |
| | Temperature (°C) | −2 | −2 | −3 | −2 | −3 |

(a)  Compute the snow density and cold content for both dates. (b) How much energy needs to be added to the snowpack before water output begins?

**\*\*5-2.**  The spreadsheet program SolarRad.xls (see Appendix E) on the CD accompanying this text computes the daily clear-sky incident solar radiation, $K_{cs}$, as a function of latitude, date, and slope steepness and orientation. Select fixed values for three of these factors and a range of values for the fourth, and use the program to compute how $K_{cs}$ varies as a function of the fourth factor. Summarize your results with appropriate tables and/or figures and a brief discussion.

**\*\*5-3.**  [Note: Before beginning this problem, please make a minor but important correction to the SNOWMELT.XLS spreadsheet. Two variables, indicated in bold, in the formula in cell D44 are transposed. Cell D44 should read: "=IF(D11>0,((1-**D6**)*(19.6*D8/1000-2.39)*D11+**D6***(10.4*D8/1000-0.7)*D11),0)"]  (a) Use the spreadsheet program SnowMelt.xls on the CD accompanying this text to compute the daily snowmelt and water output at two adjacent locations, one in a balsam-fir forest with 80% canopy closure and the other in an open

field, on two days with the contrasting weather conditions specified in the accompanying table. Both locations are nearly horizontal. The snowpack is ripe, $\rho_s = 400$ kg m$^{-3}$, and it has been 8 days since the last snowfall. All meteorological observations were made in the open area at a height of 2 m above the snow surface. Assume a roughness height of 0.002 m. Write a brief report comparing the relative importance of the various components under the differing conditions. (b) Write a paragraph comparing the melt and water output computed by the energy-balance approach, the temperature-index approach, and the hybrid approach.

| Quantity | Day 1 | Day 2 |
|----------|-------|-------|
| Clear-sky solar rad., $K_{cs}$ (MJ m$^{-2}$ day$^{-1}$) | 16.8 | 16.8 |
| Wind speed, $v_a$ (m s$^{-1}$) | 5.0 | 5.0 |
| Air temperature, $T_a$ (°C) | 4.0 | 4.0 |
| Relative humidity, $W_a$ | 0.50 | 1.00 |
| Atmospheric pressure, $P$ (kPa) | 100 | 100 |
| Rainfall, $r$ (mm day$^{-1}$) | 0 | 10 |
| Cloud cover, $C$ | 0 | 1.0 |

**\*5-4.**  Using the approach shown in Example 5-10, compute the time lag for the water output for each site and each day of Exercise 5-3.

**5-5.**  Using a snow tube, make a survey of snow depths, densities, and water equivalents under various conditions of slope, aspect, and forest cover. If a snow tube is not available, measure depths with a meter stick and collect samples to be melted to determine water equivalents. Write a brief report comparing your observations in the different environments.

**5-6.**  For one or more snowstorms, compare the catch of a rain gage with the increment of water equivalent as determined with a snow tube or by collecting and melting the snowfall. Compare the results with the information in Figure 4-16. Write a brief report discussing your findings.

# 6

# Water in Soils: Infiltration and Redistribution

This chapter deals with the movement of water in unsaturated porous earth materials, focusing on **infiltration,** which is the movement of water from the soil surface into the soil, and **redistribution,** which is the subsequent movement of infiltrated water in the unsaturated zone of a soil (Figure 6-1).

As shown in Figure 6-1, redistribution can involve **exfiltration** (evaporation from the upper layers of the soil; discussed in Section 7.4), **capillary rise** (movement from the saturated zone upward into the unsaturated zone due to surface tension; discussed in Section 8.5.4), **recharge** (the movement of percolating water from the unsaturated zone to the subjacent saturated zone; discussed in Section 8.5.1), **interflow** (flow that moves downslope; discussed in Section 9.2.3), and uptake by plant roots (transpiration; discussed in Section 7.5). **Percolation** is a general term for downward flow in the unsaturated zone.

L'vovich (1974) estimated that 76% of the world's land-area precipitation infiltrates, so infiltration is clearly of great importance quantitatively in the global hydrologic cycle. It is, of course, the process that provides all the water used by natural and cultivated plants and almost all the water that enters ground-water reservoirs.

Thus an understanding of the linked processes of infiltration and redistribution is an essential basis for many crucial aspects of water-resource management, including (1) developing strategies for crop irrigation; (2) understanding chemical processes in soils, including natural weathering and the movement of natural nutrients, fertilizers, and contaminants to ground and surface waters; and (3) estimating the timing and amounts of ground-water recharge.

In addition, much of the motivation for studying infiltration comes from the need to forecast or predict hydrologic response to rain or snowmelt events (Chapter 9). Water that does not infiltrate typically moves relatively quickly as **overland flow** toward a stream channel and contributes to short-term stream response, perhaps causing soil erosion and flooding. In contrast, infiltrated water is either retained in the soil and ultimately evapotranspired or moves, usually much more slowly, to the surface-water system via underground paths.

Infiltration and redistribution involve unsaturated flow in porous media, which can be understood and modeled on the basis of physical principles developed earlier—particularly the principle of conservation of mass (Section 2.3) and the applicable equation of motion. In order to apply these principles appropriately, we must begin by understanding the material and hydraulic properties of soils (which together are the focus of the field of **soil physics**) and the water conditions of natural soils.

**FIGURE 6-1**
Definitions of terms used to describe water movement in the unsaturated zone.

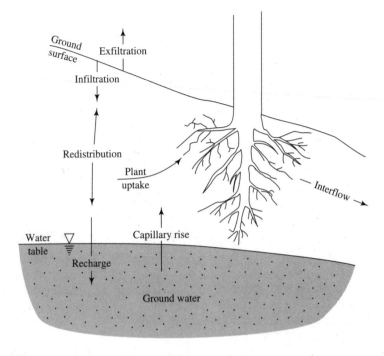

## 6.1 MATERIAL PROPERTIES OF SOIL

### 6.1.1 Distribution of Pore and Particle Sizes

#### *Definition*

Our model of a quasi-homogeneous soil consists of a matrix of individual solid grains (mineral or organic) between which are interconnected pore spaces that can contain varying proportions of water and air. To simplify discussion, we will not separately consider the organic component and will consider a soil as a three-phase system consisting of solid mineral grains, water, and air.[1]

The size of the pores through which water flow occurs is approximately equal to the grain size,[2] and the distribution of pore sizes is determined largely by the grain-size distribution. Most soils are composed of a mixture of grain sizes, and the grain-size distribution is often portrayed as a cumulative-frequency plot of grain diameter (logarithmic scale) vs. weight fraction of grains with smaller diameter (Figure 6-2). The steeper the slope of such plots, the more uniform the soil grain-size distribution.

For many purposes the particle-size distribution is characterized by the soil **texture,** which is determined by the proportions by weight of clay, silt, and sand. Figure 6-2 shows the range of grain sizes in those classes, and Figure 6-3 is the scheme developed by the U.S. Department of Agriculture for defining textures. Note that the texture is determined by the proportions of sand, silt, and clay *after* particles larger than sand (i.e., > 2 mm) are removed. If a significant proportion of the soil (> 15%) is gravel or larger, an adjective such as "gravelly" or "stony" is added to the soil-texture term.

#### *Measurement*

Weight fractions of soils of various diameters are measured by sieve analysis for particles larger than 0.05 mm and by sedimentation for the smaller grain sizes. (See Hillel 1980a.)

---

### *EXAMPLE 6-1*

A soil formed on glacial till in southwestern New Hampshire has the grain-size distribution shown in the accompanying table. What is the soil texture?

---

[1] The volume of water vapor contained in the air in soil pores is insignificant and can be neglected.
[2] For uniform spheres, it can be shown that the volume of an individual pore space ranges from 0.35 (close packed) to 0.91 (open packed) times the volume of the individual spheres.

| Diameter (mm) | 50 | 19 | 9.5 | 4.76 | 2.00 | 0.420 | 0.074 | 0.02 | 0.005 | 0.002 |
|---|---|---|---|---|---|---|---|---|---|---|
| % Finer | 100 | 95 | 90 | 84 | 75 | 64 | 42 | 20 | 7 | 2 |

**Solution** The grain-size distribution for this soil is plotted in Figure 6-2. Using the classification in Figure 6-2, we have the following proportions:

| | Total Soil | < 2 mm Fraction |
|---|---|---|
| % > 2 mm | 25 | — |
| % Sand | 40 | 53 |
| % Silt | 33 | 44 |
| % Clay | 2 | 3 |

This soil plots where the dot is shown in Figure 6-3—a sandy loam. However, since more than 15% of the particles are in the gravel range, the soil would be called a gravelly sandy loam.

### 6.1.2  Particle Density

#### Definition

**Particle density,** $\rho_m$, is the weighted average density of the mineral grains making up a soil:

$$\rho_m \equiv \frac{M_m}{V_m}, \qquad (6\text{-}1)$$

where $M_m$ is the mass and $V_m$ the volume of the mineral grains.

#### Measurement

The value of $\rho_m$ for a given soil is not usually measured, but is estimated based on the mineral composition of the soil. A value of 2650 kg m$^{-3}$, which is the density of the mineral quartz, is assumed for most soils.

### 6.1.3  Bulk Density

#### Definition

**Bulk density,** $\rho_b$, is the dry density of the soil:

$$\rho_b \equiv \frac{M_m}{V_s} = \frac{M_m}{V_a + V_w + V_m}, \qquad (6\text{-}2)$$

where $V_s$ is the total volume of the soil sample and $V_a$, $V_w$, and $V_m$ are the volumes of the air, liquid water, and mineral components of the soil, respectively. In most hydrologic problems, bulk density at any point is constant in time; however, it commonly increases with depth due to compaction by the weight of overlying soil.

#### Measurement

In practice, bulk density is defined as the weight of a volume of soil that has been dried for an extended period (16 hr or longer) at 105 °C, divided by the original volume.

### 6.1.4  Porosity

#### Definition

**Porosity,** $\phi$, is the proportion of pore spaces in a volume of soil:

$$\phi \equiv \frac{V_a + V_w}{V_s}. \qquad (6\text{-}3)$$

Like bulk density, porosity is constant over the time periods considered in most hydrologic analyses. However, in many soils, it decreases with depth due to compaction and to the development of **macropores** by biologic activity near the surface.

#### Measurement

It can be shown that

$$\phi = 1 - \frac{\rho_b}{\rho_m}, \qquad (6\text{-}4)$$

and $\phi$ is usually determined by measuring $\rho_b$ and assuming an appropriate value for $\rho_m$ (usually 2650 kg m$^{-3}$).

The range of porosities in soils is shown in Figure 6-4; in general, finer grained soils have higher porosities than coarser grained soils. This is due in part to the very open "house-of-cards" arrangement of clays, which is maintained by electrostatic forces between the roughly disk-shaped grains; in

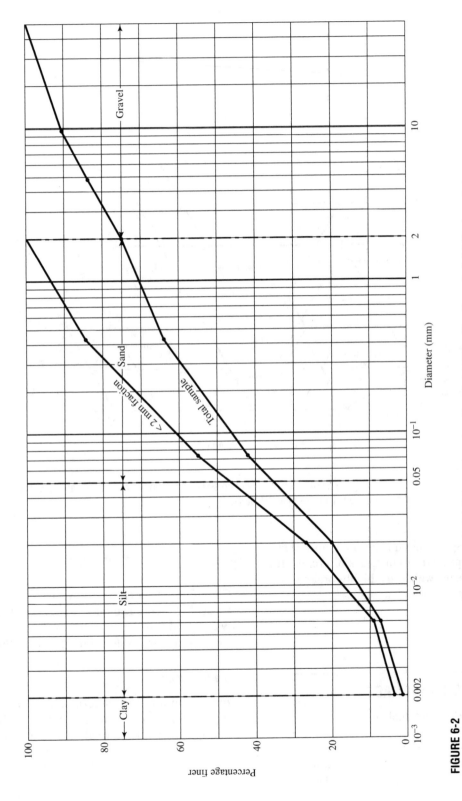

**FIGURE 6-2**

Grain-size distribution curve for a soil formed on glacial till in southwestern New Hampshire. (See Example 6-1.) The boundaries between size classes designated clay, silt, sand, and gravel are shown as vertical lines.

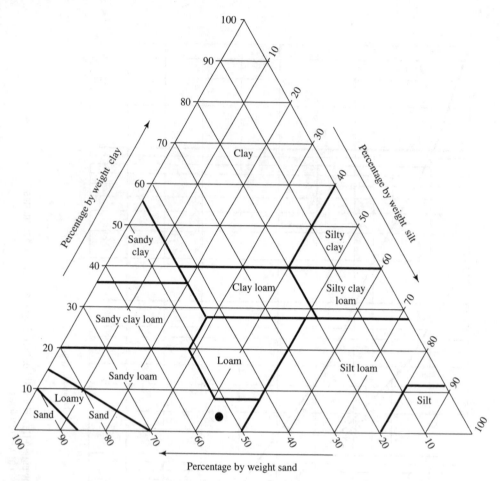

**FIGURE 6-3**
Soil-texture triangle, showing the textural terms applied to soils with various fractions of sand, silt, and clay. The dot shows where the soil in Example 6-1 plots.

contrast, quasi-spherical sand and silt grains are in a more closely packed grain-to-grain architecture (Figure 6-5). Peats, which are highly organic soils, may have porosities as high as 0.80.

## 6.2   SOIL-WATER STORAGE

### 6.2.1   Volumetric Water Content

#### Definition

**Volumetric water content,** or simply **water content,** $\theta$, is the ratio of water volume to soil volume:

$$\theta \equiv \frac{V_w}{V_s}. \tag{6-5}$$

Clearly, water content can vary in both time and space. The theoretical range of $\theta$ is from 0 (completely dry) to $\phi$ (saturation) but, as will be seen later, the range for natural soils is much less than this.

#### Measurement

In the laboratory, $\theta$ is determined by first weighing a representative soil sample of known volume, oven drying it at 105 °C, reweighing it, and calculating

$$\theta = \frac{M_{swet} - M_{sdry}}{\rho_w \cdot V_s}. \tag{6-6}$$

**FIGURE 6-4**
Ranges of porosities, field capacities, and permanent wilting points for soils of various textures. From *Water in Environmental Planning*, by Thomas Dunne and Luna B. Leopold. Copyright © 1978 by W. H. Freeman and Company. Reprinted with permission.

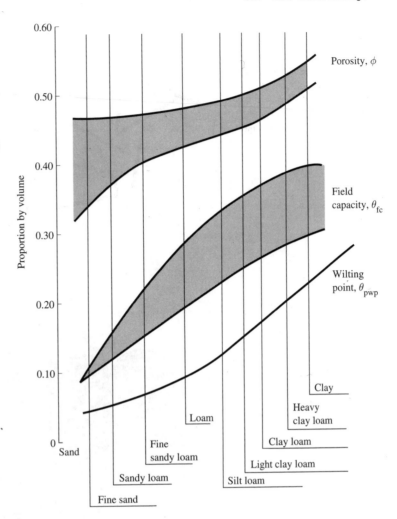

Here, $M_{swet}$ and $M_{sdry}$ are the weights before and after drying, respectively, and $\rho_w$ is the density of water.

The following methods can be used to measure water content in the field (more detailed descriptions are given in the cited references):

**Electrical resistance blocks** use the inverse relation between water content and the electrical resistance of a volume of porous material (gypsum, nylon, or fiberglass) in equilibrium with the soil (Hillel 1980a; Williams 1980).

**Neutron moisture meters** are combined sources and detectors of neutrons that are inserted into access tubes to measure the scattering of neutrons by hydrogen atoms, which is a function of water content (Hillel 1980a).

**Gamma-ray scanners** measure the absorption of gamma rays by water molecules in soil between a source and a detector at fixed locations (Hillel 1980a).

**Capacitance** (Dean et al. 1987; Bell et al. 1987; Eller and Donath 1996) and **time-domain reflectometry** (Topp et al. 1980; Zegelin et al. 1989; Roth et al. 1990; Jacobsen and Schjønning 1993; Yu et al. 1997) techniques measure the dielectric constant of a volume of soil, which increases strongly with water content.

**Nuclear magnetic resonance** techniques measure the response of hydrogen nuclei to magnetic fields. Paetzold et al. (1985) developed and tested a tractor-mounted system to measure water contents along transects.

**Microwave remote sensing** can provide information about surface soil-water content over large areas. These systems can be airplane- or satellite-borne and active or passive and can ob-

**FIGURE 6-5**
Schematic diagrams of structures
of aggregates of (a) clay (shown
as tabular particles viewed edge
on) and (b) sand (shown as
spherical particles). The architecture of clay particles is maintained by inter-grain electrostatic
forces. After Hillel (1980).

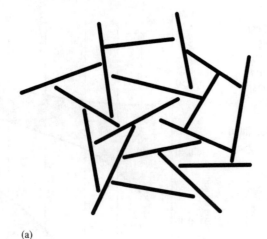

(a)

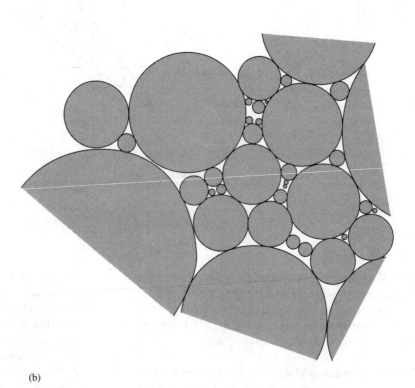

(b)

serve through clouds (Engman and Chauhan
1995; Mattikali et al. 1997). Active systems
(radar) have high resolution (400–900 m²), but
interpretation of the signal is made difficult by
surface roughness and vegetation. Passive sensors are less affected by surface conditions and
have resolution on the order of 100–1000 km²,
which can be useful for hydrologic analysis of
large areas. The depth for which water content
is determined is about 0.1 times the wave-

length and decreases with water content; it typically ranges from 0.02 to 0.1 m (Engman and
Chauhan 1995). For satellite-borne sensors, the
times of observation are dictated by the orbital
characteristics.

### 6.2.2 Degree of Saturation

The **degree of saturation,** or **wetness,** $S$, is the proportion of pores that contain water:

$$S \equiv \frac{V_w}{V_a + V_w} = \frac{\theta}{\phi}. \qquad \textbf{(6-7)}$$

Wetness is not directly measured, but is calculated using Equation (6-7). Either $S$ or $\theta$ can be used to express the amount of water in a soil.

---

## EXAMPLE 6-2

A 10-cm-long sample of the soil described in Example 6-1 is taken with a cylindrical sampling tube that has a 5-cm diameter. On removal from the tube, the sample weighs 331.8 g. After oven drying at 105 °C, the sample weighs 302.4 g. Compute the bulk density, porosity, water content, and wetness.

**Solution:** The sample volume, $V_s$, is

$$V_s = 10 \text{ cm} \times (2.5 \text{ cm})^2 \times 3.1416 = 196 \text{ cm}^3$$
$$= 1.96 \times 10^{-4} \text{ m}^3.$$

The bulk density is found from Equation (6-2) as

$$\rho_b = \frac{302.4 \text{ g}}{196.3 \text{ cm}^3} = 1.54 \text{ g cm}^{-3} = 1540 \text{ kg m}^{-3}.$$

The porosity can be calculated from Equation (6-4), assuming that $\rho_m = 2650$ kg m$^{-3}$:

$$\phi = 1 - \frac{1540 \text{ kg m}^{-3}}{2650 \text{ kg m}^{-3}} = 0.419.$$

The water content is found via Equation (6-6):

$$\theta = \frac{(0.3318 \text{ kg} - 0.3024 \text{ kg})/1000 \text{ kg m}^{-3}}{1.96 \times 10^{-4} \text{ m}^3}$$
$$= 0.150.$$

The wetness is then found using Equation (6-7):

$$S = \frac{0.150}{0.419} = 0.358.$$

---

### 6.2.3 Total Soil-Water Storage

The total amount of water stored in any layer of soil is usually expressed as a depth [L] (volume per unit area), which is the product of the volumetric water content times the thickness of the layer.

## 6.3 SOIL-WATER FLOW

### 6.3.1 Darcy's Law

Infiltration and redistribution are flows in unsaturated porous media (soils) that are described by Darcy's Law:

$$q_x = -K_h \cdot \frac{d(z + p/\gamma_w)}{dx}$$
$$= -K_h \cdot \left[ \frac{dz}{dx} + \frac{d(p/\gamma_w)}{dx} \right]. \qquad \textbf{(6-8a)}$$

Here, $q_x$ is the volumetric flow rate in the $x$-direction per unit cross-sectional area of medium [L T$^{-1}$], $z$ is the elevation above an arbitrary datum [L], $p$ is the water pressure [F L$^{-2}$], $\gamma_w$ is the weight density of water [F L$^{-3}$], and $K_h$ is the hydraulic conductivity of the medium [L T$^{-1}$].

Darcy's Law describes the flow at a "point"—actually, at a **representative elemental volume** of the soil that includes pore spaces and soil particles. Flow occurs in response to spatial gradients of mechanical potential energy, which has two components: gravitational potential energy and pressure potential energy.[3] In Equation (6-8a), $dz/dx$ is the gradient of gravitational potential energy per unit weight of flowing water, and $d(p/\gamma_w)/dx$ is the gradient of pressure potential energy per unit weight of flowing water.

In this chapter we consider only flows in the vertical ($z$) direction, so

$$q_z = -K_h \cdot \frac{d(z + p/\gamma_w)}{dz} = -K_h \cdot \left[ \frac{dz}{dz} + \frac{d(p/\gamma_w)}{dz} \right]$$
$$= -K_h \cdot \left[ 1 + \frac{d(p/\gamma_w)}{dz} \right]; \qquad \textbf{(6-8b)}$$

that is, the magnitude of the gravitational potential-energy gradient will always equal unity (+1 if directed upward, −1 if downward). The sign of $z$ depends on whether the "point" is above ($z > 0$) or below ($z < 0$) the arbitrary datum. As described in Section 6.3.2, $p \leq 0$ for the unsaturated flows considered in this chapter.

---

[3] Electrical and chemical (osmotic) forces can also induce water flow in soils, but these forces are negligible in most hydrological situations.

Since $\gamma_w$ is effectively constant for hydrologic problems that do not involve temperature or salinity gradients, it is convenient to use the **pressure head**, $\psi$, defined as

$$\psi \equiv \frac{p}{\gamma_w}. \qquad (6\text{-}9)$$

Note that $\psi$ has the dimension [L]; it is usually expressed in cm of water.[4] Since $p \le 0$, $\psi \le 0$.

In unsaturated flows, both the pressure head and the hydraulic conductivity for a given soil are functions of the soil-water content, $\theta$, so we henceforth write Darcy's Law for vertical unsaturated flow as

$$q_z = -K_h(\theta) \cdot \frac{d[z + \psi(\theta)]}{dz}$$

$$= -K_h(\theta) \cdot \left[ 1 + \frac{d\psi(\theta)}{dz} \right]. \qquad (6\text{-}10)$$

The relations between pressure and water content $[\psi(\theta)]$ and between hydraulic conductivity and water content $[K_h(\theta)]$ are crucial determinants of unsaturated flow in soils, and these relations are examined in more detail in the following sections.

### 6.3.2  Soil-Water Pressure

#### Definition

Pressure is force per unit area, $[F\ L^{-2}]$; it is a scalar quantity that acts in all directions in a fluid. It is conventional to measure pressures relative to atmospheric pressure; thus $p > 0$ and $\psi > 0$ in saturated flows, and $p < 0$ and $\psi < 0$ in unsaturated flows. The **water table** is the surface at which $p = 0$. Negative pressure is often called **tension** or **suction,** and $\psi$ is called **tension head, matric potential,** or **matric suction** when $p < 0$.

In unsaturated soils, water is held to the mineral grains by surface-tension forces; the water can be thought of as hanging suspended from menisci and as being under tension (as in the capillary tube in Figure B-9). Since infiltration can take place only into unsaturated soils, $p$ and $\psi$ will always be negative numbers in this chapter. Tension increases (i.e., pressure gets more negative) as the radii of curvature of the menisci decrease [Equation (B-9)]. Thus,

for a given soil, tension increases as the water content decreases.

### *Measurement*

The tension of soil moisture under field conditions can be directly measured by **tensiometers** (Figure 6-6). These devices consist of a hollow metal tube, of which one end is closed off by a cup of porous ceramic material and the other end is fitted with a removable airtight seal. A manometer, vacuum gage, or pressure transducer is attached to the tube. The tube is completely filled with water and inserted into the soil to the depth of measurement. Since the water in the tube is initially at a pressure somewhat above atmospheric pressure, there is a pressure-induced flow through the porous cup into the soil that continues until the tension inside the tube equals that in the soil. When equilibrium is reached, the manometer or gage gives the tension in the tube and in a roughly spherical region immediately surrounding the cup. The soil tension, $|\psi|$, is then computed as

$$|\psi| = |\psi_{gage}| - z_{ten}, \qquad (6\text{-}11)$$

where $|\psi_{gage}|$ is the reading of the tensiometer gage and $z_{ten}$ is the distance between the tensiometer bulb and the height at which the gage measures (Figure 6-6).

The practical range of tension measurable in a tensiometer is from 0 to 800 cm (0 to 78 kPa), which covers only a small part of the tension range observed in nature (Hillel 1980a). However, this practical range covers a large part of the water-content range for coarser-grained soils, and tensiometers are often very useful tools for hydrologic field studies. Most commonly they are installed in clusters of two or three, extending to different depths to provide information on vertical tension gradients. If transducers or mercury manometers are used as sensors, tensions can be recorded continuously on a chart recorder or data logger (Walkotten 1972; Williams 1978; Cooper 1980).

Other devices can be used for field determination of soil-water tension outside the range for which tensiometers are useful. Peck and Rabbidge (1966) described a method that uses osmotic principles, and Hillel (1980a) described a **thermocouple psychrometer** that can make precise determinations of soil-water tension by measuring the humidity of soil air. Gypsum electrical-resistance blocks

---

[4] Soil scientists sometimes express tension head in pF units, where $pF = \log_{10}(-\psi)$ and $\psi$ is in cm of water. [Note the analogy with pH, defined in Equation (B-1).]

**FIGURE 6-6**
Schematic diagram of two types of tensiometers, one using a vacuum gage and one using a mercury manometer to measure soil-water tension in a volume of soil surrounding the porous ceramic cup. $z_{ten}$ is the gravity-head adjustment [Equation (6-11)]. After Hillel (1980).

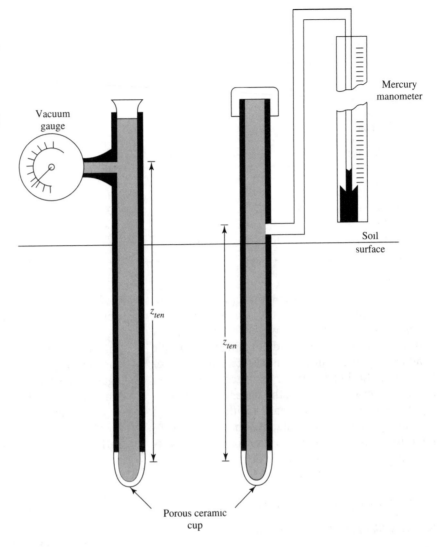

can also be used to measure tensions stronger than about 1020 cm (100 kPa) (Rawls et al. 1992).

Example 6-3 indicates how Darcy's Law [Equation (6-10)] operates in practice in near-surface unsaturated flows.

---

## EXAMPLE 6-3

Consider two adjacent tensiometers inserted into unsaturated soil to determine whether water flow is toward or away from the surface. Both tensiometers have a distance $z_{ten} = 61$ cm between the gage and the bulb. (See Figure 6-6.) The bulb of tensiometer A is installed at a depth of 20 cm; the bulb of tensiometer B is installed at a depth of 50 cm. The readings in the accompanying table (Conditions 1–5) are obtained at different times. Assuming that no maximum or minimum of tension occurs between the depths of the two sensors, which way is the water flowing in each condition?

| Condition | 1 | 2 | 3 | 4 | 5 |
|---|---|---|---|---|---|
| $\mid \psi_{gage} \mid$ at A (cm) | 93 | 76 | 151 | 217 | 71 |
| $\mid \psi_{gage} \mid$ at B (cm) | 83 | 71 | 117 | 173 | 71 |

**Solution:** The first step is to compute the value of $\mid \psi \mid$ at the bulb level by subtracting $z_{ten} = 61$ cm from the gage readings [Equation (6-11)]:

| Condition | 1 | 2 | 3 | 4 | 5 |
|---|---|---|---|---|---|
| $\mid \psi \mid$ at A (cm) | 32 | 15 | 90 | 156 | 10 |
| $\mid \psi \mid$ at B (cm) | 22 | 10 | 56 | 112 | 10 |

As indicated by Darcy's Law [Equation (6-10)], the direction of flow is given by the direction of the gradient of $(z + \psi)$ in each situation. The datum for measurement can be arbitrarily fixed at any elevation, and the most convenient choice is at the level of the lower tensiometer. With this choice, $z_A = 30$ cm and $z_B = 0$ cm. Since the soil is unsaturated, the value of $\psi$ at each depth is the negative of the value indicated in the previous table. Thus,

| Condition | 1 | | 2 | | 3 | | 4 | | 5 | |
|---|---|---|---|---|---|---|---|---|---|---|
| Tensiometer | A | B | A | B | A | B | A | B | A | B |
| $z$ (cm) | 30 | 0 | 30 | 0 | 30 | 0 | 30 | 0 | 30 | 0 |
| $\psi$ (cm) | −32 | −22 | −15 | −10 | −90 | −56 | −156 | −112 | −10 | −10 |
| $z + \psi$ (cm) | −2 | −22 | 15 | −10 | −60 | −56 | −126 | −112 | 20 | −10 |
| Gradient | A > B | | A > B | | A < B | | A < B | | A > B | |
| Direction | down | | down | | up | | up | | down | |

Note that we did not compute the actual flow rate in Example 6-3. Since there is a gradient of water content between the levels of tensiometers A and B, the product of the gradient and the hydraulic conductivity, and hence the flow rate, changes over the flow path. Furthermore, as water moves from one level toward the other, the water contents, gradients, and conductivities change with time, so that the flow is inherently unsteady. The only way to quantify such flows is by solution of the **Richards Equation,** which is discussed in Section 6.6.1.

### 6.3.3   Pressure–Water-Content Relations

#### Moisture-Characteristic Curve

The relation between pressure head, $\psi$, (often plotted on a logarithmic scale) and water content, $\theta$, for a given soil is called the **moisture-characteristic curve.** The relationship is highly nonlinear and typically has the form shown in Figure 6-7, curve a. Note that the pressure head is zero (i.e., the pressure is atmospheric) when the water content equals the porosity, and that the water content changes little as tension increases up to a point of inflection. This more-or-less distinct point represents the tension at which significant volumes of air begin to appear in the soil pores and is called the **air-entry tension,** $\psi_{ae}$.[5] The absolute value of the air-entry ten-sion head equals the height of the tension-saturated zone, or capillary fringe (discussed in Section 6.4.2).

As tension increases beyond its air-entry value, the water content begins to decrease rapidly and then more gradually. At very high tensions the curve again becomes nearly vertical, reflecting a residual water content that is very tightly held in the soil pores by capillary and electrochemical forces.

Figure 6-8 compares typical moisture-characteristic curves (shown as $\psi$ vs. $S$ rather than $\psi$ vs. $\theta$) for soils of three different textures and shows that, at a given degree of saturation, tension is much higher in finer-grained than in coarser-grained soils.

In reality the value of tension at a given water content is not unique, but depends on the soil's history of wetting and drying (Figure 6-9). While this **hysteresis** can have a significant influence on soil-moisture movement (Rubin 1967; Perrens and Watson 1977), it is difficult to model mathematically and is therefore not commonly incorporated in hydrologic models.

#### Measurement

The moisture-characteristic curve for a soil stratum can be defined, at least for the wetter portions of its range, by simultaneous field measurement of water content and tension by the methods described previously. The hanging-water-column method described by Stephens (1995) can be used in the laboratory to define curves for tensions less than about 30 kPa (310 cm). Complete curves are developed in the laboratory by successive weighing of an

---

[5] $\psi_{ae}$ is also called the **bubbling pressure.**

**FIGURE 6-7**

Typical forms of hydraulic relations $\psi(\theta) - \theta$ and $K_h(\theta) - \theta$ for unsaturated soils. For this soil, $\phi = 0.5$.

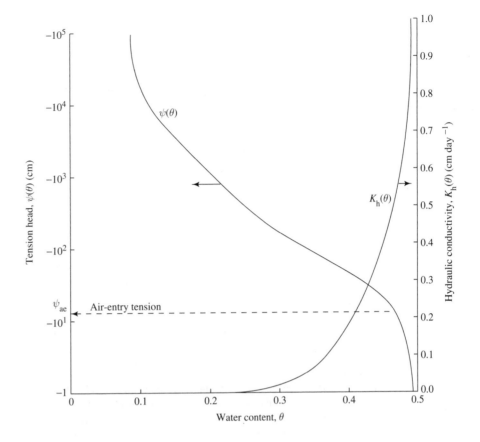

initially saturated sample after its water content has equilibrated under successively higher vacuums in a pressure-plate apparatus.

### 6.3.4 Hydraulic Conductivity

#### *Definition*

From Equation (6-10), it is clear that the hydraulic conductivity, $K_h$ [L T$^{-1}$], is the rate (volume per unit time per unit area) at which water moves through a porous medium under a unit potential-energy gradient. This rate is determined largely by the size (cross-sectional area) of the pathways available for water transmission. Under saturated conditions, this size is determined by the soil-grain size (see Figure 8-5); for unsaturated flows, it is determined by grain size and the degree of saturation.

#### *Measurement*

Hillel (1980a) and Stephens (1995) described various approaches to field and laboratory measure-

ment of unsaturated hydraulic conductivity. One such approach involves measurement of water content at several depths as the soil drains for extended periods following irrigation, with evaporation prevented. The flux of water and the vertical water-content gradients can be determined from these measurements, and the hydraulic conductivity can be calculated by substituting the measured values into Darcy's Law (Khosla 1980). The infiltrometer measurements described in Section 6.5.2 can also be used to estimate saturated hydraulic conductivity (Scotter et al. 1982; Elrick et al. 1990).

Mualem (1976) derived a method for calculating $K_h$ based on the moisture-characteristic curve and the value of $K_h^*$, which is relatively easy to measure using various laboratory and field techniques. [See, for example, Freeze and Cherry (1979).]

### 6.3.5 Hydraulic-Conductivity–Water-Content Relations

For a given soil, unsaturated hydraulic conductivity is very low at low to moderate water contents; it in-

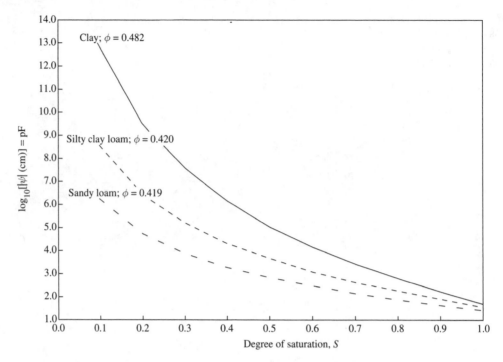

**FIGURE 6-8**

Soil-water pressure (tension), $\psi$, vs. degree of saturation, $S$, for soils of three different textures. Note that the vertical axis gives the base-10 logarithm of the absolute value of the pressure (which is negative), expressed in cm of water (pF). Curves are based on typical values given by Clapp and Hornberger (1978). The sandy-loam curve is for the soil discussed in Examples 6-1–6-3.

creases nonlinearly to its saturated value, $K_h^*$, as the water content increases to saturation (Figure 6-7). A comparison of $K_h - S$ relations for soils of different textures is shown in Figure 6-10; the form of these curves differs from that in Figure 6-7, because here the $K_h$ scale is logarithmic. Note that, for a given $S$, $K_h$ increases by several orders of magnitude in going from clay to silty clay loam to sand; also note that, for a given soil, $K_h$ increases by several orders of magnitude over the range of $S$ values. The hysteresis effect in the $K_h - \theta$ relation is less marked than in the $\psi - \theta$ relation and is usually neglected.

### 6.3.6 Analytic Approximations of $\psi - \theta$ and $K_h - \theta$ Relations

Because measurement of $\psi - \theta$ and $K_h - \theta$ relations is difficult, and because of the need to incorporate these relations into computer models of water movement, it is useful to express them in

equation form. Brooks and Corey (1964), Campbell (1974), and Van Genuchten (1980) have proposed various versions of these relations; we use the power-law equations of Campbell (1974):

$$|\psi(S)| = \frac{|\psi_{ae}|}{S^b} \qquad \textbf{(6-12a)}$$

Or, using Equation (6-7),

$$|\psi(\theta)| = |\psi_{ae}| \cdot \left(\frac{\phi}{\theta}\right)^b; \qquad \textbf{(6-12b)}$$

and

$$K_h(S) = K_h^* \cdot S^c \qquad \textbf{(6-13a)}$$

or

$$K_h(\theta) = K_h^* \cdot \left(\frac{\theta}{\phi}\right)^c. \qquad \textbf{(6-13b)}$$

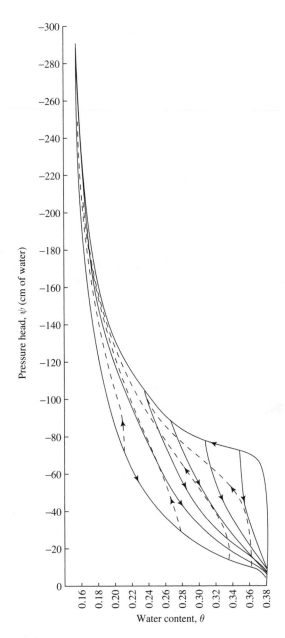

**FIGURE 6-9**
Hysteresis in the $\psi(\theta) - \theta$ relation for Rubicon sandy loam. The paths with arrows trace the relation as the soil undergoes successive cycles of wetting (arrows pointing to the right) and drying (arrows pointing to the left). From Perrens and Watson (1977); used with permission of the American Geophysical Union.

Clearly, Equations (6-12) and (6-13) ignore hysteresis. Furthermore, Equation (6-12) states that when $S = 1$ (saturation), $\psi = \psi_{ae}$; thus the equation applies only for $|\psi| \geq |\psi_{ae}|$. The moisture-characteristic curve for $|\psi| < |\psi_{ae}|$ can be crudely approximated by a vertical line from $|\psi| = 0$ to $|\psi| = |\psi_{ae}|$; Clapp and Hornberger (1978) and van Genuchten (1980) showed how the portion of the curve near $S = 1$ can be represented more accurately, if required.

The parameter $b$ (or sometimes its inverse) is often called the **pore-size distribution index**; the exponent $c$ is called the **pore-disconnectedness index**, because it is a measure of the ratio of the length of the path followed by water in the soil to a straight-line path (Eagleson 1978; Bras 1990). Approximately,

$$c = 2 \cdot b + 3. \qquad \textbf{(6-14)}$$

Values of the parameters in Equations (6-12) and (6-13) depend primarily on soil texture (Clapp and Hornberger 1978; Cosby et al. 1984; Mishra et al. 1989; Rawls et al. 1992). Typical values determined by statistical analysis of data for a large number of soils are given in Table 6-1, but one should be aware that there is considerable within-soil-type variability, as reflected in the standard deviations. Rawls et al. (1992) described how $\psi_{ae}$, $K_h^*$, and $b$ can be estimated from soil-texture information.

---

### EXAMPLE 6-4

Use the general analytic relations of Equations (6-12)–(6-14) to estimate the moisture-characteristic curve and hydraulic-conductivity–water-content relations for the soil of Examples 6-1 and 6-2.

***Solution***  The porosity of this soil was found in Example 6-2: $\phi = 0.419$. Since we have no information about the other quantities in these equations, we use the typical values for sandy loams from Table 6-1 ($|\psi_{ae}| = 21.8$ cm; $b = 4.90$; $K_h^* = 3.47 \times 10^{-3}$ cm s$^{-1}$). From Equation (6-14) we calculate

$$c = 2 \times 4.90 + 3 = 12.8.$$

Then we can calculate the following values via Equations (6-12b), (6-13b), and (6-14):

| $\theta$ | 0.042 | 0.084 | 0.126 | 0.168 | 0.210 |
|---|---|---|---|---|---|
| $\|\psi\|$ (cm) | $1.71 \times 10^6$ | $5.73 \times 10^4$ | 7860 | 1920 | 643 |
| $K_h$ (cm s$^{-1}$) | $5.67 \times 10^{-16}$ | $4.04 \times 10^{-12}$ | $7.26 \times 10^{-10}$ | $2.88 \times 10^{-8}$ | $5.02 \times 10^{-7}$ |

| $\theta$ | 0.252 | 0.294 | 0.336 | 0.378 | 0.419 |
|---|---|---|---|---|---|
| $\|\psi\|$ (cm) | 263 | 124 | 64.3 | 36.1 | 21.8 |
| $K_h$ (cm s$^{-1}$) | $5.18 \times 10^{-6}$ | $3.72 \times 10^{-5}$ | $2.06 \times 10^{-4}$ | $9.29 \times 10^{-4}$ | $3.47 \times 10^{-3}$ |

Note that the sandy-loam curves in Figures 6-8 and 6-10 show these relations in terms of $S$; it is instructive to plot them in terms of $\theta$. (Use semi-log paper for the $\psi - \theta$ relation.)

### 6.3.7 Hydraulic Diffusivity

In some situations, problems of soil-water movement can be more readily solved by defining the **hydraulic diffusivity,** $D_h(\theta)$, as

$$D_h(\theta) \equiv K_h(\theta) \cdot \frac{\partial \psi(\theta)}{\partial \theta}. \qquad \textbf{(6-15)}$$

Note that $D_h(\theta)$ has appropriate dimensions of a diffusivity, [L$^2$ T$^{-1}$]. Using Equation (6-15), we may write Darcy's Law [Equation (6-10)] as

$$q_z = -K_h(\theta) - D_h(\theta) \cdot \frac{d\theta}{dz}; \qquad \textbf{(6-16)}$$

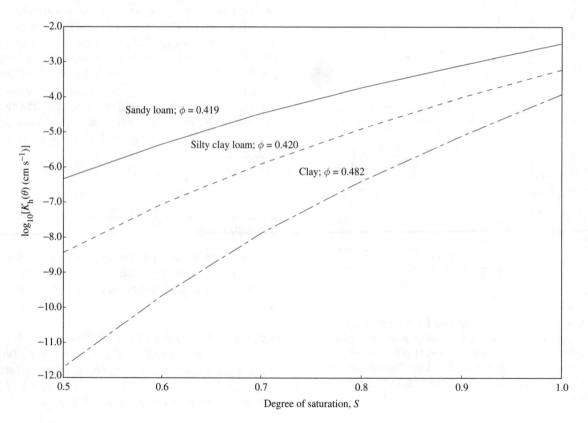

**FIGURE 6-10**

Hydraulic conductivity, $K_h$, vs. degree of saturation, $S$, for soils of three different textures. Note that the vertical axis gives the base-10 logarithm of the hydraulic conductivity, expressed in cm s$^{-1}$. Curves are based on typical values given by Clapp and Hornberger (1978). The sandy-loam curve is for the soil discussed in Examples 6-1–6-3.

**TABLE 6-1**

Representative values of parameters in Equations (6-12) and (6-13) based on analysis of 1845 soils; values in parentheses are standard deviations. Data from Clapp and Hornberger (1978).

| Soil Texture | $\phi$ | $K_h^*$ (cm s$^{-1}$) | $|\psi_{ae}|$ (cm) | $b$ |
|---|---|---|---|---|
| Sand | 0.395 (0.056) | $1.76 \times 10^{-2}$ | 12.1 (14.3) | 4.05 (1.78) |
| Loamy sand | 0.410 (0.068) | $1.56 \times 10^{-2}$ | 9.0 (12.4) | 4.38 (1.47) |
| Sandy loam | 0.435 (0.086) | $3.47 \times 10^{-3}$ | 21.8 (31.0) | 4.90 (1.75) |
| Silt loam | 0.485 (0.059) | $7.20 \times 10^{-4}$ | 78.6 (51.2) | 5.30 (1.96) |
| Loam | 0.451 (0.078) | $6.95 \times 10^{-4}$ | 47.8 (51.2) | 5.39 (1.87) |
| Sandy clay loam | 0.420 (0.059) | $6.30 \times 10^{-4}$ | 29.9 (37.8) | 7.12 (2.43) |
| Silty clay loam | 0.477 (0.057) | $1.70 \times 10^{-4}$ | 35.6 (37.8) | 7.75 (2.77) |
| Clay loam | 0.476 (0.053) | $2.45 \times 10^{-4}$ | 63.0 (51.0) | 8.52 (3.44) |
| Sandy clay | 0.426 (0.057) | $2.17 \times 10^{-4}$ | 15.3 (17.3) | 10.4 (1.64) |
| Silty clay | 0.492 (0.064) | $1.03 \times 10^{-4}$ | 49.0 (62.1) | 10.4 (4.45) |
| Clay | 0.482 (0.050) | $1.28 \times 10^{-4}$ | 40.5 (39.7) | 11.4 (3.70) |

that is, the flow due to the pressure gradient can be expressed as the product of the hydraulic diffusivity and the water-content gradient.

Equations (6-12)–(6-15) can be used to find that

$$D_h(\theta) = -b \cdot \psi_{ae} \cdot K_h^* \cdot \phi^{-b-3} \cdot \theta^{b+2}; \qquad \textbf{(6-17)}$$

thus $D_h(\theta) > 0$ (note that $\psi_{ae} < 0$) and increases with water content.

### 6.3.8 Sorptivity

In horizontal infiltration and in the earliest stages of vertical infiltration into dry soils, the pressure forces in Equations (6-10) and (6-16) are much greater than the gravity forces. The soil **sorptivity**, $S_p$, is a measure of the rate at which water will be drawn into an unsaturated soil under these conditions—that is, in the absence of gravity forces. As discussed in Section 6.6.1, sorptivity arises in formulating solutions to the Richards Equation (Box 6-1).[6] It can be related to soil properties and the initial soil wetness prior to infiltration, $\theta_0$, as

$$S_p = \left[ (\phi - \theta_0) \cdot K_h^* \cdot |\psi_{ae}| \cdot \left( \frac{2 \cdot b + 3}{b + 3} \right) \right]^{1/2} \quad \textbf{(6-18)}$$

(Rawls et al. 1992). Note that the dimensions of $S_p$ are [L T$^{-1/2}$] and that Equation (6-18) is dimensionally correct.

---

[6] Hillel (1980b, p. 16–20) provided a good explanation of the conceptual and mathematical basis for sorptivity.

## 6.4 WATER CONDITIONS IN NATURAL SOILS

### 6.4.1 Soil-Water Status

If a soil is saturated and then allowed to drain without being subject to evaporation, plant uptake, or capillary rise, its water content will decrease indefinitely in a quasi-exponential manner; note that the water content in the soil shown in Figure 6-11 is still decreasing after 156 days of drainage. However, the drainage rate also declines exponentially and typically becomes negligible (relative to, for example, the evapotranspiration rate) within at most a few days. Thus although drainage continues indefinitely, it has proven useful to define a soil's **field capacity**, designated $\theta_{fc}$, as the water content below which further decrease occurs at a "negligible" rate. The field capacity is thus an index of the water content that can be held against the force of gravity. One way of defining this arbitrary value is as the water content corresponding to a pressure head of $-340$ cm ($-33$ kPa), which can be computed from Equation (6-12b) as

$$\theta_{fc} \equiv \phi \cdot \left( \frac{|\psi_{ae}|}{340} \right)^{1/b}, \qquad \textbf{(6-19)}$$

where $\phi$ is the porosity, $\psi_{ae}$ is the air-entry tension in cm, and $b$ is the exponent describing the moisture-characteristic curve (Table 6-1). Values of $\theta_{fc}$ for sands may be as low as 0.1, while those for clays can exceed 0.3 (Figures 6-4 and 6-12).

In nature, water is removed from a soil that has reached field capacity principally by direct evaporation or by plant uptake as part of the process of transpiration (discussed more fully in Section 7.5). However, plants cannot exert suctions stronger

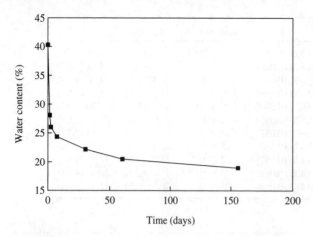

**FIGURE 6-11**
Gravity drainage of a silt-loam soil. Water content vs. time.

than about −15,000 cm (−1470 kPa) and, when the water content is reduced to the point corresponding to that value on the moisture-characteristic curve, transpiration ceases and plants wilt. Thus it is useful to define the **permanent wilting point, $\theta_{pwp}$**, as

$$\theta_{pwp} \equiv \phi \cdot \left( \frac{|\psi_{ae}|}{15,000} \right)^{1/b}, \qquad \textbf{(6-20)}$$

where the symbols are defined as in Equation (6-19). The value of $\theta_{pwp}$ ranges from about 0.05 for sands to 0.25 for clays (Figure 6-4).

The difference between the field capacity and permanent wilting point is considered to be the water available for plant use, called the **available water content, $\theta_a$**:

$$\theta_a \equiv \theta_{fc} - \theta_{pwp}. \qquad \textbf{(6-21)}$$

Figure 6-13 shows a classification of water status in soils based on pressure head. Corresponding values of water content for a particular soil can be determined from its moisture-characteristic curve. However, it should be noted that field capacity, permanent wilting point, and available water content are arbitrary values even for a particular soil and under some circumstances (e.g., when there is significant capillary rise) may have little significance. Soils in nature do not have water contents lower than that corresponding to **hygroscopic water,** since

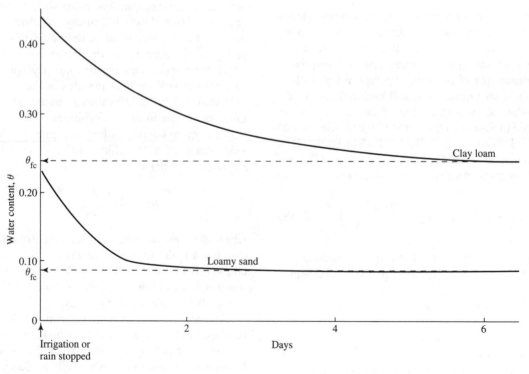

**FIGURE 6-12**
Drainage of two soils (schematic); the arrows indicate the values of field capacity. After Donahue et al. (1983).

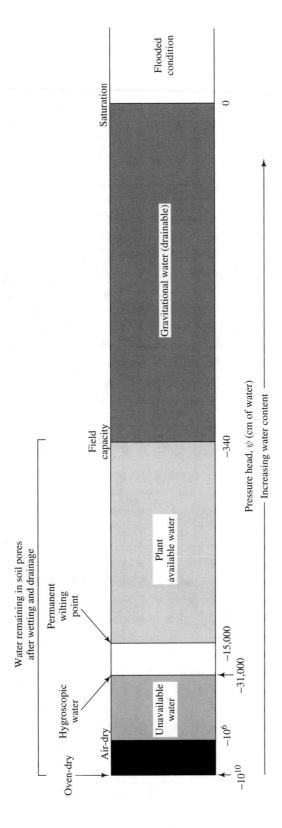

**FIGURE 6-13**

Soil-water status as a function of pressure (tension). Natural soils do not have tensions exceeding about $-31,000$ cm; in this range, water is absorbed from the air. After Donahue et al. (1983).

at this extreme dryness water is absorbed directly from the air.

---

## EXAMPLE 6-5

Estimate (a) field capacity, (b) permanent wilting point, and (c) available water content for the soil of Examples 6-1 and 6-2.

**Solution** Lacking a moisture-characteristic curve specifically for this soil, we make use of Equations (6-19)–(6-21) and the values $\phi = 0.419$, $|\psi_{ae}| = 21.8$ cm, and $b = 4.90$ previously used for this soil.

(a) Equation (6-19) yields

$$\theta_{fc} = 0.419 \times \left(\frac{21.8 \text{ cm}}{340 \text{ cm}}\right)^{1/4.90} = 0.239.$$

(b) Equation (6-20) yields

$$\theta_{pwp} = 0.419 \times \left(\frac{21.8 \text{ cm}}{15{,}000 \text{ cm}}\right)^{1/4.90} = 0.110.$$

(c) Equation (6-21) yields

$$\theta_a = 0.239 - 0.110 = 0.129.$$

---

### 6.4.2 Soil Profiles

#### Pedologic Horizons

As noted in Section 3.3.1, soils are characterized by a typical sequence of horizons that constitutes the **soil profile.** Pedologic horizons are distinguished by the proportion of organic material and the degree to which material has been removed (**eluviated**) or deposited (**illuviated**) by chemical and physical processes.

Pedologists identify soil horizons on the basis of color, texture, and structure and designate them by letters, as shown in Figure 6-14. Not all of the horizons shown in Figure 6-14 are present in all soils, and the boundaries between layers are commonly gradational. The relative development of these horizons is the basis for identifying soil orders

**FIGURE 6-14**
Designation of pedologic soil-profile horizons. **Illuviation** is the process of deposition of material leached (**eluviated**) from above. After Hillel (1980).

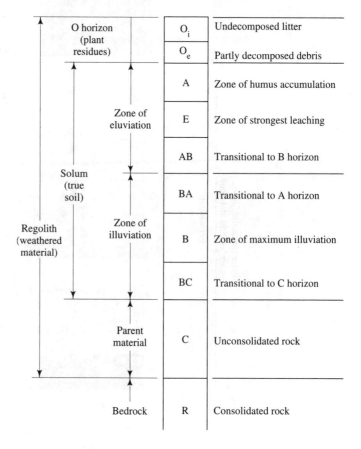

(Table 3-9) and for more detailed pedologic classification. [See for example Donahue et al. (1983).] The horizons above the C-horizon constitute the "true soil," or **solum.**

As discussed in Section 3.3.1, the development of the various pedologic horizons is determined largely by climate, but depends also on (1) local topography; (2) the effects of erosion, deposition, and other disturbances; (3) the type of parent material; and (4) the time over which soil-forming processes have been operating. The development of these horizons may vary over short distances depending on local conditions, but does not vary temporally over the time scale of most hydrologic analyses.

### Hydrologic Horizons

For purposes of describing water movement in soils, it is useful to define another set of horizons based on the normal range of water contents and the soil-water pressure, as shown in Figure 6-15. The depths and thicknesses of these hydrologic horizons vary in both time and space, and one or more of them may be absent in a given situation.

**Ground-Water Zone** The **ground-water zone** (sometimes called the **phreatic zone**[7]) is saturated and the pressure is positive. If there is no ground-water flow, the pressure will be **hydrostatic**—i.e., increasing linearly with depth according to the relation

$$p(z) = \gamma_w \cdot (z' - z_0'), \ z' > z_0', \qquad (6-22)$$

where $p$ is gage pressure, $\gamma_w$ is the weight density of water, $z'$ is the distance measured vertically downward, and $z_0'$ is the value of $z'$ at the water table. Thus the water table is at atmospheric pressure; it is the level at which water would stand in a well.

In general, the water table rises and falls in response to seasonal climatic variations and to recharge from individual storm events. In arid regions it may be at depths of many tens of meters, while elsewhere—for example, where the soil is developed in a few-meter layer of glacial till overlying dense crystalline bedrock—there may be no water table in the soil all or most of the time.

**Tension-Saturated Zone (Capillary Fringe)** The term **vadose zone**[8] is commonly used to refer to the entire zone of negative water pressures above the water table. The lowest portion of this zone is a region that is saturated or nearly saturated as a result of capillary forces. The pore spaces in a porous medium act like the capillary tube in Figure B-9, and surface-tension forces draw the water into the spaces above the water table, creating the **tension-saturated zone,** or **capillary fringe.** We designate the depth of the top of the tension-saturated zone as $z_{ts}'$.

Water in the tension-saturated zone is under tension; pressure remains hydrostatic, and head is given by

$$\psi = z' - z_0', \qquad (6-23)$$

where $z_{ts}' \leq z' \leq z_0'$ and $\psi \leq 0$. The pressure head at the top of this zone ($z' = z_{ts}'$) is the air-entry tension, $\psi_{ae}$, since at greater tensions the pores are partially filled with air. Thus $\psi_{ae}$ is equal to the height of capillary rise in the soil. This height can be approximately calculated by Equation (B-7); it ranges from about 10 mm for gravel to 1.5 m for silt to several meters for clay.

**Intermediate Zone** Water enters the intermediate zone largely as percolation from above and leaves by gravity drainage. Thus water content in this zone may rise above field capacity when water from rain or snowmelt passes through the root zone, after which it gradually returns toward field capacity following a curve like those in Figure 6-12. Tensions in the this zone are usually stronger than $\psi_{ae}$ and depend on the soil texture and the water content, as given by the moisture-characteristic curve.

The intermediate zone may occupy much of the soil profile and, in arid regions, extend over many tens of meters. In other situations, it may be absent at least seasonally—for example, in wetlands and where the root zone extends to the water table or to impermeable bedrock.

**Root Zone** The **root zone** (also called the **soil-moisture zone**) is the layer from which plant roots can extract water during transpiration; its upper boundary is the soil surface, while its lower boundary is indefinite and irregular, but effectively constant in time (except in agricultural situations). Water enters by infiltration and leaves via transpiration (or evaporation) and gravity drainage. The water content is usually above the permanent wilting point; in non-wetland soils it might occasionally approach

---

[7] "Phreatic" comes from the Greek word for "well."
[8] "Vadose" comes from the Latin word for "shallow."

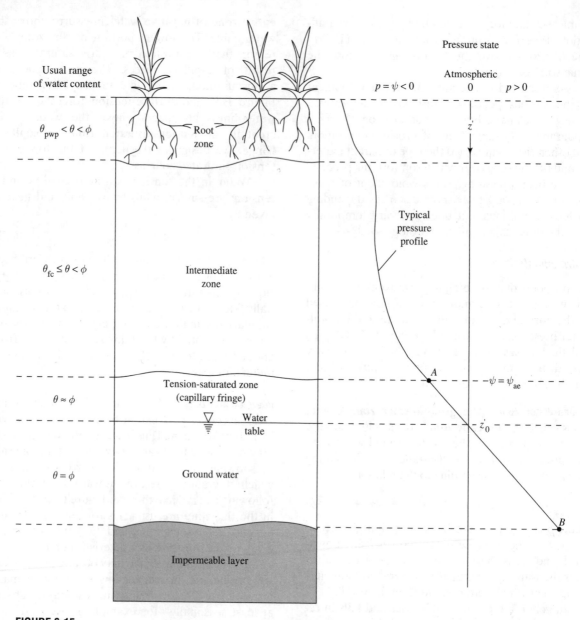

**FIGURE 6-15**
Designation of hydrologic soil-profile horizons. Note that this figure is idealized and that one or more of these horizons may be absent in a given situation. After Todd (1959).

saturation following extensive water input and infiltration, but would be below field capacity much of the time between events.

As noted previously, there are many places (including wetlands) in which the root zone extends to the water table, and it may extend throughout a thin soil lacking a subjacent water table.

**Computation of Typical Soil-Water Profiles** The equilibrium water-content profile is the water-content profile that will vertically transmit to the water table a constant flux of water equal to the local climatic-average rate of recharge, $\overline{R}$ [L T$^{-1}$] (Salvucci and Entekhabi 1995). The equilibrium profile is determined by the soil's hydraulic properties and the depth to the water table, $z_0'$, as well as by $\bar{R}$, which is equal to the climatic-average infiltration rate minus the average rate of evapotranspiration. In many regions the average infiltration rate is close to the average annual precipitation rate, so

$$\bar{R} \approx P - ET, \qquad (6\text{-}24)$$

where $P$ is the average precipitation rate and $ET$ is the average evapotranspiration rate.

The equilibrium profile is useful for portraying the typical effects of climate, soil type, and water-table depth on the hydraulic conditions of the soil and can be computed as described by Salvucci and Entekhabi (1995). First, the average depth to the top of the dynamic tension-saturated zone, $z_{ts}'$, is computed as

$$z_{ts}' = z_0' - |\psi_{ae}| \cdot (1 + \bar{R}/K_h^*); \qquad (6\text{-}25)$$

thus the effect of recharge is to increase the height of the tension-saturated zone above its static value. Below $z_{ts}'$, the soil is effectively saturated, so

$$\theta(z') = \phi, \quad z' > z_{ts}'. \qquad (6\text{-}26a)$$

For $z' \leq z_{ts}'$, Darcy's Law [Equation (6-10)] and the analytical relations for the hydraulic properties of the soil [Equations (6-12)—(6-14)] can be used to compute the equilibrium profile:

$$\theta(z') = \phi \cdot \left[ \frac{\bar{R}}{K_h^*} + \left( 1 - \frac{\bar{R}}{K_h^*} \right) \cdot \right.$$
$$\left. \left( \frac{z_0' - z'}{|\psi_{ae}| \cdot (1 + \bar{R}/K_h^*)} \right)^{-(2+3/b)} \right]^{1/(2b+3)},$$
$$z' \leq z_{ts}'. \qquad (6\text{-}26b)$$

---

### EXAMPLE 6-6

Compute the equilibrium soil-water and tension profiles for the soil of Examples 6-1 and 6-2 ($\phi = 0.419$, $|\psi_{ae}| = 21.8$ cm, $K_h^* = 12.49$ cm hr$^{-1}$, $b = 4.90$) in a location where the average recharge rate $\bar{R} = 50$ cm yr$^{-1}$ and the water-table depth $z_0' = 200$ cm.

#### Solution

1.  Convert $\bar{R}$ to cm hr$^{-1}$:

$$\bar{R} = 50 \text{ cm yr}^{-1} \times \frac{1 \text{ yr}}{365 \text{ day yr}^{-1} \times 24 \text{ hr day}^{-1}}$$
$$= 0.00571 \text{ cm hr}^{-1}.$$

2.  Compute $\bar{R}/K_h^*$:

$$\frac{\bar{R}}{K_h^*} = \frac{0.00571 \text{ cm hr}^{-1}}{12.49 \text{ cm hr}^{-1}} = 0.000457.$$

Note that this value is small enough to neglect in Equations (6-25) and (6-26).

3.  The depth to the top of the tension-saturated zone, $z_{ts}'$, is computed from Equation (6-25):

$$z_{ts}' = 200 \text{ cm} - 21.8 \text{ cm} \times 1.000457$$
$$= 178.2 \text{ cm}.$$

4.  Within the tension-saturated zone, $z_0' \leq z' \leq z_{ts}'$, $\theta(z') = \phi$ [Equation (6-26a)]. For $z' < z_{ts}'$, $\theta(z')$ is found via Equation (6-26b), and the corresponding value of $\psi(z')$ can be calculated via Equation (6-12b). Thus we find the results displayed in the following table:

| $z'$ (cm) | 0 | 20 | 40 | 60 | 80 | 100 |
|---|---|---|---|---|---|---|
| $\theta(z')$ | 0.269 | 0.275 | 0.281 | 0.288 | 0.297 | 0.308 |
| $|\psi(z')|$ (cm) | 190 | 173 | 155 | 137 | 118 | 99 |

| $z'$ (cm) | 120 | 140 | 160 | 180 | 200 |
|---|---|---|---|---|---|
| $\theta(z')$ | 0.322 | 0.341 | 0.370 | 0.419 | 0.419 |
| $|\psi(z')|$ (cm) | 80 | 60 | 40 | 20 | 0 |

---

The water-content profile for the sandy-loam soil of Example 6-6 is compared with the equilibrium profiles for a clay soil and a sand soil in Figure 6-16. Note that the finer the soil texture, the higher the water content for a given recharge rate and depth to water table.

### Relation between Pedologic and Hydrologic Horizons

Because of the variability in development of pedologic and hydrologic horizons, only a few generalizations can be made about the relations between them. In non-wetland soils, the root zone usually occupies the zone of eluviation and may extend throughout the solum. Ordinarily the zones of eluviation and illuviation develop only in the unsaturated zone above the capillary fringe, but in some soils the seasonal high water table may move into the solum. Most macropores occur in the root zone.

Bluish, grayish, and greenish mottling of subsoils, called **gleying,** usually indicates reduced aeration because the horizon is below the water table for a significant fraction of the year (Donahue et al. 1983). In wetlands, where the water table is at or

**FIGURE 6-16**
Equilibrium (climatic) water-content profiles for a clay soil, a sand soil, and the sandy loam soil of Example 6-6. The water table is arbitrarily fixed at a depth of 200 cm.

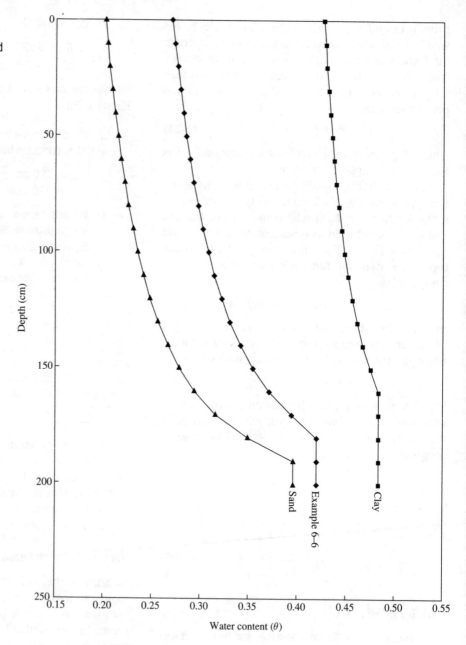

near the ground surface, the soil may consist almost entirely of organic layers or, where mineral soils are present, the solum may be absent or poorly developed.

Soil-hydraulic properties may change relatively abruptly in successive pedologic horizons. In some cases, impermeable or nearly impermeable layers called **hardpan** or **fragipan** develop at or below the B horizon. Percolating water may accumulate above these layers, forming a **perched water table** above the general regional water table. However, other soils show a more or less gradual decrease in hydraulic conductivity and porosity with depth (Beven 1984).

In general, soil-water pressure changes smoothly across the boundary between layers of contrasting hydraulic conductivity, even though there are abrupt changes in water content.

## 6.5   INFILTRATION: MEASUREMENT AND QUALITATIVE DESCRIPTION

Infiltration is the process by which water arriving at the soil surface enters the soil. At a given point, the rate of infiltration generally changes systematically with time during a given water-input event. This section and the next section of this chapter are largely concerned with understanding and modeling these within-event changes.

This section defines terms and conditions essential to understanding the infiltration process during a water-input event, describes approaches to the field measurement of infiltration, and qualitatively discusses the factors that control the process. Section 6.6 will develop quantitative descriptions of infiltration that will sharpen our understanding of the process and that can be used in modeling this critical component of the hydrologic cycle.

Conditions affecting infiltration also change between water-input events. Such changes due to hydraulic processes are discussed in Section 6.7, and those due to biologic processes (evapotranspiration) are discussed in Section 7.8.2.

### 6.5.1   Definitions

A water-input event begins at time $t = 0$ and ends at $t = t_w$. The **infiltration rate,** $f(t)$, is the rate at which water enters the soil from the surface; it has the dimensions [L T$^{-1}$]. To describe the infiltration process, we further define the following variables: (1) the **water-input rate,** $w(t)$, which is the rate at which water arrives at the surface due to rain, snowmelt, or irrigation [L T$^{-1}$]; (2) the **infiltrability** (also called **infiltration capacity**), $f^*(t)$, which is the maximum rate at which infiltration can occur [L T$^{-1}$] (note that this value is not constant, but changes during the infiltration event); and (3) the **depth of ponding,** $H(t)$, which is the depth of water standing on the surface [L].

To understand the infiltration process, it is helpful to distinguish three conditions:

1. **No ponding.** In this case, the infiltration rate equals the water-input rate and is less than or equal to the infiltrability:

$$H(t) = 0,\ f(t) = w(t) \le f^*(t).\quad \textbf{(6-27a)}$$

In this situation, infiltration is said to be **supply-controlled** (or **flux-controlled**).

2. **Saturation from above.** Ponding is present because the water-input rate exceeds the infiltrability. In this case, the infiltration rate equals the infiltrability:

$$H(t) > 0,\ f(t) = f^*(t) \le w(t).\quad \textbf{(6-27b)}$$

In this situation, the rate of infiltration is determined by the soil type and wetness and is said to be **profile-controlled** (Hillel 1980b).

3. **Saturation from below.** Ponding is present because the water table has risen to or above the surface, and the entire soil is saturated. In this case, the infiltration rate is zero:

$$H(t) \ge 0,\ f(t) = 0.\quad \textbf{(6-27c)}$$

### 6.5.2   Measurement

#### Ring Infiltrometers

A **ring infiltrometer** is a device for direct field measurement of infiltrability over a small area (0.02–1 m$^2$). The area is defined by an impermeable boundary, usually a cylindrical ring extending several centimeters above the surface, and sealed at the surface or extending several centimeters into the soil.

A condition of ponding due to saturation from above is created within the ring by direct flooding of the surface or by applying a sufficiently high rate of simulated rainfall. The rate of infiltration is obtained by (1) measuring the rate at which the level of ponded water decreases, (2) measuring the rate at which water has to be added to maintain a constant level of ponding, or (3) solving a water-balance equation for the ponded surface:

$$f(t) = \frac{W - Q - \Delta H \cdot A}{\Delta t}.\quad \textbf{(6-28)}$$

Here, $f(t)$ is the average infiltration rate over the period of measurement, $\Delta t$; $W$ is the volume of water applied during $\Delta t$; $Q$ is the volume of ponded water removed from the plot (usually by a small pump) during $\Delta t$; $\Delta H$ is the change in ponded-water level during $\Delta t$; and $A$ is the area covered by the infiltrometer.

As will be seen, infiltration rates during the early stages of the process are usually high and gradually decrease to a nearly constant value; it is this constant value that is usually taken as the infiltrability at a particular location. We will see later that this rate is approximately equal to the saturated hydraulic conductivity of the near-surface soil.

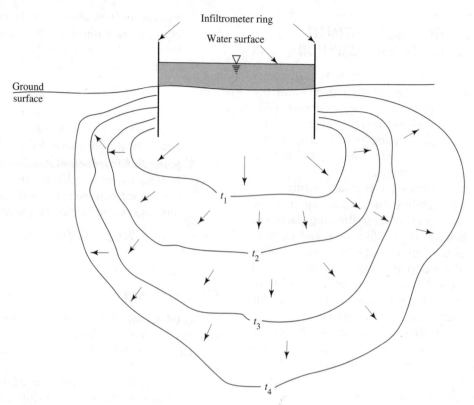

**FIGURE 6-17**
Pattern of wetting-front movement into an unsaturated soil from an infiltrometer; $t_1$, ..., $t_4$ represent successive times. Because the front moves laterally due to capillary forces, the rate of inflow at the surface exceeds the infiltration rate that would exist if the entire surface were flooded. After Hills (1971).

Because water infiltrating into an unsaturated soil is influenced by both capillary (pressure) and gravity forces, the water applied within an infiltrometer ring moves laterally as well as vertically (Figure 6-17); thus the measured infiltration rate exceeds the rate that would occur if the entire surface were ponded. One way of reducing this effect is to use a **double-ring infiltrometer** (Figure 6-18). With this device, areas within two concentric rings are ponded; the area between the two rings acts as a "buffer zone," and measurements only on the inner ring are used to calculate the infiltration rate. Swartzendruber and Olson (1961a; 1961b) suggested that the inner- and outer-ring diameters be at least 100 and 120 cm, respectively.

A more convenient approach is to use a single-ring infiltrometer and apply a correction for the capillary effect; Tricker (1978) showed that this method gave satisfactory results using a single ring of 15-cm diameter and correction factors that de-

**FIGURE 6-18**
A double-ring infiltrometer connected to an apparatus that maintains a constant water level in each ring. Photo by the author.

pend on the measured rate and the duration of the measurement. Detailed information on the design of various types of infiltrometers can be found in Johnson (1963), McQueen (1963), Tricker (1979), Wilcock and Essery (1984), Bouwer (1986), and Loague (1990).

Infiltrability tends to have considerable spatial variability, so that the value for a given soil should be the average of several measurements. Burgy and Luthin (1956) found that the average of six single-ring infiltrometer measurements of infiltrability was within 30% of the true value for a soil with uniform characteristics.

### Sprinkler-Plot Studies

Infiltration rates can also be determined by recording the rates of runoff from well-defined plots on which a known constant rate of artificial rainfall, $w$, is applied at a rate high enough to produce saturation from above. The infiltration rate, $f(t)$, is computed as

$$f(t) = w - q(t), \qquad \text{(6-29)}$$

where $q(t)$ is the rate of surface runoff from the plot (Nassif and Wilson 1975).

### Observation of Soil-Water Changes

A third method of measuring infiltration is to record changes in tension in tensiometers installed at several depths during a natural or artificial water-input event. The moisture-characteristic curve is used to relate tension to water content, and the infiltration rate is determined from the increase in soil-water content at various depths during the event.

### 6.5.3  Basic Characteristics of the Infiltration Process

#### Change in Infiltration Rate with Time during an Event

Virtually all studies of infiltration during a water-input event in which there is saturation from above [Equation (6-27b)] show high infiltration rates at the beginning of the event, followed by a relatively rapid decline and an asymptotic approach to a near-constant value. Figure 6-19 shows this typical pattern in a sprinkler-plot study.

#### Water-Content Profiles

Typical vertical profiles of water content as a function of time during infiltration are shown in Figure

| Symbol | Water-input rate, $w$ (mm h$^{-1}$) |
|--------|-------------------------------------|
| o | 312 |
| ▲ | 234 |
| × | 156 |
| ● | 78 |

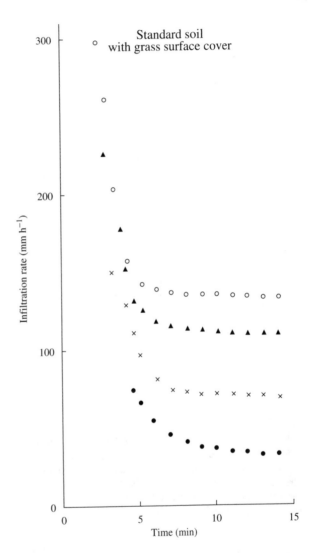

**FIGURE 6-19**

Infiltration rate into grassed loam plots as measured in laboratory studies using artificial rainstorms of 15-min duration; infiltration rate as a function of time for various water-input rates (slope = 16%). After Nassif and Wilson (1975).

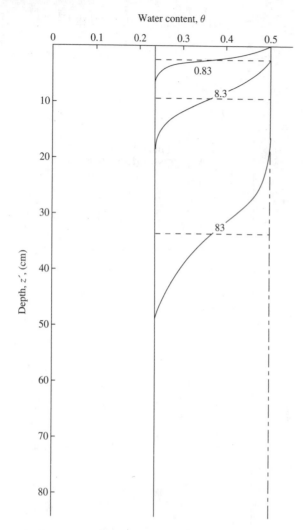

**FIGURE 6-20**
Soil-moisture profiles in a light clay soil with $\phi = 0.50$ at various times during steady infiltration with no ponding. The number on the curve is the time since infiltration began, in hours. The initial water content was 0.23. The dashed lines mark the average positions of the wetting front at the indicated times. Data from Freyberg et al. (1980).

**FIGURE 6-21**
Wetting front in a sandy soil exposed after an intense rain. Photo by the author.

wetting front may become unstable and the downward flow becomes concentrated into vertical "fingers" (Hillel 1980b).

### 6.5.4   Factors Affecting Infiltration Rate

Our ultimate objective is to estimate the infiltration rate, $f(t)$, as a function of time, $t$, during a water-input event. The value of $f(t)$ is determined by the effects of gravity and pressure forces on water arriving at the surface, which are in turn determined by the following factors:

1. (a) the rate at which the water arrives from above as rainfall, snowmelt, or irrigation; or (b) the depth of ponding on the surface;
2. the saturated hydraulic conductivity of the soil profile;
3. the degree to which soil pores are filled with water when the process begins;
4. the inclination and roughness of the soil surface;
5. the chemical characteristics of the soil surface;
6. the physical and chemical properties of water.

The model of the infiltration process that we develop in Section 6.6.2 will permit us to examine

6-20. The region of rapid downward decrease in water content is called a **wetting front,** which in this case was at about 3-cm depth after 0.83 hr and had progressed to about 33 cm after 83 hr of steady infiltration. Wetting fronts in clay soils like that in Figure 6-20 tend to be relatively diffuse due to the effects of capillarity, but are quite sharp in sandy soils (Figure 6-21). When the pressure gradient opposes the flow ($d\psi/dz' > 0$), which may occur when hydraulic conductivity increases with depth, the

quantitatively the effect of the first three of the foregoing factors on the infiltration rate during a water-input event. However, before developing that model, we will examine qualitatively the ways in which those factors are affected by environmental conditions.

### Water-Input Rate or Depth of Ponding

Equations (6-27) state the conditions under which $w(t)$ and $H(t)$ determine the range of possible values for $f(t)$.

### Hydraulic Conductivity of the Surface

Subsequent analysis will show that the minimum value of the infiltrability of a soil approximately equals its saturated hydraulic conductivity, $K_h^*$. As noted earlier, $K_h^*$ for mineral soils is determined primarily by grain size. However, there are many factors that can cause the pore size, and hence the value of $K_h^*$, at the surface to be much greater or much less than those of the underlying undisturbed mineral soil:

**Organic surface layers:** If the soil supports natural vegetation, particularly a forest, the near-surface soil will usually consist largely of leaf litter, humus, and other organic matter that has a large number of big openings, and hence a high hydraulic conductivity, regardless of the texture of the mineral soil beneath (Tricker 1981). Root growth and decay and the action of worms, soil insects, and burrowing mammals contribute to the surface porosity. This effect occurs also in grass (Hino et al. 1987) and even in arid scrubland, where infiltration was found to be nearly three times higher under individual bushes than in the open (Lyford and Qashu 1969). In deciduous broad-leaved forests, the leaf litter can produce "shingling" that can have the opposite effect, preventing infiltration at least locally.

**Frost:** If surface soil with a high water content freezes, "concrete frost" can form, making the surface nearly impermeable. However, frost action associated with lower water contents can sometimes markedly increase the surface permeability (Schumm and Lusby 1963; Dingman 1975). In some permafrost areas, seasonal freezing at the surface produces a polygonal network of cracks that can admit much of the spring meltwater.

**Swelling–Drying:** Some soils, especially Vertisols (see Section 3.3.1), contain clay minerals that swell when wet and shrink when dry. During the rainy season or during a single rainstorm, swelling can reduce effective surface porosity and permeability and limit infiltration; during dry periods, polygonal cracks develop that can accept high infiltration rates.

**Rain Compaction:** The impact of rain on bare soils causes the surface to compact, producing a soil seal that markedly decreases infiltration. The intensity of the effect is determined by the kinetic energy of the rain, modified by the physical and chemical nature of the soil (Assouline and Mualem 1997).

**Inwashing of fine sediment:** Where surface erosion occurs due to sheetflow over the surface, or where mineral grains are brought into suspension by the splashing of raindrops, fine sediment may be carried into larger pores and effectively reduce the surface pore size and permeability. These situations are most likely to occur where vegetative cover is sparse, either naturally or due to cultivation practices.

**Human modification of the soil surface:** Many human activities have a direct effect on surface porosity—most obviously, the construction of roads and parking lots. Plowing may increase or decrease surface porosity temporarily, but subsequent use of equipment for spreading fertilizers and pesticides tends to compact the soil and reduce porosity. Grazing animals also tend to compact the surface soil.

### Water Content of Surface Pores

Saturation from below occurs when local recharge and/or ground-water flow from upslope causes the local water table to rise to the surface. It can also occur in the absence of a local water table at the beginning of an event where there is a more-or-less gradual decrease of porosity and hydraulic conductivity with depth, or where there is a distinct layer with significantly reduced permeability at depth. Either situation may reduce or prevent percolation at depth, which in turn causes an accumulation of water arriving from above and may lead to creation of a saturated zone. If water input continues at a high enough rate, the saturated zone can reach the surface and prevent further infiltration, regardless of the hydraulic conductivity, the rate of input, or any other factor.

Even when surface saturation does not occur, the antecedent water content affects the infiltration rate. A higher water content increases hydraulic conductivity [Figure 6-10; Equation (6-13)] and reduces the space available for storage of infiltrating water, both of which tend to increase the speed of the wetting front. Furthermore, a soil that is relatively wet at the beginning of a water-input event will be more likely to become saturated during the event, resulting in a longer period of reduced infiltration.

The net effect of antecedent water content on infiltration thus depends on the specific conditions of the water-input rate and duration, the distribution of soil hydraulic conductivity with depth, the depth of the local water table, and the initial water content itself. The model developed in Section 6.6.2 will allow us to evaluate some of these factors quantitatively.

### Surface Slope and Roughness

As long as ponding does not occur during a water-input event, the infiltration rate is governed by the water-input rate, and the hydraulic characteristics of the soil surface have no direct effect on infiltration. However, once ponding begins, the ponding depth will increase until it is sufficient to overcome the hydraulic resistance of the surface, at which time downslope runoff or **overland flow** begins. Overland flow is usually turbulent or quasi-turbulent, and the rate of overland flow increases with increasing slope and decreases with increasing roughness (Dingman 1984).

Thus steeper slopes and smoother surfaces promote more rapid overland flow, and hence less accumulation of ponded water on the surface and, all other things equal, lower infiltration rates (Nassif and Wilson 1975).

### Chemical Characteristics of the Soil Surface

Waxy organic substances produced by vegetation and microorganisms have been found on ground surfaces under a variety of vegetation types. Since the contact angle between a water surface and waxes is negative (Table B-3), water falling on such surfaces tends to "bead up" instead of being drawn into the pores by surface-tension forces. Soils with such surfaces are called **hydrophobic** (i.e., water-repellent) soils.

The effects of hydrophobic compounds on infiltration seem to be minor under undisturbed vegetation. However, when forest or range fires occur, the organic surface layer is burned off, these substances vaporize and subsequently condense on the bare soil, and the wettability and infiltrability of the surface is significantly reduced (Branson et al. 1981).

### Physical and Chemical Properties of Water

The movement of liquid water in the unsaturated zone is affected by the surface tension, density, and viscosity of water, all of which are properties that depend on temperature (Table B-2). Viscosity is especially sensitive: Its value at 30 °C is less than half its value at 0 °C; thus the hydraulic conductivity at 30 °C is about twice as large as it is at 0 °C, all other things equal. This effect was observed by Klock (1972), who measured infiltration rates twice as large with 25 °C-water as with 0 °C-water in laboratory experiments.

## 6.6   QUANTITATIVE MODELING OF INFILTRATION AT A POINT

Our quantitative treatment of infiltration assumes that the water is moving vertically through interconnected inter-grain pores that are randomly distributed throughout a quasi-homogeneous soil, and our analysis applies to representative soil volumes that are large relative to the typical pore size.

We have seen that many factors influence infiltration in the field, and the foregoing assumptions may not hold in some situations. For example, where infiltration and vertical flow are concentrated in macropores that are significantly larger than the pores of the general soil matrix (caused by frost action, drying, animal burrowing, or decayed roots), the pipe-like flow may not be well modeled by the classical approach that we will use (Beven and Germann 1982; Wagenet and Germann 1989). We will discuss the importance of macropore flow in runoff generation in Section 9.2.3, but the complete physical modeling of such flow, to the extent it has been achieved [see Beven and Clarke (1986), Germann

(1989), and Rawls et al. (1992)], is beyond the scope of this text.

Our treatment will also assume that water moves only in the liquid state and that its movement is not significantly affected by the flow of air in the soil pores or by temperature or osmotic gradients. Air flow may be important in ponded infiltration, but can be neglected for most natural infiltration events (Youngs 1988). Water movement

in response to thermal and osmotic gradients may be important in some situations (e.g., when soil-water freezing or thawing is occurring), but is too complex for consideration here.

### 6.6.1   The Richards Equation

The Richards Equation, which is derived in Box 6-1 by combining Darcy's Law for vertical unsaturated

---

## BOX 6-1
· · · · · · ·
### Derivation of the Richards Equation for Vertical Percolation in a Soil

Figure 6-22 shows a rectangular parallelepiped of soil oriented so that one dimension is aligned with the vertical ($z$-) direction of a rectangular coordinate system. The dimensions of this volume, $\Delta x$, $\Delta y$, and $\Delta z$, are "small," but large enough to encompass a representative volume of the soil. In this derivation, flow occurs only vertically downward, which we designate as the $z'$-direction.

During a "small" time increment $\Delta t$, the conservation of water mass for this volume element is

$$\rho_w \cdot q_{z'} \cdot \Delta x \cdot \Delta y \cdot \Delta t - \rho_w$$
$$\cdot \left( q_{z'} + \frac{\partial q_{z'}}{\partial z'} \cdot \Delta z \right) \cdot \Delta x \cdot \Delta y \cdot \Delta t$$
$$= \rho_w \cdot \frac{\partial \theta}{\partial t} \cdot \Delta t \cdot \Delta x \cdot \Delta y \cdot \Delta z,$$

$$\text{(6B1-1)}$$

where $\rho_w$ is the mass density of water, $q_{z'}$ is the volumetric flow rate in the $z'$-direction at the top of the element, and $\theta$ is water content. Assuming a constant density and simplifying Equation (6B1-1) yields the continuity equation for this situation:

$$-\frac{\partial q_{z'}}{\partial z'} = \frac{\partial \theta}{\partial t}. \qquad \text{(6B1-2)}$$

From Equation (6-10), we find that Darcy's Law for unsaturated flow in the $z'$-direction is

$$q_{z'} = -K_h(\theta) \cdot \frac{\partial [z + \psi(\theta)]}{\partial z'} \qquad \text{(6B1-3)}$$

$$= -K_h(\theta) \cdot \frac{\partial z}{\partial z'} - K_h(\theta) \cdot \frac{\partial \psi(\theta)}{\partial z'}. \qquad \text{(6B1-4)}$$

But since

$$\frac{\partial z}{\partial z'} = -1, \qquad \text{(6B1-5)}$$

we have

$$q_{z'} = K_h(\theta) - K_h(\theta) \cdot \frac{\partial \psi(\theta)}{\partial z'}. \qquad \text{(6B1-6)}$$

Taking the derivative of this expression with respect to the $z'$-direction yields

$$\frac{\partial q_{z'}}{\partial z'} = \frac{\partial K_h(\theta)}{\partial z'} - \frac{\partial}{\partial z'} \left[ K_h(\theta) \cdot \frac{\partial \psi(\theta)}{\partial z'} \right]. \qquad \text{(6B1-7)}$$

Finally, substituting Equation (6B1-7) into Equation (6B1-2) gives the basic equation for vertical unsaturated porous-media flow:

$$-\frac{\partial K_h(\theta)}{\partial z'} + \frac{\partial}{\partial z'} \left[ K_h(\theta) \cdot \frac{\partial \psi(\theta)}{\partial z'} \right] = \frac{\partial \theta}{\partial t}. \qquad \text{(6B1-8)}$$

Equation (6B1-8) is commonly called the **Richards Equation,** after the American soil physicist A. L. Richards, who first derived it (Richards 1931). An alternative version of this equation can be derived using the hydraulic diffusivity defined in Equation (6-15). Darcy's Law for infiltration then becomes

$$q_{z'} = K_h(\theta) - D_h(\theta) \cdot \frac{\partial \theta}{\partial z'}. \qquad \text{(6B1-9)}$$

Taking the derivative of this expression and substituting into Equation (6B1-2) yields

$$-\frac{\partial K_h(\theta)}{\partial z'} + \frac{\partial}{\partial z'} \left[ D_h(\theta) \cdot \frac{\partial \theta}{\partial z'} \right] = \frac{\partial \theta}{\partial t}. \qquad \text{(6B1-10)}$$

**FIGURE 6-22**
Definition sketch for terms used
in deriving the Richards Equation
(Box 6-1).

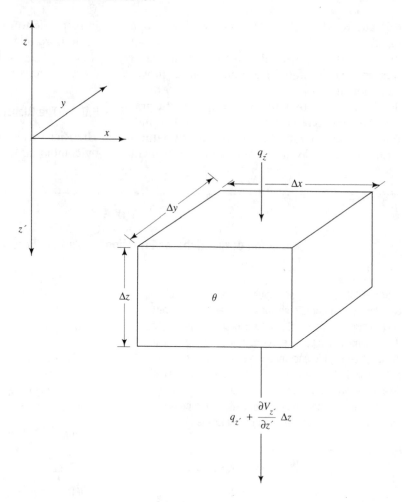

flow with the conservation of mass, is the basic theoretical equation for infiltration into a homogeneous porous medium. Because it is non-linear, there is no closed-form analytical solution except for highly simplified $\psi - \theta$ and $K_h - \theta$ relations and boundary conditions. However, the Richards Equation can be used as a basis for numerical modeling of infiltration, exfiltration, and redistribution by specifying appropriate boundary and initial conditions, dividing the soil into thin layers, and applying the equation to each layer sequentially at small increments of time. Tests have shown good agreement between the predictions of the numerically-solved Richards Equation and field and laboratory measurements (Nielsen et al. 1961; Whisler and Bouwer 1970).

Numerical solutions, however, are not very useful for providing a conceptual overview of the ways in which various factors influence infiltration, and they are generally too computationally intensive

for inclusion in operational hydrologic models. Thus, there have been many attempts to develop approximate analytical solutions to the Richards Equation for specific situations, such as infiltration (Wang and Dooge 1994). The first and best known of these attempts was developed by Philip (1957; 1969), who formulated an infinite-series solution for ponded infiltration into a indefinitely deep soil with a uniform initial water content of the form

$$f(t) = \frac{S_p}{2} \cdot t^{-1/2} + A_2 + A_3 \cdot t^{1/2} + A_4 \cdot t + \dots + A_n^{n/2-1}, \quad \textbf{(6-30)}$$

where $S_p$ is the sorptivity [Equation (6-18)]. Usually only the first two terms of this series are used, and $A_2$ is a hydraulic conductivity, designated $K_p$, so that

$$f(t) = \frac{S_p}{2} \cdot t^{-1/2} + K_p. \quad \textbf{(6-31a)}$$

[Note that for ponded infiltration, the infiltration rate, $f(t)$, equals the infiltrability, $f^*(t)$.] The cumulative infiltration, $F(t)$, is given by the time integral of Equation (6-31a):

$$F(t) = S_p \cdot t^{1/2} + K_p \cdot t. \qquad \textbf{(6-31b)}$$

Swartzendruber (1997) showed that Equation (6-31) is the exact solution of the Richards Equation only when $K_h(\theta)$ is a linear function of $\theta$ and the level of ponded water increases in proportion to $t^{1/2}$. However, that equation has often been used to model infiltration more generally, with $S_p$ and $K_p$ sometimes treated as fitting parameters determined from field infiltrometer measurements and sometimes computed from generalized soil relations [Equations (6-12)–(6-14) and Table 6-1]. These applications are discussed further in Section 6.6.4.

Note that Equations (6-31) reproduce the general behavior apparent in Figure 6-19: Infiltration decreases with time and becomes asymptotic to a constant value. However, Philip's approach has three limitations: (1) Since it applies only to ponded conditions, it is not directly applicable to rainfall infiltration before ponding occurs (a common situation discussed further later); (2) Equations (6-30) and (6-31) give $f(0) \rightarrow \infty$; and (3) The series solution of Equation (6-30) diverges as $t$ increases beyond a certain time and thus ceases to portray the actual behavior. An alternative series solution that has better convergence properties and gives results very close to numerical solutions of the Richards Equation for ponded conditions was recently formulated by Salvucci (1996).

## 6.6.2 The Green-and-Ampt Model

As noted earlier, numerical solutions of the Richards Equation are computationally intensive and require detailed soil data that are usually unavailable. Furthermore, while Equations (6-31) give some insight into how soil properties and wetness affect infiltration, they do not reveal much about the physical reasons for the behavior of the process, nor do they contain parameters that are related to the characteristics of the rainfall or snowmelt that supplies input in natural hydrologic events. Thus to better understand the essential aspects of the infiltration process, we explore the **Green-and-Ampt Model,** named for its original formulators (Green

and Ampt 1911). The development here follows that of Mein and Larson (1973).

Like the Richards Equation, the Green-and-Ampt Model applies Darcy's Law and the principle of conservation of mass, but in a finite-difference formulation that allows a more holistic and informative view of the infiltration process. The predictions of this model have been successfully tested against numerical solutions of the Richards Equation (Mein and Larson 1973).

### Idealized Conditions

As in Box 6-1, $z$ is the vertical axis (elevation), $z'$ indicates the downward direction, $f(t)$ is the infiltration rate at time $t$ [L T$^{-1}$], and $F(t)$ is the total amount of water infiltrated up to time $t$ [L].

We consider a block of soil that is homogeneous to an indefinite depth (i.e., porosity, $\phi$, and saturated hydraulic conductivity, $K_h^*$, are given parameters that are invariant throughout and there is no water table, capillary fringe, or impermeable layer) and that has a horizontal surface at which there is no evapotranspiration. The water content just prior to $t = 0$ is also invariant at an initial value $\theta_0 < \phi$.

Just before water input begins at $t = 0$, there is no water-content gradient, so the downward flux of water, $q_{z'}(z,0)$, is given by Equation (6B1-6) as

$$q_{z'}(z, 0) = K_h(\theta_0), \qquad \textbf{(6-32)}$$

because there is no vertical tension gradient. Note that these conditions are not a steady-state situation because the soil is gradually draining, but if we further assume that $\theta_0 << \theta_{fc}$, then $q_{z'}(z,0)$ can be considered negligible.

Beginning at time $t = 0$, liquid water (rain or snowmelt) begins arriving at the surface at a specified rate $w$ and continues at this rate for a specified time $t_w$. We need to consider two cases: (1) $w < K_h^*$ and (2) $w \geq K_h^*$.

### Water-Input Rate Less Than Saturated Hydraulic Conductivity

Consider a thin surface layer of soil, where Equation (6-32) applies at the instant water input begins. If we assume that $w > K_h(\theta_0)$, water will enter this layer faster than it is leaving. This water goes into storage in the layer, increasing its water content. The increase in water content causes an increase in hydraulic conductivity, diffusivity, and water-content gradient, so the flux out of the layer, $q_{z'}$, also increases [Equation (6B1-9)]. However, as long as $q_{z'}$ is less than the water-input rate, the water content will continue to increase (Figure 6-23). When the water content reaches the value $\theta_w$, at which $q_{z'} = w$, the rate of outflow from the layer equals the rate of inflow, and there is no further change in water content until water input ceases.

This process happens successively in each layer as water input continues, producing a descending wetting front at which water decreases more or less abruptly. The water content equals $\theta_w$ behind (above) the front and $\theta_0$ below it. This process results in the successive water-content profiles shown in Figure 6-24a and the corresponding graph of infiltration vs. time shown in Figure 6-24b. As the wetting front descends, the importance of the pressure (capillary) forces decreases because the denominator of the $\partial\theta/\partial z'$ term of Equation (6B1-9) increases; thus, the rate of downward flow approaches $K_h(\theta_w)$.

This analysis confirms the following model:

$$\text{If } w < K_h^*, \quad f(t) = w, \quad 0 < t \le t_w;$$
$$f(t) = 0, \quad t > t_w. \tag{6-33}$$

### Water-Input Rate Greater Than Saturated Hydraulic Conductivity

When $w > K_h^*$, the process just described will occur in the early stages of infiltration. Water will arrive at each layer faster than it can be transmitted downward and will initially go into storage, raising the water content and the hydraulic conductivity (Figure 6-23). However, the water content cannot exceed its value at saturation, $\phi$, and the hydraulic conductivity cannot increase beyond $K_h^*$. Once the surface layer reaches saturation, the wetting front begins to descend, with $\theta = \phi$ above the wetting front and $\theta = \theta_0$ below it (Figure 6-25a). As in the previous situation, the pressure force decreases as the wetting front descends, while the gravity force remains constant; thus the downward flux decreases, approaching $q_{z'} = K_h^*$ (Figure 6-25b). Since $w > K_h^*$, some rain will continue to infiltrate after the surface reaches saturation, but the excess accumulates on the surface as ponding, or **detention storage,** while the wetting front continues to descend as long as the input continues. If the ground is sloping, the excess water moves downslope as **overland flow** or **surface runoff.**

**FIGURE 6-23**
The arrows show changes in hydraulic conductivity, $K_h(\theta)$, with water content, $\theta$, during water input at a rate $w$, where $K_h(\theta_0)$ $< w = K_h(\theta_w) < K_h^*$. Water content in each layer at successive depths increases until $K_h(\theta_w) = w$.

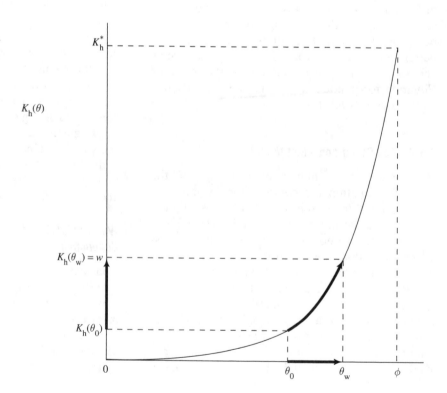

**FIGURE 6-24**
(a) Successive water-content pro-files ($t_1$, $t_2$, ... denote successive times) and (b) infiltration rate vs. time for infiltration into a deep soil when $w < K_h^*$.

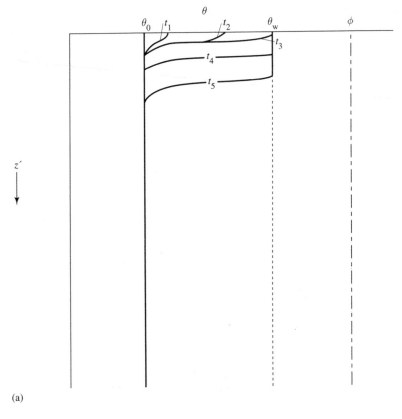

(a)

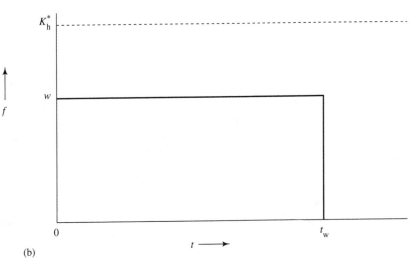

(b)

The instant when the surface layer becomes saturated is called the **time of ponding,** designated $t_p$. We can develop an equation to compute $t_p$ by approximating the wetting front as a perfectly sharp boundary (horizontal line), which is at a depth $z_f'(t_p)$ at $t = t_p$. Up to this instant, all the rain that has fallen has infiltrated, so

$$F(t_p) = w \cdot t_p. \qquad (6\text{-}34)$$

All of this water occupies the soil between the surface and $z_f'(t_p)$, so

$$F(t_p) = z_f'(t_p) \cdot (\phi - \theta_0). \qquad (6\text{-}35)$$

Combining Equations (6-34) and (6-35) yields

**FIGURE 6-25**

(a) Successive water-content profiles ($t_1$, $t_2$, ... denote successive times) and (b) infiltration rate vs. time for infiltration into a deep soil when $w > K_h^*$. The dashed lines show average wetting-front depths at the time of ponding, $z_f(t_p)$, and at some arbitrary later time $t_5$, $z_f(t_5)$.

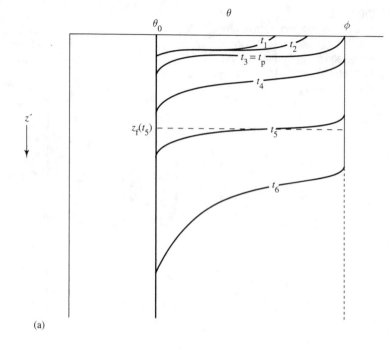

(a)

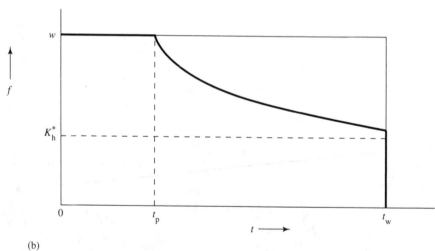

(b)

$$t_p = \frac{z_f'(t_p) \cdot (\phi - \theta_0)}{w}. \qquad (6\text{-}36)$$

In order to use Equation (6-36), we need to determine $z_f'(t_p)$. This is done by applying Darcy's Law [Equation (6B1-6)] in finite-difference form between the surface and the depth $z_f'(t_p)$:

$$q_z'(0, t_p) = f(t_p) = w = K_h^* - K_h^* \cdot \frac{\psi_f - 0}{z_f'(t_p)}. \qquad (6\text{-}37)$$

Here, $\psi_f$ is the effective tension at the wetting front (discussed further in the next section). This relation is justified because at the instant of ponding we have saturation at the surface, so the tension there is 0, the hydraulic conductivity is equal to its saturation value, and the infiltration rate is just equal to the rainfall rate. Noting that $\psi_f < 0$, we can solve Equation (6-37) for $z_f'(t_p)$:

$$z_f'(t_p) = \frac{K_h^* \cdot |\psi_f|}{w - K_h^*}. \qquad (6\text{-}38)$$

Substitution of Equation (6-38) into Equation (6-36) then yields

$$t_p = \frac{K_h^* \cdot |\psi_f| \cdot (\phi - \theta_0)}{w \cdot (w - K_h^*)} \tag{6-39}$$

as the equation for time of ponding. This expression has a logical form, in that $t_p$ increases with increasing $K_h^*$, $|\psi_f|$, and the initial soil-water deficit $(\phi - \theta_0)$ and decreases with increasing $w$.

As water input continues after the time of ponding, infiltration continues at a rate given by Darcy's Law as

$$f(t) = q_z[z_f'(t), H(t)]$$
$$= K_h^* - K_h^* \cdot \frac{\psi_f + H(t)}{z_f'(t)} = f^*(t), \tag{6-40}$$

where $H(t)$ is the depth of ponding. Thus the infiltration rate decreases with time (as in Figures 6-19 and 6-25b), because the gradient producing the pressure force (capillary suction) decreases with time as the wetting-front depth, $z_f'(t)$, increases.

Since $H(t)$ is a complicated function that depends on the amount of infiltration up to time $t$ and the surface slope and roughness, and since satisfactory results for many situations have been obtained by assuming that $H(t)$ is negligible (as will be shown later), our subsequent analysis will assume that $H(t) = 0$. Noting that

$$F(t) = z_f'(t) \cdot (\phi - \theta_0), \tag{6-41}$$

we can solve this expression for $z_f'(t)$ and substitute it into Equation (6-40) to yield

$$f(t) = f^*(t) = K_h^* \cdot \left[1 + \frac{|\psi_f| \cdot (\phi - \theta_0)}{F(t)}\right],$$
$$t_p \le t \le t_w, \tag{6-42}$$

which is the Green-and-Ampt Equation for infiltrability.

Equation (6-42) allows us to compute the infiltration rate (infiltrability) as a function of the total infiltration that has occurred. It would be most useful to a have a relation between $f(t)$ or $F(t)$ and $t$, and we can derive such a relation if we recognize that

$$f(t) = \frac{dF(t)}{dt}. \tag{6-43}$$

With Equation (6-43), we see that Equation (6-42) is actually a differential equation that can be solved

to yield a relation for $t$ as a function of $F(t)$ for $F(t) > F(t_p)$ $(t_p \le t \le t_w)$ (see Exercise 6-8):

$$t = \frac{F(t) - F(t_p)}{K_h^*} + \left[\frac{|\psi_f| \cdot (\phi - \theta_0)}{K_h^*}\right]$$
$$\cdot \ln\left[\frac{F(t_p) + |\psi_f| \cdot (\phi - \theta_0)}{F(t) + |\psi_f| \cdot (\phi - \theta_0)}\right] + t_p. \tag{6-44}$$

There is no solution for Equations (6-42) and (6-44) that gives $F(t)$ as an explicit function of $t$. Thus application of Equation (6-44) requires arbitrarily choosing values of $F(t)$ and solving for $t$. If a given value of $F(t)$ gives $t < t_p$ or $t > t_w$, it is invalid. The corresponding infiltration rate, $f(t)$, is then found by substituting valid values of $F(t)$ in Equation (6-42), and the corresponding depth of the wetting front, $z_f'(t)$, can be computed from Equation (6-41).

Note from Equation (6-42) that as time goes on and $F(t)$ increases, $f(t)$ becomes asymptotic to $K_h^*$. This behavior is the same as that given by the approximate solution to the Richards Equation in Equation (6-31a) and as observed in nature.

### Estimation of Effective Soil-Hydraulic Properties

The soil properties $\phi$, $K_h^*$, and $\psi_f$ play important roles in the Green-and-Ampt Model, and there have been several studies to determine the appropriate effective values of these parameters:

**Porosity** If measurements of $\phi$ for the soils of interest are lacking, values can be selected from Table 6-1. Or, if bulk density is known, $\phi$ can be estimated via Equation (6-4).

**Hydraulic Conductivity** Although we have developed the Green-and-Ampt Model using the actual saturated hydraulic conductivity as the asymptote for $f(t)$, there is some reason to believe that this value should be adjusted in applying the model. For example, Rawls et al. (1992) suggested adjustments to account for soil crusting in bare soils and for the presence of macropores in vegetated areas. However, Mein and Larson (1973) and other studies have found good results using the nominal value, and we will continue to use it in this text.

**Wetting-Front Suction**  Neuman (1976) has shown that the appropriate definition of $|\psi_f|$ for the Green-and-Ampt Model is

$$|\psi_f| = \int_0^{|\psi(\infty)|} \frac{K_h(|\psi|)}{K_h^*} \cdot d|\psi|. \qquad \textbf{(6-45)}$$

This integral can be evaluated if one has detailed information about the soil's hydraulic properties. Rawls et al. (1992) recommended evaluating $|\psi_f|$ as

$$|\psi_f| = \frac{2 \cdot b + 3}{2 \cdot b + 6} \cdot |\psi_{ae}|. \qquad \textbf{(6-46)}$$

---

### EXAMPLE 6-7

Use the Green-and-Ampt Model to calculate the cumulative infiltration, the infiltration rate, and the depth of the wetting front during a constant rain of $w = 5$ cm hr$^{-1}$ lasting $t_w = 2$ hr on a typical silt loam soil with an initial water content of 0.450.

**Solution**  From Table 6-1, we find that $\phi = 0.485$, $K_h^* = 7.20 \times 10^{-4}$ cm s$^{-1}$ = 2.59 cm hr$^{-1}$, $|\psi_{ae}| = 78.6$ cm, and $b = 5.3$. Then follow these steps:

1. Since $w > K_h^*$, there is potential for ponding and we proceed to the next step.
2. Compute the wetting-front suction, $|\psi_f|$, via Equation (6-46):

$$|\psi_f| = \frac{2 \times 5.3 + 3}{2 \times 5.3 + 6} \times 78.6 \text{ cm} = 64.4 \text{ cm}.$$

3. Compute the time of ponding, $t_p$, via Equation (6-39):

$$t_p = \frac{2.59 \text{ cm hr}^{-1} \times 64.4 \text{ cm} \times (0.485 - 0.450)}{5.00 \text{ cm hr}^{-1} \times (5.00 \text{ cm hr}^{-1} - 2.59 \text{ cm hr}^{-1})}$$
$$= 0.485 \text{ hr}.$$

Since $t_p < t_w$, there is ponding and we proceed with the computations.

4. The infiltration rate is constant at the rainfall rate, 5.00 cm hr$^{-1}$, from $t = 0$ to $t = 0.485$ hr, and the cumulative infiltration during this time is given by Equation (6-34) as

$$F(t_p) = 5.00 \text{ cm hr}^{-1} \times 0.485 \text{ hr} = 2.43 \text{ cm}.$$

5. Select a series of values of $F(t)$ such that 2.43 cm $\leq F(t) \leq 10$ cm, and compute the corresponding values of $t$ via Equation (6-44), values of $f(t)$ via Equation (6-40), and values of $z_f'(t)$ via Equation (6-38):

| $F(t)$ (cm) | 2.50 | 3.00 | 3.50 | 4.00 | 5.00 |
|---|---|---|---|---|---|
| $t$ (hr) | 0.50 | 0.61 | 0.72 | 0.84 | 1.10 |
| $f(t)$ (cm hr$^{-1}$) | 4.93 | 4.54 | 4.26 | 4.05 | 3.76 |
| $z_f'(t)$ (cm) | 71.4 | 85.7 | 100.0 | 114.3 | 142.9 |

| $F(t)$ (cm) | 6.00 | 7.00 | 8.00 | 9.00 |
|---|---|---|---|---|
| $t$ (hr) | 1.37 | 1.66 | 1.96 | 2.26 |
| $f(t)$ (cm hr$^{-1}$) | 3.56 | 3.42 | 3.32 | 3.24 |
| $z_f'(t)$ (cm) | 171.4 | 200.0 | 228.6 | 257.1 |

The last value of $F(t)$ selected, 9.00 cm, gives $t > t_w$, so it is not a meaningful value. By trial and error, we find that $F(t) = 8.14$ cm gives $t \approx 2.00$ hr = $t_w$, with corresponding final values of $f(t_w) = 3.31$ cm hr$^{-1}$ and $z_f'(t_w) = 232.6$ cm. Thus, for this storm under the given conditions, 8.14 cm of water infiltrates and $10.00 - 8.14 = 1.86$ cm of water runs off.

### Explicit Forms of the Green-and-Ampt Model

The need to use trial-and-error solution methods makes direct use of the Green-and-Ampt Model inconvenient for incorporation into models of land-surface hydrologic processes. To get around this problem, Salvucci and Entekhabi (1994a) developed a close approximation to the Green-and-Ampt equations that gives $f(t)$ and $F(t)$ as explicit functions of $t$. Their approach requires computation of three time parameters:

1. A "characteristic time," $T^*$, that depends on soil type and the initial water content:

$$T^* \equiv \frac{|\psi_f| \cdot (\phi - \theta_0)}{K_h^*}; \qquad \textbf{(6-47)}$$

2. A "compression time," $t_c$, which is the equivalent time to infiltrate $F_p = w \cdot t_p$ under initially ponded conditions, as given by the Green-and-Ampt Model:

$$t_c \equiv \frac{F_p}{K_h^*} - \left[ \frac{|\psi_f| \cdot (\phi - \theta_0)}{K_h^*} \right]$$
$$\cdot \ln \left[ 1 + \frac{F_p}{|\psi_f| \cdot (\phi - \theta_0)} \right]; \qquad \textbf{(6-48)}$$

3. An "effective time," $t_e$, defined as

$$t_e \equiv t - t_p + t_c, \qquad \textbf{(6-49)}$$

where $t_p$ is the time of ponding computed via Equation (6-39).

The complete explicit solution is an infinite series, retaining the first four terms of which gives sufficient accuracy for most purposes:

$$f(t) = K_h^* \cdot \left[ 0.707 \cdot \left( \frac{t_e + T^*}{t_e} \right)^{1/2} + 0.667 \right.$$
$$\left. - 0.236 \cdot \left( \frac{t_e}{t_e + T^*} \right)^{1/2} - 0.138 \cdot \left( \frac{t_e}{t_e + T^*} \right) \right];$$
(6-50)

$$F(t) = K_h^* \cdot \{ 0.529 \cdot t_e + 0.471 \cdot (T^* \cdot t_e + t_e^2)^{1/2}$$
$$+ 0.138 \cdot T^* \cdot [\ln(t_e + T^*) - \ln(T^*)]$$
$$+ 0.471 \cdot T^* \cdot \{ \ln [t_e + T^*/2 + (T^* \cdot t_e + t_e^2)^{1/2}]$$
$$- \ln (T^*/2) \} \}.$$
(6-51)

Example 6-8 shows how these explicit relations are used and compares the results with those of the application of the traditional Green-and-Ampt Model as given in Example 6-7.

---

### EXAMPLE 6-8

Calculate the cumulative infiltration, the infiltration rate, and the depth of the wetting front for the same soil, storm, and initial conditions as in Example 6-7, but using the explicit forms of the Green-and-Ampt relations.

#### Solution

1. Compute $T^*$ from Equation (6-47):

$$T^* = \frac{64.4 \text{ cm} \times (0.485 - 0.450)}{2.59 \text{ cm hr}^{-1}} = 0.870 \text{ hr.}$$

2. From Example 6-7, we have $t_p = 0.485$ hr. Thus, $F_p = 5$ cm hr$^{-1} \times 0.485$ hr = 2.43 cm. Substituting this value into Equation (6-48) along with other appropriate values yields

$$t_c = \frac{2.43 \text{ cm}}{2.59 \text{ cm hr}^{-1}} - \left[ \frac{64.4 \text{ cm} \times (0.485 - 0.450)}{2.59 \text{ cm hr}^{-1}} \right]$$
$$\times \ln \left[ 1 + \frac{2.43 \text{ cm}}{64.4 \text{ cm} \times (0.485 - 0.450)} \right]$$
$$= 0.301 \text{ hr.}$$

3. Compute $t_e$ via Equation (6-49):

$$t_e = t - 0.485 \text{ hr} + 0.301 \text{ hr} = t - 0.184 \text{ hr.}$$

4. As in Example 6-7, the infiltration rate is constant at the rainfall rate, 5.00 cm hr$^{-1}$, from $t = 0$ to $t_p = 0.485$ hr.

5. Select a series of times $t$ (0.485 hr $\leq t \leq$ 2.00 hr), and calculate the corresponding values of $t_e$ as in

Step 3; then compute the corresponding values of $f(t)$ via Equation (6-50), $F(t)$ via Equation (6-51), and wetting-front depth, $z_f'(t)$, via Equation (6-38):

| $t$ (hr) | 0.64 | 0.79 | 0.94 | 1.09 | 1.24 |
|---|---|---|---|---|---|
| $t_e$ (hr) | 0.45 | 0.60 | 0.76 | 0.91 | 1.06 |
| $f(t)$ (cm hr$^{-1}$) | 4.38 | 4.05 | 3.83 | 3.67 | 3.55 |
| $F(t)$ (cm) | 3.12 | 3.75 | 4.35 | 4.92 | 5.46 |
| $z_f'(t)$ (cm) | 89.0 | 107.2 | 124.3 | 140.5 | 156.1 |

| $t$ (hr) | 1.39 | 1.55 | 1.70 | 1.85 | 2.00 |
|---|---|---|---|---|---|
| $t_e$ (hr) | 1.21 | 1.36 | 1.51 | 1.66 | 1.82 |
| $f(t)$ (cm hr$^{-1}$) | 3.46 | 3.38 | 3.31 | 3.26 | 3.21 |
| $F(t)$ (cm) | 5.99 | 6.51 | 7.02 | 7.52 | 8.01 |
| $z_f'(t)$ (cm) | 171.3 | 186.1 | 200.5 | 214.7 | 228.8 |

The original implicit (Example 6-7) and explicit (Example 6-8) values of $f(t)$ and $F(t)$ are compared in Figure 6-26. For this case, the explicit method gives slightly lower values, but the relative difference in total infiltration is only about 1.5%.

### Summary

The system of Equations (6-34) through (6-41) provides a complete model of the infiltration process for cases where $w > K_h^*$ and the other assumptions of the idealized situation apply—most importantly, the representation of the wetting front as an abrupt "piston-like" discontinuity and a homogeneous initial water-content profile. Mein and Larson (1973) compared infiltration predictions made using this model with those based on numerical solution of the Richards Equation and found the excellent correspondence shown in Figure 6-27.

Thus in spite of its simplicity, the Green-and-Ampt Model seems to capture the essential aspects of the infiltration process, in particular the complete infiltration of rain up to the time of ponding and the quasi-exponential decline of infiltration rate (= infiltrability) thereafter. Other attractive features of the model are that its parameters are measurable physical properties of the soil and that these parameters enter the model in intuitively logical ways. The widely observed decay of the infiltration rate after the time of ponding is due to the steadily decreasing capillary gradient, $\psi_f/z_f'(t)$, as the wetting front descends, and this decay is asymptotic to $K_h^*$. Thus the minimum value of infiltrability for a soil is given by its saturated hydraulic conductivity.

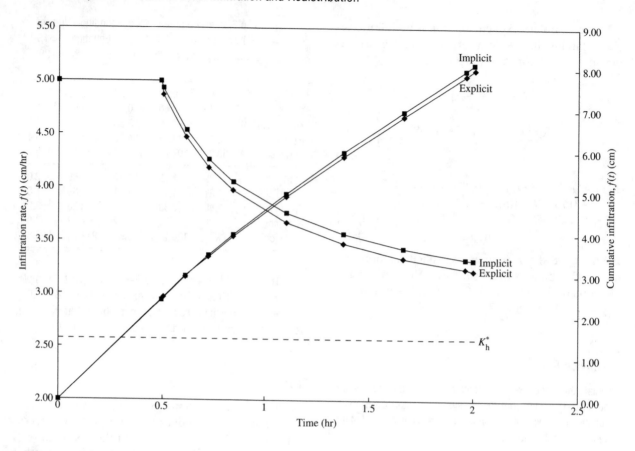

**FIGURE 6-26**

Comparison of the infiltration rate (decreasing curves) and the cumulative infiltration (increasing curves) computed by the implicit Green-and-Ampt Model (Example 6-7) and the explicit approximation of Salvucci and Entekhabi (1994a) (Example 6-8).

The assumption of a piston-like wetting front is closer to reality for sands than for clays, in which the wetting front tends to be less distinct. The computation of equilibrium water-content profiles (Figure 6-16) suggests that the assumption of homogeneous initial water content is better for clays than sands; the profile also becomes more homogeneous for deeper water tables for all soil types.

The Green-and-Ampt Model has been extended for application to conditions of unsteady rain (Chu 1978), time-varying depths of ponding (Frey-berg et al. 1980), and soils in which $K_h^*$ decreases with depth continuously (Beven 1984) or in discrete layers (Rawls et al. 1992). Rawls et al. (1992) also described an approach that adapts the Green-and-Ampt Model to situations in which macropores are important.

James et al. (1992) successfully incorporated the basic Green-and-Ampt model into a comprehensive watershed model, and Salvucci and Entekhabi's (1994a) explicit formulas appear to be even better suited to this purpose.

### 6.6.3 Green-and-Ampt Approach for Shallow Soils

As we have just seen, the Green-and-Ampt Model provides useful insight into infiltration and the phenomenon of potential runoff formation by saturation from above when $w > K_h^*$ and $t_w > t_p$ in deep homogeneous soils with a uniform initial water

**FIGURE 6-27**
Comparison of infiltration rates computed by the Green-and-Ampt Model (symbols) with those computed by numerical solution of the Richards Equation (lines). (a) shows the effects of initial water content, $\theta_0$, for an identical water-input event; (b) shows the effect of water-input rates, $w$, for identical $\theta_0$. From Mein and Larson (1973); used with permission of the American Geophysical Union.

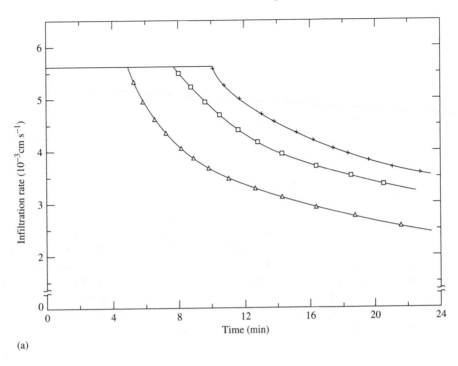

(a)

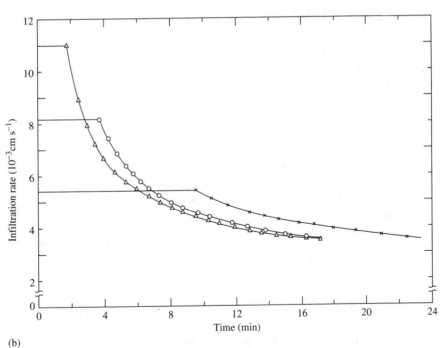

(b)

content. We can use the same approach to explore infiltration in soils underlain by an impermeable layer or by a water table at a relatively shallow depth, which can lead to saturation from below. Once such saturation occurs, infiltration ceases and detention storage builds up; as with saturation from above, this water is potential overland flow if the surface is sloping.

The analysis here is limited to a horizontal impermeable layer. If this layer or the water table is sloping, the analysis is complicated by the down-slope porous-media flow; this situation will be explored in Section 9.2.3. To get a basic understanding of the phenomenon of saturation from below and the effects of a shallow water table or impermeable layer, we first consider the case of a soil with a homogeneous initial water content, as in the deep-soil case. Following this discussion, we qualitatively explore situations in which the water content varies with depth, which is more realistic for shallow soils.

### Water-Input Rate Less Than Saturated Hydraulic Conductivity

Consider a permeable soil that is horizontal and has invariant hydraulic properties from the surface down to an effectively impermeable horizontal boundary (which could be a bedrock surface, a fragipan layer, or the top of the tension-saturated zone above a water table) at a depth $z_u'$. The initial water content is $\theta_0 << \theta_{fc}$ throughout. Water input at a rate $w$, where $K_h(\theta_0) < w < K_h^*$, begins at time $t = 0$ and stops at $t = t_w$. Infiltration will occur at the rate $w$ until the time at which the soil becomes saturated at the surface, $t_b$:

$$f(t) = w, \quad 0 < t < t_b \qquad (6\text{-}52)$$

and

$$F(t) = w \cdot t, \quad 0 < t < t_b. \qquad (6\text{-}53)$$

Here, $t_b$ is calculated as developed later in this section.

Water storage will occur at each successive depth until $q_z(\theta) = w$, at which point the water content is designated $\theta_w$. Again assuming a piston-like wetting front, above which $\theta = \theta_w$ and below which $\theta = \theta_0$, we can combine the continuity relation,

$$F(t) = (\theta_w - \theta_0) \cdot z_f'(t), \qquad (6\text{-}54)$$

with Equation (6-53) to get the position of the wetting front, $z_f'(t)$, as a function of time:

$$z_f'(t) = \frac{w \cdot t}{\theta_w - \theta_0}. \qquad (6\text{-}55)$$

From this equation, we can calculate the time it takes the wetting front to reach the effectively impermeable boundary, $t_u$:

$$t_u = \frac{(\theta_w - \theta_0) \cdot z_u'}{w}. \qquad (6\text{-}56)$$

If $t_w > t_u$, water will continue to arrive at the boundary and the water content just above the boundary will increase until it reaches saturation. Now an upward-moving wetting front will develop, below which $\theta = \phi$ and above which $\theta = \theta_w$. As long as water input continues, the position of this front, $z_f'(t)$, is given by

$$z_f'(t) = z_u' - \frac{w \cdot (t - t_u)}{\phi - \theta_w}, \quad t > t_u. \qquad (6\text{-}57)$$

By combining Equations (6-56) and (6-57), we can calculate the time, $t_b$, when the saturation front reaches the surface:

$$t_b = \frac{z_u' \cdot (\phi - \theta_0)}{w}. \qquad (6\text{-}58)$$

At the instant the saturation front reaches the surface, infiltration will decrease abruptly from $f(t) = w$ to $f(t) = 0$, and water will begin to build up on the surface.

### Water-Input Rate Greater Than Saturated Hydraulic Conductivity

If $w \geq K_h^*$ and $t_b \leq t_p$, the analysis of the section immediately preceding this one applies, and we have saturation from below.

If $t_b > t_p$, ponding occurs due to saturation from above and infiltration begins to decrease when $t = t_p$, just as in the deep-soil case. The position of the wetting front for $t > t_p$ can be found from Equations (6-41) and (6-44) (as it was in Examples 6-6 and 6-7), and the instant when $z_f'(t) = z_u'$ can be identified. Since the Green-and-Ampt Model assumes saturation above the wetting front, the entire soil layer is saturated and infiltration ceases at this instant. This situation thus involves simultaneous saturation from above and below.

### Infiltration into Soils with Depth-Varying Water Content

As shown in Figure 6-16, the presence of a water table near the surface tends to cause the water-content profile to be non-uniform. When water content increases with depth, as in typical equilibrium profiles, the descending wetting front encounters pores that are increasingly occupied with water. Thus, with a constant water-input rate, the wetting front descends more rapidly than with a homogeneous initial water content and continuously accelerates. The wetting front thus reaches the top of the tension-saturated zone more quickly than calculated in the preceding analysis for uniform initial water content. Salvucci and Entekhabi (1995) derived a Green-and-Ampt-like approach to calculating the infiltration with non-uniform (i.e., climatic-equilibrium) water-content profiles.

### 6.6.4 Application of the Philip Equation

#### Infiltration from Rain and Snowmelt

The Philip Equation [Equation (6-31)] was derived for ponded conditions. However, it can be applied to the modeling of rain and snowmelt infiltration by using generalized soil properties [Equations (6-12)–(6-14) and Table 6-1] and by applying a time-adjustment approach similar to that used for the explicit Green-and-Ampt Model.

First, $S_p$ is computed from the soil properties and initial conditions via Equation (6-18). $K_p$ is often assumed to equal the saturated hydraulic conductivity, $K_h^*$, although smaller values, in the range $K_h^*/3 \le K_p \le 2 \cdot K_h^*/3$, usually fit measured values better for short time periods (Sharma et al. 1980). Then the compression time, $t_{cp}$, is found by solving Equation (6-31a) for time when the infiltration rate equals the water-input rate, $w$:

$$t_{cp} = \left[ \frac{S_p}{2 \cdot (w - K_p)} \right]^2. \tag{6-59}$$

Next, the apparent time of ponding, $t_{pp}$, is computed as

$$t_{pp} \equiv \frac{F(t_{cp})}{w} = \frac{S_p \cdot t_{cp}^{1/2} + K_p \cdot t_{cp}}{w}. \tag{6-60}$$

Values of $f(t)$ and $F(t)$ are then computed via Equations (6-31), but with $t_{ep}$ substituted for $t$, where

$$t_{ep} = t - t_{pp} + t_{cp}. \tag{6-61}$$

### Estimation of $S_p$ and $K_p$ from Infiltrometer Measurements

In many studies, $S_p$ and $K_p$ are treated simply as empirical parameters whose values are those that best fit infiltration data measured with infiltrometers. Box 6-2 shows how such values are determined using the principle of least squares, and Example 6-9 applies this principle to the infiltration data developed in the previous examples. Because the Philip Equation applies only after ponding occurs, we define $t' \equiv t - t_p$ to simplify the notation.

---

### EXAMPLE 6-9

Find the values of $S_p$ and $K_p$ that best fit the infiltration rates after the time of ponding in Example 6-7.

**Solution** From Example 6-7, we have $N = 10$ and compute $\Sigma f(t_i') = 36.1$ cm hr$^{-1}$, $\Sigma [f(t_i')/t_i'^{1/2}] = 48.3$ cm hr$^{-1/2}$, $\Sigma (1/t_i') = 19.2$ hr$^{-1}$, and $\Sigma (1/t_i'^{1/2}) = 12.9$ hr$^{-1/2}$. Substituting these values into Equation (6B2-9) yields $S_p = 1.38$ cm hr$^{-1/2}$, about 40% of the value $S_p = 3.42$ cm hr$^{-1/2}$ given by Equation (6-18); and substituting the appropriate values into Equation (6B2-8) yields $K_p = 2.72$ cm hr$^{-1}$, which is slightly greater than $K_h^* = 2.59$ cm hr$^{-1}$.

The Philip Model is well suited for characterizing the spatial variability of infiltrometer measurements.

### 6.6.5 Infiltration over Areas

#### Spatial and Temporal Variability of Infiltration

Detailed field studies generally show a very high degree of spatial variability of infiltrability and the factors that affect infiltration rates. The experiments of Burgy and Luthin (1956) give a striking example of this variability. They first measured the infiltrability of an unvegetated clay soil by flooding a 335-m$^2$ basin and measuring the rate of decline of the water level. After the water had completely infiltrated, they made infiltrometer-ring measurements at 119 sites evenly spaced throughout the basin. The average value for the ring measurements, 16.8 cm hr$^{-1}$, was reasonably close to the basin

## BOX 6-2

. . . . . . .

### Least-Squares Estimates of Parameters of the Philip Equation

Field or laboratory experiments are often conducted to determine the values of parameters that appear in a theoretical or quasi-theoretical equation that characterizes a particular process. In such situations, one wants to find the values of the parameters that will give the "best fit" of the equation to the data.

"Best fit" is usually interpreted to mean the parameter values that minimize the sum of the squared differences between the measured values and the values given by the theoretical equation. This concept is called the "least-squares" criterion, and it is represented mathematically as

$$\text{minimize } (SS), \qquad \text{(6B2-1)}$$

where

$$SS = \sum_{i=1}^{N}(x_i - \hat{x}_i)^2, \qquad \text{(6B2-2)}$$

$x_i$ is the $i$th measured value of the process of interest, $\hat{x}_i$ is the $i$th value of the process according to the theoretical equation, and $N$ is the number of measured values.

For example, we might measure the ponded infiltration rate, $f(t')$, at a given location with an infiltrometer to determine the values of $S_p$ and $K_p$ that provide the best fit of the Philip Equation [Equation (6-31a)] to the measured values. In this case, Equation (6B2-2) becomes

$$SS = \sum_{i=1}^{N}\left[\frac{S_p}{2 \cdot t'^{1/2}} + K_p - f(t'_i)\right]^2, \qquad \text{(6B2-3)}$$

where $f(t'_i)$ are the infiltration rates measured at $N$ successive times $t'_i$. This expression can be expanded to

$$SS = \frac{S_p^2}{4} \cdot \sum\frac{1}{t'_i} + K_p \cdot S_p \cdot \sum\frac{1}{t_i'^{1/2}} - S_p \cdot \sum\frac{f(t'_i)}{t_i'^{1/2}}$$
$$+ N \cdot K_p^2 - 2 \cdot K_p \cdot \sum f(t'_i) + \sum f^2(t'_i).$$

$$\text{(6B2-4)}$$

(The summation limits are dropped to simplify the notation.)

The values of $S_p$ and $K_p$ that minimize $SS$ are found by taking the derivative of Equation (6B2-4) with respect to each of the parameters, setting the results equal to

zero, and solving for the parameters. Taking the derivatives, we find

$$\frac{\partial SS}{\partial S_p} = \frac{S_p}{2} \cdot \sum\frac{1}{t'_i} + K_p \cdot \sum\frac{1}{t_i'^{1/2}} - \sum\frac{f(t'_i)}{t_i'^{1/2}}$$

$$\text{(6B2-5)}$$

and

$$\frac{\partial SS}{\partial K_p} = S_p \cdot \sum\frac{1}{t_i'^{1/2}} + 2 \cdot N \cdot K_p - 2 \cdot \sum f(t'_i).$$

$$\text{(6B2-6)}$$

Setting these expressions equal to zero and solving yields

$$S_p = \frac{2 \cdot \sum[f(t'_i)/t_i'^{1/2}] - 2 \cdot K_p \cdot \sum(1/t_i'^{1/2})}{\sum(1/t'_i)}$$

$$\text{(6B2-7)}$$

and

$$K_p = \frac{2 \cdot \sum f(t'_i) - 2 \cdot S_p \cdot \sum(1/t_i'^{1/2})}{2 \cdot N}.$$

$$\text{(6B2-8)}$$

Equation (6B2-8) can be substituted into Equation (6B2-7) and the result solved to give the least-squares estimate of $S_p$ entirely in terms of the measured data, $t_i'$ and $f(t_i')$:

$$S_p =$$
$$\frac{2 \cdot N \cdot \sum[f(t_i')/t_i'^{1/2}] - 2 \cdot \sum f(t_i') \cdot \sum(1/t_i'^{1/2})}{N \cdot \sum(1/t_i') - [\sum(1/t_i')^{1/2}]^2}.$$

$$\text{(6B2-9)}$$

Finally, the least-squares estimate of $K_p$ is found by substituting $S_p$ from Equation (6B2-9) into Equation (6B2-8).

One can derive equivalent expressions for estimating $S_p$ and $K_p$ starting with the cumulative-infiltration form of the Philip Equation [Equation (6-31b)].

value of 10.8 cm hr$^{-1}$, but the ring values ranged from near zero to over 110 cm hr$^{-1}$.

Sharma et al. (1980) found an 11-fold and 29-fold variation in the Philip-Equation parameters $S_p$ and $K_p$, respectively, and a 7.5-fold variation in total infiltration for a 30-min storm over a 0.096-km$^2$ watershed in Oklahoma, with no consistent relation between these quantities and location or soil type. In another detailed study in that same watershed, Loague and Gander (1990) found that the scale of spatial correlation of infiltration was less than 20 m, that there was little spatial structure to the data, and that infiltration rates were strongly influenced by animal activity, vegetation, and climate rather than by soil texture. Buttle and House (1997) found similar results in Ontario, where saturated hydraulic conductivity was largely determined by macropores.

In other small-watershed-scale studies in humid areas, Tricker (1981) measured infiltrability (infiltration rates after 1 hr of ponded infiltration) over a 3.6-km$^2$ watershed in England and found values ranging from near zero to 256 cm hr$^{-1}$; these values were positively related to the thickness of the soil-litter layer and unrelated to soil texture or soil-water content. Measurements on a 15.7-km$^2$ watershed in Northern Ireland found a considerably lower range of variability in infiltrability, but a similar dependence on soil organic matter rather than soil texture (Wilcock and Essery 1984). The latter study also found a substantial seasonal variability, with average infiltrability considerably higher in summer (0.9 cm hr$^{-1}$ in June) than in winter (0.06 cm hr$^{-1}$ in January).

Springer and Gifford (1980) repeatedly measured infiltrability with infiltrometers at 20 to 25 locations on a plowed and grazed rangeland area in Idaho and found coefficients of spatial correlation [see Box 4-2, Equation (4B2-3)] in the range 0.40 to 0.68 and little seasonal variation. In another semi-arid area, Berndtsson (1987) found a very high degree of spatial variability in the coefficients of the Philip Equation over a 19.6-km$^2$ catchment in Tunisia: Values of $S_p$ and $K_p$ had coefficients of variation [Equation (C-20)] exceeding 1.00. Schumm and Lusby (1963) found wide seasonal variations in infiltrability in a sparsely vegetated badlands area.

Unfortunately, then, the dominant conclusions of these and other field studies are as follows: (1) infiltration varies greatly over short (1- to 20-m) dis-

tances, and (2) the variations are generally not related to soil textures, but are instead determined by plant and animal activity and by small-scale topographic variations. These conclusions make it difficult to transfer an understanding of the process at a point to its representation over a watershed. Some approaches to this problem are discussed in the next section.

### Infiltration in Watershed Models

In many watershed models, it is assumed that a single spatially representative value for each of the parameters of the Green-and-Ampt, Philip, or other model of infiltration can be applied to portray the process over the whole watershed. This is the approach taken in the BROOK90 model (Box 6-3). To represent spatial variability, one can divide the watershed into sub-areas, each characterized by its appropriate soil properties and initial conditions, apply the model to each sub-area, and compute an areally weighted average of infiltration at successive points in time, as done by James et al. (1992). However, the fact that soil properties vary on a scale of a few meters means that an impractically large number of sub-areas would often be required to represent the true variability using this approach.

Thus the question of how to capture the spatial variability of infiltration and runoff remains. It is clear from the non-linear forms of the equations constituting the Richards Equation and the Green-and-Ampt relations that a watershed-average value for infiltration cannot be computed from watershed-average values of the soil properties and initial water contents. This non-linearity, in fact, raises the question of whether these approaches, which are derived for virtual points on the land surface, can be applied in modeling larger areas such as watersheds or the very large grid cells used in continental or global modeling.

In an extensive review of the problem of scaling soil parameters, Kabat et al. (1997) concluded that these physically based approaches can be applied to large areas, but that at the large scales, the parameters that represent hydraulic conductivity, porosity, etc., should be treated simply as calibration parameters that do not necessarily have a physical meaning. Using numerical simulation, Kabat et al. showed that methods that represent soils over an area by the dominant soil type, or by conceptually mixing or aggregating the soils, do not successfully model the

## BOX 6-3
· · · · · · ·
### Simulation of Soil-Water Movement in the BROOK90 Model

The BROOK90 model uses Equations (6-12) and (6-13) to simulate $\psi - \theta$ and $K_h - \theta$ relations, extended to include the portion of the moisture-characteristic curve at tensions less than $|\psi_{ae}|$. However, the parameterization of these relations differs from that of Equations (6-12) and (6-13). (See the model's documentation.) For forest soils, some of the parameter values may differ from those given in Table 6-1 because of higher porosities.

Figure 2-12 is the flow chart for the model. The number of soil layers is an input parameter, *ILAYER*%; *I* is the layer counter. Each day an amount *SMLT+RNET* reaches the soil surface. A fraction of this amount is immediately allocated to streamflow, simulating impervious or permanently saturated areas. The remaining portion, *SLFL*, is allocated to soil layers by the fraction *INFRAC(I)*, where

$$INFRAC(I) = [THICKA(I)/THICKT]^{\wedge} INFEXP - [THICKA(I-1)/THICKT]^{\wedge} INFEXP. \tag{6B3-1}$$

In Equation (6B3-1), *THICKA(I)* is the depth of the bottom of layer *I*, and *THICKT* is the total soil depth. *INFEXP* is a specified parameter. When *INFEXP* = 0, all of the infiltrating water contributes to soil-water storage, *SWATI*(1), and infiltration proceeds vertically through the layers as a classic wetting front (as described later). As *INFEXP* increases, more water goes to lower layers, simulating macropore flow. Default values are *INFEXP* = 0.3 for forest, 0.1 for grass, and 0 for bare soil.

Each layer can receive input as vertical macropore flow [*INFLI(I)* = *SLFL* · *INFRAC(I)*] or via the downward movement of the wetting front, *VRFLI(I)*. *VRFLI(I)* is determined by a finite-difference form of Darcy's Law, with the area of flow reduced by the stone content of the layers [input parameters *STONE(I)*]:

$$VRFLI(I) = [GRAD(I) \cdot KKMEAN(I)/RHOWG] \cdot \tag{6B3-2}$$
$$\{1 - [STONE(I) + STONE(I+1)]/2\}.$$

*GRAD(I)* is the vertical gradient between layers *I* and *I+1*:

$$GRAD(I) = [PSIT(I) - PSIT(I+1)]/ \{[THICK(I) + THICK(I+1)]/2\}. \tag{6B3-3}$$

*PSIT(I)* is the total head:

$$PSIT(I) = PSIM(I) + PSIG(I). \tag{6B3-4}$$

*PSIM(I)* is the matric suction (in kPa), which is determined by the wetness of the layer and the soil parameters [as in Equation (6-12)]. *PSIG(I)* is the gravitational potential (kPa), which is determined by the average depth of the layer. Because potentials are given in kPa, they are divided by the density of water and gravitational acceleration (*RHOG*) to convert to length units (mm) in Equation (6B3-2).

The hydraulic conductivity [*KK(I)*] of each layer is computed via BROOK90's version of Equation (6-13) and the layer's wetness, and the geometric mean value for flow between layers *I* and *I +1*, *KKMEAN(I)*, is

$$KKMEAN(I) = exp\{\{THICK(I+1) \cdot \ln[KK(I)] + THICK(I) \cdot \ln[KK(I+1)]\}/ \{THICK(I) + THICK(I+1)\}\}. \tag{6B3-5}$$

Water may leave each layer as evapotranspiration [*SLVP* + *TRANI(I)*] (discussed in Box 7-4), as macropore flow downslope to the stream [*BYFLI(I)*], and as matrix (Darcy) flow downslope to the stream [*DSFLI(I)*] (discussed in Section 9.2.3). If the upper soil layer becomes saturated, the excess water becomes streamflow, simulating ponding due to saturation from above.

The water balance for each layer is computed for time steps that vary in length to satisfy several criteria for numerical stability. (See the program documentation for details.)

real hydrologic behavior. However, the values of the effective parameters can be fairly simply determined by using **reference curves** derived by scaling the $\psi - \theta$ and $K_h - \theta$ relations. Soils with similar hydrologic response can be identified on the basis of the dimensionless ratio, $\delta$, where

$$\delta \equiv \frac{D(\theta)}{K_h^{*2}} \qquad \textbf{(6-62)}$$

and the hydraulic diffusivity is evaluated at a particular water content for all soils.

A very simple method for incorporating the spatial variability of infiltration in a watershed model was described by Manley (1977). Based on the work of Burgy and Luthin (1956), Manley assumed a linear cumulative distribution of ultimate infiltrability ($K_h^*$) over the watershed (Figure 6-28). The intersection of a horizontal line representing the rainfall rate, $w$, with the line representing this distribution defines an area proportional to the rate of runoff, $q$; this rate can then be calculated as

$$q = \frac{w^2}{2 \cdot K_h^{*+}}, \qquad w \le K_h^{*+}; \qquad \textbf{(6-63a)}$$

$$q = \frac{3 \cdot w}{2} - K_h^{*+}, \qquad w > K_h^{*+}, \qquad \textbf{(6-63b)}$$

where $K_h^{*+}$ is the maximum value of $K_h^*$ for the watershed.

Manley's (1977) approach is similar to that used in other operational watershed models. [See Viessman et al. (1989)]. Several papers have suggested more sophisticated approaches for incorporating the variability of soil properties and initial conditions in watershed models (e.g., Sharma et al. 1980; Maller and Sharma 1981; Berndtsson and Larson 1987; Berndtsson 1987; Valdes et al. 1990).

## 6.7  REDISTRIBUTION

Following a rain or snowmelt event, infiltrated water is subject to redistribution by gravity and pressure forces and removal by evapotranspiration. This process is governed by and can be numerically modeled via the Richards Equation, with the addition of a "sink" term to represent uptake of water by plants in the root zone. Following Hillel (1980b),

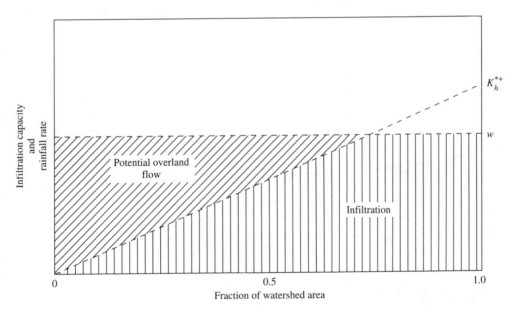

**FIGURE 6-28**

Simple approach to estimating watershed infiltration and potential overland flow. The diagonal line represents an approximation of the areal distribution of infiltration, with $K_h^{*+}$ as the maximum value for the area. The horizontal line is drawn at the rainfall rate. The diagonally shaded area is proportional to the potential overland flow; the geometry of the relation yields Equation (6-63). After Manley (1977).

we consider redistribution first in completely wetted profiles, then in partially wetted profiles.

### 6.7.1 Completely Wetted Profiles

To develop an approximate quantitative approach, we consider an initially homogeneous deep-soil profile with no water table and no evapotranspiration. To a first approximation, the water content changes equally at all depths $z'$ (Hillel and van Bavel 1976). Thus if $d\theta/dz = 0$, the vertical flux, $q_{z'}$, at any level is given by Equation (6B1-9) as

$$q_{z'} = K_h(\theta). \tag{6-64}$$

We can also reason that if $d\theta/dt$ is constant through the profile, the flux through any level $z_b'$ must equal

$$q_{z'} = -z_b' \cdot \frac{d\theta}{dt}. \tag{6-65}$$

[See Equation (6B1-2).] Equating Equations (6-64) and (6-65), using the $K_h - \theta$ relation of Equation (6-13b), and integrating yields the following relation for water content as a function of time at any level $z_b'$:

$$\theta(t) = \left[ \phi_{\cdot}^{1-c} - \frac{K_h^* \cdot (1-c)}{z_b' \cdot \phi^c} \cdot t \right]^{\frac{1}{1-c}}. \tag{6-66}$$

Figure 6-29 shows how this equation models the decrease in water content at two levels during drainage of a sand soil and a clay soil. Note that sand drains much more quickly and that the water contents do differ at the two levels after drainage begins—so that, in reality, the pressure forces ignored in the derivation of Equation (6-66) are actually present. However, the results shown are not greatly different from those obtained by including the pressure forces (Hillel and van Bavel 1976). Note also

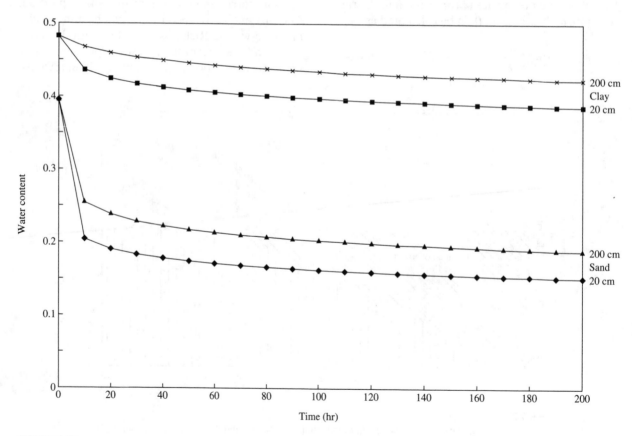

**FIGURE 6-29**
Water content as a function of time at depths of 20 and 200 cm during drainage (redistribution) for completely wetted profiles in a sand soil and a clay soil, according to Equation (6-66).

that the water contents level off after a few tens of hours, justifying the concept of field capacity.

### 6.7.2 Partially Wetted Profiles

It is more common that infiltration will cease before the wetting front has penetrated through an entire soil profile. Under these conditions, experiments have shown that in the absence of a water table there are two basic patterns of soil-water redistribution following infiltration (when no evapotranspiration occurs) (Figure 6-30). In the first, the water content always decreases monotonically with depth and the water-content gradient across the wetting front gradually decreases as the front descends. In this case, the smaller the amount of infiltrated water, $F(t_w)$, the faster the rate of redistribution. This situation occurs when the gravitational force is negligible compared with the capillary force (i.e., when the water content above the wetting front at the start of redistribution is less than field capacity). This condition occurs with fine-grained soils and small initial depths to the wetting front at the cessation of infiltration.

In the second pattern, a "bulge" in water content develops due to rapid gravitational drainage soon after infiltration ceases and persists as redistribution progresses. A sharp wetting front is maintained, but the water contents above the bulge form

a gradual "drying front" that is transitional to the field capacity. This situation occurs when the gravitational force is significant (i.e., the water content above the initial wetting front exceeds the field capacity) and hence characterizes coarse-grained soils and higher $F(t_w)$. In this case, the rate of redistribution increases with greater $F(t_w)$, and in the "final" condition, the infiltrated water is distributed over a depth $z_f'("\infty")$, given by

$$z_f'("\infty") = \frac{F(t_w)}{\theta_{fc} - \theta_0}. \qquad (6\text{-}67)$$

Once this "final" state is reached, there will be slow redistribution by pressure forces at the wetting front, as in the first pattern. Biswas et al. (1966) showed several examples of redistribution in the absence of a water table in laboratory experiments.

Using numerical solutions to the Richards Equation, Rubin (1967) showed that hysteresis has a significant effect on soil-water profiles during redistribution (again, without evapotranspiration or a water table). Interestingly, he found that profiles accounting for hysteresis were not intermediate between those using non-hysteretic wetting and non-hysteretic drying moisture-characteristic curves (Figure 6-31).

In many natural situations, extraction of water from the upper soil layers by evapotranspiration

**FIGURE 6-30**
Typical soil-water profiles showing the redistribution of water at successive times ($t_0$, ..., $t_3$) following infiltration into a deep soil with no evaporation. Time $t_0$ is when infiltration ceases and redistribution begins. (a) Redistribution when capillary force dominates gravitational force. At $t_0$ the water content above the wetting front is less than field capacity. (b) Redistribution when gravitational force dominates. At $t_0$ the water content above the wetting front is greater than field capacity. From Youngs and Poulovasisilis (1976); used with permission of the American Geophysical Union.

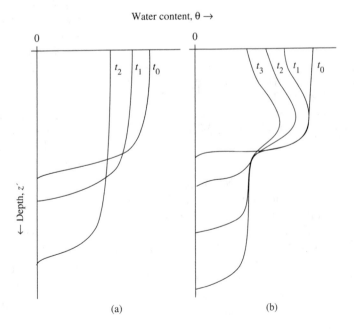

Water content, $\theta \rightarrow$

$\leftarrow$ Depth, $z'$

(a)

(b)

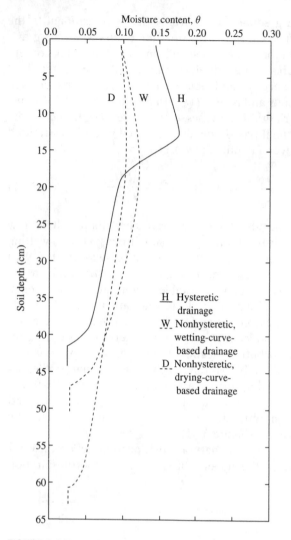

**FIGURE 6-31**
Numerically simulated water-content profiles two hours after cessation of rain on a sandy soil. Note that the profile which accounts for hysteresis in the moisture-characteristic curve (H) is not intermediate between the profiles based on non-hysteretic curves characterizing wetting (W) and drying (D). After Rubin (1967).

strongly affects water redistribution following infiltration. In general, an upward-decreasing water-content gradient similar to that of Figure 6-30b is produced, but Equation (6-67) is no longer valid. Figure 6-32 shows water-content changes following infiltration in a forest soil; most of the water loss in this case could be attributed to evapotranspiration rather than to gravitational drainage, as in Figure 6-30b. The importance of evapotranspiration to the seasonal variation of soil-water content is evident

in Figure 6-33, which shows a wide range of $\theta$ in the root zone of a silt-loam soil in an agricultural watershed and a much narrower range below this zone.

Where a water table is present, the water-content profile will fluctuate about the climatically determined equilibrium water-content profile (Figure 6-16). Salvucci and Entekhabi (1994b) showed that the equilibrium profile is a reasonable estimate of the typical antecedent conditions encountered by rain events.

### 6.7.3 Modeling

Clearly, the movement of water following infiltration is complicated: Upward- and downward-directed pressure gradients can develop in different parts of the profile due to drainage and evapotranspiration, and the effects of hysteresis in the moisture-characteristic relation can be important. Detailed modeling of the process based on the Richards Equation and accounting for hysteresis can be done analytically (Warrick et al. 1990) or via numerical methods (Rubin 1967; Musiake et al. 1988; Milly 1988). In numerical models, the soil is divided into a number of layers (generally on the order of a few centimeters thick), and the equations are solved for successive time steps (generally on the order of minutes to hours).

For many purposes, the drainage from a soil layer can be satisfactorily modeled using a simpler approach based on an approximation of Darcy's Law in which the capillary gradient is ignored, as in Equations (6-64)–(6-66), where the water content is averaged over the soil layer and the $K_h(\theta)$ relation can be represented by Equation (6-13). For example, Black et al. (1970) used Equation (6-64) in a simple water-balance procedure applied to a soil layer supporting a vegetable crop. In their application, the change in water content of the soil layer over time period $t$, $\Delta\overline{\theta}_t$, was calculated as

$$\Delta\overline{\theta}_t = \frac{-K_h(\overline{\theta}_{t-1}) \cdot \Delta t - ET_t - F_t}{\Delta z}, \quad \textbf{(6-68)}$$

where $ET_t$ and $F_t$ are the total evapotranspiration and total infiltration during period $t$, respectively; $\Delta t$ is the time step; and $\Delta z$ is the thickness of the layer. The average soil-water content at the end of period $t$, $\overline{\theta}_t$, was then found as

$$\overline{\theta}_t = \overline{\theta}_{t-1} + \Delta\overline{\theta}_t. \quad \textbf{(6-69)}$$

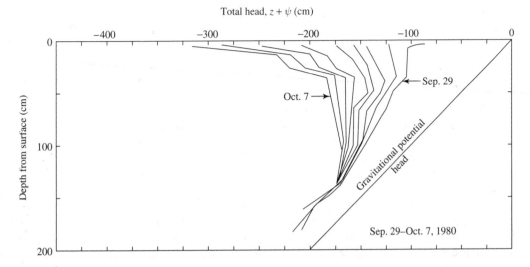

**FIGURE 6-32**

Daily profiles of total potential-energy changes due largely to evapotranspiration from a volcanic loam soil during a rainless period. The soil-water tension, which reflects the water content, is the horizontal distance between each profile and the gravitational-head line. From Musiake et al. (1988); used with permission.

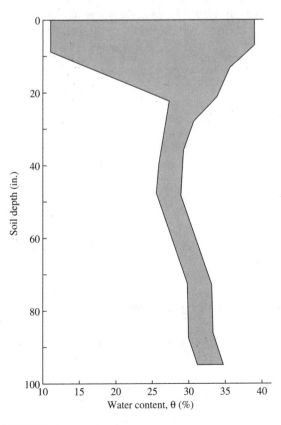

**FIGURE 6-33**

Annual range of soil-water content, $\theta$, in a silt–loam soil supporting grass and clover in Ohio. After Dreibelbis (1962).

Black et al. (1970) found that water contents estimated in this way compared well with measured values, even with $\Delta t = 1$ day and $\Delta z = 150$ cm. However, this approach can become numerically unstable unless $\Delta \bar{\theta}_t \ll \bar{\theta}_{t-1}$, so Federer and Lash (1978) found it necessary to make $\Delta t < 1$ day when applying this approach in their BROOK model. With this modification, Equations (6-68) and (6-69) appear to be sound bases for simulating percolation in general hydrologic models. [See also Groves (1989)].

Box 6-3 describes how infiltration and vertical flow are simulated in the BROOK90 watershed model using an approach similar to that of Equations (6-68)–(6-69).

## 6.8 SUMMARY

Infiltration and soil-water movement are perhaps the most important hydrologic processes, since they determine the rates and amounts of water available for surface and subsurface runoff, the amounts of water available for evapotranspiration, and the rates and amounts of recharge to ground water. These processes are also crucial in determining water quality. [See Nielsen et al. (1986) for a review of solute transport in the vadose zone.] In fact, it has been said that "hydrology is mostly soil physics" (Youngs 1988).

The classical theory of unsaturated flow based on Darcy's Law and the Richards Equation is well developed and tested and has been successfully incorporated in hydrologic models in various forms. Some of these models employ approximate analytical solutions or numerical approximations of the basic equations, while others use simpler, but physically based, representations like the Green-and-Ampt and Philip approaches.

The most critical theoretical and practical questions in the application of theory concern the representation of the typically wide variability in the conditions that determine infiltration. Because of our limited understanding of this variability and the paucity of field data, we are often led to highly simplified model representations of the process. Promising new approaches in this area include the application of various statistical methods (e.g., Berndtsson and Larson 1987) and scaling and similarity theory (e.g., Sharma et al. 1980; Wood et al. 1990; Kabat et al. 1997). There is a continuing need for field data so that we can understand what we are attempting to model, and there is indication that remote sensing of soil moisture (Jackson 1988) and other surface factors will be helpful in this regard.

We also need to be aware that the classical theory may inadequately represent soil-water processes in many situations. This is especially true where macropores are important routes of water and solute transport, and understanding and modeling flows in soils containing these conduits is an active field of research.

The redistribution of soil water can also be affected by temperature gradients and freezing and thawing, which introduce significant complications into the flow equations, and by water uptake by plants. Approaches to modeling this latter process are described in Section 7.5.

The quantitative characterizations of equilibrium soil-water profiles developed by Salvucci and Entekhabi (1994b; 1995) [Equations (6-24)–(6-26)] are important contributions to our understanding of how soil-water conditions are affected by climate, soil type, and water-table depth and appear to be very useful for large-scale water-balance modeling and for rainfall-runoff modeling. (Discussed further in Section 9.5.)

## EXERCISES

Exercises marked with ** have been programmed in EXCEL on the CD that accompanies this text. Exercises marked with * can be advantageously executed on a spreadsheet, but you will have to construct your own worksheets to do this.

*6-1. (a) Plot the grain-size distributions for the soils in the table below.
(b) Determine the texture classes of the soils of part (a).

| Diameter (mm) | 50 | 19 | 9.5 | 4.76 | 2.00 | 0.420 | 0.074 | 0.02 | 0.005 | 0.002 |
|---|---|---|---|---|---|---|---|---|---|---|
| *Soil A1* | | | | | | | | | | |
| % Finer | 100 | 100 | 100 | 100 | 100 | 100 | 97 | 79 | 45 | 16 |
| *Soil B1* | | | | | | | | | | |
| % Finer | 100 | 100 | 98 | 94 | 70 | 19 | 15 | 8 | 3 | 2 |
| *Soil C1* | | | | | | | | | | |
| % Finer | 93 | 91 | 88 | 85 | 69 | 44 | 40 | 27 | 13 | 6 |
| *Soil D1* | | | | | | | | | | |
| % Finer | 100 | 100 | 100 | 100 | 100 | 97 | 92 | 75 | 47 | 31 |

*6-2. Field and oven-dry weights of four soil samples taken with a 10-cm-long by 5-cm-diameter cylindrical tube are given in the accompanying table. Assuming $\rho_m = 2650$ kg m$^{-3}$, calculate the field water contents and degrees of saturation and the bulk densities and porosities of those soils.

| Soil | Field Weight (g) | Oven-Dry Weight (g) |
|------|-----------------|--------------------|
| A2 | 302.5 | 264.8 |
| B2 | 376.3 | 308.0 |
| C2 | 422.6 | 388.6 |
| D2 | 468.3 | 441.7 |

*6-3. Consider two adjacent tensiometers inserted into unsaturated sandy-loam soil to determine whether water flow is toward or away from the surface. Both tensiometers have $z_{ten} = 20$ cm. Tensiometer A is placed at a depth of 20 cm, tensiometer B at 60 cm. The readings shown in the accompanying table are obtained at different times. Which way is the water flowing in each condition (assuming that no maximum or minimum of tension exists between the two levels)?

| Condition | 1 | 2 | 3 | 4 | 5 |
|-----------|-----|-----|-----|-----|-----|
| $|\psi|$ at A (cm) | 123 | 106 | 39 | 211 | 20 |
| $|\psi|$ at B (cm) | 22 | 51 | 65 | 185 | 36 |

*6-4. Use Equations (6-12) and (6-13) and Table 6-1 to estimate the $|\psi| - \theta$ and $K_h - \theta$ relations for the soils in Exercise 6-1. Plot these relations. Estimate $\theta_{fc}$ and $\theta_{pwp}$ for these soils.

6-5. (a) Baver et al. (1972) suggested that field capacity be defined as the water content at which $K_h = 2$ mm day$^{-1}$. Using Equation (6-13), compute this value for the soils of Table 6-1 and compare it with the values computed via Equation (6-19). (b) Write a brief paragraph discussing the rationales for the two definitions and which appears more justified.

6-6. Obtain a soils map of your area. In the United States, soils maps are published for each county in soil-survey reports that are available from county or state offices of the Natural Resources Conservation Service (NRCS) [formerly the Soil Conservation Service (SCS)] of the U.S. Department of Agriculture. Read the descriptions of typical soil profiles and, if possible, excavate a pit to observe comparable profiles.[9] In SCS soil-survey reports, notice the information about bulk densities, permeabilities ($K_h^*$), and available water capacity ($\theta_a$) in tables of physical and chemical properties of soils.

6-7. If an infiltrometer is available, make observations of infiltration in the field. Use the data to estimate $S_p$ and $K_p$ as described in Box 6-2.

**6-8. The spreadsheet program GrnAmpt.xls on the CD that accompanies this text simulates infiltration via the Green-and-Ampt approach. Use this program to explore the effects of initial water content, $\theta_0$, and rainfall rate, $w$, on one or more of the following soils [values from Mein and Larson (1973)]:

| Soil | $K_h^*$ (cm hr$^{-1}$) | $\phi$ | $|\psi_{ae}|$ (cm) |
|------|----------------------|--------|-------------------|
| Plainfield sand | 12.4 | 0.477 | 15.4 |
| Columbia sandy loam | 5.00 | 0.518 | 31.3 |
| Guelph loam | 1.32 | 0.523 | 41.3 |
| Ida silt loam | 0.105 | 0.530 | 9.7 |
| Yolo light clay | 0.0443 | 0.499 | 29.5 |

*6-9. Compare the explicit approximations [Equations (6-47)–(6-51)] with the results for one event simulated in Exercise 6-8.

*6-10. Consider a layer of sandy-loam soil 1.0 m thick, with a porosity of 0.38 and $K_h^* = 0.0025$ cm s$^{-1}$, above an impermeable bedrock surface. Use Equation (6-58) to explore how the time to saturation from below varies as a function of (a) initial water content and (b) rainfall rate.

*6-11. Using the values of $S_p$ and $K_p$ derived in Example 6-9, compare the values of $f(t')$ vs. $t'$ estimated with the Philip Equation for that event with the values estimated by the Green-and-Ampt Model.

*6-12. Using the method described in Box 6-2, estimate $S_p$ and $K_p$ using the results for one event simulated in Exercise 6-9. Compare the estimated value of $K_p$ with $K_h^*$ for the chosen soil.

---

[9] Be sure you have the landowner's permission before entering an area and digging a pit!

# 7

# Evapotranspiration

**Evapotranspiration** is a collective term for all the processes by which water in the liquid or solid phase at or near the earth's land surfaces becomes atmospheric water vapor. The term thus includes evaporation of liquid water from rivers and lakes, bare soil, and vegetative surfaces; evaporation from within the leaves of plants (**transpiration**); and sublimation from ice and snow surfaces.

Globally, about 62% of the precipitation that falls on the continents is evapotranspired, amounting to 72,000 km$^3$ yr$^{-1}$ (Table 3-2). Of this, about 97% is evapotranspiration from land surfaces and 3% is open-water evaporation. Figures 3-23 and 3-24 show the latitudinal and global distributions of evapotranspiration, and Tables 3-3 and 3-4 give data on evapotranspiration from the continents and from major river basins. Note that evapotranspiration exceeds runoff in most of the river basins and on all the continents except Antarctica.

A quantitative understanding of evapotranspiration is of vital practical importance in several respects:

Understanding and predicting climate change requires the ability to model evapotranspiration, which is a major component of energy as well as water-vapor exchange between land surfaces and atmosphere. (See Figure 3-2.)

Over the long term, the difference between precipitation and evapotranspiration is the water available for direct human use and management. Thus quantitative assessments of water resources and the effects of climate and land-use change on those resources require a quantitative understanding of evapotranspiration.

Most of the water "lost" via evapotranspiration is used to grow the plants that form the base of the earth's land ecosystems, and understanding relations between evapotranspiration and ecosystem type is a requirement for predicting ecosystem response to climate change (Woodward 1987).

Much of the world's food supply is grown on irrigated land, and irrigation is one of the largest uses of water in the United States and many other countries. Efficient irrigation requires knowledge of crop water use (transpiration), so that water will be applied only as needed.

Evaporation has a significant influence on the yield of water-supply reservoirs, and thus on the economics of building reservoirs of various sizes.

The fraction of water falling in a given rainstorm that contributes to streamflow and to groundwater recharge is in large part determined by the "wetness" of the land; this wetness can only be assessed by determining the amount of evapotranspiration that has occurred since the previous storm.

Direct measurement of evapotranspiration is much more difficult and expensive than for precipitation and streamflow, and is usually impractical. Thus, in order to analyze these important scientific and practical issues, hydrologists have developed an array of methods that provide estimates of evapotranspiration based on measurements of more readily measured quantities. A major goal of this chapter is to show how the physics of evaporation and other basic principles such as the conservation of mass and energy are incorporated in these methods.

The basic physics of evaporation and atmospheric energy exchange are developed in Sections D.4 and D.6. These relations are involved in liquid-vapor and solid-vapor phase changes at all surfaces of hydrologic interest, including transpiring leaves, and this chapter begins by reviewing and extending those basic relations. Following this we define the basic types of evapotranspiration, and then focus on methods of estimation appropriate for each type in various situations.[1]

## 7.1   PHYSICS OF EVAPORATION AND TURBULENT ENERGY EXCHANGE

### 7.1.1   Evaporation

Evaporation is a diffusive process that follows Fick's first law [Table 2-1; Equations (D-26) and (D-38)], which can be written in finite-difference form as

$$E = K_E \cdot v_a \cdot (e_s - e_a), \qquad \text{(7-1)}$$

where $E$ is the evaporation rate [L T$^{-1}$], $e_s$ and $e_a$ are the vapor pressures of the evaporating surface and the overlying air, respectively [M L$^{-1}$ T$^{-2}$], $v_a$ is the wind speed [L T$^{-1}$], and $K_E$ is a coefficient that reflects the efficiency of vertical transport of water vapor by the turbulent eddies of the wind [L T$^2$ M$^{-1}$].

$K_E$ can be evaluated by dividing the right-hand side of Equation (D-42) by the latent heat of vaporization, $\lambda_v$, and the mass density of water, $\rho_w$. This gives an equation for evaporation rate in the same dimensions as Equation (7-1), from which we see that

$$K_E \equiv \frac{0.622 \cdot \rho_a}{P \cdot \rho_w} \cdot \frac{k^2}{\left[ \ln\left( \dfrac{z_m - z_d}{z_0} \right) \right]^2}$$

$$= \frac{0.622 \cdot \rho_a}{P \cdot \rho_w} \cdot \frac{1}{6.25 \cdot \left[ \ln\left( \dfrac{z_m - z_d}{z_0} \right) \right]^2}, \qquad \text{(7-2)}$$

where $\rho_a$ is the density of air [M L$^{-3}$], $P$ is the atmospheric pressure [M T$^{-2}$ L$^{-1}$], $k = 0.4$ [1], $z_m$ is the height at which wind speed and air vapor pressure are measured [L], $z_d$ is the **zero-plane displacement** [L], and $z_0$ is the **roughness height** of the surface [L].

Equation (7-2) is derived from the logarithmic vertical distribution of wind speed [Equation (D-22)]. As explained in Section D.6, $z_m$ is a fixed value for a given situation; $z_d$ and $z_0$ depend on the roughness of the surface, which determines the intensity of the turbulent eddies at a given wind speed; and the density of air and pressure are approximately constants ($\rho_a = 1.220$ kg m$^{-3}$; $P = 101.3$ kPa at sea level). $K_E$ can be adjusted for the stability condition of the air, which can be determined from the vertical gradients of wind speed and temperature as described in Section D.6.8 [Equations (D-53) and (D-55); Table D-5].

### 7.1.2   Vapor-Pressure Relations

The vapor pressure of an evaporating surface is equal to the saturation vapor pressure at the surface temperature, $e_s^*$, so

$$e_s = e_s^*, \qquad \text{(7-3)}$$

where $e_s^*$ is given to good approximation by Equation (D-7) with $T = T_s$:

$$e_s^* = 0.611 \cdot \exp\left( \frac{17.3 \cdot T_s}{T_s + 237.3} \right). \qquad \text{(7-4)}$$

Here, vapor pressure is in kPa and temperature in °C. The vapor pressure in the air, $e_a$, depends on the relative humidity, $W_a$, as well as the air temperature, $T_a$:

$$e_a = W_a \cdot e_a^*, \qquad \text{(7-5)}$$

where $e_a^*$ is the saturation vapor pressure at the air temperature, given by Equation (D-7) with $T = T_a$ and $W_a$ is expressed as a ratio.

Some approaches to estimating evapotranspiration make use of the slope of the relation between saturation vapor pressure and temperature, $\Delta$. Its value can be found by taking the derivative of Equation (7-4):

$$\Delta \equiv \frac{de^*}{dT} = \frac{2508.3}{(T + 237.3)^2} \cdot \exp\left(\frac{17.3 \cdot T}{T + 237.3}\right), \quad \text{(7-6)}$$

where $\Delta$ is in kPa K$^{-1}$ and $T$ is in °C. Note that, like $e^*$, $\Delta$ increases exponentially with temperature.

### 7.1.3 Latent-Heat Exchange

Evaporation is always accompanied by a transfer of latent heat from the evaporating body into the air. This heat loss tends to produce a reduction of surface temperature, which may be completely or partially compensated by heat transfer to the surface from within the evaporating body or by radiative or sensible-heat transfer from the overlying air. The rate of latent-heat transfer, $LE$ [E L$^{-2}$ T$^{-1}$], is found simply by multiplying the evaporation rate by the latent heat of vaporization, $\lambda_v$, and the mass density of water, $\rho_w$:

$$LE = \rho_w \cdot \lambda_v \cdot E = \rho_w \cdot \lambda_v \cdot K_E \cdot v_a \cdot (e_s - e_a). \quad \text{(7-7)}$$

The latent heat of vaporization decreases as the temperature of the evaporating surface increases; this relation is given approximately by

$$\lambda_v = 2.50 - 2.36 \times 10^{-3} \cdot T, \quad \text{(7-8)}$$

where $\lambda_v$ is in MJ kg$^{-1}$ and $T$ is in °C.

### 7.1.4 Sensible-Heat Exchange

The upward rate of sensible-heat exchange by turbulent transfer, $H$, is given in finite-difference form by Equation (D-52):

$$H = K_H \cdot v_a \cdot (T_s - T_a), \quad \text{(7-9)}$$

where, from Equation (D-49),

$$K_H \equiv c_a \cdot \rho_a \cdot \frac{k^2}{\left[\ln\left(\dfrac{z_m - z_d}{z_0}\right)\right]^2}$$

$$= c_a \cdot \rho_a \cdot \frac{1}{6.25 \cdot \left[\ln\left(\dfrac{z_m - z_d}{z_0}\right)\right]^2}. \quad \text{(7-10)}$$

Here $k = 0.4$, and $c_a$ is the heat capacity of air ($c_a = 1.00 \times 10^{-3}$ MJ kg$^{-1}$ K$^{-1}$), and the other symbols are as in Equation (7-2). As with $K_E$, $K_H$ can be adjusted to account for non-neutral stability conditions [Equations (D-54) and (D-55) and Table D-5].

### 7.1.5 The Bowen Ratio, the Psychrometric Constant, and the Evaporative Fraction

In developing approaches to estimating evapotranspiration based on energy balances, we will see that it is sometimes useful to incorporate the ratio of sensible-heat exchange to latent-heat exchange. This quantity, originally formulated by Bowen (1926), is called the **Bowen ratio,** $B$:

$$B \equiv \frac{H}{LE}. \quad \text{(7-11)}$$

Combining Equations (D-42) and (D-49), we have

$$B = \frac{c_a \cdot \rho_a \cdot (T_s - T_a)}{0.622 \cdot \lambda_v \cdot (e_s - e_a)} = \gamma \cdot \frac{(T_s - T_a)}{(e_s - e_a)}. \quad \text{(7-12)}$$

Thus the Bowen ratio depends on the ratio of surface-air temperature difference to surface-air vapor-pressure difference, multiplied by a factor $\gamma$, where

$$\gamma \equiv \frac{c_a \cdot P}{0.622 \cdot \lambda_v}. \quad \text{(7-13)}$$

The factor $\gamma$ enters separately into some expressions for estimating evapotranspiration, and is called the **psychrometric constant.** However, it is not strictly a constant: Pressure is a function of elevation (Figure D-2) and varies slightly over time at a given location, and latent heat varies slightly with temperature [Equation (7-8)]. Using typical values of $c_a = 1.00 \times 10^{-3}$ MJ kg$^{-1}$ K$^{-1}$, $P = 101.3$ kPa, and $\lambda_v = 2.47$ MJ kg$^{-1}$, we find $\gamma = 0.066$ kPa K$^{-1}$, and that value is commonly used. However, the decrease of pressure with elevation (Figure D-2) should be accounted for when calculating $\gamma$ for applications at high elevations.

It has also proven useful in studies of regional evapotranspiration to define the **evaporative fraction,** $EF$, as the ratio of latent-heat exchange rate to total turbulent-heat exchange rate:

$$EF \equiv \frac{LE}{LE + H} = \frac{1}{B + 1}. \quad \text{(7-14)}$$

### 7.1.6 The Energy Balance

The general energy balance for an evapotranspiring body during a time period $\Delta t$ can be written as

$$LE = K + L - G - H + A_w - \frac{\Delta Q}{\Delta t}, \quad \text{(7-15)}$$

where the first six terms represent average energy fluxes (energy per unit area of evaporating surface per unit time) via the following modes: evaporation, $LE$; net shortwave radiation input, $K$; net longwave radiation input, $L$; net output via conduction to the ground, $G$; net output of sensible-heat exchange with the atmosphere, $H$; net input associated with inflows and outflows of water (**water-advected energy**), $A_w$; and $\Delta Q$ is the change in the amount of heat stored in the body per unit area between the beginning and end of $\Delta t$.[2,3]

From Equation (7-15) we see that any evaporation occurring during $\Delta t$ must be balanced by some combination of heat inputs from radiation or sensible heat from the atmosphere or ground, and/or a loss of heat energy (i.e., a temperature reduction) in the evaporating body.

In some situations, the atmospheric conditions above the evapotranspiring region are representative of an extensive area extending beyond the region, and there is no significant horizontal transport of energy by air movement to or from the area above the region (i.e., the water-atmosphere heat exchange is in approximate local equilibrium). When such equilibrium does not exist, horizontal air flows supply **air-advected energy** to the air overlying the region to maintain the energy balance.

## 7.2   CLASSIFICATION OF EVAPOTRANSPIRATION PROCESSES

The various methods for estimating evapotranspiration have been developed for specific surface and energy-exchange situations determined by the following conditions:

*Type of surface:* open water, bare soil, leaf or leaf canopy, a specific **reference crop** (usually a complete cover of well-watered grass, as discussed later), or land region (generally including vegetated surfaces, surface-water bodies, and areas of bare soil);

*Availability of water:* unlimited water available to evaporate, or water supply to the air may be limited because water vapor must pass through plant openings or soil pores;

*Stored-energy use:* $\Delta Q$ in Equation (7-15) may be significant, negligible, or nonexistent;

*Water-advected energy use:* $A_w$ in Equation (7-15) may be significant, negligible, or nonexistent.

Table 7-1 shows how the various "types" of evapotranspiration are distinguished with respect to the above conditions, and the methods described in subsequent sections of this chapter are classified according to these types.

Additional considerations in choosing a method for use in a given circumstance are: (1) the purpose of the analysis (determination of the amount of evapotranspiration that has actually occurred in a given situation, incorporation in a hydrologic model, reservoir design, general water-resources assessment, etc.); (2) the available data (particular meteorological parameters measured and whether measurements were made at the area of interest or are estimated regional values); and (3) the time period of interest (hour, day, month, year; climatic average). We will indicate the applicability of the various methods with respect to these considerations.

## 7.3   FREE-WATER, LAKE, AND WETLAND EVAPORATION

In natural water bodies, water-advected heat and change in heat storage may play a significant role in the energy balance. The magnitude of these components in a particular case depends in large part on the area, volume, and residence time of water in the lake relative to the time period of the analysis. Because of the variable importance of these non-meteorologic factors in the energy balance, it is not generally possible to develop equations for predicting the evaporation for a particular lake from meteorologic data alone.[4]

---

[2] Where appropriate, a term for energy used in photosynthesis can be added to Equation (7-15). This use can amount to as much as 3% of $K + L$ for some crops (C.A. Federer, pers. comm.).

[3] Equation (7-15) is analogous to the energy-balance relation for a snowpack, Equation (5-26). Note, however, that the signs of $H$, $LE$, and $G$ are reversed here because we are considering *outward* latent- and sensible-heat flows to be positive. Heat input due to rain [$R$ in Equation (5-26)] is usually negligible in considering evaporation, but could be included in $A_w$ in Equation (7-15).

---

[4] Advection and heat-storage effects are often considered to balance out for annual values. However, even annual balancing of advection and storage may not occur in lakes with residence times of more than one year.

**TABLE 7-1**

Classification of Types of Evapotranspiration

| Evapotranspiration Type | Type of Surface | Availability of Water to Surface | Stored Energy Use | Water-Advected Energy Use |
|---|---|---|---|---|
| Free-water evaporation[a] | Open water | Unlimited | None | None |
| Lake evaporation | Open water | Unlimited | May be involved | May be involved |
| Bare-soil evaporation | Bare soil | Limited to unlimited | Negligible | None |
| Transpiration | Leaf or leaf canopy | Limited | Negligible | None |
| Interception loss | Leaf or leaf canopy | Unlimited | Negligible | None |
| Potential evapotranspiration | Reference crop[b] | Limited to air, unlimited to plants | None | None |
| Actual evapotranspiration | Land area[c] | Varies in space and time | Negligible | None |

[a]Also called **potential evaporation.**

[b]Usually a complete ground cover of uniform short vegetation (e.g., grass); discussed further in Section 7.7.1.

[c]May include surface-water bodies and areas of bare soil.

In order to develop general methods for estimating evaporation from surface-water bodies, hydrometeorologists have formulated the theoretical concept of **free-water evaporation:** evaporation that would occur from an open-water surface in the absence of advection and changes in heat storage (Table 7-1) and which thus depends only on regionally continuous meteorologic or climatic conditions. **Lake evaporation** is determined by adjusting mapped or computed free-water evaporation to account for the advection and heat-storage effects in a given actual water body.

In the following discussion of methods, we will indicate how each approach can be applied to estimating free-water evaporation and/or actual lake evaporation.

### 7.3.1 Water-Balance Approach

#### Theoretical Basis

The water-balance approach involves applying the water-balance equation (see Section 2.5.1) to the water body of interest over a time period $\Delta t$ and solving that equation for evaporation, $E$:

$$E = W + SW_{in} + GW_{in} - SW_{out} - GW_{out} - \Delta V. \tag{7-16}$$

Here, $W$ is precipitation on the lake; $SW_{in}$ and $SW_{out}$ are the inflows and outflows of surface water, respectively; $GW_{in}$ and $GW_{out}$ are the inflows and outflows of ground water, respectively; $\Delta V$ is the change in the amount of water stored in the lake during $\Delta t$; and all quantities have dimensions of either volume or volume per unit lake area.

#### Practical Considerations

While this approach is simple in concept, it is generally far from simple in application. In practice we do not know the true values of the quantities in Equation (7-16) and must use measured or estimated values, which are subject to uncertainty. Our computation of $E$ thus includes the net sum of all these measurement errors as well as the evaporation, as discussed in Section 2.5.2 and illustrated for watershed evapotranspiration in Example 2-1.

Measurement of all liquid-water inputs and outputs for a water body is generally a formidable task. All major streams entering the body and the outlet stream must be continuously gaged, and some method must be devised for estimating the amount of any non-channelized overland flow inputs. At best, ground-water flows are usually estimated from gradients observed in a few observation wells and assumptions about the saturated thickness and hydraulic conductivity of surrounding geologic formations, perhaps supplemented by scattered observations with seepage meters. (See Section 8.5.3.) If the lake is large and the surrounding topography irregular, it may be difficult to obtain precise measurements of precipitation. (See Sections 4.2 and 4.3.) Changes in storage can be estimated from careful observations of water levels only if one has good information on the lake's bathymetry, and if corrections are made for changes in water density.

**TABLE 7-2**

Range of Uncertainty in Precipitation and Streamflow Values Used in Computing Lake Water Balances[a]

| Time Interval | Precipitation | Streamflow Inputs[b] | Streamflow Outputs |
|---|---|---|---|
| | **General Range** | | |
| Daily | 60–75 | 5–15 (50) | 5 (15) |
| Monthly | 10–25 | 5–15 (50) | 5 (15) |
| Seasonal/annual | 5–10 | 5–15 (30) | 5 (15) |
| | **Values for Typical Lake in Northern United States** | | |
| Monthly | (26) | (31) | (12) |
| Seasonal/annual | 8 (17) | 9 (23) | 9 (12) |

[a]Values are percentages of the true values. Those without parentheses are for "best" methodology; those in parentheses are "commonly used" methodology.

[b]Does not include overland flow.

Based on analyses of Winter (1981).

Winter (1981) made a thorough analysis of the uncertainties in estimating lake water balances; even for the "readily" measurable components the results are somewhat discouraging (Table 7-2). For other components—overland flows and, especially, ground-water flows (see LaBaugh et al. 1997)—uncertainties on the order of 100% must generally be anticipated.

*Applicability*

The water-balance method is theoretically applicable in determining the amount of evaporation that has occurred in any water body during a given time period. In most situations, however, measurement problems preclude use of the method as an independent means of determination—especially where evaporation is a small fraction of the total outflow and less than the uncertainty in the overall balance (e.g., Figure 7-1). If reasonably accurate information on the significant balance components is available, the method can provide a rough—and in rare cases a quite accurate—check on evaporation estimated by other approaches. Table 7-2 indicates that the accuracy of the method should increase as $\Delta t$ increases.

Because of the difficulty in finding situations where all the significant terms on the right side of Equation (7-16) can be measured with sufficient accuracy, the water-budget approach has only rarely been used to determine lake evaporation. One of the more successful applications of the approach

was in an extensive comparison of methods for determining lake evaporation conducted by the U.S. Geological Survey in the early 1950s at Lake Hefner, OK (Harbeck et al. 1954). That lake was selected for the study because it was one of the few—out of more than 100 considered—for which a water balance could be computed with acceptable precision. Harbeck et al. (1954) concluded that daily evaporation calculated via the water balance at Lake Hefner was within 5% of the true value on 29% of the days, and within 10% on 62% of the days.

### 7.3.2 Mass-Transfer Approach

*Theoretical Basis*

The mass-transfer approach makes direct use of Equation (7-1), often written in the form

$$E = (b_0 + b_1 \cdot v_a) \cdot (e_s - e_a), \qquad \textbf{(7-17)}$$

where $b_0$ and $b_1$ are empirical constants that depend chiefly on the height at which wind speed and air vapor pressure are measured. Relations of the form of (7-17) were first formulated by the English chemist John Dalton in 1802, and are often called **Dalton-type equations.**

The theoretical analysis leading to Equations (7-1) and (7-2) indicates that $b_0 = 0$ and $b_1 = K_E$. Accepting this, we can evaluate $b_1$ by inserting appropriate numerical values for the quantities in Equation (7-2). Brutsaert (1982) suggested that $z_0 = 2.30 \times 10^{-4}$ m and $z_d = 0$ m over typical water surfaces,[5] and using these values along with standard values for air and water properties and a measurement height of 2 m, we find that

$$E = 1.26 \times 10^{-3} \cdot v_a \cdot (e_s - e_a), \qquad \textbf{(7-18a)}$$

where $E$ is in m day$^{-1}$, $v_a$ is in m s$^{-1}$, and vapor pressures are in kPa. When $v_a$ is in km day$^{-1}$, the relation becomes

$$E = 1.46 \times 10^{-5} \cdot v_a \cdot (e_s - e_a). \qquad \textbf{(7-18b)}$$

*Practical Considerations*

Use of Equation (7-18) requires measurement of wind speed; surface vapor pressure is calculated from surface temperature via Equation (7-4) and air vapor pressure from air temperature and rela-

---

[5] The values of $z_0$ and $z_d$ increase as $v_a$ increases due to the effects of waves.

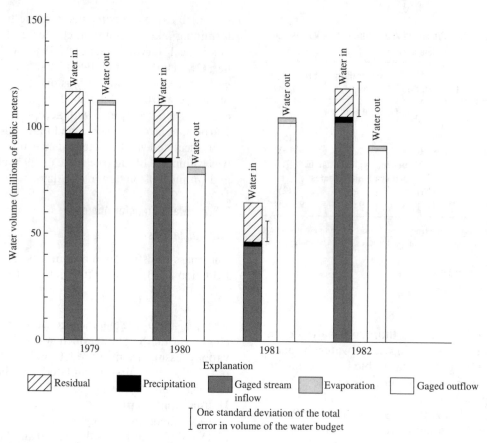

**FIGURE 7-1**

Water balance for the Williams Fork Reservoir, CO, for four years. The vertical bars represent the uncertainty in the overall budget (standard deviation of total error); note that this is several times larger than annual evaporation. From Labaugh (1984), used with permission of the American Geophysical Union.

tive humidity measurements via Equations (7-4) and (7-5). The value of $K_E$ depends on the height at which wind speed is measured [Equation (7-2)], so this height must be noted (the 2-m height is standard). Clearly all measured values should be representative of the entire water body, which may be difficult to accomplish using shore-based observation stations (Sene et al. 1991). In fact, in a comparative study, Winter et al. (1995) concluded that "the mass transfer equation is acceptable only if a raft station is used so that water temperature and wind-speed data can be obtained near the center of the lake." The air vapor pressure is usually taken to be the value that is unaffected by the lake, and thus should be measured at several meters height or on the upwind shore.

Many studies have been done to verify the appropriate values for $b_0$ and $b_1$ and—because of vari-ations in lake size, measurement height, instrument location, atmospheric stability, precision with which evaporation rate could be determined by other approaches, and time scale—each study has obtained a different set of values. The Lake Hefner project (Harbeck et al. 1954) and a study by Ficke (1972) in Indiana were among the most detailed and careful of these studies because accurate estimates of daily evaporation could be independently obtained via the water balance. In both cases, $v_a$ is the wind speed at 2-m elevation and $e_a$ represents the vapor pressure of the air unmodified by passage over the lake. Both studies confirmed that $b_0 = 0$, and Ficke (1972) found $K_E$ almost identical to the value in Equation (7-18) while Harbeck et al. (1954) found $K_E$ about 15% larger. A later study of Lake Erie found $K_E$ very close to Harbeck's Lake Hefner value (Derecki 1976).

It appears that much of the variability in $K_E$ found in various studies is a function of lake area. Harbeck (1962) found the empirical relation

$$K_E = 1.69 \times 10^{-5} \cdot A_L^{-0.05}, \qquad \textbf{(7-19)}$$

where $K_E$ is in m km$^{-1}$ kPa$^{-1}$ [i.e., for use in Equation (7-18b)] and $A_L$ is in km$^2$ (Figure 7-2). There are two reasons for this area effect: (1) since water surfaces are smoother than land surfaces, the efficiency of turbulent eddies in the vertical transport of water vapor decreases as the wind travels longer distances over a lake; and (2) the value of $e_a$ will increase with downwind distance as evaporation occurs, decreasing the effective vapor-pressure difference over the lake compared with the value measured anywhere except on the downwind shore.

The value of $K_E$ is also affected by atmospheric stability. When the water surface is warmer than the air, the air in contact with the surface is warmed by conduction and tends to rise. This tends to induce instability, so that there is turbulent transport of water vapor (and heat) away from the surface even in the absence of wind. This process, called **free convection,** is most common in situations where the water body has been artificially heated, as in a cooling pond or a river reach receiving cooling water from a power plant. However, it also occurs in large lakes because of thermal inertia: Derecki (1981) found considerable seasonal variation in $K_E$ for Lake Superior, with values much lower than those given by Equations (7-18) and (7-19) in the summer when the lake is colder than the air, and higher in the fall and winter, when the lake is warmer.

One approach to account for the effect of atmospheric stability on $K_E$ is to include a stability correction in the formulation of Equation (7-2), as described in Section D.6.8 [Equations (D-53) and (D-55)]. Alternatively, when $T_s > T_a$ the effects of free convection can be modeled by assuming that $b_0$ in Equation (7-17) has a positive value that increases with the temperature difference between the surface and the air. In a comparative study, Rasmussen et al. (1995) found that the following equation gave the best results:

$$E = \frac{[2.33 \cdot (T_s - T_a)^{1/3} + 2.68 \cdot v_a] \cdot (e_s - e_a)}{\lambda_v},$$
$$T_s > T_a. \qquad \textbf{(7-20)}$$

In this equation, $E$ is evaporation rate in mm day$^{-1}$; $T_s$ and $T_a$ are the surface and air temperatures, re-

spectively, in °C; $v_a$ is wind speed in m s$^{-1}$; $e_s$ and $e_a$ are surface and air vapor pressures, respectively, in kPa, and $\lambda_v$ is latent heat of vaporization in MJ kg$^{-1}$ [Equation (7-8)].

### Applicability

It must be emphasized that the various versions of the mass-transfer equation give the *instantaneous rate* of evaporation under given *instantaneous* values of wind speed and vapor pressure. Because the vapor pressures used in the equation are determined from measured temperatures via the nonlinear relation of Equation (7-4), and because wind speed and vapor-pressure differences may be correlated, one cannot assume that a mass-transfer equation will give the correct *time-averaged* rate of evaporation when time-averaged temperatures and wind speeds are used as independent variables. Using Lake Hefner data, Jobson (1972) determined the errors introduced in calculating average evaporation from averaged temperatures and wind speeds for averaging periods of 3 hr, 1 day, and 1 month. The results, shown in Table 7-3, indicate that little error is introduced for averaging periods up to 1 day, but that a bias is introduced with monthly averaging.

Thus the mass-transfer approach is potentially useful for determining the amount of free-water or lake evaporation that has occurred during a given time period during which the independent variables $v_a$, $T_s$, $T_a$, and $W_a$ have been measured at a representative location. Observations of $T_s$ may also be available from satellites that remotely sense thermal radiation (Croley 1989). It appears safe to use the approach when the variables are averaged over periods of up to 1 day.

The method can also be used to predict, forecast, or model free-water or lake evaporation based on predicted, forecast, or modeled values of the independent variables. It is particularly useful for modeling artificially heated water bodies, for which surface temperatures can usually be readily calculated. However, it is difficult to apply the method for naturally heated water bodies, because there are virtually no climatic data on water-surface temperatures, and modeling such temperatures is difficult. This turns out to be a serious limitation in the practical usefulness of the mass-transfer approach.

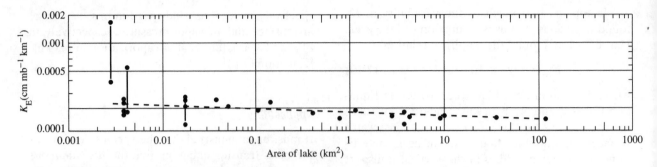

**FIGURE 7-2**
Value of mass-transfer coefficient, $K_E$, as a function of lake area. Data are for the southwestern United States, Indiana, and western Canada. The dashed line is Equation (7-19). After Dunne and Leopold (1978).

## EXAMPLE 7-1

The data shown in Table 7-4 were collected at Lake Hefner, OK, on 12 July 1951. The lake's area is 9.4 km². Compute the evaporation via Equation (7-18).

**Solution**  Substituting the lake's area into Equation (7-19) gives

$$K_E = 1.51 \times 10^{-5} \text{ m km}^{-1} \text{ kPa}^{-1}.$$

Equation (7-4) gives

$$e_s^* = 3.56 \text{ kPa}$$

and

$$e_a^* = 3.62 \text{ kPa},$$

and Equation (7-5) gives

$$e_a = 0.69 \cdot 3.62 \text{ kPa} = 2.50 \text{ kPa}.$$

Converting the wind speed to km day⁻¹ yields

$$5.81 \text{ m s}^{-1} \cdot \frac{86{,}400 \text{ s}}{1 \text{ day}} \cdot \frac{1 \text{ km}}{1000 \text{ m}} = 502 \text{ km day}^{-1}.$$

**TABLE 7-3**
Errors in Computing Time-Averaged Evaporation Using Time-Averaged Temperatures and Wind Speeds

|  | Averaging Period | | |
|---|---|---|---|
|  | **3 hr** | **1 day** | **1 month** |
| Median % error | 0 | 0 | +4 |
| % of time periods with less than 5% error | 97 | 79 | — |
| % of time periods with less than 10% error | >99 | 93 | — |

Data from Jobson (1972), based on Lake Hefner studies.

Substituting these values in Equation (7-18) gives

$$E = (1.51 \times 10^{-5} \text{ m km}^{-1} \text{ kPa}^{-1}) \cdot (502 \text{ km day}^{-1})$$

$$\cdot (3.56 \text{ kPa} - 2.50 \text{ kPa}) = 0.00803 \text{ m day}^{-1}$$
$$= 8.03 \text{ mm day}^{-1}.$$

Note that this value is considerably larger than that determined by the water-budget approach for that day.

### 7.3.3   Eddy-Correlation Approach

#### *Theoretical Basis*

As described in Section D.6.9 [Equations (D-56)–(D-58)], the rate of upward movement of water vapor near the surface (i.e., the local evaporation rate) is proportional to the time average of the product of the instantaneous fluctuations of vertical air movement, $u_a'$, and of absolute humidity, $q'$, around their respective mean values:

$$E = \frac{\rho_a}{\rho_w} \cdot \overline{u_a' \cdot q'}, \qquad (7\text{-}21)$$

where the overbar denotes time averaging.

If sensors capable of accurately recording and integrating high-frequency (on the order of 10 s⁻¹) fluctuations in humidity and vertical velocity are used, eddy-correlation measurements are often considered to be measurements of the "true" evaporation rates because the method has a sound theoretical foundation and requires no assumptions about parameter values, the shape of the velocity profile, or atmospheric stability.

**TABLE 7-4**

Daily Values Measured during Evaporation Studies at Lake Hefner, OK, 12 July 1951

| | | | Daily Average | | | |
|---|---|---|---|---|---|---|
| $T_a$ (°C) | $T_s$ (°C) | $W_a$ | $P$ (kPa) | $v_a$ (cm s$^{-1}$) | $T_{span}$ (°C) | $v_{pan}$ (cm s$^{-1}$) |
| 27.2 | 26.9 | 0.69 | 97.3 | 581 | 27.5 | 279 |

| | Daily Total | | | |
|---|---|---|---|---|
| $K_{in}$ (MJ m$^{-2}$ day$^{-1}$) | Albedo | $L_{in}$ (MJ m$^{-2}$ day$^{-1}$) | Water-Budget Evaporation (cm) | Class-A Pan Evaporation (cm) |
| 30.6 | 0.052 | 34.4 | 0.558 | 1.24 |

From U.S. Geological Survey (1954).

### Practical Considerations

The eddy-correlation approach can give measurements representative of limited (~ 1 ha) areas upwind of the sensors, with accuracies of about 10% for time periods as short as 0.5 hr (Stannard and Rosenberry 1991). However, the stringent instrumentation requirements make the approach suitable only for research applications.

### 7.3.4 Energy-Balance Approach

### Theoretical Basis

The energy-balance equation for an evaporating water body was given as Equation (7-15). Dividing that equation by the latent heat of vaporization, $\lambda_v$, and the density of water, $\rho_w$, yields

$$E = \frac{K + L - G - H + A_w - \Delta Q/\Delta t}{\rho_w \cdot \lambda_v}, \quad (7\text{-}22)$$

where $E$ has dimensions [L T$^{-1}$].

The energy-balance approach to determining the average rate of evaporation over a time period $\Delta t$ thus involves measuring or otherwise determining the rates of energy input and output by the various modes, along with the change in energy stored in the water body during $\Delta t$, and solving Equation (7-22).

Equation (7-22) can be written in a different and often more useful form by using the Bowen ratio [Equation (7-12)], so that the sensible heat-loss rate, $H$, is replaced by

$$H = B \cdot LE = B \cdot \rho_w \cdot \lambda_v \cdot E. \quad (7\text{-}23)$$

For a water body, energy can be advected in as precipitation and ground-water and surface-water inflows, and advected out by ground-water and sur-face-water outflows; all this advected energy is included in $A_w$. Then using Equation (7-23) and solving Equation (7-22) for $E$ yields a modified energy-balance equation:

$$E = \frac{K + L - G + A_w - \Delta Q/\Delta t}{\rho_w \cdot \lambda_v \cdot (1 + B)}. \quad (7\text{-}24)$$

### Practical Considerations

As with the water balance, the energy-balance approach requires precise determination of all the non-negligible quantities in the basic equation, and all measurement errors are included in the final computation of evaporation. However, in some situations the measurement problems for determining the energy balance are less severe than for the water balance because (1) it is possible to eliminate some of the terms in the energy-balance equation and (2) it may be possible to use regional climatic data to estimate the radiation components. We now consider how values for the modified energy-balance relation can be determined.

**Shortwave Radiation** Net incoming shortwave, or solar, radiation, $K$, is given by

$$K = K_{in} \cdot (1 - a), \quad (7\text{-}25)$$

where $K_{in}$ is incoming solar radiation and $a$ is the albedo (reflectivity) of the water surface.

As discussed in Section 5.4.2, incoming and reflected solar radiation can be measured with pyranometers, but this is routinely done at only scattered locations. More commonly, $K_{in}$ for a lake is estimated from the clear-sky solar radiation, $K_{cs}$, and the fraction of sky covered by cloud, $C$, where $K_{cs}$ is determined from the latitude and time of year using the relations given in Section E.1. An empirical relation that has been used in evaporation modeling (see Croley 1989) is

$$K_{in} = [0.355 + 0.68 \cdot (1 - C)] \cdot K_{cs}. \quad \textbf{(7-26)}$$

However, manual observations of $C$ are no longer available in the United States because human weather observers are being replaced by automated stations. Because of this, Lindsey and Farnsworth (1997) recommended that, where $K_{in}$ is not directly measured, it is best determined from gridded estimates of solar radiation developed from Geostationary Operational Environmental Satellite (GOES) imagery. Useful estimates of $K_{in}$ can also be developed using daily maximum and minimum temperature and precipitation data (Hunt et al. 1998).

Koberg (1964) presented an empirical relation giving the albedo of a water surface as a function of $K_{in}$:

$$a = 0.127 \cdot \exp(-0.0258 \cdot K_{in}), \quad \textbf{(7-27)}$$

where $K_{in}$ is in MJ m$^{-2}$ day$^{-1}$; other studies have assumed a constant value in the range $0.05 \leq a \leq 0.10$.

**Longwave Radiation**   Net longwave radiation input, $L$, is equal to incoming atmospheric longwave radiation flux, $L_{at}$, minus the portion of $L_{at}$ reflected and the radiation flux emitted by the water surface, $L_w$. Since the longwave reflectivity of the surface equals $1 - \varepsilon_w$, where $\varepsilon_w$ is the emissivity of the water surface, it follows that

$$L = L_{at} - (1 - \varepsilon_w) \cdot L_{at} - L_w = \varepsilon_w \cdot L_{at} - L_w. \quad \textbf{(7-28)}$$

Again, instruments exist that can measure longwave radiation fluxes, but these are even less common than pyranometers. Thus $L_{at}$ and $L_w$ are usually calculated from relations based on the Stefan-Boltzmann Equation [Equation (D-1)]. The reasoning used in Equations (5-35)–(5-37) for a snow surface can be applied to a water surface, yielding

$$L = \varepsilon_w \cdot \varepsilon_{at} \cdot \sigma \cdot (T_a + 273.2)^4 - \varepsilon_w \cdot \sigma \cdot (T_s + 273.2)^4, \quad \textbf{(7-29)}$$

where $\varepsilon_{at}$ is the effective emissivity of the atmosphere, $\sigma$ is the Stefan-Boltzmann constant ($\sigma = 4.90 \times 10^{-9}$ MJ m$^{-2}$ day$^{-1}$ K$^{-4}$), and the temperatures are in °C. The value of $\varepsilon_w$ is 0.97 (Table D-1).

As discussed in Section 5.4.2, $\varepsilon_{at}$ is largely a function of humidity and cloud cover, and can be estimated via empirical relations such as

$$\varepsilon_{at} = 1.72 \cdot \left( \frac{e_a}{T_a + 273.2} \right)^{1/7} \cdot (1 + 0.22 \cdot C^2), \quad \textbf{(7-30)}$$

where $e_a$ is in kPa, $T_a$ is in °C, and $C$ is cloud-cover fraction (Kustas et al. 1994).

**Conduction to Ground**   In most situations, the heat exchange by conduction between a lake and the underlying sediments is negligible.

**Water-Advected Energy**   Net water-advected energy is found from

$$A_w = c_w \cdot \rho_w \cdot (w \cdot T_a + SW_{in} \cdot T_{swin} - SW_{out} \cdot T_{swout} + GW_{in} \cdot T_{gwin} - GW_{out} \cdot T_{gwout}), \quad \textbf{(7-31)}$$

where $\rho_w$ is the mass density of water, $c_w$ is the specific heat of water, $w$ is average precipitation rate, $SW$ and $GW$ represent surface-water and groundwater inflows and outflows (per subscript) expressed as volumes per time per unit lake area, and the $T$s represent temperatures of the respective inflows and outflows.

**Change in Stored Energy**   The change in energy storage in a lake is found from the volumes and average temperatures of the lake water at the beginning and end of $\Delta t$:

$$\Delta Q = \frac{c_w \cdot \rho_w}{A_L} \cdot (V_2 \cdot T_{L2} - V_1 \cdot T_{L1}). \quad \textbf{(7-32)}$$

Here, $V$ is lake volume, $T_L$ is average lake temperature, the subscripts 1 and 2 designate values at the beginning and end of $\Delta t$, respectively, and $A_L$ is lake area.

Evaluation of $\Delta Q$ thus involves the same considerations and difficulties as determining $\Delta V$ in the water-balance equation [Equation (7-16)], plus the problem of computing an average lake temperature. These difficulties can be minimized by choosing $\Delta t$ so that the term in parentheses in Equation (7-32) is likely to be small. In most situations this will be true if $\Delta t = 1$ yr; in some climates lakes become isothermal at the temperature of maximum density (3.98 °C) twice a year, so those times can be used to bound $\Delta t$. It may be possible to relate average lake temperature to water-surface temperature (Rosenberry et al. 1993), which can be remotely sensed; however, the relation between the two must be established for each lake, and at least for deep lakes will vary seasonally (Figure 7-3a).

**FIGURE 7-3**

(a) Variation of total heat storage with surface temperature in Lake Superior in 1976. (b) Annual cycles of air temperature, humidity, wind speed, and evaporation in Lake Superior, 1975–1977. From Croley (1992), used with permission of the American Geophysical Union.

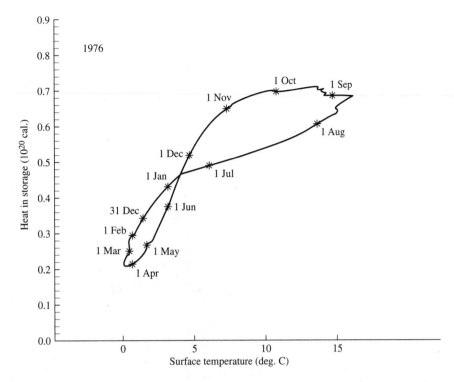

Heat storage significantly influences the timing of evaporation in large, deep lakes (Croley 1992). For example, it reaches a maximum in the fall in Lake Superior (Figure 7-3a), shifting the annual peak of evaporation such that it is out of phase with the annual cycles of air temperature, water temperature, and humidity (Figure 7-3b).

**The Bowen Ratio**  The principal advantage of using the Bowen ratio in the energy balance is to eliminate the need for wind data, which would be required if sensible-heat exchange were separately evaluated. However, one still needs data for surface and air temperatures and relative humidity, so the same considerations for time averaging apply as for the mass-transfer approach.

### Applicability

Use of Equation (7-24) to determine actual evaporation for a lake involves many of the same kinds of measurement difficulties as for the water-balance approach. L. J. Anderson, who conducted the elaborate energy-balance studies at Lake Hefner, concluded that "the energy-budget equation, when applied to periods greater than 7 days, will result in

a maximum accuracy approaching ± 5 per cent of the mean ... evaporation, providing all terms in the energy budget have been evaluated with the utmost accuracy, particularly change in energy storage" (Harbeck et al. 1954, p. 117). He found that considerable error in evaluating the change in energy storage may occur if the method is applied to periods of less than seven days. Rosenberry et al. (1993) found that the energy-budget approach gave the best estimates of lake evaporation for a small lake in Minnesota and investigated the effects of using various instrumentation and approaches to determining energy-budget components.

As with the mass-transfer method, it is difficult to apply the energy-budget approach in the prediction, forecasting, or modeling mode because of the requirement for water-surface temperatures. However, Hostetler and Bartlein (1990) have developed a model that simulates the vertical distribution of lake temperatures through time coupled to energy-balance computations, and this appears promising as an approach to estimating the effects of past and future climate change on lake evaporation.

In estimating free-water evaporation, Equation (7-24) can be simplified because the most troublesome terms, $A_w$ and $\Delta Q/\Delta t$, are small; this is also generally true in calculating annual evaporation for

**FIGURE 7-3**
(Continued)

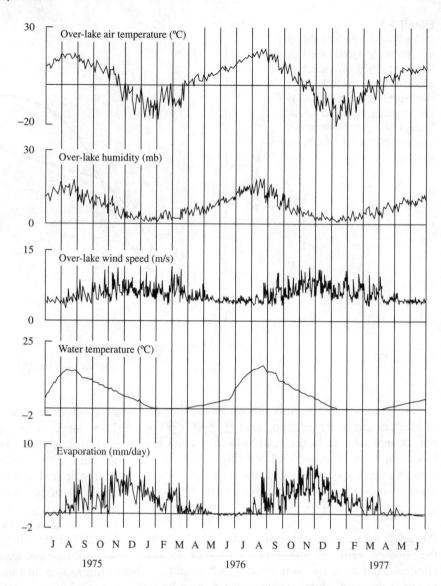

lakes if their residence times do not greatly exceed one year. However, even for these cases there is the need for water-surface temperature. As we will see in Section 7.3.5, the most useful applications of the energy-balance approach are in combination with the mass-transfer method, whereby one can eliminate the need for surface data in estimating free-water evaporation.

## EXAMPLE 7-2

Estimate the evaporation from Lake Hefner on 12 July 1951 using the energy-balance approach with the data in Table 7-4.

*Solution* Data indicate that water-advected energy and change in heat storage were negligible over this date (U.S. Geological Survey 1954). Net shortwave radiation, $K$, is

$$K = K_{in} \cdot (1 - a) = (30.6 \text{ MJ m}^{-2} \text{ day}^{-1}) \cdot (1 - 0.052)$$
$$= 29.0 \text{ MJ m}^{-2} \text{ day}^{-1}.$$

Incoming longwave radiation, $L_{at}$, was measured, so net longwave radiation is found as

$$L = L_{at} - 0.03 \cdot L_{at} - 0.97 \cdot (4.90 \times 10^{-9})$$
$$\cdot (T_s + 273.15)^4,$$

and substituting appropriate values from Table 7-4 gives $L = -5.23 \text{ MJ m}^{-2} \text{ day}^{-1}$. Thus

$$K + L = 23.8 \text{ MJ m}^{-2} \text{ day}^{-1}.$$

From Equation (7-8),

$$\lambda_v = 2.50 - (2.36 \times 10^{-3}) \cdot 26.9 = 2.44 \text{ MJ kg}^{-1}.$$

With the vapor pressures calculated in Example 7-1, the Bowen ratio, $B$, is

$$B = \frac{(1.00 \times 10^{-3} \text{ MJ kg}^{-1} \text{ K}^{-1}) \cdot (97.3 \text{ kPa}) \cdot (26.9°\text{C} - 27.2°\text{C})}{0.622 \cdot (2.44 \text{ MJ kg}^{-1}) \cdot (3.56 \text{ kPa} - 2.50 \text{ kPa})}$$
$$= -0.0183.$$

We can now substitute into Equation (7-24) as follows:

$$E = \frac{23.8 \text{ MJ m}^{-2} \text{ day}^{-1}}{(1000 \text{ kg m}^{-3}) \cdot (2.44 \text{ MJ kg}^{-1}) \cdot (1 - 0.0183)}$$
$$= 0.00959 \text{ m day}^{-1} = 9.59 \text{ mm day}^{-1}.$$

This value is about 1.6 mm day$^{-1}$ larger than that estimated via mass transfer (Example 7-1), and both estimates are considerably larger than determined from the water balance.

---

### 7.3.5 Penman or Combination Approach

#### Theoretical Basis

Following the steps shown in Box 7-1, Penman (1948) was the first to show that the mass-transfer and energy-balance approaches could be combined to arrive at an evaporation equation that did not require surface-temperature data. Van Bavel (1966) generalized Penman's original development by replacing an empirical wind function with Equation (7-2), resulting in the following theoretically sound and dimensionally homogeneous relation:

---

## BOX 7-1

. . . . . . . .

### Derivation of the Penman Combination Equation

Neglecting ground-heat conduction, water-advected energy, and change in energy storage, Equation (7-22) becomes

$$E = \frac{K + L - H}{\rho_w \cdot \lambda_v}. \qquad \textbf{(7B1-1)}$$

The sensible-heat transfer flux is given by Equation (7-9):

$$H = K_H \cdot v_a \cdot (T_s - T_a). \qquad \textbf{(7B1-2)}$$

The slope of the saturation-vapor vs. temperature curve at the air temperature can be approximated as

$$\Delta = \frac{e_s^* - e_a^*}{T_s - T_a}, \qquad \textbf{(7B1-3)}$$

from which we obtain

$$T_s - T_a = \frac{e_s^* - e_a^*}{\Delta}. \qquad \textbf{(7B1-4)}$$

Equation (7B1-4) can now be substituted into (7B1-2):

$$H = \frac{K_H \cdot v_a}{\Delta} \cdot (e_s^* - e_a^*). \qquad \textbf{(7B1-5)}$$

This relation remains true if $e_a$ is added and subtracted from each of the terms in brackets:

$$H = \frac{K_H \cdot v_a}{\Delta} \cdot (e_s^* - e_a) - \frac{K_H \cdot v_a}{\Delta} \cdot (e_a^* - e_a). \qquad \textbf{(7B1-6)}$$

Making use of Equation (7-3), we rearrange Equation (7-1) to give

$$e_s^* - e_a = \frac{E}{K_E \cdot v_a}. \qquad \textbf{(7B1-7)}$$

Substituting (7B1-7) into (7B1-6) yields

$$H = \frac{K_H \cdot v_a}{\Delta} \cdot \frac{E}{K_E \cdot v_a} - \frac{K_H \cdot v_a}{\Delta} \cdot (e_a^* - e_a). \qquad \textbf{(7B1-8)}$$

Now Equation (7B1-8) can be substituted into (7B1-1) and the result solved for $E$:

$$E = \frac{K + L + (K_H \cdot v_a/\Delta) \cdot (e_a^* - e_a)}{\rho_w \cdot \lambda_v + (K_H/K_E) \cdot \Delta}. \qquad \textbf{(7B1-9)}$$

From the definitions of $K_H$ [Equation (7-10)], $K_E$ [Equation (7-2)], and $\gamma$ [Equation (7-13)], we see that

$$K_H = \gamma \cdot \rho_w \cdot \lambda_v \cdot K_E, \qquad \textbf{(7B1-10)}$$

and substituting this relation into (7B1-9), multiplying the numerator and denominator by $\Delta$, and making use of Equation (7-5) yields Equation (7-33).

$$E =$$
$$\frac{\Delta \cdot (K + L) + \gamma \cdot K_E \cdot \rho_w \cdot \lambda_v \cdot \nu_a \cdot e_a^* \cdot (1 - W_a)}{\rho_w \cdot \lambda_v \cdot (\Delta + \gamma)}.$$

**(7-33)**

Equation (7-33) assumes no heat exchange with the ground,[6] no water-advected energy, and no change in heat storage, and makes use of one approximation [Equation (7B1-3)]. This relation has become known as the **Penman equation** or **combination equation.**

Note that the essence of the Penman equation can be represented as

$$E = \frac{\Delta \cdot \text{net radiation} + \gamma \cdot \text{"mass transfer"}}{\Delta + \gamma};$$

**(7-34)**

that is, evaporation rate is a weighted sum of a rate due to net radiation and a rate due to mass transfer. Note, however, that the mass-transfer relation now depends on the difference between the actual vapor pressure and the saturation vapor pressure at the *air* temperature, rather than at the water-surface temperature. The weighting coefficients are given by the slope of the saturation-vapor-pressure vs. temperature curve at the air temperature, $\Delta$, [Equation (7-6)], and the psychrometric constant, $\gamma$, which depends on atmospheric pressure [Equation (7-13)].

While Penman (1948) intended his approach to eliminate the need for surface-temperature data, such data are in fact required to evaluate the long-wave energy exchange, $L$ [Equation (7-29)], except in the rare instances where measured values are available. Kohler and Parmele (1967) showed that another approximation could be made to avoid this problem with little error and arrived at the following modifications of the terms in Equation (7-33):

Replace $L$ with $L'$, where

$$L' \equiv \varepsilon_w \cdot L_{at} - \varepsilon_w \cdot \sigma \cdot (T_a + 273.2)^4; \quad \textbf{(7-35)}$$

replace $\gamma$ with $\gamma'$, where

$$\gamma' \equiv \gamma + \frac{4 \cdot \varepsilon_w \cdot \sigma \cdot (T_a + 273.2)^3}{K_E \cdot \rho_w \cdot \lambda_v \cdot \nu_a}. \quad \textbf{(7-36)}$$

As noted, Penman's development did not include water-advected energy or changes in energy

---

[6] If it is significant, the ground-heat conduction term can be included by replacing $(K + L)$ with $(K + L - G)$.

storage, and hence is applicable only for free-water evaporation. Kohler and Parmele (1967) also developed a method whereby the effects of water-advected energy and changes in heat storage can be incorporated in the combination approach to provide a more generalized estimate of open-water evaporation. They reasoned that lake evaporation, $E_L$, is related to Penman evaporation, $E_P$, as

$$E_L = E_P + \alpha_{KP} \cdot \left( A_w - \frac{\Delta Q}{\Delta t} \right), \quad \textbf{(7-37)}$$

where $A_w$ and $\Delta Q$ are given by Equations (7-31) and (7-32), respectively, and $\alpha_{KP}$ is the fraction of the net addition of energy from advection and storage that was used in evaporation during $\Delta t$. Since the total net addition $(A_w - \Delta Q/\Delta t)$ is allocated among evaporation, sensible-heat transfer, and emitted radiation, it follows that

$$\alpha_{KP} =$$
$$\frac{\Delta}{\Delta + \gamma + 4 \cdot \varepsilon_w \cdot \sigma \cdot (T_w + 273.2)^3 / (\rho_w \cdot \lambda_v \cdot K_E \cdot \nu_a)}.$$

**(7-38)**

### Practical Considerations

The combination approach as originally developed for free-water evaporation requires representative data on net shortwave radiation, net longwave radiation, wind speed, air temperature, and relative humidity. As modified by Equations (7-35)–(7-36), it requires data on incoming and reflected short-wave radiation, incoming longwave radiation, wind speed, air temperature, and relative humidity. Since incoming longwave flux can be estimated from

$$L_{at} = \varepsilon_{at} \cdot \sigma \cdot (T_a + 273.2)^4, \quad \textbf{(7-39)}$$

with $\varepsilon_{at}$ estimated from relations like Equation (7-30), the data requirements reduce to incident short-wave radiation [which can be estimated from Equations (7-26) and (7-27)], wind speed, air temperature, and relative humidity.

Where humidity data are lacking, one can assume for non-arid regions that the dew-point temperature equals the daily minimum temperature (Gentilli 1955; Bristow 1992) and use that temperature in Equation (7-4) to estimate $e_a$. Linacre (1993) developed an even more simplified empirical version of the Penman equation that does not require direct measurement of radiation or humidity.

Note that Kohler and Parmele's (1967) generalization of the combination approach to give lake evaporation via Equations (7-37) and (7-38) reintroduces the need for lake water-surface temperature and the flow and temperature data needed to evaluate $A_w$ and $\Delta Q$.

### Applicability

Many studies have shown that evaporation estimates made with the combination approach compare well with those determined by other methods. For example, Van Bavel (1966) showed that free-water evaporation calculated with Equation (7-33) compared closely with actual evaporation from a shallow pan on an hourly basis (Figure 7-4), and the Penman method performed well in comparative studies of lake (Winter et al. 1995) and wetland (Souch et al. 1996) evaporation. Kohler and Parmele (1967) tested their generalized version of the combination approach against evaporation determined by water-balance methods at Lake Hefner and other lakes, and found good agreement

when evaporation was calculated daily and summed to use with weekly to monthly values of advection and storage.

Because it gives satisfactory results, because it has a theoretical foundation, and because it requires meteorological inputs that are widely available or can be reasonably well estimated from available data, the combination method has become the "standard" hydrological method for determining free-water evaporation. One of its major advantages is that it eliminates the need for surface-temperature data, so that it can be readily used both in determining evaporation from measured data and in the predictive or modeling context.

Van Bavel (1966) showed that Equation (7-33) can be used to give reliable estimates of daily evaporation using measured total daily radiation and average daily values of temperature, humidity, and wind speed. As with the mass-transfer approach, it is probably unwise to expect such reliability if averages for monthly or longer periods are used.

It must be emphasized that, as with all approaches to determining evaporation, calculated

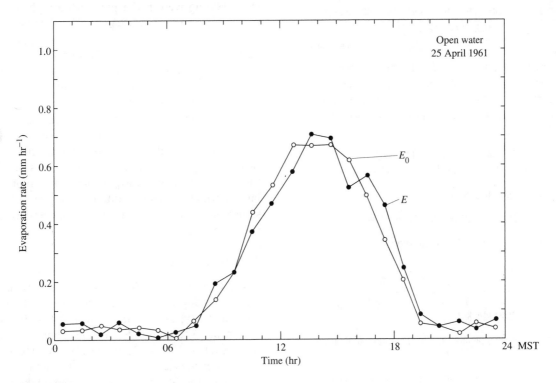

**FIGURE 7-4**
Comparison of hourly evaporation rates determined from measurements in a shallow pan (*E*) with those computed via the combination approach [*E₀*, Equation (7-33)] at Tempe, AZ. From Van Bavel (1966), used with permission of the American Geophysical Union.

values are only valid to the degree that the input data are correct and representative of the evaporating water body. In particular, Van Bavel (1966) found that significant errors can be introduced when empirical radiation relations [e.g., Equations (7-26)–(7-30)] are used in place of measured values.

---

### EXAMPLE 7-3

Estimate the evaporation from Lake Hefner on 12 July 1951 using the combination approach with the data in Table 7-4.

**Solution** The values of $\Delta$ and $\gamma$ are found from Equations (7-6) and (7-13), respectively:

$$\Delta = 0.212 \text{ kPa K}^{-1} \quad \text{and} \quad \gamma = 0.0645 \text{ kPa K}^{-1}.$$

All other values needed for Equation (7-33) are available from Table 7-4 and Examples 7-1 and 7-2. Since $K_E$ was calculated in m km$^{-1}$ kPa$^{-1}$, $v_a$ must be in km day$^{-1}$ to give consistent units. We then obtain

$$E = \frac{0.212 \cdot 23.8 + 0.0645 \cdot (1.51 \times 10^{-5}) \cdot 1000 \cdot 2.44 \cdot 502 \cdot 3.62 \cdot (1 - .69)}{1000 \cdot 2.44 \cdot (0.212 + 0.0645)}$$

$$= 0.00947 \text{ m day}^{-1} = 9.47 \text{ mm day}^{-1}.$$

This value is quite close to that calculated from the energy balance.

It is also of interest to compare this result with that obtained from the combination approach using the modifications of Kohler and Parmele (1967). To do this, we replace $L$ with $L'$, from Equation (7-35):

$$L' = 0.97 \cdot 34.4 - 0.97 \cdot (4.90 \times 10^{-9})$$
$$\cdot (27.2 + 273.15)^4 = -5.38 \text{ MJ m}^{-2} \text{ day}^{-1}.$$

This gives a net radiation term of

$$29.0 - 5.38 = 23.6 \text{ MJ m}^{-2} \text{ day}^{-1}.$$

Next we replace $\gamma$ with $\gamma'$ from Equation (7-36):

$$\gamma' = 0.0645$$
$$+ \frac{4 \cdot 0.97 \cdot (4.90 \times 10^{-9}) \cdot (27.2 + 273.2)^3}{(1.51 \times 10^{-5}) \cdot 1000 \cdot 2.44 \cdot 502}$$
$$= 0.0924 \text{ kPa K}^{-1}.$$

Substituting these quantities into the Penman Equation yields

$$E =$$
$$\frac{0.212 \cdot 23.6 + 0.0924 \cdot (1.51 \times 10^{-5}) \cdot 1000 \cdot 2.44 \cdot 502 \cdot 3.62 \cdot (1 - .69)}{1000 \cdot 2.44 \cdot (0.212 + 0.0924)}$$
$$= 0.00934 \text{ m day}^{-1} = 9.34 \text{ mm day}^{-1},$$

which is quite close to the value calculated via Equation (7-33).

Thus the values calculated from the combination method for this date are fairly close to that obtained from the energy balance, higher than given via mass transfer, and substantially higher than that determined from the water balance.

---

### 7.3.6 Pan-Evaporation Approach

#### Theoretical Basis

A direct approach to determining free-water evaporation is to expose a cylindrical pan of liquid water to the atmosphere and to solve the following simplified water-balance equation for a convenient time period, $\Delta t$ (usually one day):

$$E = W - [V_2 - V_1]. \tag{7-40}$$

Here, $W$ is precipitation during $\Delta t$ and $V_1$ and $V_2$ are the storages at the beginning and end of $\Delta t$, respectively. Precipitation is measured in an adjacent non-recording rain gage; the storage volumes are determined by measuring the water level in a small **stilling well** in the pan with a high-precision micrometer called a **hook gage.** The water surface is maintained a few centimeters below the pan rim by adding measured amounts of water as necessary.

Pans on land are placed in clearings suitable for rain gages (see Section 4.2.2), surrounded by a fence to prevent animals from drinking, and may be sunken so that the water surface is approximately in the same plane as the ground surface or raised a standard height above ground. For special studies of lake evaporation, pans are sometimes placed in the center of a floating platform with dimensions large enough to insure stability and prevent water splashing in, again with the surface either at or above the lake surface.

#### Practical Considerations

Several different standard types of pans are used by different countries and agencies. Shaw (1988) described the types used in Great Britain, the Soviet Union, and the United States, and Figure 7-5 shows the standard "Class-A" pan used by the U.S. National Weather Service (NWS) and in Canada.

An evaporation pan differs from a lake in having far less heat-storage capacity, in lacking surface- or ground-water inputs or outputs, and, with raised

**FIGURE 7-5**
Standard U.S. National Weather Service "Class-A" evaporation pan. The pan is 4 ft (1.22 m) in diameter by 10 in. (25.4 cm) high and set on a low platform. The stilling well in which water level is measured is on the right, and a floating thermometer is on the left. Typically an anemometer is installed next to the pan and a rain gage must be located nearby. Photo by author.

pans like the Class-A, in having sides exposed to the air and sun. These differences significantly affect the energy balance, elevating the warm-season average temperature and vapor pressure of the water surface of a pan relative to that of a nearby lake. The ratio of lake evaporation to pan evaporation is called the **pan coefficient**; its annual average over the United States is about 0.7. Within the evaporation season, heat-storage effects cause water temperatures in lakes to be generally lower than those of pans in the spring, and higher in the fall. Thus pan coefficients follow the same pattern, with values lower than the seasonal average in the spring and higher in the fall (e.g., Figure 7-6).

The Class-A pan is usually installed with an anemometer and a floating maximum–minimum thermometer, providing data on average daily wind speed and average water-surface temperature. Kohler et al. (1955) developed the following empirical equation to use these data along with air temperature to account for the energy exchange through the sides of a pan and thereby adjust daily pan evaporation to daily free-water evaporation:

$$E_{fw} = 0.7 \cdot [E_{pan}$$
$$\pm\ 0.064 \cdot P \cdot \alpha_{pan} \cdot (0.37 + 0.00255 \cdot v_{pan})$$
$$\cdot\ |T_{span} - T_a|^{0.88}].$$
$$(7\text{-}41)$$

In this equation, $E_{fw}$ and $E_{pan}$ are daily free-water and pan evaporation, respectively, in mm day$^{-1}$, $P$ is atmospheric pressure in kPa, $v_{pan}$ is the average wind speed at a height of 15 cm above the pan in km day$^{-1}$, $T_{span}$ is the water-surface temperature in the pan, temperatures are in °C, and the operation following $E_{pan}$ is + when $T_{span} > T_a$ and – when $T_{span} < T_a$. The factor $\alpha_{pan}$ is the proportion of energy exchanged through the sides of the pan that is used for, or lost from, evaporation; it can be estimated as

$$\alpha_{pan} = 0.34 + 0.0117 \cdot T_{span}$$
$$-\ (3.5 \times 10^{-7}) \cdot (T_{span} + 17.8)^3 + 0.0135 \cdot v_{pan}^{0.36},$$
$$(7\text{-}42)$$

using the same units as in Equation (7-41) (Linsley et al. 1982).

---

### EXAMPLE 7-4

Adjust the measured Class-A pan evaporation at Lake Hefner to give an estimate of the free-water evaporation on 12 July 1951. Data are given in Table 7-4.

**Solution** First we convert $v_{pan}$ from m s$^{-1}$ to km day$^{-1}$ as in Example 7-1, and find $v_{pan} = 241$ km day$^{-1}$. Next compute $\alpha_{pan}$ via Equation (7-42):

$$\alpha_{pan} = 0.34 + 0.0117 \cdot 27.5$$
$$-\ (3.5 \times 10^{-7}) \cdot (27.5 + 17.8)^3 + 0.0135 \cdot 241^{0.36}$$
$$= 0.726.$$

Substituting this value into Equation (7-41) yields

$$E_{fw} = 0.7 \cdot [12.4$$
$$+\ 0.064 \cdot 97.3 \cdot 0.726 \cdot (0.37 + 0.00255 \cdot 241)$$
$$\cdot\ |27.5 - 27.2|^{0.88}] = 9.76 \text{ mm day}^{-1},$$

which is close to the value given by the combination method.

---

Because inputs and outputs of energy through the sides of a pan balance out over the course of a year, adjustments via Equation (7-41) and (7-42) are not needed for annual values.

### Applicability

Pan evaporation data provide a useful basis for understanding the regional climatology of seasonal or annual free-water evaporation: The NWS publishes

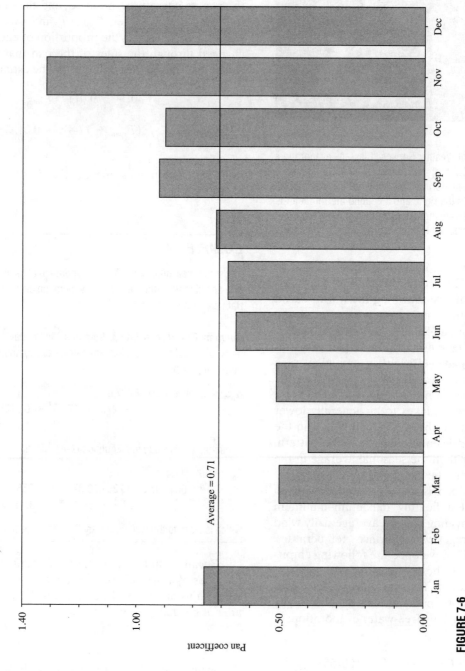

**FIGURE 7-6**
Monthly Class-A pan coefficients at Lake Hefner, OK, June 1950–May 1951. Data from Harbeck et al. (1954).

data for some 400 U.S. evaporation-pan stations in its *Climatological Data* series, and pan coefficients and annual free-water evaporation have been summarized and mapped by Farnsworth and Thompson (1982). Figure 7-7 shows free-water evaporation for the contiguous United States. Chow (1994) described an automated evaporation pan that, while relatively costly, is sufficiently accurate and reliable to make it attractive for use in an observation network.

Since year-to-year variations of pan evaporation are usually not large, observations for a few years can provide a satisfactory estimate of annual values, which can then be adjusted by the appropriate regional pan coefficient to give an estimate of free-water evaporation. While such information might be useful in planning for a water-supply reservoir, advection and storage could cause actual lake evaporation to be considerably different from pan-estimated free-water evaporation: Derecki (1981) computed 483 mm average annual evaporation for Lake Superior by the mass-transfer approach, whereas free-water–surface evaporation is about 554 mm. Based on the Lake Hefner study, M.A. Kohler concluded that "Annual lake evaporation can probably be estimated within 10 to 15% (on the average) by applying an annual coefficient to pan evaporation, provided lake depth and climatic regime are taken into account in selecting the coefficient" (Harbeck et al. 1954, p. 148).

Van Bavel's (1966) results and other studies indicate that free-water–surface evaporation calculated by the combination method closely approximates pan evaporation on an annual basis and for shorter periods if heat exchange through the pan is accounted for via Equations (7-41) and (7-42). This correspondence is seen in the results of Examples 7-3 and 7-4.

## 7.4 BARE-SOIL EVAPORATION

More than one-third of the earth's land surface consists of Entisols, Inceptisols, and Aridisols supporting little or no vegetation (Table 3-9; Figures 3-40, 3-41). In addition, most agricultural lands have meager vegetative cover much of the time. Thus evaporation from bare soil is globally significant and vitally important to farmers, especially in the management of irrigation.

Following infiltration due to rain, snowmelt, or irrigation, evaporation from bare soil (also called **exfiltration**) generally occurs in two distinct stages:

*Stage 1*: an **atmosphere-controlled stage,** in which the evaporation rate is largely determined by the surface energy balance and mass-transfer conditions (wind and humidity) and is largely independent of soil-water content. Evaporation in this stage occurs at or near the rate of free-water evaporation; and

*Stage 2*: a **soil-controlled stage,** in which the evaporation rate is determined by the rate at which water can be conducted to the surface in response to potential gradients induced by upward-decreasing soil-water contents [Equation (6-16)] rather than by atmospheric conditions. The evaporation rate in this stage is less than the free-water rate.

The transition from the first stage to the second is typically quite abrupt and can often be visually detected as an increase in the brightness (albedo) of the soil surface (Figure 7-8a).

As implied above, soil evaporation during stage 1 can be estimated by the approaches appropriate for free-water evaporation, of which the Penman Equation [Equation (7-33)] usually gives the best results (Parlange and Katul 1992a). Thus the remainder of this section will focus on evaporation during the soil-controlled stage.

Following an infiltration event, the soil dries both by drainage and by evaporation. If stage-1 evaporation has occurred at the average rate $\overline{E}_1$ for a duration $t_1$, then the total stage-1 evaporation, $F_1$, is

$$F_1 = \overline{E}_1 \cdot t_1, \qquad \textbf{(7-43)}$$

and the water content at the end of stage 1 is designated $\theta_1$ (assumed uniform with depth). Salvucci (1997) showed that the modeling of stage-2 soil evaporation can be greatly simplified by (1) invoking a relationship between the duration of stage-1 evaporation, soil properties, and soil moisture; and (2) assuming that the rate of soil drainage due to gravity is much greater than the exfiltration rate. This leads to a cumulative evaporative loss, $F_{soil}(t)$, given by

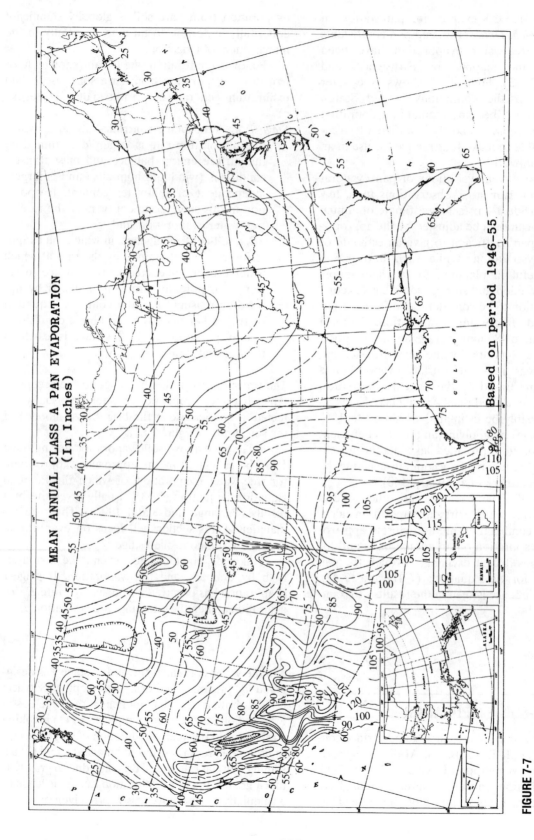

**FIGURE 7-7**

Average annual pan (= free-water-surface) evaporation (in. yr$^{-1}$) for the 48 contiguous United States based on data for 1946–1955. Map provided by U.S. National Weather Service.

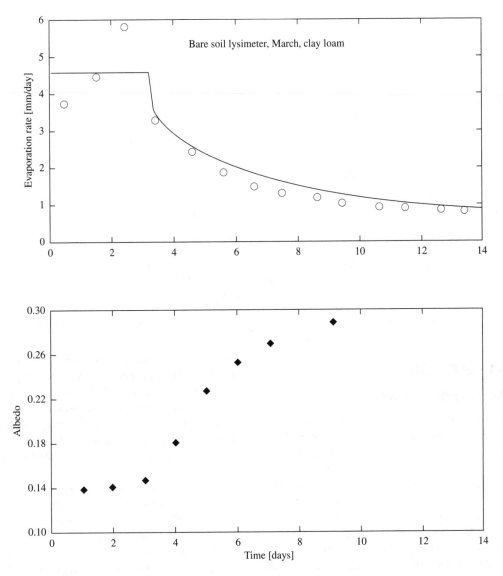

**FIGURE 7-8**
(a) Bare-soil evaporation following irrigation as measured in a lysimeter (circles) and computed via Equation (7-45) (line). (b) Concurrent soil-surface albedo. Figure provided by Guido Salvucci, Boston University.

$$F_{soil}(t) = F_1 \cdot \left[ 1 + \left( \frac{8}{\pi^{1/2}} \right) \cdot \ln\left( \frac{t}{t_1} \right) \right], \quad t \geq t_1 \quad \textbf{(7-44)}$$

and a stage-2 evaporation rate, $E_2(t)$, of

$$E_2(t) = \left( \frac{8}{\pi^{1/2}} \right) \cdot \overline{E}_1 \cdot \frac{t_1}{t} = 0.811 \cdot \frac{F_1}{t}, \quad t \geq t_1.$$

$$\textbf{(7-45)}$$

With Equations (7-44) and (7-45), one can estimate stage-2 evaporation based only on estimates of the stage-1 evaporation rate and observation of the time at which stage 1 ends—which, as noted, can usually be readily done from ground or satellite observations of albedo. This simplified approach gave estimates that compared well with observations in several field experimental situations (Figure 7-8).

## EXAMPLE 7-5

During the infiltration event of Example 6-7 a total of 81.4 mm infiltrated. Following this, stage-1 evaporation lasted for $t_1 = 3$ days and occurred at an average rate of $\bar{E}_1 = 10$ mm day$^{-1}$. Compute the stage-2 evaporation rate over the next 10 days in which no rain fell.

**Solution** From Equation (7-43), the stage-1 evaporation is

$$F_1 = 10 \text{ mm day}^{-1} \cdot 3 \text{ day} = 30 \text{ mm}.$$

For stage 2, Equation (7-44) gives the cumulative evaporation:

$$F_{\text{soil}}(t) = 30 \text{ mm} \cdot \left[ 1 + 0.811 \cdot \ln\left(\frac{t}{3 \text{ day}}\right)\right];$$

and Equation (7-45) gives the evaporation rate:

$$E_2(t) = 0.811 \cdot 10 \text{ mm day}^{-1} \cdot \frac{3 \text{ day}}{t} = \frac{24.3 \text{ mm}}{t}.$$

The results are tabulated as follows:

| $t$ (day) | 3.0 | 3.5 | 4.0 | 4.5 | 5.0 | 5.5 | 6.0 | |
|---|---|---|---|---|---|---|---|---|
| $E_2(t)$ (mm day$^{-1}$) | 8.11 | 6.95 | 6.08 | 5.40 | 4.86 | 4.42 | 4.05 | |
| $F_{\text{soil}}(t)$ (mm) | 30.0 | 33.8 | 37.0 | 39.9 | 42.4 | 44.7 | 46.9 | |

| $t$ (day) | 6.5 | 7.0 | 7.5 | 8.0 | 8.5 | 9.0 | 9.5 | 10.0 |
|---|---|---|---|---|---|---|---|---|
| $E_2(t)$ (mm day$^{-1}$) | 3.74 | 3.47 | 3.24 | 3.04 | 2.86 | 2.70 | 2.56 | 2.43 |
| $F_{\text{soil}}(t)$ (mm) | 48.8 | 50.6 | 52.3 | 53.9 | 55.3 | 56.7 | 58.0 | 59.3 |

## 7.5  TRANSPIRATION

### 7.5.1  The Transpiration Process

**Transpiration** is the evaporation of water from the vascular system of plants into the atmosphere. The entire process (Figure 7-9) involves **absorption** of soil water by plant roots; **translocation** in liquid form through the vascular system of the roots, stem, and branches to the leaves; and translocation through the vascular system of the leaf to the walls of tiny **stomatal cavities,** where evaporation takes place. The water vapor in these cavities then moves into the ambient air through openings in the leaf surface called **stomata.**

Plants live by absorbing carbon dioxide ($CO_2$) from the air to make carbohydrates, and that gas can enter the plant only when dissolved in water. The essential function of the stomatal cavities is to provide a place where $CO_2$ dissolution can occur and enter plant tissue; the evaporation of water is an unavoidable concomitant of that process. However, transpiration also performs the essential functions of maintaining the turgor of plant cells and delivering mineral nutrients from the soil to growing tissue.

Air in stomatal cavities is saturated at the temperature of the leaf, and water moves from the cavities into the air due to a vapor-pressure difference, just as in open-water evaporation. The major differ-ence between transpiration and open-water evaporation is that plants can exert some physiological control over the size of the stomatal openings, and hence the ease of vapor movement, by the action of **guard cells** (Figure 7-10). The major factors affecting the opening and closing of guard cells are (1) light (most plants open stomata during the day and close them at night), (2) humidity (stomatal openings tend to decrease as humidity decreases below its saturation value), and (3) the water content of the leaf cells (if daytime water contents become too low, stomata tend to close).[7]

It is important to emphasize that transpiration is a physical, not a metabolic, process: Water in the **transpiration stream** is pulled through the plant by potential-energy gradients that originate with the movement of water vapor into the air through the stomata in response to a vapor-pressure difference. When vapor exits through the stomata, water evaporates from the walls of the stomatal cavity to replace the loss; this loss of liquid water causes a potential-energy decrease that induces the movement of replacement water up through the vascular system; this movement ultimately produces a water-

---

[7] Several other factors are known to affect the opening and closing of stomata, including wind, $CO_2$ levels, temperature, and certain chemicals. A mathematical representation of the effects of the factors most important in hydrological modeling is given later.

**FIGURE 7-9**
Diagram of the transpiration process and enlarged views of plant cellular structure where absorption, translocation, and transpiration occur. From Hewlett (1992).

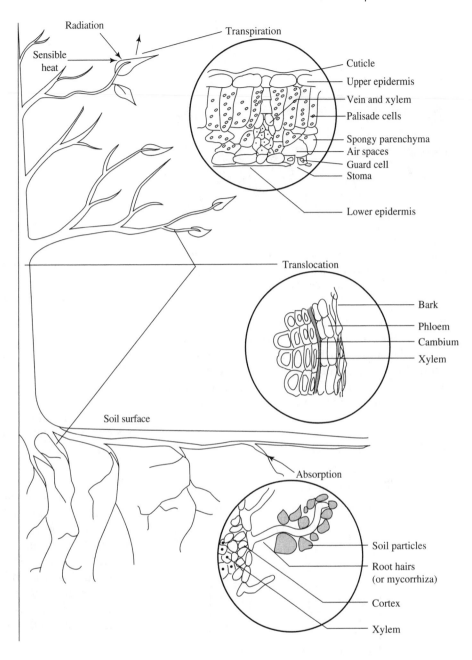

content gradient between the root and the soil; and this gradient induces movement of soil water into the root. Absorption at the root surface decreases soil-water content in the adjacent soil, inducing some flow of water toward the root following Darcy's Law [Equation (6-10)]. However, roots come into contact with soil water mostly by growing toward the water: During the growing season roots typically grow several tens of millimeters per day (Raven et al. 1976).

The great cohesive strength of water due to its intermolecular hydrogen bonds (see Section B.1.1) maintains the integrity of the transpiration stream to heights of the tallest trees.

### 7.5.2   Modeling Transpiration

Since transpiration is essentially the same physical process as open-water evaporation, it can be represented by a mass-transfer equation of the form of

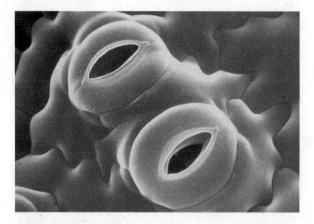

**FIGURE 7-10**
Photomicrograph of leaf surface showing stomata and the crescent-shaped guard cells that regulate their openings. Photo by Dr. Jeremy Burgess/Photo Researchers, Inc.

Equation (7-1). In order to develop an equivalent relation for transpiration, it is convenient first to recast the mass-transfer relation for evaporation using the concept of atmospheric conductance. This allows us to use some electric-circuit analogies for "scaling up" from a leaf to an entire vegetated surface.

### Atmospheric Conductance

We can represent the process of evaporation from open water by combining Equations (7-1) and (7-2) and defining an "atmospheric constant", $K_{at}$, as

$$K_{at} \equiv \frac{0.622 \cdot \rho_a}{P \cdot \rho_w},\qquad (7\text{-}46)$$

so that

$$E = K_{at} \cdot \left\{ \frac{D_V \cdot v_a}{D_M \cdot 6.25 \cdot \left[ \ln\left( \dfrac{z_m - z_d}{z_0} \right) \right]^2} \right\}$$

$$\cdot (e_s - e_a),\qquad (7\text{-}47)$$

where $D_V$ and $D_M$ are the diffusivities of water vapor and momentum, respectively [equations (D-25) and (D-30).

The term in braces in Equation (7-47) represents the efficiency of the turbulent eddies in the lower atmosphere in transporting water vapor from the surface to the ambient air. This can also be viewed as an **atmospheric conductance** for water

vapor, $C_{at}$, so that the mass-transfer equation for evaporation can be written as

$$\begin{array}{ccccc} E & = & K_{at} & \cdot & C_{at} & \cdot (e_s - e_a). \\ \text{evaporation} & & \text{atmospheric} & & \text{atmospheric} & \text{driving} \\ \text{rate} & & \text{constant} & & \text{conductance} & \text{gradient} \end{array}$$
$$(7\text{-}48)$$

If $D_V$ is assumed identical to $D_M$,[8] $C_{at}$ is given explicitly as

$$C_{at} \equiv \frac{v_a}{6.25 \cdot \left[ \ln\left( \dfrac{z_m - z_d}{z_0} \right) \right]^2}\qquad (7\text{-}49)$$

and has the dimensions $[L\ T^{-1}]$.[9]

The zero-plane displacement, $z_d$, and roughness height, $z_0$, can be approximately related to the height of vegetation, $z_{veg}$, as

$$z_d = 0.7 \cdot z_{veg}\qquad (7\text{-}50)$$

and

$$z_0 = 0.1 \cdot z_{veg}.\qquad (7\text{-}51)$$

If it is assumed that $z_m$ is 2 m above the top of the vegetation, Equations (7-49)–(7-51) can be used to generate a relation between atmospheric conductance and wind speed for various values of $z_{veg}$, as shown in Figure 7-11. Typical values of $z_{veg}$ for various land-cover types are given in Table 7-5.

Since evaporation from an open-water surface is a "one-step" process in which water molecules pass directly from the liquid surface into the atmosphere, it can be conceptually represented by the electric-circuit analogy shown in Figure 7-12a. In this analogy a current (water vapor) moves in response to a voltage (vapor-pressure difference) across a resistance (inverse of the atmospheric conductivity).

### Leaf Conductance

Transpiration is a "two-step" process, in which water molecules pass (1) from the stomatal cavity to the leaf surface, and (2) from the leaf surface into the

---

[8] This assumption is justified for short vegetation, but becomes less so for forests because $z_0$ and $z_d$ for water-vapor transfer increasingly differ from their values for momentum transfer as the surface gets rougher (C.A. Federer, pers. comm.). In the interests of simplicity, we assume $D_V = D_M$ in the developments herein.
[9] Many authors use **atmospheric resistance,** which is the inverse of atmospheric conductance.

**FIGURE 7-11**
Relation between atmospheric conductance, $C_{at}$, and wind speed, $v_a$, for vegetation of various heights, $z_{veg}$. Relation for typical water surface is also shown. It is assumed that $v_a$ is measured 2 m above the canopy and that atmospheric stability is near neutral.

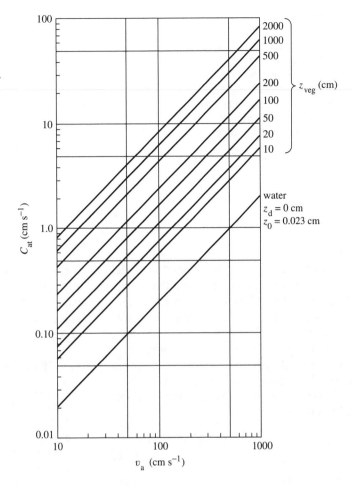

**TABLE 7-5**
Typical values of maximum leaf conductance ($C^*_{leaf}$), leaf-area index (*LAI*), albedo (*a*), and height ($z_{veg}$) for principal land-cover types.

| Land Cover | $C^*_{leaf}$ (mm s$^{-1}$) | *LAI* | *a* | $z_{veg}$ (m) |
|---|---|---|---|---|
| Conifer forest | 5.3 | 6.0 | 0.14 | 25.0 |
| Broadleaf forest | 5.3 | 6.0 | 0.18 | 25.0 |
| Savannah/shrub | 5.3 | 3.0 | 0.18 | 8.0 |
| Grassland | 8.0 | 3.0 | 0.20 | 0.5 |
| Tundra/nonforest wetland | 6.6 | 4.0 | 0.20 | 0.3 |
| Desert | 5.0 | 1.0 | 0.26 | 0.1 |
| Typical crop | 11.0 | 3.0 | 0.22 | 0.3 |

Data from Federer et al. (1996).

atmosphere. Continuing the electric-circuit analogy, the same driving force in this case operates across two resistances (inverse conductances) linked in series: leaf and atmospheric (Figure 7-12b).

**Leaf conductance** is determined by the number of stomata per unit area and the size of the stomatal openings. Stomatal densities range from 10,000 to 100,000 stomata per square centimeter of leaf surface, depending on species (Hewlett 1982). Table 7-5 lists leaf conductances at maximum stomatal opening for various land-cover types. As noted earlier, plants control the size of the stomatal openings, and hence leaf conductance, by the response of the guard cells. These cells have been found to respond to (1) light intensity, (2) ambient $CO_2$ concentration, (3) leaf-air vapor-pressure difference, (4) leaf temperature, and (5) leaf water content (Stewart 1989). We first examine how these factors affect leaf conductance, and then incorporate them in a model for the transpiration from a vegetative canopy.

**FIGURE 7-12**
Conceptualization of (a) open-water evaporation and (b) evapotranspiration in terms of atmospheric resistance ($1/C_{at}$) and leaf resistance ($1/C_{leaf}$). $\Delta e_v$ is the driving vapor-pressure difference between the evaporating surface and the atmosphere.

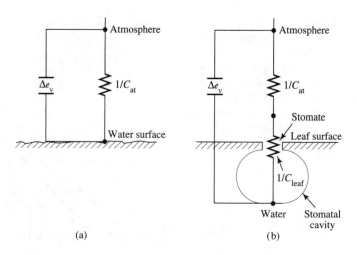

(a)                    (b)

Stewart (1988) developed and tested a model for estimating hourly evapotranspiration that incorporates four of the five factors that determine leaf conductance—the effect of $CO_2$ concentration was not included because it usually varies little with time.[10] As shown in Table 7-6, he substituted more commonly measured hydrologic variables for some of the controlling factors—in particular, soil-water deficit was used as a proxy for leaf-water content. His model has the general form

$$C_{leaf} = C_{leaf}^* \cdot f_k(K_{in}) \cdot f_\rho(\Delta\rho_v) \cdot f_T(T_a) \cdot f_\theta(\Delta\theta),$$
$$(7\text{-}52)$$

where $C_{leaf}^*$ is the maximum value of leaf conductance (i.e., values from Table 7-5), $K_{in}$ is incident short-wave radiation flux, $\Delta\rho_v$ is the humidity deficit [the difference between the saturated and actual absolute humidity of the air, calculated from vapor pressures and temperature via Equation (D-8)], $T_a$ is air temperature, and $\Delta\theta$ is the soil-moisture deficit (the difference between the field capacity and the actual water content of the root zone).

The $f$s in Equation (7-52) represent the effects of each environmental factor on relative leaf conductance; they are non-linear functions that vary from 0 to 1 as shown in Figure 7-13. While the constants in these functions have been derived from studies at only one site (a pine forest in southeast England), controlled studies indicate that their form is quite general (Jarvis 1976). An abbreviated

form of the model incorporating only $f_K(K_{in})$ and $f_\rho(\Delta\rho_v)$ (but with different constants, determined by calibration at the site) successfully modeled transpiration from prairie grasses in Kansas (Stewart and Gay 1989).

### Canopy Conductance

A vegetated surface like a grass, crop, or forest canopy can be thought of as a large number of leaf conductances in parallel. Again from the laws of electric circuits, the total conductance of a number of conductances in parallel equals the sum of the individual conductances. Thus it is possible to represent a reasonably uniform vegetated surface as a single "big leaf" whose total conductance to water vapor is proportional to the sum of the conductances of millions of individual leaves. The relative size of this big leaf is reflected in the **leaf-area index, LAI,** defined as

$$LAI \equiv \frac{\text{total area of leaf surface above ground area } A}{A}.$$
$$(7\text{-}53)$$

Canopy conductance is then given by

$$C_{can} = f_s \cdot LAI \cdot C_{leaf},  \qquad (7\text{-}54)$$

where $f_s$ is a **shelter factor** that accounts for the fact that some leaves are sheltered from the sun and wind and thus transpire at lower rates. Values of $f_s$ range from 0.5 to 1, and decrease with increasing $LAI$ (Carlson 1991); a value of $f_s = 0.5$ is probably a good estimate for a completely vegetated area (Allen et al. 1989).

---

[10] However, a long-term decrease in leaf conductance is a possible response of plants to the anthropogenic increase in atmospheric $CO_2$ concentration, as discussed in Section 3.2.9 (Rosenberg et al. 1989).

**TABLE 7-6**
Stewart's (1988) model of leaf conductance as a function of environmental factors. Functional relations are plotted in Figure 7-13.

| Factor Controlling Stomatal Opening | Quantity Representing Controlling Factor in Model | Functional Relation |
|---|---|---|
| light | incident solar radiation (MJ m$^{-2}$ day$^{-1}$) | $f_K(K_{in}) = \dfrac{12.78 \cdot K_{in}}{11.57 \cdot K_{in} + 104.4}$ <br> $0 \le K_{in} \le 86.5 \text{ MJ m}^{-2}\text{day}^{-1}$ |
| CO$_2$ concentration | (not included) | (none) |
| vapor-pressure deficit | absolute-humidity deficit, $\Delta\rho_v$ (kg m$^{-3}$) | $f_\rho(\Delta\rho_v) = 1 - 66.6 \cdot \Delta\rho_v,$ <br> $0 \le \Delta\rho_v \le 0.01152 \text{ kg m}^{-3};$ <br> $f_\rho(\Delta\rho_v) = 0.233, \ 0.01152 \text{ kg m}^{-3} \le \Delta\rho_v.$ |
| leaf temperature | air temperature, $T_a$ (°C) | $f_T(T_a) = \dfrac{T_a \cdot (40 - T_a)^{1.18}}{691}$ <br> $0 \le T_a \le 40°C$ |
| leaf water content | soil-moisture deficit, $\Delta\theta$ (cm) | $f_\theta(\Delta\theta) = 1 - 0.00119 \cdot \exp(0.81 \cdot \Delta\theta)$ <br> $0 \le \Delta\theta \le 8.4 \text{ cm}$ |

Leaf-area indices for various types of plant communities are given in Table 7-5. For deciduous forests, leaf area changes through the growing season, rising from near 0 to a maximum and back; Figure 7-14 shows the seasonal variation of *LAI* used in modeling transpiration for three types of forests.

### The Penman-Monteith Model

Making use of Equation (7-1) and the definitions of $\gamma$ [Equation (7-13)], $K_{at}$ [Equation (7-46)], and $C_{at}$ [Equation (7-49)], the Penman (combination) model for evaporation from a free-water surface [Equation (7-33)] can be written in terms of atmospheric conductance as

$$E = \frac{\Delta \cdot (K + L) + \rho_a \cdot c_a \cdot C_{at} \cdot e_a^* \cdot (1 - W_a)}{\rho_w \cdot \lambda_v \cdot (\Delta + \gamma)}. \quad \text{(7-55)}$$

[annotation: heat capacity of air → $c_a$]

Recall that the derivation of this equation assumes no water-advected energy, no ground-heat conduction, and no heat-storage effects.

Monteith (1965) showed how the Penman equation can be modified to represent the evapotranspiration rate, *ET*, from a vegetated surface by incorporating canopy conductance:

$$ET = \frac{\Delta \cdot (K + L) + \rho_a \cdot c_a \cdot C_{at} \cdot e_a^* \cdot (1 - W_a)}{\rho_w \cdot \lambda_v \cdot [\Delta + \gamma \cdot (1 + C_{at}/C_{can})]}. \quad \text{(7-56)}$$

This relationship has become known as the **Penman-Monteith Equation.** The assumptions of no water-advected energy and no heat-storage effects, which are generally not valid for natural water bodies, are usually reasonable when considering a vegetated surface. (The term for ground-heat conduction can be included with $K + L$ if it is significant.) Note that Equation (7-56) reduces to Equation (7-55) when $C_{can} \to \infty$.

The Penman-Monteith Equation has been successfully tested in many environments [see, for example, Calder (1977), (1978); Berkowicz and Prahm (1982); Lindroth (1985); Dolman et al. (1988); Stewart (1988); Stewart and Gay (1989); Allen et al. (1989); Lemeur and Zhang (1990)] and has become the most widely used approach to estimating evapotranspiration from land surfaces. The following example shows how it can be applied.

### EXAMPLE 7-6

Compare the evapotranspiration rate from the canopy for soil-moisture deficits of 0 cm and 7.0 cm at the pine forest at Thetford, England, in August. The following conditions apply (Stewart 1988): $z_{veg}$ = 16.5 m; *LAI* = 2.8; $f_s$ = 0.5; $C_{leaf}^*$ = 2.3 mm s$^{-1}$; $P$ = 101.3 kPa; $K_{in}$ = 25.1 MJ m$^{-2}$ day$^{-1}$; $a$ = 0.18; $L$ = −4.99 MJ m$^{-2}$ s$^{-1}$; $T_a$ = 19.2 °C; $W_a$ = 0.54; $v_a$ = 3.00 m s$^{-}$

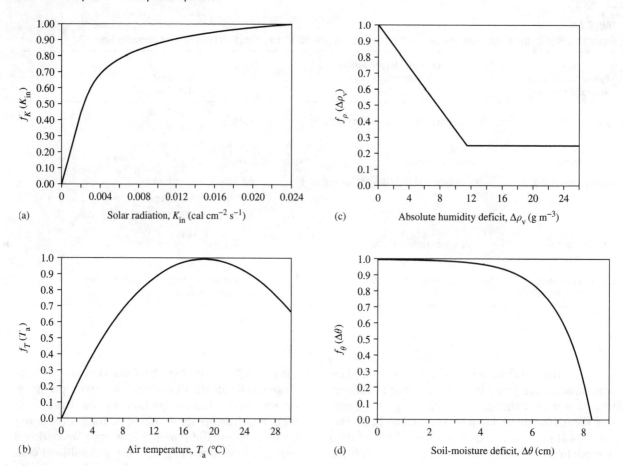

**FIGURE 7-13**
Effects of (a) solar radiation, $K_{in}$; (b) air temperature, $T_a$; (c) vapor-pressure deficit, $\Delta\rho_v$; and (d) soil-moisture deficit, $\Delta\theta$, on relative leaf conductances. (See Table 7-6.) After Stewart (1988).

(The values of $z_{veg}$, LAI, $C_{leaf}^*$, and $a$ were determined for this site, so are used in preference to values from Table 7-5.)

**Solution** Applying equations given in this chapter we have

| Quantity | Equation | Value |
|----------|----------|-------|
| $\lambda_v$ | (7-8) | 2.45 MJ kg$^{-1}$ |
| $e_a^*$ | (7-4) | 2.23 kPa |
| $\Delta$ | (7-6) | 0.139 kPa K$^{-1}$ |
| $\gamma$ | (7-13) | 0.0666 kPa K$^{-1}$ |
| $C_{at}$ | (7-49)–(7-51) | 0.232 m s$^{-1}$ |

The vapor-pressure deficit, $\Delta e_a$, is found as

$$\Delta e_a = e_a^* \cdot (1 - W_a) = 2.23 \cdot (1 - 0.54)$$
$$= 1.03 \text{ kPa};$$

using this value in Equation (D-8c) yields $\Delta\rho_v = 7.62 \times 10^{-3}$ kg m$^{-3}$. Now we can compute the values of the $f$-functions given in Table 7-6:

| Quantity | Value |
|----------|-------|
| $f_K(K_{in})$ | 0.812 |
| $f_\rho(\Delta\rho_v)$ | 0.493 |
| $f_T(T_a)$ | 0.998 |
| $f_\theta(\Delta\theta = 0 \text{ cm})$ | 1.000 |
| $f_\theta(\Delta\theta = 7 \text{ cm})$ | 0.655 |

**FIGURE 7-14**
Annual variability of transpirational leaf-area index, *LAI*, in (a) a pine forest in southeast England (Stewart 1988); (b) a hardwood forest in New Hampshire; and (c) a hardwood forest in North Carolina [both from Federer and Lash (1978)].

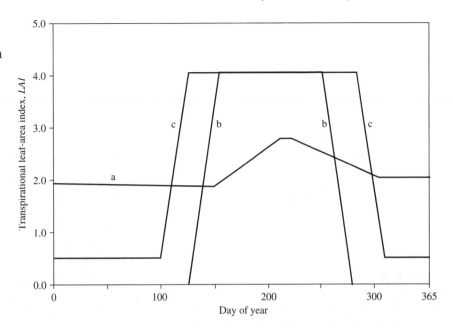

From the above *f*-values we calculate the following conductances:

| Quantity | Equation | Value (mm s$^{-1}$) |
|---|---|---|
| $C_{leaf}$ ($\Delta\theta = 0$ cm) | (7-52) | 0.918 |
| $C_{leaf}$ ($\Delta\theta = 7$ cm) | (7-52) | 0.602 |
| $C_{can}$ ($\Delta\theta = 0$ cm) | (7-54) | 1.29 |
| $C_{can}$ ($\Delta\theta = 7$ cm) | (7-54) | 0.843 |

Substituting the appropriate values from above into Equation (7-56) yields $ET = 1.04 \times 10^{-5}$ mm s$^{-1}$ (= 0.9 mm day$^{-1}$) when $\Delta\theta = 0$ cm and $ET = 6.88 \times 10^{-6}$ mm s$^{-1}$ (= 0.6 mm day$^{-1}$) when $\Delta\theta = 7$ cm.

We will consider the Penman-Monteith Equation further in discussing methods of estimating areal evapotranspiration (Section 7.8).

## 7.6  INTERCEPTION AND INTERCEPTION LOSS

**Interception** is the process by which precipitation falls on vegetative surfaces (the **canopy**), where it is subject to evaporation. As we will see, the intercepted water that is evaporated (**interception loss**) is a significant fraction of total evapotranspiration in most regions.

Interception loss depends strongly on (1) vegetation type and stage of development, which should be well characterized by leaf-area index; and (2) the intensity, duration, frequency, and form of precipitation. Vegetation type is commonly altered by human activities (e.g., deforestation), and there is concern that aspects of precipitation climatology may be altered by climate change (Waggoner 1989). A sound physical basis for understanding the interception process is essential in order to sort out and predict how these changes will affect the hydrologic cycle locally and globally. In particular, the question of whether interception loss is an addition to, as opposed to a replacement for, transpiration loss is important in evaluating the effects of vegetation changes on regional water balances.

Contact of precipitation with vegetation can significantly alter the chemical composition of water reaching the ground; thus interception can also influence weathering processes and water quality.

Under some weather conditions, vegetation can "comb" water from clouds and fog and thereby add water input to a region. This phenomenon is sometimes called "positive interception;" it is briefly discussed as "occult precipitation" in Section 4.2.2. Snow interception is treated in Section 5.3.2.

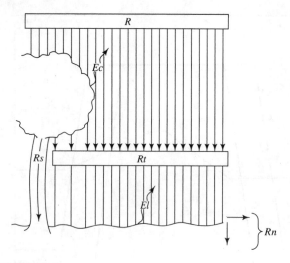

**FIGURE 7-15**
Definitions of terms used in describing the interception process. $R$ = gross rainfall; $Ec$ = canopy interception loss; $Rt$ = throughfall; $Rs$ = stemflow; $El$ = litter interception loss; $Rn$ = net rainfall.

### 7.6.1  Definitions

Figure 7-15 illustrates the following definitions used in describing and measuring interception:

**Gross rainfall,** $R$, is the rainfall measured above the vegetative canopy or in the open.

**Throughfall,** $Rt$, is rainfall that reaches the ground surface directly through spaces in the canopy and by dripping from the canopy.

**Stemflow,** $Rs$, is water that reaches the ground surface by running down trunks and stems.

**Canopy interception loss,** $Ec$, is water that evaporates from the canopy.

**Litter interception loss,** $El$, is water that evaporates from the ground surface (usually including near-ground plants and leaf litter).

**Total interception loss,** $Ei$, is the sum of canopy and litter interception losses.

**Net rainfall,** $Rn$, is the gross rainfall minus the total interception loss.

These definitions are applied over a representative area of the plant community of interest, so they take into account the typical spacing between plants. If the symbols given represent volumes of water during a given time period and have dimensions [L], we have

$$Rn = R - Ei, \qquad (7\text{-}57)$$

$$Ei = Ec + El, \qquad (7\text{-}58)$$

$$R = Rt + Rs + Ec, \qquad (7\text{-}59)$$

and

$$Rn = Rt + Rs - El. \qquad (7\text{-}60)$$

### 7.6.2  Measurement

As with other components of evapotranspiration, interception loss cannot be measured directly. The most common approach to determining the amounts of canopy interception loss in various plant communities is to measure gross rainfall, throughfall, and stemflow, and solve Equation (7-59) for $Ec$. However, this is not a simple procedure because of (1) the difficulties in accurately measuring rainfall, particularly at low rainfall intensities when interception losses are relatively large (see Section 4.2.2); (2) the large spatial variability of throughfall; and (3) the difficulty and expense of measuring stemflow.

Helvey and Patric (1965a) reviewed criteria for measuring interception quantities and concluded that averaging the catches in 20 rain gages spaced randomly over a representative portion of the community should give acceptable estimates of throughfall; typical inter-gage spacing for forest studies is on the order of 10 to 30 m (Gash et al. 1980). Large plastic sheets have also been used to get an integrated measure of net rainfall (Calder and Rosier 1976).

Stemflow is measured by attaching flexible troughs tightly around the trunks of trees and conducting the water to rain gages or collecting bottles. Helvey and Patric (1965a) stated that measuring stemflow from all trees on randomly selected plots gives the most representative results, with plot diameters at least 1.5 times the diameter of the crown of the largest trees. However, since stemflow is usually much less than throughfall, most studies have estimated stemflow less rigorously by sampling a few "typical" trees.

The few published studies of grass or litter interception have usually been done using artificial rain, either measuring the net rainfall from small isolated areas in the field (Merriam 1961) or by collecting undisturbed samples of the surface litter and setting them on recording scales in the laboratory (Pitman 1989; Putuhena and Cordery 1996).

### 7.6.3 Modeling

#### Modeling via Regression Analysis

Most of the earlier studies of interception have used results of field or laboratory measurements in particular plant communities to establish equations relating $Rt$, $Rs$, $El$, $Rn$, and/or $Ei$ to $R$ via regression analysis. These equations are usually of the form

$$Y = M_Y \cdot R + B_Y, \qquad (7\text{-}61)$$

where $Y$ is one of the components of the interception process ($Rt$, $Rs$, $Ec$, $El$, or $Ei$), $R$ is gross rainfall for an individual storm, and $M_Y$ and $B_Y$ are empirical constants determined by regression analysis. Equations in the form of (7-61) can readily be adapted to give estimates for seasonal or annual periods:

$$\Sigma Y = M_Y \cdot \Sigma R + B_Y \cdot n, \qquad (7\text{-}62)$$

where the summation sign indicates seasonal or annual totals and $n$ is the number of storms per season or year.

Table 7-7 summarizes published equations for various community types. Unfortunately, few of the studies that developed these equations made concurrent measurements of leaf-area index, so it is difficult to judge the range of applicability of their results.

Only a few studies have included estimates of litter interception loss; most of these have found $El$ in the range $0.02 \cdot R$ to $0.05 \cdot R$, though some higher values have been reported (Helvey and Patric 1965b; Helvey 1971).

As shown in Example 7-7, regression equations can be useful for showing the effects of rainfall amount and number of storms on net rainfall. However, such equations cannot usually be confidently applied to areas other than where they were developed, and they cannot be used to predict the effects of land-cover changes.

---

### EXAMPLE 7-7

Summer gross rainfall at Fairbanks, Alaska, ranges from 100 to 400 mm, typically arriving in about 45 low-intensity storms. Develop relations to estimate the effects of total summer gross rainfall and number of storms on (a) net rainfall and (b) net rainfall as a fraction of gross rainfall in birch forests in this region.

**Solution** None of the equations listed in Table 7-7 applies to central Alaska, but for purposes of illustration, we use the ones for eastern hardwoods in leaf:

$$\Sigma(Rt + Rs) = 0.941 \cdot \Sigma R - 0.092 \cdot n$$

and assume that $\Sigma El = 0.04 \cdot \Sigma R$. Then, from Equation (7-62), we have

$$\begin{aligned} \Sigma Rn &= 0.941 \cdot \Sigma R - 0.092 \cdot n - 0.04 \cdot \Sigma R \\ &= 0.901 \cdot \Sigma R - 0.092 \cdot n. \end{aligned}$$

Now we can use this relation to estimate $\Sigma Rn$ and $\Sigma Rn/\Sigma R$ for $n = 10$ to 60 storms per season over the range $100 \leq \Sigma R \leq 400$ mm, with the results shown in Figure 7-16.

---

#### Conceptual Models

Because of the uncertain transferability of regression equations and the need for simulating interception loss in predictive models, considerable effort has

**TABLE 7-7**
Regression Equations for Estimating Throughfall, $Rt$, and Stemflow, $Rs$, as Functions of Gross Rainfall, $R$, for Individual Storms. [Equation (7-61)], or for Seasons [Equation (7-62)]

| Plant Community | Quantity ($Y$, cm) | $M_Y$ | $B_Y$ (cm) |
|---|:---:|:---:|:---:|
| Eastern hardwoods full leaf | $Rt$ | 0.901 | −0.079 |
| | $Rs$ | 0.041 | −0.013 |
| | $Rt + Rs$ | 0.941 | −0.092 |
| Eastern hardwoods leafless | $Rt$ | 0.914 | −0.038 |
| | $Rs$ | 0.062 | −0.013 |
| | $Rt + Rs$ | 0.978 | −0.051 |
| Red pine | $Rt + Rs$ | 0.89 | −0.10 |
| Loblolly pine | $Rt + Rs$ | 0.88 | −0.08 |
| Shortleaf pine | $Rt + Rs$ | 0.91 | −0.10 |
| Eastern white pine | $Rt + Rs$ | 0.91 | −0.13 |
| Ponderosa pine | $Rt + Rs$ | 0.93 | −0.15 |
| Pines (average) | $Rt + Rs$ | 0.90 | −0.10 |
| Spruce-fir-hemlock | $Rt + Rs$ | 0.79 | −0.13 |

Eastern hardwood values from Helvey and Patric (1965b), conifer values from Helvey (1971).

**FIGURE 7-16**
(a) Net rainfall, *Rn*, and (b) the ratio of net rainfall to gross rainfall, *Rn/R*, as a function of seasonal gross rainfall, *R*, and number of storms per season, *n*, estimated for a central Alaskan birch forest.

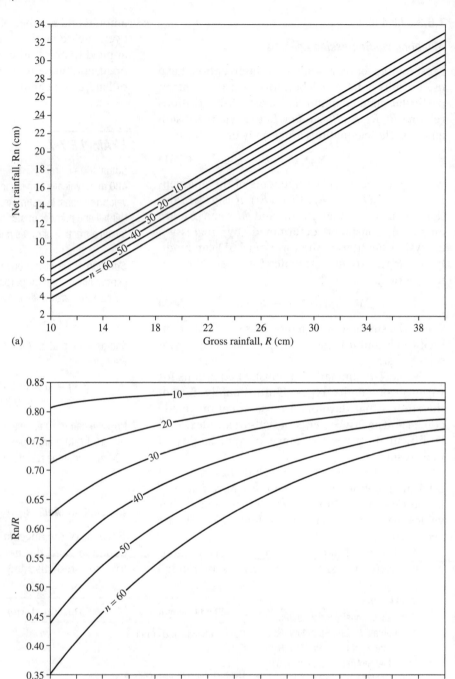

(a)

(b)

been expended to develop conceptual models of the process. The most widely used of these was originally developed by Rutter et al. (1971) and modified by Gash and Morton (1978) and Valente et al. (1997). In this model the area of interest is partitioned into the fraction covered by a forest canopy ($F$) and the open

fraction $(1 - F)$. The model computes a running water balance of the canopy (leaves and branches) and the tree trunks using the conceptual scheme shown in Figure 7-17 and a time step ($\Delta t$) of 1 day.

The canopy storage is filled by rainfall and emptied by drainage and evaporation. Drainage from the

**FIGURE 7-17**
The "sparse" Rutter conceptual model of interception as developed by Valente et al. (1997). See text for explanation.

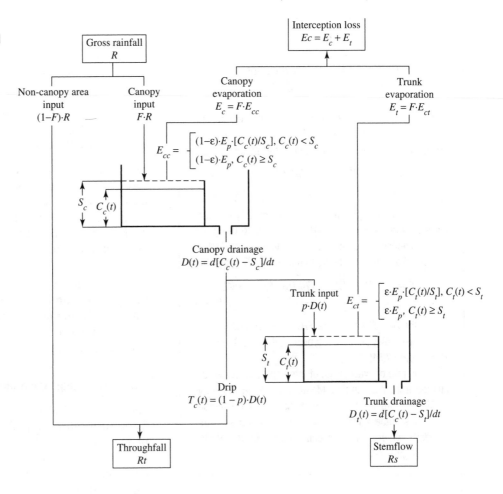

canopy occurs when the canopy storage capacity $S_c$ is exceeded. A small fraction $p$ of this drainage becomes stemflow. Throughfall ($Rt$ in Figure 7-15) is the sum of the rain falling in open areas plus the fraction of drainage that drips from the canopy ($1 - p$). When the actual amount of water on the canopy, $C_c(t)$, exceeds $S_c$, evaporation from the canopy, $E_c(t)$, occurs at the rate given by the Penman Equation [Equation (7-55)]; when $C_c(t) < S_c$ the evaporation rate equals the Penman rate times $(1 - \varepsilon) \cdot C_c(t)/S_c$, where $\varepsilon$ is the fraction of total evaporation that occurs from the trunks. The interception loss from the canopy-covered area is the sum of the evaporation from trunks and canopy. The canopy interception loss for the entire area ($Ec$ in Figure 7-15) is the sum of the evaporation from the canopy and the trunks multiplied by the canopy-covered fraction, $F$.

The structure of the vegetation is reflected in the parameters $S_c$, $S_t$, $p$, and $\varepsilon$. Valente et al. (1997) discuss how these can be determined from analysis of regression relations of the form of Equation (7-61) developed from field measurements. Typical values of $S_c$ are in the range 0.2–1.1 mm. The quantities involved in stemflow and evaporation from trunks are small: the few measured values suggest $S_t < 0.02$ mm, $p < 0.03$ and $\varepsilon \approx 0.023$.

Several studies have found good agreement between observed interception loss and that simulated by the Rutter model and its variants (e.g., Gash and Morton 1978; Gash 1979; Lloyd et al. 1988; Valente 1997). However, such models clearly require extensive field studies to establish the canopy-structure parameters. The computation of interception in the BROOK90 model uses a simplified version of the Rutter model, as described in Box 7-2.

The Penman Equation is appropriate for computing the rate of evaporation of intercepted water because the evaporation is from the leaf surface and stomata are not involved (this is equivalent to an infinite canopy conductance in the Penman–Monteith Equation). Thus canopy evaporation rates

---

## BOX 7-2

. . . . . . .

### Computation of Interception in the BROOK90 Model

The capacity of the vegetation to store intercepted rain, *INTRMX*, is related to the leaf-area index, *LAI*, and the analogously defined stem-area index, *SAI*, by the formula

$$INTRMX = CINTRL \cdot LAI + CINTRS \cdot SAI, \quad \text{(7B2-1)}$$

where the default values are *CINTRL = CINTRS* = 0.15 mm. These values give *INTRMX* = 1.0 mm at typical forest values of *LAI* = 6 and *SAI* = 0.7. Prior to filling the canopy capacity, the rate of canopy interception, *RINT*, is computed as a fraction of the rainfall rate, *RFAL*; that is,

$$RINT = (FRINTL \cdot LAI + FRINTS \cdot SAI) \cdot RFAL, \quad \text{(7B2-2)}$$

where the default values are *FRINTL = FRINTS* = 0.06. With these values and the typical *LAI* and *SAI* values, *RINT* = 0.40·*RFAL*.

The rate of evaporation of intercepted water is calculated from the Penman–Monteith Equation [Equation (7-56)] with infinite canopy conductance, as in Example 7-8. The storage of water on the canopy is tracked using standard water-balance accounting, as in the Rutter model (Figure 7-17).

---

will be larger than transpiration rates for the same conditions. For forests, the difference is considerable because of the effects of roughness on the efficiency of turbulent transfer, as illustrated by Example 7-8.

---

### *EXAMPLE 7-8*

Compare the canopy evaporation rate with the transpiration rates for the pine forest at Thetford, England, for the conditions of Example 7-6.

***Solution*** The atmospheric conductance for Example 7-6 is 0.232 m s$^{-1}$. Substituting that and the other appropriate values into Equation (7-55) yields a canopy evaporation rate of $6.21 \times 10^{-4}$ mm s$^{-1}$ or 53.6 mm day$^{-1}$, over 40 times greater than the transpiration rate with $\Delta\theta = 0$ cm. Thus if there is 1 mm of intercepted water on the canopy, this water will be

completely evaporated in (1 mm)/($6.21 \times 10^{-4}$ mm s$^{-1}$) = 1610 s = 0.45 hr.

It is of interest to compare the latent-heat flux, *LE*, associated with the evaporation of this intercepted water with the available energy (net radiation). This flux is calculated as

$$LE = \rho_w \cdot \lambda_v \cdot E = (1000 \text{ kg m}^{-3}) \cdot (2.45 \text{ MJ kg}^{-1})$$
$$\cdot (6.21 \times 10^{-7} \text{ m s}^{-1})$$
$$= 0.00152 \text{ MJ m}^{-2} \text{ s}^{-1},$$

which is about 8.4 times the net radiation. The difference between the latent-heat flux and the net radiation must be supplied by downward sensible-heat transfer from air-advected energy. Several studies have shown that such advection is commonly involved in evaporating intercepted water (e.g., Stewart 1977)—that is, interception loss can markedly cool the air.

---

The contrast between rates of evaporation of intercepted water and rates of transpiration illustrated in Example 7-8 have important bearing on understanding the hydrologic impacts of land-use changes, as discussed in Section 7.6.4.

### 7.6.4  Hydrologic Importance of Interception Loss

As shown in Table 7-8, interception loss ranges from 10 to 40% of gross precipitation in various plant communities. There has been considerable debate concerning the extent to which this loss is an addition to, as opposed to a replacement for, water loss by transpiration.

Clearly most of the interception loss that occurs in seasons when vegetation is dormant is a net addition to evapotranspiration. When intercepted water is present during the growing season, it evaporates in preference to water in stomatal cavities because it does not encounter stomatal resistance. However, in forests the evaporation of intercepted water occurs at rates several times greater than for transpiration under identical conditions. (See Example 7-8.) Thus intercepted water disappears quickly and interception loss replaces transpiration only for short periods. For example, Stewart (1977) found that annual interception loss for the forest he studied was 214 mm, and that 69 mm would have transpired during the time this loss was occurring. Thus the net additional evapotranspiration due to interception was 145 mm; this was 26% of the total annual evapotranspiration.

For short vegetation, atmospheric conductances are much lower than over forests (Figure

**TABLE 7-8**

Annual Canopy Interception Loss as a Fraction of Gross Precipitation for Various Plant Communities

| Latitude | Location | Community | Annual Pptn. (cm)[a] | *Ec/R* |
|---|---|---|---|---|
|  | Tropics | Lowland forest |  | 0.22 |
|  | Tropics | Montane forest |  | 0.18 |
| 3.0 | Manaus, Brazil | Amazonian rain forest | 281 | 0.09 |
| 4.0 | Malaysia | Lowland forest |  | 0.23 |
| 5.9 | Ivory Coast | Evergreen hardwoods |  | 0.09 |
| 6.5 | W Java | Lowland tropical rain forest |  | 0.21 |
| 7.0 | Ghana | Semideciduous moist forest |  | 0.16 |
| 10.0 | Nigeria | Forest-savannah boundary |  | 0.05 |
| 11.3 | Kottamparamba, India | Cashew | 300 | 0.31 |
| 18.3 | Mts., E Puerto Rico | Tabonuco et al. | 575 | 0.42 |
| 18.3 | Mts., E Puerto Rico | Dwarf forest |  | 0.09 |
| 34.5 | Mts., AR, U.S. | Pine-hardwood |  | 0.13 |
| 35.0 | W NC, U.S. | 60-yr-old white pine | 203 | 0.09 |
| 35.0 | W NC, U.S. | Mixed hardwoods | 203 | 0.12 |
| 35.0 | W NC, U.S. | 35-yr-old white pine | 203 | 0.19 |
| 35.0 | W NC, U.S. | 10-yr-old white pine | 203 | 0.15 |
| 42.2 | S Is. New Zealand | Mixed evergreen hardwood | 260 | 0.24 |
| 43.9 | Mts., N NH, U.S. | Mixed hardwoods | 130 | 0.13 |
| ~45.0 | NW U.S. | Douglas fir |  | 0.24 |
| ~45.0 | NW U.S. | Douglas fir et al. |  | 0.32 |
| ~45.0 | NW U.S. | Sitka spruce-hemlock et al. |  | 0.35 |
| ~45.0 | NW U.S. | Mature Douglas fir |  | 0.34 |
| ~45.0 | NW U.S. | White pine-hemlock |  | 0.21 |
| ~45.0 | NW U.S. | Douglas fir-hemlock |  | 0.24 |
| 51.4 | SE U.K. | Corsican pine | 79 | 0.35 |
| 51.4 | Hampshire, U.K. | Hornbeam |  | 0.36 |
| 51.4 | Hampshire, U.K. | Douglas fir |  | 0.39 |
| 51.4 | Hampshire, U.K. | Oak |  | 0.18 |
| 51.4 | Hampshire, U.K. | Oak—defoliated |  | 0.12 |
| 51.4 | Hampshire, U.K. | Norway spruce |  | 0.48 |
| 51.4 | Hampshire, U.K. | Corsican pine |  | 0.35 |
| 52.3 | Norfolk, U.K. | Scots & Corsican pine | 60 | 0.36 |
| 52.5 | Wales, U.K. | Sitka spruce | 187 | 0.27 |
| 52.5 | Castricum, Holland | Oak forest | 31 | 0.22 |
| 55.0 | S Scotland, U.K. | Sitka spruce | 160 | 0.30 |
| 55.0 | Northumberland, U.K. | Sitka spruce—mature | 100 | 0.49 |
| 55.0 | Northumberland, U.K. | Sitka spruce—pole timber | 100 | 0.29 |
| 55.2 | S Scotland, U.K. | Sitka spruce | 97 | 0.32 |
| 56.4 | Scotland, U.K. | Sitka spruce | 213 | 0.28 |
| 57.7 | NE Scotland, U.K. | Scots pine | 64 | 0.42 |
| 58.3 | SE AK, U.S. | Hemlock-Sitka spruce |  | 0.25 |

[a]Annual precipitation during period of measurement or climatic average, as given in source. Data from published sources.

7-11) and interception loss occurs at rates comparable to transpiration. Thus, for grasses and similar forms, interception loss is to a large extent compensated by reduction in transpiration and makes little net addition to evapotranspiration (McMillan and Burgy 1960).

Many studies have shown that tree removal by logging, forest fire, or wind damage increases the average runoff from the affected area (Hewlett and Hibbert 1961; Bosch and Hewlett 1982), and that af-

forestation decreases runoff. The magnitude of both effects is roughly proportional to the percentage change in forest cover (Figure 7-18) and is largely due to changes in evapotranspiration brought about by some combination of a change in the net additions to evaporation from interception loss and in the vertical extent of the root zone from which transpired water is extracted.

Other aspects of the effects of land-use changes on hydrology are discussed in Section 10.2.5.

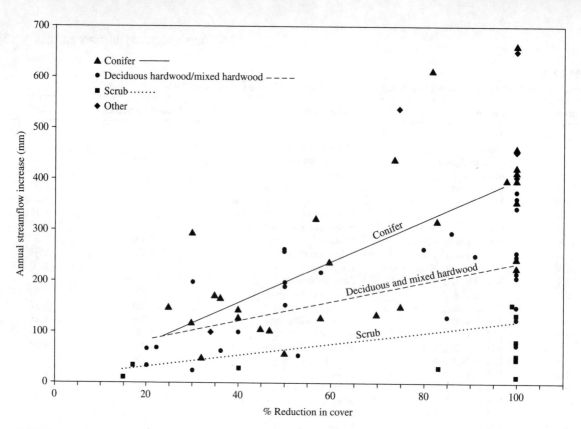

**FIGURE 7-18**

Annual streamflow increases due to reductions in vegetative cover as measured in watershed experiments. Reprinted from Bosch and Hewlett (1982) with permission of Elsevier Science.

## 7.6.5 Water-Quality Aspects

Even though intercepted water remains on plant canopies for only short times, the chemistry of throughfall is commonly significantly different from that of unintercepted precipitation (Figure 7-19). Much of the change is due to the dissolution of material that was deposited on the leaves from the air (dry deposition) and some—particularly organic carbon—represents leaching of the leaves themselves.

Lindberg and Garten (1988) found that more than 85% of the sulfate in throughfall in the southeastern United States is wash-off of dry deposition. They concluded that measurements of the sulfate content of throughfall are a good way to monitor the inputs of that constituent, which is of concern as an anthropogenic contributor to acidification of surface water.

## 7.7 POTENTIAL EVAPOTRANSPIRATION

### 7.7.1 Conceptual Definition

**Potential evapotranspiration** (PET) is the rate at which evapotranspiration would occur from a large area completely and uniformly covered with growing vegetation which has access to an unlimited supply of soil water, and without advection or heat-storage effects (Table 7-1). The concept was introduced as part of a scheme for climate classification by Thornthwaite (1948), who intended it to depend essentially on climate and to be largely independent of the surface characteristics.

However, we now know from Equation (7-56) that several characteristics of a vegetative surface have a strong influence on evapotranspiration rate even when there is no limit to the available water: (1) the albedo of the surface, which determines the net radiation (Table 7-5); (2) the maximum leaf conductance (Table 7-5); (3) the atmospheric conductance, which is largely determined by vegetation height (Table 7-5; Figure 7-11); and (4) the presence

308

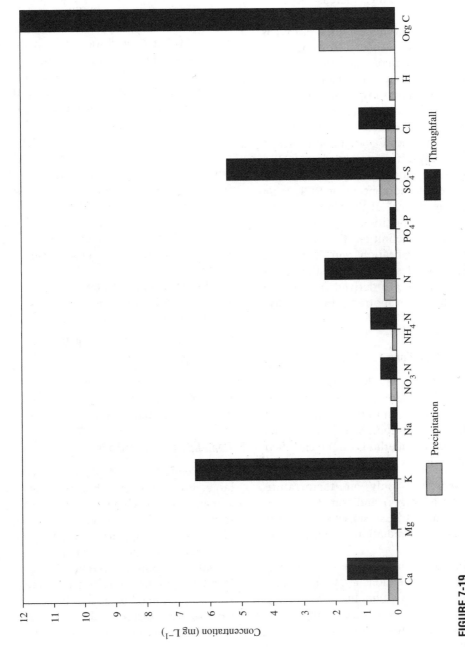

**FIGURE 7-19**
Comparison of chemical composition of incident precipitation and throughfall at Hubbard Brook Experimental Forest in the White Mountains of New Hampshire. Data from Likens et al. (1977).

or absence of intercepted water. Because of these surface effects, Penman (1956) redefined PET as "the amount of water transpired ... by a short green crop, completely shading the ground, of uniform height and never short of water," and the term **reference-crop evapotranspiration** is increasingly used as a synonym for PET.

Another concern about the definition of PET is that its magnitude is often calculated from meteorological data collected under conditions in which the actual evapotranspiration rate is less than the potential rate. However, if evapotranspiration had been occurring at the potential rate, the latent- and sensible-heat exchanges between the air and the surface, and hence the air temperature and humidity, would have been considerably different (Brutsaert 1982).

In spite of these considerable ambiguities, it has proven useful to retain the concept of PET as an index of the "drying power" of the climate or the ambient meteorological conditions, and we now examine some operational definitions that have been applied in climate classification and hydrologic modeling. Section 7.8.1 describes how estimates of actual evapotranspiration are derived from calculated values of potential evapotranspiration in hydrologic analysis.

### 7.7.2  Operational Definitions

In practice, PET is defined by the method used to calculate it, and many methods have been proffered. We limit our discussion to the methods most commonly applied in hydrologic studies. Following Jensen et al. (1990), these methods can be classified on the basis of their data requirements:

**Temperature-based:** Use only air temperature (often climatic averages) and sometimes day length (time from sunrise to sunset).

**Radiation-based:** Use net radiation and air temperature.

**Combination:** Based on the Penman combination equation; use net radiation, air temperature, wind speed, and relative humidity.

**Pan:** Use pan evaporation, sometimes with modifications depending on wind speed, temperature, and humidity.

Some of the methods do not require information about the nature of the surface and can be considered to give reference-crop evapotranspiration, others are surface-specific and require information

about albedo, vegetation height, maximum stomatal conductance, leaf-area index, and other factors.

### Temperature-Based Methods

Thornthwaite (1948) developed a complex empirical formula for calculating PET as a function of climatic average monthly temperature and day length. It turns out that Thornthwaite's temperature function has a form similar to the saturation vapor-pressure relation [Equation (7-4)], and some simplifications of his approach are based on this similarity. For example, Hamon (1963) estimated daily PET as

$$PET_H = 29.8 \cdot D \cdot \frac{e_a^*}{T_a + 273.2}, \quad \textbf{(7-63)}$$

where $PET_H$ is in mm day$^{-1}$, $D$ is day length in hr [calculated via Equations (E-5)], and $e_a^*$ is the saturation vapor pressure at the mean daily temperature, $T_a$ (°C), in kPa. Equation (7-63) gives values close to those given by the original Thornthwaite formulation and has been used in several hydrologic models.

Malmstrom (1969) used an approach similar to Hamon's and claimed improved climate classification by estimating monthly climatic PET, $PET_M$, as

$$PET_M = 40.9 \cdot e_a^*, \quad \textbf{(7-64)}$$

where $PET_M$ is in mm month$^{-1}$, $e_a^*$ is in kPa, and the temperature used to compute $e_a^*$ is the climatic mean monthly air temperature in °C for months when average temperature exceeds 0 °C.

### Radiation-Based Methods

Slatyer and McIlroy (1961) reasoned that air moving large distances over a homogeneous well-watered surface would become saturated, so that the mass-transfer term in the Penman Equation [Equation (7-55)] would disappear. They defined the evapotranspiration under these conditions as the **equilibrium potential evapotranspiration,** $PET_{eq}$. Subsequently, Priestley and Taylor (1972) compared $PET_{eq}$ with values determined by energy-balance methods over well-watered surfaces and found a close fit if $PET_{eq}$ was multiplied by a factor $\alpha_{PT}$ to give

$$PET_{PT} = \frac{\alpha_{PT} \cdot \Delta \cdot (K + L)}{\rho_w \cdot \lambda_v \cdot (\Delta + \gamma)}. \quad \textbf{(7-65)}$$

A number of field studies of evapotranspiration in humid regions have found $\alpha_{PT} = 1.26$, and theoretical examination has shown that that value in fact represents equilibrium evapotranspiration over well-watered surfaces under a wide range of conditions (Eichinger et al. 1996). Thus $PET_{PT}$ is often referred to as the equilibrium potential evapotranspiration, and Equation (7-65) gives an estimate of PET that depends only on net radiation and air temperature. This relationship has proven useful in hydrologic analysis.

### Combination Methods

If the required data are available, the Penman–Monteith Equation [Equation (7-56)], using a $C_{leaf}$ value calculated from Equation (7-52) with $f_\theta(\Delta\theta) = 1$, can be used to estimate PET for a specified vegetated surface. Shuttleworth (1994) defined the reference crop as grass with a height ($z_{veg}$) of 120 mm, an albedo ($a$) of 0.23, and a canopy conductance ($C_{can}$) of 14.5 mm s$^{-1}$.

### Pan-Based Methods

The potential evapotranspiration for short vegetation is commonly very similar to free-water evaporation (Linsley et al. 1982; Brutsaert 1982). This may be because lower canopy conductance over the vegetation fortuitously compensates for the lower atmospheric conductance over the pan. In any case, annual values of pan evaporation (= free-water evaporation as shown in Figure 7-7) are essentially equal to annual PET, and pan evaporation corrected via Equations (7-41) and (7-42) can be used to estimate PET for shorter periods.

Potential evapotranspiration can also be directly measured by various forms of **atmometers** in which evaporation occurs from porous surfaces (Giambelluca et al. 1992) or flat plates (Fontaine and Todd 1993).

### 7.7.3 Comparison of PET Estimation Methods

Jensen et al. (1990) compared PET computed by 19 different approaches with measured reference-crop evapotranspiration in weighing lysimeters[11] at 11 locations covering a range of latitudes and elevations. The Penman–Monteith method gave the best overall results (Figure 7-20a). Equilibrium evapotranspi-

---

[11] A **weighing lysimeter** is an enclosed volume of soil for which the inflows and outflows of liquid water can be measured and changes in storage can be monitored by weighing. They are described further in Section 7.8.2.

ration [Equation (7-65) with $\alpha_{PT} = 1.26$] gave reasonable agreement up to rates of 4 mm day$^{-1}$ but considerable underestimation at higher rates (Figure 7-20b). Monthly Class-A pan evaporation correlated well with measured PET, but with considerable scatter presumably due to variability of heat exchange through the pan walls (Figure 7-20c).

For short vegetation, the Penman Equation [Equation (7-55)] gives nearly the same estimates as the Penman–Monteith Equation, and Van Bavel (1966) found a close correspondence for hourly and daily evapotranspiration computed by the Penman Equation and that measured for well-watered alfalfa growing in a lysimeter (Figure 7-21).

Vörösmarty et al. (1998) compared the estimates of annual PET given by nine different methods in a global-scale water-balance model. The various methods were used as a basis for estimating actual evapotranspiration in a monthly water-balance model (as discussed in Section 7.8.1), and these estimates were compared with the difference between measured precipitation and measured streamflow. Considering the differences in conceptual basis and data requirements, the various methods gave surprisingly similar results overall. Interestingly, the Hamon method [Equation (7-63)], which is based only on temperature and day length, performed the best. Several variations of the Penman–Monteith method [Equation (7-56)] performed well, while the equilibrium method [Equation (7-65)] tended to overestimate in regions with higher ET rates and the Penman Equation [Equation (7-55)] overestimated for all locations.

---

## 7.8 ACTUAL EVAPOTRANSPIRATION

### 7.8.1 Potential-Evapotranspiration Approaches

#### Relation to Precipitation/Potential Evapotranspiration Ratio

In hot arid regions, potential evapotranspiration greatly exceeds precipitation so that average actual evapotranspiration is water-limited, and is essentially equal to average precipitation. In regions with abundant rainfall in all seasons, evapotranspiration is limited by the available energy, so that average evapotranspiration is essentially equal to average potential evapotranspiration. Pike (1964) thus reasoned that annual evapotranspiration, $ET$, is determined by

(a)

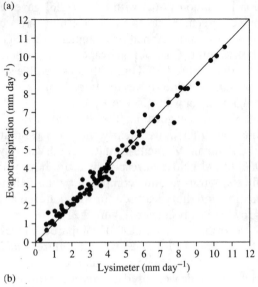

(b)

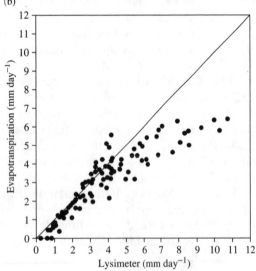

(c)

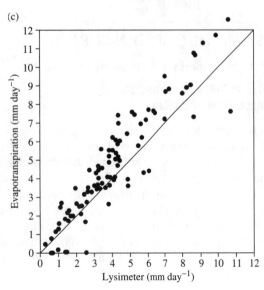

the ratio of average precipitation, $W$, to potential evapotranspiration, $PET$, and proposed the following relation for estimating annual evapotranspiration:

$$ET = \frac{W}{\left[1 + \left(\dfrac{W}{PET}\right)^2\right]^{1/2}}. \qquad \textbf{(7-66)}$$

In spite of its purely empirical nature, Equation (7-66) gives reasonably good "first-cut" estimates of climatic average evapotranspiration (Figure 7-22).

### Monthly Water-Balance Models

Thornthwaite and Mather (1955) developed a water-balance model that estimates monthly actual evapotranspiration from monthly potential evapotranspiration. PET is calculated from monthly average temperature using the Hamon [Equation (7-63)], Malmstrom [Equation (7-64)], or other temperature-based approach. Monthly precipitation is input to a simple model of soil-moisture storage, which computes actual ET and updates soil moisture via a "bookkeeping" procedure. One version of this approach is described in detail in Box 7-3, and Table 7-9 and Figure 7-23 give an example of its application.

"Thornthwaite-type" monthly ET models can also be extended to estimate monthly ground-water recharge and runoff, and one can verify ET estimates by comparing estimated and measured runoff. In spite of their extremely simple structure, models of the Thornthwaite type generally estimate monthly runoff values reasonably well (Alley 1984; Calvo 1986), and this correspondence suggests that their estimates of actual evapotranspiration are also generally reasonable. Somewhat more elaborate versions of the basic monthly water-balance model described in Box 7-3 are used to simulate land-surface hydrology in many of the general circulation models used to forecast the impacts of climate change (e.g., Vörösmarty et al. 1998).

**FIGURE 7-20**
Comparison of average monthly potential evapotranspiration computed by (a) the Penman–Monteith equation [Equation (7-56)], (b) the Priestley–Taylor equilibrium method [Equation (7-65)], and (c) uncorrected Class-A pan evaporation, with values determined in lysimeters containing well-watered alfalfa at 11 locations. Overestimation by pan evaporation is probably due largely to heat exchange through the sides of the pan. From Jensen et al. (1990).

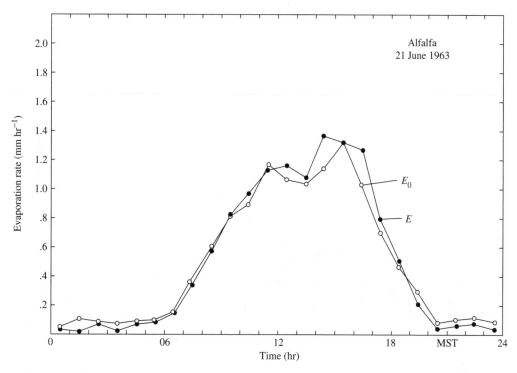

**FIGURE 7-21**
Comparison of observed hourly evapotranspiration for well-watered alfalfa (closed circles) and that calculated via the Penman equation (open circles). From Van Bavel (1966), used with permission of the American Geophysical Union.

### Use of Soil-Moisture Functions

One of the most widely used methods for estimating actual evapotranspiration makes use of meteorologic data to estimate potential evapotranspiration by relations like those discussed earlier, and then computes actual evapotranspiration as

$$ET = F(\theta_{rel}) \cdot PET. \qquad (7\text{-}67)$$

Here $\theta_{rel}$ is the **relative water content,** defined as

$$\theta_{rel} \equiv \frac{\theta - \theta_{pwp}}{\theta_{fc} - \theta_{pwp}}, \qquad (7\text{-}68)$$

in which $\theta$ is the current water content, $\theta_{fc}$ is the field-capacity, and $\theta_{pwp}$ is the permanent wilting point of the root-zone soil (Section 6.4.1). The relation between $ET/PET$ and $\theta_{rel}$ usually has a form like that shown in Figure 7-24: $ET/PET$ increases quasi-linearly as $\theta_{rel}$ increases, and reaches 1 at some water content $\theta_{crit}$ (e.g., Davies and Allen 1973; Federer 1979, 1982; Spittlehouse and Black 1981). Typically $0.5 \cdot \theta_{fc} \leq \theta_{crit} \leq 0.8 \cdot \theta_{fc}$.

The Stewart (1988) model of canopy conductance [Equations (7-52) and (7-54)] includes a function $f_\theta(\Delta\theta)$ (Table 7-6) that can be used with the Penman–Monteith Equation [Equation (7-56)] to give

$$\frac{ET}{PET} = \frac{\Delta + \gamma \cdot \left\{ 1 + \dfrac{C_{at}}{C_{can}[f_\theta(\Delta\theta) = 1]} \right\}}{\Delta + \gamma \cdot \left\{ 1 + \dfrac{C_{at}}{C_{can}[f_\theta(\Delta\theta)]} \right\}}.$$

$$(7\text{-}69)$$

The form of this expression can be explored in Exercise 7-12.

A third approach to estimating $ET$ from $PET$ relates the value of $\alpha_{PT}$ in Equation (7-65) to some measure of soil-water content (e.g., Mukammal and Neumann 1977).

Methods of estimating $ET$ from $PET$ and soil moisture are well suited to use in "real-time" estimation, where $\theta$ is measured every few days, and in hydrologic models like the BROOK90 model, where $\theta$ is estimated by a "bookkeeping" algorithm

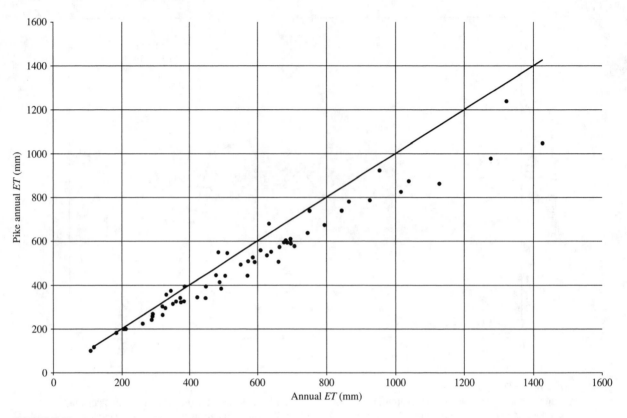

**FIGURE 7-22**
Comparison of annual evapotranspiration computed by the Pike Equation [Equation (7-66)] with that computed via a monthly water-balance model for selected North American stations.

along with equations for infiltration and deep drainage (Box 6-3). Box 7-4 describes the approach used by the BROOK90 model.

### Complementary (Advection-Aridity) Approach

Following Bouchet (1963), consider a uniform surface of 1 to 100 km$^2$ area evapotranspiring at the potential rate $ET = PET_0$ under a steady set of meteorological conditions. If these conditions remained constant, eventually the soil moisture would fall below field capacity and $ET$ would be less than $PET_0$. A flux of energy, $Q$, [E L$^{-2}$ T$^{-1}$] equivalent to the difference between $PET_0$ and $ET$ would then not be used for evapotranspiration and would become available to warm the atmosphere. Thus,

$$PET_0 - ET = \frac{Q}{\rho_w \cdot \lambda_v}. \qquad (7\text{-}70)$$

The reduced evapotranspiration decreases the humidity, and the warming increases the air temperature. Under these circumstances, one would calculate a new potential evapotranspiration $PET$ that is larger than $PET_0$ by the amount $Q/(\rho_w \cdot \lambda_v)$:

$$PET - PET_0 = \frac{Q}{\rho_w \cdot \lambda_v}. \qquad (7\text{-}71)$$

Combining Equations (7-70) and (7-71) yields the **complementary relationship** between $ET$ and $PET$:

$$ET = 2 \cdot PET_0 - PET. \qquad (7\text{-}72)$$

(See Figure 7-25.)

Brutsaert and Stricker (1979) reasoned that $PET_0$ is the PET under equilibrium conditions [Equation (7-65)] and $PET$ is the "actual" PET given by the Penman Equation using measured current values of the meteorological variables [Equation (7-33)]. Substituting those relationships into Equation (7-72) yields

$$ET = \frac{(2 \cdot \alpha_{PT} - 1) \cdot \Delta \cdot (K + L) - \gamma \cdot K_E \cdot \rho_w \cdot \lambda_v \cdot v_a \cdot e_a^* \cdot (1 - W_a)}{\rho_w \cdot \lambda_v \cdot (\Delta + \gamma)},$$

$$(7\text{-}73)$$

# BOX 7-3

· · · · · · ·

## Thornthwaite-Type Monthly Water-Balance Model

Referring to Table 2-3, Thornthwaite-type monthly water-balance models are lumped conceptual models that can be used to simulate steady-state seasonal (climatic average) or continuous values of watershed or regional water input, snowpack, soil moisture, and evapotranspiration. Input for such models consists of monthly values of precipitation, $P_m$, and temperature, $T_m$, representative of the region of interest. For steady-state applications, these values are monthly climatic averages, in which case $m = 1, 2, ..., 12$; for continuous simulations they are actual monthly averages, in which case $m = 1, 2, ..., 12 \cdot N$, where $N$ is the number of years of record. Such models typically have a single parameter, the soil-water storage capacity of the soil in the region, $SOIL_{max}$, which is defined as

$$SOIL_{max} = (\theta_{fc} - \theta_{pwp}) \cdot Z_{rz}, \qquad \text{(7B3-1)}$$

where $\theta_{fc}$ is the field capacity and $Z_{rz}$ the vertical extent of the root zone. Typically $SOIL_{max} = 100$ or $150$ mm. For continuous applications an initial value of soil moisture, $SOIL_0$, must also be specified.

All water quantities in the model represent depths (volumes per unit area) of liquid water; inputs and outputs are monthly totals and snowpack and soil storage are end-of-month values.

### Snowpack, Snowmelt, and Water Input

Monthly precipitation is divided into rain, $RAIN_m$, and snow, $SNOW_m$, where

$$RAIN_m = F_m \cdot P_m \qquad \text{(7B3-2)}$$

and

$$SNOW_m = (1 - F_m) \cdot P_m, \qquad \text{(7B3-3)}$$

in which $F_m$ is the **melt factor**. Following Figure 4-14, $F_m$ is computed as follows:

If $T_m \leq 0°C$:   $F_m = 0$;
if $0°C < T_m < 6°C$:   $F_m = 0.167 \cdot T_m$;
if $T_m \geq 6°C$:   $F_m = 1$.      **(7B3-4)**

The melt factor is also used in a temperature-index snowmelt model [Equation (5-57)] to determine the monthly snowmelt, $MELT_m$, as

$$MELT_m = F_m \cdot (PACK_{m-1} + SNOW_m), \qquad \text{(7B3-5)}$$

where $PACK_{m-1}$ is the snowpack water equivalent at the end of month $m - 1$. The snowpack at the end of month $m$ is then computed as

$$PACK_m = (1 - F_m)^2 \cdot P_m + (1 - F_m) \cdot PACK_{m-1}. \qquad \text{(7B3-6)}$$

By definition, the water input $W_m$, is

$$W_m = RAIN_m + MELT_m. \qquad \text{(7B3-7)}$$

### Evapotranspiration and Soil Moisture

Following Alley (1984), if $W_m \geq PET_m$, ET takes place at the potential rate, i.e.,

$$ET_m = PET_m \qquad \text{(7B3-8)}$$

and soil moisture increases or, if already at $SOIL_{max}$, remains constant. Thus

$$SOIL_m = \min\{[(W_m - PET_m) + SOIL_{m-1}], SOIL_{max}\}, \qquad \text{(7B3-9)}$$

where $\min\{...\}$ indicates the smaller of the quantities in the braces. In the original formulation, Thornthwaite (1948) used an empirical function that depends on the climatic average monthly temperature to calculate $PET_m$. Most current applications of the method use a simpler temperature-based method [e.g., Equations (7-63) or (7-64)] or, if data are available, one of the other approaches for estimating $PET_m$.

If $W_m < PET_m$, $ET_m$ is the sum of water input and an increment removed from soil storage; that is,

$$ET_m = W_m + SOIL_{m-1} - SOIL_m, \qquad \text{(7B3-10)}$$

where the decrease in soil storage is modeled via the following conceptualization:

$$SOIL_{m-1} - SOIL_m = SOIL_{m-1} \cdot \left[1 - \exp\left(-\frac{PET_m - W_m}{SOIL_{max}}\right)\right]. \qquad \text{(7B3-11)}$$

### Computation

If the model is used with climatic monthly averages, the computations in Equations (7B3-5), (-9), (-10), and (-11) are "wrapped around" from $m = 12$ to $m = 1$ so that

---

### BOX 7-3 (CONTINUED)
· · · · · · · ·
### Thornthwaite-Type Monthly Water-Balance Model

$m − 1 = 12$ when $m = 1$. Thus the computations are circular and must be iterated until all the monthly quantities converge to constant values. This iteration is automatically carried out in EXCEL spreadsheets when the "Iterations" box in the "Tools Option Calculation" menu is checked. Otherwise you will get an error message "Cannot resolve circular references."

#### Overall Water Balance

The model output is a table of monthly values that can be graphed to give a concise picture of the annual cycle of inputs, soil and snowpack storage, evaporation, and water available for ground-water recharge and streamflow at any location. Table 7-9 is a completed water-balance spreadsheet for Omaha, NE, at 41.3° N latitude; the annual values are the sums of the monthly values. Note

that the annual precipitation and water input are equal, as must be the case. As shown in Figure 7-23, the snowpack begins to build up in December, reaches a peak in February, and melts in March and April. $ET = PET$ for the months when $W > PET$ (March–June). Soil-water storage is recharged by rain and snowmelt beginning in February and reaches its capacity in April. It declines in the period from July to October because $PET > P$ and some of the evaporative demand is satisfied by withdrawal of water from soil storage. The last line, $W − ET − \Delta SOIL$, is the average monthly "water surplus" (i.e., the water available for recharge and runoff).

This model has been programmed in a spreadsheet in file "ThornEx.xls" on the CD accompanying this text, and Exercise 7-13 suggests ways to experiment with the model.

---

**TABLE 7-9**

Thornthwaite-type Monthly Water Balance (Box 7-3) for Omaha, NE, Computed by ThornEx.xls Spreadsheet Model on Text CD. Temperatures in °C, water-balance terms in mm. $SOIL_{max} = 100$ mm.

|  | J | F | M | A | M | J | J | A | S | O | N | D | Year |
|---|---|---|---|---|---|---|---|---|---|---|---|---|---|
| *P* | 21 | 24 | 37 | 65 | 88 | 115 | 86 | 101 | 67 | 44 | 32 | 20 | 700 |
| *T* | −5.4 | −3.1 | 2.7 | 10.9 | 17.2 | 22.8 | 25.8 | 24.6 | 19.4 | 13.2 | 3.8 | −2.1 | |
| *F* | 0 | 0 | 0.45 | 1 | 1 | 1 | 1 | 1 | 1 | 1 | 0.63 | 0 | |
| *RAIN* | 0 | 0 | 17 | 65 | 88 | 115 | 86 | 101 | 67 | 44 | 20 | 0 | 603 |
| *SNOW* | 21 | 24 | 20 | 0 | 0 | 0 | 0 | 0 | 0 | 0 | 12 | 20 | 97 |
| *PACK* | 45 | 69 | 49 | 0 | 0 | 0 | 0 | 0 | 0 | 0 | 4 | 24 | |
| *MELT* | 0 | 0 | 40 | 49 | 0 | 0 | 0 | 0 | 0 | 0 | 7 | 0 | 97 |
| *W* | 0 | 0 | 57 | 114 | 88 | 115 | 86 | 101 | 67 | 44 | 28 | 0 | 700 |
| *PET* | 0 | 0 | 29 | 56 | 90 | 130 | 151 | 131 | 88 | 54 | 26 | 0 | 755 |
| *W-PET* | 0 | 0 | 27 | 61 | 8 | 1 | −50 | −26 | −25 | −18 | −5 | 0 | |
| *SOIL* | 29 | 29 | 55 | 100 | 100 | 100 | 60 | 47 | 36 | 30 | 29 | 29 | |
| Δ*SOIL* | 0 | 0 | 27 | 45 | 0 | 0 | −40 | −14 | −10 | −6 | −2 | 0 | |
| *ET* | 0 | 0 | 29 | 56 | 90 | 129 | 126 | 113 | 73 | 46 | 26 | 0 | 688 |
| *W-ET-* Δ*SOIL* | 0 | 0 | 0 | 12 | 0 | 0 | 0 | 0 | 0 | 0 | 0 | 0 | 12 |

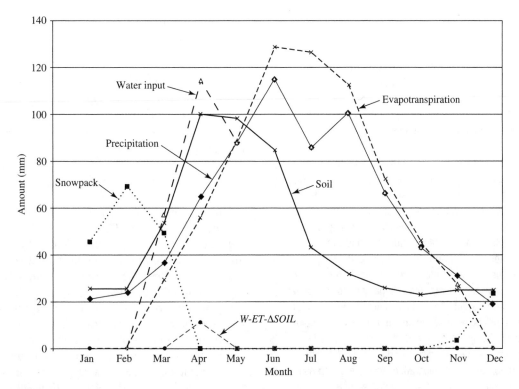

**FIGURE 7-23**
Annual cycle of water-balance components as computed by Thornthwaite-type monthly water-balance model for Omaha, NE (Table 7-9).

**FIGURE 7-24**
General form of relations between *ET/PET* and soil-water content, $\theta$, used to estimate *ET*. Different studies have used different functions to express soil wetness. When the soil-water content variable is less than the critical value $\theta_{crit}$, *ET* is less than *PET* and plants are considered under water stress.

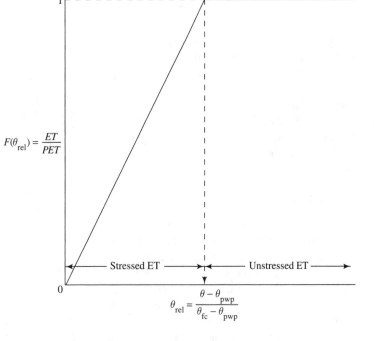

---

**BOX 7-4**

• • • • • • •

**Computation of Evapotranspiration in the BROOK90 Model**

BROOK90 uses the Shuttleworth and Wallace (1985) approach to modeling transpiration and soil-water evaporation. Briefly, this approach applies the Penman–Monteith Equation [Equation (7-56)] twice, once to compute potential transpiration for the vegetated fraction of the region (as in Example 7-6) and again for the ground-surface fraction. In computing potential soil evaporation, atmospheric conductance is modified by appropriate adjustments of the surface roughness and zero-plane displacement height, and the surface conductance decreases as surface soil-water content decreases. BROOK90 computes evaporation separately for the daylight and non-daylight hours using different values for air temperature and wind speed and, of course, solar radiation for the two periods. Total daily evapotranspiration is an appropriately weighted sum of daytime and nighttime transpiration and soil evaporation. Actual evapotranspiration and soil evaporation are then determined by multiplying potential values by factors that depend on internal plant resistance to water transport, the maximum potential plants can exert on soil water, and water contents in the various soil layers (Federer 1995).

---

which Brutsaert and Stricker (1979) called the **advection-aridity** interpretation of the complementary approach. Its main advantage is that it uses readily available meteorological data and does not require calibration to a specific site. It has been found to give estimates of daily ET that compare well with those using other approaches (Figure 7-26; see also Parlange and Katul 1992b).

### 7.8.2 Water-Balance Approaches

Actual evapotranspiration from a region over a time period $\Delta t$ can in principle be determined by measuring water inputs and outputs and changes in storage and solving the water-balance equation, just as for open-water evaporation. The application of this principle to various types of regions is discussed in the following sections; in all cases the precision of the determination is dictated by the precision with which all the other water-balance components can be measured.

### Land-Area Water Balance

As discussed in Section 2.5.2, the most common method of estimating actual evapotranspiration from a land area is the application of the water-balance equation in the form

$$ET = W - Q - G_{out}, \qquad \textbf{(7-74)}$$

where $W$ is precipitation, $Q$ is streamflow, and $G_{out}$ is ground-water outflow. The major problems in applying Equation (7-74) were discussed in Section 2.5.2, and include obtaining a reliable estimate of regional precipitation (see also Sections 4.2 and 4.3), obtaining reliable measurements of liquid outflows, especially where ground-water flow is significant (see also Section 8.5), and assuring that changes in storage over the period of measurement are negligible (or are measured and included in the equation).

In regions where most of the storage is in the form of soil water, the assumption of negligible change in storage typically leads to only small errors in estimating ET using data from time periods as short as a few years (Box 2-1; Hudson 1988). Such errors can be further minimized by selecting a water year that begins and ends during the season when soil moisture is near its maximum, as discussed in Section 2.5.2.

Apparently no studies have examined the validity of negligible storage change where other storage components, such as ground water and large lakes, are important. However, the levels of the Great Lakes and Great Salt Lake, U.S., show periods of several decades of steadily declining or rising levels (U.S. Geological Survey 1984). These trends suggest that significant errors are possible in estimating ET from Equation (7-74) for some large drainage basins, even when quantities are averaged over long periods.

### Lysimeter Measurement

A **lysimeter** is an artificially enclosed volume of soil for which the inflows and outflows of liquid water can be measured and, commonly, changes in storage can be monitored by weighing. Lysimeters range from 1 m$^3$ or less to over 150 m$^3$ in size and are usu-

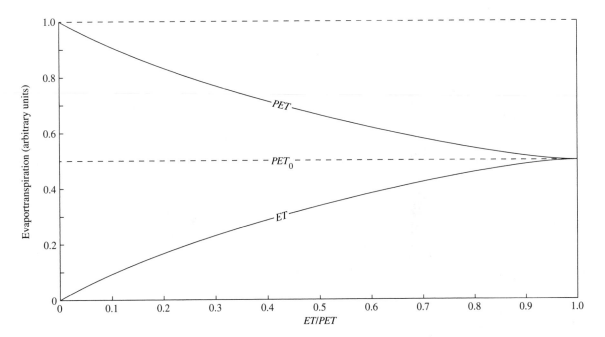

**FIGURE 7-25**
Bouchet's (1963) complementary relationship: $PET + ET = 2 \cdot PET_0$ [Equation (7-72)]. After Brutsaert (1982).

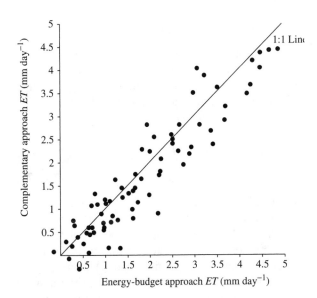

**FIGURE 7-26**
Comparison between estimates of daily $ET$ obtained by the complementary (advection-aridity) approach [Equation (7-73)] and an energy-budget method. From Brutsaert (1982) used with permission of Kluwer Academic Publishers.

ally designed so that their soil and vegetation are as closely identical as possible to those of the surrounding area. Details of standard lysimeter construction can be found in Dunne and Leopold (1978), Brutsaert (1982), and Shaw (1988); Grimmond et al. (1992) describe a portable mini-lysimeter ($< 0.2$ m$^2$ area).

Carefully obtained lysimeter measurements are usually considered to give the best determinations of actual evapotranspiration during a time period, and are often taken as standards against which other methods are compared (e.g., Jensen et al. 1990). However, lysimeters must have provision for drainage, and the introduction of atmospheric pressure at depth can result in a water-content profile, and hence an evapotranspiration rate, different from those in the surrounding soil. Unfortunately, it is virtually impossible to use the technique for forest vegetation.

### Soil-Moisture Balance

One can estimate the total evapotranspiration in a rain-free time period $\Delta t$ by carefully monitoring soil-water content profiles [$\theta(z')$] at spatially representative locations. As shown in Figure 7-27, the

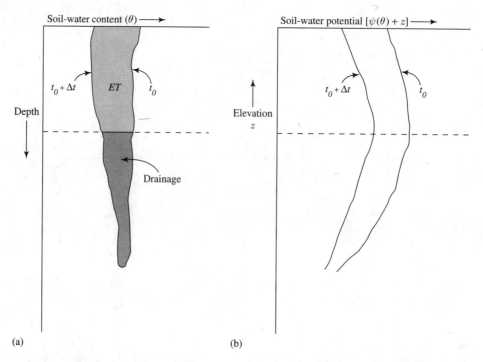

**FIGURE 7-27**
Conceptual basis for estimating evapotranspiration from the soil-water balance. (a) Change in soil-water content with depth during time period $\Delta t$. (b) Profiles of soil-water potential defining the zero-flux plane (dashed line) that divides water lost to evapotranspiration (ET) from that lost to drainage. After Shuttleworth (1992).

total soil-water loss is the difference in water content through the soil profile between times $t_0$ and $t_0 + \Delta t$. The portion of this loss due to evapotranspiration is determined by identifying the "zero-flux plane," which is the boundary between upward-directed water movement due to evapotranspiration and downward-directed movement due to drainage. The average location of the zero-flux plane is found by plotting profiles of the vertical soil-water potential $\psi(\theta) + z$, which is determined from the $\theta(z')$ values and the moisture-characteristic curve for the soil (Section 6.3.3).

This method essentially creates a "lysimeter without walls" that does not distort the soil water-content profile and can be especially useful in forests. However, obtaining representative values of $\theta(z)$ is not easy, and the method will not give good results if the water table is near the surface, if there is horizontal water movement, or if soil properties are highly variable. Rouse and Wilson (1972) found that the minimum length of $\Delta t$ for reliable results is 4 days, and is considerably longer under many conditions.

### Atmospheric Water Balance

Evaporation can also be estimated by applying the water-balance equation to a volume of the lower atmosphere. For a control volume of height $z_a$ and perimeter $X$ above an area $A$ and a time interval $\Delta t$, this equation becomes

$$ET = W - \frac{1}{A \cdot \rho_w} \cdot \int_0^{z_a} \int_X (\overline{\rho_v \cdot \overline{v}_n}) \cdot dx \cdot dz - M_2 + M_1 ,$$

**(7-75)**

where $ET$ and $W$ are the net evapotranspiration into and precipitation out of the volume (per unit area), $(\overline{\rho_v \cdot \overline{v}_n})$ is the time-averaged product of the outward-directed wind velocity normal to the perimeter and the absolute humidity, and $M_1$ and $M_2$ are the total water content per unit area of the control volume at the beginning and end of $\Delta t$ respectively.

As summarized by Brutsaert (1982), this method has been applied in several studies using both routine and specially collected atmospheric data. Typically, $7 < z_a < 8$ km. The spatial and temporal coarseness of network upper-air observations limit its routine application to areas of $2.5 \times 10^5$ km$^2$ or more to provide estimates of monthly evaporation. Munley and Hipps (1991) showed the importance of vertical resolution in obtaining accurate estimates.

Several recent research efforts have applied the atmospheric water balance to estimate evapotranspiration from large areas of land, and it appears that the method will play an increasing role in expanding understanding of global-scale hydrology (Shuttleworth 1988; Brutsaert 1988). Kuznetsova (1990) summarized several applications of the approach at the large-river-basin, subcontinental, and continental scales.

### 7.8.3 Turbulent-Transfer/Energy-Balance Methods

#### Penman–Monteith Approach

The Penman-Monteith Equation [Equation (7-56)], with the vegetative canopy treated as a "big leaf" [Equations (7-52)–(7-54)], is commonly used to estimate land-area evapotranspiration. This approach can be refined by treating the vegetated and unvegetated portions of a given area separately, using Equations (7-43)–(7-45) for the bare soil areas. A detailed methodology combining canopy and bare-soil evapotranspiration was developed by Shuttleworth and Wallace (1985); this is the basis for modeling evapotranspiration in the BROOK90 model (Box 7-4).

#### Bowen-Ratio Approach

Direct application of the mass-transfer equation [Equation (7-17)] to estimating actual evapotranspiration from a land surface is generally infeasible because of the absence of surface-temperature data and, most of the time, the absence of a surface that is at saturation.

However, in principle evapotranspiration can be evaluated by applying the mass-transfer equation in the form that makes use of measurements of wind speed and humidity at two levels in the air near the surface [Equation (D-41)]. Dividing both sides of Equation (D-41) by $\lambda_v$ and $\rho_w$ gives

$$ET = \frac{0.622 \cdot \rho_a \cdot (v_2 - v_1) \cdot (e_1 - e_2)}{P \cdot \rho_w \cdot 6.25 \cdot \left[\ln\left(\dfrac{z_2 - z_d}{z_1 - z_d}\right)\right]^2}, \quad \textbf{(7-76)}$$

where the subscripts 1 and 2 refer to measurements at the lower and upper levels, respectively.

However, rather than apply Equation (7-76) directly, we can eliminate the need for wind-speed data and for estimates of the roughness height by making use of the Bowen-ratio approach [as in the development of Equation (7-24)] and an energy-balance relation. This is done using Equation (D-48), which gives the sensible-heat transfer rate, $H$, as

$$H = \frac{c_a \cdot \rho_a \cdot (v_2 - v_1) \cdot (T_1 - T_2)}{6.25 \cdot \left[\ln\left(\dfrac{z_2 - z_d}{z_1 - z_d}\right)\right]^2}. \quad \textbf{(7-77)}$$

From Equations (7-76) and (7-77), the Bowen ratio [Equation (7-11)] is

$$B \equiv \frac{H}{\rho_w \cdot \lambda_v \cdot ET} = \frac{c_a \cdot P \cdot (T_1 - T_2)}{0.622 \cdot \lambda_v \cdot (e_1 - e_2)}. \quad \textbf{(7-78)}$$

The energy balance, assuming that the change in heat storage and other terms are negligible, is

$$K + L - \rho_w \cdot \lambda_v \cdot ET - H = 0. \quad \textbf{(7-79)}$$

Then making use of the Bowen ratio and solving for $ET$ yields

$$ET = \frac{K + L}{\rho_w \cdot \lambda_v \cdot (1 + B)}, \quad \textbf{(7-80)}$$

where $K + L$ is the net radiation and $B$ is calculated from Equation (7-78).

Because of the need for measurements at two levels, Equation (7-80) is useful only in an elaborately instrumented research setting. Furthermore, the approach may not give good results for forest evapotranspiration because the diffusivities of momentum and water vapor may differ significantly over rough surfaces.

#### Eddy-Correlation Approach

The principles of the **eddy-correlation approach** are developed in Section D.6.9, leading to Equation (D-58), namely,

$$ET = \frac{\rho_a}{\rho_w} \cdot \overline{u_a' \cdot q'}, \quad \textbf{(7-81)}$$

where $u_a'$ and $q'$ are concurrent instantaneous variations in the *vertical* component of wind speed and specific humidity, respectively, and the overbar represents a time average.

Eddy-correlation measurements are often considered to be measurements of the "true" evaporation rate, because the method has a sound theoretical foundation and requires no assumptions about parameter values, the shape of the vertical velocity profile, or atmospheric stability. However, sensors capable of recording high-frequency (on the order of $10 \text{ s}^{-1}$) fluctuations in humidity and vertical air velocity are required, and because of these stringent instrumentation requirements the method is feasible only in a research facility. A full theoretical development and discussion of instrumentation was given by Brutsaert (1982).

### 7.8.4    Methods Based on Water-Quality Analyses

Approaches based on chemical and isotopic composition can provide very useful estimates of space- and time-integrated evapotranspiration. However, their application appears limited to situations where the hydrology is fairly simple.

#### Methods Based on Dissolved-Solids Concentrations

Water evaporates as individual $H_2O$ molecules (Section D.4), and any dissolved solids remain in the liquid water. Thus there is a tendency for the concentration of dissolved solids to increase in proportion to the amount of water that has evaporated, and this tendency can be used to estimate the amount of evaporation if other complicating factors (e.g., sporadic additions of water of varying concentration, dissolution of new materials) can be accounted for.

Margaritz et al. (1990) were able to use isotopes to determine the source of dissolved chloride, and then examined the increase in concentration of the chloride that came with precipitation to estimate evapotranspiration in the Jordan River basin. Claasen and Halm (1996) found that annual evapotranspiration in a number of Rocky Mountain watersheds could be well estimated from the chloride concentration in a single sample of stream water where the chloride concentrations of precipitation were well known.

#### Methods Based on Isotopic Composition

Isotopically lighter water molecules are more likely to evaporate than heavier ones, so the liquid water left behind tends to become enriched in the heavier isotopes of hydrogen and oxygen. (See Section B.1.5.) Several studies have taken advantage of this enrichment to estimate evapotranspiration from land surfaces (e.g., Allison and Barnes 1983; Walker and Brunel 1990).

---

## EXERCISES

Exercises marked with ** have been programmed in EXCEL on the CD that accompanies this text. Exercises marked with * can be advantageously executed on a spreadsheet, but you will have to construct your own worksheets to do so.

*7-1.    Measurements of water-balance components have been made for a one-year period on a lake with an area of $4.2 \text{ km}^2$ and a drainage basin of $52.1 \text{ km}^2$ (including the lake), with the following results: $W = 1083 \text{ mm}$; $SW_{in} = 2.33 \times 10^7 \text{ m}^3$; $GW_{in} = 2.2 \times 10^5 \text{ m}^3$; $GW_{out} = 0.6 \times 10^5 \text{ m}^3$; $SW_{out} = 2.70 \times 10^7 \text{ m}^3$. The lake-surface elevation, $h$, at the end of the year was 108 mm higher than at the beginning. (a) What is the water-balance estimate of the lake evaporation for that year [Equation (7-16)]? (b) Referring to Table 7-2, give a qualitative evaluation of the uncertainty of this estimate.

**7-2.    The table that follows gives the hourly air temperature, $T_a$ (°C), relative humidity, $W_a$, wind speed, $v_a$ (m s$^{-1}$), and water-surface temperature, $T_s$ (°C), for Lake Hefner, OK, on 3 May 1951. (These data are also in file HefnrHly.xls on the CD accompanying this text.) The lake area is $9.4 \text{ km}^2$. Compare the average evaporation rate for that day via the mass-transfer approach by (a) calculating the evaporation rate for each hour and averaging the results; and (b) averaging the values in the table and using those averages in the mass-transfer equation.

| Hour | 1 | 2 | 3 | 4 | 5 | 6 | 7 | 8 |
|------|------|------|------|------|------|------|------|------|
| $T_a$ | 20.8 | 21.0 | 21.3 | 21.9 | 22.6 | 23.5 | 26.4 | 28.2 |
| $W_a$ | 0.92 | 0.92 | 0.91 | 0.90 | 0.88 | 0.87 | 0.72 | 0.62 |
| $v_a$ | 8.14 | 7.83 | 7.62 | 7.94 | 8.53 | 9.47 | 9.68 | 9.84 |
| $T_s$ | 16.2 | 16.2 | 16.2 | 16.3 | 16.8 | 17.4 | 18.0 | 18.9 |

| Hour | 9 | 10 | 11 | 12 | 13 | 14 | 15 | 16 |
|------|------|------|------|------|------|------|------|------|
| $T_a$ | 28.2 | 29.1 | 30.9 | 32.6 | 32.4 | 32.0 | 31.9 | 30.3 |
| $W_a$ | 0.58 | 0.49 | 0.41 | 0.30 | 0.33 | 0.38 | 0.42 | 0.57 |
| $v_a$ | 10.71 | 10.05 | 9.46 | 9.37 | 9.02 | 8.88 | 8.55 | 8.34 |
| $T_s$ | 19.5 | 19.7 | 19.8 | 20.2 | 20.1 | 20.0 | 19.8 | 19.5 |

| Hour | 17 | 18 | 19 | 20 | 21 | 22 | 23 | 24 |
|------|------|------|------|------|------|------|------|------|
| $T_a$ | 29.6 | 26.2 | 25.8 | 24.9 | 23.6 | 23.3 | 22.8 | 22.5 |
| $W_a$ | 0.64 | 0.75 | 0.79 | 0.81 | 0.84 | 0.86 | 0.88 | 0.92 |
| $v_a$ | 8.01 | 8.03 | 7.99 | 8.15 | 8.24 | 8.00 | 7.84 | 7.62 |
| $T_s$ | 19.1 | 18.9 | 18.7 | 18.4 | 18.3 | 18.0 | 17.6 | 17.4 |

**\*7-3.** Given the following meteorological conditions (average or total daily values for 10 September 1950 at Lake Hefner, OK) compare the open-water evaporation rates given by (a) the mass-transfer, (b) energy-balance, and (c) combination approaches:

| $T_a (°C)$ | $W_a$ | $v_a (m\ s^{-1})$ | $T_s (°C)$ | $P (kPa)$ |
|------|------|------|------|------|
| 22.3 | 0.68 | 2.16 | 23.7 | 97.3 |

| $K_{in} (MJ\ m^{-2}\ day^{-1})$ | $a$ | $L_{in} (MJ\ m^{-2}\ day^{-1})$ |
|------|------|------|
| 16.2 | 0.057 | 30.6 |

**\*7-4.** For the conditions given in Exercise 7-3, compute the open-water evaporation rate given by the combination approach with the Kohler–Parmele modifications [Equations (7-35)–(7-36)].

**\*7-5.** Given the following conditions at the Class-A pan at Lake Hefner on 10 September 1950, compute the free-water evaporation [pan evaporation adjusted via Equations (7-41) and (7-42)] and compare it with the results of Exercises 7-3 and 7-4:

| Pan Evaporation (mm) | $T_{span} (°C)$ | $v_{pan} (m\ s^{-1})$ |
|------|------|------|
| 6.1 | 25.3 | 0.89 |

**\*\*7-6.** Use the spreadsheet program PenMontx.xls on the CD accompanying this text to explore the effects on the evapotranspiration rate of any two of the Input-Data variables, as computed via the Penman–Monteith Equation. Write a paragraph or two, supplemented with appropriate graphs, describing the sensitivity of ET to the variables you selected.

**\*7-7.** In central Alaska, spruce forests typically occupy north-facing slopes, with birch forests on the south-facing slopes. Redo Example 7-7 using the regression equation for spruce-fir-hemlock from Table 7-7, compare the results with Figure 7-16, and write a paragraph comparing the runoff-producing potential of north- vs. south-facing slopes in this region.

**\*\*7-8.** Pick a set of meteorological conditions and vegetation characteristics typical of the growing season in the region in which you live, and use the spreadsheet program PenMontx.xls to compare the rate of transpiration with the rate of evaporation of intercepted water, as in Example 7-8. How long would it take to evaporate 1 mm of intercepted water at the computed rate? Does the evaporation rate exceed the net radiation? If so, what are the implications for energy supply?

**\*7-9.** For the following conditions, compare the potential evapotranspiration as given by Equations (7-56), (7-63), (7-64), and (7-65):

| $T_a (°C)$ | $W_a$ | $K + L (MJ\ m^{-2}\ day^{-1})$ | $z_m (m)$ | $z_{veg} (m)$ |
|------|------|------|------|------|
| 20.7 | 0.49 | 13.9 | 3.66 | 0.50 |

| $v_a (m\ s^{-1})$ | $P (kPa)$ | $D (hr)$ | $C_{leaf}^* (m\ s^{-1})$ | $LAI$ |
|------|------|------|------|------|
| 2.48 | 101.3 | 14.8 | 0.0050 | 4.5 |

**7-10.** Obtain Class-A pan-evaporation data for your region (published in monthly Climatological Bulletins of the U.S. National Weather Service) and examine the seasonal and year-to-year variability of selected stations.

**\*7-11.** Use the information in Table 6-1 and the method of Example 6-5 to estimate $\theta_{fc}$ and $\theta_{pwp}$ for a silt-loam soil.

Construct a graph like Figure 7-24 for this soil, with actual values of $\theta_{rel}$ on the abscissa, where $\theta_{rel}$ has the form of Equation (7-68) and $ET = PET$ when $\theta_{rel} > 0.3$.

**\*7-12.** Use Equation (7-69) and the $f_\theta(\Delta\theta)$ relation of Table 7-6 to develop a relation between $ET/PET$ and $\theta$ at air temperatures of 0, 10, and 20 °C.

**\*\*7-13.** File ThornEx.xls is a spreadsheet implementing a Thornthwaite-type monthly water-balance model for climatic-average data. File ThornDatax.xls contains monthly average precipitation and temperature for many locations throughout the world. These can be used to compute actual evapotranspiration and to explore a range of hydrologic questions, such as the following:

1) How does $ET$ and the ratio $ET/PET$ vary with latitude on each continent? (Compare with Figures 3-23 and 3-24.)
2) How does $ET/PET$ vary with annual precipitation?
3) How does annual runoff vary with latitude? (Compare with Figures 3-25 and 3-26.)
4) How does annual $ET$ computed via ThornEx.xls compare with that given by the Pike Equation [Equation (7-66)] for selected stations?
5) What proportion of annual precipitation falls as snow as a function of latitude for selected continents? (Compare with Figure 5-11.)
6) How would a 2 °C average temperature increase (as projected due to increased atmospheric $CO_2$) affect annual $ET$ and runoff for selected stations? Is the relative response of $ET$ or runoff more sensitive to a temperature increase? (Compare with Box 3-4 and Figure 3-37.) What are the implications for water resources?
7) How would a 2 °C average temperature increase affect the timing of water "surplus" (monthly $W - ET - \Delta SOIL$) for selected stations? What are the implications for water resources?

# 8

# Ground Water in the Hydrologic Cycle

**Ground water** is water under positive (i.e., greater than atmospheric) pressure in the saturated zone of earth materials. Most water enters the ground-water reservoir when infiltrated water arrives at the **water table** as **recharge** (Figure 8-1). The water table is the fluctuating upper boundary of the ground-water zone at which pressure is atmospheric (denoted on diagrams by the **hydrat** symbol, ▽). Recharge can also occur by horizontal or vertical seepage from surface-water bodies. Under natural conditions ground water eventually discharges into rivers or lakes or, in coastal areas, directly into the ocean; water can also leave the ground-water reservoir by moving upward from the water table into the capillary fringe. (See Section 6.4.2.)

As shown in Table 3-1, ground water constitutes about 30% of the world's total fresh water and 99% of its total stock of liquid fresh water. As with all hydrologic stocks, ground water is in continual motion, albeit slow (typically less than 1 m day$^{-1}$). Using the values from Figure 3-16 in Equation (2-27), one can calculate that the overall residence time for the global ground-water reservoir is about 235 yr; for moderate- to large-scale regional flow systems in various parts of the world, residence time varies from a few years to 1000 years or more. In spite of its slow pace, ground water is a crucial link in the hydrologic cycle because it is the source of most of the water in rivers and lakes.

Ground water is, of course, also important as the direct source of water withdrawn for domestic water use, irrigation, and industrial uses worldwide. In the United States, about one-fourth of the water used for these purposes comes from ground water, and the proportion is much higher for many regions; concern about the quantity and quality of ground water is one of the major water-resource issues there and in many other parts of the world.

Our focus in this chapter is on ground water as a link in the hydrologic cycle. Thus, after reviewing the basic physical relations of ground-water flow and storage, we will examine the effects of topography and geology on natural ground-water flows in drainage basins; how ground water interacts with streams, lakes, wetlands, and the ocean; the role of ground water in the drainage-basin water balance; and approaches to the quantitative evaluation of components of the ground-water budget. We will also briefly examine the hydraulics of wells, how human use of ground water affects the basin water balance, and the concept of "safe yield." We do not explore the chemical evolution of ground water or its relation to geotechnical problems or geologic processes; these topics are well covered in many texts devoted exclusively to ground-water hydrology (e.g., Freeze and Cherry 1979; Fetter 1994).

We begin by reviewing some of the basic physics of ground-water movement and the properties of earth materials that determine the nature of that movement.

**325**

**FIGURE 8-1**
Generalized recharge, discharge, and flow relations in a porous deposit overlying impermeable rocks in a coastal region. Water can also move upward from the ground-water reservoir into the capillary fringe, which occupies a thin layer immediately above the water table.

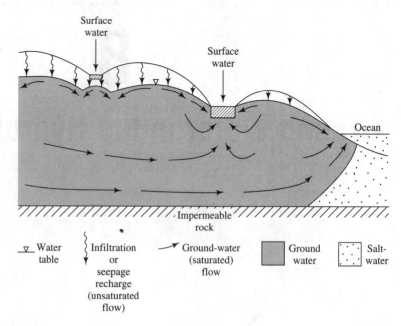

▽ Water table    ⟨ Infiltration or seepage recharge (unsaturated flow)    ⟶ Ground-water (saturated) flow    ▨ Ground water    ⦂ Salt-water

## 8.1 BASIC PRINCIPLES OF GROUND-WATER FLOW

### 8.1.1 Darcy's Law

The bulk flow of water in a saturated porous medium is governed by Darcy's Law for saturated flow:[1]

$$V_x \equiv \frac{Q}{A_x} = -K_{hx} \cdot \frac{dh}{dx}. \qquad \textbf{(8-1)}$$

The terms in Darcy's Law can be visualized by reference to Figure 8-2: $V_x$ [L T$^{-1}$] is the **specific discharge,** which is defined as the volume rate of flow, $Q$ [L$^3$ T$^{-1}$], per unit area, $A_x$ [L$^2$], of porous medium at right angles to the $x$-direction; $K_{hx}$ [L T$^{-1}$] is the **saturated hydraulic conductivity** of the medium[2] in the $x$-direction; and $h$ [L] is the **total hydraulic head** (usually called simply "head") of the fluid.

Head is mechanical energy per unit weight of fluid, so the hydraulic-head gradient represents the gradient of potential energy that induces the flow. The total potential energy of the fluid is the sum of its **gravitational head,** given by elevation above an

---

[1] Darcy's Law for unsaturated flow was developed in Equations (6-8)–(6-10).
[2] The symbol $K_h^*$ was used for this quantity in Chapter 6. Because the present chapter deals only with saturated flow, we drop the asterisk.

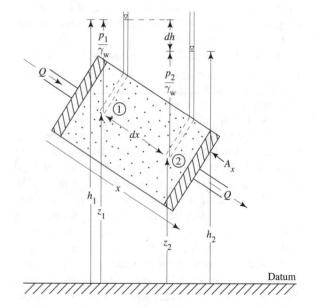

**FIGURE 8-2**
A simple experimental device illustrating the terms in Darcy's Law [Equation (8-1)]. The subscripts 1 and 2 refer to "upstream" and "downstream" points, respectively.

arbitrary horizontal datum, $z$ [L], and its **pressure head,** $p/\gamma_w$ [L]; thus

$$h = z + \frac{p}{\gamma_w}, \qquad (8\text{-}2)$$

where $p$ is the pressure [F L$^{-2}$] and $\gamma_w$ is the weight density of water [F L$^{-3}$]. Pressure is conventionally measured as **gage pressure,** which is the difference between actual pressure and atmospheric pressure; thus $p \geq 0$ for ground-water flows.

Note that Darcy's Law describes flow in a particular direction through a small volume that is large enough to be representative of the local medium and large enough to permit averaging of inter-pore and intra-pore variations of velocity. In diagrams of ground-water flow systems, such a volume is represented as a point. For a fluid of constant density (which we assume throughout this chapter), the head at any "point" in a ground-water flow can be measured as the height above the selected arbitrary datum to which water rises in a tube connecting the "point" to the atmosphere. (See Figure 8-2.) Such a tube is called a **piezometer.**

In Equation (8-1) the specific discharge, $V_x$, has the dimensions of a velocity and is sometimes called the "Darcy velocity." However, the actual average velocity of a ground-water flow is found by dividing the discharge by the cross-sectional area occupied by the flowing water. In a saturated porous medium, that area is given by $\phi \cdot A_x$, where $\phi$ is porosity, so the average velocity of a ground-water flow, $U_x$, is given by

$$U_x = \frac{Q}{\phi \cdot A_x} = \frac{V_x}{\phi}. \qquad (8\text{-}3)$$

Darcy's Law applies only to laminar porous-media flows in which flow velocities are low enough that the inertial forces are negligible. In such flows the **Reynolds number,** $\Re_{pm}$, has a value less than unity, where

$$\Re_{pm} \equiv \frac{V_x \cdot d}{\nu}, \qquad (8\text{-}4)$$

and $V_x$ is the specific discharge, $d$ is the average grain size of the medium, and $\nu$ is the kinematic viscosity of the fluid. (See Section B.2.3.) Most natural ground-water flows meet this criterion, but non-Darcy flow can occur in very large soil pores ("macropores," see Section 9.2.3), in large openings

produced by fracturing or dissolution of earth materials, and in the immediate vicinity of pumped wells.

### 8.1.2 Classification of Ground-Water Flows

For our purposes, an **aquifer** is a geologic unit that can store enough water and transmit it at a rate fast enough to be hydrologically significant.[3] A given aquifer has reasonably uniform water-storage and water-transmitting properties, and the ground-water movement within it can be considered to be the flow field induced by a single coherent potential-energy field.

Figure 8-3 illustrates the two major classes of ground-water flows, which are distinguished by the nature of the upper boundary of the flow.

#### *Unconfined Aquifers*

If the upper boundary of a ground-water flow is a water surface at atmospheric pressure (the water table), the flow and the aquifer in which it occurs are said to be **unconfined.** Since the pressure at the water table is atmospheric, $p = 0$ and the total head at the water table is equal to its elevation above the selected datum, $z$. The elevation of the water table can be determined as that of the water surface in a well open along its length and extending just deep enough to encounter standing water (Freeze and Cherry 1979).

Recharge to unconfined aquifers usually occurs from water percolating vertically to the water table over a significant portion of the upper surface. The elevation of the water table changes as the flow through the aquifer varies, thus flow in unconfined aquifers is analogous to free-surface flow in streams.

#### *Confined Aquifers*

If an aquifer is saturated throughout and bounded above and below by formations with significantly lower hydraulic conductivity (called **confining layers** or **aquicludes**), the flow and the aquifer are **confined.** The water level in an observation well penetrating a confined aquifer will rise above the upper boundary of the aquifer to coincide with the

---

[3] In the water-resource context, an aquifer is "a geologic unit that can store and transmit enough water to be a significant water resource."

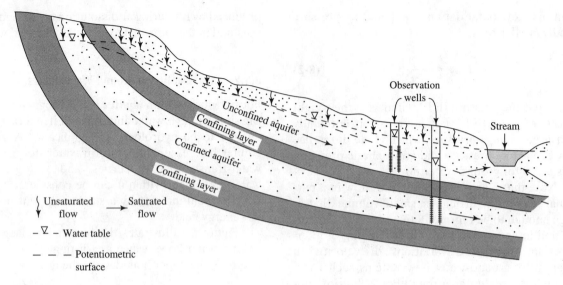

**FIGURE 8-3**
Unconfined and confined aquifers in a region of folded sedimentary rocks. Note relations of water levels in wells penetrating unconfined and confined aquifers.

**potentiometric surface,** an imaginary surface analogous to the water table.

Major recharge to confined aquifers typically occurs from water infiltrating at the "upstream" end of confined aquifers, where the flow is not confined and a water table is present. The boundary of the flow in a confined aquifer does not change with time, thus flow in confined aquifers is analogous to flow in pipes.

It is important to recognize that these classes represent idealizations, and that the spatial variability of geologic conditions produces flows that may have some of the characteristics of both types. In fact, most confining layers can transmit some ground water, and if the amount transmitted is significant such layers are called **leaky aquitards.**

### 8.1.3   Storage Properties of Porous Media

#### Basic Definitions

The degree of saturation [defined in Equation (6-7)] for ground-water flows is always unity. The volume of water stored in a saturated porous medium per unit volume of medium is the **porosity,** $\phi$, as defined in Equation (6-3).[4]

Consider a small unit area (say 1 m$^2$) on the earth's surface above an aquifer. The saturated thickness of the aquifer beneath that area is designated $H$ [L]. When the hydraulic head beneath the area decreases (increases), water is released from (taken into) storage. The volume of water that a unit volume of aquifer beneath the area releases (takes up) in response to a unit decrease (increase) in head is called the **specific storage,** $S_s$, of the aquifer:

$$S_s \ [\mathrm{L}^{-1}] \equiv$$

$$\frac{\text{Volume of water} \left\{ \dfrac{\text{taken into}}{\text{released from}} \right\} \text{storage } [\mathrm{L}^3]}{\text{Volume of aquifer } [\mathrm{L}^3] \cdot \left\{ \dfrac{\text{increase}}{\text{decrease}} \right\} \text{in head } [\mathrm{L}]}.$$

$$(8\text{-}5)$$

The **storage coefficient,** $S$, of an aquifer is defined as

$$S \ [1] \equiv$$

$$\frac{\text{Volume of water} \left\{ \dfrac{\text{taken into}}{\text{released from}} \right\} \text{storage } [\mathrm{L}^3]}{\text{Surface area of aquifer } [\mathrm{L}^2] \cdot \left\{ \dfrac{\text{increase}}{\text{decrease}} \right\} \text{in head } [\mathrm{L}]}.$$

$$(8\text{-}6)$$

---

[4] Ground-water hydrologists often distinguish between the **primary porosity,** which is the original intergranular porosity of a soil or sedimentary deposit, and the **secondary porosity,** which is

due to void spaces developed subsequently by fracturing and/or dissolution.

Thus storage coefficient [1] and specific storage [L$^{-1}$] are related as

$$S = H \cdot S_s. \qquad (8\text{-}7)$$

Note that the storage coefficient is defined only for the case of two-dimensional flow in the horizontal plane.

The mechanisms relating changes in head and changes in storage, and the relative magnitudes of these changes, differ between unconfined and confined aquifers, as described in the following sections. (See Figure 8-4.)

### *Unconfined Aquifers*

In an unconfined aquifer, a decrease (increase) in head is reflected in the lowering (raising) of the water table and the concomitant decrease (increase) in water content of the portion of the aquifer through which the water table descends (ascends). The amount of water-content change is characterized by the **specific yield,** $S_y$, defined as the volume of stored ground water released (taken up) per unit surface area per unit decline (increase) of water table (Figure 8-4a). Thus for an unconfined aquifer,

$$S = S_y. \qquad (8\text{-}8)$$

Typical values of $S_y$ for sedimentary deposits are in the range 0.05–0.35 (Table 8-1).

The relative volume of water retained in the portion of the aquifer experiencing a water-table decline is the **specific retention,** $S_r$. Thus

$$\phi = S_y + S_r. \qquad (8\text{-}9)$$

Note that specific retention is essentially identical to field capacity, defined in Section 6.4.1. Note also that soil drainage is not instantaneous, and many days may be required for water content to decline to $S_r$ in a draining aquifer (Figure 6-11).

### *Confined Aquifers*

In a confined aquifer, a unit decrease (increase) in head is reflected in a lowering (raising) of the piezometric surface, but the aquifer remains saturated (Figure 8-4b). The decrease (increase) of storage accompanying the head decrease (increase) is due to (1) compression (expansion) of the aquifer as the weight of the overlying material is transferred from (to) the liquid to (from) the solid grains, resulting in

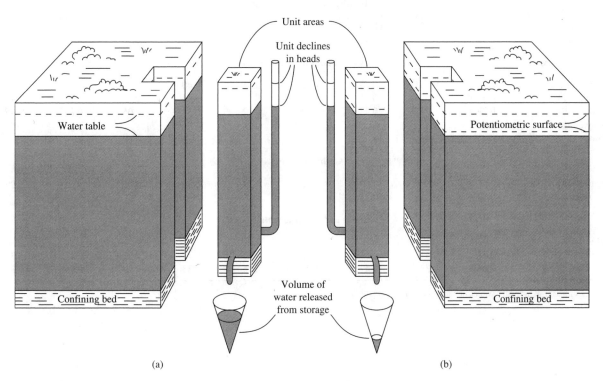

(a)                (b)

**FIGURE 8-4**
Definition of storage coefficient, *S*, in (a) unconfined and (b) confined aquifers. [See Equation (8-6).] From Heath (1982).

**TABLE 8-1**

Porosities, Specific Yields, and Specific Retentions of Geologic Materials[a]

| Material | Porosity | | | Specific Yield | | | Specific Retention | | |
|---|---|---|---|---|---|---|---|---|---|
| | Min. | Avg. | Max. | Min. | Avg. | Max. | Min. | Avg. | Max. |
| **Unconsolidated Alluvial Deposits** | | | | | | | | | |
| Clay | 0.34 | 0.42 | 0.57 | 0.01 | 0.06 | 0.18 | 0.25 | 0.38 | 0.47 |
| Silt | 0.34 | 0.46 | 0.61 | 0.01 | 0.20 | 0.39 | 0.03 | 0.28 | 0.45 |
| Fine sand | 0.26 | 0.43 | 0.53 | 0.01 | 0.33 | 0.46 | 0.03 | 0.08 | 0.43 |
| Medium sand | 0.29 | 0.39 | 0.49 | 0.16 | 0.32 | 0.46 | 0.01 | 0.04 | 0.18 |
| Coarse sand | 0.31 | 0.39 | 0.46 | 0.18 | 0.30 | 0.43 | 0.05 | 0.07 | 0.18 |
| Fine gravel | 0.25 | 0.34 | 0.39 | 0.13 | 0.28 | 0.40 | 0.00 | 0.07 | 0.17 |
| Medium gravel | 0.24 | 0.32 | 0.44 | 0.17 | 0.24 | 0.44 | 0.01 | 0.07 | 0.15 |
| Coarse gravel | 0.24 | 0.28 | 0.37 | 0.13 | 0.21 | 0.25 | 0.03 | 0.09 | 0.14 |
| **Unconsolidated Glacial Deposits** | | | | | | | | | |
| Silty till | 0.30 | 0.34 | 0.41 | 0.01 | 0.06 | 0.13 | 0.23 | 0.28 | 0.33 |
| Sandy till | 0.22 | 0.31 | 0.37 | 0.02 | 0.16 | 0.31 | 0.03 | 0.14 | 0.29 |
| Gravelly till | 0.22 | 0.26 | 0.30 | 0.05 | 0.16 | 0.34 | 0.01 | 0.12 | 0.25 |
| **Unconsolidated Aeolian Deposits** | | | | | | | | | |
| Loess | 0.44 | 0.49 | 0.57 | 0.14 | 0.18 | 0.22 | 0.22 | 0.27 | 0.30 |
| Aeolian sand | 0.40 | 0.45 | 0.51 | 0.32 | 0.38 | 0.47 | 0.01 | 0.03 | 0.06 |
| **Unconsolidated Biogenic Deposits** | | | | | | | | | |
| Peat | | 0.92 | | | 0.44 | | | 0.44 | |
| **Weathered Rock (Saprolites)** | | | | | | | | | |
| Granite | 0.34 | 0.45 | 0.57 | | | | | | |
| Gabbro | 0.42 | 0.43 | 0.45 | | | | | | |
| **Clastic Sedimentary Rocks** | | | | | | | | | |
| Fine sandstone | 0.14 | 0.33 | 0.49 | 0.02 | 0.21 | 0.40 | 0.01 | 0.13 | 0.31 |
| Med. sandstone | 0.30 | 0.37 | 0.44 | 0.12 | 0.27 | 0.41 | 0.05 | 0.10 | 0.19 |
| Siltstone | 0.29 | 0.35 | 0.48 | 0.01 | 0.12 | 0.33 | 0.05 | 0.29 | 0.45 |
| Claystone | 0.41 | 0.43 | 0.45 | | | | | | |
| Shale | 0.01 | 0.06 | 0.10 | | | | | | |
| **Carbonate Rocks** | | | | | | | | | |
| Limestone | 0.07 | 0.30 | 0.56 | 0.02 | 0.14 | 0.36 | 0.05 | 0.13 | 0.29 |
| Dolomite | 0.19 | 0.26 | 0.33 | | | | | | |
| **Igneous and Metamorphic Rocks** | | | | | | | | | |
| Basalt | 0.03 | 0.17 | 0.35 | | | | | | |
| Volcanic tuff | 0.07 | 0.41 | 0.55 | 0.02 | 0.21 | 0.47 | 0.06 | 0.21 | 0.38 |
| Schist | 0.04 | 0.38 | 0.49 | 0.22 | 0.26 | 0.33 | 0.22 | 0.26 | 0.33 |

[a]Values measured in small samples by Morris and Johnson (1967).

a slight decrease (increase) in porosity; and (2) a slight expansion (compression) of the water due to the lowered (increased) pressure. A more detailed discussion of these mechanisms can be found in Freeze and Cherry (1979).

The magnitude of the storage coefficient for confined aquifers is usually in the range $5 \times 10^{-5}$ to $5 \times 10^{-3}$, that is, at least an order of magnitude less than the specific yield for unconfined aquifers.

### 8.1.4    Transmission Properties of Porous Media

The **intrinsic permeability,** $k_I$ [L²] of a porous medium is a function of the size and configuration of its water-transmitting pathways. For a granular medium, $k_I$ is related to its effective ($\approx$ average) grain size, $d$ [L], as

$$k_I = C \cdot d^2, \qquad \textbf{(8-10)}$$

where $C$ is a dimensionless constant that reflects the configuration of flow paths. The hydraulic conductivity, $K_h$ [L T⁻¹] is determined by properties of both the medium and the fluid; specifically,

$$K_h = \frac{\gamma_w}{\mu} \cdot k_I = \frac{\gamma_w}{\mu} \cdot C \cdot d^2, \qquad \textbf{(8-11)}$$

where $\mu$ is the dynamic viscosity [F T L⁻²] and $\gamma_w$ the weight density [F L⁻³] of the fluid (water). $K_h$ varies over a $10^{12}$-fold range for earth materials (Figure 8-5), and most of this variability is due to the $10^6$-fold range of $d$; $\gamma_w/\mu$ varies by only a factor of 2 or so under typical near-surface conditions.

Values of saturated hydraulic conductivity for small samples can be readily determined in **permeameters** similar to the device shown in Figure 8-2, where $A_x$ and $dx$ are known, $Q$ and $dh$ are measured, and $K_{hx}$ is calculated via Equation (8-1). However, conductivities measured in permeameters may differ greatly from field values because (1) it is often difficult to avoid disturbing the structure of materials when taking samples for a permeameter and (2) conductivities measured in small samples do not include the effects of fractures and larger-scale geologic variability, and hence are generally smaller than effective regional-scale values.

Values of $K_h$ for undisturbed materials over much larger volumes can be measured by removing

**FIGURE 8-5**

Ranges of saturated hydraulic conductivities for various geologic materials. From Heath (1982).

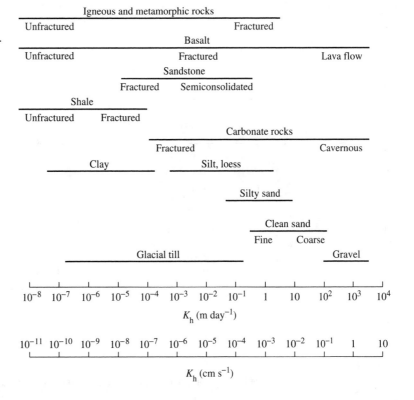

water from or adding it to a well and observing the rate at which water level changes in the well or at one or more nearby observation wells. Details of these methods can be found in most ground-water hydrology texts (e.g., Freeze and Cherry 1979).

In general, hydraulic conductivity has significant spatial variability even within a given aquifer, and conductivity at a "point" may be different in different directions. If $K_h$ at a "point" is the same for all directions, the medium is **isotropic**; if it differs for different flow directions, it is **anisotropic.** If the conductivity in all directions is the same at all "points," the medium is **homogeneous**; otherwise it is **heterogeneous.** Figure 8-6 illustrates the four possible combinations of anisotropy and heterogeneity.

When dealing with flows in which the saturated thickness, $H$, is only slightly variable and the flow

paths are approximately horizontal, ground-water hydrologists often use the concept of aquifer **transmissivity,** $T$ [$L^2 T^{-1}$], defined as

$$T \equiv H \cdot K_h. \qquad (8\text{-}12)$$

### 8.1.5 Response Characteristics of Porous Media

Aquifers are reservoirs in which outflow is related to storage [Equation (2-25)], and storage is related to inflow and outflow. A relative measure of the time scale at which aquifer storage and outflow vary, which appears in various contexts later in this chapter, is given by

$$\tau \equiv \frac{L^2 \cdot S}{H \cdot K_h} = \frac{L^2 \cdot S}{T}, \qquad (8\text{-}13)$$

Homogeneous, isotropic

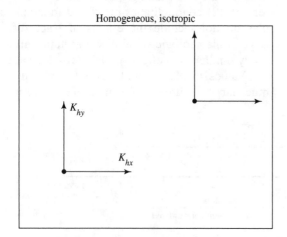

Homogeneous, anisotropic

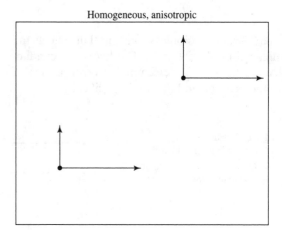

Heterogeneous, isotropic

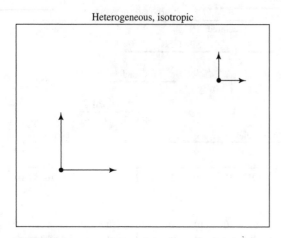

Heterogeneous, anisotropic

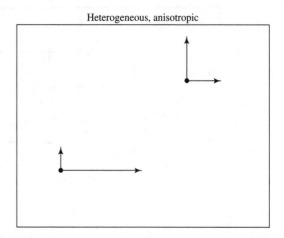

**FIGURE 8-6**
Four possible combinations of heterogeneity and anisotropy in saturated hydraulic conductivity. After Freeze and Cherry (1979).

where $\tau$ is **aquifer response time** [T], $L$ is a measure of the horizontal extent of the aquifer [L], and the other symbols are as defined earlier. The inverse of $\tau$ is often called the **aquifer response rate** (Erskine and Papaioannou 1997).

### 8.1.6   General Ground-Water Flow Equation

The general equation for flow in a ground-water flow system is developed by combining Darcy's Law with expressions of the law of conservation of fluid mass (continuity). The three-dimensional continuity

equation for porous-media flow is derived in Box 8-1. When Darcy's Law in the form of Equation (8-1) for each of the three directions is inserted into Equation (8B1-5), we have the equation that must be satisfied for all points in a time-varying flow in a saturated aquifer, assuming constant density:

$$\frac{\partial}{\partial x}\left(K_{hx} \cdot \frac{\partial h}{\partial x}\right) + \frac{\partial}{\partial y}\left(K_{hy} \cdot \frac{\partial h}{\partial y}\right)$$
$$+ \frac{\partial}{\partial z}\left(K_{hz} \cdot \frac{\partial h}{\partial z}\right) = S_s \cdot \frac{\partial h}{\partial t}.$$

(8-14)

---

## BOX 8-1

· · · · · · ·

### Derivation of Equation of Continuity for Ground-Water Flow

We apply the general conservation equation [Equation (2-1)] to water entering a rectangular parallelipiped control volume $dx \cdot dy \cdot dz$ (Figure 8-7) in time interval $dt$. The volume must be large enough to contain many typical pore spaces but small enough to allow us to use continuous mathematics. For this volume over $dt$,

$$M_{in} - M_{out} = \Delta M, \qquad \text{(8B1-1)}$$

where $M_{in}$ is the mass of water flowing into the volume, $M_{out}$ is the mass of water flowing out of the volume, and $\Delta M$ is the change in mass storage.

The flow can be in any direction, and the specific-discharge vector can be resolved into vectors parallel to the sides of the volume, $V_x$, $V_y$, and $V_z$. Then $M_{in}$ (entering the volume through faces 1, 2, and 3) is

$$M_{in} = \rho \cdot V_x \cdot dy \cdot dz \cdot dt + \rho \cdot V_y \cdot dx \cdot dz \cdot dt$$
$$+ \rho \cdot V_z \cdot dx \cdot dy \cdot dt,$$

(8B1-2)

where $\rho$ is the mass density of water.

In general, the rate of flow changes in space, so that $M_{out}$ (leaving through faces 4, 5, and 6) is given by

$$M_{out} = \left[\rho \cdot V_x + \frac{\partial(\rho \cdot V_x)}{\partial x} \cdot dx\right] \cdot dy \cdot dz \cdot dt$$
$$+ \left[\rho \cdot V_y + \frac{\partial(\rho \cdot V_y)}{\partial y} \cdot dy\right] \cdot dx \cdot dz \cdot dt$$

$$+ \left[\rho \cdot V_z + \frac{\partial(\rho \cdot V_z)}{\partial z} \cdot dz\right] \cdot dx \cdot dy \cdot dt.$$

(8B1-3)

From Equation (8-5),

$$\Delta M = \rho \cdot S_s \cdot \frac{\partial h}{\partial t} \cdot dx \cdot dy \cdot dz \cdot dt, \qquad \text{(8B1-4)}$$

where $S_s$ is specific storage.

Substituting Equations (8B1-2)–(8B1-4) into (8B1-1), assuming constant density, and simplifying yields the continuity equation for three-dimensional time-varying ground-water flow:

$$-\frac{\partial V_x}{\partial x} - \frac{\partial V_y}{\partial y} - \frac{\partial V_z}{\partial z} = -S_s \cdot \frac{\partial h}{\partial t}. \qquad \text{(8B1-5)}$$

If water is being added to the control volume by recharge, or extracted by pumping or evaporation, a source or sink term, $r_s$, is added, yielding

$$-\frac{\partial V_x}{\partial x} - \frac{\partial V_y}{\partial y} - \frac{\partial V_z}{\partial z} + r_s = -S_s \cdot \frac{\partial h}{\partial t}, \qquad \text{(8B1-6)}$$

where $r_s$ is the volume of water added ($r_s > 0$) or extracted ($r_s < 0$) per unit aquifer volume and unit time [T$^{-1}$].

**FIGURE 8-7**
Definitions of terms for derivation of the continuity equation for ground-water flow (Box 8-1).

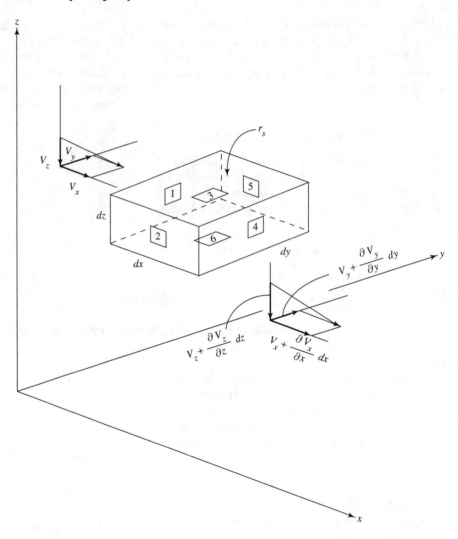

Since in general the conductivities, $h$, and specific storage, $S_s$, can vary from point to point in the flow system, Equation (8-14) can describe time-varying ground-water flow in a heterogeneous and anisotropic porous medium.

For steady-state flow $\partial h / \partial t = 0$ and the general equation simplifies to

$$\frac{\partial}{\partial x}\left(K_{hx} \cdot \frac{\partial h}{\partial x}\right) + \frac{\partial}{\partial y}\left(K_{hy} \cdot \frac{\partial h}{\partial y}\right)$$
$$+ \frac{\partial}{\partial z}\left(K_{hz} \cdot \frac{\partial h}{\partial z}\right) = 0. \quad \textbf{(8-15)}$$

For an isotropic and homogeneous aquifer, $K_{hx} = K_{hy} = K_{hz} = K_h$ and $K_h$ does not vary with location, so that Equation (8-14) becomes

$$\frac{\partial^2 h}{\partial x^2} + \frac{\partial^2 h}{\partial y^2} + \frac{\partial^2 h}{\partial z^2} = \frac{S_s}{K_h} \cdot \frac{\partial h}{\partial t}. \quad \textbf{(8-16)}$$

Finally, for steady-state flows in a homogeneous and isotropic aquifer, Equation (8-14) reduces to

$$\frac{\partial^2 h}{\partial x^2} + \frac{\partial^2 h}{\partial y^2} + \frac{\partial^2 h}{\partial z^2} = 0. \quad \textbf{(8-17)}$$

Equation (8-17) is known as the **Laplace Equation.** As we shall see, solutions to this equation are widely used to develop understanding of regional ground-water flow systems.

Solutions to Equation (8-14) and its various simplifications consist of a coherent potential-energy ($h$) field in one or more contiguous aquifers, and are largely determined by the **boundary conditions** of the flow system. For steady-state flows, these boundary conditions are specified in terms of (1) the configuration of the system's boundaries, often expressed as the coordinates of points on the boundaries, $(x_b, y_b, z_b)$, and (2) at boundary points, either the head $h(x_b, y_b, z_b)$ or their designation as impermeable barriers. In general, we must also specify as parameters the values of hydraulic conductivity and storage coefficient at all points and, for time-varying flows, the initial conditions, $h(x, y, z, 0)$. For Equation (8-17), only the boundary conditions are needed.

If the flow is steady and two-dimensional and the boundary conditions have a simple configuration—for example, if the region is a rectangle with flow across only one or two sides—the two-dimensional form of the Laplace Equation,

$$\frac{\partial^2 h}{\partial x^2} + \frac{\partial^2 h}{\partial z^2} = 0, \qquad \textbf{(8-18)}$$

can be solved analytically (Tóth 1962). However, for most situations, solutions must be numerically approximated by transforming the partial-differential equations to a system of algebraic equations and solving for $h$ at selected points throughout the flow system and at successive times.[5] The points are the nodes of a grid which are separated by finite distances, and the equations are solved at successive finite time steps. The process is ended when the successive head values change by less than a prescribed amount.

Once the $h$ values are found at all points for the steady-state solution, those values can be plotted and contoured. Lines of equal head are called **equipotential lines,** and the directions of ground-water flow are given by **streamlines** (also called **flowlines**) drawn at right angles to these lines. The space between each pair of adjacent streamlines is called a **streamtube.**

A diagram showing equipotentials and streamlines is called a **flow net.** Principles of graphical flow-net construction are developed in most ground-water texts [e.g., Freeze and Cherry (1979); Fetter (1994)] and are discussed in detail by Cedergren (1989). For a homogeneous and isotropic aquifer, the configuration of a flow net is determined only by the boundary conditions; it is independent of the hydraulic conductivity. If the aquifer is anisotropic or if the conductivity is a function of location, the flow-net shape is determined by the boundary conditions and the *relative* values of the various conductivities.

Qualitatively, the relative intensity of the circulation in various portions of the flow net is inversely proportional to the width of the streamtube. The rate of flow in the direction of the streamline at any point can be calculated by determining the head gradient in that direction and multiplying it by the local value of hydraulic conductivity. As discussed later in this chapter, properly constructed flow nets can be used to compute the total discharge (per unit distance at right angles to the flow) by applying Darcy's Law to each streamtube and adding the flows for all streamtubes.

Figure 8-8 shows the effects of boundary conditions on solutions to Equation (8-18), where $x$ represents the horizontal coordinate and $z$ the vertical coordinate. Where the aquifer is deep relative to its length (Figure 8-8b), the vertical components of the streamlines are much greater than where the aquifer is relatively shallow (Figure 8-8a). The following section makes extensive use of such two-dimensional solutions to depict the effects of topographic and geologic boundary conditions on ground-water flow systems.

## 8.2    REGIONAL GROUND-WATER FLOW[6]

To illustrate the essential features of regional ground-water flow systems, we will use models that consist of Equation (8-18) solved for boundary conditions that represent various configurations of topography and geology. The solutions are shown

---

[5] Wang and Anderson (1982) provided an excellent introduction to the two approaches for finding numerical solutions to the ground-water flow equation: the finite-difference approach and the finite-element method.

[6] This section follows the development in Freeze and Cherry (1979).

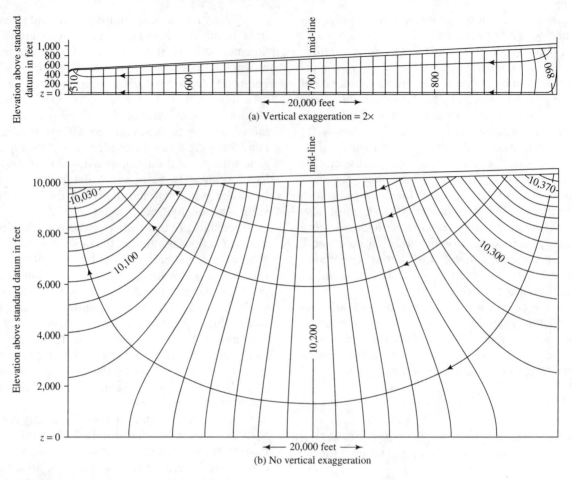

**FIGURE 8-8**
Flow-net configurations computed by applying the Laplace Equation [Equation (8-18)] to idealized approximations of unconfined ground-water flow to a stream where (a) the aquifer is relatively shallow (note flow is essentially horizontal) and (b) the aquifer is relatively deep (note flow has significant vertical component except near the mid-line). Contours are equipotentials; lines with arrows are streamlines. From Tóth (1962), used with permission of the American Geophysical Union.

as flow nets for two-dimensional vertical slices through unconfined aquifers (like Figures 8-1 and 8-8).

Note that the flow-net diagrams represent the pattern of equipotentials and streamlines that is consistent with the selected configuration of boundary conditions, including that of the water table, and the specified distribution of conductivity values. The processes of recharge and capillary rise are not included in the modeling; instead, water-table configurations typical of those observed in nature are specified. However, in nature there is an interactive relation in which topography, conductivity, and climate influence the water-table configuration, and

the water-table configuration influences the distribution of recharge and discharge conditions.

The steady-state condition represented by Equation (8-18) never strictly occurs in nature. However, if the general configuration of the water table does not greatly change through the year, and if the fluctuations of the water table at any point are small relative to the thickness of the saturated zone, Equation (8-18) provides a picture of the essential features of regional ground-water flow under various boundary conditions. We will modify these boundary conditions to illustrate the effects of particular topographic and geologic conditions on regional ground-water flows.

### 8.2.1 General Features

Figure 8-9 shows a cross section through an idealized hilly upland area in a humid region, underlain by permeable deposits with homogeneous and isotropic conductivity resting on an impermeable base. The balance between a steady net vertical recharge and lateral ground-water flow typically results in a water table that is a subdued replica of the ground surface, as shown in Figure 8-9. The higher the hydraulic conductivity, the more subdued the replica.

With such a water table as a boundary condition, the solution of Equation (8-18) results in a general head gradient from the surface divides toward the streams, with the flow moving downward near the divide and then upward and converging toward the streams, where ground-water discharge occurs. The symmetry of the situation produces one streamline that flows vertically downward at the divides, horizontally along the impermeable base, and vertically upward under the streams; thus the surface divides and the streams mark ground-water divides as well. (However, in regions of natural non-symmetric topography or heterogeneous geology, ground-water divides may not coincide with surface-water divides and surface-water bodies.)

A **recharge area** is a portion of a drainage basin in which the ground-water flow is directed away from the water table, and a **discharge area** is a region in which the ground-water flow is directed toward the water table. The water table is usually at or near the surface in discharge areas, and such areas are usually the sites of surface-water bodies: streams, lakes, wetlands, or, if highly localized, springs. The line separating recharge and discharge areas is called a **hinge line.**

Note that basic water-balance considerations dictate that the average rate of discharge from a regional aquifer must be equal to the average rate of recharge. Since recharge rate is determined largely by climate ($\approx$ precipitation less evapotranspiration), topography and geology determine only the spatial distributions of discharge and recharge, not the absolute rates.

### 8.2.2 Effects of Topography

Figure 8-10 shows the flow nets for two situations of identical geology, depth to impermeable base, and lateral dimension. Both have a major valley on the left and an upland area toward the right, but the detailed topography as represented by the steady-state water table differs. The water table in Figure 8-10a is a gently sloping plane, such as might be found beneath a region of lake deposits or undeformed coastal-plain sediments; the water table in Figure 8-10b reflects a hilly upland superimposed on the general leftward-sloping topographic trend, such as might be found in a region of glacial deposits.

The plane water table of Figure 8-10a results in a single flow system, with the recharge area extend-

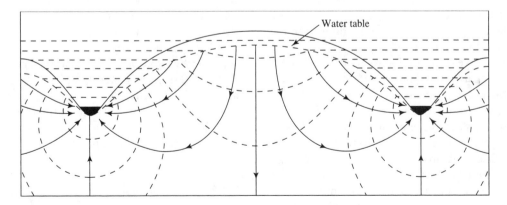

**FIGURE 8-9**
Ground-water flow net as given by solving the Laplace Equation [Equation (8-18)] for a vertical section through idealized hills and valleys in a permeable material resting on an impermeable base. Dashed lines are equipotentials; arrows are streamlines. Streams, lakes, or wetlands are present in valleys where the water table intersects the land surface. After Hubbert (1940).

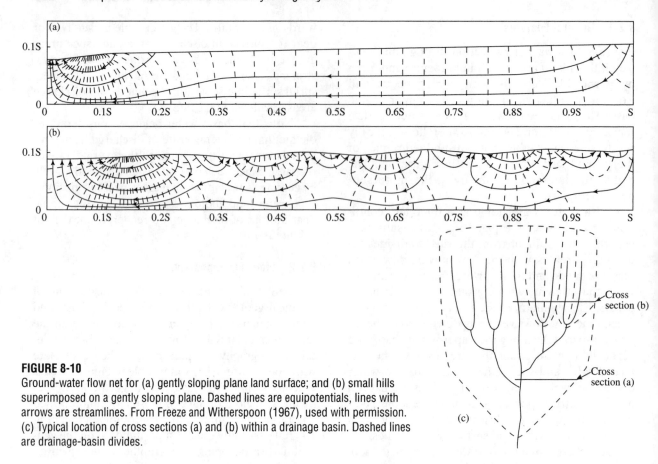

**FIGURE 8-10**
Ground-water flow net for (a) gently sloping plane land surface; and (b) small hills superimposed on a gently sloping plane. Dashed lines are equipotentials, lines with arrows are streamlines. From Freeze and Witherspoon (1967), used with permission. (c) Typical location of cross sections (a) and (b) within a drainage basin. Dashed lines are drainage-basin divides.

ing down to the central valley. In Figure 8-10b, on the other hand, each hill in the water table produces a small-scale recharge-discharge system that circulates above an intermediate-scale flow in the left half of the system, and both these systems circulate above a large-scale flow system extending from the major divide to the major valley.

Patterns like that in Figure 8-10b led Tóth (1963) to conclude that, in many situations, one can identify **local flow systems,** in which water moves from a recharge area to the next adjacent discharge area; **regional flow systems,** in which the flow is from the recharge area farthest from the main valley to the discharge area in the main valley; and **intermediate flow systems,** in which the flow path is longer than local but shorter than regional.

Regions with little local relief typically have only regional systems, and regions with pronounced local relief typically have only local systems. However, the pattern of development of flow systems of various scales is affected also by the overall system geometry: Development of local flow systems is fa-

vored where the depth to the impermeable layer is small relative to the distance from the main valley to the divide, and the more pronounced the relief, the deeper the local systems extend; regional systems are favored where the impermeable layer is deep relative to that distance.

Thus, as noted by Freeze and Cherry (1979), topography alone can create complex ground-water flow patterns. In general, uplands are recharge areas and lowlands are discharge areas; hinge lines are usually closer to the valleys than to the divides and discharge areas typically constitute less than 30% of a given drainage basin.

### 8.2.3 Effects of Geology

The most important geologic factors controlling the directions and relative rates of ground-water movement are as follows:

1. **Lithology:** The mineral composition, grain-size distribution, and grain-shape character-

istics of rocks and unconsolidated geologic materials, which control the hydraulic conductivity distribution.

2. **Stratigraphy:** The geometrical and age relations among the various formations, which, except for intrusive igneous rocks, are typically layered.

3. **Structure:** The general disposition, attitude, arrangement, or relative positions of formations, especially as modified by deformational processes such as folding, faulting, and intrusion of igneous rocks.

A regionally extensive mappable geologic unit with reasonably constant lithology is called a **formation**; an aquifer may comprise one or more formations, or may consist of particular high-conductivity beds within a formation.

As noted by Freeze and Cherry (1979), an understanding of the lithology, structure, and stratigraphy of a region usually leads directly to an understanding of the distribution of aquifers and confining beds, and hence to a qualitative understanding of at least the major characteristics of ground-water movement. Complete understanding of the role of ground water in a region's hydrology must be based on detailed geologic mapping and subsurface exploration.

Table 8-1 gives some typical values of porosity, specific yield, and specific retention and Figure 8-5 shows the ranges of saturated hydraulic conductivity for various geologic materials. Figure 8-11 shows the distribution of the principal types of aquifers in the conterminous United States. More detailed information about regional ground-water geology and the ground-water characteristics of various types of geologic materials can be found in McGuinness (1963), U.S. Geological Survey (1985), Freeze and Cherry (1979), and Fetter (1994).

Geologic conditions display infinite variability, so we can only suggest the types and magnitudes of geologic effects on ground-water flow patterns by examining a few idealized situations. In these depictions, variations in geology are represented by differences in the relative magnitudes of saturated hydraulic conductivity.

Figure 8-12 shows the effects on the flow net for a sloping-plane topography with a buried layer having hydraulic conductivity 10, 100, and 1000 times greater than the overlying layer. Comparing these nets with Figure 8-10a, we see that the flow in the high-conductivity layers is essentially horizontal; the flow in the upper layer gets increasingly vertical as the conductivity contrast increases; and the hinge line moves upslope.

The effect of a buried high-conductivity layer in a region of hilly topography can be seen by comparing Figure 8-13 with Figure 8-10b. Here the buried layer changes the local flow systems near the divide to intermediate systems, reduces the flow intensities in the remaining local systems, and increases the intensity and extent of the discharge area in the main valley.

Figure 8-14 shows the effect of a basal lens of high-conductivity material in various positions on the plane-topography flow net. Where the lens is in the upper half of the basin (Figure 8-14a), the flow net is much altered from that in Figure 8-10a, with a discharge zone in mid-basin; the effects for lenses in the other positions are less profound but still substantial.

Figure 8-15 shows the effect of a sloping high-conductivity bed; note that it produces an extensive discharge area just to the left of where that bed outcrops in addition to the discharge area in the main valley.

Figure 8-16 illustrates the effects of anisotropy on regional flow. For the vertical/horizontal conductivity ratios shown, the flow nets are not greatly different from that of Figure 8-10a.

Figure 8-17 is a cross section of an actual flow system in glacial deposits in Saskatchewan, Canada, that illustrates the complexity of flow patterns that can exist in nature. Note that recharge occurs over almost the entire section, and that Notukeu Creek is a site of recharge, not discharge. Discharge occurs only to Wiwa Creek, but there is also flow beneath that stream and the northern upland to some other regional discharge site to the northeast.

Freeze and Witherspoon (1967) presented flow nets for combinations of idealized topography and geology other than those shown here. All these examples show clearly that variations in geology, which might not be apparent from surface observation, can produce a wide range of possible flow-net configurations consistent with a given water-table configuration. Thus information about subsurface geologic conditions and hydraulic head is necessary for formulating an accurate picture of regional ground-water flow.

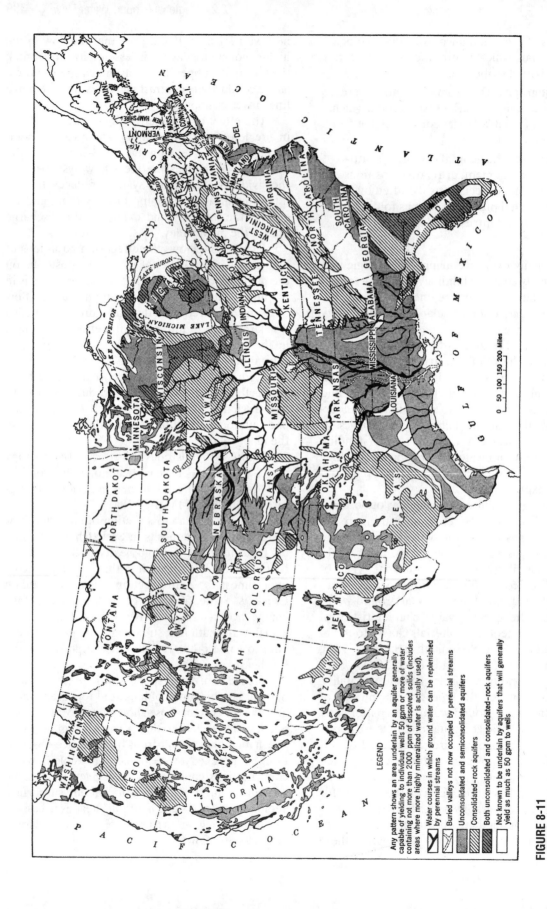

**LEGEND**

Any pattern shows an area underlain by an aquifer generally capable of yielding to individual wells 50 gpm or more of water containing not more than 2000 ppm of dissolved solids (includes areas where more highly mineralized water is actually used).

Water courses in which ground water can be replenished by perennial streams

Buried valleys not now occupied by perennial streams

Unconsolidated and semiconsolidated aquifers

Consolidated-rock aquifers

Both unconsolidated and consolidated-rock aquifers

Not known to be underlain by aquifers that will generally yield as much as 50 gpm to wells

**FIGURE 8-11**

Major aquifers of the coterminous United States. From McGuinness (1963).

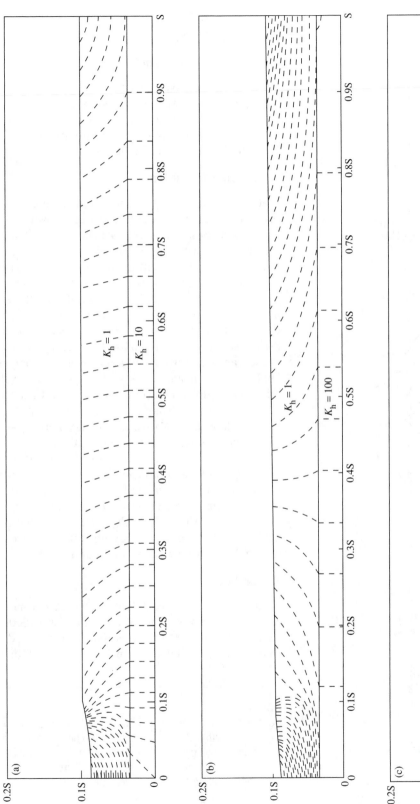
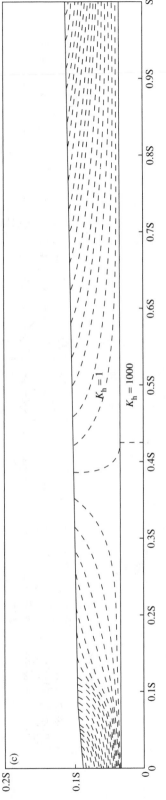

**FIGURE 8-12**
Equipotentials for topography of Figure 8-10a, but with a buried layer with hydraulic conductivity (a) 10, (b) 100, and (c) 1000 times greater than that of the overlying material. From Freeze and Witherspoon (1967), used with permission of the American Geophysical Union.

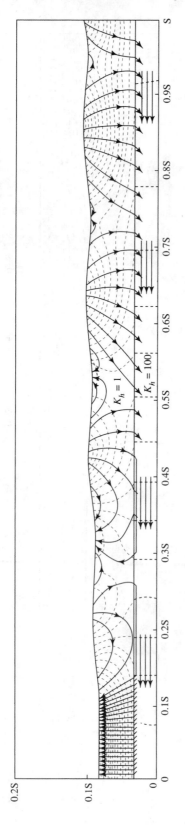

## 8.3 GROUND-WATER–SURFACE-WATER RELATIONS

### 8.3.1 Ground Water and Streams

Here we qualitatively examine the types of relations that can exist between streams and ground water and define some useful terms. Approaches to quantifying ground-water contributions to streamflow are discussed later in this chapter.

#### Event Flow and Base Flow

Water that enters streams promptly in response to individual water-input events (rain or snowmelt) is called **event flow, direct flow, storm flow,** or **quick flow.** This is distinguished from **base flow,** which is water that enters from persistent, slowly varying sources and maintains streamflow between water-input events.

It is usually assumed that most, if not all, base flow is supplied by ground-water circulation in the drainage basin as depicted in Figures 8-8–8-10 and 8-12–8-17. However, streamflow between water-input events can also derive from drainage of lakes or wetlands, or even from the slow drainage of relatively thin soils on upland hillslopes (Hewlett and Hibbert 1963). Conversely, ground water can also contribute to quick flow; these aspects of ground-water–surface-water relations are examined in Chapter 9.

Streams that receive large proportions of their flow as ground-water base flow tend to have relatively low temporal flow variability, and hence provide a more reliable source of water for various water-resource purposes (water supply, waste-water dilution, navigation, hydropower generation, etc.; see discussion of flow-duration curves in Section 10.2.5).

#### Stream Types

A stream or stream reach that occurs in a discharge area and receives ground-water flow is called a **gaining** (or **effluent**) stream or reach because its discharge increases downstream. The surface of gain-

**FIGURE 8-13**

Flow net for hilly topography of Figure 8-10b, but with a buried layer with hydraulic conductivity 100 times greater than that of the overlying material. Dashed lines are equipotentials, arrows are streamlines. From Freeze and Witherspoon (1967), used with permission of the American Geophysical Union.

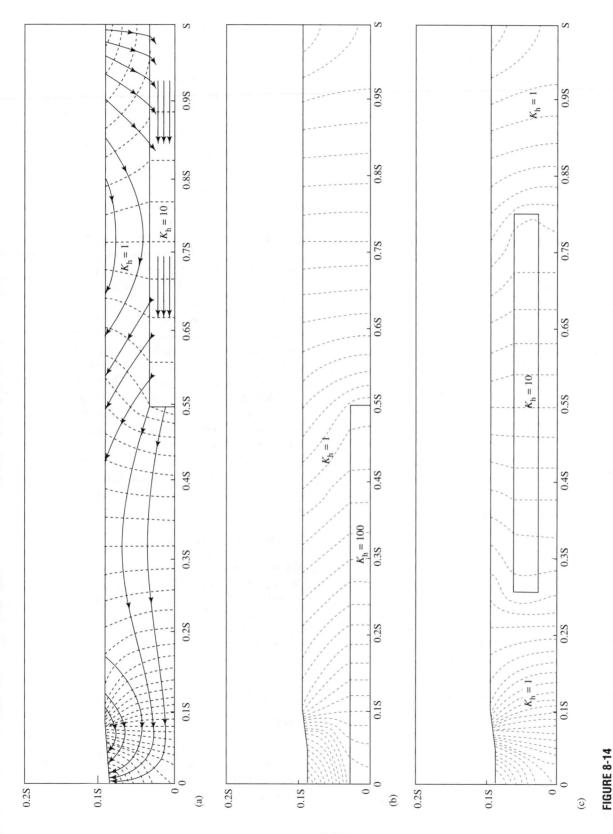

**FIGURE 8-14**

Flow nets for topography of Figure 8-10a but with a buried lens with hydraulic conductivity 10 times greater than that of the overlying material in different positions. From Freeze and Witherspoon (1967), used with permission of the American Geophysical Union.

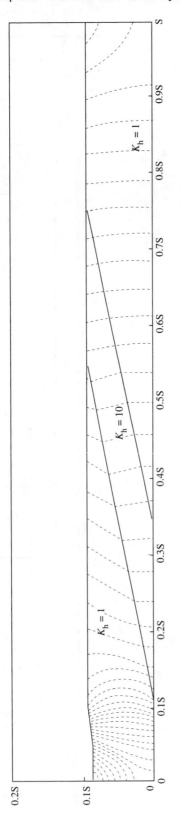

ing streams is generally very slightly below the outcropping of the water table, with a thin **seepage face** between the two surfaces (Figure 8-18a).

A **losing** (or **influent**) **stream** is one in which discharge decreases downstream; such a stream may occur in a recharge zone and be connected to (Figure 8-18b) or "perched" above (Figure 8-18c) the general ground-water flow. A **flow-through stream** is one that simultaneously receives and loses ground water (Figure 8-18d).

Figure 8-19 shows an idealized relation between regional water-table contours and a river. At any point the regional ground-water flow vector, $V$, is perpendicular to the contours, but may be resolved into a down-valley, or **underflow component, $V_u$,** and a riverward, or **base-flow component, $V_b$.** Larkin and Sharp (1992) found that rivers can be classified as base-flow–dominated ($V_b > V_u$), underflow-dominated ($V_u > V_b$), or mixed-flow ($V_b \approx V_u$), and that these types can be distinguished on the basis of river characteristics that can be readily determined from maps (Table 8-2). Streams of each type may be gaining or losing, depending on the relation between river level and water-table level. Figure 8-20 shows examples of underflow- and base-flow–dominated rivers.

Streams that flow all year are **perennial streams,** and those that flow only during wet seasons are **intermittent streams;** these are almost always gaining streams that are sustained by ground-water flow between water-input events. **Ephemeral streams** flow only in response to a water-input event and are usually losing.

### Hyporheic Flow and Bank Storage

At a more local scale, a river bed typically is at least locally permeable, and river water may exchange between the river and its bed and banks. The zone of down-river ground-water flow in the bed is called the **hyporheic zone,** and the importance of this zone to aquatic organisms, including fish spawning, is increasingly being recognized (e.g., Hakenkamp et al. 1993).

---

**FIGURE 8-15**
Equipotentials for topography of Figure 8-10a, but with a sloping bed with hydraulic conductivity 10 times greater than that of the overlying and underlying material. From Freeze and Witherspoon (1967), used with permission of the American Geophysical Union.

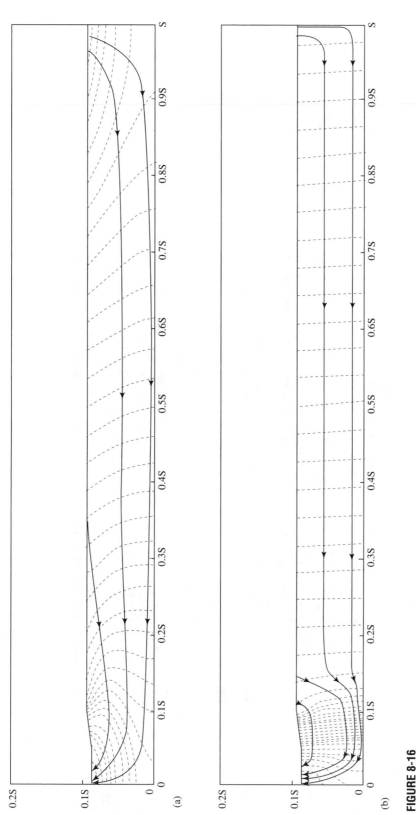

**FIGURE 8-16**
Flow nets for topography of Figure 8-10a, but with horizontal conductivity (a) 10 and (b) 0.1 times vertical. Dashed lines are equipotentials, arrows are streamlines. From Freeze and Witherspoon (1967), used with permission of the American Geophysical Union.

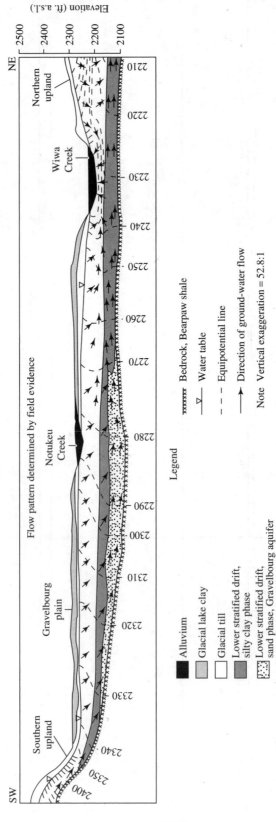

**FIGURE 8-17**

Geologic cross section and ground-water flow pattern in glacial deposits in the Wiwa Creek drainage basin, Saskatchewan, Canada, as determined from field measurements. From Freeze and Witherspoon (1968), used with permission of the American Geophysical Union.

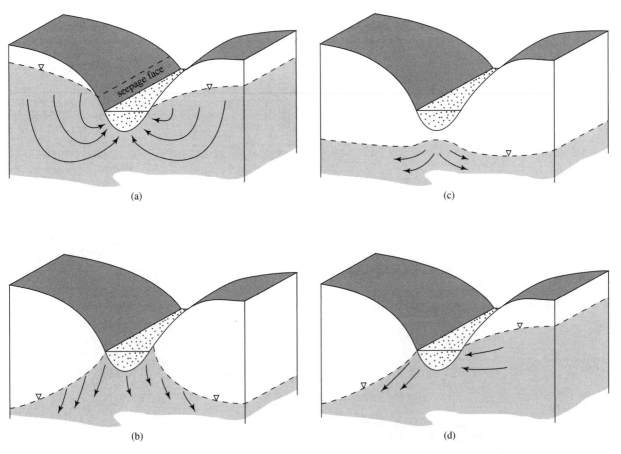

**FIGURE 8-18**
Stream-aquifer relations. (a) Gaining stream receiving water from local, intermediate, or regional ground-water flow; (b) losing stream connected to water table; (c) losing stream perched above water table; (d) flow-through stream.

The lateral exchange of water between the channel and banks is commonly significant during high flows, and is termed **bank storage.** When an event flow enters a gaining stream, a flood wave forms and travels downstream (Section 9.1.1). As the leading edge of the wave passes any cross section, the stream-water level rises above the water table in the bank, reversing the head gradient and inducing flow from the stream into the bank (Figure 8-21b). After the peak of the wave passes the section, the stream level declines and a streamward gradient is once again established (Figure 8-21c). Now the wedge of stream-water storage created by the rapid rise drains in both directions, but ultimately all returns to the stream.

By temporarily removing water from the channel, bank storage reduces the magnitude and delays the peak of the flood wave that would otherwise have occurred in response to the water-input event

(Figure 8-22). The importance of this natural flood-control process varies depending on the channel configuration and material, the extent of the aquifer, and the rate of rise, magnitude, and duration of the flood wave. Approaches to quantitative modeling of bank storage have been developed by Rorabaugh (1964), Moench et al. (1974), Hunt (1990), and Whiting and Pomeranets (1997).

### 8.3.2   Ground Water and Lakes and Wetlands

As with regional flow systems, much of our understanding of ground-water–lake interactions is based on mathematical simulations of idealized situations (McBride and Pfannkuch 1975; Winter 1976; 1978; 1983; Cheng and Anderson 1994), supplemented increasingly by field studies (e.g., Crowe and Schwartz 1985; Cherkauer and Zager 1989; Shaw

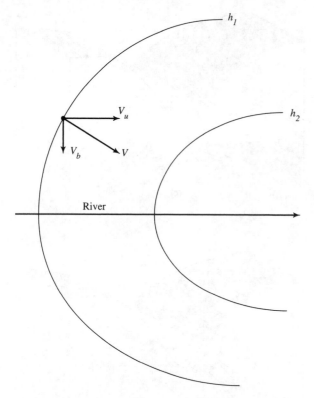

**FIGURE 8-19**

Idealized river–ground-water relations. $h_1$ and $h_2$ are elevations of water table; $h_1 > h_2$. $V_u$ and $V_b$ are underflow and base-flow components of regional ground-water flow, respectively. Modified after Larkin and Sharp (1992).

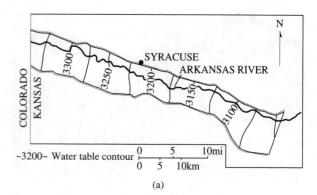

(a)

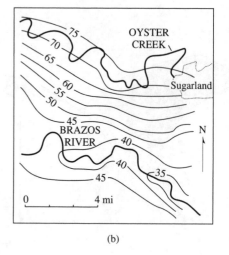

(b)

**FIGURE 8-20**

(a) An underflow-dominated river: the Upper Arkansas River and its aquifer, Kansas. (b) A base-flow–dominated river: The Brazos River and its aquifer, Texas. Contours are water-table elevations in ft above sea level. From Larkin and Sharp (1992), used with permission of the Geological Society of America, Boulder, CO; copyright © 1992 Geological Society of America.

and Prepas 1990). Both types of studies have shown that, like perennial streams, permanent lakes are almost always sites of ground-water discharge; smaller lakes for local flow systems and larger lakes for intermediate or regional systems (Figure 8-23a).

These studies have also shown that, where the surrounding geologic materials are homogeneous,

**TABLE 8-2**

Relations Between River–Ground-Water Interaction and River Type (Larkin and Sharp 1992)

| Dominant Ground-Water Flow Direction | Channel Slope | Sinuosity[1] | Width/Depth Ratio | Penetration[2] | Sediment Load |
|---|---|---|---|---|---|
| Underflow | high (> 0.0008) | low (< 1.3) | high (> 60) | low (< 20%) | mixed-bedload |
| Base flow | low (< 0.0008) | high (> 1.3) | low (< 60) | high (> 20%) | suspended load |
| Mixed | ≈ valley slope; lateral valley slope flat | | | | |

[1]Ratio of stream length to down-valley distance.

[2]Degree of incision into valley fill.

**FIGURE 8-21**
The bank-storage process. In (a) the stream is receiving base flow only. In (b) a flood peak is passing, and flow is induced into the banks. In (c) the peak has passed and the bank-storage wedge is draining.

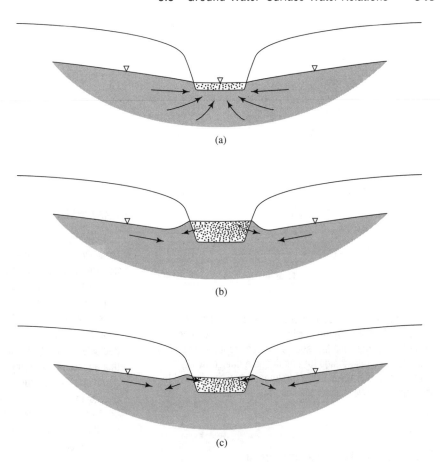

(a)

(b)

(c)

ground-water inflow to lakes is concentrated in the littoral zone, whether or not there are relatively impermeable sediments present in the deeper part of the lake. However, geologic heterogeneity can significantly alter the spatial distribution of inflows (Cherkauer and Nader 1989).

Under some water-table and geologic configurations, a lake can be the locus of both recharge and discharge. For example, Figure 8-23b shows a situation topographically identical to Figure 8-23a, but where lenses of high hydraulic conductivity at depth change the flow net so that the highest lake contributes recharge through its bottom while receiving discharge around its perimeter. **Flow-through lakes,** which receive discharge on one side and contribute recharge on the other, can exist where a lake is the outcropping of a continuously sloping water table (Figure 8-24).

In some situations, lakes can be recharge sites exclusively, at least at certain seasons. Meyboom (1966) showed that, in southern Saskatchewan, Canada, spring snowmelt runoff accumulates in small depressions (prairie potholes) that are above the recharge zone of a regional flow system. The subsequent leaking of these small ponds produces localized ground-water mounds representing recharge to this system. This local ground-water–flow pattern reverses in the summer, when trees growing around the edges of the potholes extract ground water and thereby make the potholes sites of discharge for temporary local flow systems. Studies by Winter (1983), Sacks et al. (1992), and Anderson and Cheng (1993) also showed significant seasonal changes in local ground-water flow systems adjacent to lakes.

Most freshwater wetlands are on the borders of streams or lakes, or are former lakes that have been largely or wholly filled with mineral and organic soil and vegetation in various proportions. Thus they are hydrologically similar to lakes, and the above generalizations can also be applied to bogs, swamps, and marshes. Dooge (1975) and LaBaugh (1986) provided useful reviews of wetland hydrology and Doss (1993), Hunt et al. (1996), and Rosenberry and Winter (1997) described field investigations.

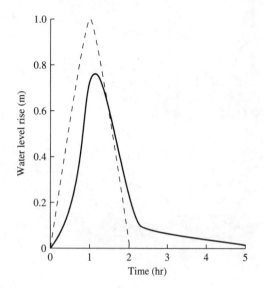

**FIGURE 8-22**
Flood-control effect of bank storage. The dashed line shows the magnitude of a hypothetical flood wave in the absence of bank storage; the solid line shows the peak reduction and delay due to bank storage for conditions modeled by Hunt (1990). Bank storage is filling (draining) when the dashed hydrograph is above (below) the solid-line hydrograph.

### 8.3.3   Ground Water and the Ocean

Figure 8-25 depicts the relations between fresh and salt ground water in a simple coastal aquifer. The Ghyben–Herzberg relation (Box 8-2) indicates that the interface lies at a depth below sea level equal to 40 times the height of the water table above sea level, and the water-table elevation decreases to zero at the coast. Equation (8B2-5) gives the depth of the interface as a function of distance from the coast.

In real situations the position of the interface differs somewhat from the Ghyben–Herzberg relation because the aquifer receives recharge from infiltration, and water-balance considerations dictate that the average net recharge rate must be balanced by an equal average discharge to the ocean. This discharge occurs through an **outflow face** that extends seaward from the coast, the width of which is given by Equation (8B2-7). The average discharge, $q(0)$, per unit length of coastline is given by

$$q(0) = R \cdot X, \qquad \textbf{(8-19)}$$

where $R$ is the net recharge rate $[\text{L T}^{-1}]$ and $X$ is the distance inland to the ground-water divide. Equation (8B2-8) gives the depth of the interface when the outflow face is accounted for.

As Example 8-1 shows, the width of the outflow face is usually small compared to the scale of the flow system, and the Ghyben–Herzberg relation gives a useful approximation except very near the coast in many situations.

---

### *EXAMPLE 8-1*

Long Island extends eastward from New York, NY, into the Atlantic Ocean. It is underlain by relatively homogeneous glacially deposited sands and gravels up to 120 m thick (the "Upper Glacial Aquifer") overlying coastal-plain sedimentary rocks. The distance from the central ground-water divide to the south coast is about 16 km. The average annual recharge is estimated at about 0.57 m yr$^{-1}$. (See Example 8-2.) The hydraulic conductivity of the Upper Glacial Aquifer is about 50 m day$^{-1}$. For this aquifer (a) Compute the distance $x^*$ at which the interface depth equals the aquifer thickness; (b) For $x \leq x^*$, compare the depths to the interface computed via Equations (8B2-5) and (8B2-8) at increasing distances from the coast; and (c) Calculate the width of the outflow face, $x_0$, extending seaward from the south coast.

**Solution**  (a) From Equation (8-19), compute

$$q(0) = 0.57 \text{ m yr}^{-1} \cdot 16{,}000 \text{ m} = 9120 \text{ m}^2 \text{ yr}^{-1}.$$

Convert

$$K_h = 50 \text{ m day}^{-1} \cdot 365 \text{ day yr}^{-1} = 18{,}250 \text{ m yr}^{-1}.$$

To compute the distance $x^*$ at which the interface depth equals the aquifer thickness $z_s(x^*)$, we rearrange Equation (8B2-5):

$$x^* = \frac{K_h \cdot z_s(x^*)^2}{2 \cdot \Gamma \cdot q(0)} = \frac{(18{,}250 \text{ m yr}^{-1}) \cdot (120 \text{ m})^2}{2 \cdot 40 \cdot (9120 \text{ m}^2 \text{ yr}^{-1})}$$
$$= 360 \text{ m}.$$

A similar rearrangement of Equation (8B2-8) yields

$$x^* = \frac{K_h \cdot \left( z_s(x^*) - \dfrac{\Gamma \cdot q(0)}{K_h} \right)^2}{2 \cdot \Gamma \cdot q(0)}$$

$$= \frac{(18{,}250 \text{ m yr}^{-1}) \cdot \left( 120 \text{ m} - \dfrac{40 \cdot 9120 \text{ m}^2 \text{ yr}^{-1}}{18{,}250 \text{ m yr}^{-1}} \right)^2}{2 \cdot 40 \cdot (9120 \text{ m}^2 \text{ yr}^{-1})}$$

$$= 250 \text{ m}.$$

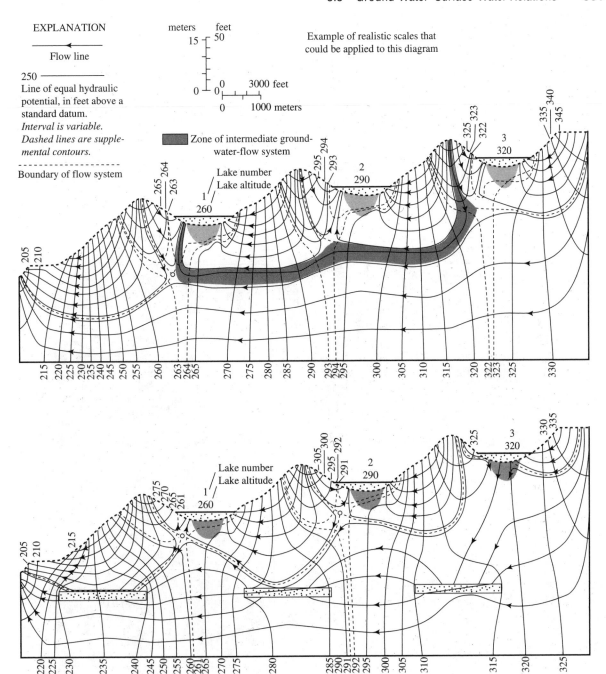

**FIGURE 8-23**

Flow nets for a hypothetical system of three lakes above a main stream. (a) With a homogeneous aquifer there are local, intermediate, and regional flow systems and the lakes are zones of discharge for local systems. (b) With three high-conductivity lenses at depth the intermediate system disappears and the highest lake receives discharge near its edge and contributes recharge in its center. Note vertical exaggeration. From Winter (1976).

**FIGURE 8-24**
The relation of Mirror Lake, NH, to the regional ground-water table shows that it is a flow-through lake. From Winter (1984).

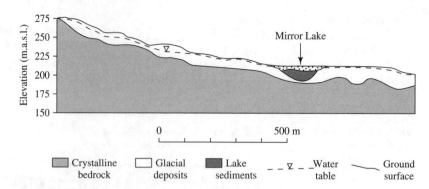

**FIGURE 8-25**
Definition sketch for deriving the Ghyben–Herzberg relation (Box 8-2). The salt-water–fresh-water interface position predicted by this hydrostatic relation differs somewhat from the actual position near the coast due to the dynamics of the flow, which produces an outflow face of width $x_o$. $x_o$ can be estimated via Equation (8B2-7).

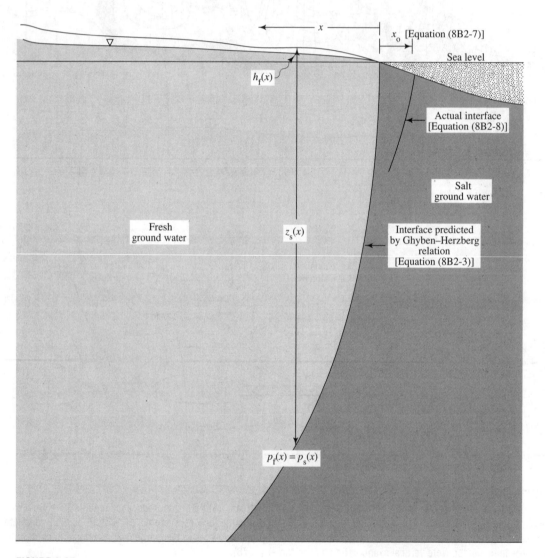

## BOX 8-2

$\cdot$ $\cdot$ $\cdot$ $\cdot$ $\cdot$ $\cdot$ $\cdot$

### The Fresh-Salt Ground-Water Interface in Coastal Aquifers

Figure 8-25 shows the interface between fresh and salt ground water in a homogeneous unconfined aquifer at a coastline. If we assume that this interface is a sharp boundary, the hydrostatic pressure on the freshwater side of the fresh-salt interface, $p_f(x)$, is

$$p_f(x) = \gamma_f \cdot [h_f(x) + z_s(x)], \qquad \textbf{(8B2-1)}$$

where $x$ is the distance inland from the coast, $\gamma_f$ is the weight density of freshwater, $h_f(x)$ is the elevation above sea level of the water table at $x$, and $z_s(x)$ is the elevation below sea level of the interface.

The hydrostatic pressure on the seawater side of the interface, $p_s(x)$, is

$$p_s(x) = \gamma_s \cdot z_s(x), \qquad \textbf{(8B2-2)}$$

where $\gamma_s$ is the weight density of seawater.

At hydrostatic equilibrium $p_f(x) = p_s(x)$, so equating Equations (8B2-1) and (8B2-2) and solving for $z_s(x)$ yields

$$z_s(x) = \Gamma \cdot h_f(x), \qquad \textbf{(8B2-3)}$$

where

$$\Gamma \equiv \frac{\gamma_f}{\gamma_s - \gamma_f} = \frac{9{,}800 \text{ N m}^{-3}}{10{,}045 \text{ N m}^{-3} - 9{,}800 \text{ N m}^{-3}}$$
$$= 40. \qquad \textbf{(8B2-4)}$$

This analysis of the fresh-salt ground-water interface was formulated over 100 years ago, and is known as the **Ghyben–Herzberg relation.** It states that at any distance inland, the fresh-salt interface is at a depth below sea level that is 40 times the height of the water table above sea level.

Using the Ghyben–Herzberg assumptions and the Dupuit Equation for flow in an unconfined aquifer (see

Box 8-4), Todd (1953) found the following relation for the depth to the interface $z_s(x)$:

$$z_s(x) = \left( \frac{2 \cdot \Gamma \cdot q(0) \cdot x}{K_h} \right)^{1/2}. \qquad \textbf{(8B2-5)}$$

Here, $q(0)$ is the ground-water discharge per unit length of coastline at the coast $[\text{L}^2 \text{ T}^{-1}]$. Equating (8B2-3) and (8B2-5) gives relations for the height of the water table as a function of distance from the coast:

$$h_f(x) = \left( \frac{2 \cdot q(0) \cdot x}{\Gamma \cdot K_h} \right)^{1/2} = \left( \frac{q(0) \cdot x}{20 \cdot K_h} \right)^{1/2}.$$
$$\textbf{(8B2-6)}$$

As noted in the text, the actual position of the interface must allow for an outflow face to discharge the fresh ground water. Glover (1954) developed a simple model that accounts for this; it gives the seaward extent of the outflow face, $x_0$, as

$$x_0 = \frac{\Gamma \cdot q(0)}{2 \cdot K_h} = 20 \cdot \frac{q(0)}{K_h} \qquad \textbf{(8B2-7)}$$

and an additional term in the relation for depth of the interface:

$$z_s(x) = \left( \frac{2 \cdot \Gamma \cdot q(0) \cdot x}{K_h} \right)^{1/2} + \frac{\Gamma \cdot q(0)}{K_h}. \qquad \textbf{(8B2-8)}$$

The relative importance of this additional term increases toward the coast, and at the coast,

$$z_s(0) = \frac{\Gamma \cdot q(0)}{K_h}. \qquad \textbf{(8B2-9)}$$

(b) Comparing Equations (8B2-5) and (8B2-8), we obtain the following table:

| Distance $x$ (m) | Using Equation (8B2-5) | | Using Equation (8B2-8) | |
| | Depth $z_s$ (m) | Height $h_f$ (m) | Depth $z_s$ (m) | Height $h_f$ (m) |
| --- | --- | --- | --- | --- |
| 0 | 0 | 0.0 | 20 | 0.5 |
| 10 | 20 | 0.5 | 40 | 1.0 |
| 20 | 28 | 0.7 | 48 | 1.2 |
| 50 | 45 | 1.1 | 65 | 1.6 |
| 100 | 63 | 1.6 | 83 | 2.1 |
| 200 | 89 | 2.2 | 109 | 2.7 |
| 250 | 100 | 2.5 | 120 | 3.0 |
| 300 | 110 | 2.7 | 120 | --- |
| 360 | 120 | 3.0 | 120 | --- |

(c) From Equation (8B2-7),

$$x_o = \frac{40 \cdot 9120 \text{ m}^2 \text{ day}^{-1}}{2 \cdot 18{,}250 \text{ m yr}^{-1}} = 10.0 \text{ m.}$$

## 8.4  GROUND WATER IN THE REGIONAL WATER BALANCE

### 8.4.1  Basic Water-Balance Relations

Figures 8-8–8-10 and 8-12–8-16 depict the flow of ground water in one-half of hypothetical symmetrical drainage basins. In all these situations, at least some of the flow is in a regional system that would typically contribute to the flow of the major stream draining the basin. In some of the cases, especially those with hilly uplands, some of the flow is in local or intermediate systems that would typically discharge to lower-order streams within the larger basin (Figure 8-10c). These diagrams could apply to basins of any scale; that is, the major stream could be a small brook draining a basin of a few square kilometers area or a large river draining many thousands of square kilometers.

Under natural (i.e., no-pumping) conditions[7], the water balance for a drainage basin defined by topographic divides and extending to a depth where ground-water flow is negligible can be written as

---

[7] The effect of pumping on the ground-water balance is considered in Sections 8.6.2 and 10.2.4.

$$P + G_{in} = Q + ET + G_{out}, \qquad \text{(8-20)}$$

where $P$ is precipitation, $Q$ is streamflow leaving the basin, $ET$ is evapotranspiration, $G_{in}$ is water entering the basin as ground water, $G_{out}$ is water leaving the basin as ground water, and all quantities are long-term average values (i.e., storage changes are assumed zero) (Figure 8-26).

Under the same conditions, the long-term average water balance for the aggregated ground-water reservoir in the basin is

$$R_I + R_{SW} + G_{in} = CR + Q_{GW} + G_{out}, \qquad \text{(8-21)}$$

where $R_I$ is recharge from infiltration (percolation), $R_{SW}$ is recharge from surface-water bodies, $CR$ is the movement of water from ground water into the capillary fringe (**capillary rise**)[8], and $Q_{GW}$ is the ground-water contribution to streamflow. We can then define **net recharge**, $R$, as

$$R \equiv R_I + R_{SW} - CR. \qquad \text{(8-22)}$$

(Note that the terms in these water-balance relations can be expressed as volumes [L$^3$] or volumes per unit area [L] during a specific time period, or as average rates [L$^3$ T$^{-1}$] or [L T$^{-1}$].)

Equations (8-20)–(8-22) can be combined to give

$$P - Q - ET = R_I + R_{SW} - CR - Q_{GW} = R - Q_{GW}, \qquad \text{(8-23a)}$$

or

$$P - Q_{SW} - ET = R_I + R_{SW} - CR = R, \qquad \text{(8-23b)}$$

where $Q_{SW}$ is the non-ground-water contribution to streamflow. The relative hydrologic importance of ground water in a drainage basin, $I_{GW}$, can be expressed as

$$I_{GW} = \frac{Q_{GW} + G_{out} - G_{in}}{Q + G_{out} - G_{in}} = \frac{R}{P - ET}. \qquad \text{(8-24)}$$

### 8.4.2  Ground-Water Residence Time

As discussed in Section 2.8.3, the residence time of water in a reservoir is the average length of time a

---

[8] This water movement occurs in response to capillary gradients created as transpiring plants or freezing reduce the liquid water content in the vadose zone. Capillary rise can be a significant portion of basin evapotranspiration in many semiarid and arid regions where a class of vascular plants called **phreatophytes** grows near streams and obtains water via roots that extend to the capillary fringe. Saltcedar, arrowweed, cottonwood, cattails, and willows are common phreatophytes.

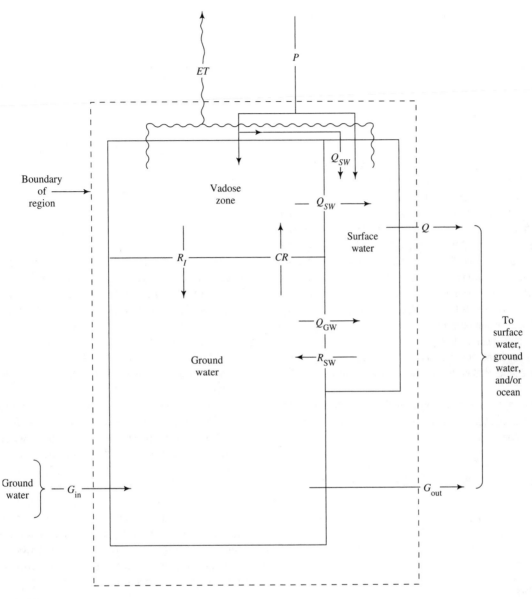

**FIGURE 8-26**
Schematic water balance for a drainage basin.

"parcel" of water spends in the reservoir. Aquifer residence time is an important parameter in characterizing the timing of stream responses to recharge and the fate of contaminants that are subject to chemical and biological processes as they move through aquifers to surface-water bodies (Haitjema 1995).

From Equation (2-27), the general definition of residence time, $T_R$, is

$$T_R \equiv \frac{\text{average storage in reservoir [L}^3\text{]}}{\text{average rate of input or output [L}^3\,\text{T}^{-1}\text{]}}.$$
(8-25)

For an unconfined aquifer of average saturated thickness $H$, this becomes

$$T_R \equiv \frac{S_y \cdot H}{R},$$
(8-26)

where $R$ is recharge rate per unit area $[\text{L}\,\text{T}^{-1}]$.

Example 8-2 summarizes one of many studies in which a form of the water-balance equation has been applied in an attempt to estimate basin recharge and other terms characterizing the role of ground water in basin hydrology.

---

### EXAMPLE 8-2

Cohen et al. (1968) developed a water balance for central Long Island, NY, (area = 760 mi$^2$) based on standard network meteorologic and hydrologic data collected over the period 1940–1965. Data are given in Table 8-3. $P$ estimates were based on data collected at five stations on the island and $Q$ estimates on records at five gaging stations. $ET$ was estimated based on average monthly temperatures via the Thornthwaite approach. (See Section 7.8.1.) Estimates of $CR$ (here assumed equal to direct evapotranspiration from ground water), $Q_{GW}$, $R_{SW}$, and $G_{in}$ were based on hydrologic judgment and the configuration of the water table. (a) Estimate $G_{out}$, $R_I$, $R$, and $I_{GW}$. (b) Assuming that most of the ground-water flow occurs in the surficial sand and gravel aquifer, which has an average thickness of 200 ft, what is the average residence time of water?

***Solution*** (a) Estimates are found by direct substitution of the appropriate values from Table 8-3 into the appropriate water-balance equations:

| Quantity | Equation | million gallons per day | in. yr$^{-1}$ |
|----------|----------|-------------------------|----------------|
| $G_{out}$ | (8-20) | 470 | 12.9 |
| $R_I$ | (8-21) | 805 | 22.1 |
| $R$ | (8-22) | 790 | 21.7 |

**TABLE 8-3**
Long-Term Average Water-Balance Components for Central Long Island, New York

| Quantity | mgd[a] | in. yr$^{-1}$ | Source[b] |
|----------|--------|----------------|-----------|
| $P$ | 1600 | 44.0 | m |
| $ET$ | 775 | 21.3 | em, eh |
| $Q$ | 355 | 9.8 | m |
| $CR$ | 15 | 0.4 | eh |
| $Q_{GW}$ | 320 | 8.8 | eq |
| $R_{SW}$ | 0 | 0 | eh |
| $G_{in}$ | 0 | 0 | eh |

[a]Millions of gallons per day

[b]m = measured; em = estimated from meteorological data; eh = estimated via hydrologic judgement based on water-table configuration; eq = estimated via base-flow analysis as discussed later in this chapter.

From Equation (8-24) $I_{GW}$ = 790/825 = 0.96. Note that $G_{out}$ in this case represents direct ground-water flow to the ocean [estimated by Cohen et al. (1968) as 340 mgd] plus outflow across the drainage-basin boundaries (130 mgd).

(b) From Table 8-1, the specific yield of unconsolidated sand and gravel $S_y \approx 0.30$. To use Equation (8-26) we first convert the units of $R$:

$$R = \frac{21.7 \text{ in. yr}^{-1}}{12 \text{ in. ft}^{-1}} \text{ in. ft}^{-1} = 1.81 \text{ ft yr}^{-1};$$

then

$$T_R = \frac{200 \text{ ft} \cdot 0.30}{1.81 \text{ ft yr}^{-1}} \approx 33 \text{ yr.}$$

---

Example 8-2 illustrates a common approach to water-balance studies, in which standard network hydroclimatological data and general knowledge of basin geology and water-table configuration are used to estimate various ground-water balance terms. In general, this approach has severe limitations: Even if we make the assumptions that long-term average values of $P$, $Q$, and $ET$ are well known, we can at best use Equation (8-20) to estimate $(G_{out} - G_{in})$, and Equations (8-21) and (8-22) to estimate $(R_I + R_{SW} - CR - Q_{GW}) = (R - Q_{GW})$.

Thus even under ideal conditions where standard network data provide good estimates of $P$, $ET$, and $Q$, firm knowledge of a basin's water balance usually requires evaluating at least some of the terms in Equation (8-21). Before proceeding to discussion of methods for obtaining quantitative estimates of these terms, we next describe an important conceptual framework for modeling flows in unconfined aquifers and understanding relations among major water-balance components.

### 8.4.3 The Dupuit Approximation for Modeling Flow in Unconfined Aquifers

In gaining streams, which are typical of humid regions, streamflow consists largely of drainage from unconfined aquifers [Figures 8-9 and 8-18a; $Q_{GW}$ in Figure 8-26 and Equations (8-23)–(8-25)]. Flow in unconfined aquifers is inherently difficult to characterize because the position of the upper flow boundary (the water table) is not fixed, but changes with time. However, useful approximate analytical solutions to many unconfined flow problems can be developed using assumptions formulated by the French engineer A. J. E. J. Dupuit in 1863. The basic

---

## BOX 8-3

. . . . . . .

## Dupuit Approximation for Unconfined Flow

The Dupuit formulation follows the reasoning in Box 8-1, but with the following simplifying assumptions (see Figure 8-27): (1) the control volume extends from a horizontal impermeable base in the $x$-$y$ plane up to the water table; (2) at any point in the $x$-$y$ plane the total head, $h$, is constant in the vertical ($z$-) direction so that $V_z = 0$; and (3) the head gradients are assumed equal to the slope of the water table. These assumptions do not introduce significant errors for water-table slopes $< 0.18$ (Smith and Wheatcraft 1992).

Under these conditions, the mass inflow, $M_{in}$, (through faces 1, 2, and 3) during time period $dt$ is

$$M_{in} = \rho \cdot V_x \cdot h \cdot dy \cdot dt + \rho \cdot V_y \cdot h \cdot dx \cdot dt$$
$$+ \rho \cdot R_l \cdot dx \cdot dy \cdot dt, \qquad \text{(8B3-1)}$$

where $R_l$ is the net rate of recharge [L T$^{-1}$] and $V_x$ and $V_y$ are the Darcy velocities [L T$^{-1}$] in the $x$- and $y$-directions, respectively. Since both $V$ and $h$ may change in the $x$- and $y$-directions, the outflow (through faces 4 and 5 only, since face 6 is impermeable) in $dt$, $M_{out}$, is

$$M_{out} = \rho \cdot \left[ V_x \cdot h + \frac{\partial(V_x \cdot h)}{\partial x} \cdot dx \right] \cdot dy \cdot dt$$
$$+ \rho \cdot \left[ V_y \cdot h + \frac{\partial(V_y \cdot h)}{\partial y} \cdot dy \right] \cdot dx \cdot dt. \qquad \text{(8B3-2)}$$

The change in storage during $dt$ is

$$\rho \cdot S_Y \cdot \frac{\partial h}{\partial t} \cdot dx \cdot dy \cdot dt, \qquad \text{(8B3-3)}$$

where $S_y$ is the aquifer specific yield. Thus

$$M_{in} - M_{out} = R_l - \frac{\partial(V_x \cdot h)}{\partial x} - \frac{\partial(V_y \cdot h)}{\partial y}$$
$$= S_y \cdot \frac{\partial h}{\partial t}. \qquad \text{(8B3-4)}$$

Replacing $V_x$ and $V_y$ with the flow rates given by Darcy's Law [Equation (8-1)] then yields

$$R_l - \frac{\partial}{\partial x}\left(-K_{hx} \cdot \frac{\partial h}{\partial x} \cdot h\right) - \frac{\partial}{\partial y}\left(-K_{hy} \cdot \frac{\partial h}{\partial y} \cdot h\right)$$
$$= S_y \cdot \frac{\partial h}{\partial t}. \qquad \text{(8B3-5)}$$

If the aquifer is homogeneous and isotropic ($K_{hx} = K_{hy} = K_h$) this is further simplified to

$$\frac{R_l}{K_h} + \frac{\partial}{\partial x}\left(h \cdot \frac{\partial h}{\partial x}\right) + \frac{\partial}{\partial y}\left(h \cdot \frac{\partial h}{\partial y}\right) = \frac{S_y}{K_h} \cdot \frac{\partial h}{\partial t}. \qquad \text{(8B3-6)}$$

We can make use of the mathematical identities

$$h \cdot \frac{\partial h}{\partial x} = \frac{1}{2} \cdot \frac{\partial(h^2)}{\partial x} \text{ and } h \cdot \frac{\partial h}{\partial y} = \frac{1}{2} \cdot \frac{\partial(h^2)}{\partial y} \qquad \text{(8B3-7)}$$

to rewrite Equation (8B3-6) as

$$\frac{2 \cdot R_l}{K_h} + \frac{\partial^2(h^2)}{\partial x^2} + \frac{\partial^2(h^2)}{\partial y^2} = \frac{2 \cdot S_y}{K_h} \cdot \frac{\partial h}{\partial t}, \qquad \text{(8B3-8)}$$

which is the Dupuit Equation for time-varying flow in a homogeneous isotropic aquifer. For steady flow in a homogeneous isotropic aquifer, Equation (8B3-8) becomes

$$\frac{2 \cdot R_l}{K_h} + \frac{\partial^2(h^2)}{\partial x^2} + \frac{\partial^2(h^2)}{\partial y^2} = 0. \qquad \text{(8B3-9)}$$

---

**Dupuit Equations** are derived in Box 8-3, and are applied to a simplified steady-state model of an aquifer draining to streams in Box 8-4.

Equation (8B4-9) gives significant insight into the water-balance relations between ground-water and surface-water flows: It shows that average ground-water discharge to streams depends only on the average rate of recharge from infiltration, which is essentially climatically determined, and the stream spacing, $X$, which is a function of drainage-basin size. The hydraulic conductivity determines the configuration of the water table required to

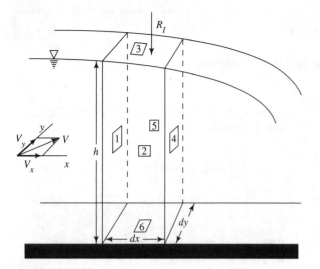

**FIGURE 8-27**
Definition diagram for derivation of the Dupuit Equation (Box 8-3).

transmit the recharge to streams [Equation (8B4-5)], but not the flow rate.

Note that $q_{GW}$ in Equation (8B4-9) is the ground-water contribution to streamflow per unit length of stream from one-half of the drainage basin. Thus for a drainage basin,

$$q_{GW} = \frac{Q_{GW}}{2 \cdot L}, \qquad \textbf{(8-27)}$$

where $L$ is the length of the main stream. An important implication of Equation (8B4-9) is that one can estimate recharge from infiltration as

$$R_I = \frac{Q_{GW}}{X \cdot L}. \qquad \textbf{(8-28)}$$

Note that Equation (8-28) is a restatement of Equation (8-23a) for the common situation in which $R_{SW}$ and $CR$ are small. Methods for evaluating $Q_{GW}$ are discussed in the following section.

## 8.5   EVALUATION OF GROUND-WATER–BALANCE COMPONENTS

This section reviews approaches to quantitative evaluation of the components of the ground-water balance [Equation (8-23), Figure 8-26] using field

observations. Most of these approaches are based on application of basic water-balance concepts and Darcy's Law, in some cases employing various water-quality constituents as tracers. A still-useful and historically interesting review of early (1686–1931) approaches was compiled by Meinzer (1932).

### 8.5.1   Recharge from Infiltration, $R_I$

Infiltrated water can carry contaminants from agriculture, industries, and waste-disposal sites to the ground-water reservoir, so quantification of $R_I$ has important implications for the study of water quality as well as quantity.

The water-table configurations in the regional ground-water flows depicted in Figures 8-8–8-10 and 8-12–8-16 were specified as boundary conditions to illustrate typical situations, and these conditions determined the locations of recharge and discharge zones. Since recharge from infiltration is the principal source of water to unconfined aquifers and is the ultimate source of most streamflow, it is of interest to explore further the natural factors that determine the regional distribution of recharge and discharge.

Direct measurement of recharge requires elaborate instrumentation and is feasible only in a research setting. In one of the few such studies, Wu et al. (1996) installed lysimeters consisting of 60-cm diameter soil-filled cylinders in which the water table was kept at depths ranging from 1.5 to 5 m. At the shallowest depth almost every water-input event caused a separate recharge event reflected in a rise, peak, and decline of the water table; but at 5 m individual events were not identifiable and there was a single annual peak. At intermediate depths, individual recharge events were associated with input events separated by a "critical" time interval that increased with depth. The relationships between recharge, precipitation ($P$), and evapotranspiration ($ET$) for individual events were

$$R_I = 0.87 \cdot (P - 5.25) \qquad \textbf{(8-29a)}$$

and

$$R_I = 0.87 \cdot (P - ET - 27.4) \qquad \textbf{(8-29b)}$$

for the 1.5-m depth and 4.5-m depth, respectively, and all quantities are in cm.

## BOX 8-4

· · · · · · ·

### Dupuit Approximation for Unconfined Aquifer Draining to Streams

Figure 8-28 is a simplified version of Figure 8-9, in which the land surface between streams is horizontal and the streams extend downward to the basal impermeable layer. We consider long-term average (i.e., steady-state) conditions, and to make the development more general we show an asymmetrical situation, with $h_0 \neq h_x$. Since flow is in the x-direction only, Equation (8B3-9) becomes

$$\frac{2 \cdot R_I}{K_h} + \frac{\partial^2 (h^2)}{\partial x^2} = 0. \qquad \text{(8B4-1)}$$

Separating variables and integrating twice yields

$$h^2 = -\frac{R_I}{K_h} \cdot x^2 + C_1 \cdot x + C_2. \qquad \text{(8B4-2)}$$

The constants of integration, $C_1$ and $C_2$, are evaluated by noting that

$$h = h_0 \text{ at } x = 0 \qquad \text{(8B4-3)}$$

and

$$h = h_x \text{ at } x = X, \qquad \text{(8B4-4)}$$

so that

$$h^2 = -\frac{R_I}{K_h} \cdot x^2 + \left( \frac{h_x^2 - h_0^2}{X} + \frac{R_I \cdot X}{K_h} \right) \cdot x + h_0^2. \qquad \text{(8B4-5)}$$

Equation (8B4-5) states that the water table in this situation is a curved surface whose shape is determined by the stream spacing ($X$), the hydraulic conductivity ($K_h$), the recharge rate ($R_I$), and the stream elevations $h_0$ and $h_x$. By manipulation of this relation, we can show that the maximum water-table elevation (i.e., the ground-water divide), $h_{max}$, occurs at $x = X_d$, where

$$X_d = \left[ \frac{X}{2} - \frac{K_h \cdot (h_0^2 - h_x^2)}{2 \cdot R_I \cdot X} \right] \qquad \text{(8B4-6)}$$

and has a value

$$h_{max} = \left[ h_0^2 - \frac{(h_0^2 - h_x^2) \cdot X_d}{X} - \frac{R_I \cdot (X - X_d) \cdot X_d}{K_h} \right]^{1/2}. \qquad \text{(8B4-7)}$$

Note that for $h_0 = h_x$ the configuration becomes symmetrical, analogous to Figure 8-9. Under these conditions, the discharge to the streams (per unit stream length), $q_{GW}$, is

$$q_{GW} = q_{GW0} = q_{GWx} = -K_h \cdot h_x \cdot \left. \frac{dh}{dx} \right|_x. \qquad \text{(8B4-8)}$$

Evaluating $dh/dx$ from Equation (8B4-5) and substituting into Equation (8B4-8) then leads to

$$q_{GW} = \frac{R_I \cdot X}{2}. \qquad \text{(8B4-9)}$$

In nature the depth of the water table at a given location is determined by feedbacks among precipitation, infiltration, runoff, and evapotranspiration at that location along with the regional flow, which integrates those quantities throughout the drainage basin under the influence of topography and geology. On average, the water table is at an equilibrium depth such that the net recharge (i.e., percolation minus capillary rise to supply evapotranspiration) from above is just balanced by the net ground-water flow away from (recharge areas) or toward (discharge areas) the water table. When the water table is above this depth, losses to evapotranspiration exceed recharge and a discharge zone exists; when below this depth, recharge exceeds evapotranspiration and a recharge zone exists.

Levine and Salvucci (1999) quantitatively modeled these relations and showed how water-table depth is related to net recharge for various soil types. They found that net recharge increases and evapotranspiration and surface runoff generally decrease as water-table depth increases (Figure 8-29).

Figure 8-30 shows regional ground-water flow and recharge and discharge zones modeled using

**FIGURE 8-28**
Definition sketch for Dupuit flow to streams (Box 8-4). Under the Dupuit assumptions (Box 8-3) equipotential lines are vertical and streamlines are horizontal, in contrast to Figure 8-9.

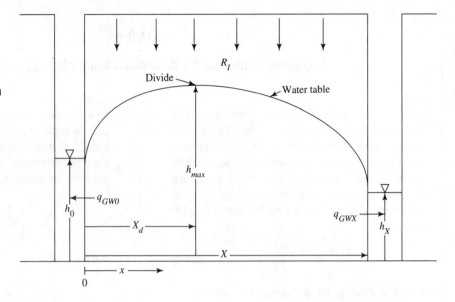

these more realistic relationships in climatic, topographic, and geologic conditions similar to those assumed in Figure 8-10b. Note that local, intermediate, and regional circulations occur in both, but that discharge zones are more concentrated in Figure 8-30. Salama et al. (1994) showed that aerial photographs and satellite imagery can be used to map regional recharge and discharge areas.

Because of the difficulty in direct measurement of recharge, hydrologists have attempted to evaluate $R_I$ by applying various combinations of water-balance concepts, applications of Darcy's Law, soil-physics principles, mathematical systems models, and water-quality measurements. The major methodological approaches are briefly described here, following in part the review of Van Tonder and

**FIGURE 8-29**
Variations of evapotranspiration, net recharge, and surface runoff as a function of water-table depth for a silt-loam soil. From Levine and Salvucci (1999), used with permission of the American Geophysical Union.

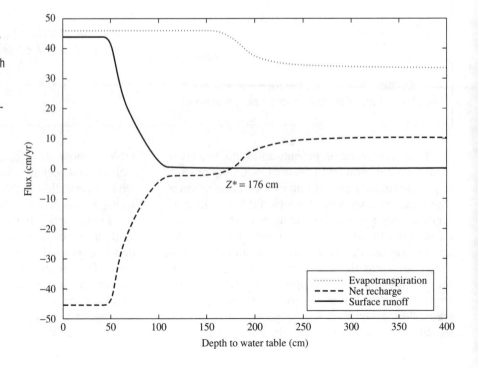

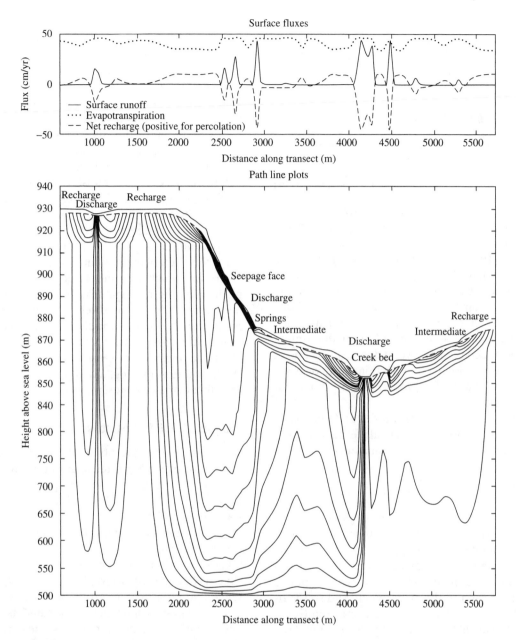

**FIGURE 8-30**

Recharge areas, discharge areas, and streamlines for a cross section of a drainage basin in Alberta, Canada, as determined by coupling recharge determined via unsaturated flow modeling with a ground-water flow model. Note that the discharge areas are more concentrated than predicted from ground-water flow modeling with imposed water-table configuration. (Compare Figure 8-10.) From Levine and Salvucci (1999) used with permission of the American Geophysical Union.

Kirchner (1990). Most of these methods require elaborate and careful data collection, and Sophocleous and Perry (1984) provided a useful overview of considerations in selecting instrumentation for recharge studies.

### Soil-Water-Balance Method

The water balance for the root zone of a soil for a time period $\Delta t$ can be written as

$$R_I = P - ET - Q_{SW} - R_{SW} + CR + \Delta S, \quad \text{(8-30)}$$

where $Q_{SW}$ represents overland flow, $\Delta S$ is the change in soil-water storage during $\Delta t$, and the other symbols are as defined for Equation (8-20) and (8-21). Equation (8-30) is usually applied to a small plot where $P$ and $Q_{SW}$ are directly measured, $ET$ is determined by measuring meteorological parameters and applying one of the approaches discussed in Section 7.8, $R_{SW}$ and $CR$ are assumed negligible, and $\Delta S$ is calculated from determinations of water content at several depths using one of the approaches discussed in Sections 6.2.1 or 6.3.3. Rushton and Ward (1979) found that $\Delta t$ should not exceed 1 day.

In one application, Steenhuis et al. (1985) used Equation (8-30) assuming that all terms except $P$ and $ET$ were negligible and estimated $R_I$ on Long Island, NY, as simply

$$R_I = P - ET, \quad \text{(8-31)}$$

where $P$ was measured and $ET$ was calculated from detailed energy-balance measurements at the site.

One limitation of this approach is that to characterize recharge for a drainage basin, the measurements must be made at a number of points sufficient to capture the climatologic and soil variability. Note too that, at a vegetated site, $P$ in Equation (8-30) should be replaced by throughfall. (See Section 7.6.) Another limitation arises from the uncertainties in estimating actual evapotranspiration, which is usually done using an empirical relation like Equation (7-67). Finch (1998) found that recharge estimates are highly sensitive to soil characteristics, and Rushton and Ward (1979) concluded that uncertainties of $\pm$ 15% should be expected with this approach.

However, the basic method—with varying approaches to modeling $ET$ and soil-water storage—has been successful in estimating $R_I$ in a small coastal-plain drainage basin in Maryland (Ras-

mussen and Andreasen 1959), chalk and sandstone aquifers of England (Wellings 1984; Ragab et al. 1997) and France (Thiery 1988); glacial deposits on Long Island, NY (Steenhuis et al. 1985; Steenhuis and Van Der Molen 1986) and Sweden (Johansson 1987); and in a large drainage basin in Australia (Chiew and McMahon 1990).

A variation of this approach makes use of the "zero-flux plane" (Figure 7-27), in which soil-moisture measurements are used to identify a level below which changes in moisture content are attributed to percolation and $R_I$ is estimated as the difference between soil-water profiles graphically integrated over the soil layer at successive times. Major limitations in applying this approach are that (1) a zero-flux plane may not always be present in layers for which $\theta$ is measured, (2) the zero-flux plane may migrate significantly between measurement times, and (3) rapid recharge through macropores could occur without being reflected as a change in measured values of $\theta$. Wellings (1984) applied the method successfully to estimate $R_I$ in chalk aquifers in England for periods when the zero-flux plane was present.

### Analysis of Well Hydrographs

Well hydrographs are observations of water levels in monitoring wells plotted against time; to evaluate recharge it is useful to superimpose on the plots hyetographs showing the timing and amount of water input (Figure 8-31). Recharge is reflected in the rise of the hydrograph following a water-input event, following which drainage to surface-water bodies (usually streams) is reflected in a gradual decline, called the **hydrograph recession.** There are several approaches to estimating recharge from well hydrographs; most of these assume that unconfined aquifers can be modeled as linear reservoirs (Box 8-5; Figure 8-32), which is often approximately true (Box 8-6; Brutsaert and Lopez 1998). Note also from Equations (8B6-6), (8B6-7), and (8B6-11) that ground-water discharge to streamflow decreases at a rate that is proportional to conductivity and inversely related to size and storage coefficient. [See Equation (8-13).]

If recharge from streams and capillary rise are negligible, the ground-water–balance equation, accounting for storage changes, is

$$R_I = Q_{GW} + \Delta S, \quad \text{(8-32)}$$

**FIGURE 8-31**
Typical plot for estimation of recharge via well-hydrograph analysis. (a) Rainfall hyetograph. (b) Well hydrograph plotted on arithmetic scale. (c) Well hydrograph plotted on logarithmic scale. Dashed lines are the extensions of hydrograph recessions (assumed exponential decay). $R_{I2}$ and $R_{I3}$ are recharge from storm events 2 and 3, respectively estimated as the vertical distance between the extended recession and the hydrograph peak times the specific yield. [See Equation (8-34).]

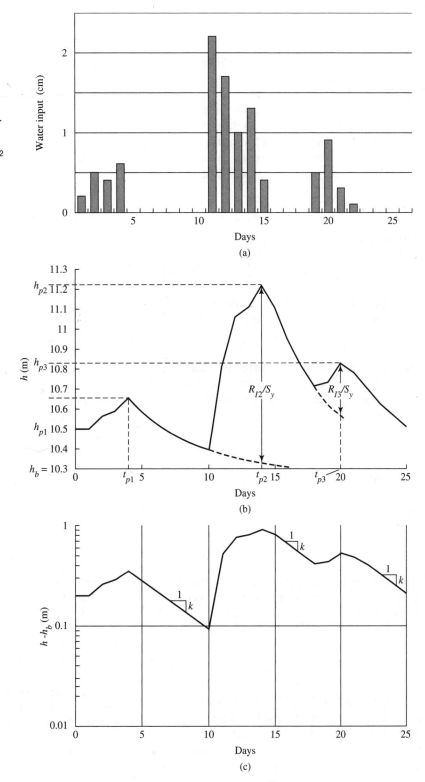

## BOX 8-5

. . . . . . . .

## Drainage of a Linear-Reservoir Aquifer

Figure 8-32 shows an idealized unconfined aquifer of area $A$ and specific yield $S_y$ receiving recharge from infiltration, $R_I$ [L T$^{-1}$] and discharging to a stream at a rate $Q_{GW}$ [L$^3$ T$^{-1}$]. Assuming constant density, the conservation of mass equation for a time period $dt$ for this situation is

$$A \cdot R_I \cdot dt - Q_{GW} \cdot dt = A \cdot S_y \cdot dh. \quad \text{(8B5-1)}$$

Defining $q_{GW} \equiv Q_{GW}/A$, this becomes

$$R_I - q_{GW} = S_y \cdot \frac{dh}{dt}. \quad \text{(8B5-2)}$$

If the aquifer behaves as a linear reservoir, outflow rate is proportional to storage [see Equation (2-26)]; that is,

$$q_{GW} = k \cdot S_y \cdot (h - h_b), \quad \text{(8B5-3)}$$

where $h_b$ is a level below which no discharge occurs and $k$ [T$^{-1}$] is the inverse of the residence time [Equation (8-26)] of the aquifer. With Equation (8B5-3), (8B5-2) becomes

$$R_I - k \cdot S_y \cdot (h - h_b) = S_y \cdot \frac{dh}{dt}. \quad \text{(8B5-4)}$$

If there is no recharge or capillary rise and the aquifer is draining, $R_I = 0$ and Equation (8B5-4) can be written as

$$-k \cdot dt = \frac{dh}{h - h_b}. \quad \text{(8B5-5)}$$

Integrating (8B5-5) yields

$$-k \cdot t = \ln(h - h_b) + C, \quad \text{(8B5-6)}$$

and evaluating the constant of integration, $C$, from the initial condition $h = h_0$ when $t = 0$ leads to

$$h = h_b + (h_0 - h_b) \cdot \exp(-k \cdot t). \quad \text{(8B5-7)}$$

Equation (8B5-7) shows that drainage of a linear aquifer follows an exponential decay asymptotic to $h_b$ with decay constant $k$. Substituting Equation (8B5-7) into Equation (8B5-3) yields

$$q_{GW} = k \cdot S_y \cdot h_0 \cdot \exp(-k \cdot t), \quad \text{(8B5-8)}$$

and we see that a linear aquifer produces ground-water outflow (base flow) that also follows an exponential decay with the same decay constant. The analysis in Box 8-6 provides theoretical relations between the decay constant $k$ and aquifer properties. However, for well-hydrograph analysis, $k$ is usually evaluated empirically as the slope of the hydrograph when plotted on a semi-logarithmic graph as in Figure 8-31.

where $\Delta S$ now denotes the change in storage in the aquifer. If the aquifer behaves as a linear reservoir, the analysis in Box 8-5 shows that Equation (8-32) becomes

$$R_I = k \cdot S_y \cdot (h - h_b) + S_y \cdot \frac{dh}{dt}, \quad \text{(8-33)}$$

where $R_I$ has units of [L T$^{-1}$], $h$ is the water level in the aquifer, and $h_b$ is the level at which $Q_{GW}$ becomes negligible. As shown in Figure 8-31, the constant $k$ is evaluated as the slope of the straight line that best defines the hydrograph recessions when plotted on a semi-logarithmic graph, and $h_b$ can be evaluated as the level to which the recessions become asymptotic during extended periods of no

recharge. The seasonal variation of $R_I$ can then be estimated by observing water-table elevations in monitoring wells and using Equation (8-33). Figure 8-34 is an example of this approach as applied to a glacial aquifer in Sweden.

Another approach to well-hydrograph analysis applies the linear-reservoir model in a somewhat different way. Again assuming an exponential recession and referring to Figure 8-31, recharge for event $i$, $R_{I,i}$ [L], becomes

$$R_{I,i} = \{h_{p,i} - h_{p,i-1} \cdot \exp[-k \cdot (t_{p,i} - t_{p,i-1})]\} \cdot S_y, \quad \text{(8-34)}$$

where $h_{p,i}$ and $h_{p,i-1}$ are the peak water levels associated with events $i$ and $i-1$, respectively; $t_{p,i}$ and $t_{p,i-1}$

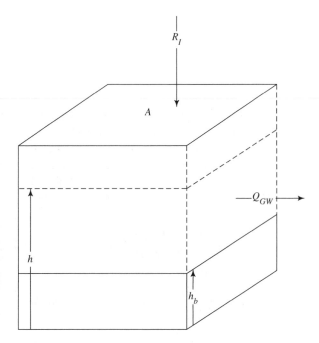

**FIGURE 8-32**
Definition diagram for analysis of drainage of a horizontal aquifer (Box 8-5).

are the times of occurrence of the successive peak water levels; and $S_y$ is the specific yield. Rasmussen and Andreasen (1959) obtained reasonable results with this method in Maryland, with a constant $S_y$ determined via successive approximations as the value most consistent with weekly water-balance data.

A third approach to estimating recharge from well hydrographs applies a simplified water-balance equation to the entire aquifer of area $A$:

$$R_I = S_y \cdot \Delta h_+ \cdot A - (G_{in} - G_{out}), \quad \textbf{(8-35)}$$

where $\Delta h_+$ is the spatial average increase in aquifer water level in response to a water-input event and $G_{in}$ and $G_{out}$ are the ground-water inflows and outflows, respectively, to the aquifer. Van Tonder and Kirchner (1990) suggested estimating the $G$ terms by approximating Darcy's Law as

$$G = T \cdot L \cdot \frac{i_1 + i_2}{2}, \quad \textbf{(8-36)}$$

where $G$ is the appropriate flow rate, $T$ is the transmissivity [Equation (8-12)] for the inflow or outflow boundary, $L$ is the width of the boundary, and $i_1$ and $i_2$ are the hydraulic gradients at the boundary at the beginning and end of the observation period, respectively. Van Tonder and Kirchner (1990) found

that this method was the only one to give reliable estimates of recharge in clastic sedimentary rock aquifers in South Africa. Das Gupta and Pudyal (1988) showed how an approximate analytical solution of the one-dimensional ground-water flow equation could be used to account for the $G$ terms when Equation (8-35) is applied to an aquifer.

Estimates based on analyses of well hydrographs are subject to two limitations: (1) the difficulty of determining the appropriate value of specific yield; and (2) the uncertainty that a given increase in water level actually represents an increment of recharge. A characteristic value of $S_y$ for the aquifer material is often assumed, but this can range widely (Table 8-1). Furthermore, the specific yield cannot in general be assumed constant in the near-surface zones of aquifers. This is especially true if the water table is within a few meters of the surface, where water contents are a function of depth due to the complex interplay of infiltration, percolation, and capillary rise. In general, the equilibrium water-content profile varies with the water-table elevation [Equation (6-26b)], so a given water-table rise represents different amounts of recharge depending on the elevation range covered. Another cause of inconstancy in $S_y$ is related to the wetting/drying hysteresis in soils (Figure 6-9): In a wetting soil, air bubbles are typically trapped in the pores, so that $S_y$ for a rising water table is generally less than $S_y$ for a falling water table. Sophocleous (1985) showed that recharge estimates based on the assumption of constant $S_y$ can be seriously in error.

Increases in water-table elevation unrelated to recharge may occur due to (1) fluctuations in atmospheric pressure caused by the expansion and contraction of air trapped beneath the water table; (2) thermal effects, including freezing and thawing; and (3) pressurization of the capillary fringe. This latter phenomenon is especially likely to occur where the capillary fringe extends up to the soil surface; infiltrating water can then cause an almost instantaneous rise of the water table to the surface with virtually no change in ground-water storage (Novakowski and Gillham 1988; this is discussed further in Section 9.2.3).

### Direct Application of Darcy's Law

This approach requires careful determination of the relation between soil-water content, $\theta$, and hydraulic conductivity, $K_h(\theta)$, for the soil of interest

## BOX 8-6
• • • • • • • •
### Drainage of an Unconfined Sloping Aquifer

Box 8-4 describes the Dupuit approximation for a horizontal aquifer draining to streams. Brutsaert (1994) showed how the Dupuit approach can be generalized to apply to aquifers with a sloping impermeable base, such as might be found where a relatively thin permeable soil overlies a bedrock base in hilly or mountainous terrain. We consider a one-dimensional case of "pure drainage"—that is, there is no recharge or evapotranspiration—and the base has a constant slope $\beta$ (Figure 8-33).

Using an approach similar to Dupuit's, Boussinesq (1877) derived the following version of Darcy's Law for this situation:

$$q_{GW} = -K_h \cdot h \cdot \frac{\partial h}{\partial x} \cdot \cos \beta - K_h \cdot h \cdot \sin\beta. \quad \textbf{(8B6-1)}$$

Boussinesq then combined this equation with the continuity relation to give

$$\frac{\partial}{\partial x}\left(h \cdot \frac{\partial h}{\partial x}\right) \cdot \cos \beta + \frac{\partial h}{\partial x} \cdot \sin \beta = \frac{S_y}{K_h} \cdot \frac{\partial h}{\partial t}, \quad \textbf{(8B6-2)}$$

where the symbols are as in Box 8-4 and Figure 8-33. Equation (8B6-2) is known as the "Boussinesq Equation." Note that for a horizontal slope ($\beta = 0$) and no recharge, this is identical to Equation (8B3-6).

To find an analytical solution to Equation (8B6-2), Brutsaert (1994) linearized Equation (8B6-1) to

$$q_{GW} = -\alpha \cdot K_h \cdot h_0 \cdot \frac{\partial h}{\partial x} \cdot \cos \beta - K_h \cdot h \cdot \sin \beta, \quad \textbf{(8B6-3)}$$

where $\alpha$ is a coefficient $\approx 0.5$ that adjusts for the use of the constant $h_0$ instead of the variable $h$. With this simplification, Equation (8B6-2) becomes

$$\overset{\text{diffusion term}}{\left(\frac{\alpha \cdot K_h \cdot h_0 \cdot \cos\beta}{S_y}\right) \cdot \frac{\partial^2(h_0 - h)}{\partial x^2}}$$
diffusion coefficient

$$+ \overset{\text{advection term}}{\left(\frac{K_h \cdot \sin\beta}{S_y}\right) \cdot \frac{\partial(h_0 - h)}{\partial x}} = \overset{\text{storage term}}{\frac{\partial h}{\partial t}}$$
advection coefficient

$$\textbf{(8B6-4)}$$

Equation (8B6-4) has the form of an "advection-diffusion equation." The diffusion term represents the effect of the slope of the water table relative to the aquifer base, and the advection term represents the effect of the sloping base. A dimensionless number, $J$, that gives a measure of the relative importance of the two terms can be formed as

$$J \equiv \frac{\text{advection coefficient} \cdot \text{aquifer length}}{\text{diffusion coefficient}}$$

$$= \frac{\dfrac{K_h \cdot \sin\beta}{S_Y} \cdot X}{\dfrac{\alpha \cdot K_h \cdot h_0 \cdot \cos \beta}{S_Y}} = \frac{\tan \beta \cdot X}{\alpha \cdot h_0}. \quad \textbf{(8B6-5)}$$

When $J$ is large (steep slope and relatively thin aquifer), the advection term dominates; when small (flat slope and relatively thick aquifer), the diffusion term dominates.

Brutsaert (1994) gives the general solution to Equation (8B6-4), which involves the sum of a series of exponential-decay terms. For a horizontal base ($\beta = 0$; the Dupuit conditions), the solution simplifies to

$$q_{GW} = \frac{h_0^2 \cdot K_h}{X}$$

$$\cdot \sum_{i=1}^{\infty} \exp\left[-\frac{(2 \cdot i - 1)^2 \cdot \pi^2 \cdot K_h \cdot h_0}{8 \cdot X^2 \cdot S_Y} \cdot t\right]. \quad \textbf{(8B6-6)}$$

The importance of the successive terms diminishes rapidly, and if only the first is retained, we have

$$q_{GW} = \frac{h_0^2 \cdot K_h}{X} \cdot \exp\left[\frac{1.23 \cdot K_h \cdot h_0}{X^2 \cdot S_Y} \cdot t\right] \quad \textbf{(8B6-7)}$$

as the equation for the drainage of a Dupuit aquifer.

For a very steep, thin aquifer (large $J$) we can simplify the problem even further by assuming that the water table and flow lines remain parallel to the slope during drainage. Then the volume of gravity-drainable water stored in the aquifer at any time, $B$, is

$$B = h \cdot S_y \cdot X. \quad \textbf{(8B6-8)}$$

Because the streamlines are assumed parallel to the slope, the discharge from the aquifer is given by Darcy's Law as

---

**BOX 8-6 (*continued*)**

· · · · · · ·

**Drainage of an Unconfined Sloping Aquifer**

$$q_{GW} = h \cdot K_h \cdot \tan \beta. \qquad \textbf{(8B6-9)}$$

Since $q_{GW}$ represents the only input or output from the aquifer,

$$q_{GW} = -\frac{dB}{dt}. \qquad \textbf{(8B6-10)}$$

Substituting Equations (8B6-7)–(8B6-9) into (8B6-10), separating variables, and integrating gives

$$q_{GW} = K_h \cdot h_0 \cdot \tan \beta \cdot \exp\left( -\frac{K_h \cdot \tan \beta}{S_Y \cdot X} \cdot t \right). \qquad \textbf{(8B6-11)}$$

Equations (8B6-6), (8B6-7), and (8B6-11) suggest that the drainage of unconfined aquifers approximates an exponential decay under a wide range of conditions. This implies that such aquifers function approximately as linear reservoirs (Box 8-5).

---

[Equation (6-13)] and periodic measurement of the vertical water-content gradient in the unsaturated zone of that soil. Recharge is then computed as the flux across the base of the root zone as given by Darcy's Law for vertical unsaturated flow [Equation (6B1-6)].

Sophocleous and Perry (1985) and Stephens and Knowlton (1986) obtained reasonable recharge estimates using this approach in humid and semi-arid environments, respectively. Steenhuis et al. (1985) also successfully applied this method in a humid region, using bi-weekly observations of soil-water content at depths of 90 and 120 cm in a sandy loam soil. They found that the hydraulic gradient across the soil layer varied linearly with time following a rainstorm, and the hydraulic conductivity varied linearly with the square root of time. Thus the average rate of recharge between two observation times $t_1$ and $t_2$ was calculated from a modified version of Darcy's Law for unsaturated flow [Equation (6-10)]:

$$R_I = -[K_h(\theta_{t1}) \cdot K_h(\theta_{t2})]^{1/2}$$
$$\cdot \left(\frac{1}{2}\right) \cdot \left[ \left(\frac{dh}{dz}\right)_{t1} + \left(\frac{dh}{dz}\right)_{t2} \right] \cdot (t_2 - t_1).$$

$$\textbf{(8-37)}$$

Here, $h$ is the hydraulic head $(= [\psi(\theta) + z])$ and the subscripts indicate the time of measurement.

A variation on this method involves ignoring the capillary-force gradient in the one-dimensional form of Darcy's Law and estimating recharge as a function of water content:

$$R_I = K_h(\theta). \qquad \textbf{(8-38)}$$

Stephens and Knowlton (1986) found good agreement between results obtained with Equation (8-38) and those using the more complete version of Darcy's Law.

Principal limitations of this approach are the spatial variability of soils in a typical drainage basin, the difficulty in establishing precise $K_h(\theta)$–$\theta$ relations for a given soil, and the possibility that recharge can occur via macropores or fissures in

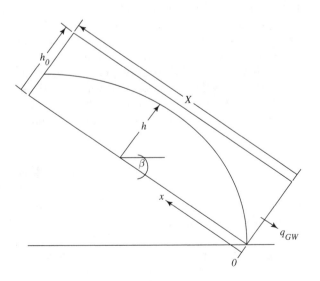

**FIGURE 8-33**
Definition diagram for analysis of drainage of a sloping aquifer (Box 8-6).

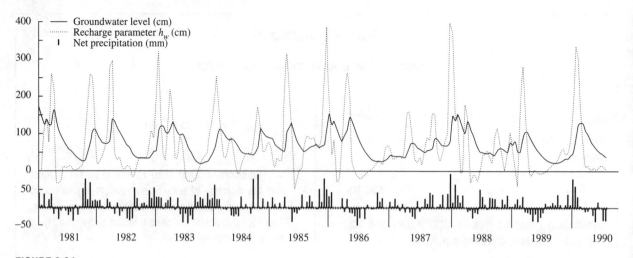

**FIGURE 8-34**
Hyetograph (histogram), well hydrograph (solid line), and recharge as estimated via Equation (8-33) (dashed line) in Sweden over a 10-yr period. From Olin and Svensson (1992).

which flow is not well modeled by Darcy's Law (Van Tonder and Kirchner 1990).

### Inverse Application of Ground-Water Flow Equation

Equation (8-15) can be incorporated into a model that accounts for an assumed regional distribution of recharge. The inverse application of such a model involves finding the values of recharge by calibration: The values of recharge in the model are changed until the computed head distribution corresponds to the values observed in piezometers in the region.

The inherent problem of the inverse approach is the non-uniqueness of solutions: There may be many distributions of recharge that give head distributions matching the field observations within the uncertainty of those observations. The problem is exacerbated by the difficulties in obtaining a precise picture of the distribution of hydraulic conductivities.

However, a number of studies have shown that useful estimates of recharge can be obtained via the inverse method, including those of Smith and Wikramaratna (1981), Wikramaratna and Reeve (1984), Chiew et al. (1992), Boonstra and Bhutta (1996), and Levine and Salvucci (1999).

### Impulse-Response Analysis

The objective of impulse-response analysis is to determine recharge from the relation between water inputs (the impulse) and water-table rises (the response) using the principles of the mathematical analysis of time series. Given a series of daily observations of water level in an observation well and water input, the level on a given day, $h_i$, is assumed to depend on the previous day's level and the previous days' water inputs as

$$h_i = a_0 + a_1 \cdot h_{i-1} + b_0 \cdot W_i + b_1 \cdot W_{i-1} + \ldots + b_n \cdot W_{i-n}, \quad (8\text{-}39)$$

where the $a$ and $b$ values are constants, the $W$ values are water inputs, the subscripts are day counters, and $n$ is the maximum number of days required for water to percolate to the water table.

The values of the constants in Equation (8-39) are found by mathematical techniques that minimize the differences between estimated and observed $h_i$ values over some period of observation of the particular wells of interest (Viswanathan 1984).

Viswanathan (1984) also showed that the average fraction of water input that becomes recharge is given by

$$\frac{R_I}{W} = S_y \cdot \sum_{j=0}^{n} b_j, \quad (8\text{-}40)$$

so that the method can be used to estimate long-term average recharge from long-term average water input (= average precipitation). Note, though, that this relation is also based on the uncertain assumption of a constant specific yield.

Viswanathan (1984) showed that the best results from time-series analysis were found when the "constants" in Equation (8-39) were functions of the time of year, and showed how to derive these functions from observations.

### Methods Based on Water Quality

Under some circumstances, recharge rates can be estimated from the concentrations of certain chemicals and stable and radioactive isotopes that function as natural tracers of water movement.

**Chemical Tracers**   A chemical tracer suitable for estimating recharge must be (1) present in measurable amounts in precipitation or, if deposited in solid form from the atmosphere, highly soluble; (2) not taken up or released in the vadose zone; and (3) not taken up or released by vegetation. For a column of the vadose zone in which no horizontal flow occurs above a water table, the balance for such a tracer is

$$C_w \cdot W = C_{GW} \cdot R_I, \qquad (8\text{-}41)$$

where $W$ is water input (or throughfall if measurements are made beneath a vegetative canopy), $C_w$ is the concentration of the tracer in the water input as it infiltrates, $C_{GW}$ is the concentration in ground water at the water table, and all terms represent long-term averages. Recharge can thus be directly calculated if all other terms are determined from observations.

Application of this approach is limited by the difficulty of finding tracers that satisfy all the necessary assumptions for application of Equation (8-41). One candidate is chloride ion ($Cl^{-1}$), which is commonly present in precipitation, is not used by plants, and is not commonly involved in soil-chemical reactions. Recharge has been estimated from $Cl^{-1}$ concentrations in Colorado (Claassen et al. 1986), Australia (Thorburn et al. 1991; Walker et al. 1991), and Niger (Bromley et al. 1997). Detailed aspects of using $Cl^{-1}$ to estimate recharge were discussed by Allison et al. (1984) and Taniguchi and Sharma (1990).

Chlorofluorocarbons (CFCs) are chemically stable man-made compounds that have been accumulating in the atmosphere since their introduction as propellants and refrigerants in the 1930s. Detectable concentrations of CFCs are present in ground water that fell as precipitation since 1945, and their concentration can be used to compute recharge rates. Busenberg and Plummer (1992) and Dunkle et al. (1993) discuss the methodology in detail.

**Stable Isotopes**   A small fraction of water molecules contain the heavy oxygen isotope, $^{18}O$, or the heavy hydrogen isotope $^2H$ (deuterium). (See Section B.1.5.) The concentrations of these isotopes in precipitation tend to vary seasonally at a given location, but they are not affected by soil-chemical reactions or by plant uptake. Thus comparisons of concentrations in ground water with those in precipitation at various times of the year can provide qualitative information on the seasonality of recharge.

It is usually difficult to use stable isotopes for quantitative estimates of $R_I$ without additional information or assumptions about the percolation mechanism—i.e., whether a piston-like wetting front occurs, whether macropores are important, or whether isotopic equilibration takes place between percolating and immobile water. Darling and Bath (1988) cited several studies in which $^{18}O$ and $^2H$ were used to obtain information about the seasonality and rate of recharge.

**Radioactive Isotopes**   Atmospheric testing of nuclear weapons in the 1950s and 1960s increased concentrations of tritium ($^3H$) in precipitation by orders of magnitude above its natural levels (Figure B-7). In subsequent decades, $^3H$ proved very useful for recharge studies, as it is part of the water molecule and its concentration is not affected by chemical reactions.

In one application, Larson et al. (1987) used $^3H$ concentrations to estimate recharge rate in the sandy glacial aquifer of Cape Cod, Massachusetts. This aquifer is relatively homogeneous, and since the water samples were taken near a ground-water divide where the flow is vertically downward, they assumed that recharge occurred essentially as a piston-like movement. For this situation, the actual velocity of the infiltrating water, $U_I$, could be calculated as

$$U_I = \frac{Z_B}{t_B}, \qquad (8\text{-}42)$$

where $Z_B$ is the maximum depth at which water with high (bomb-produced) $^3H$ was found and $t_B$ is the time since bomb-$^3H$ was first produced. Then the recharge rate can be estimated as the vertical flux,

$$R_I = \phi \cdot U_I, \qquad (8\text{-}43)$$

where $\phi$ is porosity. [See Equation (8-3).]

Larson et al.'s (1987) results ($R_I$ = 34 to 41 cm yr$^{-1}$) agreed well with those based on water-balance computations. Their report cited several other studies in which $^3$H was used in recharge estimation. However, because of the short half-life of $^3$H (Section B.1.5), its levels in ground water have now decreased to the point where this approach is no longer feasible.

### 8.5.2 Recharge from Surface Water, $R_{SW}$

Although there are few published studies in which $R_{SW}$ has been estimated, we can outline the principal approaches that can be applied to evaluation of this quantity.

#### Ground-Water Balance Computation

In arid regions, average evapotranspiration is nearly equal to average precipitation [Equation (7-66)], so most infiltrating water evaporates and most recharge is from losing streams ($R_{SW}$ ; Figure 8-18b,c) rather than percolation ($R_I$). For example, Osterkamp et al. (1994) applied water-balance computations along with streamflow measurements (see later section) and ground-water modeling techniques in a 20,000-km$^2$ river basin in Nevada

and California to estimate that only 1.6% of basin precipitation became recharge and that 90% of that was from streams.

#### Direct Measurement of Ground-Water Potentials or Flux

The gradient of ground-water flow out of (or into) a stream can be measured by comparing (1) the water level of the stream with the head measured in a piezometer inserted into the subjacent bed, or (2) the levels in two piezometers inserted to different distances below the bed (Figure 8-35). If the gradient is directed downward, the ground-water flux out of the stream can be calculated directly via Darcy's Law using measured or estimated hydraulic conductivities. However, there is likely to be considerable spatial and temporal variability in local values (and even direction) of gradients as well as conductivities, so such measured values must be extrapolated with caution. Winter et al. (1988) gave details on the construction of piezometers used for this purpose.

Workman and Serrano (1999) combined measurements of near-stream water levels with a simple ground-water flow model to estimate that 65% of the recharge to alluvial aquifers in Ohio came

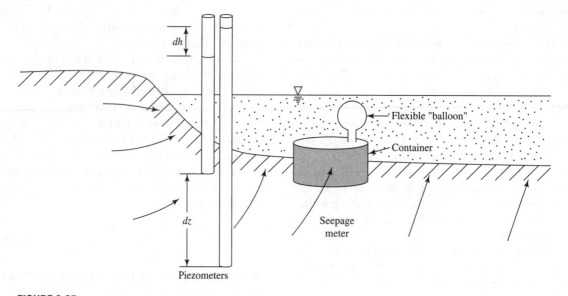

**FIGURE 8-35**
Sketch showing installation of piezometers and seepage meter to measure ground-water flux to a stream. Arrows indicate flow direction. *dh* is the head difference and *dz* is the elevation difference between the two piezometers; vertical flow is calculated via Darcy's Law. Water flowing into the container of the seepage meter is collected in the flexible "balloon."

from overbank flow due to floods, and hence was highly sporadic.

### Direct Measurement of Streamflow Increments

The difference between stream discharge at the upstream and downstream limits of a stream reach that receives no overland or tributary flow can be attributed to ground-water inflow or outflow. Techniques for measuring streamflow are discussed in Appendix F; typically "velocity-area" methods, in which flow depth and average flow velocity are measured at a number of points across the stream width, are most applicable since they require no fixed installations.

The accuracy of an estimate of $R_{SW}$ or $Q_{GW}$ using this approach depends on the precision of the streamflow measurements (see Section F.2.4) and is increased by maximizing the difference between the upstream and downstream discharges (i.e., by making the measurement reach as long as possible). Here again one must be aware that rates and directions of ground-water/surface-water exchange may be highly variable in space and time.

### Ground-Water Temperature Measurements

Stream temperatures are temporally highly variable due to daily and seasonal cycles. Where streams are losing water to subjacent ground water, this variability becomes progressively damped with distance below the stream bed. The outflow flux can be evaluated by measuring temperature fluctuations a short distance below the channel bottom, measuring the thermal properties of the bed sediment, and applying basic heat-conductivity relations (Silliman et al. 1995).

### 8.5.3 Ground-Water Contributions to Streamflow, $Q_{GW}$

Qualitative relations between ground water and surface water were discussed in Section 8.3. Here we outline approaches to obtaining quantitative estimates of the contribution of ground water to streamflow, $Q_{GW}$. Additional aspects of this topic are discussed in Section 9.1.2.

### Flow Equation for Horizontal Unconfined Flow

As described in Boxes 8-3 and 8-4, flow in horizontal unconfined aquifers draining to streams can be usefully described by the Dupuit approximations. If significant water is not being pumped from such aquifers (see Section 8.6), and if capillary rise is negligible, these equations state that $Q_{GW} = R_I$ on a per-unit-area basis [Equation (8-28)]. Thus, under these conditions, the techniques described above for estimating $R_I$ also provide information about $Q_{GW}$.

### Use of Flow Nets

If enough information on the spatial distribution of hydraulic conductivities and average water-table elevations is available, one can develop a numerical or graphical solution to Equation (8-18) and sketch one or more flow nets that represent the major ground-water flow features of the basin along various stream reaches. Referring to Figure 8-36, the discharge per unit length of stream $\Delta q_{GW}$ into a stream for each streamtube can then be calculated as

$$\Delta q_{GW} = K_h \cdot \Delta h, \qquad (8\text{-}44)$$

where $K_h$ is the appropriate conductivity and $\Delta h$ is the head difference between the stream and the next up-gradient equipotential line; the total discharge per unit stream length for a reach, $q_{GW}$, is given by

$$q_{GW} = n \cdot \Delta q_{GW}, \qquad (8\text{-}45)$$

where $n$ is the number of streamtubes discharging to the stream. The total flow into the stream for the reach, $Q_{GW}$, is then found as

$$Q_{GW} = L \cdot q_{GW}, \qquad (8\text{-}46)$$

where $L$ is the length of the reach.

### EXAMPLE 8-3

Figure 8-36 shows a portion of a flow net adjacent to one-half a symmetrical stream, representative of a reach of length $L = 1500$ m. The aquifer is homogeneous and isotropic, with a hydraulic conductivity $K_h = 10^{-5}$ m s$^{-1}$. The equipotentials are drawn at $\Delta h = 1$-m intervals. What is $Q_{GW}$ for this case?

**Solution** Equation (8-44) gives

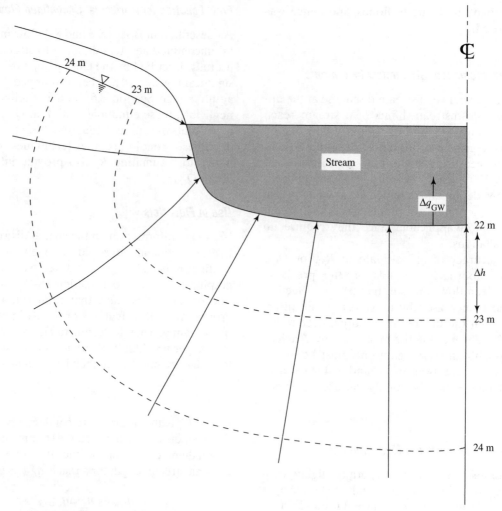

**FIGURE 8-36**
Flow net in the vicinity of a stream receiving ground water, defining terms in Equations (8-44)–(8-45).

$$\Delta q_{GW} = (10^{-5}\ \text{m s}^{-1}) \cdot (1.00\ \text{m}) = 10^{-5}\ \text{m}^2\ \text{s}^{-1}.$$

Since $n = 12$ increments of $\Delta q_{GW}$ enter the stream (only six of which are shown), the total discharge into the stream per unit length is found from Equation (8-45) as

$$q_{GW} = 12 \cdot (10^{-5}\ \text{m}^2\ \text{s}^{-1}) = 1.2 \times 10^{-4}\ \text{m}^2\ \text{s}^{-1}.$$

Finally, Equation (8-46) gives

$$Q_{GW} = (1500\ \text{m}) \cdot (1.2 \times 10^{-4}\ \text{m}^2\ \text{s}^{-1}) = 0.18\ \text{m}^3\ \text{s}^{-1}$$

for the reach.

---

Flow nets constructed in the horizontal plane can also be used to estimate ground-water inflow to streams. In practice, however, it is usually difficult to obtain enough information about subsurface geology and conductivities to warrant computations based on a flow net, and there are only a few published studies that have used this approach (Freeze and Witherspoon 1968; Freeze 1968; Ophori and Tóth 1990).

### Direct Measurement of Ground-Water Potentials or Flux

As for $R_{SW}$, $Q_{GW}$ can be evaluated by measuring the heads in adjacent piezometers in the stream bed or banks or comparing the stream water level with the sub-bed hydraulic head (Winter et al. 1988) and using those values along with information about conductivities to calculate the ground-water flux into the stream.

Direct local measurement of $Q_{GW}$ can be made using seepage meters, which are devices that capture water flowing upward into a portion of a water body over some time period (Figure 8-35). The flux into the container is simply the volume collected divided by the time period. Lee (1977) described the construction and use of seepage meters.

As noted earlier, the main concern in using piezometer or seepage-meter measurements is whether the sampling is spatially and temporally representative. Piezometer observations in stream beds in relatively homogeneous glacial deposits on Long Island, New York, were consistent with mathematical simulations (Prince et al. 1989). However, Lee and Hynes (1978) found very large spatial variability of ground-water input to a small stream in Ontario, Canada, and concluded that determination of average rates of ground-water input to streams from point measurements of seepage flux was not generally possible.

### Direct Measurement of Streamflow Increments

As described for evaluating $R_{SW}$, direct measurement of streamflow via velocity-area stream gaging (Section F.2) at the upstream and downstream ends of a stream reach in which no tributaries enter provides a direct measurement of $Q_{GW}$. Cey et al. (1998) found that this approach gave the most reliable results in their study of a small Canadian stream.

### Methods Based on Water Quality

If streamflow, $Q$, at any instant is assumed to be a mixture of water from two sources, surface water, $Q_{SW}$, and ground water, $Q_{GW}$, each of which has a characteristic concentration of some chemical or isotope, $C_{SW}$ and $C_{GW}$, then

$$Q = Q_{SW} + Q_{GW} \qquad (8\text{-}47)$$

and

$$C \cdot Q = C_{SW} \cdot Q_{SW} + C_{GW} \cdot Q_{GW}, \qquad (8\text{-}48)$$

where $C$ is the concentration in the streamflow. If $Q$ and $C$ are measured and $C_{SW}$ and $C_{GW}$ determined from sampling the respective sources in the watershed, $Q_{GW}$ can be found by combining Equations (8-47) and (8-48):

$$Q_{GW} = Q \cdot \left( \frac{C - C_{SW}}{C_{GW} - C_{SW}} \right). \qquad (8\text{-}49)$$

Use of Equation (8-49) assumes that there are only two sources of streamflow, that each source has a constant concentration of tracer, and that these concentrations are significantly different. To relax some of these constraints, Pilgrim et al. (1979) showed how the relation could be modified when the concentration of one of the components is a function of time since the beginning of the event, and Swistock et al. (1989) derived a version of Equation (8-49) for use when there are three runoff components.

Equation (8-49) has been used with apparent success in several studies, using as tracers various anions and cations [e.g., Newbury et al. (1969); Pinder and Jones (1969)], stable isotopes $^{18}$O and $^{2}$H [*e.g.*, Sklash and Farvolden (1979); Space et al. (1991)], and the radioactive gas radon ($^{222}$Rn) [e.g., Ellins et al. (1990)]. Figure 8-37 shows the estimated ground-water contribution to streamflow in a case where $SO_4^{-2}$ was used as the tracer in Equation (8-49).

### Base-Flow Analysis

As explained more fully in Section 9.1, a stream responds to a significant water-input event with a relatively rapid flow increase to a peak, which usually occurs soon after the input ceases, followed by a more gradual decline, called the **recession,** that continues until the next event. (See Figure 9-5.) Flow ($Q$) during the recession represents drainage from watershed storage and declines as storage decreases; thus we can infer that

$$Q = f(B), \qquad (8\text{-}50)$$

where $B$ is the volume of water in storage and the form of $f$ is determined largely by watershed characteristics (e.g., size, geology, slope, etc.). [See Equation (2-25).]

As noted earlier, base flow is the portion of streamflow that is derived from persistent, slowly varying sources; thus at some point recession flow becomes coincident with base flow. Base flow is usually assumed to be from unconfined aquifers draining to the stream network, and the attempt to identify ground-water contributions to streamflow by analysis of stream hydrographs, usually called

**FIGURE 8-37**
Total streamflow and streamflow attributed to ground-water base flow for four runoff events in Wilson Creek, Manitoba, as estimated using Equation (8-49) with $SO_4^{-2}$ as a tracer. From Newbury et al. (1969).

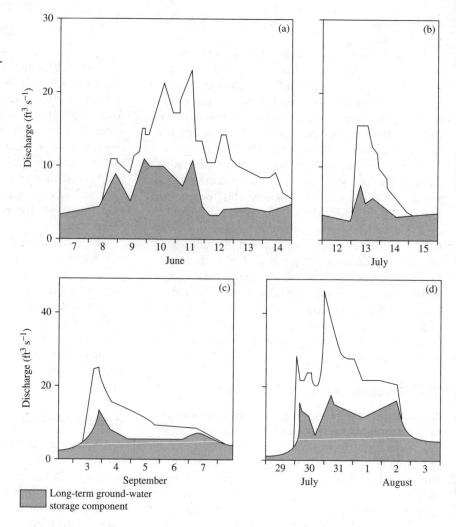

Long-term ground-water storage component

**base-flow analysis, base-flow separation,** or **recession analysis,** has a long history [Hall (1968); Tallaksen (1995)] and remains a common practice (Chapman 1999).

In general, base-flow analysis involves plotting the measured stream hydrograph, usually on semi-logarithmic paper, and constructing a line coincident with or below the hydrograph that purports to represent the ground-water component. Construction of the line requires arbitrary answers to the following questions: (1) How does base flow behave while the stream is responding to water input? and (2) At what point does base flow become equal to total flow?

The analyses of Boxes 8-4, 8-5, and 8-6 provide theoretical analyses of aquifer drainage suggesting that such drainage approximates an exponential

decay, and this concept has been the justification and basis for most approaches to base-flow separation. In practice, however, the relation between base flow and storage [Equation (8-50)] is often assumed to be of simpler form:

$$Q_{GW} = k \cdot B^b; \qquad \textbf{(8-51)}$$

if $b = 1$, this leads to

$$Q_{GW} = Q_{GW0} \cdot \exp\left[-k \cdot (t - t_0)\right] \qquad \textbf{(8-52)}$$

(the linear reservoir of Box 8-5); if $b > 1$, then

$$Q_{GW} = Q_{GW0} \cdot \left[1 + d \cdot (t - t_0)\right]^{b/(1-b)}, \qquad \textbf{(8-53)}$$

where $k, b,$ and $d$ are constants, $t$ is time, and $Q_{GW0}$ is the flow rate at the initial time $t_0$. Values of the constants are found empirically by choosing $t_0$ and finding values of $b$ and $d$ that give a line that best fits

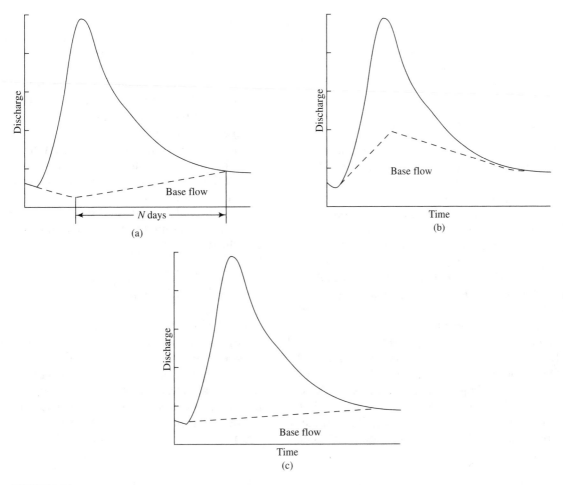

**FIGURE 8-38**

Methods of graphical base-flow separation. (a) The pre-event flow trend is projected until the time of peak, after which the base-flow hydrograph is connected by a straight line that intersects the total-flow hydrograph $N$ days after the peak, where $N$ (days) $= A^{0.2}$, and $A$ is drainage area in mi². (b) The hydrograph is plotted on semi-logarithmic paper (log[$Q(t)$] vs. $t$). A straight line is fitted to the end of the hydrograph recession on this graph and projected backward in time under the peak. This projected line is transferred onto arithmetic graph paper and a smooth line is sketched connecting it to the end of the preceding recession. (c) From the point of initial hydrograph rise, a line that slopes upward at a rate of 0.05 ft³ s⁻¹·$A$ (mi²) per hour is drawn and extended until it intercepts the hydrograph ($A < 20$ mi²). From *Water in Environmental Planning* by Thomas Dunne and Luna B. Leopold. Copyright © 1978 by W. H. Freeman and Co. Reprinted with permission.

the subsequent recession. Alternatively, the problem of choosing $t_0$ can be avoided by plotting $dQ/dt$ vs. $Q$ on logarithmic graph paper and determining $b$ and $d$ from the slope and intercept of the line that best fits the plot (Brutsaert and Nieber 1977).

Figure 8-38 illustrates some of the approaches used in base-flow separation; others are described by Tallaksen (1995), Arnold and Allen (1999), and Chapman (1999). Note that the total volume of base flow varies greatly depending on the method

used. This illustrates the crucial point: If a consistent method is applied, base-flow separation can be a useful tool for comparing the relative contributions of ground water to streamflow in different watersheds; however, all such methods require arbitrary decisions and thus cannot be used to identify the actual ground-water component of streamflow [Freeze (1972a); Anderson and Burt (1980)]. In applying these methods, one must also be aware that measurements of streamflow when flows are very low are commonly very imprecise, and the true shape of a recession hydrograph may be highly uncertain.

The topic of base-flow separation is examined further in Section 9.1.2.

### 8.5.4  Capillary Rise, *CR*

Capillary rise is induced by extraction of water from the vadose zone and capillary fringe by evaporation from the soil and transpiration, and by migration of water to a freezing front. Some of the water that freezes may return to the water table by percolation after thawing occurs, but some may ultimately be lost to evapotranspiration.

Capillary rise is usually considered to be a minor to negligible component of the water budget in humid regions, but may be a significant proportion of evapotranspiration in semi-arid and arid areas. In general, net capillary rise may be difficult to estimate as a separate component, and it is often tacitly included as part of basin evapotranspiration.

One approach to estimating capillary rise is to identify those portions of the drainage basin where the water table is close enough to the surface that plants can obtain water from the capillary fringe. The presence of wetland vegetation and plants that

are known phreatophytes can be used in this identification. One can then assume that evapotranspiration from these areas will always be at the potential rate and use one of the methods described in Section 7.7 to estimate potential evapotranspiration. Nichols (1993; 1994) applied such methods in estimating *CR* in the western United States.

In areas where plants are extracting water from the capillary fringe, the water table may show a diurnal fluctuation (Figure 8-39). Johansson (1986) showed that transpiration could produce such fluctuations even with water tables at depths of 2 m or more. White (1932) suggested that evapotranspiration could be estimated from such diurnal water-table fluctuations as

$$ET = \left[ 24 \cdot \left( \frac{\Delta h}{\Delta t} \right)_r + \Delta h_{24} \right] \cdot S_y, \quad \textbf{(8-54)}$$

where *ET* is the daily evapotranspiration [L], $(\Delta h / \Delta t)_r$ is the rate of rise of the water table during the period midnight to 4:00 A.M., and $\Delta h_{24}$ is the net fall in water level over the 24-hr period. This method was successfully used by Meyboom [as cited in Freeze and Cherry (1979)], who suggested that the appropriate value of $S_y$ used in Equation (8-54) is one-half the conventional value.

**FIGURE 8-39**
Estimation of evapotranspiration from the capillary fringe (capillary rise) using diurnal water-table fluctuations [Equation (8-54)]. See text for explanation. After Freeze and Cherry (1979).

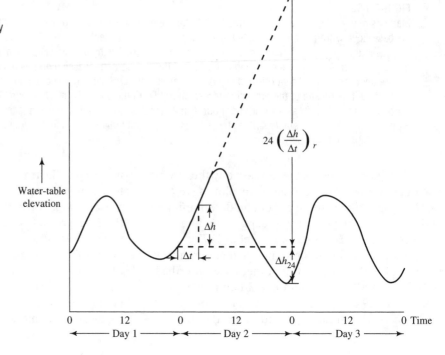

Daniel (1976) developed a method for estimating evaporative extractions of ground water based on theoretical aquifer-drainage relations, and applied it successfully in Alabama.

### 8.5.5 Deep Seepage, $G_{in}$ and $G_{out}$

In the context of the water-balance relations developed earlier [Equations (8-20)–(8-24)], **deep seepage** refers to the ground-water inflow and outflow terms $G_{in}$ and $G_{out}$. The magnitudes of these terms are very difficult to determine, and they are often assumed to be negligible or to cancel (i.e., $G_{out} = G_{in}$). However, the earlier discussion of regional ground-water flow suggests that it is often unwise to cavalierly adopt such assumptions.

#### Use of Flow Nets

At least a qualitative indication of the importance of deep seepage can be obtained if enough information about basin geology and water-table configurations is available to determine boundary conditions and develop flow nets via solutions to Equation (8-18) or via graphical methods, as described earlier. The studies by Tóth (1962; 1963), Freeze and Witherspoon (1968), and Ophori and Tóth (1990) are examples of this approach.

#### Use of Piezometer Measurements

Installation of piezometers and observation wells at strategic locations gives the most definitive information about the magnitude of deep seepage. This information is most effectively used in combination with flow-net construction, as by Freeze and Witherspoon (1968). Winter et al. (1989) combined piezometer observations with hydroclimatologic observations to develop information on the magnitude of deep seepage in upland New England.

#### Water-Balance Relations for Basin Segments

Figure 8-40 shows a hypothetical drainage basin in which the main stream is gaged at $N$ successive downstream locations. Starting at each of these locations, divides can be delineated that define $N$ sub-basin segments, numbered $i = 1, 2, ..., N$ from upstream to downstream. If we assume that there is no flow across the main basin divide, and that

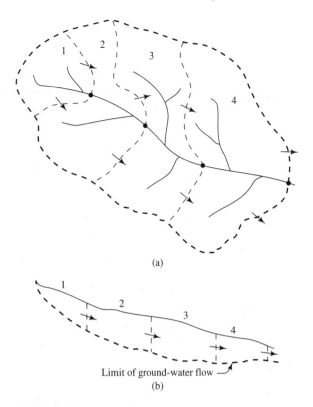

(a)

Limit of ground-water flow
(b)

**FIGURE 8-40**
Schematic diagram defining terms for derivation of water-balance estimates of deep seepage [Equations (8-55)–(8-56)]. (a) Plan view of basin showing segments. (b) Longitudinal cross section. Arrows indicate general direction of ground-water flow, dots are gaging stations.

ground water as well as surface water moves only in the down-basin direction, the long-term average water balance for the $i$th segment is

$$Q_i = (P_i - ET_i) \cdot A_i + G_{i-1} - G_i + Q_{i-1}, \quad \text{(8-55)}$$

where $Q$ is stream outflow $[L^3 \, T^{-1}]$, $P$ and $ET$ are areal average precipitation and evapotranspiration rates, respectively $[L \, T^{-1}]$, $A$ is the area of the segment $[L^2]$, and $G$ is ground-water outflow rate $[L^3 \, T^{-1}]$.

If we assume that all the quantities in Equation (8-55) can be determined with reasonable precision except the $G$ terms, there are $N$ equations of the form of (8-55) with $N + 1$ unknowns (including $G_0$). However, if $G_0$ as well as $Q_0 = 0$ as assumed, there are actually only $N$ unknowns, and the ground-water outflow for the $i$th segment can be found as

$$G_i = (P_i - ET_i) \cdot A_i + G_{i-1} + Q_{i-1} - Q_i. \quad \text{(8-56)}$$

### EXAMPLE 8-4

The table that follows gives areas and long-term average values of precipitation, evapotranspiration, and streamflow for four segments of the Contoocook River basin in central New Hampshire. Estimate the ground-water outflow from each segment and compare it with the stream outflow.

| i | Stream Gage at | $A_i$ (km²) | $P_i$ (mm yr⁻¹) | $ET_i$ (mm yr⁻¹) | $Q_i$ (m³ s⁻¹) |
|---|----------------|-------------|------------------|-------------------|----------------|
| 1 | Peterboro      | 176 | 1300 | 500 | 3.31 |
| 2 | Henniker       | 777 | 1180 | 500 | 18.0 |
| 3 | W. Hopkinton   | 153 | 1020 | 550 | 19.7 |
| 4 | Penacook       | 878 | 1070 | 550 | 35.5 |

**Solution** For the first segment, Equation (8-56) gives

$$G_1 = [(1300 - 500) \text{ mm yr}^{-1}] \cdot (176 \text{ km}^2)$$
$$\cdot (3.17 \times 10^{-5} \text{ m}^3 \text{ yr mm}^{-1} \text{ km}^{-2} \text{ s}^{-1})$$
$$+ 0 \text{ m}^3 \text{ s}^{-1} + 0 \text{ m}^3 \text{ s}^{-1} - 3.31 \text{ m}^3 \text{ s}^{-1} = 1.15 \text{ m}^3 \text{ s}^{-1}.$$

For the second segment, we have

$$G_2 = [(1180 - 500) \text{ mm yr}^{-1})] \cdot (777 \text{ km}^2)$$
$$\cdot (3.17 \times 10^{-5} \text{ m}^3 \text{ yr mm}^{-1} \text{ km}^{-2} \text{ s}^{-1})$$
$$+ 1.15 \text{ m}^3 \text{ s}^{-1} + 3.31 \text{ m}^3 \text{ s}^{-1} - 18.0 \text{ m}^3 \text{ s}^{-1}$$
$$= 3.21 \text{ m}^3 \text{ s}^{-1}.$$

Similarly, for the third and fourth segments, the equation yields $G_3 = 3.76 \text{ m}^3 \text{ s}^{-1}$ and $G_4 = 2.50 \text{ m}^3 \text{ s}^{-1}$. The ratios of ground-water outflow to stream outflow are thus as follows:

| i | $G_i/Q_i$ |
|---|-----------|
| 1 | 0.35 |
| 2 | 0.18 |
| 3 | 0.19 |
| 4 | 0.07 |

As with all water-balance approaches, the uncertainty in all the "known" terms is hidden in the terms found by subtraction. However, the above computations suggest that deep seepage is a significant component of the water balance in the Contoocook River basin.[9] Exercise 8-10 gives you an

opportunity to explore the effect of uncertainties in the values of $P_i$, $ET_i$, and $Q_i$ on the estimated $G_i$.

### Water Balance as a Function of Basin Elevation

Where evapotranspiration decreases with elevation, ground-water outflow may be estimated from measurements of long-term average precipitation, $P$, long-term average streamflow, $Q$, and estimates of actual or potential evapotranspiration (Sections 7.7 and 7.8) as follows: For each gaged basin, plot (1) $P - Q$, where $Q$ is converted to units of $[\text{L T}^{-1}]$; and (2) actual or potential evapotranspiration, vs. mean basin elevation (Figure 8-41). For basins that plot to the right of the line for $ET$, $P - Q > ET$, suggesting the presence of ground-water outflow via deep seepage. Again, the strength of this inference depends on the precision with which $P$, $Q$, and $ET$ can be determined.

### Areal Discharge Relations

Consider a region in which long-term average annual precipitation, $P$ $[\text{L T}^{-1}]$, and evapotranspiration, $ET$ $[\text{L T}^{-1}]$, do not vary spatially and in which there is no deep seepage. Then for all gaging stations in the region,

$$Q \propto A, \qquad \textbf{(8-57)}$$

where $Q$ is long-term average streamflow, $[\text{L}^3 \text{ T}^{-1}]$, and $A$ is drainage area, $[\text{L}^2]$.

If however, there is deep seepage, the relationship between streamflow and area will be

$$Q \propto A^y, \qquad \textbf{(8-58)}$$

where $y > 1$ because larger basins should be sites of discharge of deep seepage from smaller basins. This reasoning suggests that if the relation between $Q$ and $A$ for a region has the form of Equation (8-58) and $y > 1$,[10] it is qualitative evidence for deep seepage. Such evidence would be all the stronger if, as is often the case, precipitation increases and evapotranspiration decreases with elevation, because in

---

[9] If there are random positive and negative uncertainties in the "known" values in Example 8-4, there should be an equal chance that each value of $G_i$ is positive or negative. The probability of getting four positive values under these circumstances is thus $0.5^4 = 0.0625$. This low probability further supports the idea that there is ground-water outflow in this region.

[10] The value of $y$ can be estimated as the slope in the regression analysis of $\log (Q)$ vs. $\log (A)$. Statistics books describe the methodology and tell how to determine the uncertainty in the estimate of that slope.

**FIGURE 8-41**
Long-term average precipitation (*P*) minus streamflow (*Q*) vs. mean basin elevation for 19 small gaged basins in New Hampshire and Vermont. The lines are estimates of average evapotranspiration (*ET*) using the Hamon and equilibrium estimates (see Section 7.7.2), with the value determined at Hubbard Brook Experimental Forest (HBEF) as control. The existence of deep seepage is suggested for basins that plot to the right of the *ET* lines. From Dingman (1981).

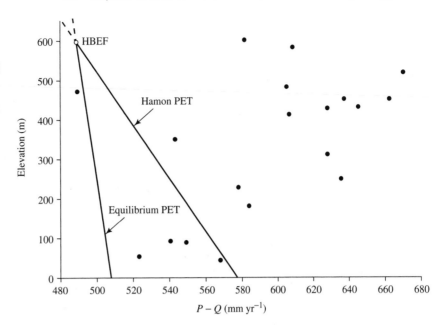

that case *y* would be less than 1 in the absence of deep seepage.

# 8.6   IMPACTS OF GROUND-WATER DEVELOPMENT ON BASIN HYDROLOGY

Our consideration of the impacts of ground-water development focuses on unconfined aquifers, because they have the most direct connections with other portions of the land phase of the hydrologic cycle and their exploitation as water sources usually has the most direct impacts on regional hydrology. However, we will use the mathematically simpler but essentially similar behavior of confined aquifers to illustrate the most basic features of the effects of the extraction of ground water on drainage-basin hydrology. This is justified because the behavior of unconfined aquifers is nearly identical to that of confined aquifers as long as the changes in water-table elevation are small relative to the saturated thickness.

To further simplify the discussion we consider only homogeneous, isotropic aquifers, simple aquifer configurations, and fully penetrating wells. General ground-water texts such as Bear (1979), Freeze and Cherry (1979), and Fetter (1994) should be consulted to explore more exact models of

ground-water development in unconfined flows and in more complex boundary conditions.

## 8.6.1   Hydraulics of Ground-Water Development

### Radial Flow to a Well

Consider the highly idealized case of a well completely penetrating a homogeneous unconfined aquifer of infinite extent resting on a horizontal impermeable base (Figure 8-42). The water table is initially horizontal everywhere at a height $h_0$ above the base, and there is no recharge or capillary rise.

When the well is pumped at a constant rate $Q_w$, water is withdrawn from aquifer storage, the water table declines toward the well, and flow is induced toward the well from all directions. Thus the flow has radial symmetry, and if we approximate the unconfined case by equivalent confined conditions (i.e., assume negligible water-table decline and horizontal streamlines and ignore a transient initial period prior to the establishment of gravity drainage), it can be described by transforming the two-dimensional form of Equation (8-16) to polar coordinates (Freeze and Cherry 1979):

$$\frac{\partial^2 h}{\partial r^2} + \frac{1}{r} \cdot \frac{\partial h}{\partial r} = \frac{S_y}{K_h \cdot h_0} \cdot \frac{\partial h}{\partial t}. \quad \textbf{(8-59)}$$

Here, *r* is the radial distance measured outward from the well.

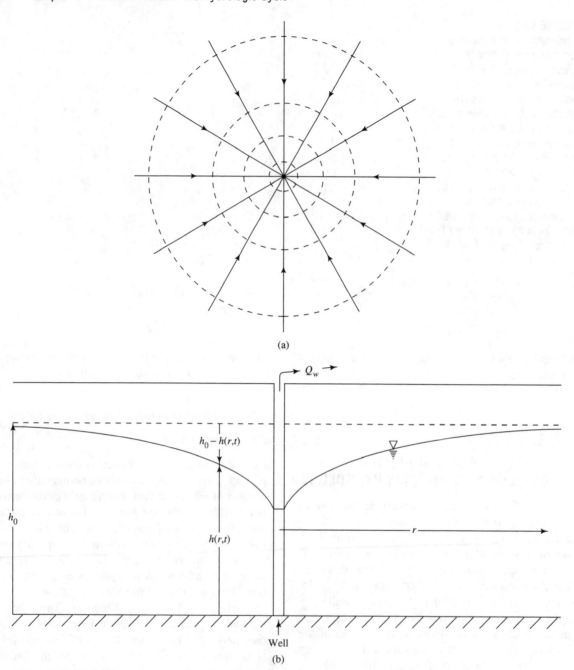

(a)

(b)

**FIGURE 8-42**
Definitions of terms for equations describing radial flow to a well in an unconfined aquifer [Equations (8-59)–(8-62)]. (a) Plan view; dashed lines are equipotentials, arrows show flow directions. (b) Cross section.

Theis (1935) showed that an analytical solution for Equation (8-59) exists:

$$h_0 - h(r, t) = \frac{Q_w}{4 \cdot \pi \cdot K_h \cdot h_0} \cdot W[u(r, t)], \quad \textbf{(8-60)}$$

where

$$W[u(r, t)] \equiv \int_{u(r, t)}^{\infty} \frac{\exp[-u(r, t)]}{u(r, t)} \cdot du(r, t) \quad \textbf{(8-61)}$$

and

$$u(r, t) \equiv \frac{S_y \cdot r^2}{4 \cdot K_h \cdot h_0 \cdot t}. \quad \textbf{(8-62)}$$

The quantity $u(r,t)$ is a measure of the aquifer response time [Equation (8-13)].

The function $W[u(r,t)]$ is known as the **well function,** and its values are tabulated in Table 8-4.[11] The quantity $[h_0 - h(r,t)]$ is the **drawdown,** and when its values are plotted as a function of distance at any time they define a **drawdown curve** or, in three dimensions, a **cone of depression** that is asymptotic to $h_0$. Example 8-5 shows how this relation is applied.

### EXAMPLE 8-5

For an ideal aquifer with $K_h = 10^{-5}$ m s$^{-1}$, $h_0 = 20$ m, and $S_y = 0.20$, compute the drawdown at distances of 1, 5, 10, 50, and 100 m from the well at times 1, 2, 40, and 80 hr after the start of pumping at a constant rate of 0.001 m$^3$ s$^{-1}$.

**Solution** For the first values of time and distance we use Equation (8-62) to find $u(r,t)$ as

$$u(1 \text{ m}, 3600 \text{ s})$$
$$= \frac{0.2 \cdot (1 \text{ m})^2}{4 \cdot (10^{-5} \text{ m s}^{-1}) \cdot (20 \text{ m}) \cdot (3600 \text{ s})} = 0.07.$$

Repeating this for all combinations of time and distance gives the following values of $u(r,t)$:

| $t$ (s) \ $r$ (m) | 1 | 5 | 10 | 50 | 100 |
|---|---|---|---|---|---|
| 3600 | 0.07 | 1.74 | 6.94 | 173.61 | 694.44 |
| 7200 | 0.03 | 0.87 | 3.47 | 86.81 | 347.22 |
| 144,000 | 0.00 | 0.04 | 0.17 | 4.34 | 17.36 |
| 288,000 | 0.00 | 0.02 | 0.09 | 2.17 | 8.68 |

[11] A formula that approximates $W[u(r,t)]$ is used in the spreadsheet program WellFunc.xls on the CD accompanying the text.

From Table 8-4, we find the values of $W[u(r,t)]$ that correspond to the foregoing values of $u(r,t)$:

| $t$ (s) \ $r$ (m) | 1 | 5 | 10 | 50 | 100 |
|---|---|---|---|---|---|
| 3600 | 2.16 | 0.07 | 0.00 | 0.00 | 0.00 |
| 7200 | 2.82 | 0.28 | 0.01 | 0.00 | 0.00 |
| 144,000 | 5.78 | 2.60 | 1.34 | 0.00 | 0.00 |
| 288,000 | 6.47 | 3.27 | 1.95 | 0.04 | 0.00 |

Finally, the drawdown is calculated by multiplying the preceding values of $W[u(r,t)]$ by

$$\frac{Q_w}{4 \cdot \pi \cdot K_h \cdot h_0} = \frac{0.001 \text{ m}^3 \text{ s}^{-1}}{4 \cdot 3.14 \cdot (10^{-5} \text{ m s}^{-1}) \cdot (20 \text{ m})}$$
$$= 0.398 \text{ m}$$

to give $h_0 - h(r,t)$ in m:

| $t$ (s) \ $r$ (m) | 1 | 5 | 10 | 50 | 100 |
|---|---|---|---|---|---|
| 3600 | 0.86 | 0.03 | 0.00 | 0.00 | 0.00 |
| 7200 | 1.12 | 0.11 | 0.00 | 0.00 | 0.00 |
| 144,000 | 2.30 | 1.04 | 0.53 | 0.00 | 0.00 |
| 288,000 | 2.58 | 1.30 | 0.78 | 0.02 | 0.00 |

The area over which the pumping causes drawdown is called the **area of influence.** For the idealized situation of Figure 8-42, the lines of equal drawdown and the extent of the cone of depression are circular in plan view and the area of influence coincides with the projected area of the cone.

Clearly, the drawdown is proportional to pumping rate, and it decreases with distance at any time and increases with time at any distance. For a given pumping rate, the rates of change are controlled by the aquifer properties: Other factors equal,

lower $K_h \rightarrow$ larger drawdown spread over a smaller area;
higher $K_h \rightarrow$ smaller drawdown spread over a greater area;

lower $S_y \rightarrow$ larger drawdown spread over a greater area;
higher $S_y \rightarrow$ smaller drawdown spread over a smaller area.

An interesting and useful property of the solution to Equation (8-59) is that the drawdown at any location in an aquifer due to the pumping of more than one well is equal to the sum of the drawdowns that would be produced at that location by each of the wells individually.

**TABLE 8-4**
Values of $W[u(r, t)]$ for Various Values of $u(r, t)$

| $u(r, t)$ | 1.0 | 2.0 | 3.0 | 4.0 | 5.0 | 6.0 | 7.0 | 8.0 | 9.0 |
|---|---|---|---|---|---|---|---|---|---|
| $\times 1$ | 0.219 | 0.049 | 0.013 | 0.0038 | 0.0011 | 0.00036 | 0.00012 | 0.000038 | 0.000012 |
| $\times 10^{-1}$ | 1.82 | 1.22 | 0.91 | 0.70 | 0.56 | 0.45 | 0.37 | 0.31 | 0.26 |
| $\times 10^{-2}$ | 4.04 | 3.35 | 2.96 | 2.68 | 2.47 | 2.30 | 2.15 | 2.03 | 1.92 |
| $\times 10^{-3}$ | 6.33 | 5.64 | 5.23 | 4.95 | 4.73 | 4.54 | 4.39 | 4.26 | 4.14 |
| $\times 10^{-4}$ | 8.63 | 7.94 | 7.53 | 7.25 | 7.02 | 6.84 | 6.69 | 6.55 | 6.44 |
| $\times 10^{-5}$ | 10.94 | 10.24 | 9.84 | 9.55 | 9.33 | 9.14 | 8.99 | 8.86 | 8.74 |
| $\times 10^{-6}$ | 13.24 | 12.55 | 12.14 | 11.85 | 11.63 | 11.45 | 11.29 | 11.16 | 11.04 |
| $\times 10^{-7}$ | 15.54 | 14.85 | 14.44 | 14.15 | 13.93 | 13.75 | 13.60 | 13.46 | 13.34 |
| $\times 10^{-8}$ | 17.84 | 17.15 | 16.74 | 16.46 | 16.23 | 16.05 | 15.90 | 15.76 | 15.65 |
| $\times 10^{-9}$ | 20.15 | 19.45 | 19.05 | 18.76 | 18.54 | 18.35 | 18.20 | 18.07 | 17.95 |
| $\times 10^{-10}$ | 22.45 | 21.76 | 21.35 | 21.06 | 20.84 | 20.66 | 20.50 | 20.37 | 20.25 |
| $\times 10^{-11}$ | 24.75 | 24.06 | 23.65 | 23.36 | 23.14 | 22.96 | 22.81 | 22.67 | 22.55 |
| $\times 10^{-12}$ | 27.05 | 26.36 | 25.96 | 25.67 | 25.44 | 25.26 | 25.11 | 24.97 | 24.86 |
| $\times 10^{-13}$ | 29.36 | 28.66 | 28.26 | 27.97 | 27.75 | 27.56 | 27.41 | 27.28 | 27.16 |
| $\times 10^{-14}$ | 31.66 | 30.97 | 30.56 | 30.27 | 30.05 | 29.87 | 29.71 | 29.58 | 29.46 |
| $\times 10^{-15}$ | 33.96 | 33.27 | 32.86 | 32.58 | 32.35 | 32.17 | 32.02 | 31.88 | 31.76 |

*Source*: Wenzel (1942).

### Contributing Areas

The **contributing area** of a well is the area on the land surface above the volume of aquifer from which water is flowing to the well. Identification of this area for unconfined aquifers receiving recharge from infiltration or surface-water bodies is important because any water-contaminating substances introduced into the contributing area will eventually reach the well. The delineation of these areas by analytical and numerical methods was reviewed by Morrissey (1987).

For the ideal, infinite, homogeneous aquifer with an initially horizontal water table described in the preceding section, all the water extracted from the cone of depression eventually arrives at the well, and the contributing area at any time is identical to the area of influence. However, actual aquifers do not have horizontal water tables (which would imply no flow) and do not extend infinitely. If the water table is initially sloping, the cone of depression is no longer circular and the contributing area does not coincide with the area of influence (Figure 8-43). If the aquifer is in a river valley, the contributing area may extend to or even beyond the river (Figure 8-44).

### 8.6.2 Effects of Ground-Water Extraction

#### Effects on Natural Recharge and Discharge

Consider the ground-water system of a drainage basin in which there is no deep seepage in or out. Under natural (no-development) conditions, the long-term average recharge and discharge for this system must be in balance, and from Equations (8-23),

$$R_{nat} - Q_{GWnat} = 0, \qquad \text{(8-63)}$$

where the subscript *nat* denotes the natural recharge and discharge rates.

If one or more wells begins pumping ground water from the system, water will be removed from aquifer storage as the cones of depression develop. In addition, the natural rates of recharge and discharge will in general be changed as the water-table configuration is altered by the pumping. Thus, during development, the water-balance for the system becomes

$$(R_{nat} + \Delta R) - (Q_{GWnat} + \Delta Q_{GW}) - \Sigma Q_w = \frac{\Delta S}{\Delta t}, \qquad \text{(8-64)}$$

where $\Delta R$ and $\Delta Q_{GW}$ are the induced changes in recharge and discharge, respectively; $\Sigma Q_w$ is the total pumping rate; and $\Delta S/\Delta t$ is the rate at which water is withdrawn from aquifer storage (i.e., $\Delta S/\Delta t < 0$). Combining Equations (8-63) and (8-64) yields

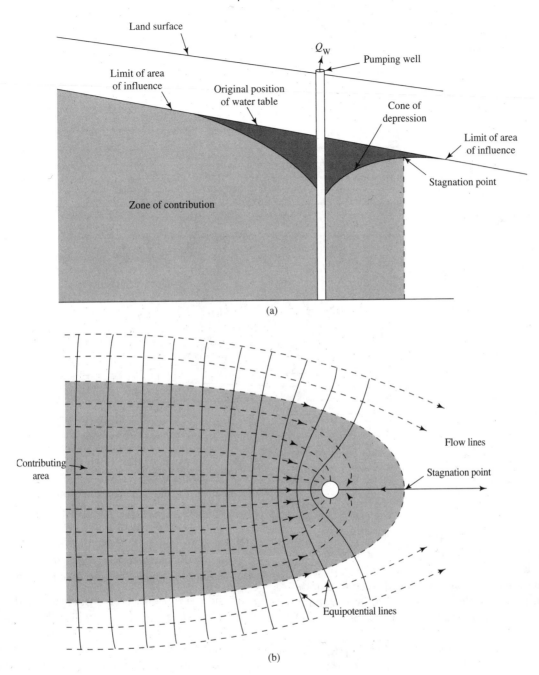

**FIGURE 8-43**
The cone of depression, area of influence, and contributing area for an aquifer with a sloping water table. (a) Cross section. (b) Plan view. After Morrissey (1987).

**FIGURE 8-44**

Contributing areas and water-table contours for wells near a river. (a) Natural condition before pumping. (b) Well intercepts water that was flowing to the river. (c) Well intercepts water and extracts flow from the river. (d) Well intercepts water from both sides of the valley and extracts water from the river. Existence of conditions (b), (c), or (d) depends on pumping rate and aquifer configuration and properties. From Morrissey (1987).

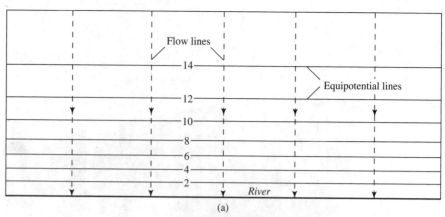

(a)

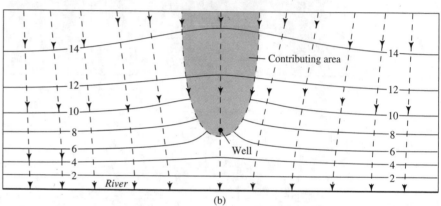

(b)

$$\Sigma Q_w = \Delta R - \Delta Q_{GW} - \frac{\Delta S}{\Delta t}$$

$$= \Delta R - \Delta Q_{GW} + \left| \frac{\Delta S}{\Delta t} \right|. \qquad \textbf{(8-65)}$$

Equation (8-65) states that ground-water development must be balanced by a decrease in storage (which always occurs) and, in general, by some combination of increased (induced) recharge or decreased ground-water discharge. We now examine how lowered water tables due to pumping affect $\Delta R$ and $\Delta Q_{GW}$.

From the definition of recharge [Equation (8-22)], we see that $\Delta R$ must be due to some combination of (1) increased recharge from infiltration, $R_I$; (2) increased recharge from surface-water bodies, $R_{SW}$; and (3) decreased capillary rise, $CR$. Lowered water tables due to pumping affect these components in the following ways:

$R_I$: As shown in Figure 8-29, net recharge from infiltration tends to increase with water-table depth up to a point, beyond which there is little change.

$R_{SW}$: As shown in Figure 8-44, the cone of depression from wells near streams can extend to the stream, locally reverse the potential gradient, and induce recharge from the stream. (This effect is discussed further in the next section.)

$CR$: Again from Figure 8-29, lowered water tables tend to decrease capillary rise by lowering the capillary fringe beyond the reach of plant roots.

Thus the overall effect of ground-water development tends to be a net increase in recharge; however, the magnitude of the effect will be highly dependent on the drainage-basin geology, topography, and climate and the placement and pumping rates of wells.

It should be clear from the preceding discussion that some "mining"—that is, removal of water from aquifer storage—is required to reach any extraction rate. However, if a constant rate of ground-water extraction is imposed on a basin for a sufficient time, a new equilibrium state may eventually be reached in which there is no further change in storage ($\Delta S/\Delta t = 0$) and the extraction rate ($\Sigma Q_w$) is supplied by increased recharge ($\Delta R_I, \Delta R_{SW} > 0$) or

**FIGURE 8-44**
(continued)

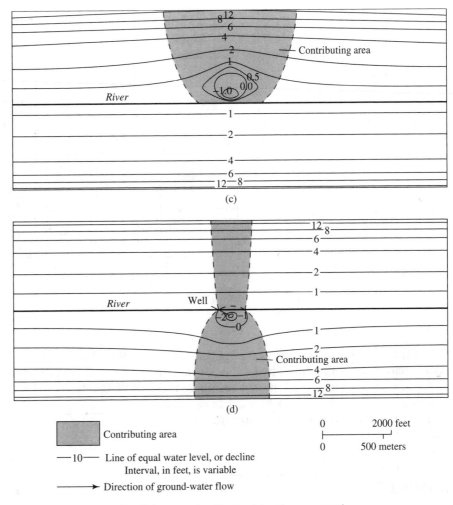

(c)

(d)

Contributing area

—10— Line of equal water level, or decline
Interval, in feet, is variable

⟶ Direction of ground-water flow

For all figures, units of head and drawdown expressed
in feet relative to river stage

0        2000 feet
├───────┼───────┤
0        500 meters

reduced discharge ($\Delta CR, \Delta Q_{GW} < 0$). Bredehoeft et al. (1982) pointed out that the time required to reach this equilibrium may be very long indeed (hundreds of years), depending on the basin size, hydrology, and geology and on the locations and pumping rates of wells. In some situations, it is not possible to reach an equilibrium state before drawdown at wells equals its maximum value, $h_0$.

### Effects on Streams

As shown in Figure 8-44, ground-water extraction can reduce streamflow by inducing local recharge from a gaining stream ($\Delta R_{SW} > 0$) and intercepting water that would naturally discharge to streams ($\Delta Q_{GW} < 0$). The net of these two effects is called **stream depletion**, $D_w$. Thus,

$$D_w \equiv \Delta R_{SW} - \Delta Q_{GW}. \qquad \textbf{(8-66)}$$

Jenkins (1968) showed that, under the same idealized aquifer conditions used to solve Equation (8-59), the ratio of depletion rate to a constant rate of pumping, $Q_w$, from a well located a distance $x$ from a stream is given by a **depletion function**, $D$, where

$$\frac{D_w(t)}{Q_w} = D\left(\frac{K_h \cdot h_0 \cdot t}{x^2 \cdot S_y}\right). \qquad \textbf{(8-67)}$$

(Note again that the argument of $D$ is a measure of the aquifer response time similar to $u(r,t)$ [Equation (8-13); Equation (8-62)].)

Figure 8-45 gives the form of this depletion function. Note that it is asymptotic to $D_w(t)/Q_w = 1$; thus ultimately all the water withdrawn by the well depletes the streamflow.

**FIGURE 8-45**
The stream-depletion function, $D_w(t)/Q_w$, as a function of $K_h \cdot h_0 \cdot t/(x^2 \cdot S_y)$. See Example 8-6. From Jenkins (1968).

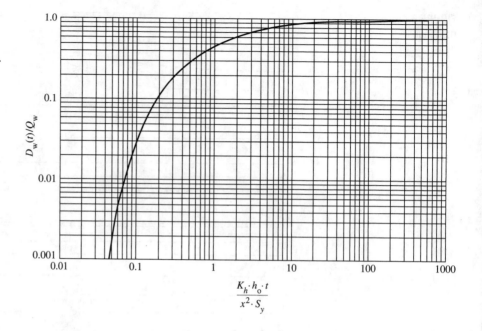

Example 8-6 is an example of the application of this relation to estimating streamflow depletion rates. Jenkins (1968) gave additional examples, and showed how the relation can be used to estimate depletion rates and volumes due to continual and intermittent pumping.

## EXAMPLE 8-6

If the well of Example 8-5 is located 20 m from a stream, calculate the streamflow depletion rate at 1, 10, 30, 60, 180, and 365 days of continuous pumping.

**Solution**  We first calculate

$$\frac{K_h \cdot h_0}{x^2 \cdot S_y} = \frac{(10^{-5} \text{ m s}^{-1}) \times (20 \text{ m})}{(20 \text{ m})^2 \times (0.20)} = 2.5 \times 10^{-6} \text{ s}^{-1}.$$

Next, we find the argument of $D(\ )$ by multiplying the preceding value by the times of interest:

| time (day) | 1 | 10 | 30 |
|---|---|---|---|
| time (s) | $8.64 \times 10^4$ | $8.64 \times 10^5$ | $2.59 \times 10^6$ |
| argument | 0.216 | 2.16 | 6.48 |

| time (day) | 60 | 180 | 365 |
|---|---|---|---|
| time (s) | $5.18 \times 10^6$ | $1.56 \times 10^7$ | $3.15 \times 10^8$ |
| argument | 13.0 | 38.9 | 78.8 |

The value of $D_w(t)/Q_w$ is now found from the curve in Figure 8-45, and this ratio is multiplied by $Q_w$ (= 0.001 m³ s⁻¹) to give the actual stream depletion rate in m³ s⁻¹:

| time (day) | 1 | 10 | 30 |
|---|---|---|---|
| $D_w(t)/Q_w$ | 0.11 | 0.62 | 0.77 |
| $D_w$ (m³ s⁻¹) | $1.1 \times 10^{-4}$ | $6.2 \times 10^{-4}$ | $7.7 \times 10^{-4}$ |

| time (day) | 60 | 180 | 365 |
|---|---|---|---|
| $D_w(t)/Q_w$ | 0.82 | 0.90 | 0.92 |
| $D_w$ (m³ s⁻¹) | $8.2 \times 10^{-4}$ | $9.0 \times 10^{-4}$ | $9.2 \times 10^{-4}$ |

(Note that, if the well in Example 8-5 were only 20 m from a stream, it would not be in a quasi-infinite aquifer and the drawdown calculations in that example would not be good approximations of the actual drawdown.)

### Saltwater Intrusion

Consider a well located above the fresh-salt interface in a coastal aquifer like that shown in Figure 8-25. By the Ghyben-Herzberg Principle (Box 8-2), the elevation of the interface will increase by 40 m for every 1 m in drawdown caused by pumping. Thus if the drawdown at the well approaches 1/40th of the vertical distance between the bottom of the well and the interface, the well is likely to pump saltwater. (See Figure 10-6.)

To the extent that the idealized conditions assumed in deriving them are not satisfied, Equations (8-60)–(8-62) will not exactly predict the drawdown in a coastal aquifer. This may occur when the aquifer is not quasi-infinite, the well is not fully penetrating, or the interface between freshwater and saltwater is not sharp. Modified versions of those equations are available to account for these complicating conditions (Freeze and Cherry 1979).

### 8.6.3 "Safe Yield"

It is widely believed, even by many hydrologists and water-resource managers, that the sustainable rate of extraction—or "**safe yield**"—of ground water from a basin equals the rate of natural recharge, $R_{nat}$. It should be clear from the preceding discussion that this is not true: Equation (8-65) shows that the rate of extraction is supplied by a decrease in storage and, in general, by *changes* in recharge and/or discharge. $R_{nat}$ itself does not enter into that equation, and in fact it is relevant to the determination of "safe yield" only to the extent that it should be accounted for in ground-water models.

Instead, "safe yield" is best defined as "the rate at which ground water can be withdrawn without producing undesirable effects." The preceding discussion has identified the most important hydrologic impacts of ground-water extraction, and most of these have potentially undesirable effects:

- Lowered water table may cause land subsidence as some of the overburden stresses formerly supported by ground water are transferred to the mineral grains.
- Costs of pumping, which are proportional to depth of water table, rise.
- Water tables lowered by one developer may fall below depths of nearby wells belonging to others, perhaps resulting in legal action.
- Reductions of streamflow may seriously reduce surface water available for instream and withdrawal uses.
- Levels or extents of lakes and wetlands may be reduced, with consequent loss of valued habitat.
- Extent of areas where water is available to plants that exploit the capillary fringe

(phreatophytes) may be reduced, with consequent loss of habitat.[12]
- Ground-water outflow to the ocean may be reduced, with consequent effect on coastal wetlands and near-shore benthic marine habitats.
- The fresh-salt interface may be raised, increasing the likelihood of saltwater intrusion.

Due to the varying importance of all these hydrologic effects and their economic, social, environmental, and legal consequences in different regions and within a given region, there is no general formula for computing "safe yield." Acceptable rates of development can only be determined by (1) determining the likely hydrologic effects of various combinations of rates, timing, and location of ground-water extraction; (2) assessing the environmental, economic, legal, and social impacts of these effects; and (3) balancing the benefits afforded by the ground water provided against the undesirable consequences of the various schemes. These considerations are discussed more fully in Section 10.2.4.

## EXERCISES

Exercises marked with ** have been programmed in EXCEL on the CD that accompanies this text. Exercises marked with * can be advantageously executed on a spreadsheet, but you will have to construct your own worksheets to do this.

**8-1.** Obtain data on the configuration and properties of the coastal aquifers closest to your region and compute (a) the depth to the fresh-salt interface as a function of distance from the coast and (b) the width of the seepage face through which fresh ground water discharges to the ocean. (c) Find out from marine biologists if it is possible that discharging fresh ground water has any effects on benthic marine life.

**\*\*8-2.** (a) Using the WellFunc.xls spreadsheet on the CD accompanying the text, compute the drawdown for a well pumping 100,000 gal day$^{-1}$ from the Long Island aquifer described in Example 8-1 for distances up to 3 km and times up to 10 yr. Assume the aquifer has a specific

---

[12] Destruction of phreatophytes has sometimes been advanced as a strategy for reducing evapotranspiration losses and increasing the water available from basins in semi-arid and arid regions.

yield of 0.20 and that the well extends to 30 m below sea level. (b) Calculate the position of the fresh-salt interface at those times—is there a danger of saltwater intrusion?

**\*8-3.**    (a) Confirm the computations in Example 8-2 for the water balance of Long Island, NY. (b) Calculate the range of possible values for $G_{out}$, $R_I$, and $R$ if we accept that $R_{SW}$ and $G_{in} = 0$ and the values in Table 8-2 have the following relative uncertainties: $P$, $Q$: $\pm$ 5 %; $ET$, $CR$, $Q_{GW}$: $\pm$ 10 %.

**\*8-4.**    Develop a water balance like that of Example 8-2 for a drainage basin in your region, including estimates of uncertainty in measured and estimated quantities.

**8-5.**    Find reports estimating recharge from infiltration in your region. (a) What methods were used? (b) Do various methods and reports give consistent results? (c) How are spatial and temporal variability accounted for? (d) What is your overall assessment of the state of knowledge of $R_I$ in your region? (e) What type of studies would you recommend to improve knowledge of this quantity? (f) Do any of these studies assert that $R_I$ is the value of safe or sustainable yield for your region?

**8-6.**    (a) What are the most important potential adverse impacts of ground-water development in your region? (b) Is there evidence that any of these impacts have occurred? (c) What studies would you recommend for determining "safe yield" in your region?

**\*8-7.**    Redo Example 8-3 with $L = 1000$ m; $K_h = 10^{-4}$ m s$^{-1}$; $\Delta h = 0.5$ m.

**8-8.**    Locate a gaining stream reach at least 10 stream widths long, which receives no tributaries and which can be waded to make accurate velocity-area discharge measurements (see Section F.2) at the upstream and downstream ends. (a) Estimate $Q_{GW}$ for the reach by measuring streamflow at these two locations during a period when only base flow is present. (b) Estimate the uncertainty in your measurement. (c) Is your estimate consistent with those given by other lines of reasoning?

**8-9.**    If $a_0 = 0$ in Equation (8-39), show by comparing Equations (8-34) and (8-39) that $a_I = \exp(-k)$ for a period when $W_i = 0$ for more than $n$ days.

**\*\*8-10.**    (a) Use the spreadsheet program DeepSeep.xls to explore the effects of various degrees of relative uncertainty in estimates of precipitation and evapotranspiration on the $G$ terms as estimated by Equation (8-56) for Example 8-4. (b) Does your analysis tend to confirm or cast doubt on the hypothesis that deep seepage is important in this region?

**\*\*8-11.**    (a) Find a drainage basin in your region in which an analysis like that of Example 8-4 can be done, and carry out the analysis. (b) Test for the effects of uncertainty on your estimates of deep seepage. (c) What do you conclude?

**\*8-12.**    Plot the drawdown curves (water-table elevations) in Example 8-5 as a function of distance. Show the curves for the various times on the same graph.

**\*\*8-13.**    (a) Use the spreadsheet program WellFunc.xls to explore the effects of varying at least two of the variables $Q_w$, $K_h$, $S_y$, and $h_0$ on drawdown. (b) Write a paragraph describing your results.

**\*\*8-14.**    (a) Use the spreadsheet program WellFunc.xls to compute drawdowns as a function of distance and time for a typical well in your region. (b) How well do the assumptions involved in deriving the well function apply for this case?

# 9

# Stream Response to Water-Input Events

Streams are the routes by which the precipitation excess (i.e., the portion of precipitation not evaporated) on the continents is returned to the oceans, completing the global hydrologic cycle (Tables 3-1, 3-2; Figure 3-16). There are three principal scientific and practical motivations for studying stream response to water-input events:

*Water Supply* The precipitation excess moving through the stream network constitutes the water resources available for human use and management.

*Flood Prediction and Forecasting*[1] Flood predictions are the basis for the design of bridges, dams, and levees and the formulation of floodplain land-use plans and regulations. (See Section 10.4.) Flood forecasts are estimates of the streamflow response to an actual event that is occurring or is forecast to occur; these are used to guide the operation of reservoir systems and to provide flood warnings to floodplain occupants.

*Water Quality* Water quality is strongly influenced by chemical and biological reactions that occur as water moves over and through the land surface toward streams.

The principal focuses of this chapter are (1) the processes and routes by which precipitation excess on a drainage basin enters streams and moves through the stream network in response to individual rainstorm and snowmelt events and (2) some of the simpler and more commonly used approaches to forecasting and predicting such responses ("rainfall–runoff modeling").

In order to develop a sound understanding of these topics, we begin with a description of the basic aspects of stream response, including definitions of the terms used to quantify response and the ways in which response is affected by characteristics of the input and the watershed. With this background, we then focus on the physical processes that produce event response and the watershed conditions under which the various processes operate. This is followed by an introduction to the physics of open-channel flows and streamflow routing, and to the quantitative description of the stream networks through which these flows occur. The chapter concludes with a discussion of some of the fundamental considerations in rainfall–runoff modeling and an introduction to some simple conceptual rainfall–runoff models that are widely applied in flood prediction and forecasting.

## 9.1 BASIC ASPECTS OF STREAM RESPONSE

### 9.1.1 The Phenomenon of Stream Response

Figure 9-1 shows possible flow paths in a small upland watershed during a water-input event. Input rates $[L\,T^{-1}]$ are measured at one or more points on the watershed and spatially averaged (see Section 4.3); a graph of water input vs. time is called a

[1] The distinction between predictions and forecasts is introduced in Section 2.9.

**389**

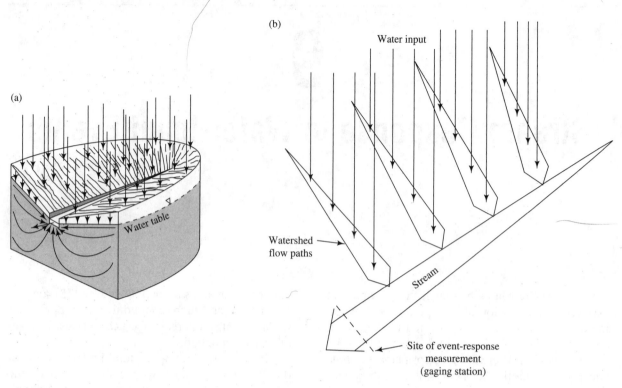

**FIGURE 9-1**

(a) Schematic flow paths in a small upland watershed receiving water input. (b) The essence of watershed response as the integrated result of flow with lateral inflows.

**hyetograph.** Watershed response to the event (output) is characterized by measuring the stream discharge (i.e., the volume rate of flow $[L^3 \, T^{-1}]$) at a single "point" in the stream network—a stream cross section whose location determines the extent of the watershed. (See Section 2.4.) A graph of stream discharge vs. time is a **streamflow hydrograph;** for direct comparison with the input it may be expressed as flow rate per unit watershed area $[L \, T^{-1}]$.

The hydrograph of a stream responding to an isolated period of rain or snowmelt of significant magnitude and areal extent usually has the characteristic form shown in Figure 9-2: Some time after the beginning of the event, the flow rate begins to increase relatively rapidly from the pre-event rate to a well-defined **peak discharge** (the **hydrograph rise**), then it declines more slowly (the **hydrograph recession**) to a rate near its pre-event value.

Figure 9-1 shows that streamflow is a spatially and temporally integrated response determined by (1) the spatially and temporally varying input rates; (2) the time required for each drop of the water to travel from where it strikes the watershed surface

to the stream network (determined by the length, slope, vegetative cover, soils, and geology of the watershed hillslopes); and (3) the time required for water to travel from its entrance into the channel to the point of measurement (determined by the length and nature of the channel network). In small watersheds (less than about 50 km² area), the travel time to the watershed outlet is determined mostly by the hillslope travel time; for larger watersheds the travel time in the stream network becomes increasingly important (Kirkby 1993).

Once the flood wave leaves the portion of the stream network that has been affected by a given input event, its shape is affected solely by bank-storage effects (see Section 8.3.1) and channel hydraulics.

Figure 9-3 shows a typical example of how the effects of hillslope-response mechanisms are gradually superseded by channel-hydraulic effects through a stream network. The hydrograph shape for the smallest watershed is strongly influenced by the form of the hyetograph. Subsequently, the hydrograph is increasingly affected by tributary inputs and by the storage effects of the stream channels, and the net

**FIGURE 9-2**
A storm hydrograph is the time trace made by an observer at a fixed point of a flood wave moving downstream.

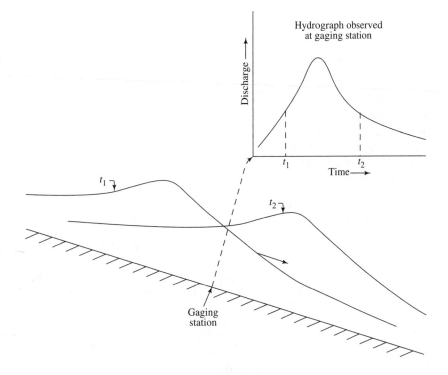

result is an increase in the lag time between the rainfall inputs and the peaks and a decrease in hydrograph ordinates (when scaled by drainage area). The hydrograph also becomes smoother, and at the lowest two gages the formerly multiple-peaked hydrograph has become single-peaked.

In analyzing and modeling these event-response processes, one should keep in mind the following points:

### On the hillslopes

- Water moves in an infinite number of surface and/or subsurface flow paths of varying length and character.
- During the water-input event, each flow path is an accumulation of **lateral inflows** of precipitation and/or snowmelt that vary in space and time (Figure 9-1b).
- Flow in each path can in principle be described by invoking the conservation of mass and an equation of motion appropriate to the mode of flow. The principles of unsaturated and saturated subsurface flow were discussed in Sections 6.3.1 and 8.1, respectively; the principles of surface overland flow are discussed in Section B.3.2.

### In the stream network

- During the event, and while the land surface is draining, flow is an accumulation of temporally and spatially varying lateral inflows from the hillslope flow paths distributed along the channel length (Figure 9-1b).
- Movement of water in the stream can in principle be described by invoking the conservation of mass and the equation of motion for open-channel flow. These principles are introduced in Section 9.3.1 and elaborated in Section B.3.2.
- Flow in the stream takes the form of a **flood wave**[2] that moves downstream through the stream network (Figure 9-2).
- The observed hydrograph records the movement of the flood wave past the fixed point of measurement (Figure 9-2).

### In the overall watershed response

- The volume of water appearing in the response hydrograph for a given event is only a fraction (often a very small fraction) of the

---

[2] This term is used even if no overbank flooding occurs.

**FIGURE 9-3**
Changes in hydrograph shape at a series of gaging stations along the Sleepers River in Danville, VT, in response to an intense rainstorm (hyetograph at top). Note that the left-hand hydrograph ordinates show flow rates per unit drainage area, while right-hand ordinates show actual flow rates. From *Water in Environmental Planning*, by Thomas Dunne and Luna B. Leopold. Copyright © 1978 by W.H. Freeman and Company. Reprinted by permission.

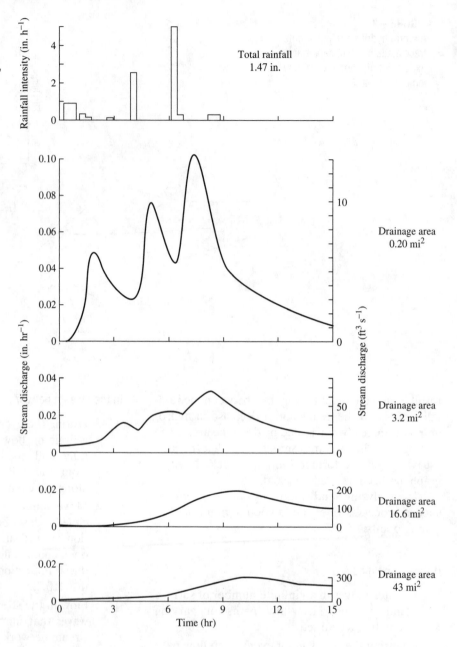

total input. The remainder of the water input ultimately leaves the watershed as (1) evapotranspiration, (2) streamflow that occurs so long after the event that it cannot be associated with that event ("base flow"), or (3) ground-water outflow.

- The water identified as the response to a given event may originate on only a fraction of the watershed; this fraction is called the **contributing area.**

- The extent of the contributing area may vary from event to event and during an event.
- At least some of the water identified as the response to a given event may be "old water" that entered the watershed in a previous event.

These points are elaborated in subsequent sections of this chapter.

**FIGURE 9-4**
Hyetograph and response hydro-graph for an isolated storm on Watershed W-3, Sleepers River Research Watershed, Danville, VT, September 1973. Numbers on time axis indicate noon on indicated dates.

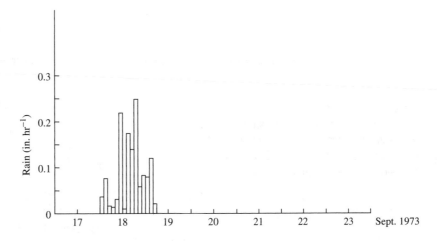

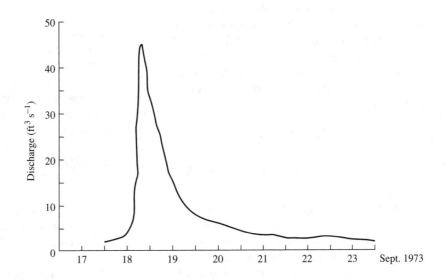

## 9.1.2   Hydrograph Separation

Figure 9-4 shows the response of a watershed to an isolated rainfall event lasting about 24 hr. As in most humid-region streams, there is non-zero flow before the event and very gradually declining flow persisting long after the obvious hydrograph response. Clearly, some of the flow during the hydrograph rise and early recession must have entered the watershed in previous events, and some of the water that entered during the current event will flow out of the watershed well after the identifiable hydrograph response. This typical behavior raises a question that is central to the analysis of stream response to water inputs: How do we identify the por-tion of stream response that is associated with a given event?

This question can be answered with confidence only if water entering the stream from specific sources (e.g., direct rain onto the channel, flow from the watershed surface, ground water) has a charac-teristic water-quality "signature" and if the compo-nents of this signature in streamflow are monitored during and following an event. Box 9-1 describes how chemical and isotopic tracers can be used to identify the various sources of streamflow, and Fig-ure 8-37 shows an example of the application of these methods.

## BOX 9-1

. . . . . . .

### Chemical and Isotope Separation of Streamflow Components

If streamflow comes from two sources (e.g., overland flow and ground water), each of which has a constant known characteristic concentration of a particular chemical or isotopic constituent, we can write mass-balance equations for water and for the constituent as

$$q(t) = q_1(t) + q_2(t) \tag{9B1-1}$$

and

$$c(t) \cdot q(t) = c_1 \cdot q_1(t) + c_2 \cdot q_2(t), \tag{9B1-2}$$

where the $c$s represent concentrations, the $q$s represent discharges, the subscripts denote the two components, and $(t)$ denotes variability with time. Then if the total flow rate, $q(t)$, and its concentration, $c(t)$, are measured, one can combine Equations (9B1-1) and (9B1-2) to calculate the discharge contributed by each component:

$$q_2(t) = \frac{c(t) - c_1}{c_2 - c_1} \cdot q(t); \tag{9B1-3}$$

$$q_1(t) = q(t) - q_2(t). \tag{9B1-4}$$

Contributions of three streamflow components can be determined, if one can measure the flow rate contributed by one of the components as well as the total flow rate and concentration. For example, Swistock et al. (1989) used rainfall measurements to determine the contribution of channel precipitation, $q_{cp}(t)$, and were able to determine the contributions of soil water, $q_s(t)$, and ground water, $q_{gw}(t)$, as

$$q_s(t) = q(t) \cdot \frac{c(t) - c_{gw}}{c_s - c_{gw}} - q_{cp}(t) \cdot \frac{c_{cp} - c_{gw}}{c_s - c_{gw}} \tag{9B1-5}$$

and

$$q_{gw}(t) = q(t) - q_{cp}(t) - q_s(t), \tag{9B1-6}$$

where the subscripts $cp$, $s$, and $gw$ denote channel precipitation, soil, and ground water, respectively, and $^{18}O$ was the tracer.

These relations have been applied using various chemical and isotopic tracers. The concentrations of chemical tracers can generally be measured more readily than those of isotopes and may be more distinctive than signatures of the stable isotopes $^{18}O$ and $^2H$. (See Section B.1.5.) However, concentrations of chemical constituents may be modified by chemical reactions with mineral or organic components of soils and thus violate the assumption of constant values. Only physical processes such as evaporation and condensation affect variations in isotopic concentrations.

It is important to note that very precise measurements of concentrations are usually needed to get meaningful hydrograph separations, because small errors in concentrations can lead to large errors in computed flow rates (Rice and Hornberger 1998).

Special issues of the *Journal of Hydrology* (v. 116, 1990) and *Water Resources Research* (v. 26, no. 12, 1990) are dedicated to this topic, and Bonell (1993) provided a useful review.

However, knowledge of actual sources of streamflow as provided by chemical or isotopic data is unavailable except in specialized research situations. Thus, almost always, the only data available are measurements of total water input and total streamflow. Since this flow is the sum of responses to the event of interest plus an indeterminate number of preceding events, the problem of identifying particular responses arises. Hydrologists faced with the practical problems of forecasting and predicting streamflow response have filled this knowledge vacuum by developing graphical techniques to separate streamflow into two operationally defined components:

**event flow:** flow that is considered to be the direct response to a given water-input event. This may also be called **direct runoff, storm runoff, quick flow,** or **storm flow.**

**base flow:** flow that is not associated with a specific event. As discussed in Section 8.3.1, this is commonly, though not always correctly, assumed to be due to ground-water inputs.

Three commonly used methods of graphical separation are described in Figure 8-38. As discussed there, the volume of water represented by the area between the measured hydrograph and the separation line is the event-flow volume, $Q_{ef}$. At any instant, we have

$$q_{ef} = q - q_{bf}, \qquad (9\text{-}1)$$

where $q$ is the tota flow rate, $q_{bf}$ is the base-flow rate, and $q_{ef}$ is the event-flow rate. Example 9-1 applies one of these methods to the event shown in Figure 9-5.

---

## EXAMPLE 9-1

Figure 9-5 shows the hydrograph of Watershed W-3 (drainage area = 3.23 mi²) in the Sleepers River Research Watershed in northeastern Vermont in response to a rainstorm with a total area-averaged storm depth $W = 1.59$ in. Find the volume of event flow, $Q_{ef}$, and the ratio of event flow to total rainfall, $Q_{ef}/W$.

**Solution** We use the graphical separation method described in Figure 8-38c. For this case, the separation line (dashed line in Figure 9-5) slopes upward from the time of initial hydrograph rise at a rate of

$$(0.05 \text{ ft}^3 \text{ s}^{-1} \text{ mi}^{-2} \text{ hr}^{-1}) \times (3.23 \text{ mi}^2) = 0.162 \text{ ft}^3 \text{ s}^{-1} \text{ hr}^{-1}$$

and is extended until it intersects the measured hydrograph at about 0500 on 19 September. Using a digitizer or planimeter, we measure the area between the total hydrograph and the separation line and find that this represents a volume of $1.10 \times 10^6$ ft³. Dividing by watershed area, we find that

$$Q_{ef} = \frac{(1.10 \times 10^6 \text{ ft}^3)}{(3.23 \text{ mi}^2)} \times \frac{(1 \text{ mi}^2)}{(5280^2 \text{ ft}^2)} \times \frac{(12 \text{ in.})}{(1 \text{ ft})} = 0.147 \text{ in.}$$

Thus for this storm, $Q_{ef}/W = 0.147$ in./1.59 in. $= 0.092$.

---

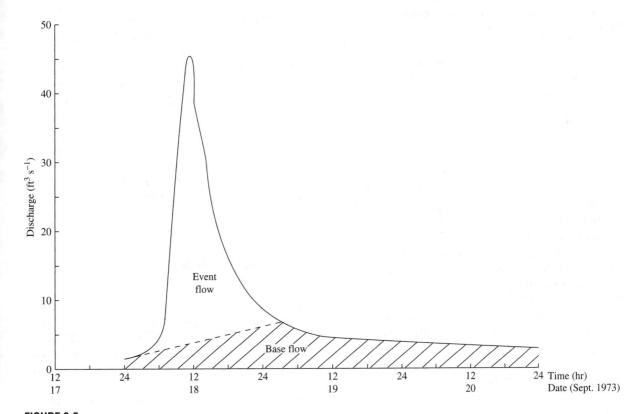

**FIGURE 9-5**
Hydrograph separation applied to the hydrograph of Watershed W-3, Sleepers River Research Watershed, Danville, VT, to rain of 18 September 1973. See Example 9-1.

Unfortunately, there are two serious problems with graphical hydrograph separation: (1) We do not know how to draw the lines separating "event flow" from "base flow" unless we have conducted detailed observations involving tracers that allow us to identify "old" vs. "new" water flowing in the stream (as discussed in Box 9-1). (2) As discussed further in Sections 9.2.3 and 9.2.4, even if old and new water can be distinguished, some of the old water may have been released from storage and entered the stream as a result of the new event. Thus there may be no clear association of "new" water with the most recent event or "old" water with preceding events.

Graphical hydrograph separation must therefore be regarded as a "convenient fiction" that must often be invoked in order to analyze and model the ways in which event response is influenced by permanent and temporary watershed characteristics and by the spatial and temporal variability of water input. In applying such methods in a given study, it is most important to be consistent: there may be a reasonably constant (if unknown) relation between the true event flow and the event flow identified by a particular method on a particular watershed. However, one must not fall into the trap of thinking that graphical separation actually identifies flow from different sources.

### 9.1.3   Event-Flow Volume

#### Spatial Variability

The results of Example 9-1 are quite typical for most regions: The ratio $Q_{ef}/W$ for individual events is usually considerably less than 0.5 and commonly less than 0.1.

Woodruff and Hewlett (1970) investigated the long-term average values of $Q_{ef}/W$ for a large sample of small to moderate-sized watersheds in the southeastern United States (where snow is absent). As in Example 9-1, they used the method of Figure 8-38c to identify event-flow volumes. Their results, summarized in Figure 9-6, reinforce the conclusion that the amount of water involved in event response is a small fraction of inputs and illustrate that average values of $Q_{ef}/W$ are highly variable in space. Colonell and Higgins (1973) found similar ratios in New England.

Woodruff and Hewlett's (1970) results can also be used to show that for the region they studied, the ratio of effective to total runoff, $Q_{ef}/Q$, is consider-

ably less than 0.5 (Figure 9-7). This suggests that more than half of streamflow typically travels to streams via delayed routes as base flow, presumably in large part as the regional ground-water flow discussed in Section 8.2.

The implication of these results is that forecasts and predictions of runoff response are very sensitive to the estimate of total event flow—a small error in estimating the fraction of water input appearing in the response could lead to a large relative error in the predicted runoff peak and volume.

#### Temporal Variability

Figure 9-8 shows $Q_{ef}$ as a function of $W$ for 16 rainstorms during four summers on a research watershed in central Alaska. (The separation in this case was done using the method of Figure 8-38b.) For this period, the $Q_{ef}/W$ ratio varied over an order of magnitude, from 0.03 to 0.42. A similarly high degree of temporal variability in $Q_{ef}/W$ in a single watershed is common in most regions.

An even more striking example of the range of variability of $Q_{ef}/W$ for a given watershed is shown in Figure 9-9. Figure 9-9a is the response of this watershed to an isolated storm in August 1963 during a very wet summer in northern Alaska, $Q_{ef}/W = 0.63$. In the very dry summer of 1965, a storm of similar intensity and total volume resulted in the hydrograph of Figure 9-9b, in which $Q_{ef}/W = 0.01$—less than 1/50th of the earlier value!

The implication of these results is that failure to account for time-varying watershed conditions may result in significant forecasting or prediction errors.

### 9.1.4   Quantitative Description of Response Hydrographs

The terms used to analyze and model event hyetographs and response hydrographs are defined on Figure 9-10 and in Table 9-1 and are discussed below. Note that we are focusing only on the portion of streamflow identified as event flow, $Q_{ef}$, and the fraction of water input that appears as event flow, called the **effective water input** (or **effective rainfall** or **excess rainfall**), $W_{eff}$. Note also that

$$W_{eff} = Q_{ef}. \tag{9-2}$$

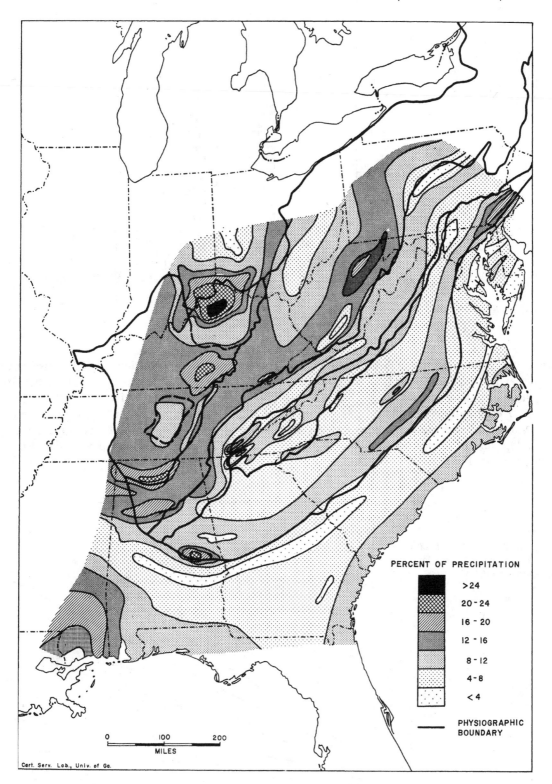

**FIGURE 9-6**
Spatial distribution of average annual event-flow volume for watersheds less than 200 mi$^2$ (500 km$^2$) area in the southeastern United States expressed as ratio to average annual precipitation. From Woodruff and Hewlett (1970), used with permission of the American Geophysical Union.

**FIGURE 9-7**

Frequency distribution of average annual $Q_{ef}/Q$ for 59 watersheds in the southeastern United States. $Q_{ef}$ data from Woodruff and Hewlett (1970).

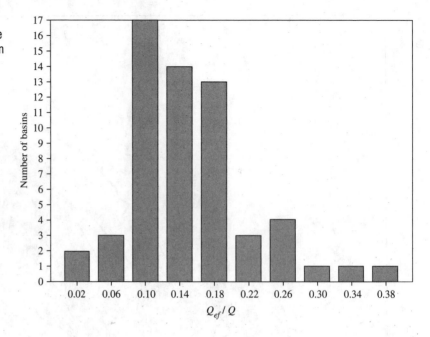

In Figure 9-10, water input begins at time $t_{w0}$ and ends at $t_{we}$; the input duration (often called **storm duration**), is $T_W \equiv t_{we} - t_{w0}$. Stream response, or **hydrograph rise,** begins at $t_{q0}$ and usually continues more or less steadily to a peak flow, $q_{pk}$, at time $t_{pk}$. Typically the peak occurs when or soon after input ceases. The delay between the beginning of input and the beginning of the rise is the **response lag,** $T_{LR} \equiv t_{q0} - t_{w0}$. The time between the beginning of input and peak is the **lag-to-peak,** $T_{LP} \equiv t_{pk} - t_{w0}$. The duration of the hydrograph rise is called the **time of rise,** or **time to peak,** $T_r \equiv t_{pk} - t_{q0}$. Following the peak, the response declines quasi-exponentially as water input drains

**FIGURE 9-8**

Effective precipitation, $W_{eff}$, vs. total precipitation, $W$, for 16 summer rainstorms on a 1.8-km$^2$ watershed in central Alaska. Data from Dingman (1970).

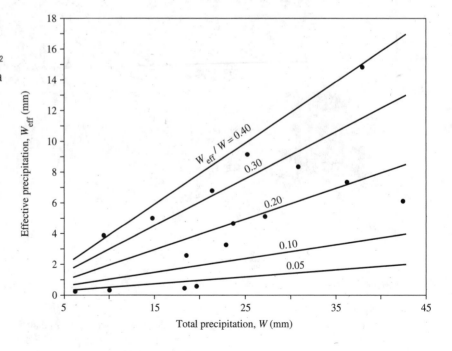

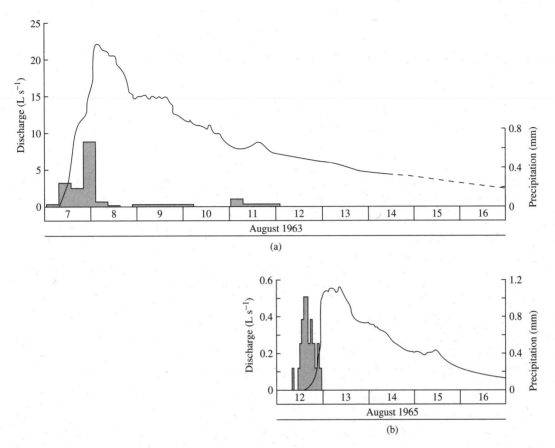

**FIGURE 9-9**
Hydrographs and hyetographs (histograms) for two similar storms on a 1.6-km² watershed at Barrow, Alaska. Note different scales. (a) August 1963: $W$ = 7.9 mm, $T_w$ = 24 hr; $Q$ = 5.0 mm, $Q/W$ = 0.63. (a) August 1965: $W$ = 6.9 mm, $T_w$ = 13.5 hr; $Q$ = 0.08 mm, $Q/W$ = 0.012.

from watershed storage; response ceases at time $t_{qe}$. The total duration of the response hydrograph is called the **time base,** $T_b \equiv t_{qe} - t_{q0}$.

The **centroid,** or **center of mass,** of the input, $t_{wc}$, is the weighted-average time of occurrence of the input hyetograph.[3] If we have values of input, $W_i$, measured for $i$ = 1, 2, ..., $n$ time periods of equal length, the centroid, $t_{wc}$, is calculated as

$$t_{wc} = \frac{\sum\limits_{i=1}^{n} W_i \cdot t_i}{\sum\limits_{i=1}^{n} W_i}, \qquad \textbf{(9-3)}$$

where $t_i$ is the time halfway through period $i$.

---

### EXAMPLE 9-2

The hyetograph for the rain storm recorded at Watershed W-3 of the Sleepers River Research Watershed, Vermont, on 18 September 1973 is given in the following table. Calculate the centroid.

---

[3] If the hyetograph or hydrograph is imagined to represent the distribution of masses on a bar, the centroid is the balance point.

| Clock Time (hr) | 0 | 1 | 2 | 3 | 4 | 5 | 6 | 7 | 8 |
|---|---|---|---|---|---|---|---|---|---|
| Rain $W_i$ (in.) | 0.04 | 0.08 | 0.04 | 0.00 | 0.04 | 0.22 | 0.23 | 0.22 | |
| Time $t_i$ (hr) | 0.5 | 1.5 | 2.5 | 3.5 | 4.5 | 5.5 | 6.5 | 7.5 | |

| Clock Time (hr) | 9 | 10 | 11 | 12 | 13 | 14 | 15 | 16 | 17 |
|---|---|---|---|---|---|---|---|---|---|
| Rain $W_i$ (in.) | 0.11 | 0.21 | 0.09 | 0.09 | 0.09 | 0.02 | 0.07 | 0.04 | |
| Time $t_i$ (hr) | 8.5 | 9.5 | 10.5 | 11.5 | 12.5 | 13.5 | 14.5 | 15.5 | |

**Solution** Calculate the sums in Equation (9-3) using the data from the preceeding table. The sum of the $W_i \cdot t_i$ products is

$$(0.04 \text{ in.} \times 0.5 \text{ hr}) + (0.08 \text{ in.} \times 1.5 \text{ hr}) + \dots$$
$$+ (0.04 \text{ in.} \times 15.5 \text{ hr}) = 12.715 \text{ in.} \times \text{ hr.}$$

The total rainfall is

$$0.04 \text{ in.} + 0.08 \text{ in.} + \dots + 0.04 \text{ in.} = 1.59 \text{ in.}$$

Then, from Equation (9-3), the centroid is

$$t_{wc} = \frac{12.715 \text{ in. hr}}{1.59 \text{ in.}} = 8.00 \text{ hr.}$$

Since $t_0 = 00{:}00$ clock time on 18 September, the centroid of this storm occurred at $00{:}00 + 8.00 \text{ hr} = 08{:}00$ clock time.

**FIGURE 9-10**
Definitions of terms used to describe hyetographs and response hydrographs. See Table 9-1.

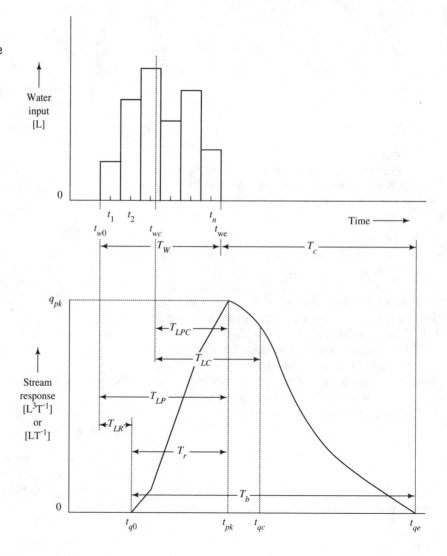

**TABLE 9-1**

Definitions of Terms Used to Describe Hyetographs and Response Hydrographs. See Figure 9-10.

| Time Instants | Time Durations |
|---|---|
| $t_{w0} \equiv$ beginning of effective water input | $T_w \equiv$ duration of effective water input $= t_{w0} - t_{we}$ |
| $t_{wc} \equiv$ centroid of effective water input | $T_{LR} \equiv$ response lag $= t_{q0} - t_{w0}$ |
| $t_{we} \equiv$ end of effective water input | $T_r \equiv$ time of rise $= t_{pk} - t_{q0}$ |
| $t_{q0} \equiv$ beginning of hydrograph rise | $T_{LP} \equiv$ lag-to-peak $= t_{pk} - t_{w0}$ |
| $t_{pk} \equiv$ time of peak discharge | $T_{LPC} \equiv$ centroid lag-to-peak $= t_{pk} - t_{wc}$ |
| $t_{qc} \equiv$ centroid of response hydrograph | $T_{LC} \equiv$ centroid lag $= t_{qc} - t_{wc}$ |
| $t_{qe} \equiv$ end of response | $T_b \equiv$ time base $= t_{qe} - t_{q0}$ |
| | $T_c \equiv$ time of concentration $= t_{qe} - t_{we}$ |
| | $T_{eq} \equiv$ time to equilibrium $\approx T_c$ |

The centroid of a response hydrograph, $t_{qc}$, can be estimated by approximating the continuous curve with a histogram and substituting the appropriate values into Equation (9-3).

The **centroid lag,** $T_{LC}$, is defined as the time between the centroids of input and response: $T_{LC} \equiv t_{qc} - t_{wc}$. As will be discussed more fully later, the centroid lag is a theoretically useful value characterizing the response time of a watershed. However, a more commonly used measure of watershed response time is the time between the centroid of input and the peak, called the **centroid lag-to-peak,** $T_{LPC} \equiv t_{pk} - t_{wc}$.

Another measure of watershed response time is the **time of concentration,** $T_c$, defined as the time it takes water to travel from the hydraulically most distant part of the contributing area to the outlet. Thus, $T_c = t_{qe} - t_{we}$. It is difficult to define the time of concentration for an entire watershed because of uncertainties about the nature of the actual flow paths. Formulas for estimating $T_c$ from watershed and storm characteristics are discussed in Section 9.6. It is commonly assumed that $T_{LPC} \approx 0.60 \cdot T_c$.

In spite of the difficulty in obtaining a precise operational a priori definition of time of concentration in most natural situations, the concept is useful in visualizing hydrologic response: If effective water input continues at a constant rate for a duration equal to $T_c$, eventually the outflow rate will become equal to the input rate and the hydrograph peak will be effectively constant at that rate until input ceases (Figure 9-11). This condition is called **equilibrium runoff,** and the time at which the input and output rates become effectively equal is an apparent time of concentration, called the **time to equilibrium,** $T_{eq}$. When equilibrium runoff is occurring, all parts of the contributing area provide flow, and we can write

$$q(t) = q_{pk} = w \cdot A_c, \quad T_{eq} < t - t_{w0} < T_w, \quad \text{(9-4)}$$

where $q_{pk}$ is the peak flow rate [L$^3$ T$^{-1}$], $w$ is the effective water-input rate [L T$^{-1}$], and $A_c$ is the contributing area [L$^2$].

Equilibrium runoff almost never occurs in natural watersheds, but it can occur from small areas with short times of concentration, such as parking lots. Thus most hydrographs from natural areas have an instantaneous peak, as in Figure 9-4, and the peak flow rate will be less than the input rate. This implies that the entire contributing area does not produce runoff simultaneously and that

$$q_{pk} < w \cdot A_c. \quad \text{(9-5)}$$

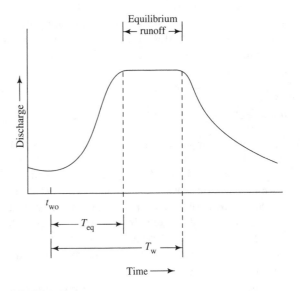

**FIGURE 9-11**

Schematic hydrograph for equilibrium runoff resulting when input duration, $T_w$, exceeds the watershed's time of equilibrium, $T_{eq}$.

### 9.1.5 Effects of Input and Basin Characteristics on the Hydrograph

The objective of this section is to explore in a general way how storm size and timing and drainage-basin characteristics affect the hydrograph characteristics defined in Table 9-1 and Figure 9-4. To do this we use the simple linear-watershed model developed in Box 9-2. This model is based on the principle of continuity [Equation (9B2-1)] and on a very simple but reasonable concept of watershed function—the linear-reservoir model [Equation (9B2-2)].

A linear watershed is characterized by a single parameter, $T^*$, called the **response time**—the larger the value of $T^*$, the more sluggish the response of the watershed. Thus $T^*$ integrates the effects of all the factors that affect the travel time of water input to the basin outlet: watershed size, shape, slope, and the nature of the various flow paths as determined by land-surface conditions, soils, and geology. The discussion in Section 9.2 will provide more detailed insight into these relationships.

In the examples used here to illustrate basic hydrograph features, the input, output, and time values are arbitrary. Thus although $T^*$ is given in "hr," this is really only an arbitrary measure of watershed response time, and all times are relative. Inputs and outputs are also in arbitrary units with dimensions $[L^3 T^{-1}]$ or $[L T^{-1}]$. Inputs to the model represent effective water input, and in all the simulations examined here the total volume of input, $W$, is the same: 1 unit. Because the model preserves continuity, the total volume of outflow, $Q$, is also always 1 unit.

#### Basic Hydrograph Shape

Figure 9-12 shows the response of the linear watershed with $T^* = 1$ "hr" to a constant unit input for 1 "hr." The general form of this hydrograph is similar to that for natural watersheds, suggesting that the typical pattern of quick rise to an instantaneous peak followed by gradual recession is inherent to the general form of storage–outflow relations for watersheds in which $T_w < T_{eq}$, as expressed in Equations (2-26) and (9B2-2).

Figure 9-13 shows the same input volume applied to the same watershed, but with $T_w$ equal to $T_{eq}$ as given by Equation (9B2-8). This produces equilibrium runoff as described earlier.

Figure 9-14 shows the same input amount and timing as in Figure 9-12, but to a watershed with $T^* = 5$ "hr." Note that in this case the peak is much lower and the recession much longer.

### Response Time, $T^*$, and Centroid Lag, $T_{LC}$

As noted earlier, $T^*$ is related to the time required for water to travel to the watershed outlet, so that its value is determined largely by the following factors:

*Watershed Size* Larger watersheds have larger $T^*$, other things equal. Analyses of large floods by Holtan and Overton (1963) indicated that $T^*$ is strongly related to drainage area; however, the relationship varied from region to region, presumably largely because of differences in watershed geology, soils, and topography.

*Soils and Geology* Water moves fastest toward streams as sheet flow and slowest as subsurface flow (Section 9.2). Thus, watersheds with low surface hydraulic conductivities (e.g., clay soils), where less infiltration and more overland flow occur, should have smaller $T^*$ than those with higher conductivities (e.g., sandy soils). However, if flow paths are predominantly in the subsurface, watersheds with higher conductivities will have smaller values of $T^*$.

*Slope* Steeper slopes should be associated with faster surface and subsurface water movement and hence smaller $T^*$.

*Land Use* Watersheds with intensive urbanization (more impermeable areas and storm sewers) should have faster water movement and smaller $T^*$.

However, determining $T^*$ for real watersheds is difficult because of the lack of information on the timing and amounts of effective, as opposed to total, water inputs and outputs. Furthermore, $T^*$ generally varies temporally on a given watershed due to the variability of contributing area and flow rates, both of which are dependent on the antecedent watershed wetness and other seasonal factors (Figures 9-8, 9-9). However, it may be less important to account for temporal variability if

## BOX 9-2

· · · · · · · ·

### Linear-Reservoir Model of Watershed Response

We can develop a very simple conceptual model of the response of a watershed to water input based on (1) the principle of conservation of mass,

$$w - q = \frac{dV}{dt} \qquad \text{(9B2-1)}$$

[see Equation (2-9)], and (2) the linear-reservoir conceptual model of watershed behavior given in Equation (2-26):

$$q = \frac{1}{T^*} \cdot V. \qquad \text{(9B2-2)}$$

Here $w$ is water-input rate ([L$^3$ T$^{-1}$] or [L T$^{-1}$]), $q$ is event flow ([L$^3$ T$^{-1}$] or [L T$^{-1}$]), $V$ is storage of event-flow water ([L$^3$] or [L]), $t$ is time, and $T^*$ is a time constant that characterizes the watershed response [T]. Real watersheds are not strictly linear reservoirs, but Equation (9B2-2) captures the most basic aspects of watershed response and is mathematically tractable.

Combining Equations (9B2-1) and (9B2-2) yields

$$w - q = T^* \cdot \frac{dq}{dt}, \qquad \text{(9B2-3)}$$

which, for constant $w$, has the solution

$$q = w + (q_0 - w) \cdot \exp(-t/T^*), \qquad \text{(9B2-4)}$$

where $q_0$ is the outflow at $t = 0$. To model an isolated hydrograph rise in response to a constant input beginning at $t = 0$ and lasting until time $t_w$, we can set $q_0 = 0$ at $t = 0$ so that Equation (9B2-4) becomes

$$q = w \cdot [1 - \exp(-t/T^*)], \ t \leq t_w. \qquad \text{(9B2-5)}$$

The peak discharge, $q_{pk}$, is then given by

$$q_{pk} = w \cdot [1 - \exp(-t_{pk}/T^*)]. \qquad \text{(9B2-6)}$$

For the recession, $w = 0$, the "initial" discharge is $q_{pk}$ at time $t = t_{pk} = t_w$, and Equation (9B2-4) becomes

$$q = q_{pk} \cdot \exp[-(t - t_{pk})/T^*], \ t \geq t_{pk}. \qquad \text{(9B2-7)}$$

The properties of this model are as follows:

1. The model preserves continuity (i.e., the volume of event response equals the volume of effective water input $= w \cdot T_w$).
2. $T^*$ can be shown to be equal to the centroid lag of the watershed (i.e., $t_{qc} - t_{wc} = T^*$).
3. The hydrograph rise begins as soon as input begins ($T_{LR} = 0$) and peaks exactly at $t_{pk} = t_{we}$. (However, it would be a simple matter to add a time delay to the model.)
4. Because outflow decreases exponentially after input ceases and approaches zero asymptotically, time of concentration is infinite. However, if we define the time to equilibrium, $T_{eq}$, as the time it takes for the outflow rate to reach 1% of a constant inflow rate, it can be shown that

$$T_{eq} = -\ln(0.01) \cdot T^* = 4.605 \cdot T^*. \qquad \text{(9B2-8)}$$

---

one restricts consideration to large floods, when the watershed is generally wet and contributing area may be fairly constant (Askew 1970; Pilgrim 1976).

In the linear watershed model the centroid lag, $T_{LC}$, is equal to $T^*$, regardless of the timing of inputs. Thus, if the timing and amounts of effective water input and output can be determined, $T_{LC}$ would appear to be the best estimate of the characteristic response time for a given watershed.

### Response Lag, $T_{LR}$

In the linear watershed model response begins as soon as input begins. The time between the beginning of water input and the beginning of measurable response in real watersheds is largely determined by the time required to fill storage (canopy interception, infiltration, and surface depressions) plus the average travel time to the basin outlet. Very short response times can result when water inputs cause a

**FIGURE 9-12**
Response, $q$, of linear watershed with $T^* = 1$ hr to 1 unit of input, $W$, in 1 hr. Input rate $w = 1$ unit/hr.

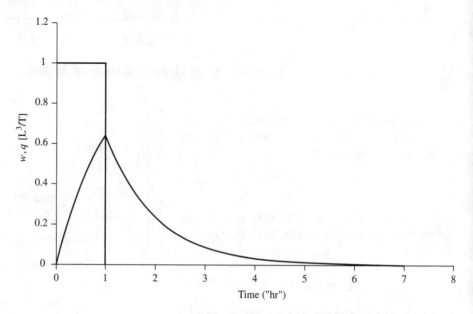

nearly immediate change in hydraulic conditions that causes channelward flow of "old" water held in soil storage near streams (Section 9.2.3).

### Time of Rise, $T_r$

Experiments with the linear watershed model show that, for a constant input rate, the time of rise equals the duration of effective water input (compare Figures 9-12 and 9-13). Observations on real watersheds also show a close correspondence between $T_r$ and storm duration for large storms (Holtan and Overton 1963; 1964). If effective input rates vary

markedly during an event, the time of peak is usually determined by the timing of the highest input rates, as illustrated in Figures 9-15 and 9-16 (and, for a real watershed, Figure 9-3).

### Lag-to-Peak, $T_{LP}$, and Centroid Lag-to-Peak, $T_{LPC}$

Some hydrologists have considered $T_{LP}$ or $T_{LPC}$ to be a characteristic time of a watershed, determined by the same factors that control $T^*$. However, experiments with the linear watershed model show that the timing of the inputs is at least as important in determining these lags: For a constant input rate, the lag-

**FIGURE 9-13**
Response, $q$, of linear watershed with $T^* = 1$ hr to 1 unit of input, $W$, in 4.5 hr. Input rate $w = 0.222$ unit/hr.

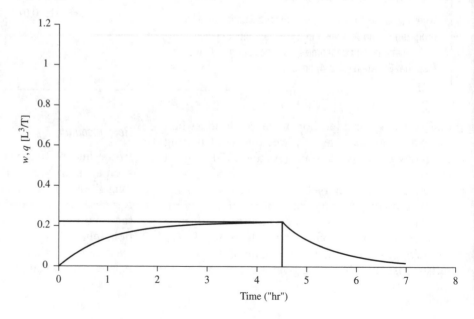

**FIGURE 9-14**
Response, $q$, of linear watershed with $T^* = 5$ hr to 1 unit of input, $W$, in 1 hr. Input rate $w = 1$ unit/hr.

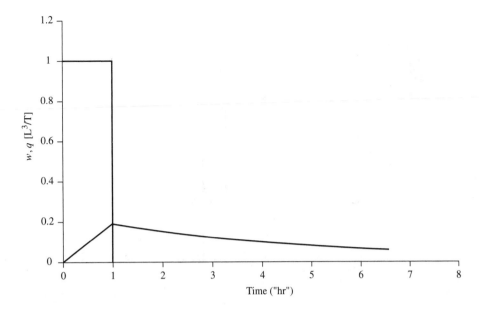

to-peak and the centroid lag-to-peak of the linear watershed are equal to one-half the input duration; when input rates vary, these times are determined by both the exact timing of the inputs and the watershed characteristics as reflected in the value of $T^*$.

### Time Base, $T_b$, and Recession

The recession of the linear watershed follows an exponential decay with decay constant equal to $T^*$ [Equation (9B2-7)]; thus the theoretical time base is infinite. However, defining the end of runoff when $q = 0.01 \cdot w$ as in Box 9-2, we see that $T_b = T_w + 4.605 \cdot T^*$ and is therefore determined by both the duration of input and by watershed characteristics.

To the extent the recession for a given stream approximates exponential decay, the decay constant can be used to estimate $T^*$ [Equation (9B2-7)]. However, recessions for a given watershed often appear to follow different decay "constants" (apparent values of $T^*$) in different discharge ranges and at different seasons, making it

**FIGURE 9-15**
Response, $q$, of linear watershed with $T^* = 1$ hr to 1 unit of input, $W$, in 1 hr. Input rate $w = 0.75$ unit/hr for first 0.5 hr and 0.25 unit/hr for second 0.5 hr.

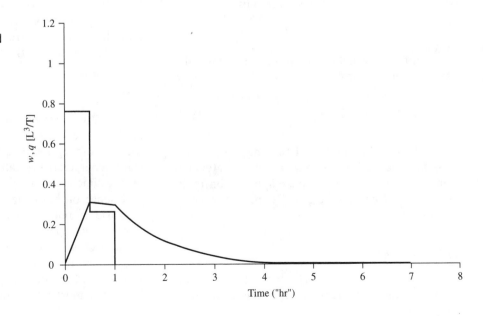

**FIGURE 9-16**
Response, $q$, of linear watershed with $T^* = 1$ hr to 1 unit of input, $W$, in 1 hr. Input rate $w = 0.25$ unit/hr for first 0.5 hr and 0.75 unit/hr for second 0.5 hr.

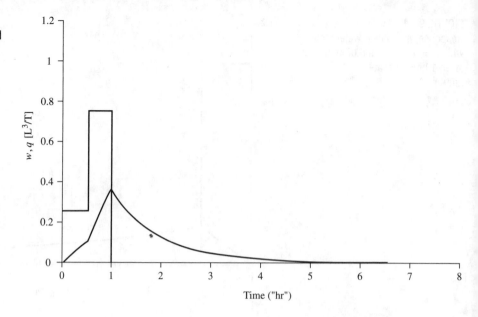

difficult to identify a "fundamental" time constant for a watershed. As noted in Section 8.5.3, the exponential-decay model has been used as a basis for graphical hydrograph separation (Figure 8-38b).

### Time of Concentration, $T_c$

Conceptually, the time of concentration is a constant characteristic time for a given watershed, given by the time between the end of input and the end of response. Usually this time will be approximately equal to the duration of the recession, and from Box 9-2 we see that this is given by

$$T_c = 4.605 \cdot T^*. \qquad \textbf{(9-6)}$$

Empirical formulas for estimating $T_c$ typically have the general form

$$T_c \propto \frac{L^a}{S^b}, \qquad \textbf{(9-7)}$$

where $L$ is some measure of the length of watershed flow paths, $S$ is some measure of watershed slope, and $a$ and $b$ are positive empirical constants. Several such formulas are given later in this chapter (Table 9-9).

### Peak Discharge, $q_{pk}$

Equation (9B2-6) shows that, for a constant water-input rate, the peak discharge rate is determined by the rate and duration of input and the basin characteristics as reflected in $T^*$. Interestingly, Holtan and Overton (1963) found that peaks of large floods on actual watersheds could be well estimated by a relation almost identical to that of Equation (9B2-6).

The effect of $T^*$ on $q_{pk}$ can be seen by comparing Figure 9-14 with Figure 9-12: For a watershed with $T^* = 1$ "hr" the peak rises to more than 60% of the input rate, then recedes to near zero 5.5 "hr" after input ceases. When the same input is applied to a basin with $T^* = 5$ "hr," the peak is about 20% of the input rate and the recession is much slower—outflow is still well above zero 7 "hr" after input stops.

Comparison of Figures 9-15 and 9-16 shows that the temporal variability of input affects the magnitude as well as the timing of peaks.

### Summary

Table 9-2 summarizes the conclusions gleaned from applying the linear reservoir model. Conceptually, the centroid lag and time of concentration are "constant" characteristic watershed times that depend on the time of travel of water to the basin outlet, and hence on basin size, topography, geology, and land use. In real watersheds, these times may vary temporally depending on the amount of water in storage. Other time characteristics of the response hydrograph depend on both the watershed characteristics and the duration or timing of input. Peak flow—the most important hydrograph feature for

**TABLE 9-2**

Dependence of Response Hydrograph Characteristics on Watershed and Input Characteristics as Predicted by the Linear Watershed Model (Box 9-2).

| Hydrograph Characteristic | Watershed Characteristics, $T^*$ | Input Duration, $T_w$ | Input Timing |
|---|:---:|:---:|:---:|
| Centroid Lag, $T_{LC}$ | X | | |
| Response Lag, $T_{LR}$ | | | |
| Time of Rise, $T_r$ | | X | X |
| Lag-to-Peak, $T_{LP}$ | X | | X |
| Centroid Lag-to-Peak, $T_{LPC}$ | X | | X |
| Time Base, $T_b$ | X | X | |
| Time of Concentration, $T_c$ | X | | |
| Peak discharge, $q_{pk}$ | X | X | X |

flood forecasting and prediction—depends on watershed characteristics and both the duration and timing of inputs.

## 9.2 MECHANISMS PRODUCING EVENT RESPONSE

In this section we explore the actual physical mechanisms by which water produced at the ground surface by rain or snowmelt can travel to a stream to produce an event hydrograph. These mechanisms are classified in Table 9-3, and the climatic, topographic, and pedologic-geologic conditions in which they occur will be described later in this section. It

**TABLE 9-3**

Classification of Flow Mechanisms That Produce Event Responses[a]

I. Channel precipitation
II. Overland flow (surface runoff)
  A. Hortonian overland flow
  B. Saturation overland flow
III. Subsurface flow
  A. Flow in the saturated zone
    1. Flow from local ground-water mounds
      a. Gradual mound development
      b. Sudden mound development by pressurization of capillary fringe
    2. Flow from perched saturated zones
      a. Matrix (Darcian) flow
      b. Macropore flow
  B. Flow in the unsaturated zone
    1. Matrix (Darcian) flow
    2. Macropore flow

[a]See Figures 9-17–9-31 and text for description.

should be noted that some or all of these mechanisms may operate simultaneously on a given watershed, and that their relative importances may fluctuate seasonally or even during a single water-input event.

Much of our understanding of the importance of these sources in various climatic, geologic, and topographic contexts has come from studies using various chemical and/or isotopic tracers, as explained in Box 9-1.

### 9.2.1 Channel Precipitation

**Channel precipitation** (also called **channel interception**) is the rain that falls directly on the stream to become incorporated in channel flow. Runoff from this source occurs in all rainstorms and in all regions and, although the total area of stream channels is almost always less than 1% of total drainage area, it can be a significant component of peak flow and total event flow.

The volume of channel precipitation, $W_{cp}$, for an event can be readily calculated as

$$W_{cp} = W \cdot A_{cn}, \tag{9-8}$$

where $W$ is total water input (rain) and $A_{cn}$ is the surface area of stream channels above the point of measurement. Referring to Figure 9-17, we see that the hydrograph of runoff from channel precipitation can be approximated by a triangle such that

$$W_{cp} = 0.5 \cdot q_{pkc} \cdot (T_w + T_{cn}), \tag{9-9}$$

where $q_{pkc}$ is the peak discharge from channel precipitation, $T_w$ is storm duration, and $T_{cn}$ is the time of concentration for the channel network. Setting Equations (9-8) and (9-9) equal to each other leads to an expression for $q_{pkc}$:

**FIGURE 9-17**
Approximation of the hydrograph of channel precipitation, $q_c$, as a triangle. The total area of the triangle is the volume of channel precipitation, $W \cdot A_{cn}$ [Equations (9-8) -(9-10)].

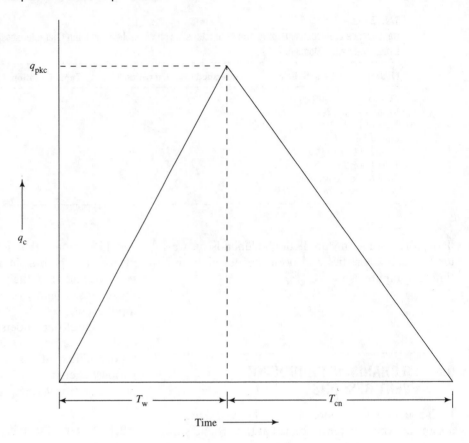

$$q_{pkc} = \frac{2 \cdot W \cdot A_{cn}}{T_w + T_{cn}}. \qquad \textbf{(9-10)}$$

Using Equations (9-9) and (9-10), Dingman (1970) calculated that up to 5% of total event flow and 40% of the peak discharge were due to channel interception for a small Alaskan watershed. Beven and Wood (1993) cited several studies in which channel precipitation was a significant part of the response hydrograph, and Crayosky et al. (1999) found that channel precipitation accounted for up to 29% of total storm flow and for all of the rising limb of storm hydrographs during the dry season for a small stream in Pennsylvania.

### 9.2.2 Overland Flow

**Overland flow** (or **surface runoff**) occurs on a sloping surface that is either (1) saturated from above (**Hortonian overland flow**) or (2) saturated from below (**saturation overland flow**). Emmett (1978) and Abrahams et al. (1986) described methods of measuring overland flow on hillslopes. The hydraulics of overland flows is discussed in Section B.3.2.

### Hortonian Overland Flow

Hortonian overland flow (Figure 9-18) is overland flow that results from saturation from above (including that which occurs on impermeable surfaces). The mechanism is named for Robert E. Horton, who described the process in a series of papers in the 1930s and 1940s (Horton 1933; 1945). Horton portrayed this mechanism as the principal, if not the only, process responsible for event response. This belief became inculcated in a generation of hydrologists and was incorporated as the basis for many event–response models. As will be seen subsequently, a number of field studies beginning in the 1960s showed that Hortonian overland flow is in fact far from ubiquitous.

As developed at length in Section 6.6.2, saturation from above occurs where water-input rate, $w(t)$, exceeds the saturated hydraulic conductivity, $K_h^*$, of the surface layer for a duration exceeding the time of ponding, $t_p$, which is given by Equation (6-39). The infiltration rate, $f(t)$, under these conditions is given by Equation (6-42), and the rate at which water becomes available for Hortonian overland flow, $q_{ho}(t)$, is

**FIGURE 9-18**
Hortonian overland flow over an entire slope of length $X_s$. $q_{ho}(X_s, t)$ is the lateral inflow to the stream at time $t$.

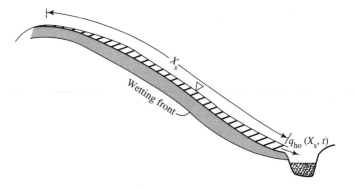

$$q_{ho}(t) = w(t) - f(t) \qquad \textbf{(9-11)}$$

(Figure 9-19).

The development in Section 6.6.2 showed that $f(t)$ decreases asymptotically to $K_h^*$ as input continues after the time of ponding. Thus Hortonian overland flow can occur only when and where $w(t) > K_h^*$ and $t_w > t_p$. Figure 9-20 compares the range of natural rainfall intensities[4] with $K_h^*$ values of various soils and geologic materials, and indicates that $K_h^* < w(t)$ only for relatively intense rains on relatively fine-grained soils. Thus Hortonian overland flow is an important response mechanism only in (1) semi-arid to arid regions, where rainfalls tend to be intense and natural surface conductivities low; (2) areas where soil frost or human or animal activity has reduced surface conductivity (see Section 6.5.4); and (3) impermeable areas.

In fact, actual occurrences of $w(t) > K_h^*$ may be even rarer than Figure 9-20 implies, because the $K_h^*$ values indicated there are for the mineral soil horizons. The surface horizons of many soils have considerably larger conductivities because of organic matter and biological activity.

Horton (1933; 1945) postulated that overland flow due to saturation from above would occur from virtually an entire upland watershed. This view was modified by Betson (1964), who proposed the **partial-area concept:** Event response may originate as Hortonian overland flow on a limited contributing area that varies from basin to basin but, except for the possibility of some expansion during extreme events, remains fairly constant on a given basin. Betson (1964) found that the stable contributing area ranged from 4.6% to 46% on agricultural watersheds in the southern Appalachian Mountains.

Eagleson (1970) and Stephenson and Meadows (1986) presented approaches for computing overland-flow hydrographs as a function of slope conditions and rainfall rates.

### Saturation Overland Flow

**Saturation overland flow** (Figure 9-21) is overland flow that occurs due to saturation from below; it consists of direct water input to the saturated area plus the **return flow** contributed by the "break-out" of ground water from upslope. The process of saturation from below is discussed in Section 6.6.3.

The importance of near-stream saturation overland flow as a stream-response mechanism in humid regions was first established by the intensive field studies of Dunne and Black (1970). As described in Section 8.3.1, streams in humid areas are typically gaining streams; the water table is usually coincident with the stream surface and not far below the ground surface in near-stream areas. When water input occurs over the drainage basin, all or part of it infiltrates and some of this infiltration percolates to recharge ground water and raise the water table. Since the water table is close to the surface near the stream, it commonly rises to the surface there. Once this happens, all further water input on the saturated zone travels as overland flow to the stream regardless of water-input rate.

Geometrical considerations dictate that near-stream saturated zones will be most extensive in drainage basins with concave hillslope profiles and wide, flat valleys. However, a number of studies (see Ward 1984) have shown that saturation overland flow is not restricted to near-stream areas. Saturation from below can also occur (1) where subsurface flow lines converge in slope concavities ("hillslope hollows") and water arrives faster than it can be transmitted downslope as subsurface flow

---

[4] Snowmelt rates seldom exceed 0.5 cm hr⁻¹ (0.2 in. hr⁻¹).

**FIGURE FIGURE 9-19**
Hortonian overland flow, $q_{ho}(t)$, at a point on a hillslope during a water-input event. $w(t)$ is water-input rate, shown as constant, $f(t)$ is infiltration rate given by Equation (6-42).

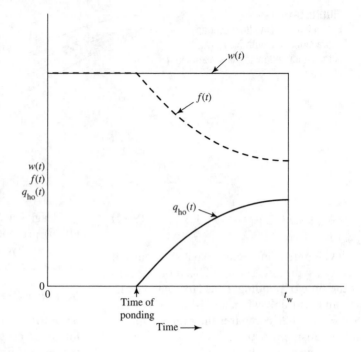

**FIGURE 9-20**
Saturated hydraulic conductivities for several soils and geologic materials compared to usual range of rainfall intensities. From Freeze (1972b), used with permission of the American Geophysical Union.

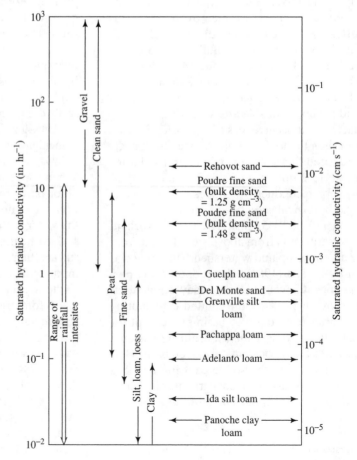

**FIGURE 9-21**
Saturation overland flow and sub-surface event flow due to near-stream ground-water mounding. (a) Early stages of event; overland flow absent and only regional ground-water flow (base flow) occurring. (b) Later, water table has risen to the surface in near-stream areas due to local and up-slope recharge; infiltration ceases and saturation overland flow results along with subsurface event flow. Return flow is the portion of saturation overland flow contributed by "breakout" of ground water. Flow contributing to mounding results from both vertical recharge and downslope flow in the saturated zone. After Ward (1984).

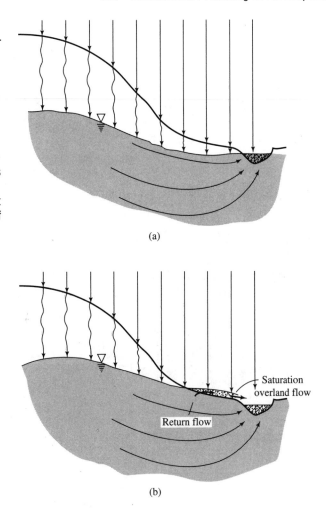

(a)

(b)

(Figure 9-22a); (2) at concave slope breaks, where the hydraulic gradient inducing subsurface flow from upslope is greater than that inducing down-slope transmission (Figure 9-22b); (3) where soil layers conducting subsurface flow are locally thin (Figure 9-22c); and (4) where hydraulic conductivity decreases abruptly or gradually with depth, and percolating water accumulates above the low-conductivity layers to form "perched" zones of saturation that reach the surface (Figure 9-22d). If any of these conditions occurs in areas that have an effective hydraulic connection to the stream, they can contribute significantly to event response.

During an event, return flow is usually much less important than direct water input on the saturated zone, especially in areas distant from the stream. However, return flow can be a significant contributor to streamflow when it persists after input ceases (Whipkey and Kirkby 1978).

A number of field (Ragan 1966a; Dunne and Black 1970; Dunne 1978; Ward 1984) and modeling (Freeze 1972b; 1974) studies indicate that saturation overland flow is a, if not the, major mechanism producing event response in humid regions. These studies have established the **variable source-area concept:** Within a given watershed, the extent of areas saturated from below varies widely with time, reflecting the overall watershed wetness (Figure 9-23). This variability is in large part responsible for the tremendous temporal variability of storm runoff that is observed in many regions (e.g., Figures 9-8 and 9-9), and thus has extremely important implications for understanding and modeling event response.

Dunne et al. (1975) have shown how the extent of source areas for saturation overland flow can be identified in the field and Van de Greind and Engman (1985) reviewed approaches for identifying

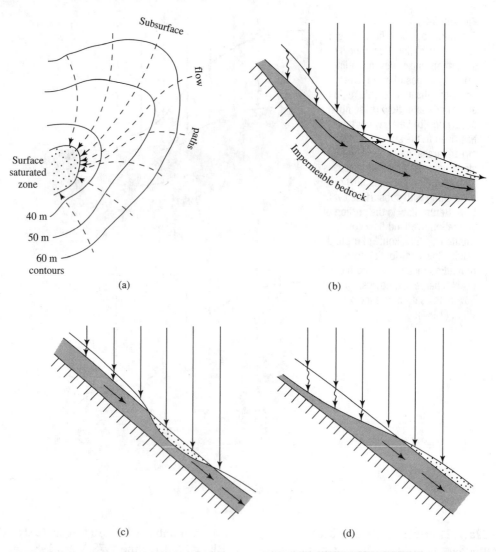

(a)

(b)

(c)

(d)

**FIGURE 9-22**

Situations in which saturation overland flow may arise on hillslopes outside of near-stream areas. (a) Plan view showing convergence of subsurface flow paths. (b) Cross section showing downslope reduction in hydraulic gradient associated with slope break. (c) Cross section showing local area of thin soil. (d) Cross section showing formation of perched saturated zone above low-conductivity layer with constant slope and soil thickness. After Ward (1984).

such areas via remote sensing. O'Loughlin (1981) developed generalized relations that give the extent of saturated zones as a function of hillslope gradient, planform geometry (diverging, planar, or converging), hydraulic conductivity, depth, and flow rate. A conceptually similar approach, called "TOP-MODEL," is widely used to simulate saturated overland flow in hydrologic models; this is described in Box 9-3.

### 9.2.3 Subsurface Event Flow

Discussion in Section 8.3.1 makes it clear that regional ground-water flow is usually the source of most streamflow between event responses (base flow). Such flow enters the drainage basin in the same rain and snowmelt events that produce event responses, but travels to the stream through ground-water reservoirs with residence times that

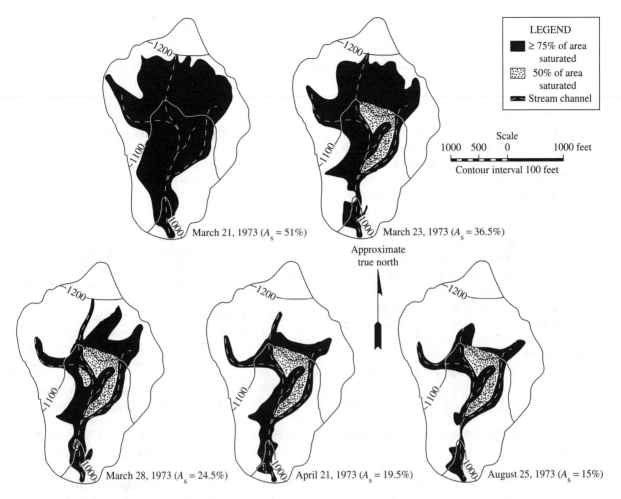

**FIGURE 9-23**
Seasonal variation of extent of areas saturated from below in a drainage basin with gentle slopes and moderately to poorly drained soils in northeastern Vermont. $A_s$ is percent of total area that is saturated. From Dunne et al. (1975).

are so large that short-term pulses of input are damped out. Here we examine mechanisms by which subsurface flow enters streams quickly enough to contribute to the event response. Such flow is called **subsurface event flow** or **subsurface storm flow,** and we consider situations under which such flow can occur under saturated and unsaturated conditions.

Dunne and Black (1970), Anderson and Burt (1978), and Atkinson (1978) have described field methods for measuring subsurface flow and for instrumenting slopes with tensiometers (Section 6.3.2), piezometers (Section 8.1.1), and other instruments to detect soil–water states.

### Flow in the Saturated Zone

Model studies incorporating the ground-water flow equations lead to the conclusion that basin-wide regional ground-water flow (as depicted, for example, in Figures 8-8 and 8-9) cannot respond quickly enough to contribute to such response (Freeze 1974). However, studies using various tracers have indicated that ground water can be a significant component of event response under some circumstances (see Figure 8-37). These ground-water contributions to event flow must arise from mechanisms that quickly produce steep hydraulic gradients in materials of high conductivity in near-stream areas. Such mechanisms are described here.

---

**BOX 9-3**

· · · · · · · ·

## The TOPMODEL Approach to Distributed Hydrologic Modeling of Saturation Overland Flow

As depicted in Figures 9-22 and 9-23, the locus of areas that generate saturation overland flow is determined essentially by local topography. TOPMODEL (Beven and Kirkby 1979; Wood et al. 1990; Beven and Wood 1993) has proven to be a useful approach to modeling runoff in humid regions by identifying the time-varying portions of a watershed that can produce saturation overland flow.

In this approach, the watershed is assumed to be covered by a uniform, relatively thin soil layer through which downslope saturated flow occurs beneath a water table that is parallel to the soil surface (the "sloping slab"). At each point in the watershed, the propensity to produce saturation overland flow is directly proportional to the tendency to collect subsurface flow from upslope areas and inversely proportional to the tendency to transmit that flow downstream. These opposing tendencies are quantified in a **topographic index,** $TI_i$, where

$$TI_i \equiv \ln\left(\frac{a_i}{S_i}\right), \qquad \text{(9B3-1)}$$

$a_i$ is the area draining to point $i$ per unit contour length, and $S_i$ is the local slope tangent. Thus points that have large drainage areas and flat slopes have high $TI$ values (e.g., the near-stream areas in Figure 9-21 and the situations shown in Figure 9-22a and 9-22b). Typically, $TI$ is computed for a grid of points such as provided by a digital elevation model (DEM).

Saturated hydraulic conductivity is assumed to decrease exponentially with depth to near zero at the base of the soil. This relation can be expressed in terms of the local value of the difference between the current soil-water content and the maximum (saturated) value, $d_i$ [$L^3$ $L^{-2}$]. Downslope subsurface flow per unit width through the $i$th point, $q_i$ [$L^2 T^{-1}$], is then given by a modified version of Darcy's Law as

$$q_i = T_0 \cdot S_i \cdot \exp(-d_i/M), \qquad \text{(9B3-2)}$$

where $T_0$ is the transmissivity [$L^2 T^{-1}$] of the soil when saturated to the surface and $M$ characterizes the rate at which conductivity decreases with depth. Equation (9B3-2) assumes that the local hydraulic gradient is parallel to the local slope.

The local storage deficit is linked to the watershed mean storage deficit, $\bar{d}$, by the relation

$$d_i = \bar{d} + M \cdot (\overline{TI} - TI_i), \qquad \text{(9B3-3)}$$

where $\overline{TI}$ is the watershed mean value of $TI$. The value of $\bar{d}$ is computed at successive time steps by keeping track of the watershed water balance (precipitation, evapotranspiration, and outflow). Then at each time step, the points capable of generating saturation overland flow are those at which $d_i = 0$. Given the spatial pattern of values of $d_i$ and the percentage of the watershed that is saturated, the spatial pattern of saturated points can be determined and the amount of saturation overland flow computed.

Alternative forms of the topographic index that may be more applicable than Equation (9B3-1) in particular watersheds have been developed (Ambroise et al. 1996).

---

**Flow from Local Ground-Water Mounds** The same process that produces saturation overland flow induces subsurface stormflow due to mounding: Because the water table adjacent to streams in humid regions is near the surface, percolation recharges ground water in near-stream areas before it does so at upslope locations. This recharge produces a mound, or ridge, that steepens the hydraulic gradient both toward and away from the stream (Figures 9-21b, 9-25). The steepened streamward gradient can produce a prompt and sustained contribution to streamflow.

In some situations, the pressurization of the tension-saturated zone, or capillary fringe (Section 6.4.2), plays a role in the rapid formation of ground-water mounds. This zone extends above the water table a distance that is inversely related to soil-pore size. Laboratory and field studies (Gillham 1984; Abdul and Gillham 1984; 1989; Jayatilaka et al. 1996) showed that when even a small amount of water percolates to the top of this zone, the menisci that maintain the tension are obliterated and the pressure state of the water is immediately changed from negative (tension) to positive (pressure) (Figure 9-24). This phenomenon can thus produce an almost instantaneous rise in the near-stream water table (Figure 9-25), and the streamflow contribution induced thereby may greatly exceed the quantity of water input that induced it. Note also that streamflow produced in this way is "old" water—water that fell in previous events.

As with saturation overland flow, it is clear that subsurface event flow due to near-stream ground-water mounding is most likely to occur in watersheds with concave slopes and wide, flat valleys.

**Flow from Perched Saturated Zones** In many regions, hillslopes consist of a thin layer of permeable soil overlying relatively impermeable materials, or soils in which the hydraulic conductivity decreases markedly with depth. In these situations, infiltration and percolation of water input commonly produce a more or less temporary thin saturated zone that is not connected to a regional ground-water flow, and downslope flow in this zone can be a significant contributor to event response (Figure 9-26). This situation is often referred to as a **sloping slab.**

The response time for the sloping slab can be obtained by applying the same reasoning used to develop the Green-and-Ampt approach to modeling infiltration. We consider an initially unsaturated soil layer of uniform depth, $z_u$, hydraulic properties, and initial water content, $\theta_0$, overlying a uniformly sloping impermeable base. For these conditions, the time, $T_u$, required for a wetting front due to the complete infiltration of input at a constant water-input rate, $w$, to reach the base can be determined from Equation (6-56) as

$$T_u = \frac{(\theta_w - \theta_0) \cdot z_u}{w \cdot \cos(\beta_s)}. \qquad \textbf{(9-12)}$$

Here, $\theta_w$ is the soil–water content at which the hydraulic conductivity equals $w$, $\beta_s$ is the slope angle, and the cosine accounts for the fact that infiltration is vertical. Beven (1982a) showed how Equation (9-12) can be modified to give $T_u$ when hydraulic conductivity decreases with depth.

Sloan and Moore (1984) and Ormsbee and Khan (1989) have shown that a simple conceptualization can be used as the basis for modeling flow from the base of a steeply sloping slab. In this model, the entire saturated zone that forms beginning at $t = T_u$ is taken as the control volume for formulating the conservation-of-mass relation, and this zone is assumed to be a triangular wedge that

thickens from zero at the top of the slope to a maximum value at the base (Figure 9-26a).[5] This model is developed in Box 9-4 and applied in Example 9-3.

---

### EXAMPLE 9-3

Consider a hillslope of length $X_s = 50$ m and inclination angle $\beta_s = 22.9°$ and consisting of a 1-m layer of uniform sandy soil with hydraulic conductivity $K_h^* = 1.00$ m hr$^{-1}$, porosity $\phi = 0.40$, and field capacity $\theta_{fc} = 0.10$ overlying an impermeable bedrock base. The initial water content is $\theta_0 = 0.08$. Compute (a) the time to beginning of runoff, $T_u$, for a steady storm of $w = 25$ mm hr$^{-1}$; (b) the centroid lag, $T_{ss}^*$, for the saturated wedge; and (c) the time to equilibrium, $T_{eqss}$, for the saturated wedge and for the slope.

**Solution:**
(a) By definition, $K_h(\theta_w) = w$. From Equation (6-13b),

$$w = K_h(\theta_w) = K_h^* \cdot \phi^{-c} \cdot \theta_w^c.$$

Using a value of $c = 11$ for the soil (Table 6-1), we find that

$$0.025 \text{ m hr}^{-1} = (1.00 \text{ m hr}^{-1}) \times (0.40^{-11}) \times \theta_w^{11},$$

from which it follows that $\theta_w = 0.29$. Then, from Equation (9-12),

$$T_u = \frac{(0.29 - 0.08) \times (1 \text{ m})}{(0.025 \text{ m hr}^{-1}) \times \cos(22.9°)} = 9.1 \text{ hr}.$$

Thus runoff from the slope would not begin until 9.1 hr after rain began.
(b) First we calculate $\theta_{av}$ via Equation (9B4-2a) as

$$\theta_{av} = 0.40 - 0.29 = 0.11.$$

$T_{ss}^*$ is then found from Equation (9B4-8):

$$T_{ss}^* = \frac{(50 \text{ m}) \times 0.11}{2 \times (1 \text{ m hr}^{-1}) \times 0.40} = 6.9 \text{ hr}.$$

(c) $T_{eqss}$ for the saturated wedge can be found from Equation (9B2-8):

$$T_{eqss} = 4.605 \times 6.9 \text{ hr} = 31.8 \text{ hr}.$$

The total time to equilibrium for the slope equals $T_{eqss} + T_u = 40.9$ hr.

---

[5] Experimental evidence shows that the saturated wedge does not usually extend the full length of the slope, but is present only near the slope base (Nutter 1975; Anderson and Burt 1977; see Figure 9-30). Nonetheless, the model described here has been successfully used to simulate sloping-slab response and even appeared superior to a model that accounted for the less extensive saturated wedge (Sloan and Moore 1984).

**FIGURE 9-24**
(a) Field instrumentation used to record rapid pressurization of the tension-saturated zone in a sandy soil. (b) Response of pressure head and total hydraulic head to sudden application of 3 mm of water to the surface in (a). From Gillham (1984), used with permission.

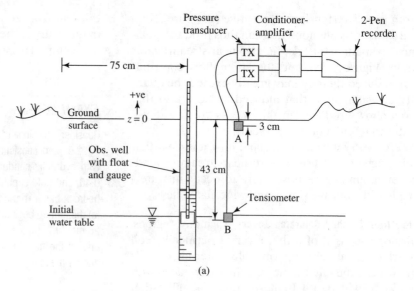

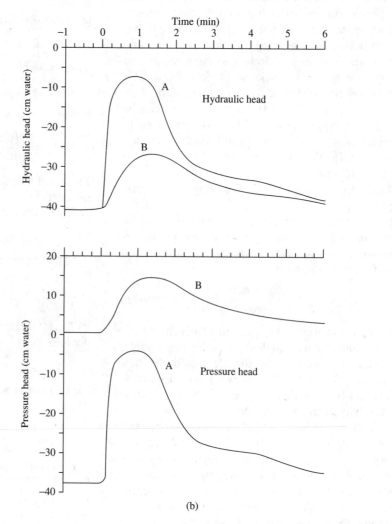

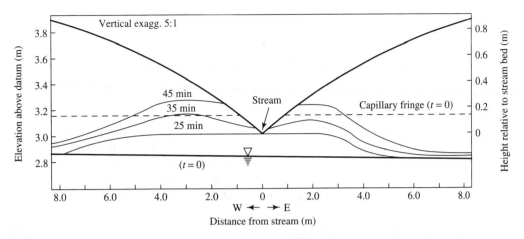

**FIGURE 9-25**
Response of near-stream water table due to pressurization of the capillary fringe during a simulated rain of $w = 2.0$ cm hr$^{-1}$, $T_w = 50$ min, in a sandy soil. Lines show position of water table at successive times after onset of rain. From Abdul and Gillham (1989), used with permission.

**FIGURE 9-26**
The "sloping slab." Formation of a perched saturated zone on a hillslope in which a relatively conductive soil overlies a nonconductive layer. (a) Subsurface stormflow from basal saturated zone at slope base. (b) Sloping slab with "breakout" near the stream, producing saturation overland flow along with subsurface stormflow.

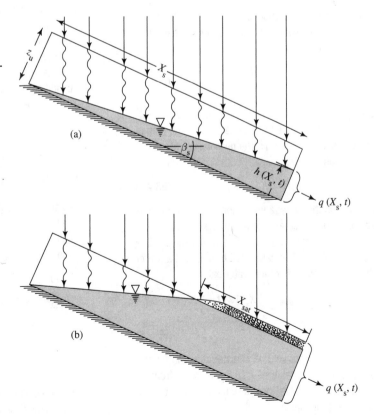

---

**BOX 9-4**
. . . . . . .
**Linear-Reservoir Model of Response of a Sloping Slab**

Referring to Figure 9-26a, we find that the storage per unit width in the saturated wedge, $V_s(t)$ at any time $t > t_u$ is

$$V_s(t) = \frac{h(t) \cdot X_s \cdot \theta_{av}}{2}, \qquad \textbf{(9B4-1)}$$

where $\theta_{av}$ is the pore space available for storage. When this wedge is filling,

$$\theta_{av} = \phi - \theta_w, \qquad \textbf{(9B4-2a)}$$

and when it is draining,

$$\theta_{av} = \phi - \theta_{fc}, \qquad \textbf{(9B4-2b)}$$

where $\phi$ is the porosity and $\theta_{fc}$ the field capacity of the soil. If $X_s \gg z_u$, the slope of the water table (i.e., the top surface of the saturated wedge) is essentially equal to that of the slope, $S_s$, and Darcy's Law gives the drainage (per unit width) from the slab, $q(t)$, as

$$q(t) = K_h^* \cdot h(t) \cdot S_s. \qquad \textbf{(9B4-3)}$$

(See Box 8-6.)

If no evapotranspiration is occurring, the finite-difference form of the continuity equation for the saturated wedge can be written as

$$p(t_i) \cdot X_s - q(t_i) = \frac{V_s(t_{i+1}) - V_s(t_i)}{\Delta t}, \qquad \textbf{(9B4-4)}$$

where $p(t)$ is the rate of percolation from the unsaturated zone into the saturated wedge, $\Delta t$ is the time step of the model, and $t_{i+1} = t_i + \Delta t$. The percolation rate is approximated as

$$p(t) = K_h[\theta_u(t)], \qquad \textbf{(9B4-5)}$$

where $\theta_u(t)$ is the average water content of the unsaturated zone. This assumes that percolation is driven only by gravity (i.e., pressure gradients are negligible). [See Equation (6-64).] $\theta_u(t)$ is calculated from an initial water content and a finite-difference water balance:

$$\theta_u(t_{i+1}) = \frac{\theta_u(t_i) \cdot V_u(t_i) + X_s \cdot [w(t_i) - p(t_i)] \cdot \Delta t}{V_u(t_i)}, \qquad \textbf{(9B4-6)}$$

where $V_u(t)$ is the total volume of water stored in the unsaturated zone. The value of $K_h[\theta_u(t)]$ is found from an empirical relation of the form of Equation (6-13b) with constants appropriate to the soil.

Equation (9B4-4) is exactly analogous to the convex model [Equation (9B5-3)], and substituting Equations (9B4-1) and (9B4-3) into Equation (9B4-4) yields a convex routing equation for the drainage from the slab for a sequence of percolation inputs calculated from Equation (9B4-5) for each time step:

$$q(t_{i+1}) = \frac{\Delta t}{T_{ss}^*} \cdot p(t_i) \cdot V_s + \left[1 + \frac{\Delta t}{T_{ss}^*}\right] \cdot q(t_i). \qquad \textbf{(9B4-7)}$$

Using the convex analog and Equations (9B4-1) and (9B4-3) allows computation of the centroid lag for the saturated wedge, $T_{ss}^*$:

$$T_{ss}^* = \frac{X_s \cdot \theta_{av}}{2 \cdot K_h^* \cdot S_s}. \qquad \textbf{(9B4-8)}$$

With this value the time to equilibrium, $T_{eqss}$, can be obtained via Equation (9B2-8). Note that these quantities apply to the saturated wedge only, not to the coupled unsaturated–saturated flow.

Sloan and Moore (1984) also showed how the convex sloping-slab model can be modified to simulate saturation overland flow that results when the saturated wedge reaches the surface (Figure 9-26b). Under these conditions, Equation (9B4-1) becomes

$$V_s(t) = \left[X_{sat}(t) + \frac{X_s - X_{sat}(t)}{2}\right] \cdot z_u \cdot \theta_{av}, \qquad \textbf{(9B4-9)}$$

and Equation (9B4-3) becomes

$$q(t) = w(t) \cdot X_{sat}(t) + K_h^* \cdot z_u \cdot S_s, \qquad \textbf{(9B4-10)}$$

where $X_{sat}(t)$ is the length of slope over which "breakout" and saturation overland flow occur. Substituting Equations (9B4-9) and (9B4-10) into Equation (9B4-4) allows computation $X_{sat}(t)$ at successive times:

$$X_{sat}(t_{i+1}) = \frac{\dfrac{X_{sat}(t_i) \cdot z_u \cdot \theta_{av}}{\Delta t} - w(t_i) + 2 \cdot [w(t_i) \cdot X_s - z_u \cdot K_h^* \cdot S_s]}{\dfrac{z_u \cdot \theta_{av}}{\Delta t} + w(t_i)}. \qquad \textbf{(9B4-11)}$$

Once $X_{sat}(t_i)$ values are known, we can compute storage and outflow via Equations (9B4-9) and (9B4-10).

Example 9-3 and studies by Freeze (1972) and Beven (1982b) suggest that the response of even a relatively shallow, permeable, and steeply sloping slab to a relatively intense water-input rate is usually quite sluggish where percolation and downslope flow occur via Darcy's Law through the soil matrix. However, rapid subsurface response to water input can occur on slopes containing **macropores** produced by roots, soil fauna, or desiccation cracking. Macropores are typically on the order of 3 to 100 mm in diameter and are interconnected to varying degrees (Figure 9-27); thus they can allow water to bypass the soil matrix and move rapidly to a basal saturated zone and/or move downslope as pipe flow at speeds much greater than predicted by Darcy's Law (Kirkby 1988).

Beven and Germann (1982) reviewed studies of the origins and distributions of macropores and their effects on infiltration and storm-runoff generation. More recent studies that document the importance of macropore flow in event response in a wide range of environments are those of Roberge and Plamondon (1987) in northern Quebec, McDonnell (1990) in New Zealand, Chapman et al. (1993) in Wales, Villholth et al. (1998) in Denmark, and Newman et al. (1998) in New Mexico. In a detailed study in a New Hampshire forest, Stresky (1991) found that more than 60% of the macropores were in the upper 0.15 m of the soil and were created by live roots, decayed roots, and animal burrows. More than 70% of the macropores were less than 20 mm in diameter—though conduits exceeding 25 mm were present. Macropore networks were generally oriented downslope and were interconnected over distances of at least tens of meters.

It is clear that macropores play an important role in delivering water input to streams at least at some times in many regions, particularly forested areas. Many studies indicate that their importance increases with the amount of rain or snowmelt in an event. Ormsbee and Khan (1989) achieved improved simulations of hydrologic responses of small watersheds in Kentucky and Virginia by modifying the sloping-slab model of Box 9-4 to account for preferential flow through macropores. However, it is generally difficult to assess the importance of macropores or to simulate their effects in models because their number, orientation, size, and interconnectedness are highly dependent on local geology, soils, vegetation, and fauna.

Flow through macropores has also been shown to affect the chemical quality of runoff water (Chapman et al. 1993; Villholth et al. 1998).

### Flow in the Unsaturated Zone

Downslope flow occurring between the ground surface and a perched or regional water table is called **interflow** or **throughflow**. Many writers have referred to such flow, but have often been ambiguous as to its exact mechanism: It is sometimes described as unsaturated Darcian flow in the soil matrix, sometimes as pipe flow in macropores that largely bypasses the unsaturated soil matrix, and sometimes as flow in saturated zones of very limited vertical extent caused by soil horizons that impede vertical percolation. Since the last of these mechanisms is essentially identical to the sloping slab just discussed, this section focuses on the other two.

**Matrix (Darcian) Flow**  Water in a stream is under positive pressure, while water in the unsaturated zone is under tension (negative pressure). Thus unsaturated flow governed by Darcy's Law [Equation (6-8a)] cannot enter a stream without passing through a saturated zone that provides a streamward hydraulic gradient. Here we examine the direction and rate of flow in the unsaturated soil matrix upslope from this zone during and following an infiltration event.

Figure 9-28 shows the hydraulic-head contours during and following infiltration on a slope underlain by a homogeneous soil. During water input, the wetting front remains parallel to the surface (Figure 9-28a,b). However, the near-surface hydraulic-head contours gradually rotate toward the horizontal as infiltration continues (Figure 9-28b). Thus initial infiltration is normal to the slope, but becomes nearly vertical if the event continues long enough. After water input ceases, the near-surface head contours rotate further, becoming nearly normal to the surface (Figure 9-28c,d), indicating downslope flow.

The simulations shown in Figure 9-28 indicate that downslope Darcian unsaturated flow does not occur during a storm or snowmelt event, although there is a downslope component of the drainage following the event at velocities not exceeding a few centimeters per hour. Thus Darcian unsaturated

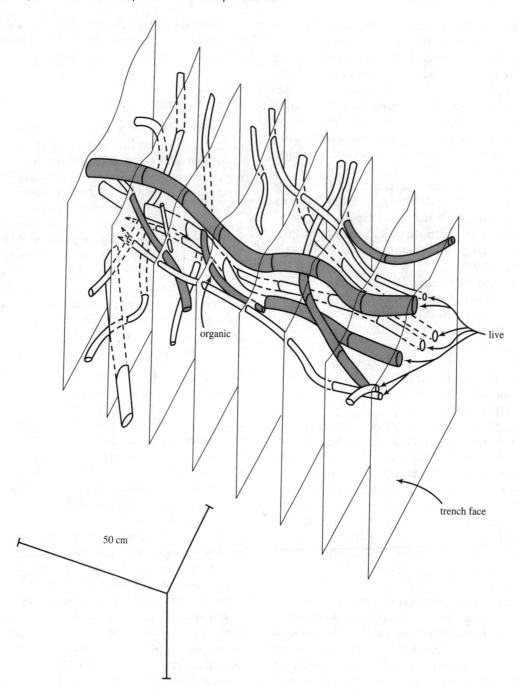

50 cm

trench face

live

organic

**FIGURE 9-27**
Macropores in an approximately 1-m portion of hillslope in a New Hampshire forest floor. Live-root and organic macropores are indicated; other macropores are decayed roots. For live root macropores, the entire root is shown; flow occurs in the space between the roots and the soil matrix. Dark portions show dye paths. From Stresky (1991).

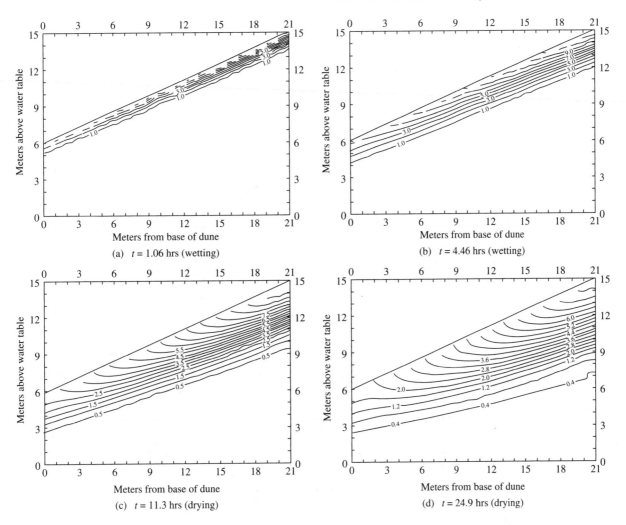

**FIGURE 9-28**
Modeled total-head contours in an isotropic soil draining after rainfall at a rate of 1 cm hr$^{-1}$ for $0 < t < 5$ hr. (a) $t = 1.06$ hr; (b) $t = 4.46$ hr; (c) $t = 11.3$ hr; (d) $t = 24.9$ hr. From Jackson (1992), used with permission of the American Geophysical Union.

flow does not appear to be a likely source of event flow under most circumstances.

However, it is clear from several studies that unsaturated matrix flow can contribute to, and may be the major source of, streamflow recessions and baseflow in upland watersheds in many regions (Hewlett and Hibbert 1963; Weyman 1970; Nutter 1975; Anderson and Burt 1977). This is strikingly illustrated in Figure 9-29, which shows the drainage of a confined 1 m × 1 m column of homogeneous sandy–clay–loam soil extending 15 m on a 40° slope, with an outlet at the slope base. The soil was initially saturated and covered to prevent evaporation or water input. After 1.5 days the saturated zone had retreated to within a few centimeters of

the outlet, and flow into that zone came from unsaturated downslope flow. Note that discharge rates declined very rapidly in the first few days, but drainage persisted until the experiment was terminated at 145 days. The total change in water content over this period was relatively small, from about 0.50 at saturation to 0.38 at 145 days.

Figure 9-30 shows the head contours measured after more than 30 days of drainage in another sloping-slab experiment, confirming the existence of a downslope-oriented gradient and showing the development of a saturated wedge at the slope base.

These experiments show that base flow to upland streams can be supplied by unsaturated drainage for extended periods.

(a)

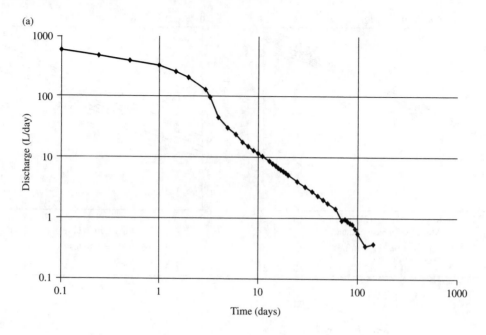

(b)

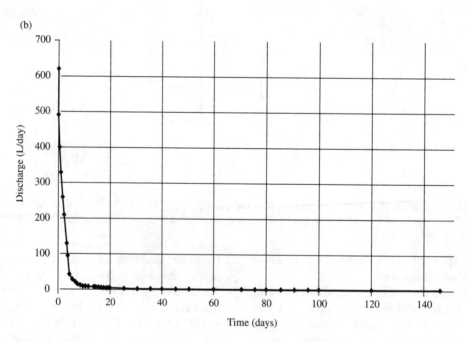

**FIGURE 9-29**
Drainage from a 1 × 1 column of sandy clay loam soil extending 15 m down a 40° slope for 145 days. Soil was initially saturated and covered to prevent evaporation. (a) Log–log plot. (b) Arithmetic plot. Data from Hewlett and Hibbert (1963).

**FIGURE 9-30**
(a) Water content and (b) hydraulic head distribution in a sloping slab after 749 hr of drainage. Slope = 15°. Note the saturated wedge at the slope base. From Nutter (1975).

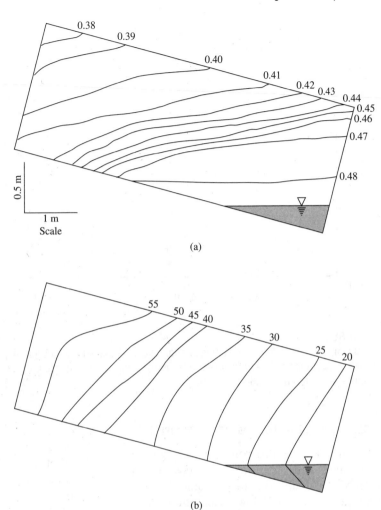

(a)

(b)

**Macropore Flow** As noted earlier, macropores can play a significant role as runoff conduits on hillslopes. In many cases it is unclear whether the soil matrix surrounding the macropores was saturated or unsaturated. However, there is ample evidence that macropores can conduct water downslope considerable distances through otherwise unsaturated soils at velocities of several millimeters per second (Mosley 1979; 1982; Beven and Germann 1982). This phenomenon is called **bypass flow;** in it the potential gradient causing the macropore flow is not in equilibrium with the gradient in the soil matrix. Kirkby (1988) presented a general relation between the minimum pore diameter that will cause bypassing and the pore (grain) size of the soil matrix (Figure 9-31).

As noted by Ward (1984, p. 181), the functioning of a thatched roof provides an apt analogy for efficacy of the preferential flow paths provided by macropores in an unsaturated matrix:

No hydrologist, having measured the infiltration characteristics of bundles of straw, would recommend their use as roofing material. And yet, in even the heaviest rain, the building remains dry, no water runs over the thatch as "overland" flow, there is no "groundwater" and no evidence of zones of "temporary saturation," i.e. all the rainfall is evacuated along the narrow layer of the thatch itself. The thatched roof works because the alignment of straw imparts a preferential permeability along the stems and because the roof slopes.

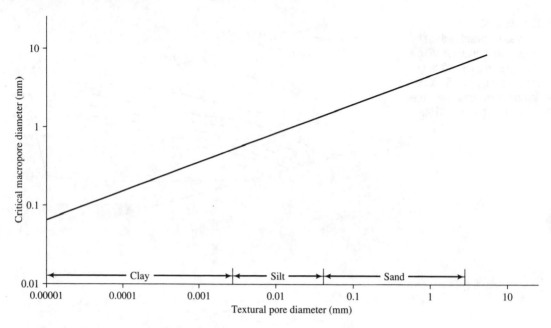

**FIGURE 9-31**
Relation between critical minimum macropore diameter that will allow bypassing and pore (grain) diameter of soil matrix. Thus for a fine sand (diameter = 0.1 mm), bypassing would occur in macropores exceeding about 1 mm diameter. After Kirkby (1988).

### 9.2.4 Overview of Event-Response Mechanisms

Understanding of event-response mechanisms has been evolving rapidly since the mid-1960s, and this evolution will no doubt continue as studies based on chemical and isotopic flow separations (Box 9-1) and detailed measurements of soil and ground water are conducted in new environments. Bonell (1993) provided a comprehensive review of studies investigating runoff mechanisms in forests, and Table 9-4 summarizes a recent sample of such studies.

Table 9-5 summarizes the current understanding of the soil/geologic, topographic, vegetative, and water-input conditions that favor the various mechanisms. The Hortonian overland-flow mechanism that dominated hydrologic thinking and modeling beginning in the 1930s is now known to be common only on natural hillslopes in semi-arid to arid regions and on human-disturbed and impermeable areas, where rainfall rates commonly exceed surface infiltration capacities. Hortonian overland flow can also occur on ground rendered impermeable by soil frost.

In most humid forested watersheds, a very large proportion of rain and snowmelt usually infiltrates, and response hydrographs are dominated by subsurface flow supplemented to varying extents by saturation overland flow. The relative importance of the various subsurface mechanisms is determined largely by watershed geology, soils, topography, and the amount, intensity, and spatial and temporal distribution of water input, but the exact mechanism that contributes the subsurface flow is often unclear.

Whatever the mechanism, it is widely accepted that the water that gives rise to response hydrographs usually comes from only a limited portion of the topographically defined watershed, that this contributing area generally varies strongly as a function of watershed wetness, and that in many cases much of the event response consists of "old" water.

### 9.3 OPEN-CHANNEL FLOW AND STREAMFLOW ROUTING

We have seen that response hydrographs are increasingly modified as the flood wave travels through the stream network—peak flows per unit

**TABLE 9-4.**

A Sampling of Recent Field Studies of Runoff Mechanisms.

| Location | Mechanisms | Separation Basis | Source |
|---|---|---|---|
| Upland forest, Pennsylvania | Ground-water mounding; pressurization of capillary fringe; minor channel precipitation | $^{18}O$ | Swistock et al. (1989) |
| Upland forest, Georgia | Sloping slab (mineral soil); sloping slab (organic soil); ground-water mounding | Six chemical constituents | Hooper et al. (1990) |
| Gently sloping forested hillside, Australia | Sloping slab (macropores) | $^{2}H$; $Cl^{-1}$ | Leaney et al. (1993) |
| Coastal Plain, Virginia | Saturation overland flow | $Cl^{-1}$ | Eshleman et al. (1993) |
| Forested swamp, Ontario, Canada | Saturation overland flow; ground water (macropore flow) | $^{18}O$, $Cl^{-1}$, $Li^{+1}$ | Waddington et al. (1993) |
| Forested upland, deep soils, Tennessee | Bedrock (dolomite) ground water; ground-water mounding; sloping slab | Flow measurement; Ca, $SO_4$ | Mulholland (1993) |
| Upland unforested watershed, Scotland | Hortonian overland flow; sloping slab; ground water | ANC[a] | Giusti and Neal (1993) |
| Forest and pasture watershed, Switzerland | Saturation overland flow; ground water; Hortonian overland flow | $^{18}O$ | Jordan (1994) |
| Upland forest, Virginia | Saturation overland flow; subsurface flow | $^{18}O$, $Cl^{-1}$ | Bazemore et al. (1994) |
| Tropical rain forest, Australia | Saturation overland flow; sloping slab; ground-water mounding | $K^{+1}$, ANC[a], $^{18}O$, DOC[b] | Elsenbeer et al. (1995) |
| Shallow-soil forest, Canadian Shield | Saturation overland flow; sloping slab | | Peters et al. (1995) |
| Mixed-forest, New Brunswick, Canada | Ground water | Conductivity, Alkalinity, pH, $Na^{+1}$, $Mg^{+2}$, $Ca^{+2}$ | Caissie et al. (1996) |
| Unforested permafrost watershed, northern Alaska | Water tracks[c] | Conductivity, $^{18}O$ | McNamara et al. (1997) |
| Steep forested slope, Japan | Sloping slab | Flow measurement; tensiometers | Tani (1997) |
| Catskill Mts., New York | Sloping slab; ground-water mounding | Several solutes | Evans et al. (1998) |
| Steep, forested watershed, Maryland | Subsurface flow; channel precipitation | $^{2}H$, $^{18}O$, $Cl^{-1}$, $SiO_2$, $Na^{+1}$ | Rice and Hornberger (1998) |

[a] Acid-neutralizing capacity.

[b] Dissolved organic carbon.

[c] Subsurface "channels" of enhanced soil moisture that conduct flow directly downslope to streams.

watershed area tend to decrease and become delayed (Figure 9-3). **Streamflow routing** is the computational procedure for modeling this process: The **outflow hydrograph** at the downstream end of a stream reach is predicted, given an **inflow hydrograph** for the upstream end of the reach, the hydraulic characteristics of the reach, and a **lateral-inflow hydrograph.**

The most general approaches to this problem are based on the basic equations of open-channel flow, which are developed in Section B.3.2. However, such approaches are generally time-consuming and data-intensive, and engineers have developed simpler routing methods that can usefully model the movement of flood waves under many condi-

tions. In order to understand the hydraulic factors that affect response hydrographs, we explore one of these simple methods, the **convex method.** This method is based on elementary relations of open-channel flow, which are introduced in the next section. Gupta et al. (1979), Weinman and Laurenson (1979), and Nwaogazie and Tyagi (1984) discussed more elaborate routing methods.

### 9.3.1   Basic Relations of Open-Channel Flow

#### Conservation of Mass (Continuity)

The volumetric flow rate through a stream cross section (discharge), $Q$ [$L^3 T^{-1}$], is the product of the

**TABLE 9-5**

Environmental Factors Favoring Hillslope Event-Response Mechanisms. $K_h^*$ is saturated hydraulic conductivity.

| Mechanism | Soils/Geology | Water Table | Topography | Vegetation | Water-Input Rate[a] |
|---|---|---|---|---|---|
| Hortonian overland flow | Low surface $K_h^*$ | Deep | Steep slopes | Absent to sparse | High |
| Saturation overland flow | Slopes: High surface $K_h^*$, decreasing gradually or abruptly at shallow depth; Conditions of Figure 9-22. Valley bottoms: Low to high $K_h^*$ | Near surface | Concave, convergent slopes; Wide valleys | Absent to abundant | Low to high |
| Ground-water mounding | Slopes: Deep soils with high surface $K_h^*$. Valley bottoms: High $K_h^*$. Silty soils enhance flow from pressurized capillary fringe | Slopes: Deep; Valley bottoms: Near surface | Concave slopes, wide valleys | Absent to abundant | Low to moderate |
| Perched ground water (sloping slab) | Slopes: High surface $K_h^*$, decreasing gradually or abruptly at shallow depth; Macropores present | Absent to present in high-$K_h^*$ layer | Steep slopes; straight to convex | Absent to abundant | Low to moderate |

[a]Relative to $K_h^*$.

average flow velocity, $U$ [L T$^{-1}$], the average depth of the flow, $Y$ [L], and the water-surface width, $B$ [L]; that is,

$$Q = U \cdot Y \cdot B \qquad (9\text{-}13)$$

or

$$Q = U \cdot A, \qquad (9\text{-}14)$$

where $A$ [L$^2$] is the cross-sectional area of the flow. Equation (9-13) is the basis for the velocity–area method of discharge measurement, described in Section F.2.

The general equation of continuity for an open-channel flow is

$$q_L - \frac{\partial Q}{\partial x} = \frac{\partial A}{\partial t}, \qquad (9\text{-}15)$$

where $q_L$ is the rate of lateral inflow per unit channel length [L$^2$ T$^{-1}$], $x$ [L] is downstream distance,

and $t$ [T] is time. Usually $q_L$ is positive due to lateral inflows from ground water and/or overland flow and tributaries, but it can be negative in a losing stream or due to bank storage (Section 8.3.1). If the rate of change with time is small ("quasi–steady flow"), Equation (9-15) shows that

$$\frac{\partial Q}{\partial x} = q_L; \qquad (9\text{-}16)$$

that is, the downstream change in discharge is due to lateral inflow or outflow only.

### Equation of Motion

The basic equation for the velocity of one-dimensional steady open-channel flows is developed from force–balance considerations that involve the **hydraulic radius**, $R$, defined as

$$R \equiv \frac{A}{P}, \qquad \textbf{(9-17)}$$

where $A$ is the cross-sectional area of the flow and $P$ is the length of the wetted portion of the flow boundary (the "wetted perimeter"). In most channels, $R$ is virtually identical to the average depth, $Y$, so we will use $Y$ instead of $R$ henceforth.

The average open-channel flow velocity may be given by either the **Chézy Equation,**

$$U = u_c \cdot C \cdot R^{1/2} \cdot S^{1/2} \approx u_c \cdot C \cdot Y^{1/2} \cdot S^{1/2}, \qquad \textbf{(9-18)}$$

or the **Manning Equation,**

$$U = \frac{u_m \cdot R^{2/3} \cdot S^{1/2}}{n} \approx \frac{u_m \cdot Y^{2/3} \cdot S^{1/2}}{n}. \qquad \textbf{(9-19)}$$

Here $C$ ("Chézy's $C$") and $n$ ("Manning's $n$") are factors characterizing channel conductance/resistance, $S$ is the water-surface slope, and $u_c$ and $u_m$ are unit-conversion factors. The conversion factors are required because the equations are dimensionally inhomogeneous (see Section A.4); their values in the various unit systems are given in Table 9-6. From Equations (9-18) and (9-19), we see that

$$C = \frac{u_m \cdot R^{1/6}}{u_c \cdot n} = \frac{u_m \cdot Y^{1/6}}{u_c \cdot n}. \qquad \textbf{(9-20)}$$

Values of $C$ or $n$ depend on channel roughness and irregularity. They can be estimated from empirical formulas (Cowan 1956; Arcement and Schneider 1989), comparison with photographs of stream reaches with known resistance (Barnes 1967; Hicks and Mason 1991), or from tables like Table 9-6.

### Flood-Wave Velocity

It can be shown (Dingman 1984) that the velocity of a flood wave, $U_F$, is given by

$$U_F = \frac{1}{B} \cdot \frac{\partial Q}{\partial Y}. \qquad \textbf{(9-21)}$$

Combining Equations (9-13) and (9-19), we see that

$$Q = \frac{u_m \cdot Y^{5/3} \cdot S^{1/2} \cdot B}{n}; \qquad \textbf{(9-22)}$$

therefore,

$$\frac{\partial Q}{\partial Y} = \frac{5}{3} \cdot \left( \frac{u_m \cdot Y^{2/3} \cdot S^{1/2} \cdot B}{n} \right). \qquad \textbf{(9-23)}$$

Substituting Equations (9-19) and (9-23) into (9-21) yields

$$U_F = \frac{5}{3} \cdot U. \qquad \textbf{(9-24)}$$

Thus we have the interesting result that a flood wave travels with a velocity about 1.67 times that of the water itself.

Equations (9-22)–(9-24) are valid for flood waves that remain within the channel and can be considered to have a single representative average velocity. However, the relation between flow velocity and flood-wave velocity may be altered when a flood overtops the channel banks and inundates the floodplain.

Typically, the velocities of the overbank portions of a flow are much lower than in-channel flows because they are shallower and encounter much higher resistance due to vegetation. Referring to Figure 9-32, we can rewrite Equation (9-24) as

$$U_F = \frac{5}{3} \cdot \frac{U_{LB} \cdot B_{LB} + U_C \cdot B_C + U_{RB} \cdot B_{RB}}{B_{LB} + B_C + B_{RB}}, \qquad \textbf{(9-25)}$$

where the subscripts $LB$ and $RB$ refer, respectively, to the left and right overbank portions of the flow and $C$ to the central channel. If the floodplain velocities $U_{LB}$ and $U_{RB}$ are negligible, Equation (9-25) becomes

$$U_F = \frac{5}{3} \cdot \frac{U_C \cdot B_C}{B_{LB} + B_C + B_{RB}} = \frac{5}{3} \cdot \frac{U_C \cdot B_C}{B}. \qquad \textbf{(9-26)}$$

Because the ratio $B_C/B$ is less than one, Equation (9-26) shows that the flood wave velocity $U_F$ will be less than 1.67 times the channel velocity; if $B_C/B < 3/5$, the flood-wave velocity will be less than the central channel velocity. Thus by providing areas for storage of water as a flood wave moves downstream, floodplains reduce the velocity (and the height) of the flood peak.

### 9.3.2 The Convex Routing Method

The **convex routing method** is a simple model of the hydraulic behavior of a stream reach of a specified length ($X$) and specified constant slope ($S$), width ($B$), and resistance ($n$). A hydrograph of inflow at the upstream end of the reach, $QI(t)$, as a function of time, $t$, is also specified. A lateral-inflow hydrograph may be specified as well.

In addition to the hydraulic relations of Equations (9-19)–(9-24), the convex model is based on the following relations:

**TABLE 9-6**

Values of Manning's *n* for Channels
of Various Types[a]

| Type of Channel and Description | n Minimum | n Normal | n Maximum |
|---|---|---|---|
| Minor streams (top width at flood stage <100 ft) | | | |
| Streams on plain | | | |
| 1. Clean, straight, full stage, no riffles or deep pools | 0.025 | 0.030 | 0.033 |
| 2. Same as above, but more stones and weeds | 0.030 | 0.035 | 0.040 |
| 3. Clean, winding, some pools and shoals | 0.033 | 0.040 | 0.045 |
| 4. Same as above, but some weeds and stones | 0.035 | 0.045 | 0.050 |
| 5. Same as above, but lower stages, more ineffective slopes and sections | 0.040 | 0.048 | 0.055 |
| 6. Same as 4, but more stones | 0.045 | 0.050 | 0.060 |
| 7. Sluggish reaches, weedy, deep pools | 0.050 | 0.070 | 0.080 |
| 8. Very weedy reaches, deep pools, or floodways with heavy stand of timber and underbrush | 0.075 | 0.100 | 0.150 |
| Mountain streams, no vegetation in channel, banks usually steep, trees and brush along banks submerged at high stages | | | |
| 1. Bottom: gravels, cobbles, and few boulders | 0.030 | 0.040 | 0.050 |
| 2. Bottom: cobbles with large boulders | 0.040 | 0.050 | 0.070 |
| Floodplains | | | |
| Pasture, no brush | | | |
| 1. Short grass | 0.025 | 0.030 | 0.035 |
| 2. High grass | 0.030 | 0.035 | 0.050 |
| Cultivated areas | | | |
| 1. No crop | 0.020 | 0.030 | 0.040 |
| 2. Mature row crops | 0.025 | 0.035 | 0.045 |
| 3. Mature field crops | 0.030 | 0.040 | 0.050 |
| Brush | | | |
| 1. Scattered brush, heavy weeds | 0.035 | 0.050 | 0.070 |
| 2. Light brush and trees, in winter | 0.035 | 0.050 | 0.060 |
| 3. Light brush and trees, in summer | 0.040 | 0.060 | 0.080 |
| 4. Medium to dense brush, in winter | 0.045 | 0.070 | 0.110 |
| 5. Medium to dense brush, in summer | 0.070 | 0.100 | 0.160 |
| Trees | | | |
| 1. Dense willows, summer, straight | 0.110 | 0.150 | 0.200 |
| 2. Cleared land with tree stumps, no sprouts | 0.030 | 0.040 | 0.050 |
| 3. Same as above, but with heavy growth of sprouts | 0.050 | 0.060 | 0.080 |
| 4. Heavy stand of timber, a few down trees, little undergrowth, flood stage below branches | 0.080 | 0.100 | 0.120 |
| 5. Same as above, but with flood stage reaching branches | 0.100 | 0.120 | 0.160 |
| Major streams (top width at flood stage >100 ft)[b] | | | |
| Regular section with no boulders or brush | 0.025 | | 0.060 |
| Irregular and rough section | 0.035 | | 0.100 |

[a]The values of $u_c$ and $u_m$ for the common unit systems are

$$u_c = \text{SI}: 0.552; \quad \text{cgs}: 5.52; \quad \text{English}: 1.00;$$
$$u_m = \text{SI}: 1.00; \quad \text{cgs}: 4.64; \quad \text{English}: 1.49.$$

[b]The *n* value is less than that for minor streams of similar description, because banks offer less effective resistance.

Data from Chow (1959).

**FIGURE 9-32**
Definitions of terms used in deriving effects of overbank flows on flood-wave velocity [Equations (9-25) and (9-26)]. After Gray and Wigham (1970).

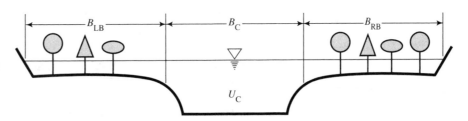

*Conservation of mass* [see Equation (2-6)]
Conservation of mass is expressed as

$$QI(t) - QO(t) = \frac{dV(t)}{dt}, \qquad (9\text{-}27)$$

where $QO(t)$ is the rate of outflow from the reach and $V(t)$ is the volume of water stored in the reach.

*Outflow–storage relation*
The outflow–storage relation is expressed as

$$QO(t) = \frac{1}{T^*} \cdot V(t), \qquad (9\text{-}28)$$

where $T^*$ is the time it takes a flood wave to travel through the reach; that is,

$$T^* = \frac{X}{U_F}. \qquad (9\text{-}29)$$

Equation (9-28) portrays the reach as a **linear reservoir,** because outflow is proportional to the first power of storage. Real stream reaches are not strictly linear reservoirs, but Equation (9-28) captures the most basic aspects of the storage–outflow relation and is mathematically tractable.

The derivation of the convex method is given in Box 9-5. It involves transforming the continuous relation of Equation (9-27) into discrete form by selecting a **routing time step,** $\Delta t \le T^*$, and defining a **routing coefficient,** $CX$, where

$$CX \equiv \frac{\Delta t}{T^*}. \qquad (9\text{-}30)$$

---

## BOX 9-5

· · · · · · ·

### Derivation of the Convex Routing Model

The conservation-of-mass relation can be expressed in finite-difference form as

$$QI_i - QO_i = \frac{V_{i+1} - V_i}{\Delta t}, \qquad (9\text{B}5\text{-}1)$$

where the subscripts denote values at successive instants, each separated by $\Delta t$. $\Delta t$ is the "time step" for the routing procedure, which is selected by the analyst (see Box 9-6) and must be less than or equal to the reach travel time, $T^*$.

From Equation (9-28), we can write

$$V_i = T^* \cdot QO_i, \qquad (9\text{B}5\text{-}2)$$

and putting Equation (9B5-2) into (9B5-1) yields

$$QI_i - QO_i = \frac{T^* \cdot (QO_{i+1} - QO_i)}{\Delta t}. \qquad (9\text{B}5\text{-}3)$$

If we assume that $T^*$ is known for a reach, that $QI_i$ is known for all *i* during an event (the inflow hydrograph), and that an initial outflow value $QO_0$ is known, Equation (9B5-3) can be rearranged into a form useful for routing, namely,

$$QO_{i+1} = CX \cdot QI_i + (1 - CX) \cdot QO_i, \qquad (9\text{B}5\text{-}4)$$

where $CX\, (\equiv \Delta t/T^*)$ is the routing coefficient [Equation (9-30)]. Equation (9B5-4) is known as the **convex routing equation.**

It can be shown that the convex routing model preserves continuity (i.e., total output always equals total input). From Equation (9B5-4), we see that the behavior of the model is controlled by $CX$, which must be in the range $0 < CX \le 1$. As explained in the text, the smaller the value of $CX$, the more the flood wave is modified in moving through the reach.

## BOX 9-6

· · · · · · · ·

### Application of the Convex Routing Method

A spreadsheet program for applying the method is given on the accompanying computer CD as GoConvex.xls; the steps described here follow that program and the methods described in U.S. Soil Conservation Service (1964). SI units are assumed. The hydrograph of inflow to the channel reach is given, either as values of $QI_i$ or as a graph from which values of $QI_i$ can be determined. The development here does not consider lateral inflows; when required, these can be included as described in the reference mentioned earlier. Perform the following steps:

1. In the spreadsheet, enter the following reach properties where indicated: reach length, $X$ (m); width, $B$ (m); Manning's $n$; and slope, $S$.
2. Enter the peak discharge, $QI_{pk}$, and the time of rise, $T_r$, of the inflow hydrograph.
3. The velocities used in routing are those associated with three-fourths of the peak-inflow discharge, $QI_{3/4} \equiv 0.75 \cdot QI_{pk}$. To compute this velocity, designated $U_{3/4}$, we first find the corresponding flow depth, $Y_{3/4}$, from the Manning Equation [Equation (9-22)]:

$$Y_{3/4} = \left[ \frac{QI_{3/4} \cdot n}{S^{1/2} \cdot B \cdot u_m} \right]^{3/5}. \qquad \text{(9B6-1)}$$

The Manning Equation [Equation (9-19)] is then used to find $U_{3/4}$:

$$U_{3/4} = \frac{u_m \cdot Y_{3/4}^{2/3} \cdot S^{1/2}}{n}. \qquad \text{(9B6-2)}$$

The flood-wave velocity, $U_F$, is then computed via Equation (9-24):

$$U_F = 1.67 \cdot U_{3/4}. \qquad \text{(9B6-3)}$$

The travel time for the reach, $T^*$, is found via Equation (9-29):

$$T^* = \frac{X}{U_F}. \qquad \text{(9B6-4)}$$

4. Select a convenient routing time step, $\Delta t$, less than $T^*$ and less than one-fifth the time of rise of the inflow hydrograph. The routing coefficient is found from Equation (9-30).
5. From the inflow hydrograph, enter the values of $QI_i$ at successive time steps in the appropriate column. GoConvex.xls computes the outflows using Equation (9B5-4).

**FIGURE 9-33**

Effect of routing coefficient, $CX$, in the convex method on the outflow hydrograph from a hypothetical channel reach. Curve labels are values of $CX$. When $CX = 1$ there is no diffusive effect and the floodwave is purely translatory. Successively smaller values of $CX$ (longer travel times for a given reach) successively flatten and delay the outflow peak.

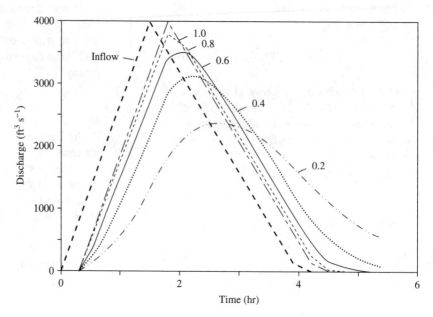

The value of $CX$ determines the degree to which flow through the reach reduces and delays the peak (Figure 9-33): If $CX = 1$, the outflow at the end of a time step equals the inflow at the end of the preceding step, and the flood wave travels through the reach at the velocity $U_F$ but does not change shape during transit. Such motion is called **purely translatory.** From Equations (9-18) or (9-19) and (9-26) we see that flatter slopes, lower flow depth, higher resistance (smaller $C$ or larger $n$), and the presence of slow-moving overbank flow reduce the speed of the flood wave and decrease the value of $CX$, which increases peak-flattening and delay.

Box 9-6 shows how the convex routing process is carried out in practice, and Example 9-4 shows its application to a specific hypothetical case.

---

### EXAMPLE 9-4

Consider a stream reach for which $X = 1000$ m, $B = 30$ m, Manning's $n = 0.05$, and $S = 0.0005$. Compute the outflow hydrograph for this reach given a triangular inflow hydrograph that begins at $QI(0) = 0$ m$^3$ s$^{-1}$; reaches a peak, $QI_{pk}$, of 35.0 m$^3$ s$^{-1}$ at 1.0 hr; and declines to 0 m$^3$ s$^{-1}$ at 2.8 hr. Assume no lateral inflow or overbank flow.

**Solution** It is convenient to use the spreadsheet program Go-Convex.xls on the CD accompanying this text and follow the steps in Box 9-6:

1. Enter the reach characteristics $X$, $B$, $S$, and $n$.
2. Enter peak, $QI_{pk}$, and duration of rise, $T_r$, of the inflow hydrograph.

3. $QI_{3/4}$ is calculated as

$$QI_{3/4} = 0.75 \times (35.0 \text{ m}^3 \text{ s}^{-1}) = 26.3 \text{ m}^3 \text{ s}^{-1}.$$

From Equation (9B6-1),

$$Y_{3/4} = \left[ \frac{(26.3 \text{ m}^3 \text{ s}^{-1}) \times 0.05}{0.0005^{1/2} \times (30 \text{ m}) \times 1} \right]^{3/5} = 1.50 \text{ m}.$$

From Equation (9B6-2),

$$U_{3/4} = \frac{1 \times (1.50 \text{ m})^{2/3} \times 0.0005^{1/2}}{0.05} = 0.58 \text{ m s}^{-1}.$$

From Equation (9B6-3),

$$U_F = 1.67 \times (0.58 \text{ m s}^{-1}) = 0.97 \text{ m s}^{-1}.$$

From Equation (9B6-4),

$$T^* = \frac{1000 \text{ m}}{0.97 \text{ m s}^{-1}} = 1026 \text{ s} = 0.285 \text{ hr}.$$

4. Selecting $\Delta t = 0.10$ hr, which satisfies the criteria given in Box 9-6, we use Equation (9-30) to give the routing coefficient,

$$CX = \frac{0.1 \text{ hr}}{0.285 \text{ hr}} = 0.35.$$

5. Enter the inflow hydrograph values $QI_i$ at 0.1-hr increments. Equation (9B5-4) yields the following:

| Time Step $i$ | Time (hr) | Inflow, $QI_i$ (m$^3$ s$^{-1}$) | Outflow, $QO_i$ (m$^3$ s$^{-1}$) | |
|---|---|---|---|---|
| 0 | 0.0 | 0.00 | | 0.00 |
| 1 | 0.1 | 3.50 | $0.35 \times 0.00 + (1 - 0.35) \times 0.00 =$ | 0.00 |
| 2 | 0.2 | 7.00 | $0.35 \times 3.50 + (1 - 0.35) \times 0.00 =$ | 1.23 |
| 3 | 0.3 | 10.50 | $0.35 \times 7.00 + (1 - 0.35) \times 1.23 =$ | 3.25 |
| . | . | . | . | |
| 12 | 1.2 | 31.11 | $0.35 \times 33.06 + (1 - 0.35) \times 28.61 = 30.17$ | |
| 13 | 1.3 | 29.17 | $0.35 \times 31.11 + (1 - 0.35) \times 30.17 = 30.50$ | |
| 14 | 1.4 | 27.22 | $0.35 \times 29.17 + (1 - 0.35) \times 30.50 = 30.03$ | |
| . | . | . | . | |
| 28 | 2.8 | 0.00 | $0.35 \times 1.94 + (1 - 0.35) \times 7.47 =$ | 5.53 |
| . | . | . | . | |
| 44 | 4.4 | 0.00 | $0.35 \times 0.00 + (1 - 0.35) \times 0.01 =$ | 0.00 |

Figure 9-34 shows the inflow and outflow hydrographs for this example.

## 9.4    THE STREAM NETWORK

We have seen that streamflow response to a water-input event is a complicated process involving spatial and temporal integration of lateral inflows and the generation of a flood wave that changes as it travels through the stream network. These changes are due to the time and space variations of the lateral inflows and tributary inflows and to the hydraulics of open-channel flow. This section describes some of the basic characteristics of stream networks and their relations to stream response.

### 9.4.1    Quantitative Description of Stream Networks

#### Stream Orders and the Laws of Network Composition

Figure 9-35 shows the most common approach to quantitatively describing stream networks (Strahler 1952): Streams with no tributaries are designated **first-order streams,** the confluence of two first-order streams is the beginning of a **second-order stream,** the confluence of two second-order streams produces a **third-order stream,** etc. When a stream of a given order receives a tributary of lower order, its order does not change. The order of a drainage basin is the order of the stream at the basin outlet.

The actual size of the streams designated a particular order depends on the scale of the map or image used and the conventions used in designating stream channels. A scale of about 1:25,000 is standard; at that scale the Mississippi River is a 12th-order stream at its mouth (Leopold et al. 1964).

Within a given drainage basin, the numbers, average lengths, and average drainage areas of streams of successive orders usually show consistent relations of the form shown in Figure 9-36. These relations are called the **laws of drainage-network composition** and are summarized in Table 9-7.

#### Links, Nodes, and Magnitude

A stream network can also be quantitatively described by designating the junctions of streams as **nodes** and the channel segments between nodes as **links.** Links connecting to only one node (i.e., first-order streams) are called **exterior links,** the others are **interior links.** The **magnitude** of a drainage-basin network is the total number of exterior links it contains; thus the network of Figure 9-35 is of magnitude 43. Typically the number of links of a given order is about half the number for the next lowest order (Kirkby 1993).

**FIGURE 9-34**
Inflow (*QI*) and outflow (*QO*) hydrographs for Example 9-4.

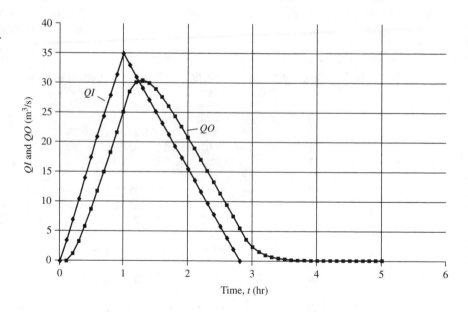

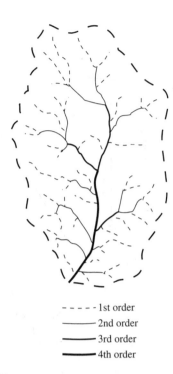

- - - - - 1st order
——— 2nd order
——— 3rd order
■■■ 4th order

**FIGURE 9-35**
Strahler (1952) system of designating stream orders.

## 9.4.2 Drainage Density

The **drainage density,** $D_d$, of an area, $A_D$, is the total length of streams draining that area, $\Sigma L$, divided by the area:

$$D_d \equiv \frac{\Sigma L}{A_D}. \qquad (9\text{-}31)$$

Drainage density thus has dimensions $[L^{-1}]$.

Drainage density is usually measured on maps or aerial photographs, and this can now be done relatively easily using digitizers or geographic information systems (GIS). Sellmann and Dingman (1970) showed that drainage density can also be readily measured by constructing a number of straight lines over the region of interest and counting the number of times the lines cross streams. Then

$$D_d = \frac{\pi}{2} \cdot \frac{N_x}{\Sigma L_x}, \qquad (9\text{-}32)$$

where $N_x$ is the total number of crossings and $\Sigma L_x$ is the total length of lines.

The value of $D_d$ for a given region will increase as the scale of the map on which measurements are made increases. Sellmann and Dingman (1970) found that drainage densities measured on standard U.S. Geological Survey topographic maps at a scale of 1:24,000 were close to true values observed in the field.

The drainage density of an area of similar geology in a given climatic region tends to have a characteristic value; values ranging from less than 2 km$^{-1}$ to over 100 km$^{-1}$ have been reported. Drainage density has been found to be related to average precipitation, with low values in arid and humid areas and the largest values in semi-arid regions. In a given climate, higher values of $D_d$ are generally found on less permeable soils, where channel incision by overland flow is more common, with lower values on more permeable materials.

### 9.4.3 Relations between Network Characteristics and Stream Response

Understanding hydrologic response requires knowledge of both hillslope processes and flow in channel networks, and the values of the bifurcation, length, and drainage-area ratios can be used to calculate some hydrologically important characteristics of watersheds, such as (1) the fraction of a basin draining directly to streams of a given order and (2) the fraction of streams of any order that are tributary directly to streams of any higher order. Formulas relating those characteristics to $R_B$, $R_A$, and $R_L$ were derived by Rodriguez-Iturbe and Valdes (1979); for example, if the bifurcation and drainage-area ratios are constant for all orders, the fraction of a watershed draining directly to first-order streams, $\theta_1$, is given by

$$\theta_1 = \frac{R_B^{\Omega-1}}{R_A^{\Omega-1}}, \qquad (9\text{-}33)$$

where $R_B$ is the bifurcation ratio, $R_A$ is the drainage-area ratio, and $\Omega$ is the order of the largest stream in the drainage basin.

**FIGURE 9-36**
(a) Numbers, $N_\omega$, (b) average lengths, $L_\omega$, and (c) average drainage areas, $A_\omega$, of streams of order $\omega = 1, 2, ..., 5$ in a drainage basin in England. See Table 9-7 and Example 9-5. From Knighton (1984).

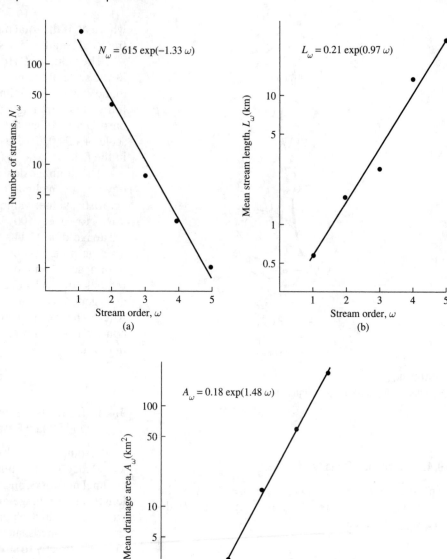

## EXAMPLE 9-5

For the watershed depicted in Figure 9-36 estimate (a) the bifurcation, drainage-area, and length ratios and (b) the fraction of the watershed that drains directly to first-order streams.

**Solution** (a) From the definitions given in Table 9-7 and the equations given in Figure 9-36 relating the number, $N_\omega$, average

length, $L_\omega$, and average drainage area, $A_\omega$, of streams to order, we find that

$$R_B = \exp(1.33) = 3.78,$$
$$R_L = \exp(0.97) = 2.64,$$

and

**TABLE 9-7**
The Laws of Drainage-Network Composition.

| Law | Relation[a] | Usual Range | Author |
|---|---|---|---|
| Law of stream numbers | $R_B \equiv \dfrac{N_\omega}{N_{\omega+1}}$ | $3 < R_B < 5$ | Horton (1945) |
| Law of stream lengths | $R_L \equiv \dfrac{L_{\omega+1}}{L_\omega}$ | $1.5 < R_L < 3.5$ | Horton (1945) |
| Law of drainage areas | $R_A \equiv \dfrac{A_{\omega+1}}{A_\omega}$ | $3 < R_A < 6$ | Schumm (1956) |

[a]$R_B \equiv$ bifurcation ratio, $R_L \equiv$ length ratio, $R_A \equiv$ drainage-area ratio, $N_\omega \equiv$ number of streams of order $\omega$, $L_\omega \equiv$ average length of streams of order $\omega$, and $A_\omega \equiv$ average drainage area of streams of order $\omega$; order: $\omega = 1, 2, 3, \dots \Omega$.

$$R_A = \exp(1.48) = 4.39.$$

(b) The drainage basin is of order $\Omega = 5$, so from Equation (9-33) we have

$$\theta_1 = \frac{3.78^4}{4.39^4} = 0.55.$$

These results are approximate, because there is some variation in the bifurcation, length, and drainage-area ratios as a function of order. For example, the actual fraction of the basin depicted in Figure 9-36 that is drained by first-order streams is 0.44.

---

It can be shown from straightforward geometry that the average distance from basin divide to stream channel, $X_d$, in a region is

$$X_d = \frac{1}{2 \cdot D_d}. \tag{9-34}$$

Therefore the average distance that a drop of water travels to a stream, $X_h$, is approximately

$$X_h = \frac{X_d}{2} = \frac{1}{4 \cdot D_d}. \tag{9-35}$$

Thus drainage density is an indicator of the efficiency of a stream network in draining an area, especially if its definition is modified to eliminate the effects of stream sinuosity:

$$D_e \equiv \frac{\Sigma L_v}{A_D}, \tag{9-36}$$

where $\Sigma L_v$ is the total length of stream *valleys* and $D_e$ is called the **effective drainage density** (Dingman 1978).

Because of its apparent relation to drainage efficiency, many studies have attempted to use drainage density as a predictor of hydrologic characteristics, such as the magnitudes of flood flows or low flows. However, Dingman (1978) cautioned that apparent relations between hydrology and drainage density may be spurious because of the varied routes by which water travels to streams and the many factors other than distance that affect drainage efficiency. (See also Harlin 1984.)

The **network width** is the number of links as a function of distance upstream from the basin outlet. For watersheds in which the time of travel in the stream network dominates over that in the hillslopes (typically watersheds with areas exceeding about 50 km[2]), the shape of the network-width function is related to the time between the occurrence of water input and the occurrence of the peak of the response hydrograph: The closer the maximum in the width function is to the watershed outlet, the shorter the time between input and peak response (Kirkby 1976).

Quantitative relations between event-flow hydrographs and the bifurcation, length, and drainage-area ratios are incorporated in the "geomorphologic instantaneous unit hydrograph" (Section 9.6.3).

## 9.5 RAINFALL–RUNOFF MODELING

One of the principal applications of hydrology is in the forecasting and predicting of flood peaks and runoff volumes due to large rain and snowmelt events. To do this, hydrologists use models that simulate the stream response to a water-input event of a given magnitude and spatial and temporal distribution on a drainage basin; these are usually referred to as **rainfall–runoff models.** (Some have been adapted for snowmelt as well.)

Ideally, rainfall–runoff models should simulate the physical processes by which water moves from the land surface to streams. This is especially important for predicting streamflow responses under conditions other than those for which we have recorded experience, including (1) extreme flood-producing rainfalls; (2) major land-use changes, such as deforestation and urbanization; and (3) altered climatic regimes. The most general way to model event response is to use spatially distributed numerical solutions to the complete equations of saturated–unsaturated subsurface and open-channel flow. Freeze (1972a; 1972b; 1974) used this approach to elucidate the conditions under which certain response mechanisms are important, but it is too computationally intensive and requires too much field data to be a practical approach to runoff modeling.

More practical physically based models commonly represent a watershed as a collection of hillslope strips on which simplified representations of the appropriate mechanisms operate, for example, the kinematic-wave model of overland flow (Eagleson 1970; Stephenson and Meadows 1986) or the sloping-slab/saturated overland flow model (Box 9-4). And, as discussed earlier, a new generation of event models incorporates saturation overland flow on variable source areas whose extent depends on basin topography and watershed wetness through the TOPMODEL approach (Box 9-3) (Hornberger et al. 1985; Wood et al. 1990). The coupling of such models with remotely sensed data and geographic information systems is a very promising area of current research (Van de Griend and Engman 1985; Shuttleworth 1988).

However, some of the most commonly used rainfall–runoff models are based on conceptual and empirical relations that appear to give "reasonable" results. Development of such expedient models has been a natural consequence of the widespread need for predictions and forecasts coupled with (1) the complexity and spatial and temporal variability of hydrologic processes (as we have seen in the preceding chapters); (2) the limited availability of spatially and temporally distributed hydrologic, climatologic, geologic, pedologic, and land-use data; and (3) the limited resources available to collect data and to develop, calibrate, and validate models.

This section introduces some of the most important considerations in rainfall–runoff modeling and some of the most commonly used simple conceptual rainfall–runoff models. More elaborate models used for flood forecasting and for design of large structures were reviewed by DeVries and Hromadka (1992) and Lettenmaier and Wood (1992).

Because of the importance of simulation models in understanding and predicting streamflow, you should review the discussion of modeling in Section 2.9 (particularly Table 2-3 and Figure 2-13) as background for the material in this section. In particular, the modeler must be aware of the need for calibration and validation and the dangers in relying on expedient empirical models to simulate events and conditions that lie outside the ranges for which they have been developed and tested (Klemeš 1986).

### 9.5.1 Basic Approach: The Systems View

In this section we view a watershed as a system, or "black box," that produces outputs (hydrographs) in response to inputs (hyetographs), without detailed consideration of the physical processes that are producing that response. In this view, a portion of the total water input in an event (the shaded part of the hyetograph in Figure 9-37a) produces an identifiable response (the shaded part of the hydrograph).

In any application of the systems approach, it is important to keep the following points in mind:

1. The hyetograph represents only the portion of total input that appears in the associated response hydrograph, called the **effective precipitation** (also called **effective rainfall** or **rainfall excess**). It does not include the generally larger portion of the input that ultimately evapotranspires or the portion that appears in the stream after the apparent current response has ended.

2. The hydrograph represents only the identifiable response to the current input hyetograph, called the **event flow** (also called **storm runoff, direct runoff,** or **quick flow**). It does not include concurrent streamflow resulting from earlier input events (**base flow**) or the portion of the streamflow from the

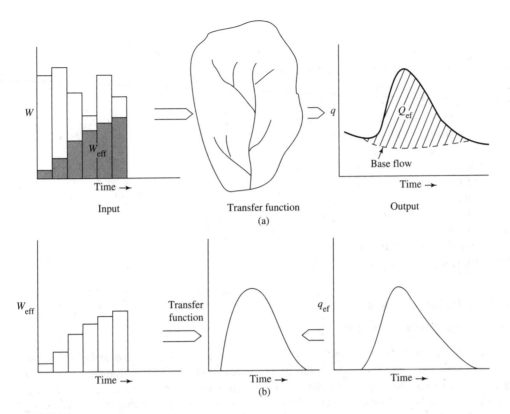

**FIGURE 9-37**

In the systems view, watershed response consists only of event flow, $Q_{ef}$, which is volumetrically equal to effective water input, $W_{eff}$. (a) In practice, the systems approach requires a priori determination of $W_{eff}$ from a measured or specified time distribution of $W$ (see Figure 9-40), routing through a specified transfer function (such as the unit hydrograph), and addition of the predicted response to a specified base flow. (b) The inverse problem involves identification of the transfer function from observed time distributions of input and response. Once the transfer function for a given watershed is determined, it is used to forecast or predict responses to actual or design storms.

current event that appears after the apparent response has ended.

3. Determination of the event-flow volume is usually based on graphical hydrograph separation, as discussed in Sections 8.5.3 (Figure 8-38) and 9.1.2.

4. The volume of effective water input equals the volume of event flow:

$$W_{eff} = Q_{ef}. \qquad \textbf{(9-37)}$$

5. In general, systems models are based on "reasonable" assumptions about, rather than a real understanding of, hydrologic processes. Unwarranted belief that these models portray hydrologic reality has not only hampered the development of hydro-logic science (Klemeš 1986), but has also led to inaccurate runoff forecasts (e.g., Loague and Freeze 1985).

An implicit assumption of the systems approach and underlying Equation (9-37) is that the water that appears as output, $Q_{ef}$, is the same water identified as input, $W_{eff}$. However, many of the studies of actual runoff sources and mechanisms described earlier (Table 9-4) have found that a substantial portion of the water appearing as stream response to a given event is "old" water—water that entered the watershed in previous rain or snowmelt events and was hydraulically displaced by the "new" water. Recalling this fact will remind the hydrologist that systems-based models are expedient conceptual devices rather than models of actual hydrologic processes.

Application of the systems approach requires (1) identification of an appropriate **transfer function** for the situation to be modeled and (2) estimation of $W_{eff}$ (and for some methods the time-distribution of $W_{eff}$) (Figure 9-37). The problem of identifying the model structure and/or parameters from measured values of input and output is called **system identification** (Figure 9-37b). Approaches to these problems are considered in the following sections.

Further detailed discussion of rainfall–runoff models can be found in the review by Pilgrim and Cordery (1992) and in the many textbooks on engineering hydrology (e.g., Ponce 1989; McCuen 1998).

### 9.5.2 Fundamental Considerations

#### Runoff Processes and Rainfall–Runoff Models

Here we summarize how the understanding of the physical mechanisms of runoff response developed earlier in this chapter can be applied to guide the application of the models discussed in Section 9.6.

In general, the processes with the longest residence times control the shape of the hydrograph, and Kirkby (1988) suggested that satisfactory event models can be developed considering only the two processes with the longest residence times on the watershed of interest (Table 9-8). Thus a small watershed might be efficiently modeled by taking account of infiltration, percolation, and the appropriate hillslope process. For a larger watershed, the response model might be simplified to include only the hillslope process combined with a channel-routing procedure. Of course, whichever processes are modeled, it is critical that the appropriate parameter values are determined through calibration and validation (Davis et al. 1999).

Following Woods and Sivaplan (1999), the residence time for watershed runoff for a given event, $T_{Rq}$, can be expressed as the sum of the residence times of water input, $T_{Rw}$, hillslope runoff production, $T_{Rh}$, and travel through the stream network, $T_{Rn}$:

$$T_{Rq} = T_{Rw} + T_{Rh} + T_{Rn}. \qquad (9\text{-}38)$$

$T_{Rw}$ can be approximated as one-half the storm duration, $T_W$; that is,

$$T_{Rw} = T_W/2. \qquad (9\text{-}39)$$

**TABLE 9-8**

Ranges of Residence Times Associated with Event-Response Processes on Hillslopes and in Channels.

| Process | Residence Times (hr) |
|---|---|
| Hillslope processes | |
|   Surface detention | 0.1–1 |
|   Infiltration | 1–20 |
|   Percolation | 1–50 |
|   Downslope flow | 1–12 |
| Channel flow | |
|   Watershed area = 1 km$^2$ | 0.5 |
|   Watershed area = 100 km$^2$ | 7 |
|   Watershed area = 10,000 km$^2$ | 100 |

From Kirkby (1988).

$T_{Rn}$ can be approximated as

$$T_{Rn} = \frac{A_D^{1/2}}{V_n}, \qquad (9\text{-}40)$$

where $A_D$ is drainage area and $V_n$ is the average flow velocity in the channel network. For rough estimates, $V_n$ can be taken as 1 m s$^{-1}$; if the appropriate information is available, more refined estimates can be obtained via Equation (9-19).

Hillslope residence times might be estimated more precisely using knowledge of local topography and geology and the operative runoff mechanism. For example, for runoff due to flow in a sloping slab,

$$T_{Rh} = \frac{1}{4 \cdot D_d \cdot V_h}, \qquad (9\text{-}41)$$

where $D_d$ is drainage density [see Equations (9-34) and (9-35)] and $V_h$ is the typical flow velocity. $V_h$ could be calculated by approximating Darcy's Law as

$$V_h = K_h^* \cdot \sin(\beta_s), \qquad (9\text{-}42)$$

where $K_h^*$ is hydraulic conductivity and $\beta_s$ is the slope angle [as in Equation (9B4-3)].

Figure 9-38 shows the ranges of lag-to-peak times and peak flow rates measured in field studies for watersheds in which responses were due to Hortonian overland flow, saturation overland flow, and "throughflow." The exact nature of the throughflow represented in the data is not clear; it probably refers largely to macropore-dominated flow in sloping slabs, but may include other subsurface mechanisms as well. Clearly, overland flow yields lag times

**FIGURE 9-38**

(a) Ranges of lag-to-peak and (b) peak flow rates associated with various response mechanisms. "Throughflow" probably includes all subsurface mechanisms in Table 9-3 except ground-water mounding. From Kirkby (1988).

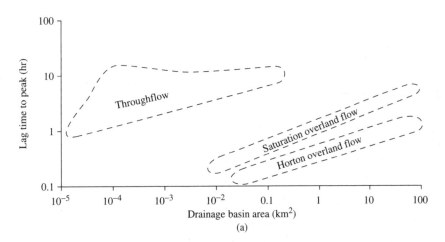

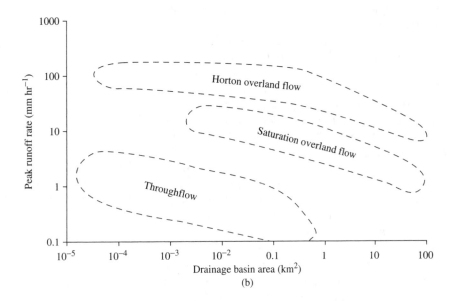

an order of magnitude shorter and peak flows at least an order of magnitude larger than subsurface flow. Channel hydraulics significantly affects the peaks and lag times for watersheds larger than a few tens of square kilometers.

The work of Wood et al. (1990) indicates that actual patterns of soil and water-input variability do not have to be modeled for regions less than about 1 km$^2$ in area; instead the spatial variability can be accounted for more simply through the areal means and variances of those quantities. And, because channel processes tend to dominate the response hydrographs of larger watersheds, much of the variability and non-linearity of hillslope event-response

mechanisms can be ignored or averaged when modeling watersheds over about 100 km$^2$ in area. For these watersheds, simple conceptual approaches like the unit-hydrograph model described in Section 9.6.3 may provide a satisfactory model.

### Conceptual Model of Effective Water Input

Effective water input is conventionally considered to be

$$W_{eff} = W - \text{"losses,"} \qquad \textbf{(9-43)}$$

where $W$ is total water input during an event. In Equation (9-43),

$$\text{losses} = ET + \Delta S_c + \Delta D + \Delta\theta, \quad \textbf{(9-44)}$$

where $ET$ is the portion of the event water evapotranspired during the event, $\Delta S_c$ is the net addition to storage on the vegetative canopy (Figure 7-17), $\Delta D$ is the net addition to **depression storage** (i.e., the water added to lakes, ponds, wetlands, puddles, and smaller depressions), and $\Delta\theta$ is the net addition to soil-water storage during the event.

Because rainfall events are usually of short duration and are accompanied by high humidity and low solar radiation, $ET$ is usually small. As discussed in Section 7.6.3, canopy-storage capacity is on the order of 1 mm × leaf-area index (Figure 7-17). Thus it is usually filled quickly, and $\Delta S_c$ is also usually negligible for storms that generate significant responses.

Depression storage is spatially variable and difficult to estimate, and only very meager data on depression-storage volumes have been published. Values on the order of 5–10 mm have been estimated for turf (Bras 1990). Evaporation and infiltration usually empty small surface depressions between rainstorms (Kirkby 1988); however, they may be maintained full during active snowmelt.

Because of the difficulty of evaluating depression storage, it is usually treated conceptually in combination with soil-water storage. These combined storage components are typically modeled as filling in the same way that infiltration occurs (Figures 6-19 and 6-25). Thus the ratio $W_{eff}/W$ is largely determined by the degree to which the available near-surface storage capacity is already filled, that is, by the antecedent soil-water content, $\theta_0$ (Section 6.5.4). Operational methods for relating effective water input to antecedent conditions of watershed wetness are discussed later in this section.

### Design Floods vs. Floods From Actual Storms

Rainfall–runoff models are used both to generate design floods and floods from actual storms (Figure 9-39). A **design flood** is a flood of a specified return period[6] that is used in the design of culverts,

---

[6] As explained in Section C.3, the return period of a rainfall event or flood is the inverse of the probability that it will be exceeded in any year. Thus there is a probability of 0.1 that the "10-year flood" discharge will be exceeded in any year, and a probability of 0.01 that the "100-year flood" discharge will be exceeded in any year.

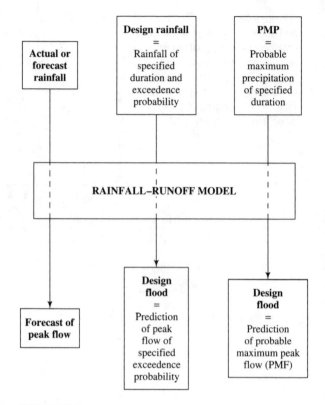

**FIGURE 9-39**
Use of systems models for forecasting and prediction. In addition to flood peaks, other characteristics of response such as total flood volume or duration may also be forecast or predicted.

bridges, flood-retention basins, levees, dam spillways, or floodplain-management plans. Floods from actual storms are generated in two contexts: (1) to forecast flooding from an in-progress storm and (2) to calibrate rainfall–runoff models using historical storms as inputs.

**Design Floods**    By far the most common use of rainfall–runoff models is for generating design floods. For these situations, the goal is to estimate the design flood using an appropriate design rainfall as the input. The design return period depends on the consequences of exceedence: small culverts may be designed to pass a low return-period flood (e.g., the 10-year flood), bridges a somewhat larger flood (e.g., the 25-year flood), and spillways of large dams the 100-year flood. If the failure of a dam would cause great economic damage and loss of life, the "probable maximum flood" (PMF) is used for the spillway design.

Estimates of design rainfalls are developed using depth–duration–frequency analysis methods or, for the PMF, by using the probable maximum precipitation (PMP), as described in Section 4.4.4.

A crucial question in selection of the design rainfall is the determination of the storm duration that will generate the required design flood for a particular watershed. For a given region, the critical duration increases with watershed area; it is generally of the order of the watershed's time of concentration, which can be estimated by formulas like those summarized in Table 9-9.

The critical duration for a given watershed can also be determined by trial and error. To do this,

rainfalls of a given exceedence probability but varying durations are used as input to the rainfall–runoff model, and the duration that gives the largest peak is selected.

Another important question in generating design floods is the relationship between the return period of the rainfall used as model input and the return period of the resulting flood. Approaches to developing answers to this question are described in Box 9-7.

**Floods from Actual Storms** Flood forecasting is usually done via complex hydrologic models that are to varying degrees physically based, rather than the simple conceptual models discussed in Section 9.6. These complex models typically divide a larger

**TABLE 9-9**

Formulas for Estimating Time of Concentration, $T_c$, from Watershed Characteristics and Rainfall Intensity. Symbols defined after table.

| Equation[a] | Remarks | Source |
|---|---|---|
| $T_c = 0.0278 \cdot \dfrac{B^{0.6}}{k^{0.4} \cdot (K_h^* \cdot i_{eff} \cdot S_s)^{0.2}}$ | Sloping slab, steep forested watershed | Loukas and Quick (1996) |
| $T_c = 0.0664 \cdot \left(\dfrac{L}{S_c^{0.5}}\right)^{0.77}$ | Agricultural watersheds in TN; $0.003 \leq A_D \leq 0.5$ km$^2$ | Kirpich (1940)[b] |
| $T_c = 0.161 \cdot \left(\dfrac{L}{S_c^{0.5}}\right)^{0.64}$ | Midwestern U.S.; $0.012 \leq A_D \leq 18.5$ km$^2$ | Chow (1962)[b] |
| $T_c = 0.927 \cdot \left(\dfrac{L}{S_c^{0.5}}\right)^{0.47}$ | United Kingdom | NERC (1975)[b] |
| $T_c = 0.700 \cdot \left(\dfrac{L \cdot L_c}{S_c^{0.5}}\right)^{0.38}$ | Appalachian Mountains | Snyder (1938)[b] |
| $T_c = 0.128 \cdot \left(\dfrac{L}{S_c^{0.5}}\right)^{0.79}$ | U.S and Canada; $0.01 \leq A_D \leq 5840$ km$^2$; $0.00121 \leq S_c \leq 0.0978$ | Watt and Chow (1985)[b] |
| $T_c = 0.817 \cdot \dfrac{L \cdot n^{0.75}}{A_D^{0.25} \cdot k^{0.5} \cdot i_{eff}^{0.25} \cdot S_c}$ | Overland flow | Aron et al. (1991)[b] |
| $T_c = 2.15 \cdot \dfrac{L^{0.5} \cdot n^{0.52}}{S_c^{0.31} \cdot i_{eff}^{0.38}}$ | Rural watersheds; $A_D < 5$ km$^2$ | Papadakis and Kazan (1987)[b] |

[a]Assumes $T_c = 1.67 \cdot T_{LPC}$.

[b]See Loukas and Quick (1996) for original reference.

$A_D$ = drainage area (km$^2$)
$B$ = factor integrating travel times (dimensionless)
$i_{eff}$ = effective rainfall intensity (mm hr$^{-1}$)
$k$ = channel shape factor (dimensionless)
$K_h^*$ = hydraulic conductivity of slope soil (mm hr$^{-1}$)
$L$ = length of main stream (km)
$L_c$ = stream distance from basin outlet to point opposite watershed centroid (km)
$n$ = Manning's resistance factor for channel (Table 9-6)
$S_c$ = sine of channel slope angle (dimensionless)
$S_s$ = sine of hillslope angle (dimensionless)
$T_c$ = time of concentration (hr)
$T_{LPC}$ = centroid lag-to-peak (Table 9-1) (hr)

---

**BOX 9-7**

. . . . . . .

**Approaches to Developing Relations Between Rainfall Return Period and Flood Return Period**

As noted by Pilgrim and Cordery (1992), there are four general approaches to this problem:

1. The simplest and most direct approach requires (a) frequency analysis of floods at gaging stations in the region of interest, using the methods described in Box C-1 or methods that apply a specific probability distribution (Interagency Committee on Water Data 1982) and (b) frequency analysis of rainfalls of appropriate durations for the gaged watersheds (Section 4.4.4). By trial and error, determine the model parameters that convert a rainfall of a given return period to a flood peak of the same return period.

2. For a given watershed, determine the model parameters that best reproduce the measured floods for each of a number of events. Select the median values of these parameters for use; this should result in a close correspondence between the return periods of rainfall and the resulting flood.

3. Calibrate the selected model for the watershed of interest. Then use the model to generate a continuous record of simulated streamflows from a long record of historical rainfall data. Finally, conduct a frequency analysis of the rainfall and flow data, and relate the return periods.

4. Conduct a joint probability analysis of parameters of an appropriate model and inputs.

---

watershed into sub-watersheds that are modeled separately. The models also contain representations of the physical processes discussed in previous chapters, particularly snowmelt and infiltration. Such models may simulate the specific runoff processes discussed in Section 9.2 and generally contain procedures for streamflow routing. Excellent reviews of these models and their use in forecasting have been prepared by DeVries and Hromadka (1992) and Lettenmaier and Wood (1992).

Meaningful predictions and forecasts are possible only with models that have been calibrated and validated for situations similar to those for which they will be applied. As discussed in Section 2.9, calibration (or, more accurately, "parameter estimation") involves use of actual rainfall data as model input, comparison of model outputs with actual streamflow data, and selection of the model parameters that give the "best" model performance. Validation (or, more accurately, "acceptance testing") involves comparing model outputs with actual flows for events not used in calibration. One of the drawbacks of complex models with many parameters is that significantly differing sets of parameters may give equally good results, reducing confidence in model validity.

Since systems models relate *effective* rainfall to storm runoff, calibration and validation of such models requires separation of hyetographs; this is discussed in the following section.

### Forecasting or Predicting Effective Water Input

However the transfer function is developed, it is important to emphasize that the validity of the estimate of $W_{eff}$ is at least as important as the exact nature of the transfer function in determining the accuracy of the forecast response (Figures 9-8 and 9-9).

Unless unusually detailed observations of storage components—especially soil moisture—are available, there is very little basis for physically based forecasts or predictions of the quantity of effective water input. The need for such forecasts and predictions has thus led to the development of empirical methods for estimating $W_{eff}$.

Most of these methods are based on one of the conceptual relations shown in Figure 9-40. When attempting to solve the system identification problem for a given watershed, one can determine the total losses for various events a posteriori by hydrograph separation and then estimate their time distribution via one of the models in Figure 9-40.

**FIGURE 9-40**
Conceptual models for estimating effective water input, $W_{eff}$, from hyetograph of water input, $W$. Losses are shaded portions, and $W_{eff}$ is unshaded. (a) Losses equal a constant fraction of the water input for each time period. (b) Losses equal a constant rate throughout the event. (c) Losses are given by an **initial abstraction** (which may be a specified amount or all input over an initial time period) followed by a constant rate (which may be zero). (d) Losses are given by an approximation to an infiltration-type curve, such as given by the Green-and-Ampt (Section 6.6.2) or Phillips approach (Section 6.6.4) (dotted line). After Pilgrim and Cordery (1992).

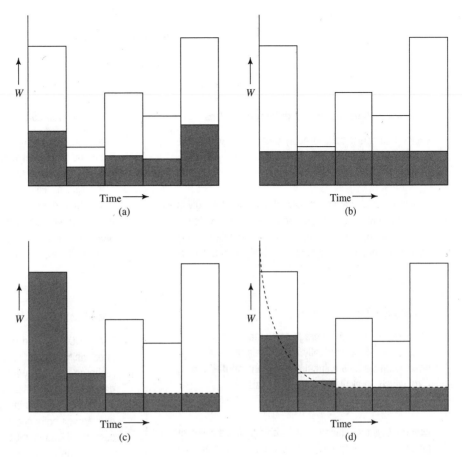

For future applications in predicting the response of the watershed, one might then attempt to relate $W_{eff}$ to a measurable quantity that reflects basin storage via one of the approaches described in Box 9-8.

## 9.6 RAINFALL–RUNOFF MODELS

In this section, we introduce three simple rainfall–runoff models. The "rational method" and the "SCS method" are commonly used for generating design flows from small watersheds for simple, relatively inexpensive structures such as culverts, small bridges, surface-drainage systems, and runoff-detention basins. The rational method is most often applied to urban areas, and the SCS method to suburban and rural areas. The "unit-hydrograph method" is often applied to generate design flows from larger watersheds.

### 9.6.1 The Rational Method

The **rational method** postulates a proportionality between peak discharge, $q_{pk}$, and rainfall intensity, $i_{eff}$; that is,

$$q_{pk} = u_R \cdot C_R \cdot i_{eff} \cdot A_D, \qquad (9\text{-}45)$$

where $u_R$ is a unit-conversion factor,[7] $A_D$ is drainage area, and $C_R$ is the **runoff coefficient,** which depends on watershed land use.

Equation (9-45) was derived from a simplified conceptual model of travel times on basins with negligible surface storage and is widely used for drainage design for small suburban and urban watersheds. The duration of the rainfall to be used in Equation (9-45) is taken as the time of concentration of the watershed, for which values can be esti-

[7]

| $q_{pk}$ | $i_{eff}$ | $A_D$ | $u_R$ |
|---|---|---|---|
| m³ s⁻¹ | mm hr⁻¹ | km² | 0.278 |
| ft³ s⁻¹ | in. hr⁻¹ | acre | 1.008 |

**BOX 9-8**

. . . . . . . .

**Estimating $W_{eff}$**

### Empirical Relations with Storm Characteristics

Lacking sufficient data to apply other approaches, one can estimate

$$W_{eff} = a_0 + a_1 \cdot W, \qquad \text{(9B8-1)}$$

where $a_0$ and $a_1$ were empirically determined for a given watershed. Although Loague and Freeze (1985) found that the approach did not yield precise estimates, it may have some predictive value and be useful if data limitations preclude other approaches.

### Antecedent Rainfall Indices

Following a storm, the storage components enumerated in Equation (9-44) gradually empty. This process is approximated via an empirical **antecedent rainfall index, $I_a(d)$**, which is calculated on a daily basis as

$$I_a(d) = I_a(0) \cdot k^d, \qquad \text{(9B8-2)}$$

where $I_a(0)$ is the value for a day with rain, $k$ is a constant (usually $0.80 < k < 0.98$), and $d$ is the number of days since the last rainfall. The values of $I_a(0)$ and $k$ are empirically determined for a particular watershed. Conceptually, $I_a(0)$ represents the total watershed near-surface storage (usually expressed as a depth of water), and $I_a(d)$ is the amount of water from the previous storm that remains in that storage on day $d$. One then identifies the empirical relation between $W_{eff}$ and $I_a(d)$ for past storms. Linsley and Kohler (1951) discussed this approach in detail.

### Indices Related to Ground-Water and/or Soil-Water Levels

Forecasts of $W_{eff}$ for a particular watershed or region can be empirically related to water levels in observation wells or soil-water content measured at one or more index sites (Dunne et al. 1975; Troch et al. 1993).

### Empirical Relations with Antecedent Discharge

Assuming that the rate of outflow from a watershed reflects the amount of water stored in it [Equation (2-25)], it is reasonable to attempt to relate $W_{eff}$ to antecedent discharge, $q_0$. Figure 9-41 shows the relation between $W_{eff}/W$ and $q_0$ for the same watershed portrayed in Figure 9-8 and indicates that it has useful predictive ability there. Gburek (1990) reported similar results for a watershed in Pennsylvania, but with considerably greater scatter.

### Use of Continuous Watershed Models

The BROOK model described in previous chapters simulates daily values of evapotranspiration, percolation, recharge, and soil-water storage. If calibrated and validated for a particular watershed, such models can be used to keep a running account of the state of the various storage components. When a water-input event occurs, the filling of those components can be estimated using models of the interception, infiltration, and percolation processes as discussed in earlier chapters; $W_{eff}$ is then estimated as the residual. Several such models were described by Viessman et al. (1989).

### Probability Models

Beran and Sutcliffe (1972) described how the probability distribution of $W_{eff}$ can be determined from the distribution of $W$ in Britain.

### SCS Method

In the absence of specific information on antecedent conditions, the SCS curve-number method described in Box 9-9 is widely used in the U.S. to estimate $W_{eff}$ given total rainfall $W$ and information on watershed soils.

**FIGURE 9-41**
Relation between $Q_{ef}/W$ and antecedent discharge, $q_0$, for a small watershed in central Alaska for the same 16 storms shown in Figure 9-8. From Dingman (1970).

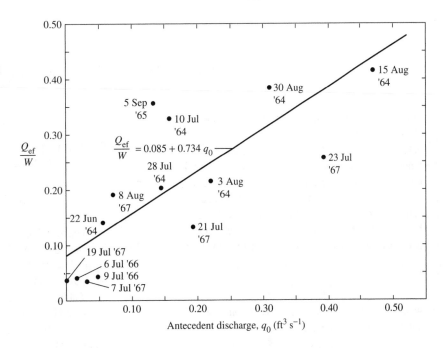

mated by formulas like those in Table 9-9. For design purposes, the return period of the design peak flow is equal to the return period of the rainfall.

One can see from Equation (9-45) that the sole model parameter, $C_R$, is the ratio of peak streamflow per unit area to rainfall intensity. Obviously, the results obtained with the method are highly sensitive to the value chosen for $C_R$; values range from 0.05 for gently sloping lawns up to 0.95 for highly urbanized areas of roofs and pavement (Table 9-10).

**TABLE 9-10.**
Runoff Coefficients, $C_R$, for the Rational Method [Equation (9-45)]. Values apply to storms with return periods of 5 to 10 yr; higher values should be used for higher return-period storms.[a]

| Type of Surface | $C_R$ |
|---|---|
| Pavement: Asphalt and Concrete | 0.70–0.95 |
| Pavement: Brick | 0.70–0.85 |
| Roofs | 0.75–0.95 |
| Lawns (sandy soil) :slope < 2% | 0.05–0.10 |
| Lawns (sandy soil) :slope 2%–7% | 0.10–0.15 |
| Lawns (sandy soil) :slope > 7% | 0.15–0.20 |
| Lawns (heavy soil) :slope < 2% | 0.13–0.17 |
| Lawns (heavy soil) :slope 2%–7% | 0.18–0.22 |
| Lawns (heavy soil) :slope > 7% | 0.25–0.35 |
| Parks and Cemeteries | 0.10–0.25 |
| Playgrounds | 0.20–0.35 |
| Railroad yards | 0.20–0.35 |
| Unimproved | 0.10–0.30 |

[a]Values from McCuen (1989).

The rational method is widely used in urban drainage design, but Pilgrim and Cordery (1992) caution that there are typically few data available to guide the selection of $C_R$, and that $C_R$ for a given watershed may vary widely from storm to storm due to differing antecedent conditions.

### 9.6.2 SCS Curve-Number Method

The most widely used rainfall–runoff model for routine design purposes in the United States is the **SCS method.** This method was developed by the U.S. Soil Conservation Service (now the U.S. Natural Resources Conservation Service, NRCS) and makes direct use of soils information routinely mapped by that organization. Given a watershed in which the hydrologic characteristics of the soils (discussed further in the following sections) have been mapped and a design rainfall volume, $W$, there are two basic computations in the method: (1) estimation of the effective rainfall, $W_{eff}$ (= the event-flow volume, $Q_{ef}$), and (2) estimation of the peak discharge, $q_{pk}$, and, if desired, the entire runoff hydrograph.

#### *Estimation of Effective Rainfall, $W_{eff}$*

The SCS method relates $W_{eff}$ to total rainfall, $W$, and watershed storage capacity, $V_{max}$, for small watersheds via the empirical relation

---

## BOX 9-9

. . . . . . .

### SCS Approach to Estimating $W_{eff}$ for Rainstorms

Figure 9-42 shows a schematic hyetograph for a rainstorm. The SCS approach assumes that the total rainfall volume $W$ is allocated to (1) **initial abstraction,** $V_I$, the amount of storage that must be satisfied before event flow can begin; (2) **retention,** $V_R$, the amount of rain that falls after the initial abstraction is satisfied, but which does not contribute to event flow; and (3) **event flow,** $Q_{ef}$. It is further assumed that the watershed has a maximum retention capacity $V_{max}$ and that the following relation obtains:

$$\frac{V_R}{V_{max}} = \frac{W_{eff}}{W - V_I}. \qquad \text{(9B9-1)}$$

All the quantities are volumes expressed on a per-watershed-area basis as lengths.

The actual retention is

$$V_R = W - V_I - W_{eff}. \qquad \text{(9B9-2)}$$

Combining Equations (9B9-1) and (9B9-2) and solving for $W_{eff}$, we find that

$$W_{eff} = \frac{(W - V_I)^2}{W - V_I + V_{max}}. \qquad \text{(9B9-3)}$$

Examination of actual response hydrographs led to the generalization that $V_I = 0.2 \cdot V_{max}$ under conditions of "normal" watershed wetness. Substitution of that relation into Equation (9B9-3) gives Equation (9-46).

Although Equation (9-46) is widely used, there are concerns about inconsistencies in the definitions of the quantities (Chen 1982; Sabol and Ward 1983; Boughton 1989; 1994).

---

$$W_{eff} = \frac{(W - 0.2 \cdot V_{max})^2}{W + 0.8 \cdot V_{max}}. \qquad \text{(9-46)}$$

The rationale for the SCS approach is given in Box 9-9. Note from Figure 9-42 that this rationale coincides closely with the behavior of infiltration when $w$ exceeds the saturated hydraulic conductivity of the surface soil, $K_h^*$, as modeled by the Green and Ampt approach in Section 6.6.2. However, the model is not explicitly an infiltration model, and $V_{max}$ supposedly incorporates all watershed storage.

The NRCS classifies and maps at various scales the soils in the United States, and the wide acceptance of the SCS method is largely because values of $V_{max}$ can be determined from such maps, along with land-cover information. Each mapped soil type is assigned to one of the hydrologic soils groups described in Table 9-11. These groups are based largely on minimum infiltration capacity, which should be approximately equal to $K_h^*$ of the surface (Section 6.6.2).

Table 9-12 gives the SCS **curve numbers** assigned to each hydrologic soil group under various land uses. $V_{max}$ is determined from these curve numbers, $CN$, as

$$V_{max} = \frac{1000}{CN} - 10, \qquad \text{(9-47)}$$

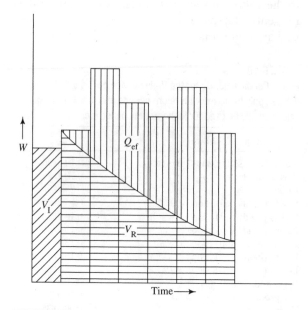

**FIGURE 9-42**
Definitions of initial abstraction, $V_I$, retention, $V_R$, and event flow, $Q_{ef}$, in the SCS method.

**TABLE 9-11**

Hydrologic Soils Groups as Defined by U.S. Soil Conservation Service (1964)[a]

| Soil Group | Characteristics |
|---|---|
| A | Low overland-flow potential; high minimum infiltration capacity even when thoroughly wetted (>0.30 in. h$^{-1}$ = 0.76 cm h$^{-1}$). Deep, well- to excessively drained sands and gravels. |
| B | Moderate minimum infiltration capacity when thoroughly wetted (0.15 to 0.30 in. h$^{-1}$ = 0.38 to 0.76 cm h$^{-1}$). Moderately deep to deep, moderately to well-drained, moderately fine- to moderately coarse-grained (e.g., sandy loam). |
| C | Low minimum infiltration capacity when thoroughly wetted (0.05 to 0.15 in. h$^{-1}$ = 0.13 to 0.38 cm h$^{-1}$). Moderately fine- to fine-grained soils or soils with an impeding layer (fragipan). |
| D | High overland-flow potential; very low minimum infiltration capacity when thoroughly wetted (<0.05 in. h$^{-1}$ = 0.13 cm h$^{-1}$). Clay soils with high swelling potential, soils with permanent high-water table, soils with a clay layer near the surface, shallow soils over impervious bedrock. |

[a]Minimum infiltration capacities given should approximate saturated hydraulic conductivities. (See Section 6.6.2.)

where $V_{max}$ is in inches. Note that this is an empirical dimensionally inhomogeneous equation; the constants differ when other units are used (see Section A.4).

Curve numbers may be further adjusted to reflect the antecedent wetness of the watershed, as prescribed in Table 9-13; however, the values given in Table 9-12 are used for most design purposes.

When a watershed consists of more than one soil/land-use complex, the standard approach is to compute a weighted average curve number as in Example 9-6. [See Grove et al. (1998) for alternative approaches.]

## EXAMPLE 9-6

Using the SCS method, compute $W_{eff}$ for a rain event of 4.2 in. in 3.4 hr for antecedent wetness conditions I, II, and III on a 1.24-mi$^2$ watershed with a mainstream length of 0.84 mi, a main-channel slope of 0.08, and the following land-cover characteristics:

| Land Cover | Soil Group | Area (mi$^2$) | Fraction of Total Area | Condition II Curve Number |
|---|---|---|---|---|
| Forest | B | 0.72 | 0.58 | 58 |
| Forest | C | 0.15 | 0.12 | 72 |
| Meadow | A | 0.26 | 0.21 | 30 |
| Meadow | B | 0.11 | 0.09 | 58 |

**Solution** The curve numbers for Condition II were found from Table 9-12. The weighted-average curve number for Condition II is found as

$$0.58 \times 58 + 0.12 \times 72 + 0.21 \times 30 + 0.09 \times 58$$
$$= 53.8 \rightarrow 54.$$

Then from Equation (9-47),

$$V_{max} = \frac{1000}{54} - 10 = 8.52 \text{ in.},$$

and using this value in Equation (9-46) gives

$$W_{eff} = \frac{(4.2 \text{ in.} - 0.2 \times 8.52 \text{ in.})^2}{(4.2 \text{ in.} + 0.8 \times 8.52 \text{ in.})} = 0.57 \text{ in.}$$

Repeating the computations using the curve numbers adjusted for Conditions I and III yields the following data:

| Condition | Weighted Curve Number | $V_{max}$ (in.) | $W_{eff}$ (in.) |
|---|---|---|---|
| I | 35 | 18.6 | 0.012 |
| II | 54 | 8.5 | 0.57 |
| III | 72 | 3.9 | 1.60 |

### Estimation of Peak Discharge, $q_{pk}$

Peak discharge is computed by assuming that the runoff hydrograph is a triangle, with a time of rise, $T_r$, (see Table 9-1) given by

$$T_r = 0.5 \cdot T_W + 0.6 \cdot T_c, \qquad (9\text{-}48)$$

where $T_W$ is the duration of excess rainfall and $T_c$ is the watershed time of concentration estimated from an appropriate formula from Table 9-9. The time base of the hydrograph, $T_b$, is then given by

$$T_b = 2.67 \cdot T_r. \qquad (9\text{-}49)$$

The total runoff volume is set equal to the triangular area, so that

$$Q_{ef} = W_{eff} = 0.5 \cdot q_{pk} \cdot T_b. \qquad (9\text{-}50)$$

Combining Equations (9-48)–(9-50), solving for $q_{pk}$, and adjusting for units then yields

$$q_{pk} = \frac{484 \cdot W_{eff} \cdot A_D}{T_r}, \qquad (9\text{-}51)$$

where $q_{pk}$ is in ft$^3$ s$^{-1}$, $W_{eff}$ is in in., $A_D$ is in mi$^2$, and $T_r$ is in hr.

## EXAMPLE 9-7

Compute the peak discharge for the event of Example 9-6 for antecedent conditions I, II, and III.

**TABLE 9-12**

SCS Curve Numbers for Various Soils/Land-Cover Complexes, Antecedent Wetness Condition II ("average").

| Land Use or Cover | Treatment or Practice | Hydrologic Condition | Hydrologic Soil Group A | B | C | D |
|---|---|---|---|---|---|---|
| Fallow | Straight row | Poor | 77 | 86 | 91 | 94 |
| Row crops | Straight row | Poor | 72 | 81 | 88 | 81 |
| | Straight row | Good | 67 | 78 | 85 | 89 |
| | Contoured | Poor | 70 | 79 | 84 | 88 |
| | Contoured | Good | 65 | 75 | 82 | 86 |
| | Contoured and terraced | Poor | 66 | 74 | 80 | 82 |
| | Contoured and terraced | Good | 62 | 71 | 78 | 81 |
| Small grain | Straight row | Poor | 65 | 76 | 84 | 88 |
| | Straight row | Good | 63 | 75 | 83 | 87 |
| | Contoured | Poor | 63 | 74 | 82 | 85 |
| | Contoured | Good | 61 | 73 | 81 | 84 |
| | Contoured and terraced | Poor | 61 | 72 | 79 | 82 |
| | Contoured and terraced | Good | 59 | 70 | 78 | 81 |
| Close-seeded legumes or rotation meadow | Straight row | Poor | 66 | 77 | 85 | 89 |
| | Straight row | Good | 58 | 72 | 81 | 85 |
| | Contoured | Poor | 64 | 75 | 83 | 85 |
| | Contoured | Good | 55 | 69 | 78 | 83 |
| | Contoured and terraced | Poor | 63 | 73 | 80 | 83 |
| | Contoured and terraced | Good | 51 | 67 | 76 | 80 |
| Pasture or range | | Poor | 68 | 79 | 86 | 89 |
| | | Fair | 49 | 69 | 79 | 84 |
| | | Good | 39 | 61 | 74 | 80 |
| | Contoured | Poor | 47 | 67 | 81 | 88 |
| | Contoured | Fair | 25 | 59 | 75 | 83 |
| | Contoured | Good | 6 | 35 | 70 | 79 |
| Meadow (permanent) | | Good | 30 | 58 | 71 | 78 |
| Woodlands (farm woodlots) | | Poor | 45 | 66 | 77 | 83 |
| | | Fair | 36 | 60 | 73 | 79 |
| | | Good | 25 | 55 | 70 | 77 |
| Farmsteads | | | 59 | 74 | 82 | 86 |
| Roads, dirt | | | 72 | 82 | 87 | 89 |
| Roads, hard-surface | | | 74 | 84 | 90 | 92 |

From U.S. Soil Conservation Service (1964).

**TABLE 9-13**
Antecedent Wetness Conditions and Curve-Number Adjustments for the SCS Method[a]

| Curve Numbers | | |
|---|---|---|
| **Condition II** | **Condition I** | **Condition III** |
| 100 | 100 | 100 |
| 95 | 87 | 98 |
| 90 | 78 | 96 |
| 85 | 70 | 94 |
| 80 | 63 | 91 |
| 75 | 56 | 88 |
| 70 | 51 | 85 |
| 65 | 45 | 82 |
| 60 | 40 | 78 |
| 55 | 35 | 74 |
| 50 | 31 | 70 |
| 45 | 26 | 65 |
| 40 | 22 | 60 |
| 35 | 18 | 55 |
| 30 | 15 | 50 |
| 25 | 12 | 43 |
| 20 | 9 | 37 |
| 15 | 6 | 30 |
| 10 | 4 | 22 |
| 5 | 2 | 13 |

| | | Total Rain 5 Previous Days (in.) | |
|---|---|---|---|
| Condition | Soil Wetness | Dormant Season | Growing Season |
| I | Dry but above wilting point | <0.5 | <1.4 |
| II | Average | 0.5–1.1 | 1.4–2.1 |
| III | Near saturation | >1.1 | >2.1 |

From U.S. Soil Conservation Service (1964).

**Solution**  To calculate the time of concentration, $T_c$, select an appropriate formula from Table 9-9. For this example, we select the Watt and Chow formula. To use this formula, we must convert mainstream length from mi to km:

$$L = 0.84 \text{ mi} \times 1.609 \frac{\text{km}}{\text{mi}} = 1.35 \text{ km.}$$

Then, with $S_c = 0.08$, we find that

$$T_c = 0.128 \cdot \left( \frac{1.35}{0.08^{0.5}} \right)^{0.79} = 0.44 \text{ hr.}$$

From Equation (9-48),

$$T_r = 0.5 \times 3.4 \text{ hr} + 0.6 \times 0.44 \text{ hr} = 1.96 \text{ hr.}$$

Finally, from Equation (9-51), we derive the following result for Condition II:

$$q_{pk} = \frac{484 \times 1.24 \times 0.57}{1.96} = 175 \text{ ft}^3 \text{ s}^{-1}.$$

The results for Conditions I and III are

$$q_{pk} = 3.1 \text{ ft}^3 \text{ s}^{-1}$$

and

$$q_{pk} = 490 \text{ ft}^3 \text{ s}^{-1},$$

respectively.

### Estimation of Runoff Hydrograph

The analysis of a large number of hydrographs developed for watersheds over a range of sizes and locations led to the formulation of the SCS generalized dimensionless synthetic hydrograph (Mockus 1957). In this approach the time of rise is estimated via Equation (9-48) and the peak discharge via Equation (9-51). These values are then used to scale the time and discharge axes as indicated in Table 9-14.

### Application and Validity

The results of studies comparing SCS-method predictions with measured data have been mixed. For example, Kumar and Jain (1982) applied the method to estimate $W_{eff}$ for 11 storms on a research watershed in Iowa and found poor agreement with values determined by hydrograph separation (Figure 9-43). Wood and Blackburn (1984) also found discrepancies for rangelands; however, others have found satisfactory agreement (e.g., Mostaghimi and Mitchell 1982).

Results like those shown in Figure 9-43—for a region in which the method should be well suited—suggest that it is unwise to accept uncritically the predictions of the SCS method. The user of the SCS relations must bear in mind that they are generalized and may not be very accurate for a specific watershed. The NRCS now recognizes that the relation of a given set of soil- and land-use conditions to curve number may vary regionally (Miller and Cronshey 1989). Thus field observations are al-

**TABLE 9-14**
Ordinates of the SCS Dimensionless Unit Hydrograph.

| $t/t_{pk}$ | $q/q_{pk}$ | $t/t_{pk}$ | $q/q_{pk}$ |
|------|-------|------|-------|
| 0.0 | 0.0 | 1.4 | 0.75 |
| 0.1 | 0.015 | 1.5 | 0.66 |
| 0.2 | 0.075 | 1.6 | 0.56 |
| 0.3 | 0.16 | 1.8 | 0.42 |
| 0.4 | 0.28 | 2.0 | 0.32 |
| 0.5 | 0.43 | 2.2 | 0.24 |
| 0.6 | 0.60 | 2.4 | 0.18 |
| 0.7 | 0.77 | 2.6 | 0.13 |
| 0.8 | 0.89 | 2.8 | 0.098 |
| 0.9 | 0.97 | 3.0 | 0.075 |
| 1.0 | 1.00 | 3.5 | 0.036 |
| 1.1 | 0.98 | 4.0 | 0.018 |
| 1.2 | 0.92 | 4.5 | 0.009 |
| 1.3 | 0.84 | 5.0 | 0.004 |

From Viessman et al. (1989).

ways advisable; one can make observations of rise times simply by measuring water levels at the basin outlet during a water-input event.

However, the SCS approach is seductive and will no doubt continue in use because (1) it is computationally simple; (2) it uses readily available watershed information; (3) it has been "packaged" in readily available tables, graphs, and computer pro-

grams; (4) it appears to give "reasonable" results under many conditions; and (5) in the absence of detailed watershed information, there are few other practicable methodologies for obtaining a priori estimates of $W_{eff}$ that are known to be better (Ponce and Hawkins 1996). One must be careful, though, not to confuse the use and manipulation of curve numbers and related devices with the science of hydrology (Klemeš 1986).

### 9.6.3 The Unit Hydrograph

#### Definition

The most venerable and widely used transfer function for systems modeling of hydrologic response is the **unit hydrograph** (also called the **unit graph**). A "$T_W$-hour unit hydrograph" is the characteristic response of a given watershed to a unit volume (e.g., 1 in. or 1 cm) of *effective* water input (usually rain) applied at a constant rate for $T_W$ hours.

The central hypothesis of the unit-hydrograph approach is that watershed response is *linear;* that is, the ordinates of the hydrograph responding to a steady input of $W_{eff}$ units for a duration $T_W$ are

**FIGURE 9-43**
$W_{eff}$ estimated a priori by the SCS method vs. $W_{eff}$ determined by hydrograph separation for 11 storms on Ralston Creek research watershed, Iowa. Data from Kumar and Jain (1982).

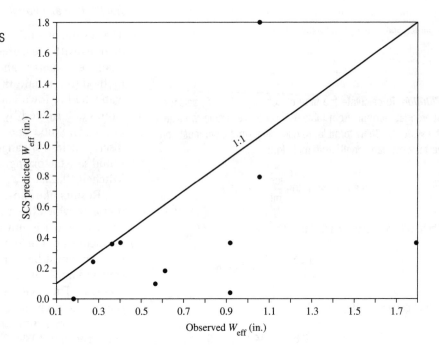

equal to $W_{eff}$ times the ordinates of the $T_W$-hour unit hydrograph. This means that the time base of the hydrograph remains constant for all inputs of duration $T_W$.

Example 9-8 shows how the unit hydrograph approach is applied; the derivation of unit hydrographs for different durations is discussed subsequently.

### EXAMPLE 9-8

Figure 9-44 shows the 2.5-hr unit (1-in.) hydrograph for a 7.2-mi² agricultural watershed in Ohio. The peak flow is 2980 ft³ s⁻¹ (0.17 in. hr⁻¹). Find the response hydrograph for an effective rainfall of 2.4 in. for the same duration.

**Solution** The hydrograph ordinates for the 2.4-in. storm are found by multiplying the unit-hydrograph ordinates by 2.4, yielding the larger hydrograph in Figure 9-44. The predicted peak flow is thus 7160 ft³ s⁻¹ (0.41 in. hr⁻¹).

### Determination of the Unit Hydrograph from Observations

Unit hydrographs for a watershed can be constructed from observations of input and response for several significant storms of approximately equal

duration. Dunne and Leopold (1978) recommended the following steps:

1. Choose four or five hydrographs from intense storms of approximately equal duration and at least moderately uniform distribution.
2. Plot each hydrograph and separate event response from base flow using one of the methods described in Figure 8-38.
3. For each hydrograph, determine $W_{eff}$ by measuring the area between the measured flow and the separation line, and convert the result to in. hr⁻¹ (or cm hr⁻¹ or other convenient unit).
4. Multiply the ordinates of each hydrograph by the corresponding value of $1/W_{eff}$ to give the unit hydrograph ordinates for each storm. Note that these ordinates have units of hr⁻¹.
5. Plot the unit hydrographs on the same graph, each beginning at the same time (Figure 9-45a).
6. Determine the peak of the composite unit graph as the average of all the peaks and plot the average peak at the average time of occurrence of all the peaks.
7. Sketch the composite unit graph to conform to the average shape of the plotted unit graphs, with the peak as determined in Step 6.

**FIGURE 9-44**
The 2.5-hr unit hydrograph for a 27.2-mi² watershed in Ohio, as derived by Dunne and Leopold (1978), and the hydrograph given by the unit-hydrograph approach for 2.4 in. of effective rain applied for 2.5 hr on that watershed.

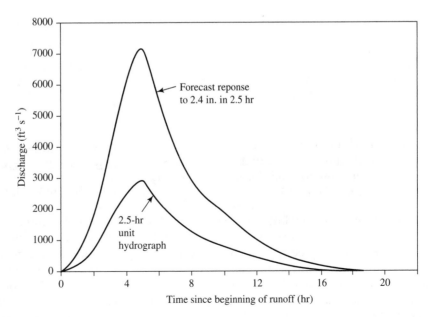

**FIGURE 9-45**
(a) Unit hydrographs for the watershed of Example 9-8 derived from four storms of $T_w \approx 2.5$ hr. (b) Composite "average" 2.5-hr unit hydrograph derived from (a) as described in the text. This is the same unit hydrograph shown in Figure 9-44; Multiply in. hr$^{-1}$ by 17,533 to get ft$^3$ s$^{-1}$. From *Water in Environmental Planning*, by Thomas Dunne and Luna B. Leopold. Copyright © 1978 by W.H. Freeman and Company. Reprinted by permission.

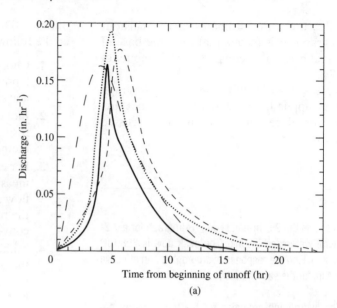

(a)

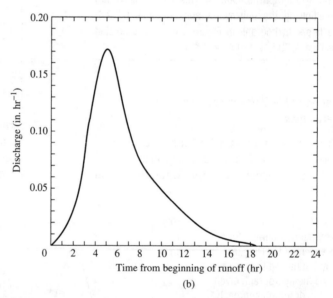

(b)

**8.** Measure the area under the sketched curve and adjust the curve until this area is satisfactorily close to 1 unit (inch or centimeter) of runoff (Figure 9-45b).

Theoretically, unit hydrographs can also be determined from observations of rainfall and runoff using matrix algebra. [See, for example, Viessman et al. (1989).] However, application of the matrix approach is severely limited in practice because it requires that the time distribution of effective water input be known.

### Relations between Unit Hydrographs of Different Durations

Once the unit graph for a given duration of excess water input, say $T_w'$, is obtained, the unit graph for any other duration can be readily derived.

To see this note that, if an event producing 1 unit of effective input over the period 0 to $T_w'$ hr is followed by a second identical event over the period $T_w'$ to $2 \cdot T_w'$ hr, the response hydrograph will be given by the sum of the ordinates of two $T_w'$-hr unit hydrographs, one beginning at 0 hr and the other at

$T_W'$ hr. This new hydrograph thus represents the response to 2 units (inches) of effective input in $2 \cdot T_W'$ hr. Thus if we divide the ordinates of this new hydrograph by 2, we have the $2 \cdot T_W'$-hr unit hydrograph.

---

### EXAMPLE 9-9

Find the 5-hr unit graph for the watershed in Example 9-8.

**Solution** Figure 9-46a shows the 2.5-hr unit graph from Example 9-8 lagged by 2.5 hr and summed; this gives the response to 2 in. of effective rainfall in 5 hr. Figure 9-46b shows the 5-hr unit graph obtained by halving the ordinates of the summed hydrograph of Figure 9-46a.

---

To generalize the procedure just described: Given a $T_W'$-hr unit graph, we can obtain the $n \cdot T_W'$-hr unit graph for $n = 2, 3, 4,...$ by adding the $n$ unit graphs, each lagged by $T_W'$, and dividing the resulting ordinates by $n$.

Unit graphs for a duration $T_W$ less than $T_W'$ can be obtained from the $T_W'$-hr unit graph via the construction of the **S-hydrograph** according to the following steps (Figure 9-47):

1. A series of $T_W'$-hr unit hydrographs is plotted, beginning at successive intervals of $T_W'$ hr.
2. The ordinates of all the unit graphs are added to give the ordinates of the S-hydrograph. Successive unit graphs are added until the S-hydrograph ordinate becomes effectively constant.
3. Plot the S-hydrograph twice, lagged by $T_W$, and subtract the ordinates for the lagged curve from those of the first curve.
4. Multiply the ordinates found in Step 3 by $T_W'$ to give the ordinates for the $T_W$-hr unit graph.

The S-hydrograph thus represents the hydrograph of a storm of infinite duration at an intensity of 1 unit/$T_W'$. Each watershed is characterized by a single S-hydrograph, from which the unit hydrograph for any duration of input can be obtained.

---

### EXAMPLE 9-10

Derive the 1-hr unit hydrograph for the watershed of Example 9-8.

**Solution** Figure 9-47a shows the S-hydrograph derived by summing the 2.5-hr unit graph (Figure 9-44) lagged by 2.5 hr.

Figure 9-47b shows that S-hydrograph lagged by 1 hr. Figure 9-47c shows the difference between the two S-hydrographs of Figure 9-47b; this represents the response to 1 in./2.5 hr = 0.4 in. hr$^{-1}$ effective rain in 1 hr. Figure 9-47d is the 1-hr unit graph obtained by multiplying the ordinates in Figure 9-47c by 2.5.

---

### The Instantaneous Unit Hydrograph

If the input duration $T_W$ used to define the unit hydrograph is allowed to become infinitesimal, the resulting response function is the **instantaneous unit hydrograph**—the response of the watershed to a unit volume of input applied instantaneously.

One advantage of the instantaneous unit hydrograph (IUH) concept is that it permits the use of continuous mathematics in developing the transfer function from measurements of the watershed response hydrograph, $q_{ef}(t)$, to a continuous input, $w_{eff}(t)$. Bras (1990) reviewed several approaches to doing this.

### Synthetic Unit Hydrographs

Several methods have been devised for estimating **synthetic unit hydrographs** on the basis of watershed characteristics, for use where observations of input and response are lacking. Here we describe a few of these approaches; others were discussed by Viessman et al. (1989), McCuen (1998), and Wilson and Brown (1992).

**Linear Watershed Instantaneous Unit Hydrograph**   One approach to developing a synthetic unit hydrograph can be based on the linear watershed model (Box 9-2). It can be shown that the linear watershed model is equivalent to an IUH having ordinates, $h(t)$, given by the simple mathematical form

$$h(t) = \frac{1}{T^*} \cdot \exp\left(-\frac{t}{T^*}\right) \qquad \textbf{(9-52)}$$

(Bras 1990). If $T^*$ can be estimated from relations like those in Table 9-9 and Equation (9-6), then the response of the linear model to a unit input for the duration of interest is the unit graph for that duration. Equivalently, once $T^*$ is determined, the linear model can be used directly to estimate the response to any sequence of inputs without explicitly invoking the unit hydrograph.

**FIGURE 9-46**
(a) The 2.5-hr unit hydrograph of Figure 9-45 lagged by 2.5 hr and summed to give the response to 2 in. of effective rainfall in 5 hr.
(b) The 5-hr unit hydrograph derived by halving the ordinates of the hydrograph of (a).

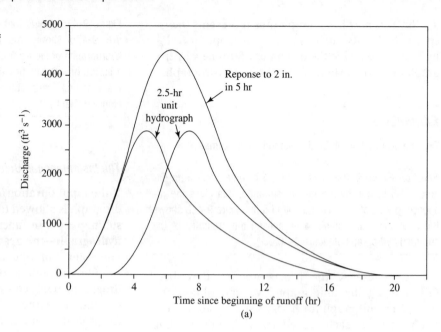

(a)

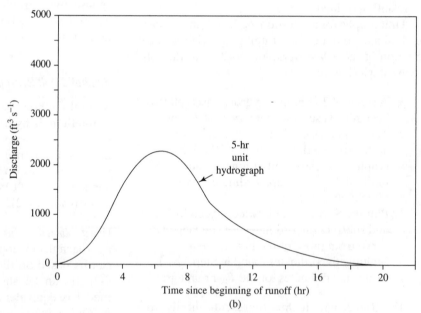

(b)

**Geomorphologic Instantaneous Unit Hydrograph** The concept of the **geomorphologic instantaneous unit hydrograph** (GIUH) was introduced in a series of papers by Rodriguez-Iturbe and Valdes (1979), Valdes et al. (1979), and Rodriguez-Iturbe et al. (1979) and developed further by Gupta et al. (1980) and Rodriguez-Iturbe (1993). Very briefly, they used statistical concepts and extensions of the laws of drainage composition (Table 9-7) to develop the following relations for estimating the peak, $q_{pk}$, and time-of-rise, $T_r$, of the IUH:

$$\frac{q_{pk}}{Q_{ef}} = 1.31 \cdot \frac{R_L^{0.43} \cdot U_{pk}}{L_\Omega} \qquad \textbf{(9-53)}$$

and

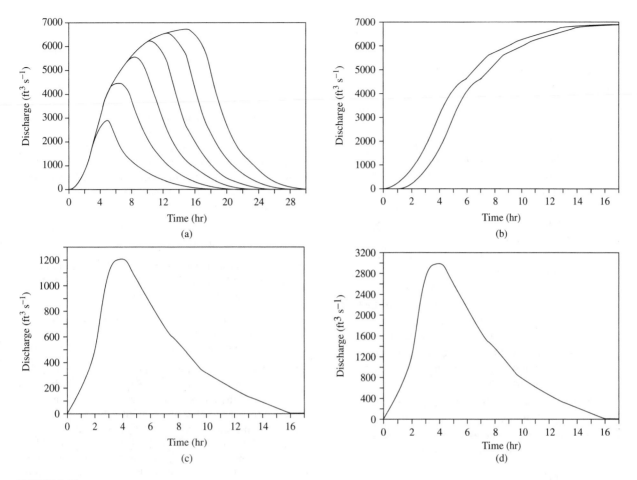

**FIGURE 9-47**
(a) The S-hydrograph for the watershed of Example 9-8 derived by summing the 2.5-hr unit hydrograph (Figure 9-45b) lagged by 2.5 hr. (b) The S-Hydrograph lagged by 1 hr. (c) The difference between the two S-hydrographs of (b). (d) The 1-hr unit hydrograph obtained by multiplying the ordinates of (c) by 2.5.

$$T_r = 0.44 \cdot \frac{L_\Omega}{U_{pk}} \cdot \left(\frac{R_B}{R_A}\right)^{0.55} \cdot \left(\frac{1}{R_L}\right)^{0.38}. \qquad \textbf{(9-54)}$$

In these equations, $q_{pk}/Q_{ef}$ is in hr$^{-1}$; $T_r$ is in hr; $L_\Omega$ is the length of the highest-order stream in km; $R_B$, $R_A$, and $R_L$ are defined in Table 9-7; and $U_{pk}$ is the velocity of flow in the channel network in m s$^{-1}$, which can be estimated via Equation (9-19).

### Application of the Unit-Hydrograph Approach

Freeze (1972a) concluded from a series of experiments with physically based models that "no physical reason seems to exist why watersheds should respond in a linear fashion," and it is clear that the linear-response assumptions of unit-hydrograph theory are at best only approximated by actual watersheds. [See, for example, Valdes et al. (1979).]

However, the unit-hydrograph is often the only feasible way to approach practical problems. Dunne and Leopold (1978) stated that the unit-hydrograph method gives estimates of flood peaks that are usually within 25% of their true value; this is close enough for most planning purposes and about as close as can be expected given the usual lack of detailed information about watershed processes and states. Errors larger than 25% can be expected if synthetic unit graphs are used without verification for the region of application. The assumption of linearity that underlies the unit-hydrograph approach seems most valid for relatively large watersheds ($A_D > 100$ km$^2$).

The most important difficulties in applying the unit-hydrograph approach are as follows:

1. The "theoretical" relationships are between *effective* input and response, and it is not clear how to make reliable a priori estimates of effective input. There is usually little information about the magnitudes of the quantities in Equation (9-44), and small relative errors in estimating those quantities can lead to large relative errors in estimating event flow.

2. Because of the temporal variability of contributing area and rates of water movement, it is difficult to identify characteristic response times for real watersheds. In fact, unless the contributing area in a given watershed follows exactly the same patterns of temporal and spatial change in every storm, the watershed cannot have a single unit hydrograph.

3. In general, the characteristic response of a given drainage basin depends on antecedent conditions, so the assumption of linearity of response may be grossly in error.

However, if the unit hydrograph is developed and applied for large runoff events only, the effects of variable contributing areas and antecedent conditions are generally less significant.

# EXERCISES

Exercises marked with ** have been programmed in EXCEL on the CD that accompanies this text.

**\*\*9-1.** Consider a stream reach for which $X = 2500$ m, $B = 10$ m, Manning's $n = 0.042$, and $S = 0.0005$. Compute the outflow hydrograph for this reach given a triangular in-flow hydrograph that begins at $QI(0) = 0$ m$^3$ s$^{-1}$; reaches a peak, $QI_{pk}$, of 100.0 m$^3$ s$^{-1}$ at 3.0 hr; and declines to 0 m$^3$ s$^{-1}$ at 10.0 hr. Use the spreadsheet program GoConvex.xls.

**9-2.** Obtain topographic maps of a region of interest containing the drainage basin of a fourth- or fifth-order stream. (a) Designate stream orders as indicated in Figure 9-35 and determine whether the law of stream numbers is followed. (b) Determine the average lengths and areas of streams of each order (you need measure only a selected sample of the low-order streams) and see if the laws of stream lengths and areas is followed. (c) Compute the fraction of the largest drainage basin that drains directly into first-order streams.

**9-3.** Use the method of Equation (9-32) to determine the drainage density of a region.

**9-4.** Obtain the hydrograph of a stream responding to a known rainfall and use one of the methods depicted in Figure 8-38 to determine $W_{eff}$ and $W_{eff}/W$.

**9-5.** If soils and land-use information are available for the watershed used in Exercise 9-4, compare the "actual" $W_{eff}$ with that given by the SCS curve-number method.

**9-6.** Obtain the hyetograph and hydrograph for a rainfall event on a watershed. (a) Separate the hydrograph using one of the methods of Figure 8-38 and estimate the hyetograph of the effective rainfall. (You may make the assumption used in Example 9-2.) (b) Estimate the centroid of the effective rain and calculate the centroid lag. (c) Compare the "actual" values of centroid lag, peak flow rate, and response time with $T_c$ given by one or more of the formulas in Table 9-9, $T_r$ given by Equation (9-48), and $q_{pk}$ given by Equation (9-51).

**9-7.** Compare the hydrograph of event flow developed in Exercise 9-6 with that obtained via the linear watershed model (Box 9-2) for the same effective input hyetograph, assuming $T^* =$ the "actual" centroid lag.

**9-8.** Compare the hydrograph of event flow developed in Exercise 9-6 with that obtained via the SCS dimensionless unit hydrograph (Table 9-14) for the same effective input hyetograph.

**9-9.** From the geologic, soils, topographic, and vegetative conditions in your region, which mechanism(s) are likely to contribute water to event responses? Have any studies that shed light on this question been done in the region?

# 10

# Hydrology and Water-Resource Management

Previous chapters of this text have emphasized the science of physical hydrology. To a large extent, that science has grown out of the need to solve practical water-resource-management problems, and most applications of hydrologic science continue to be motivated by some issue of water-resource management. To give some perspective on these motivating issues, this chapter begins with an overview of the goals and processes of water-resource management. The following four sections discuss some of the ways hydrologic analysis is applied in the process of water-resource management for water-supply and -demand issues, water-quality issues, analysis of flood hazards, and analysis of low streamflows and droughts. The treatment in these sections attempts to provide a conceptual framework as well as to introduce the application of statistical and other quantitative tools to these problems. The chapter concludes with an overview of current and projected water use in the United States and the world.

Although the science of hydrology is an essential basis for the practice of water-resource management, it is well for the hydrologist to keep in mind that political, legal, economic, and social factors are the ultimate determinants of water-resource decisions. The comprehensive and very readable history of water issues in the western United States by Reisner (1986) makes this lesson indelibly clear.

## 10.1 WATER-RESOURCE MANAGEMENT

Water-resource management is the process by which governments, businesses, and/or individuals reach and implement decisions that are intended to affect the future availability and/or quality of water for beneficial uses or the risk of water-related hazards that threaten beneficial activities. The term "beneficial" means uses or activities that are perceived to increase individual, business, or societal well-being.

### 10.1.1 Water-Resource Management Goals and Objectives

Table 10-1 summarizes broad policy goals commonly motivating national or international water-resource management. Achievement of such goals requires the development of operational definitions and specific criteria by which the degree of advancement toward those goals can be measured.

In the United States, articulation of public water-resource-management goals began with the adoption of the Constitution in 1789 and has been evolving ever since (Holmes 1972; 1979). The requirement that the economic benefits of federally supported water-resource management projects exceed their economic costs was explicitly stated in

**TABLE 10-1**
Policy Goals for National and International Water-Resource Management.

| Goal | Operational Definition | Assessment Tool |
|------|----------------------|-----------------|
| Economic development | Increase the value of the output of goods and services | Benefit–cost analysis (Box 10-1) |
| Environmental quality | Management, conservation, preservation, creation, restoration, or improvement of the quality of natural and cultural resources and ecological systems | Environmental impact statement |
| Social well-being | More equable income distribution, increased employment, increased self-reliance, improvement of health, and enhancement of educational and cultural opportunities | Various types of social analyses |
| Sustainability | Meet the needs of the present without compromising the ability of future generations to meet their needs | Benefit–cost, environmental-impact, and social analyses (Table 10-2) |

the Flood Control Act of 1936. Thus enhancing **national economic development** became the overriding goal of water-resource activities, and decision-makers relied on **benefit–cost analyses** to justify water-management projects (Box 10-1).

In 1969 the National Environmental Quality Act made the preparation of environmental impact statements (EIS) a requirement for projects involving significant federal investment, and environmental concerns were explicitly incorporated as a goal of federal water-resources management in the "Principles and Standards for Water and Related Land Resources Planning" (U.S. Water Resources Council 1973). That document stated that the federal government would participate in such planning to enhance environmental quality by

the management, conservation, preservation, creation, restoration, or improvement of the quality of certain natural and cultural resources and ecological systems.

Although subsequently somewhat modified (U.S. Water Resources Council 1982), economic development and environmental quality remain the essential goals of federal water-resources activities in the United States, and similar policy goals underlie the activities of states and other levels of government in the United States.

In other countries, various aspects of **social well-being** are commonly included as water-resource-management goals (Goodman 1984; Petersen 1984). Specific components of this goal may include income redistribution, increase in employment levels, increase in self-reliance, health and safety, and the enhancement of educational and cultural opportunities.

In recent decades, the concept of **sustainability** has emerged as an overall guiding principle for development. Sustainability may be broadly defined as

development that meets the needs of the present without compromising the ability of future generations to meet their own needs (World Commission on Environment and Development 1987).

Newson (1992) suggested that, as applied to water-resource management, the concept would include

1. use of resources at space and time scales appropriate to the optimum functioning of river basins,
2. assessment of the impacts of both technical and policy measures, and
3. monitoring the states of processes in both pristine and developed watersheds.

While there is much debate about operational definitions of sustainability and considerable doubt about whether it can actually be achieved, the emergence of the concept has engendered more serious and comprehensive examination of the long-term economic, environmental, and social impacts of proposed water-resources development in many instances. Of course, the ability of economists, scientists, and social scientists to foresee such impacts is limited, and this has led to the formulation of the **precautionary principle** as a companion to the principle of sustainability:

Where there are threats of serious or irreversible environmental damage, lack of full scientific certainty should not be used as a reason for postponing mea-

**BOX 10-1**

• • • • • • •

**Benefit–Cost Analysis**

The objective of benefit–cost (B–C) analysis is to determine the difference between the economic benefits (B) and the economic costs (C) of an undertaking (alternative plan or project). The boundaries of a B–C analysis must always be specified at the outset; these are usually political boundaries (country, state), but may also be a corporate entity. If B > C, the undertaking results in a net increase in the value of goods and services *within the boundaries* and hence represents an economically efficient use of resources for that entity. In choosing among alternative plans, the one with the largest value of B − C represents the most efficient use of those resources that can be economically evaluated.

For most water-resources projects the economic costs involve an initial one-time outlay of money for construction and land purchase, while the benefits (e.g., increased agricultural income, decreased flood damages) accrue over subsequent years. In economic theory, the present value of an amount A made available t yr from now, $PV_t(A)$, is

$$PV_t(A) = \frac{A}{(1 + r)^t} \approx A \cdot \exp(-r \cdot t), \quad \textbf{(10B1-1)}$$

where r is the prevailing interest rate (also called the **discount rate**). Thus the present value of a benefit stream of A dollars in each year over a T-yr period, $PV_T(A)$ is

$$PV_T(A) = A \cdot \sum_{t=1}^{T} \frac{1}{(1 + r)^t} = A \cdot \left[ \frac{1 - (1 + r)^{-T}}{r} \right]$$
$$\approx \frac{A}{r} \cdot [1 - \exp(-r \cdot T)]. \quad \textbf{(10B1-2)}$$

From Equation (10B1-2), we see that as T increases, $PV_T(A)$ approaches A/r. Thus if r = 0.1, the initial (year 0) cost of a project generating an indefinitely long stream of annual benefits of $1 million must be less than $10 million to be economically justifiable.

Note that the smaller the discount rate, r, the larger the present value of the benefits, other things equal. For governments, the discount rate to be used in B–C analysis is determined by the political process, but is usually related to current prevailing interest rates. Specification of the discount rate is one important way that political, social, and economic factors determine the evaluation of alternative plans and hence the outcome of the water-resource management process.

Estimating the construction costs of a water-resource management plan is usually a relatively straightforward, but not simple, task. However estimating other costs, such as the economic value of white-water canoeing lost when a dam is constructed, can be conceptually challenging. Similarly, estimating future economic benefits of increased recreation, flood-warning systems, or improved water quality can be very difficult and subject to considerable uncertainty.

sures to prevent environmental degradation, [and development] decisions should be guided by (i) careful evaluation to avoid, wherever practicable, serious or irreversible damage to the environment; and (ii) an assessment of the risk-weighted consequences of various options (Dovers and Handmer 1995).

Marchand and Toornstra (1986) suggested guidelines for river-basin management that are consistent in spirit with the principle of sustainability and the precautionary principle; these are summarized in Table 10-2.

The specific objectives of water-resource planning are sometimes called **purposes.** Table 10-3 relates common purposes of water-resources planning and management to the general goals of economic development, environmental quality, and social well-being. Usually only a limited number of purposes are included in a given planning situation.

**TABLE 10-2**

Guidelines for Sustainable River Basin Management.[a]

1. Preserve or improve spontaneous river-basin functions by
   a. minimal alteration of natural erosion and sedimentation processes;
   b. preserving genetic diversity;
   c. preserving the self-purifying capacity of the river via wastewater treatment plants and at-source anti-pollution measures.
2. Conserve the natural values of the river basin by
   a. legislating to prevent deterioration of natural resources;
   b. establishing reserves in vulnerable ecosystems;
   c. establishing environmental education programs;
   d. initiating programs to promote sustainable exploitation of ecosystems (fisheries, herding, forestry).
3. Conserve "extensive exploitation functions" of the river basin by
   a. protecting productive zones such as floodplains, estuaries, and lakes;
   b. implementing reforestation schemes.
4. Conserve "intensive exploitation functions" of the river basin by
   a. developing a basin-wide water-allocation plan consistent with items 1–3;
   b. developing small-scale projects for irrigation, aquaculture, and forestry;
   c. improving the use of rivers for transportation.
5. Improve public health of river-basin inhabitants by
   a. combating water-borne diseases;
   b. improving the food situation quantitatively and qualitatively;
   c. making clean drinking water available for all.

[a]After Marchand and Toornstra (1986) as summarized by Newson (1992).

### 10.1.2 The Geographical Unit for Water-Resources Management

As we have seen in previous chapters, the geologic, pedologic, topographic, biologic, and land-use characteristics of the drainage basin (along with its climate), determine the magnitude, timing, and quality of the basin's surface and ground waters. Thus the drainage basin (also called river basin, watershed, or catchment) at the appropriate scale is generally the most logical geographical unit for

**TABLE 10-3**

Relations among Water-Resource Management Goals, Purposes, and Types of Hydrologic Analyses

| Purpose | Goals | | | Types of Hydrologic Analysis[a] |
| --- | --- | --- | --- | --- |
| | Economic Development | Environmental Quality | Social Well-Being | |
| Public water supply | X | | X | WS, D, Q |
| Industrial water supply | X | | X | WS, D, Q |
| Irrigation | X | | X | WS, D, Q |
| Hydroelectric power | X | | X | WS |
| Navigation | X | | X | WS |
| Waste transport and treatment | X | X | X | WS, Q |
| Recreation | X | | X | WS, Q |
| Wildlife habitat | | X | X | WS |
| Reduction of flood damages | X | | X | F |

[a]WS = water supply; D = drought; Q = water quality; F = flood magnitude-frequency.

water-resources planning. However, in some regions, particularly those with semi-arid or arid climates where ground water is the principal water source, aquifer boundaries may determine the more appropriate unit.

Of course, political and land-ownership boundaries do not in general conform to drainage divides or aquifer boundaries, so it is usually difficult to develop and implement plans strictly on the basis of such natural units. Many international disputes have arisen because of this (McCaffrey 1993); on the other hand, mutual interest in the resources of a shared watershed or aquifer can be the basis for political accord and cooperation.

In the United States, the federal government conducts various assessments of water-resources availability, quality, and use based on 16 major **water-resource regions** and 106 **sub-regions** whose boundaries largely follow drainage divides (Table 10-4; Figure 10-1). For several large river basins, the sharing states have formed compacts or commissions to deal with water-resource problems. Within individual states, resource-planning agencies and private conservation groups are commonly organized on a watershed basis.

### 10.1.3  The Management Process

The process of water-resource management is usually initiated when a government entity, company, or group of citizens with political influence perceives that their or their constituents' interests would be advanced by one or more of the purposes listed in Table 10-3. The management process can be viewed as involving the following components and "mapped" as shown in Figure 10-2:

1. Establish specific objectives (purposes) for current planning process consistent with overall policy goals.
2. Assess resource capabilities and project conditions expected if no action is taken (problem identification).
3. Formulate alternative plans for accomplishing objectives.
4. Evaluate each plan with respect to (a) its achievement of objectives and other beneficial economic, environmental, and social effects and (b) economic costs and adverse environmental and social impacts. All evaluations must be made relative to the "no-action" projections of Step 2.
5. Select the "best" plan (or no action) and identify entities to implement plan components.
6. Implement decisions.

Although the six steps just outlined are a logical sequence for arriving at a plan and a useful framework for viewing the management process, they are seldom strictly followed. In practice, there is considerable overlap and feedback between one phase and another. For example, the evaluations in Step 4 may lead to formulation of new or modified alternative plans (Step 3) or even to identification of new objectives that may be achieved by modifying plans (Step 1). (See Figure 10-2.)

Steps 1, 5, and 6 of the planning process are essentially political activities, but it must be anticipated that political considerations may be injected at any step and that in many situations, political, legal, or other external considerations will short-circuit the process.

One of the most critical determinants of the outcome of the planning process is usually the eco-

**TABLE 10-4**
U. S. Water-Resources Regions.
(See Figure 10-1.)

| Number | Name | Number | Name |
|--------|------|--------|------|
| 1 | New England | 12 | Texas-Gulf |
| 2 | Mid-Atlantic | 13 | Rio Grande |
| 3 | South Atlantic-Gulf | 14 | Upper Colorado |
| 4 | Great Lakes | 15 | Lower Colorado |
| 5 | Ohio | 16 | Great Basin |
| 6 | Tennessee | 17 | Pacific Northwest |
| 7 | Upper Mississippi | 18 | California |
| 8 | Lower Mississippi | 19 | Alaska |
| 9 | Souris-Red-Rainey | 20 | Hawaii |
| 10 | Missouri | 21 | Puerto Rico-Caribbean |
| 11 | Arkansas-White-Red | | |

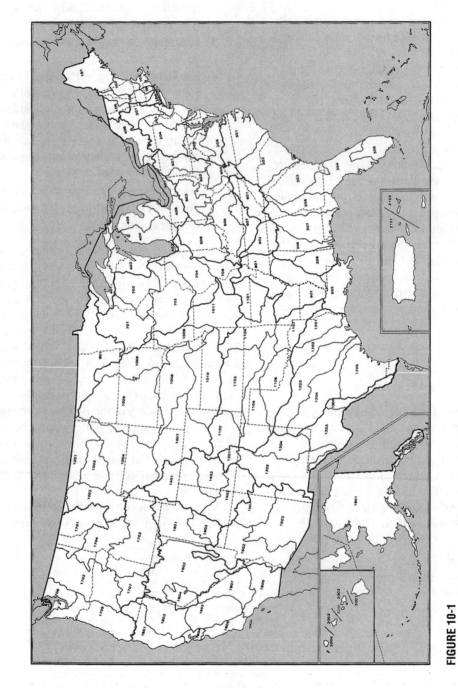

**FIGURE 10-1**
United States water-resource regions (see Table 10-4) and subregions.

**FIGURE 10-2**
Flow chart of the water-resource
management planning process.

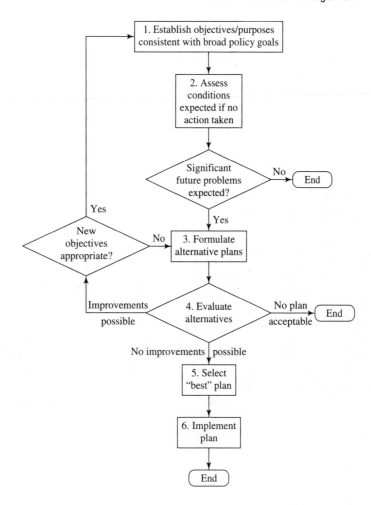

nomic- and population-growth forecasts used in
Step 2 (James et al. 1969). It is sobering to note that
such forecasts are usually seriously in error (Osborn
et al. 1986; Brown 2000). Furthermore, to the extent
that forecasts are correct, it is often because they are
self-fulfilling: If, for example, plans are implemented
to provide more water for irrigation in a region,
more development of irrigation will usually occur.

### 10.1.4  The Role of Hydrologic Analysis

Hydrologic analyses are generally involved in projecting the "no-action" conditions in Step 2, are usually required in the formulation of the specific
components of alternative plans in Step 3, and form
the basis for evaluating the beneficial and adverse
economic, environmental, and social impacts of alternative plans in Step 4.

Hydrologic analysis for water-resources management can be categorized as

1. assessment of the present and future *supply*
   of water available from surface- and/or
   ground-water sources,
2. assessment of the present and future *quality*
   of surface and/or ground water,
3. assessment of the present and future frequencies with which human activities will
   be subject to *floods*, and
4. assessment of the present and future frequencies of *low streamflows* and *drought*.

The relevance of each of these kinds of analyses to each purpose is shown in Table 10-3, and the
examination of the application of hydrology to
water-resource issues in this chapter follows this

classification. For each set of issues, we first define the terms that are necessary for understanding the problems and methods of analysis, then introduce some of the types of hydrologic analyses that contribute to assessing the problems and solutions.

## 10.2 HYDROLOGIC ANALYSIS: WATER SUPPLY AND DEMAND

Issues of water-resource supply can be understood only if we have clear definitions of the meanings of demand, use, and supply in the water-resource context. Here we develop those definitions and introduce the concepts needed as a basis for hydrologic analyses. Following this, we explore the evaluation of supply—"safe yield"—for ground- and surface-water supplies in more depth. Although not considered in detail here, one must keep in mind that considerations of water quality can also enter into the evaluation of supply and demand.

Note that demand, use, and supply are all measured as flow rates, that is, as volumes per unit time $[L^3 T^{-1}]$.

### 10.2.1 Classification of Water Uses

The first nine purposes listed in Table 10-3 constitute the principal types of water use. There is a fundamental distinction between (1) **offstream** (or **withdrawal**) **uses,** in which the water is withdrawn or diverted from a surface- or ground-water source, and (2) **instream uses,** in which benefits are derived from the water flowing in a stream channel (Figure 10-3).

For offstream uses, an important distinction must be made between consumptive and nonconsumptive uses:

**Consumptive use** is the portion of withdrawn water that is either evaporated, transpired, or incorporated into a product or crop and is consequently unavailable for subsequent downstream use. It may include the portion of withdrawn water lost due to evaporation and leakage in transit, called **conveyance loss.**

```
                    WATER USE
              /                  \
INSTREAM USE              WITHDRAWAL (OFFSTREAM) USE
(nonconsumptive)          (nonconsumptive/consumptive)

Hydropower                Thermoelectric cooling

Waste transport           Irrigation
and treatment

Fish and wildlife habitat Domestic

Navigation                Commercial

Recreation                Livestock

Esthetic                  Mining
```

**FIGURE 10-3**
Classification of water uses and types of uses in each category.

**Non-consumptive use** is the portion of withdrawn water that remains available for subsequent use. The portion of withdrawn water that is discharged to a surface- or ground-water body is called **return flow.** Strictly speaking, the conveyance losses due to leakage are part of return flow. The amount of water that evaporates from streams is negligible, so all instream use is non-consumptive.

### 10.2.2 Water Use, Demand, and Shortage

**Use** is the rate at which water is actually applied to one or more purposes, while **demand** is the rate at which water would be used at a given time and place if available at a given price. Water use generally responds to price changes, and the **price elasticity of demand** is the rate of change of demand with price. Some uses of water, like domestic use in a given culture, are relatively inelastic; others, like some industrial and agricultural uses, vary relatively strongly with price.

In general, use is less than or equal to the demand at the existing price. When use is less than demand, a **shortage** exists. It is important to distinguish between the different situations that can lead to a shortage (Table 10-5):

A **resource shortage** exists when demand exceeds the sustainable supply ("safe yield"). This situation can be addressed by developing additional

**TABLE 10-5**
Types of Water-Resource Shortages.

| Supply "Safe Yield" > Demand? | Treatment/ Distribution System Adequate? | Type of Shortage | Appropriate Strategy |
|---|---|---|---|
| No | Yes or No | Resource | Reduce demand or develop additional supplies |
| Yes | Yes or No | Drought | Temporary water-use restrictions |
| Yes | No | Infrastructure | Construct additional treatment/ distribution facilities |

sources (increasing supply) or by reducing demand by increasing price, instituting conservation measures, increasing water recycling, or increasing water or wastewater treatment.[1]

A **drought shortage** exists when the sustainable supply exceeds the demand, but the *current* availability of water is less than the "safe yield" due to drought. This situation is temporary and may be addressed by imposing voluntary or mandatory water-use restrictions for the duration of the drought.

An **infrastructure shortage** exists when the sustainable supply is greater than or equal to demand, but the capacity of the infrastructure for treating and/or distributing water is less than the demand. This type of shortage is addressed by construction of pipelines and treatment plants.

### 10.2.3   Water Supply and "Safe Yield": Basic Concepts

It should be clear from the preceding chapters that all the water in a watershed is connected in a spatially and temporally integrated network of flows. Thus it is not possible for humans to withdraw any amount of water from this network without affecting the quantity, timing, and/or quality of "downstream" flows to some extent. The role of hydrologic analysis is to determine the location, nature, and degree of the hydrologic impacts of an existing or proposed water use. However, any definition of available water supply, or "safe yield," must depend essentially on non-hydrologic factors: Are the ecological, human-health, economic, legal, or other consequences of those impacts acceptable?

These considerations lead to the following definition:

**Water supply** is the maximum sustainable rate at which water can be withdrawn from existing sources without causing undesirable ecological, human-health, economic, legal, or other consequences.

Thus water supply is identical to "safe yield" (also "reliable yield" or "sustainable yield"), the term often used to specify a rate of water utilization (withdrawal) that can be maintained more or less indefinitely and relied upon as a measure of the available water resources of a region or a specific water source. The impression is often given that "safe yield" is a well-defined quantity whose value can be established by hydrologic analysis. However, because its definition refers to a range of undesirable non-hydrologic consequences, this impression is incorrect. To reflect its misleading connotation, we will always enclose the term in quotation marks.

As discussed in Chapters 2, 8, and 9, virtually all ground water that is not withdrawn directly for human use eventually enters a stream or lake (except near the coast, where some may flow directly into the ocean) and, conversely, a large proportion of the flow of streams consists of ground-water inflow. Thus this basic hydrologic consideration must guide any discussion of the concept of "safe yield" of surface- and ground-water supplies:

Ground water and surface water are intimately connected and constitute a single resource.

However, in spite of the intimate connections between ground and surface water, there are some distinctive characteristics of each that affect the standard approaches to evaluating "safe yield." Typically, the ground-water reservoir of a region has a residence time [Equations (2-27) and (8-26); Table 9-8] that is orders of magnitude larger than

---

[1] Increased treatment may reduce the amount of instream flow needed to achieve required water-quality levels.

that of the stream network and has a correspondingly smaller time variability. Thus discussion of the "safe yield" of aquifers or individual wells has tended to ignore time variability and to focus instead on the impacts of a given level of withdrawal on the water table (unconfined aquifers) or the piezometric surface (confined aquifers). In contrast, specification of the "safe yield" of surface-water supplies usually accounts for its significant time variability.

There is another difference between ground-water and surface-water supplies that plays a large role in the perception and management of those resources: Ground water is invisible. The depletion of a surface-water reservoir can make a stark image (Figure 10-4) that can evoke action to develop alternate sources or limit demand through conservation; however, a ground-water reservoir can be depleted gradually over a long period of time before serious and often irreversible impacts become apparent.

### 10.2.4  Water Supply and "Safe Yield": Ground Water

#### Basic Considerations

As described in Section 8.6.3, water-balance considerations along with an understanding of the effects of ground-water extraction from an aquifer are the basis for determining "safe yield." Equation (8-65) shows that the rate of pumping from an aquifer, $\Sigma Q_w$, is always supplied by a decrease in storage,

**FIGURE 10-4**
A dry reservoir is a stark indicator of a drought or resource shortage. *Photo by Didier Dorval/Explorer/Photo Researchers, Inc.*

$\Delta S$, and, in general, by induced increases in the rate of ground-water recharge, $\Delta R$, or induced decreases in the rate of ground-water discharge to surface water, $\Delta Q_{GW}$:

$$\Sigma Q_w = \Delta R - \Delta Q_{GW} - \frac{\Delta S}{\Delta t}. \qquad \textbf{(10-1)}$$

As discussed in Section 8.4.1, recharge $R$ has three components, given by

$$R = R_I + R_{SW} - CR, \qquad \textbf{(10-2)}$$

where $R_I$ is recharge from infiltration, $R_{SW}$ is recharge from surface water, and $CR$ is capillary rise (usually assumed to be induced by transpiration of phreatophytic plants). Combining Equations (10-1) and (10-2), we obtain

$$\Sigma Q_w = \Delta R_I + \Delta R_{SW} - \Delta CR - \Delta Q_{GW} - \frac{\Delta S}{\Delta t}.$$
$$\textbf{(10-3)}$$

Equation (10-3) states that the pumped water comes only from the water withdrawn from storage ($\Delta S/\Delta t < 0$), plus any *induced* increases in recharge ($\Delta R_I$, $\Delta R_{SW} > 0$) and/or *induced* decreases in discharge ($\Delta CR$, $\Delta Q_{GW} < 0$). The natural recharge rate does not enter into this equation, and one should never make the common mistake of equating "safe yield" to the natural recharge rate. The natural recharge rate is relevant to the determination of "safe yield" only insofar as it should be included in ground-water models used to determine the effects of pumping.

It should be clear from Section 8.6.2 that some "mining" (i.e., removal of water from aquifer storage and a negative value of $\Delta S/\Delta t$) is required to reach any extraction rate. However, if a constant rate of ground-water extraction is imposed on a basin for a sufficient time, a new equilibrium state may eventually be reached in which there is no further change in storage ($\Delta S/\Delta t = 0$). After this the extraction rate ($\Sigma Q_w$) is supplied only by some combination of increased recharge from infiltration and/or surface water and reduced discharge via evapotranspiration by phreatophytes and/or inflow to streams.

Bredehoeft et al. (1982) pointed out that the time required to reach this equilibrium may be very long indeed (hundreds of years), depending on the basin size, hydrology, and geology and on the locations, depths, and pumping rates of wells. In some situations, it is not possible to reach an equilibrium state before drawdown at wells equals its maximum value (i.e., the water table has declined to the bottom of the deepest well). At this point, no more extraction is possible at the existing well locations and pumping rates. As noted earlier, such a situation could give the impression that ground water was being utilized at a sustainable rate for many years before the consequences became apparent.

Thus, rather than a specific flow quantity, "safe yield" is best defined as "the rate at which ground water can be withdrawn without producing undesirable effects" (Lohman 1979). The discussion in Section 8.6.3 identified the most important hydrologic impacts of ground-water extraction, and most of these have potentially undesirable effects:

- Reductions of streamflow may seriously reduce surface water available for instream and withdrawal uses.
- Levels and/or extents of lakes and wetlands may be reduced, with consequent loss of valued habitat.
- Extent of areas where water is available to plants that exploit the capillary fringe (phreatophytes) may be reduced, with consequent loss of habitat.[2]
- Ground-water outflow to the ocean may be reduced, with consequent effect on coastal wetlands and/or nearshore benthic marine habitats.
- The fresh-salt interface in the aquifer may be raised, increasing the likelihood of salt-water intrusion into wells.
- Lowered water table may cause land subsidence as some of the overburden stresses formerly supported by ground water are transferred to the mineral grains.
- Costs of pumping, which are proportional to depth of water table, rise.

---

[2] Destruction of phreatophytes has sometimes been advocated as a strategy for reducing evapotranspiration and increasing the water available from basins in semi-arid and arid regions.

- Water tables lowered by one developer may fall below depths of nearby wells belonging to others, perhaps resulting in legal action.

Because of the varying importance of all these hydrologic effects and their economic, social, environmental, and legal consequences, there is no general formula for computing "safe yield." Acceptable rates of development can be determined only by (1) finding the likely hydrologic effects of various combinations of rates, timing, and location of ground-water extraction, usually by means of ground-water modeling; (2) assessing the environmental, economic, legal, and social impacts of these effects; and (3) balancing the benefits afforded by the ground water provided against the undesirable consequences of the various schemes (Table 10-1).

Lohman (1979) has shown how application of Equation (10-3) provides insight to the factors determining "safe yield" in various typical aquifer settings (Figure 10-5). The following discussion and Table 10-6 are based on his analysis and on an additional study of a heavily utilized island aquifer.

### Valley of Large Perennial Stream in a Humid Region

Figure 10-5a is a cross-section of a river valley cut into relatively impermeable bedrock and filled with thick permeable alluvium (sand and gravel). Such geologic configurations are the major aquifers in the northeastern United States (Randall and Johnson 1988). Pumping may induce some additional recharge from infiltration ($\Delta R_I$) and reduce evapotranspiration by riparian vegetation ($\Delta CR$), but the major sources of ground water are the induced recharge from streams ($\Delta R_{SW}$) and the intercepted ground-water discharge to streams ($\Delta Q_{GW}$) (Figures 8-43 and 8-44).

Equilibrium ($\Delta S/\Delta t = 0$) is possible; wells placed relatively far from the river or pumping at lower rates intercept ground water discharging to the river ($\Delta Q_{GW}$), while those nearer the river or pumping at higher rates may induce direct recharge from it ($\Delta R_{SW}$). In any case, streamflow is reduced [Equation (8-67)], and the major limitations to development are conflicts with requirements to maintain streamflows at levels necessary to provide for downstream withdrawals of stream water and/or maintenance of acceptable water quality and aquatic habitat.

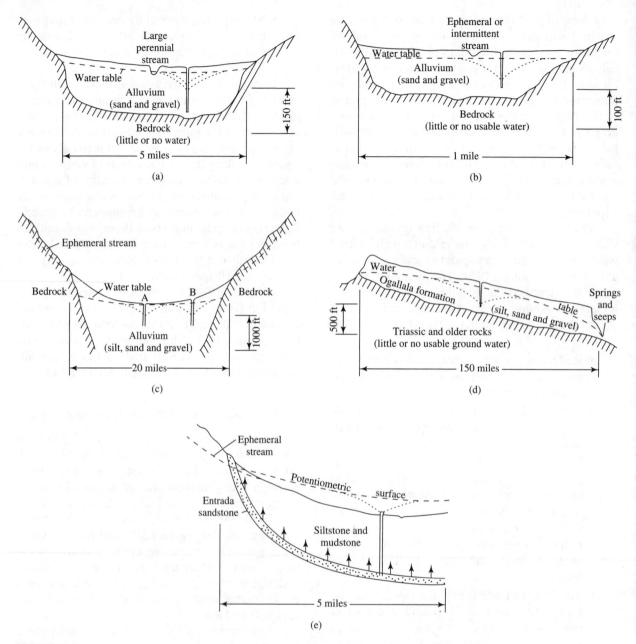

**FIGURE 10-5**

Ground-water development in (a) valley of a large perennial stream in a humid region; (b) valley of an ephemeral stream in a semi-arid region; (c) closed desert basin; (d) escarpment-bounded aquifer in an arid region; (e) confined aquifer in a semi-arid region. See text for discussion. From Lohman (1979).

### Valley of Ephemeral Stream in a Semi-Arid Region

The geologic setting here (Figure 10-5b) is similar to that of Figure 10-5a, except that the stream is ephemeral (losing) and the water table lies below the channel. Since the water table is at considerable depth, pumping does not affect recharge from infiltration ($\Delta R_I$) or capillary rise ($\Delta CR$). Under natural conditions, ground-water flow has a down-valley component (Figures 8-19 and 8-20) and eventually discharges to surface water, so there is some ground-water discharge ($\Delta Q_{GW}$) to intercept. The

**TABLE 10-6**

Sources of Water for Ground-Water Development in Various Aquifer Settings (Figures 10-5 and 10-6). Note that $\Delta S / \Delta t < 0$ for all situations until equilibrium is reached. See text for discussion.

| Aquifer Setting (Figure) | $\Delta R_I$ (>0) | $\Delta R_{SW}$ (>0) | $\Delta CR$ (<0) | $\Delta Q_{GW}$ (<0) | At Equilibrium ($\Delta S / \Delta t = 0$) $\Sigma Q_W =$ |
|---|---|---|---|---|---|
| River valley alluvium, humid region (northeastern U.S.; Figure 10-5a) | Some | Major source if well near stream | Some | Major source if well distant from stream | $\Delta R_{SW} - \Delta Q_{GW}$ |
| Ephemeral stream valley, semi-arid region (eastern Colorado; Figure 10-5b) | None | May be large from floods | None | Minor | Equilibrium not possible: $\Delta R_{SW} - \Delta S / \Delta t - \Delta Q_{GW}$ |
| Closed basin, desert region (Nevada; Figure 10-5c) | Minor near center; none near edge | Occurs near edge | May be large near center; none near edge | May be possible near edge | $\Delta R_{SW} - \Delta CR - \Delta Q_{GW}$; Equilibrium not possible for large pumping: $- \Delta S / \Delta t$ |
| Escarpment-bounded aquifer, Arid region (High Plains of Texas and New Mexico; Figure 10-5d) | None | None | None | Very small | Equilibrium not possible: $- \Delta S / \Delta t$ |
| Confined aquifer, semi-arid region (western Colorado; Figure 10-5e) | None | None | None | None | Equilibrium not possible: $- \Delta S / \Delta t$ |
| Island sand-and-gravel aquifer, humid region (Nantucket Island, Massachusetts; Figure 10-6) | None | None | None | None | Equilibrium not possible: $- \Delta S / \Delta t$ |

principal recharge is from floods, and if pumping keeps the water table low, additional recharge can be induced ($\Delta R_{SW}$). Otherwise, water is supplied largely from storage and to a lesser extent from intercepted ground-water discharge.

The major limitations for such a system are that (1) flooding must be frequent enough to supply water needs and (2) the water table must be kept low enough by pumping to provide adequate storage for flood waters but high enough for economical well operation. If these conditions are not met, equilibrium will not be attained and storage will be continuously depleted.

### Closed Desert Basin

In the Basin and Range region of Nevada, the aquifer consists of alluvial material eroded from the surrounding mountains (Figure 10-5c). Near the mountain front these deposits are coarse and permeable, but grade into finer and less permeable material toward the basin centers. The water table is near the surface in the center, where a **playa** (intermittent shallow lake) and phreatophytes are typical-

ly present. Basin precipitation is only 75–125 mm/yr, though it may be up to 750 mm in the mountains.

Ground-water development near the basin centers is limited by the low permeabilities and by the fact that water there is usually too highly mineralized for most uses. Better quality and higher permeability are present near the mountain front, but the water table is deeper and more costly to pump. Under natural conditions the pumped water comes from a reduction in water use by phreatophytes ($\Delta CR$) plus a reduction in flow to the playa ($\Delta Q_{GW}$), plus withdrawal from storage. For large pumping rates, equilibrium cannot be achieved and storage will be depleted over time.

### Escarpment-Bounded Aquifer in an Arid Region

This aquifer consists of gently sloping, moderately permeable rocks resting on a relatively impermeable base and bounded by escarpments (Figure 10-5d). This is the situation in the Ogalalla aquifer of the southern High Plains in Texas and New Mexico. The only recharge is infiltration from precipitation, which is scanty (1 to 15 mm yr$^{-1}$). Natural discharge is from seeps and springs at the downslope (eastern) boundary and equals the natural recharge.

The Ogalalla aquifer has been highly exploited for irrigation. Because the natural water table was deep (> 20 m) and there are no significant surface-water bodies, the development does not induce recharge ($\Delta R_I$ or $\Delta R_{SW}$) or affect capillary rise ($\Delta CR$). To date, it has not significantly affected the water-table gradient near the edges, so there has been no capture of discharge ($\Delta Q_{GW}$) either. Even if ultimately captured, this discharge could only supply a few percent of the current withdrawal rates. Thus equilibrium cannot be established and the sole source of the pumped water is storage.

Without remedial measures, the Ogalalla aquifer will eventually be depleted to the point that pumping is uneconomical. Lohman (1979) discussed some of the schemes proposed for artificially recharging the aquifer, but none so far appear feasible. It is estimated that the extent of irrigation on this aquifer will be reduced by about one-half by 2030.

### Confined (Artesian) Aquifer in a Semi-Arid Region

Figure 10-5e shows a sandstone aquifer confined between less permeable formations. Such a situation exists in the Entrada aquifer of western Colorado, where recharge occurs largely from ephemeral (losing) streams where the aquifer outcrops. Natural discharge consists only of minor leakage into the confining beds. Prior to development, the piezometric surface was as much as 50 m above the ground surface.

Pumping occurs a few kilometers "downstream" where the aquifer is at a depth of 150 to 300 m and has significantly lowered the piezometric surface. However, the aquifer has remained saturated. Thus the pumping has not induced any recharge ($\Delta R_I$ or $\Delta R_{SW}$) or captured any discharge ($\Delta CR$ or $\Delta Q_{GW}$), and all the pumped water has come from storage.

In contrast to the Ogallala aquifer, which has a storativity [Equation (8-6)] of about 0.15, the Entrada aquifer has a storage coefficient of $5 \times 10^{-5}$. Thus over 1 m² of aquifer surface, a 1-m decline in the water table yields 0.15 m³ of water from the Ogalalla, while a 1-m decline in the piezometric surface yields only $5 \times 10^{-5}$ m³ from the Entrada. Increasing development will continue to lower the piezometric surface and increase the costs of pumping. If continued indefinitely, the head could fall below the aquifer surface and deplete the storage at a more rapid rate.

### Island Sand-and-Gravel Aquifer in a Humid Region

Nantucket is a 123-km² island located about 45 km off the Massachusetts coast. It consists of glacial moraine and outwash deposits (largely sand and gravel) constituting an aquifer that is recharged solely by infiltration and which discharges to the ocean. Figure 10-6a shows the water-table contours. As explained in Box 8-2, a fresh water–salt water interface lies at a depth below sea level equal to about 40 times the water-table elevation.

Nantucket is an increasingly popular summer-vacation destination, and much of its population is served by a municipal well field in the center of the island. Person et al. (1998) have modeled the aquifer response to projected increasing seasonal pumping. As shown in Figure 10-6b (top), their model indicates that by 2020 this pumping will produce an approximately 1-m drawdown at the well field. The water-table gradient at the coasts is not affected, so discharge to the ocean is unchanged and $\Delta Q_{SW} = 0$. Because $\Delta R_I$, $\Delta R_{SW}$, and $\Delta CR$ are also negligible, virtually all the pumped water comes from aquifer storage.

Limitations on development are due to the rise in the fresh–salt interface that accompanies the drawdown, inducing salt-water intrusion by 2014 (Figure 10-6b and 10-6c). The average rate of pumping at that time is only about 1% of the natural recharge—underscoring the point that "safe yield" is not equal to the natural recharge rate.

### Summary Comment

A major difficulty in determining "safe yield" for ground-water supplies is due to the very long times over which storage may be depleted. Often the undesirable effects of a given level of exploitation become apparent only gradually, after the pumping has continued for many years with few serious impacts. This may be the case in many of the parts of the United States shown in Figure 10-7 and in many regions of the world where irrigated agriculture supplied by ground water is providing significant portions of the food supply. The only way to predict these long-term impacts is through detailed ground-water modeling based on a sound understanding of local geology and hydrology.

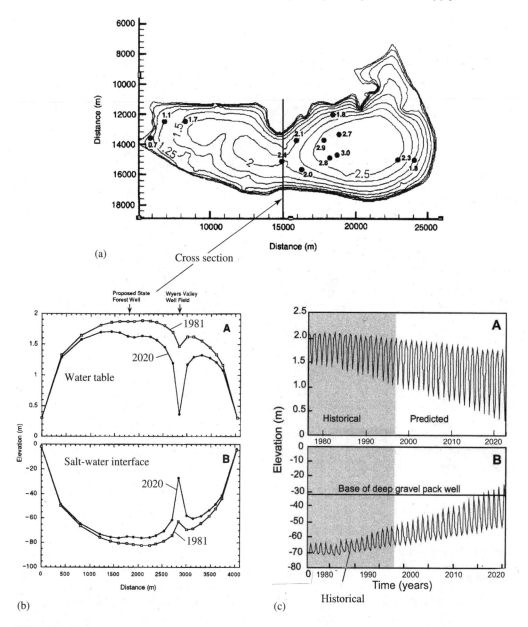

**FIGURE 10-6**

Present (1981) and projected (2020) ground-water development on Nantucket Island, MA. (a) Current water-table elevations. (b) Cross sections [vertical line in (a)] showing present and projected elevations of water table (top) and freshwater-saltwater interface. (c) Time series of elevations of water table (top) and freshwater-saltwater interface. Shaded portion is historical data, unshaded is model projection. From Person et al. (1998).

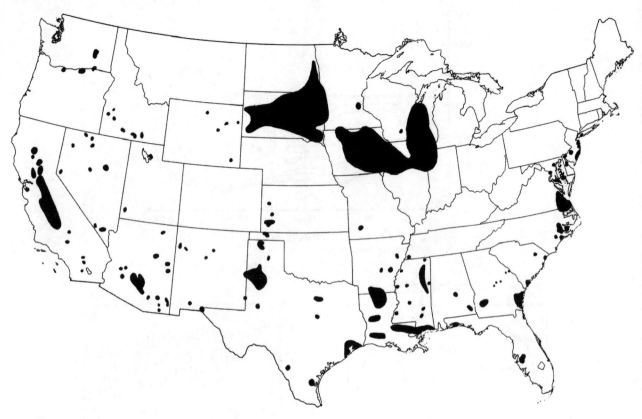

**FIGURE 10-7**
Areas of the United States where water table or piezometric surface has declined more than 40 ft since development began. From U.S. Geological Survey *Water-Supply Paper* 2325 (1988).

### 10.2.5    Water Supply and "Safe Yield": Surface Water

#### River Flow Regimes

In contrast with the usual approach for ground water, analysis of the "safe yield" of surface-water supplies must account for the typically significant variability of river flows on various time scales. The **flow regime** of a river reach can be characterized by its typical seasonal (intra-annual) pattern of flow variability, its year-to-year (inter-annual) flow variability, and various quantitative and qualitative descriptors of the time series of low flows, average flows, and flood flows (Section 2.7.1; Appendix C).

The inter-annual flow regime can be characterized by the mean and standard deviation of annual streamflows. Table 10-7 gives equations that can be used to estimate those quantities in the water-re-

source regions of the conterminous United States based on drainage area, mean annual precipitation, and mean annual temperature.

Figure 10-8 shows the typical seasonal variability of river flows in North America. More detailed examples of inter-annual and intra-annual variability are illustrated in Figure 10-9: (a) little variability due to relatively constant precipitation inputs and large ground-water contributions; (b) high variability where snow is absent, ground-water contribution is small, and storms occur in all seasons; (c) relatively constant pattern of seasonal variability due to accumulation and melt of significant snowpack; and (d) a pronounced low-flow season due to high summer evapotranspiration, with random distribution of rainstorms in other seasons.

Figure 10-10 illustrates the connections between the components of flow regime—magnitude,

**TABLE 10-7**

Regression Equations for Estimating Mean ($\mu_Q$, m³ s⁻¹) and Standard Deviation ($\sigma_Q$, m³ s⁻¹) of Annual Streamflows as Functions of Drainage Area ($A$, km²), Mean Annual Precipitation ($P$, mm yr⁻¹), and Mean Annual Temperature ($T$, °F ×10) for Water-Resources Regions of the Conterminous United States. $r^2$ is the fraction of variability of $\mu_Q$ and $\sigma_Q$ explained by the equation. From Vogel et al. (1999b); see that paper for further details.

| Region | Equations | $r^{2\ a}$ |
|---|---|---|
| 1 | $\mu_Q = 8.03 \times 10^{-5} \cdot A^{1.012} \cdot P^{1.213} \cdot T^{-0.512}$ | 0.997 |
|  | $\sigma_Q = 1.86 \times 10^{-6} \cdot A^{0.989} \cdot P^{0.724} \cdot T^{0.473}$ | 0.992 |
| 2 | $\mu_Q = 6.67 \times 10^{-2} \cdot A^{0.979} \cdot P^{1.625} \cdot T^{-2.051}$ | 0.994 |
|  | $\sigma_Q = 1.00 \cdot A^{0.935} \cdot P^{0.209} \cdot T^{-1.053}$ | 0.992 |
| 3 | $\mu_Q = 4.10 \times 10^{-5} \cdot A^{0.984} \cdot P^{2.260} \cdot T^{-1.607}$ | 0.989 |
|  | $\sigma_Q = 7.93 \times 10^{-10} \cdot A^{0.993} \cdot P^{1.677} \cdot T^{0.561}$ | 0.986 |
| 4 | $\mu_Q = 3.42 \times 10^{-3} \cdot A^{0.965} \cdot P^{2.289} \cdot T^{-2.319}$ | 0.971 |
|  | $\sigma_Q = 1.00 \cdot A^{0.956} \cdot P^{0.932} \cdot T^{-1.950}$ | 0.939 |
| 5 | $\mu_Q = 7.51 \times 10^{-3} \cdot A^{0.993} \cdot P^{2.325} \cdot T^{-2.509}$ | 0.994 |
|  | $\sigma_Q = 1.96 \times 10^{-10} \cdot A^{0.995} \cdot P^{1.374} \cdot T^{1.169}$ | 0.992 |
| 6 | $\mu_Q = 1.48 \times 10^{-4} \cdot A^{0.964} \cdot P^{1.358} \cdot T^{-0.748}$ | 0.987 |
|  | $\sigma_Q = 1.00 \cdot A^{1.011} \cdot P^{1.023} \cdot T^{-2.027}$ | 0.984 |
| 7 | $\mu_Q = 7.06 \times 10^{-6} \cdot A^{1.002} \cdot P^{4.560} \cdot T^{-3.898}$ | 0.988 |
|  | $\sigma_Q = 2.28 \times 10^{-10} \cdot A^{0.975} \cdot P^{1.665} \cdot T^{0.863}$ | 0.965 |
| 8 | $\mu_Q = 1.00 \cdot A^{0.984} \cdot P^{3.157} \cdot T^{-4.190}$ | 0.985 |
|  | $\sigma_Q = 1.00 \cdot A^{0.983} \cdot P^{-0.670}$ | 0.989 |
| 9 | $\mu_Q = 1.00 \cdot A^{0.816} \cdot P^{6.422} \cdot T^{-7.655}$ | 0.957 |
|  | $\sigma_Q = 1.00 \cdot A^{0.827} \cdot P^{4.405} \cdot T^{-5.615}$ | 0.922 |
| 10 | $\mu_Q = 1.80 \times 10^{-5} \cdot A^{0.894} \cdot P^{3.200} \cdot T^{-2.452}$ | 0.864 |
|  | $\sigma_Q = 1.00 \cdot A^{0.914} \cdot P^{1.926} \cdot T^{-3.032}$ | 0.853 |
| 11 | $\mu_Q = 8.13 \times 10^{-9} \cdot A^{0.965} \cdot P^{3.815} \cdot T^{-1.967}$ | 0.962 |
|  | $\sigma_Q = 7.57 \times 10^{-9} \cdot A^{0.949} \cdot P^{2.960} \cdot T^{-1.099}$ | 0.932 |
| 12 | $\mu_Q = 1.00 \cdot A^{0.847} \cdot P^{3.834} \cdot T^{-4.715}$ | 0.869 |
|  | $\sigma_Q = 1.00 \cdot A^{0.802} \cdot P^{2.886} \cdot T^{-3.724}$ | 0.893 |
| 13 | $\mu_Q = 1.00 \cdot A^{0.772} \cdot P^{1.964} \cdot T^{-2.828}$ | 0.874 |
|  | $\sigma_Q = 1.00 \cdot A^{0.757} \cdot P^{1.499} \cdot T^{-2.419}$ | 0.867 |
| 14 | $\mu_Q = 5.24 \times 10^{-5} \cdot A^{0.987} \cdot P^{2.469} \cdot T^{-1.877}$ | 0.935 |
|  | $\sigma_Q = 1.00 \cdot A^{0.892} \cdot P^{0.991} \cdot T^{-1.991}$ | 0.953 |
| 15 | $\mu_Q = 1.00 \cdot A^{0.866} \cdot P^{2.507} \cdot T^{-3.427}$ | 0.852 |
|  | $\sigma_Q = 1.00 \cdot A^{0.864} \cdot P^{2.539} \cdot T^{-3.523}$ | 0.863 |
| 16 | $\mu_Q = 1.00 \cdot A^{0.837} \cdot P^{2.167} \cdot T^{-3.054}$ | 0.896 |
|  | $\sigma_Q = 1.00 \cdot A^{0.863} \cdot P^{1.556} \cdot T^{-2.539}$ | 0.952 |
| 17 | $\mu_Q = 3.79 \times 10^{-5} \cdot A^{1.003} \cdot P^{1.864} \cdot T^{-1.158}$ | 0.956 |
|  | $\sigma_Q = 1.00 \cdot A^{0.958} \cdot P^{1.430} \cdot T^{-2.503}$ | 0.961 |
| 18 | $\mu_Q = 2.16 \times 10^{-4} \cdot A^{0.974} \cdot P^{1.999} \cdot T^{-1.532}$ | 0.947 |
|  | $\sigma_Q = 1.00 \cdot A^{0.853} \cdot P^{1.048} \cdot T^{-1.814}$ | 0.927 |

[a]$r^2$ values indicated for $\sigma_Q$ equations are actually for equations estimating $\sigma_Q^2$.

frequency, duration, timing, and rate of change—and the major factors affecting ecological integrity. Table 10-8 summarizes the principal ways in which humans alter the components of flow regime; Poff et al. (1997) provided a comprehensive review of how such alterations impact the ecological integrity of riverine systems. They argued that the natural flow regime is essential to ecosystem function and natural biodiversity and that maintenance and restoration of natural regimes is a desirable and realistic management goal. However, as noted previously, any withdrawal of ground or surface water has some impact on the timing and magnitude of downstream flows, so conflicts between human water use and ecosystem function are inevitable.

In the remainder of this section we focus on the flow-duration curve, which depicts the magnitude and frequency components of flow regime, as a basic tool for analyzing surface-water supplies.

### Flow-Duration Curves

A **flow-duration curve** (FDC) is the relation between the magnitudes of streamflow, $q$, at a point and the frequency (probability) with which those magnitudes are exceeded over an extended time period (i.e., many years). Almost always, FDCs are constructed for *daily average streamflows*, and we will assume that that is the case here.

The concept of the FDC as a concise picture of the temporal variability of streamflow was introduced in Section 2.7.2. Here we introduce the use of the FDC in assessing the "safe yield" of the surface-water supply available at a point on a stream.

As noted in Section 2.5.1, the long-term average streamflow rate, $\mu_Q$, represents the maximum rate at which water is potentially available for human use and management in a watershed or region. However, the actual flow rate varies greatly with time, and the FDC depicts this variability by showing the relation between a given flow rate, $q_p$, and the fraction of time, $p$, that that flow is equaled or exceeded. In the language of statistics and probability (Section C.3), $p$ is the **exceedence probability** of the flow $q_p$, $EP_Q(q_p)$; that is,

$$p = \Pr\{Q > q_p\} = EP_Q(q_p) = 1 - F_Q(q_p),$$
$$(10\text{-}4)$$

where $F_Q(q)$ is the cumulative distribution function of daily average streamflow.

Under natural conditions in most regions, the average flow rate $\mu_Q$ is available less than half of the time. Thus a more realistic expression of the usable water resources in a watershed or region is the flow rate available most of the time—say the flow exceeded 95% of the time, $q_{.95}$. In most regions, the natural $q_{.95}$ is only a few percent of $\mu_Q$. This rate

**FIGURE 10-8**

Flow regimes of North American rivers. Height of histogram bar is proportional to fraction of annual streamflow occurring in month. Numbers under histograms are average flow in $m^3 s^{-1}$ (left) and $mm\ yr^{-1}$. From Riggs and Harvey (1992), used with permission.

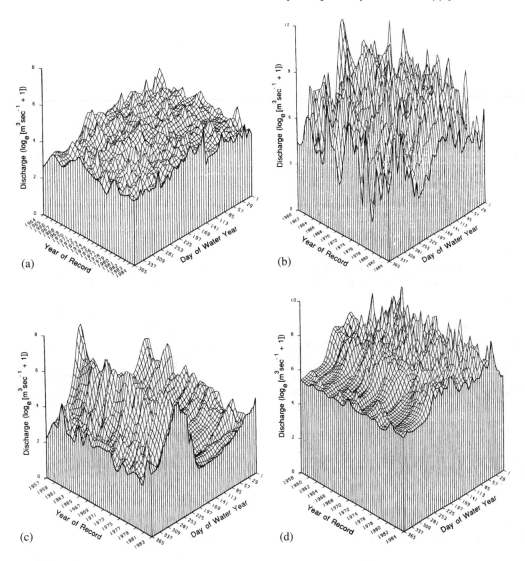

**FIGURE 10-9**
Intra-annual flow variability of four streams. (a) Augusta Creek, MI; (b) Satilla River, GA; (c) upper Colorado River, CO; (d) South Fork of McKenzie River, OR. From Poff et al (1997), used with permission.

could be termed the "safe yield" of the watershed; however, it must be remembered that $q < q_{.95}$ on 18 days out of a typical year and that the minimum flow rate from any watershed is 0 (i.e., the quantity available with 100% reliability is 0; $q_1 = 0$).

Box 10-2 describes how FDCs are constructed at sites where long-term gaging records exist and the following example applies the methods described there.

### EXAMPLE 10-1

Figure 10-11 shows two FDCs for average daily flows for the White River at West Hartford, VT, in the humid northeastern United States for 10 water years 1960–1969. The data were obtained by downloading the daily average flows for that period from the USGS Internet page as described in Box 2-2 and following the methods described in Box 10-2. The period-of-record FDC is based on the ranking of all 3,653 flows; the median FDC is based

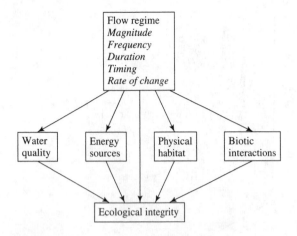

**FIGURE 10-10**
Elements of river flow regime showing their influence on the components of ecological integrity. From Poff et al. (1997), used with permission.

on the 365 medians for each rank. As found by Fennessey and Vogel (1990), the median FDC plots above the period-of-record FDC in the low range and reflects more typical behavior of the stream. For the 10-yr period the median flow ($q_{50}$) is 457 ft$^3$ s$^{-1}$, while the average flow, $\mu_Q$, is 890 ft$^3$ s$^{-1}$, which is exceeded about 28% of the time (i.e., the flow is less than $\mu_Q$ 72% of the time). The $q_{95} = 124$ ft$^3$ s$^{-1}$, only 14% of $\mu_Q$.

Approaches to estimating FDCs for reaches where long-term streamflow records are lacking are reviewed in Box 10-3.

Because FDCs are constructed by ranking all flows in the period of record regardless of their season or sequence of occurrence, they do not reflect some characteristics of daily streamflows that are very important for some purposes. Streamflows generally follow marked patterns of seasonality (Figure 10-8) that are critical when considering virtually all offstream and instream water uses. In addition, daily and even annual streamflow time series usually have significant autocorrelation (i.e., the flow for a given day or year tends to be similar in magnitude to the flow for the preceding day or year; see Sections 2.8.2 and C.9). This autocorrelation is not revealed by the FDC, but must be accounted for when analyzing the amount of reservoir storage required to provide a given amount of water with a given reliability, as will be discussed later.

However, there are many purposes for which the overall relation between flow magnitude and frequency depicted by the FDC provides a very useful basis for water-resource analysis, and the following sections introduce such applications, largely following the review by Vogel and Fennessey (1995).

### Water-Resources-Index Duration Curves

Many quantities essential to the water-resource management purposes listed in Table 10-3 and Figure 10-3 are directly related to streamflow. Such a

**TABLE 10-8**
Major Impacts of Water-Resource-Management Activities on River-Flow Regimes. [See Poff et al. (1997).]

| Activity | Magnitude-Frequency | Timing | Duration | Rate of Change |
|---|---|---|---|---|
| Damming[a] | Reduced variability (WS, FC); Reduced peak flows (FC) | Altered (WS, FC, HP) | Reduced periods of inundation (FC) | Rapid fluctuations (HP) |
| Diversion | Reduced flows; Reduced variability | Altered | | |
| Urbanization and drainage | Increased variability; Increased peak flows | | Reduced periods of floodplain inundation due to stream entrenchment | |
| Levees and channelization | May increase downstream peak flows | | Reduced periods of floodplain inundation | |
| Ground-water pumping | Reduced low flows | | | |
| Deforestation | Increased variability; Increased peak flows; Reduced low flows | | | |

[a]WS = water supply; FC = flood control; HP = hydropower.

---

## BOX 10-2

· · · · · · · ·

### Construction of Flow–Duration Curves (FDCs) at Long-Term Gaging Stations

The FDC is a plot of the daily average flow magnitude, $q$, vs. exceedence probability, $EP_0(q)$, where

$$EP_0(q) = 1 - F_0(q), \qquad \textbf{(10B2-1)}$$

and $F_0(q)$ is the cumulative distribution function (non-exceedence probability) of the value $q$. Hence the problem is that of estimating quantiles of daily average flows using plotting-position formulas, as explained in Box C-1.

There are two approaches to construction of FDCs at gaging stations where daily average flows have been measured for long periods (10 yr or more): (1) period-of-record FDCs (the conventional method) and (2) median-annual FDCs (the preferred method). To discuss these approaches, we will assume we have $N$ years of 365 daily average streamflows, all of which are greater than 0. (The flows of 29 February in leap years can be discarded without affecting the analysis).

#### Period-of-Record FDCs

Rank all the $365 \cdot N$ daily flows from lowest (rank $i = 1$) to highest (rank $i = 365 \cdot N$) and designate the $i$th-ranked flow as $q_{(i)}$. (Ties are ignored in the ranking.) Estimate the non-exceedence frequency of each flow as

$$F_0(q_{(i)}) = \frac{i}{365 \cdot N + 1} \qquad \textbf{(10B2-2)}$$

(Vogel and Fennessey 1994), and compute the corresponding exceedence probability via Equation (10B2-1). The curve is then constructed by plotting the $q_{(i)}$ values

(vertical axis) vs. the $EP_0(q_{(i)})$ values, usually with a logarithmic scale for $q$ and a probability scale for $EP_0(q)$.

Period-of-record FDCs depict the historical variability of measured streamflows, but give no information about the year-to-year variability of flows or the inherent uncertainty of the estimated exceedence frequencies due to the finite record length. Typically, the low-flow end of the period-of-record FDC is significantly influenced by the particular years over which flow was measured.

#### Median-Annual FDCs

In order to construct FDCs less influenced by the particular period of record, and for estimating the inter-annual variability and uncertainty of FDCs, Vogel and Fennessey (1994) recommended first applying Equation (10B2-2) separately for each complete water year of record, that is, compute

$$F_0(q_{(i)}) = \frac{i}{365 + 1} \qquad \textbf{(10B2-3)}$$

and the corresponding $EP_0(q_{(i)})$ values for the flows of each year. Then compute the median (see Box C-1) of the $N$ values of $q_{(i)}$ associated with each exceedence probability and plot these as the FDC. Vogel and Fennessey (1994) also showed how confidence intervals for the estimated FDC can be constructed using the annual FDCs.

LeBoutillier and Waylen (1993) described an alternative statistical approach for estimating the FDC and its uncertainty from analysis and synthesis of the annual FDCs.

---

quantity is called a **water-resource index.** By exploiting such relationships, FDCs provide an important tool of hydrologic analysis for those purposes.

The general relation between a water-resource index, $I$, and discharge, $q$, is

$$I = I(q). \qquad \textbf{(10-5)}$$

Such a relation can have four possible forms: (1) monotonic increasing (Figure 10-12a); (2) monotonic decreasing (Figure 10-12b); (3) non-monotonic convex (Figure 10-12c); or (4) non-monotonic con-

cave (Figure 10-12d). Given the FDC and $I(q)$, one can construct a water-resource-index duration curve (WRIDC) that gives the fraction of time that a given index value is equaled or exceeded. Such a relation can be used to design various water-resource-management projects and to assess the impacts of such projects.

For any of the four forms, the WRIDC can be constructed using **Monte-Carlo analysis,** as explained in Box 10-4. If the relationship is monotonic increasing or decreasing, the WRIDC can also be

**FIGURE 10-11**
Period-of-record and median annual flow-duration curves (FDCs) for the White River at West Hartford, VT. The two curves coincide above $q_{.50,}$ but the median FDC gives higher low flows. Note that the mean flow (890 ft$^3$ s$^{-1}$) is exceeded only 28% of the time and that $q_{.95}$ (124 ft$^3$ s$^{-1}$) is only about 14% of average flow.

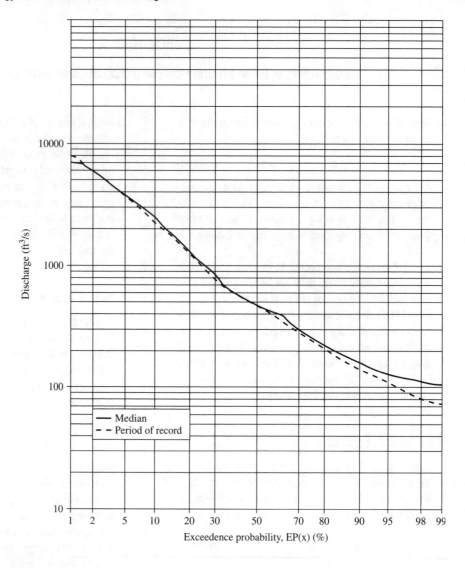

constructed by the simpler graphical methods described in Box 10-5 and Figure 10-13.

The following sections describe how various types of WRIDCs can be used in water-resource planning and management.

**Wetland Inundation** The temporal variation of water level, or **hydroperiod,** of a wetland is an important part of its ecological signature (Mitsch and Gosselink 1993; Fretwell et al. 1996). Several types of wetlands are adjacent to streams, and the depth of standing water on such wetlands is closely related to streamflow. Figure 10-14a shows a hydroperiod diagram for a typical alluvial wetland. Since river stage (water-surface elevation) is a monotonically increasing function of dis-

charge, the stage–discharge relation is a water-resources index that can be used to develop a WRIDC (Figure 10-14b) that characterizes the magnitude and frequency of water levels in alluvial wetlands.

**River, Reservoir, and Lake Sedimentation** The concentration of suspended sediment, $C$ [M L$^{-3}$], transported by a river is typically a strong function of discharge that can generally be characterized by a power-law relation:

$$C = a \cdot q^b, \qquad \textbf{(10-6)}$$

where $a$ and $b$ are empirical constants that are established by statistical (regression) analysis of concurrent measurements of $C$ and $q$ for the river

---

## BOX 10-3
. . . . . . .
### Construction of Flow–Duration Curves at Stream Reaches Without Long-Term Gaging Stations

There are two approaches to estimating FDCs where long-term records of daily streamflows are lacking: (1) identification and application of regional FDC characteristics and (2) correlation of concurrent flows. In the following, $\mu_Q$ is the mean flow; $q_p$ is the flow exceeded $100 \cdot p$ percent of the time; $A$ is drainage area; and $a, b, c$, and $d$ are empirical constants that differ in each study.

### Regional FDC Characteristics

A number of approaches have been used in various regions:

*Illinois*: Singh (1971) found that $q_p/\mu_Q = a \cdot A^b$, where $a$ and $b$ depend on $p$ and on the hydrologic region. $\mu_Q$ is estimated from a map and $a$ and $b$ from graphs.

*New Hampshire*: Dingman (1978b) identified the following consistencies among daily-flow FDCs: (1) $\mu_Q$ is exceeded 27% of the time; (2) $q_{.02}/\mu_Q = 6.0$; (3) $q_{.05}/\mu_Q = 3.9$; (4) $q_{.30}/\mu_Q = 0.88$; and (5) $q_{.95}/A$ can be estimated as a function of mean drainage-basin elevation. The estimated FDC is constructed by estimating mean basin elevation; estimating $\mu_Q/A$ as a function of mean basin elevation and each of three geographic regions; plotting $q_{.02}$, $q_{.05}$, $\mu_Q$, $q_{.30}$, and $q_{.95}$ on logarithmic-probability paper; and sketching the FDC through the points.

*Philippines*: Quimpo et al. (1983) used the equation

$$q_p = q_A \cdot \exp(-c \cdot p)$$

to approximate the form of daily-flow FDCs, where $q_A$ is a power-law function of $A$, and $c$ is determined by climate and physiography and could be contoured for the region.

*Greece*: Mimikou and Kaemaki (1985) used the equation

$$q_p = a - b \cdot p + c \cdot p^2 - d \cdot p^3$$

to approximate the form of monthly-flow FDCs. The parameters $a, b, c$, and $d$ can be estimated from power-law regression relations using mean-annual precipitation, drainage area, drainage-basin relief (difference between highest and lowest elevations), and main-stem river length.

*Massachusetts:* Fennessey and Vogel (1990) modeled the lower half of the daily-flow FDC ($p \geq 0.50$) as a log-normal distribution, for which the mean can be estimated from the drainage area and the standard deviation from the drainage-basin relief.

### Correlation of Concurrent Flows

Searcy (1959) described (1) how a series of spot discharge measurements at an otherwise ungaged stream reach can be related to concurrent flows at a nearby stream gage with a long-term record and (2) the relationship used to estimate the FDC for the ungaged reach. The flows at the ungaged reach are plotted against concurrent flows at the long-term gage on log-log paper, and a line of relation established. Since the low ends of FDCs are most dependent on local geologic conditions, it is most critical to obtain measurements during periods of low flow. To do this, the measurements on the ungaged reach should be made well after any significant rain, but individual measurements should be separated by periods of higher flow so that each measurement is independent. In general, the period of spot gaging will extend over two or more years before an adequate sample (say, 10 or more) of independent low flows is obtained.

---

reach of interest. The rate of sediment discharge, or load, $L$ [M T$^{-1}$], is the product of concentration and discharge [Equation (3-1)], so

$$L = a \cdot q^{b+1}. \qquad \textbf{(10-7)}$$

Thus Equation (10-7) becomes a monotonic-increasing water-resource index for sediment transport. Coupling this with an FDC thus produces a WRIDC for sediment, which can be integrated to estimate the long-term average sediment transport. Such information can then be used to estimate the rate of erosion of the watershed or the rate at which a reservoir fed by the stream is accumulating sediment.

**FIGURE 10-12**
General forms of relations between water-resource indices and discharge: (a) monotonic increasing; (b) monotonic decreasing; (c) non-monotonic convex; (d) non-monotonic concave.

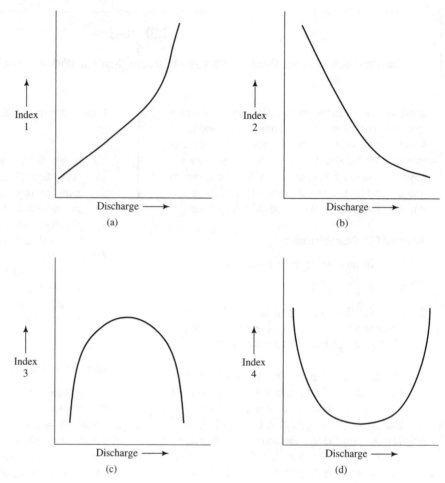

**Stream Habitat** The effects of water withdrawals on stream habitats are becoming a major water-resource issue. For many aquatic organisms the critical physical habitat components are bed-material composition, velocity, and depth. Since the latter two are monotonic-increasing functions of discharge, the depth– and velocity–discharge relations can be used as water-resource indexes for a stream reach.

The U.S. Fish and Wildlife Service has developed a methodology for relating habitat suitability for particular fish species to discharge, called the Physical Habitat Simulation System ("PHABSIM"; Milhous et al. 1990). Typically the relation between habitat suitability and discharge is convex (Figure 10-13c); thus the suitability-index duration curve is developed by Monte-Carlo methods. Figure 10-15 illustrates the procedure, using a fictitious situation: A typical FDC is shown in Figure 10-15a, and a convex relation between a habitat index and discharge

in Figure 10-15b. The Monte-Carlo simulation procedure described in Box 10-4 then leads to the WRIDC shown in Figure 10-15c.

**Hydropower** The economic feasibility of a hydropower installation depends on the long-term average amount of power that can be generated. The power, $P$ [F L T$^{-1}$], that can be generated at a hydropower dam is given by

$$P = u \cdot e \cdot \gamma \cdot h \cdot q, \qquad (10\text{-}8)$$

where $u$ is a unit-conversion factor [1], $e$ is the efficiency of the system [1], $\gamma$ is the weight density of water [F L$^{-3}$], $h$ is the drop in water level through the turbines ("head") [L], and $q$ is the discharge through the turbines [L$^3$ T$^{-1}$] (McMahon 1992). The head usually decreases with discharge, so that the relation between power (the water-resource index) and discharge is often convex. Thus the Monte-

The principle of Monte-Carlo analysis is based on generating a large number (say, $N = 10,000$) of synthetic-streamflow values that have a probability distribution that mimics the distribution (i.e., the FDC) of actual flows at the site of interest. This can be done by using the random-number generator found in most spreadsheets to generate $N$ numbers uniformly distributed between 0 and 1. These numbers represent exceedence probabilities for streamflow, $EP_Q(q)$. The FDC relates a streamflow value, $q$, to each value of $EP_Q(q)$, and the $N$ values of $q$ corresponding to each $EP_Q(q)$ can be determined either by fitting a mathematical function to the entire FDC, or by fitting several functions piecewise to portions of the FDC.

Once the $N$ values of $q$ are generated, the corresponding values of the water-resource index, $I(q)$, can be computed using the appropriate version of Equation (10-5). The $N$ values of $I(q)$ can then be ranked from lowest (rank = 1) to highest (rank = $N$) and their exceedence probability estimated using a plotting-position formula as in Box 10-2.

Carlo method similar to that shown in Figure 10-15 can be used to generate the power–duration curve; that curve can be integrated to compute the long-term-average power generation.

**Water Quality** In the United States, water-quality management efforts are now emphasizing total daily maximum loads (TDMLs) of critical water-quality constituents. The concentrations of many of these, especially nutrients and dissolved oxygen, which is essential for aquatic life, are usually strongly dependent on discharge. For soluble pollutants, the relation is usually monotonic-decreasing, that is, the concentration decreases as discharge increases. Dissolved-oxygen concentration typically increases as discharge increases.

If the relation between critical pollutants and/or dissolved oxygen is established by measurement, these concentrations become water-resource indexes, and their WRIDCs can be computed from the FDC as described earlier. The WRIDC gives the percentage of time that various concentrations are exceeded and thus provides critical information about the frequency with which allowable total daily maximum loads (TDMLs) are violated. Such applications are described more fully in Section 10.3.

**Monotonic-increasing $I(q)$**

In Figure 10-13a, the graph in the upper right-hand quadrant is the FDC, and that in the upper left-hand quadrant is the relation between $I(q)$ and $q$. The lower-left quadrant is simply a 45°, or 1:1, line. To construct the WRIDC, a number of points on the FDC are selected covering the entire curve. From each point, a vertical line is projected into the lower right quadrant, and a horizontal line is projected into the upper left quadrant to its intersection with the $I(q)$ relation. A vertical line is projected from each intersection to intersect with the 1:1 line in the lower left quadrant. Finally, horizontal lines are extended from those points to intersect with the vertical lines in the lower right quadrant. Those intersections define the

relation between values of the water-resource index, $I(q)$, and the corresponding exceedence probability, which gives the WRIDC.

**Monotonic-decreasing $I(q)$**

If $I(q)$ is monotonic-decreasing (Figure 10-13b), the procedure for constructing the curve is exactly the same as for the monotonic-increasing case except that the exceedence-probability scale for $I(q)$ must be reversed; that is,

$$EP_I[I(q)] = 1 - EP_Q(q). \quad \text{(10B5-1)}$$

The reversed scale is shown as the lower horizontal scale in the lower right quadrant of Figure 10-13b.

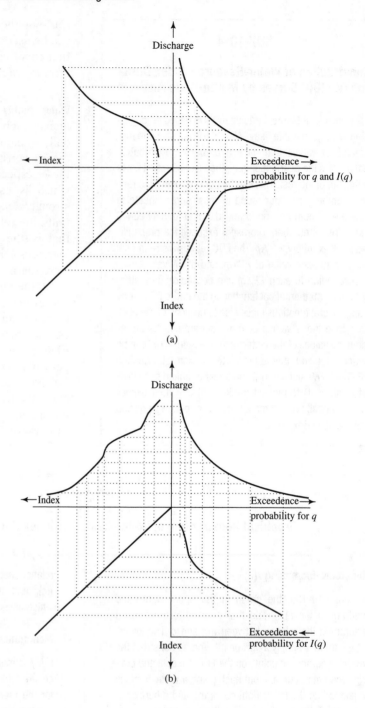

**FIGURE 10-13**
Graphical construction of a water-resource-index duration curve for (a) monotonic-increasing index; (b) monotonic-decreasing index. See text.

### Reservoir Storage

Water-supply reservoirs store water when natural flow rates are high so that it can be made available when flow rates are low. Reservoirs in humid regions tend to refill annually and function principally to smooth out intra-annual (seasonal) fluctuations

in flows; this is called **within-year storage.** In more arid regions, additional **carry-over storage** is required to smooth out inter-annual variations; reservoirs in these regions fill only rarely.

In either case, reservoirs reduce the variability of downstream flows and affect FDCs as shown in Figure 10-16: Neglecting evaporation, the mean

**FIGURE 10-14**
(a) Typical hydroperiod diagram for alluvial wetland. After Lewis et al. (1995). (b) Corresponding stage-duration curve.

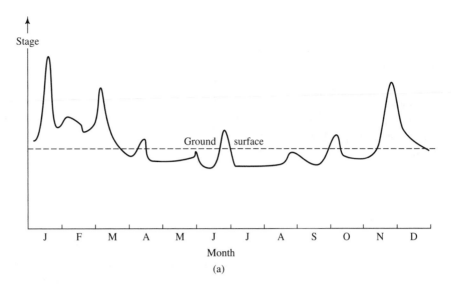

(a)

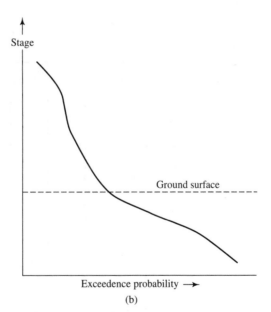

(b)

flow is not changed, but the FDC is flattened. The area between the natural and altered FDCs above and below their point of intersection are equal and equal to the average rate of release from the reservoir, called the **yield**,[3] $Y$ [$L^3 T^{-1}$]. In Figure 10-16 $\Delta Y_p$ is the increase in the flow rate available $100 \cdot p\%$ of the time due to reservoir regulation, and

$$Y_p = q_p + \Delta Y_p, \qquad \textbf{(10-9)}$$

where $Y_p$ is the new "safe yield" with reliability $p$.

To develop general relations describing reservoir behavior, it is useful to express the relevant quantities as ratios to the average annual flow at the reservoir site, $\mu_Q$ [$L^3 T^{-1}$]. Thus for a given level of reliability, $p$, we define the relative yield or **draft**, $D_p$, in dimensionless terms as

$$D_p \equiv \frac{Y_p}{\mu_Q}. \qquad \textbf{(10-10)}$$

The storage is also expressed in relative terms as the **residence time,** $T_R$ [T] of the reservoir (see Section 2.8.3); that is,

---

[3] Also called regulation, release, draft, or withdrawal.

**FIGURE 10-15**
Example illustrating construction of a WRIDC for a non-monotonic (convex) stream-habitat water-resource index via the Monte Carlo method (Box 10-4). (a) Typical FDC (arithmetic scales); (b) relation between habitat-suitability index and discharge [$I(q)$]; (c) WRIDC.

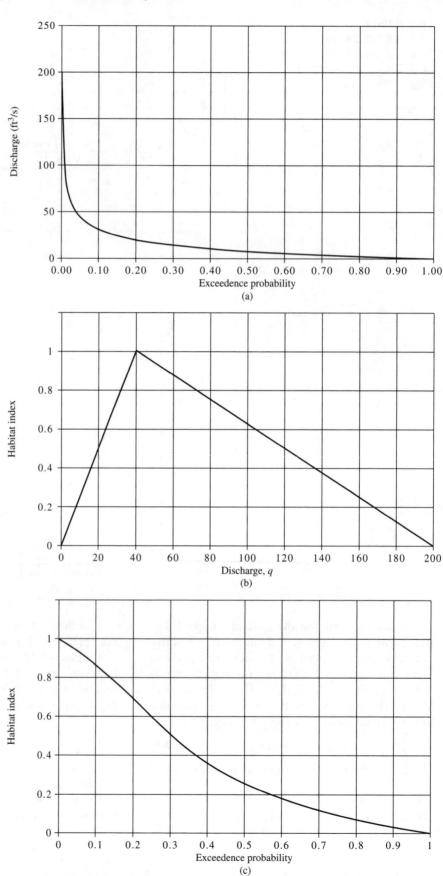

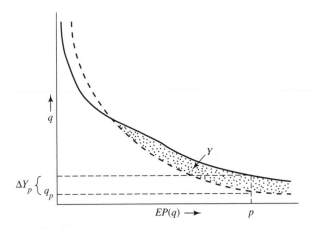

**FIGURE 10-16**
Effect of reservoir regulation on flow-duration curve. Dashed curve is natural (unregulated) FDC; regulation raises low end of curve and lowers high end, giving solid-curve FDC. The area under the two curves is equal (i.e., mean flow is not changed). $Y$ is the yield (or regulation) provided by the reservoir; $\Delta Y_p$ is the increase in yield available with probability $p$ due to regulation.

$$T_R \equiv \frac{S}{\mu_Q}, \qquad (10\text{-}11)$$

where $S$ is the storage volume [L$^3$].

A useful index of reservoir behavior is the **standardized inflow**, $M$, defined as

$$M \equiv \frac{\mu_Q - Y_p}{\sigma_Q} = \frac{(1 - D_p) \cdot \mu_Q}{\sigma_Q} = \frac{1 - D_p}{CV_Q}, \qquad (10\text{-}12)$$

where $\sigma_Q$ is the standard deviation and $CV_Q$ the coefficient of variation ($\equiv \sigma_Q/\mu_Q$) of annual inflows, and $D_p$ and $Y_p$ are the draft and yield, respectively (McMahon 1992).

Vogel et al. (1999a) showed that one can distinguish between reservoirs dominated by within-year storage vs. carry-over storage on the basis of their $M$ and $CV_Q$ values: if $M > CV_Q$ the reservoir is dominated by within-year variation, otherwise it is dominated by carry-over variation (Figure 10-17).

Storage-yield relations are the basis for reservoir design and evaluation. McMahon (1992) stated that, for reservoirs in which carry-over storage dominates, this relation can be estimated as

$$D_p = 1 - \frac{z_p^2 \cdot CV_Q^2}{4 \cdot (T_R + g_p \cdot CV_Q^2)}, \qquad (10\text{-}13)$$

where $z_p$ is the standard normal variate corresponding to a $100 \cdot p\%$ reliability [i.e., $100 \cdot (1 - p)\%$ risk of failure], $T_R$ is the residence time of the reservoir, and $g_p$ is an adjustment factor used if annual flows are not normally distributed[4] (Table 10-9). Note that Equation (10-13) is approximate, and that local site-specific data are required to develop storage–yield relations where within-year storage dominates (McMahon 1992; Vogel and McMahon 1996).

Adjustments for evaporation losses should be made to the storage–yield relation given by Equation (10-13). This can be done by increasing the required storage by an amount $\Delta S$, where

$$\Delta S = [PET_A - (\mu_{PA} - \mu_{QA})] \cdot A, \qquad (10\text{-}14)$$

where $PET_A$ is the potential evapotranspiration for the reservoir area; $\mu_{PA}$ and $\mu_{QA}$ are the pre-reservoir average precipitation and runoff, respectively; and $A$ is reservoir area (McMahon 1992). $PET_A$ represents the open-water evaporation and can be estimated from pan-evaporation or other methods described in Section 7.3.

Figures 10-18 and 10-19 are graphs constructed from Equation (10-13); in all cases, yield increases with storage and each curve is asymptotic to $D_p = 1$ (i.e., to a value of draft equal to the mean annual discharge). Figure 10-18 shows the effect of flow variability on "safe yield": The higher the relative variability ($CV_Q$), the smaller the fraction of the mean flow that can be made available for a given storage residence time. Figure 10-19 shows how draft is related to storage at different levels of reliability for a given flow variability: A storage residence time of 1 yr provides about 0.3 times the mean flow with 99.5% reliability, and about 0.9 times the mean flow with 90% reliability.

Note that the equations in Table 10-7 can be used to estimate $CV_Q$ for sites in the conterminous United States, and those values can be used in Equation (10-13) to estimate storage–yield–reliability relations (Exercise 10-5).

### Effects of Land-Use Changes on Streamflow

Major land-use changes such as afforestation/ deforestation, agricultural development, wetland

---

[4] Methods for determining whether annual flows can be represented by a normal distribution are described in Section C.5.

**FIGURE 10-17**
Relation between standardized reservoir inflow ($M$) and coefficient of variation of annual streamflow ($CV$) for selected values of draft ($D$) [Equation (10-12)]. Straight line is $M = CV$. Within-year storage dominates for $M > CV$ (typical of humid regions); carry-over storage dominates for $M < CV$ (typical of arid and semi-arid regions).

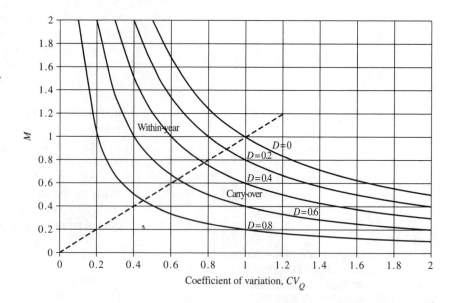

drainage, and urbanization can cause significant changes in the magnitude, frequency, duration, and timing of streamflow. Some of these effects are outlined in the following sections; more comprehensive reviews were given by Hewlett and Hibbert (1961), Bosch and Hewlett (1982), Calder (1992), and McCulloch and Robinson (1993). Only effects on streamflow magnitude and timing are discussed here; typically there are also major effects on water quality. In all cases, the effects of land-use changes on streamflow are proportional to the ratio of the area affected by the change to the total watershed area.

**Afforestation/Deforestation**  Observational and experimental studies in many parts of the world have established that forests consume more water in evapotranspiration than do any other forms of

**TABLE 10-9**
Standard Normal Variate, $z_p$, and Adjustment Factor, $g_p$, for Calculating the Draft, $D_p$, Available with Probability, $p$, as a Function of Reservoir Residence Time, $T_R$, and Coefficient of Variation of Annual Inflows, $CV_Q$, via Equation (10-13). From McMahon (1992).

| Reliability, $p$ | $z_p$ | $g_p$ |
|---|---|---|
| 0.995 | 3.30 | * |
| 0.99 | 2.33 | 1.5 |
| 0.98 | 2.05 | 1.1 |
| 0.95 | 1.64 | 0.6 |
| 0.90 | 1.28 | * |

* Method not applicable for non-normal distributions in these ranges.

vegetation or land uses. The principal reasons for this are that (1) interception-loss rates increase with vegetation height such that interception loss is largely an addition to, rather than a replacement for, transpiration for tall vegetation (see Section 7.6.4) and (2) trees have deeper roots than most shorter vegetation has, and hence can transpire at close to the potential rate for more of the time.

Thus afforestation almost always decreases average and dry-season streamflow, and deforestation by logging, forest fire, or wind damage increases streamflow. As shown in Figure 7-18, the magnitude of either effect is approximately proportional to the percentage change in forest cover.

Afforestation may also reduce flood peaks, at least for smaller floods. This is because of increased interception loss and because forest floors tend to have highly permeable surfaces that induce infiltration in preference to Hortonian overland flow (Sections 6.5.4 and 9.2.2). However, larger floods tend to be produced by rainfall and/or snowmelt events of exceptional magnitude, intensity, areal extent, and/or duration that tend to overwhelm any effects of land use or vegetation.

The discussion in Section 4.1.5 indicates that evapotranspiration from land surfaces may be an important source of precipitation in some regions (Table 4-2). Where this occurs, forest removal could have significant feedback effects such that the resulting streamflow reductions might reduce or even reverse the effects of reduced evapotranspiration.

**FIGURE 10-18**
Draft available with probability $p$ = 0.95 ($D_{.95}$) as a function of reservoir residence time ($T_R$) for regions with various values of coefficient of annual inflows ($CV_Q$) [Equation (10-13)].

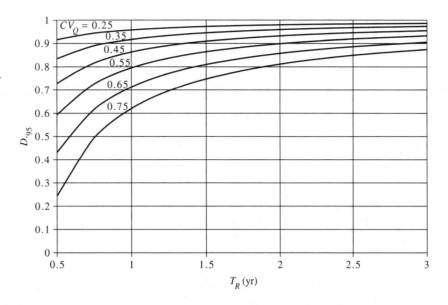

**Agricultural Development**   To the extent that agricultural development involves deforestation, it tends to increase average flow, dry-season streamflow, and peak flows for smaller floods. Development of accompanying drainage systems also tends to increase flow at least initially; however, the long-term effects may be higher wet-season flows and lower dry-season flows as well as increased flood frequency.

The withdrawal of irrigation water from streams is also likely to reduce average flows and dry-season flows. A similar effect can be expected where ground water is exploited for irrigation, though the effect may be felt some distance away, in the region where the ground water formerly discharged to streams.

**Wetland Drainage**   Wetlands are characterized by significant periods of standing or near-surface water induced either by flooding (saturation from above) or by ground-water discharge (saturation from below). Where flooding occurs, the wetland may be a locus of ground-water recharge as well as a natural flood-control reservoir.

**FIGURE 10-19**
Draft available with various probabilities $p$ ($D_p$) as a function of reservoir residence time ($T_R$) for a region with coefficient of annual inflows ($CV_Q$) = 0.5 [Equation (10-13)].

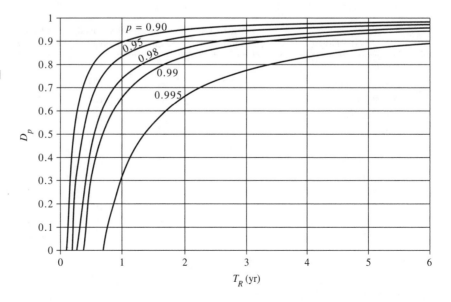

The long-term hydrologic effect of wetland drainage depends largely on the factors that produced the wetland and the type of land use that replaces it. If flooded wetlands are replaced by structures and/or fill, an increase in downstream flood flows proportional to the volume affected will result. Wetland drainage will tend to increase dry-season flows if vegetation is replaced by pavement, buildings, or fill. However, if the vegetation is replaced by forest, both dry-season and average flows will be decreased.

**Urbanization**  Urbanization involves (1) replacement of soil and vegetation by pavement and buildings, reducing infiltration and evapotranspiration; and (2) replacement of natural stream networks by streets, storm sewers, and other artificial drainage systems, increasing hydrograph peak discharges and velocities. As a result, urbanization significantly increases the frequency, volume, and peak discharges of floods (Urbonas and Roesner 1992). This is reflected in the tabulations of runoff coefficients (Table 9-10) and runoff curve numbers (Table 9-12).

## 10.3  HYDROLOGIC ANALYSIS: WATER QUALITY

Since water-quality constituents travel with water, an understanding of the movement of water is a necessary, though not sufficient, condition for an understanding of the movement of pollutants in the environment. This text has focused on water movement; detailed discussion of water chemistry and other aspects of water quality is beyond its scope. However, water quality is a vitally important societal concern and an important motivation for the study of hydrology. Thus this section introduces basic water-quality concepts and the relations between water quality and physical hydrology in the context of water-resource management.

### 10.3.1  Definitions and Basic Concepts

#### Water Quality

Water quality refers to the concentration of dissolved solids and gases, suspended solids, hydrogen ions, pathogenic organisms, and heat in a given quantity of water. The concentration of dissolved and suspended solids and dissolved gases is expressed as the mass of constituent per unit volume of solution or suspension $[\mathrm{M\,L^{-3}}]$ or as the mass of constituent per unit mass of solution or suspension $[\mathrm{M\,M^{-1}}] = [1]$. The concentration of pathogenic organisms is expressed as the number of organisms per unit volume of water $[\mathrm{L^{-3}}]$, and the concentration of heat as the temperature. As indicated in Table 10-10, most uses of water require that the concentrations of at least some of these constituents be within certain ranges, and most uses affect the concentration of one or more constituents.

**TABLE 10-10**
Relations of Types of Water-Quality Constituents to Types of Water Uses.[a]

| Water Use | pH | Dissolved Solids | Suspended Solids | Dissolved Oxygen | Organics and Petroleum Compounds | Pathogenic Organisms | Excess Heat |
|---|---|---|---|---|---|---|---|
| Public | R | R | R | R | R | R | |
|   Water Supply | | A | A | A | | A | |
| Industrial | R | R | R | R | R | R | |
|   Water Supply | | A | A | A | | A | A |
| Irrigation | | R | R | | R | R | |
| | | A | | | | | |
| Hydroelectric Power | | | R | | | | |
| Navigation | | | | | | | |
| Waste | R | | R | R | | | |
|   Transport and Treatment | A | A | A | A | A | A | |
| Recreation | R | | R | R | R | R | R |
| Wildlife Habitat | R | R | R | R | R | R | R |

[a] "R" indicates that a given use usually requires a restricted range of concentrations of the constituent. "A" indicates that a given use usually significantly affects the concentration of the constituent.

**TABLE 10-11**

Relation of Type of Water-Quality Constituent to Type of Water Resource.[a]

| Water Resource | pH | Dissolved Solids | Suspended Solids | Dissolved Oxygen | Organics and Petroleum Compounds | Pathogenic Organisms | Excess Heat |
|---|---|---|---|---|---|---|---|
| Precipitation | X | X | | | | | |
| Ground Water | | X | | | X | X | |
| Streams | X | X | X | X | X | X | X |
| Lakes | X | X | | X | | | X |

[a]"X" indicates that a given type of water-quality constituent is typically of concern in a given type of water resource.

## Pollutants

The transport and treatment of human-generated wastes are among the principal instream uses of water. When the concentration of a constituent reaches a level that adversely affects the suitability of water for habitats or human uses, the constituent is a pollutant. Dissolved and suspended solids, miscible and immiscible liquids, pathogenic organisms, heat, and the dissolved radioactive gas radon are considered pollutants when their concentrations exceed certain levels. Because the dissolved gas oxygen is a necessary constituent for most aquatic life and human uses, the **biochemical oxygen demand** (BOD), which reflects the concentration of organic wastes that consume dissolved oxygen, is also considered a pollutant. Thus water in which the concentration of oxygen is less than a certain level may be considered polluted. For pH, either excessively low or excessively high values can adversely affect many uses and hence be classified as pollution.

Table 10-11 shows the types of water-quality constituents that are of particular concern in precipitation, ground water, streams, and lakes.

## Mechanisms of Ground-Water Contamination[5]

The most common mechanisms by which pollutants enter ground water are downward percolation from surface or near-surface sources, leakage from sources below the water table, and "upconing" of naturally occurring saline water into pumped wells (salt-water intrusion) (Figure 10-20). **Point sources** are those of less than a few acres area and include landfills, surface impoundments (lagoons, pits, and ponds), spills, and injection wells. **Nonpoint sources** range in size from several acres to hundreds of

square miles and include pesticides and fertilizers applied in agriculture and silviculture, organic wastes and nutrients from septic tanks and cesspools, road salt, organic wastes and nutrients from animal-feed lots, and dissolved and suspended pollutants from mining processes. Table 10-12 summarizes the principal factors affecting aquifer susceptibility to contamination.

Ground-water contaminants are advected with ground water as it moves down the prevailing hydraulic gradient, creating a **contaminant plume** (Figure 10-21). Variations in hydraulic conductivity cause the leading edge of the plume to be irregular, with pollutants extending farther in higher conductivity zones. On a smaller scale, variations in velocity and flow-path length around individual grains and in fractures cause diffusion that further "smears out" the plume boundaries.

## Mechanisms of Surface-Water Contamination

Pollutants enter streams and lakes from atmospheric sources as well as from the land. Aspects of the quality of precipitation, throughfall, and snowmelt water—particularly acidification—were described in Sections 4.5, 5.6, and 7.6.5. Acidification arises because sulfur and nitrogen compounds produced by the burning of fossil fuels in power plants, factories, and automobiles are incorporated in raindrops and snowflakes and deposited as dry particles. Subsequent chemical reactions can produce strong acids that, if not neutralized by reaction with watershed soils and rocks, can affect soil–chemical processes and eventually enter surface water. The resulting lowered pH can adversely affect—even eliminate—aquatic biota, reduce forest productivity, and threaten human health through its effects on drinking-water supplies (Glass et al. 1982; Schindler 1988).

---

[5] This discussion follows the summary by Johnston (1988).

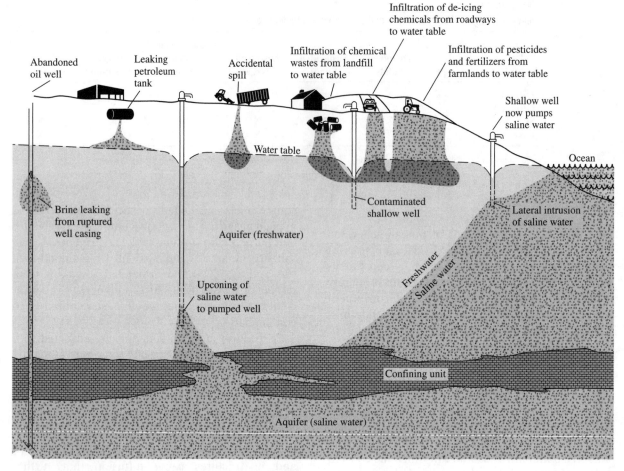

**FIGURE 10-20**
Mechanisms of ground-water contamination. From Johnston (1988).

**TABLE 10-12**

Principal Geologic and Hydrologic Features that Influence an Aquifer's Vulnerability to Contamination. From Johnston (1988).

| Feature Determining Aquifer Vulnerability to Contamination | Low Vulnerability | High Vulnerability |
|---|---|---|
| Unsaturated zone | Thick unsaturated zone containing clay and organic materials | Thin unsaturated zone in sand, gravel, limestone, and basalt |
| Confining unit | Thick confining unit of clay or shale above aquifer | No confining unit |
| Aquifer properties | Silty sandstone or shaly limestone of low permeability | Cavernous limestone, sand and gravel, or basalt of high permeability |
| Recharge rate | Negligible recharge rate, as in arid regions | High recharge rate, as in humid regions |
| Proximity to recharge or discharge area | Locations in deep, slow-flowing part of regional flow system | Locations in recharge area or within cone of depression of pumped well |

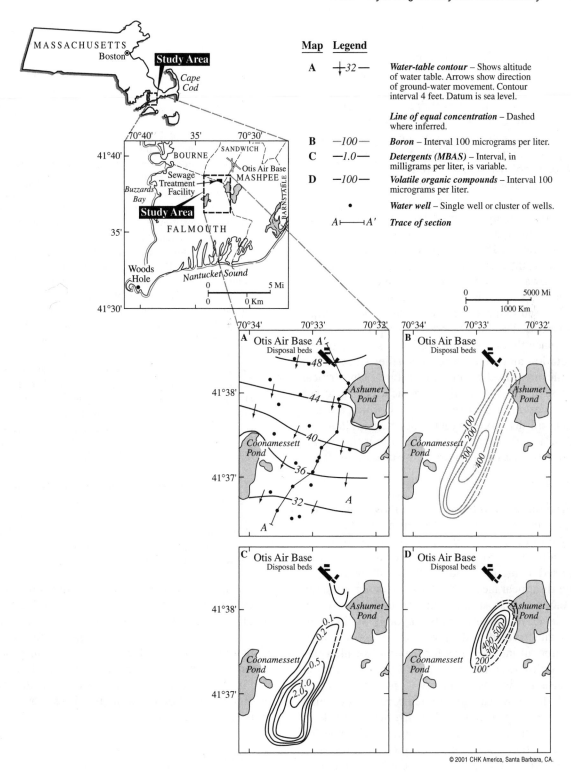

**FIGURE 10-21**

Contaminant (sewage) plume in ground water, Otis Air Force Base, Cape Cod, MA. (a) Water-table elevations (ft) and directions of ground-water flow (arrows). (b) Boron concentration ($\mu$g L$^{-1}$). (c) Detergent concentration (mg L$^{-1}$). (d) Volatile-organic-compound concentration ($\mu$g L$^{-1}$). From Hess (1988).

As with ground water, surface-water pollution sources are classified as **point sources,** which are discrete pipes or canals, or **nonpoint sources,** which include ground-water discharges to streams and diffuse flows from the land surface. Point sources are usually the returns of withdrawal uses and may carry dissolved organic or inorganic solids, BOD, pathogenic organisms, and elevated temperatures. Nonpoint sources of pollution usually represent surface runoff from urban areas or ground- or surface-water runoff from agricultural sites, construction sites, areas of timber harvest, or mining activities. Nonpoint sources typically introduce inorganic and organic suspended and dissolved solids, and can include BOD and pathogenic organisms.

After introduction from point and nonpoint sources, water quality in streams can change due to in-channel processes. Natural or artificially induced erosion and sedimentation will alter concentrations of suspended sediment, and chemical reactions can take place that affect dissolved solids. The types of vegetation along the banks of smaller streams can significantly affect water temperatures. One critical process in streams carrying organic wastes involves the oxidation of the organic matter, which reduces BOD and dissolved oxygen, and the simultaneous dissolution of oxygen from the atmosphere; this process is discussed in more detail in Section 10.3.3.

### Eutrophication

**Eutrophication** is the process whereby a lake or estuary accumulates essential plant nutrients, principally phosphorus and nitrogen. This process occurs naturally at varying rates in every lake and, along with sedimentation, leads to the in-filling and ultimate disappearance of the lake. However, in many developed areas, streams and ground water carry high levels of these nutrients that originate in fertilizers and treated and untreated sewage effluent. This human-induced over-fertilization is called **cultural eutrophication,** which is a form of pollution because it induces accelerated growth of algae, which die seasonally and in the process of decaying consume oxygen dissolved in the water. Lack of oxygen makes a lake or estuary uninhabitable for fish and other aquatic animals. Increased levels of phosphorus are usually responsible for accelerated eutrophication of lakes, while excess nitrogen is often the cause in estuaries. Application of hydrologic analysis to assessing the threat of eutrophication is discussed in Section 10.3.3.

### 10.3.2 Overview of Major Water-Quality Issues

#### Ground Water

An overview of ground-water quality in the United States (U.S. Geological Survey *Water-Supply Paper* 2325, 1988, p. 137) may be paraphrased as follows:

> Overall, the quality of the nation's ground water is good. Virtually all the states note that most of their ground water meets drinking-water standards and that the quality of the ground water is suitable for most uses. However, this assessment is based largely on analyses of inorganic chemicals. Although analyses of toxic constituents and organic chemicals are relatively scarce, available data indicate that these chemicals have contaminated shallow aquifers in many parts of the country. Reports of contamination are likely to increase as the search for contaminants intensifies and as more sophisticated analytical techniques are used to detect trace amounts of these chemicals.
>
> Standards for many toxic and organic chemicals have not been set because the effects of small concentrations of these chemicals on human health and wildlife are unknown. The presence of such chemicals does not necessarily imply a health or environmental threat, but it does raise questions about the source of the chemicals and the possibility that concentrations might increase over time to toxic levels.
>
> Given the very large number of sources of contamination (see Table 10-13), the relatively high susceptibility of shallow aquifers to the effects of waste disposal and other land uses, and the great difficulties in removing contamination once it enters an aquifer, the potential for more ground-water contamination is a very real and serious problem. Fortunately, much can be done to prevent future contamination of ground water as well as mitigate existing contamination.

As indicated in Table 10-13, ground-water contamination from pesticides, petroleum products, and organic industrial compounds is a significant problem in virtually every region of the United States. High concentrations of dissolved solids in ground water due to natural processes, salt-water intrusion due to pumping in coastal regions, and salinization of water applied in irrigation are also serious problems in many regions.

#### Surface Water

A study of water-quality trends in the nation's rivers from 1980 to 1989 (Smith et al. 1993) found that discharges of BOD from industrial and municipal point sources decreased dramatically since the

**TABLE 10-13**

Summary of a State-by-State Survey of Activities Contributing to Ground-Water Contamination. From U.S. Geological Survey *Water-Supply Paper* 2325 (1988).

| Activity | Number of States Reporting | Estimated Number of Sites | Contaminants Frequently Associated |
|---|---|---|---|
| **Waste disposal** | | | |
| Septic systems | 41 | 22 million | Pathogens, nutrients, chloride, organics |
| Landfills | 51 | 16,400 | Dissolved solids, iron, manganese, trace metals, acids, organics, pesticides |
| Impoundments | 32 | 191,800 | Brines, acids, feedlot wastes, trace metals, organics |
| Injection wells | 10 | 280,800 | Dissolved solids, bacteria, sodium, chloride, nutrients, organics, pesticides, acids |
| Land application of wastes | 12 | 19,000 | Bacteria, nutrients, trace metals, organics |
| **Material storage and handling** | | | |
| Underground tanks | 39 | 2.4–4.8 million | Organics and petroleum products |
| Above-ground tanks | 16 | Unknown | Organics, acids, metals, petroleum products |
| Handling and transfers | 29 | 10,000–16,000 spills/yr | Petroleum products, aluminum, iron, sulfate, trace metals |
| **Mining activities** | 23 | 15,000 active; 67,000 inactive | Acids, metals, sulfate, radioactive material |
| **Oil and gas activities** | 20 | 550,000 active; 1.2 million abandoned | Brines |
| **Agriculture** | | | |
| Fertilizer and pesticide application | 44 | 363 million acres | Nutrients, pesticides |
| Irrigation | 22 | 49 million acres | Dissolved solids, nutrients, pesticides |
| Feedlots | 17 | 1,900 | Nutrients, pathogens |
| **Urban activities** | | | |
| Runoff | 15 | 47 million acres | Bacteria, hydrocarbons, dissolved solids, lead, trace metals |
| De-icing chemicals | 14 | Not reported | Sodium chloride, ferrocyanides, phosphate, chromate |
| **Saline intrusion** | 29 | Not reported | Dissolved solids, brines |

passage of the Clean Water Act in 1972. However, there was evidence of only marginal improvement in dissolved-oxygen contents in the rivers receiving those discharges, probably because of an increase in oxygen-demanding wastes from nonpoint sources (especially agricultural lands).

Nonpoint source pollution from agricultural lands is an important problem in virtually every area of the country. Smith et al. (1987a,b) reported that suspended sediment and phosphorus concentrations have been increasing in rivers draining croplands. Agricultural lands are also the principal source of widespread increases in stream nitrate concentrations, but atmospheric deposition of nitrogen compounds from automobile exhausts and industrial combustion is also significant—and increasing. Runoff from urban areas, which often contains high levels of sediment, toxic metals, nutrients, and BOD, is also a significant nonpoint-pollution source in many areas of the country.

Excess nutrients, contributed largely from agricultural and urban nonpoint sources, have made cultural eutrophication a problem in many areas of the United States, including Lake Erie and a portion of Lake Superior. Nitrogen loadings to coastal areas and the Great Lakes have been increasing dramatically, while phosphorus loadings have increased in some areas and decreased in others. The increase in nitrogen compounds is of particular concern in coastal areas, because of the tendency of elevated nitrogen to induce eutrophication in estuaries.

As noted earlier, acid precipitation caused by emissions from industries and automobiles is an increasing threat to water quality in the United States and other regions of the world. Precipitation pH values below about 5.0 are considered abnormally acidic. Where acid precipitation falls on land with geology and soils that have limited capacity for acid neutralization, there is high potential for acid surface waters. Rates of acidic deposition have decreased over large portions of North America and Europe since about 1980, due largely to decreases in sulfur emissions from power plants and other industries (Stoddard et al. 1999). Nitrate deposition has generally leveled off, but not declined. Surface water alkalinity has increased in response to the decline in acid deposition in many regions, but not in large parts of mid-western and eastern United States and southern Canada. A state-by-state summary of surface-water quality problems can be found in U.S. Geological Survey *Water-Supply Paper* 2400 (1993).

### 10.3.3  Examples of Hydrologic Analysis

Here we illustrate how basic hydrologic concepts and relations are applied to the analysis of two important types of surface-water-quality problems: (1) stream dissolved oxygen and (2) lake eutrophication.

#### Stream Dissolved Oxygen

Historically, one of the principal water-quality problems in the United States and other countries has been the low dissolved-oxygen concentrations induced in rivers by the decay of organic wastes, especially sewage and runoff from agricultural lands. When such wastes are introduced into a stream, bacteria begin to oxidize the organic matter, depleting the oxygen that is essential for the waste-purification process and for a healthy stream habitat.

The standard approach to modeling the process of oxygen depletion and reaeration is to use the "Streeter–Phelps Equations" (Box 10-6). These equations define an **oxygen-sag curve** (Figure 10-22) in which dissolved oxygen downstream of a point source of oxygen-demanding wastes at first declines (reflecting the net consumption of oxygen by bacteria) and then rises (reflecting the net reaeration of the stream by dissolution of oxygen from the atmosphere).

Although the dissolved-oxygen model that produces the oxygen-sag curve is not based on hydrologic principles, hydrologic considerations are vital for assessing stream dissolved oxygen. As indicated in

Box 10-6, important variables in the Streeter–Phelps Model are functions of streamflow rate:

- The initial values of dissolved oxygen and BOD depend on the degree to which the wastewater is diluted by the streamflow—the larger the discharge, the greater the dilution and the higher the initial mixed dissolved-oxygen concentration [Equation (10B6-1)].
- The oxidation rate $k_1$ may be related to discharge [Equation (10B6-8)].
- The reaeration rate $k_2$ depends directly on stream velocity and inversely on depth, both of which increase as discharge increases [Equation (10B6-9)].

Of critical interest are the location and magnitude of the point of minimum dissolved oxygen, which can be calculated via the Streeter–Phelps Model. The maximum concentration of oxygen that can be dissolved decreases with water temperature as indicated in Figure 10-23. Usually, at least 4 to 5 mg L$^{-1}$ of dissolved oxygen is required to support a reasonably healthy aquatic habitat. If the concentration drops to zero, the stream becomes anaerobic and incapable of supporting aquatic animals.

Thus dissolved oxygen is a critical water-resource index for water quality, and the Streeter–Phelps Model can be used to develop a relation between minimum dissolved-oxygen concentration and discharge. When coupled with the flow-duration curve, a WRIDC can be developed (as in Figure 10-12a) which can be used to determine the percentage of time the dissolved-oxygen will be less than the minimum allowable level specified by regulation.

#### Lake Eutrophication

As noted previously, cultural eutrophication may occur when a lake receives excess inputs of plant nutrients derived from human activities in the lake's drainage basin. Box 10-7 develops a simple model based on conservation equations (Section 2.3) that gives a picture of how the nutrient concentration in the lake responds to such inputs.

Equation (10B7-6) shows how the concentration of phosphorus in the lake changes with time; Figure 10-24 illustrates the situation. At time $t = 0$, $C_L(t) = C_L(0)$, the initial concentration. At very large times ($t \rightarrow \infty$), $C_L(t)$ approaches a constant

# BOX 10-6
. . . . . . .
## The Streeter-Phelps[1] Analysis of Stream Dissolved Oxygen Response to Biochemical Oxygen Demand

Consider a point source of oxygen-demanding waste-water (e.g., a sewage-treatment plant) discharging to a river (Figure 10-22a). Assuming steady flow in the river and the waste stream and immediate complete mixing of the wastewater, the concentration of BOD in the river at the point of discharge, $B(0)$, is

$$B(0) = \frac{B_u \cdot q_u + B_w \cdot q_w}{q_u + q_w}, \qquad \textbf{(10B6-1)}$$

where the $B$s are BOD concentrations [M L$^{-3}$], the $q$s are discharges [L$^3$ T$^{-1}$], subscript $u$ refers to the river just upstream of the discharge point, and subscript $w$ refers to the waste source. The introduced BOD is oxidized by bacteria making use of (depleting) the dissolved oxygen. This rate of depletion is proportional to the BOD concentration, which leads to

$$B(x) = B(0) \cdot \exp(-k_1 \cdot t) = B(0) \cdot \exp\left(-\frac{k_1}{U} \cdot x\right), \qquad \textbf{(10B6-2)}$$

where $x$ is river distance below the waste source [L], $t$ is time [T], $U$ is the average river velocity [L T$^{-1}$], and $k_1$ is the proportionality between BOD oxygenation rate and concentration (**deoxygenation coefficient**) [T$^{-1}$].

However, as the dissolved oxygen is depleted by the bacteria, it is replaced by reaeration from the air at a rate given by

$$\frac{dc(x)}{dt} = k_2 \cdot [c_s - c(x)], \qquad \textbf{(10B6-3)}$$

where $c(x)$ is the dissolved-oxygen concentration [M L$^{-3}$] at $x$, $c_s$ is the dissolved-oxygen concentration at saturation (a function of temperature), and $k_2$ is the **reaeration coefficient** [T$^{-1}$].

Combining Equations (10B6-2) and (10B6-3) yields

$$c(x) = c_s - \left(\frac{k_1 \cdot B(0)}{k_2 - k_1}\right)$$
$$\cdot \left[\exp\left(-\frac{k_1}{U} \cdot x\right) - \exp\left(-\frac{k_2}{U} \cdot x\right)\right]$$
$$- [c_s - c(0)] \cdot \exp\left(-\frac{k_2}{U} \cdot x\right). \qquad \textbf{(10B6-4)}$$

Equation (10B6-4) gives the concentration of dissolved oxygen downstream from the waste source; this is called the **oxygen-sag curve** (Figure 10-22b). The point of minimum dissolved oxygen, $x_{min}$, and the minimum value of dissolved oxygen at that point, $c_{min}$, are found as

$$x_{min} = \left(\frac{U}{k_2 - k_1}\right)$$
$$\cdot \ln\left\{\left(\frac{k_2}{k_1}\right) \cdot \left[1 - \frac{(k_2 - k_1) \cdot [c_s - c(0)]}{k_1 \cdot B(0)}\right]\right\} \qquad \textbf{(10B6-5)}$$

and

$$c_{min} = c_s - \left(\frac{k_1 \cdot B(0)}{k_2}\right) \cdot \exp\left(-\frac{k_1}{U} \cdot x_{min}\right). \qquad \textbf{(10B6-6)}$$

Both $k_1$ and $k_2$ are functions of temperature. The values for 20° C, designated $k_{1\text{-}20}$ and $k_{2\text{-}20}$, are usually given, and adjustments to a given temperature, $T$, made as

$$k_1 = k_{1-20} \cdot 1.047^{T-20} \text{ and } k_2 = k_{2-20} \cdot 1.024^{T-20}. \qquad \textbf{(10B6-7)}$$

The value of $k_{1\text{-}20}$ can be determined from laboratory studies of the wastewater or can be estimated as

$$k_{1-20} = 1.80 \cdot q_w^{-0.49}, \qquad \textbf{(10B6-8)}$$

where $k_{1\text{-}20}$ is in day$^{-1}$ and $q_w$ is in m$^3$ s$^{-1}$. The value of $k_{2\text{-}20}$ depends on the efficiency of turbulent mixing in the river and can be estimated as

$$k_{2-20} = 4.55 \cdot \frac{U^{0.703}}{Y^{1.054}}, \qquad \textbf{(10B6-9)}$$

where $k_{2\text{-}20}$ is in day$^{-1}$, $U$ is in m s$^{-1}$, and $Y$ is average river depth in m (Huber 1992).

These equations are included in the spreadsheet model DOSag.xls on the CD included with the text.

---

[1] This model was originally developed by Streeter and Phelps (1925). More complete discussions of the model, including incorporating other sources and sinks of oxygen, can be found in Tchobanoglous and Schroeder (1985) and Huber (1992).

**FIGURE 10-22**
Typical dissolved oxygen (DO) relations at a point source of oxygen-demanding wastes. (a) Sewage-treatment plant (STP) discharges wastewater to river at $x = 0$. (b) Plot of DO vs. distance (oxygen-sag curve): Mixed wastewater and river water reduce dissolved oxygen (DO) at $x = 0$ to $c(0)$. The balance between oxygenation of BOD and reaeration causes further DO reduction until a minimum ($c_{min}$) is reached, after which net reaeration gradually increases DO toward saturation. Symbols defined in Box 10-6.

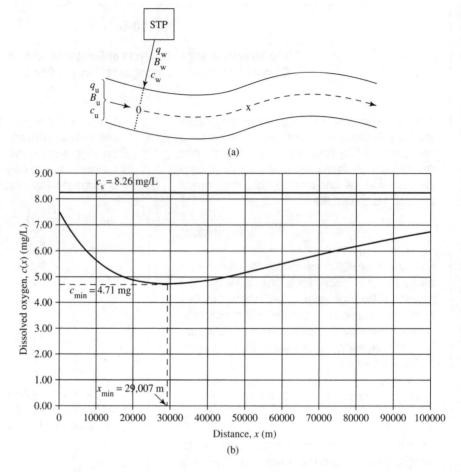

**FIGURE 10-23**
Solubility of oxygen as a function of water temperature.

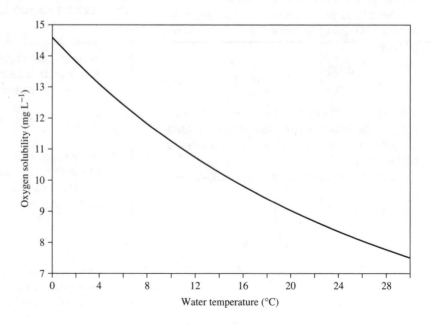

**BOX 10-7**

. . . . . . .

## Simple Lake Eutrophication Model

The water-balance equation for a lake is

$$P + R - E - Q = 0, \qquad \text{(10B7-1)}$$

where $P$ is precipitation rate, $R$ is rate of runoff (stream-flow plus ground water) into the lake, $E$ is evaporation rate, and $Q$ is stream outflow rate. All quantities are expressed as $[L^3 T^{-1}]$ and are long-term averages, so that the rate of change of storage is 0.

The mass-balance equation for a dissolved constituent such as phosphorus can be written in terms of the products of the average concentration $[M\ L^{-3}]$ in each water-balance component and the flow rate $[L^3 T^{-1}]$ of the component. However, since we want to model the change in concentration in the lake, the change-in-storage term is not zero and the balance equation for a constituent is

$$C_P \cdot P + C_R \cdot R - C_Q \cdot Q = \frac{d[C_L(t) \cdot V]}{dt}. \qquad \text{(10B7-2)}$$

Here the $C$-terms represent concentrations for each component, $C_L(t)$ is the average concentration in the lake at time $t$, and $V$ is the volume of the lake. (Evaporation does not appear in Equation (10B7-2) because the evaporating water does not contain any dissolved constituents.)

Equation (10B7-2) can be simplified in two ways: (1) from the rules of derivatives,

$$\frac{d[C_L(t) \cdot V]}{dt} = C_L(t) \cdot \frac{dV}{dt} + V \cdot \frac{dC_L(t)}{dt} \qquad \text{(10B7-3)}$$

and, since we assumed in Equation (10B7-2) that there was no change of storage in the lake, $dV/dt$ must equal 0; and (2) if the lake is reasonably well mixed, we can

assume that the concentration in the water leaving the lake, $C_O$, is the same as that in the lake water, $C_L(t)$. Thus Equation (10B7-2) becomes

$$C_P \cdot P + C_R \cdot R - C_L(t) \cdot Q = V \cdot \frac{dC_L(t)}{dt}. \qquad \text{(10B7-4)}$$

Solving Equation (10B7-1) for $Q$ and substituting the result into Equation (10B7-4) gives

$$C_P \cdot P + C_R \cdot R - C_L(t) \cdot (P + R - E) = V \cdot \frac{dC_L(t)}{dt}. $$

$$\text{(10B7-5)}$$

Equation (10B7-5) can be integrated to give the relation between $C_L(t)$ and $t$ as

$$C_L(t) = C_{eq} + [C_L(0) - C_{eq}] \cdot \exp(-t/T_L), $$

$$\text{(10B7-6)}$$

where we specify the initial concentration in the lake, $C_L(0)$, and define the **equilibrium concentration,** $C_{eq}$ as

$$C_{eq} \equiv \frac{C_P \cdot P + C_R \cdot R}{P + R - E} = \frac{C_P \cdot P + C_R \cdot R}{Q} \qquad \text{(10B7-7)}$$

and the **hydraulic residence time,** $T_L$, as

$$T_L \equiv \frac{V}{P + R - E} = \frac{V}{Q}. \qquad \text{(10B7-8)}$$

These equations are included in the spreadsheet model PolPot.xls on the CD included with the text.

A fuller development of this model, including cases where the concentration in the inputs grows linearly and exponentially with time, was given by Dingman and Johnson (1971).

value, termed the **equilibrium concentration,** $C_{eq}$ [Equation (10B7-7)].

The rate at which the concentration approaches the equilibrium concentration is determined by $T_L$ [Equation (10B7-8)]. This quantity is very similar to the residence time [Equation (2-26)]; it differs only in that the volume of the lake is divided by the

average rate of inflow minus the average rate of evaporation, rather than the average rate of inflow. $T_L$ is called the **hydraulic residence time** and is the average length of time that the water that leaves the lake in liquid form spends in the lake (rather than the average time for all water). It is therefore also the average residence time for dissolved consti-

**FIGURE 10-24**
Increase of pollutant concentration in a lake following an increase in the rate at which the pollutant enters the lake. $C_L(0)$ is the initial concentration and $C_{eq}$ is the equilibrium concentration (See Box 10-7.)

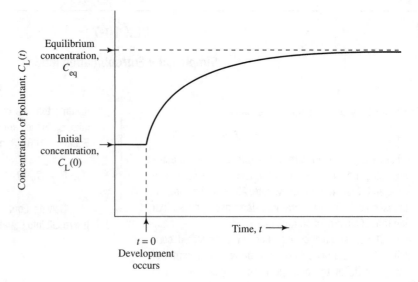

tuents like phosphorus [assuming that Equation (10B7-2) contains all the components of the mass balance for the constituent].

The model of Box 10-7 can also be used to predict how pollutant concentrations in a lake will decrease once action has been taken to reduce the rate at which pollutants are entering. In this case $C_{eq} < C_L(0)$, and the concentration decreases exponentially with time (Figure 10-25). Rainey (1967) used a simplified version of this model to calculate the response of the Great Lakes to a cessation of pollutant inputs. Because of their short hydraulic residence times, he found that the quality of the lower lakes (Erie and Ontario) would improve

markedly within 20 years. However, the quality of Lake Superior, which has a hydraulic residence time of 189 yr, would take centuries to recover.

Example 10-2 applies the model to a temperate lake where, as is often the case, phosphorus is the critical limiting nutrient for eutrophication.

---

### EXAMPLE 10-2

To provide a realistic case, we apply the model to Squam Lake in north-central New Hampshire. Table 10-14 lists the basic hydrologic parameters for the lake and its drainage basin, which we assume is initially undisturbed by human activity. The lake

**FIGURE 10-25**
Decrease of pollutant concentration in a lake following a decrease in the rate at which the pollutant enters the lake. $C_L(0)$ is the initial concentration and $C_{eq}$ is the equilibrium concentration (See Box 10-7.)

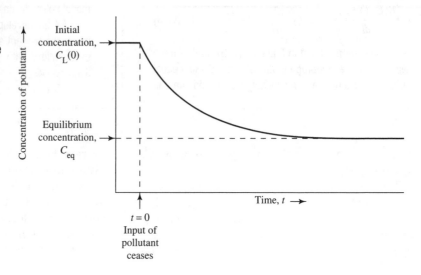

**TABLE 10-14**
Data for Computing Response of Squam Lake, New Hampshire, to Increase in Phosphorus Concentration Due to Development (Example 10-2, Box 10-7).

| Lake Drainage Area ($km^2$) | Lake Area ($km^2$) | Average Depth (m) | $P$ (mm yr$^{-1}$) | $E$ (mm yr$^{-1}$) | $R$ ($m^3$ s$^{-1}$) | $C_P$ (g m$^{-3}$) | $C_L(0)$ (g m$^{-3}$) |
|---|---|---|---|---|---|---|---|
| 84.2 | 27.3 | 11.0 | 1160 | 580 | 1.35 | 0.000 | 0.00399 |

has an initial natural concentration of inorganic phosphorus (as orthophosphate, PO$_4$), of $C_L(0) = 0.00399$ g m$^{-3}$. The critical level of phosphorus at which a lake becomes eutrophic is generally taken to be 0.01 g m$^{-3}$ (Hammer and MacKichan 1981). Suppose a developer proposes to remove the forest and construct an extensive resort community on 70% of the watershed. Will the proposed development elevate the concentration of phosphorus to the point where cultural eutrophication will occur? (Exercise 10-7 asks you to experiment with this model, using Squam Lake as an example.)

**Solution** To calculate the concentration of phosphorus in runoff, $C_R$, resulting from the proposed development, we make use of empirical data in Table 10-15, which indicates that the average export of dissolved phosphorus in runoff from forested land is 4,100 g km$^{-2}$ yr$^{-1}$ and from urban land is 15,000 g km$^{-2}$ yr$^{-1}$. Thus the predicted rate of loading of phosphorus is

$$0.3 \times (84.2 - 27.3) \text{ km}^2 \times 4100 \text{ g km}^{-2} \text{ yr}^{-1} + 0.7$$
$$\times (84.2 - 27.3) \times 15{,}000 \text{ g km}^{-2} \text{ yr}^{-1} = 6.67 \times 10^5 \text{ g yr}^{-1},$$

and the predicted concentration in runoff is found by dividing the loading rate by the average rate of runoff and converting units:

$$C_R = \frac{6.67 \times 10^5 \text{ g yr}^{-1}}{1.35 \text{ m}^3 \text{ s}^{-1}} \times \frac{1 \text{ yr}}{365 \text{ day}} \times \frac{1 \text{ day}}{86{,}400 \text{ s}}$$
$$= 1.56 \times 10^{-2} \text{ g m}^{-3}.$$

To calculate $C_{eq}$ and $T_L$, we need to express $P$ and $E$ in terms of m$^3$ s$^{-1}$ by multiplying the per-unit-area values times the area of the lake:

$$P = 27.3 \text{ km}^2 \times 1.16 \text{ m yr}^{-1} \times$$
$$\frac{10^6 \text{ m}^2}{1 \text{ km}^2} \times \frac{1 \text{ yr}}{365 \text{ day}} \times \frac{1 \text{ day}}{86{,}400 \text{ s}}$$
$$= 1.00 \text{ m}^3 \text{ s}^{-1};$$

and

$$E = 27.3 \text{ km}^2 \times 0.58 \text{ m yr}^{-1} \times$$
$$\frac{10^6 \text{ m}^2}{1 \text{ km}^2} \times \frac{1 \text{ yr}}{365 \text{ day}} \times \frac{1 \text{ day}}{86{,}400 \text{ s}}$$
$$= 0.502 \text{ m}^3 \text{ s}^{-1}.$$

$C_{eq}$ is now found via Equation (10B7-7):

$$C_{eq} = \frac{0 \text{ g m}^{-3} \times 1.00 \text{ m}^3 \text{ s}^{-1} + 0.0156 \text{ g m}^{-3} \times 1.35 \text{ m}^3 \text{ s}^{-1}}{1.86 \text{ m}^3 \text{ s}^{-1}}$$
$$= 0.0114 \text{ g m}^{-3}.$$

Since this value is above the concentration generally assumed to cause eutrophication (0.01 g m$^{-3}$), it appears that the proposed development presents a danger to the lake.

The rate of increase of phosphorus concentration can be calculated from Equation (10B7-6) after computing $T_L$ via Equation (10B7-8). The lake volume, $V$, is the product of the lake area and the average depth:

$$V = 27.3 \text{ km}^2 \times \frac{10^6 \text{ m}^2}{1 \text{ km}^2} \times 11.0 \text{ m} = 3.00 \times 10^8 \text{ m}^3.$$

Then

$$T_L = \frac{3.00 \times 10^8 \text{ m}^3}{(1.00 + 1.35 - 0.502) \text{m}^3 \text{ s}^{-1}} \times \frac{1 \text{ yr}}{365 \text{ day}} \times \frac{1 \text{ day}}{86{,}400 \text{ s}}$$
$$= 5.13 \text{ yr}.$$

**TABLE 10-15**
Average Export of Phosphorus as a Function of Land Use. Data from Omernik (1976).

| Land Use | Average Phosphorus Export Rate as PO$_4$ (g km$^{-2}$ yr$^{-1}$) |
|---|---|
| Forest | 4,100 |
| Forest with minor agriculture and urban | 7,000 |
| Mixed | 7,900 |
| Urban with minor forest and agriculture | 15,000 |
| Agriculture with minor urban and forest | 9,200 |
| Agriculture | 13,300 |

The model developed in Box 10-7 shows that the timing and magnitude of the response of a lake to changes in the inputs of nutrients is determined by the lake's physical hydrology (i.e., its water balance and residence time) as well as by the quality (concentrations) of the inputs. This indicates that climate as well as land-use changes could have direct effects on lake water quality, and these effects could also be explored via this model.

## 10.4 HYDROLOGIC ANALYSIS: FLOODS

The major types of noncoastal water-related hazards are river floods, high lake levels, river and lakeshore erosion, sedimentation, landslides, and ground subsidence. With the exception of subsidence, the magnitude and frequency of most of these hazards are largely determined by regional or local geological conditions and meteorologic events, although human activity can have significant exacerbating or moderating effects. Ground subsidence is usually caused by human activity: the extraction of ground water, which increases the stresses on soil and rock.

Floods are the most destructive natural hazard in the United States (Changnon 1985), and a significant proportion of the hydrologic effort in this country and elsewhere (e.g., National Environment Research Council 1975) has been devoted to their study. Thus although hydrologic concepts and analyses are essential tools in dealing with all water-related hazards, our discussion will focus largely on river flooding.

### 10.4.1 Definitions and Basic Concepts

Most stream channels are incised in floodplains that are relatively flat areas of varying width constructed by the alluvial processes of lateral accretion and overbank deposition. Hydrologically, a flood occurs when the drainage basin experiences an unusually intense or prolonged water-input event and the resulting streamflow rate exceeds the channel capacity. It is important to understand that floods are natural events that occur fairly frequently on virtually all streams—a stream that is unaffected by dams or other hydrologic modifications will typically overflow its banks every one to three years.

From the human viewpoint, however, floods are relatively rare events, and the gentle terrains of floodplains are commonly attractive sites for various types of development. Flooding becomes a management and planning issue when it threatens the human life and property associated with such development. Thus it is important to distinguish between the assessment of the magnitude and frequency of floods (expressed in terms of river discharges or levels) and the magnitude and frequency of the human impacts of floods (usually expressed in terms of dollar damage and lives lost).

**Floodplain management** is a broad term referring to any action intended to reduce future flood damages. In the context of benefit–cost analysis (Box 10-1), the economic benefits attributable to a proposed floodplain-management plan are equal to the difference between flood damages with the plan and those without the plan. These benefits are weighed against the economic costs of implementing the plan and any adverse environmental and social effects. Ideally, the plan with the largest

**TABLE 10-16**
Flood-Damage-Reduction Measures and Their Impacts on Existing and Future Structures.

| Measure | Reduces Damages to | | May Induce Damage-Prone Development |
| --- | --- | --- | --- |
| | **Existing Structures** | **Future Structures** | |
| **Flood Control** | | | |
| Flood-control dams | X | X | X |
| Dikes and levees | X | X | X |
| Channelization | X | X | X |
| **Damage Control** | | | |
| Floodproofing | X | X | |
| Removal of structures | X | | |
| Flood warning | X | X | |
| Floodplain zoning | | X | |

difference between benefits and costs, including noneconomic adverse effects, would be selected for implementation.

The list of alternative measures to be considered in a floodplain-management plan may include some or all of the measures listed in Table 10-16. In evaluating the benefits of a plan, the distinction between damages to existing or future structures is important. The flood-control measures listed in Table 10-16 can reduce damages to existing flood-prone development, but they may actually increase damage to future development. A prime example of this effect occurred along the Mississippi River, where an extensive system of privately and federally funded levees was built beginning in the early 18th century. Many of these levees failed during the extreme flooding of 1993, causing extensive damage to development built under the assumption that it was protected from flooding. Similarly, when a flood-control dam is built the downstream reduction in flood threat often makes the floodplain more attractive to developers. This can lead to more development, so that when a large flood does occur, damages are higher than they would have been without the dam (White 1975). Furthermore, while dikes and levees protect the areas immediately behind them, they can exacerbate flood problems by increasing flood depths and velocities downstream.

Thus in considering floodplain-management alternatives, it is well to keep the following principles in mind:

- Flood-control measures can reduce flood frequencies at specific locations, but they cannot eliminate flooding altogether.
- Flood-control measures generally trade reduced flood frequencies at some locations for increased flood frequencies at others. With a dam, flood frequency is increased in the impoundment of the dam; with levees and channelization, flood frequency is generally increased downstream.
- The effects of flood-control dams on flood frequencies decrease rapidly downstream of the dam (Leopold and Maddock 1954).

The United States has a long history of attempting to deal with flood damages by structural means, and it is now widely accepted that these have seldom been cost-effective. Most planners now encourage some form of land-use control on

the floodplain to prevent damages to future structures, coupled with floodproofing or other local measures to protect existing properties if economically warranted.

### 10.4.2  Overview of Major Flood Issues

In the United States, floods cause on average about $4 billion in property damage and claim about 100 lives annually. They also cause considerable psychic trauma (Changnon 1985).

Table 10-17 lists the major floods in the United States since 1889, and Figure 10-26 shows areas in which flooding causes particularly heavy damages. Flood damages are divided approximately equally between agriculture and urban development (U.S. Water Resources Council 1978), but urban damages are increasing more rapidly. As long ago as 1970, some 53% of developed urban land in the United States was in floodplains (Schneider and Goddard 1974), and that proportion has probably increased since that time. However, as illustrated in Figure 10-27, flood damages have been a problem for many years in communities of all sizes.

In addition to river flooding, excessively high lake levels cause damages in several regions of the United States. At the end of 1985, most of the Great Lakes were at record levels due to a period of above-average precipitation and below-average air temperatures (which lead to lower evaporation) (U.S. Geological Survey 1986). Similar climatic trends produced a rapid rise in the level of the Great Salt Lake in Utah, where 1983 levels reached record highs (U.S. Geological Survey 1985). As a result, increased shoreline erosion and extensive damages to structures occurred in both regions.

A state-by-state summary of damaging historical floods and floodplain-management activities can be found in U.S. Geological Survey *Water-Supply Paper* 2375 (1991).

### 10.4.3  Framework for Analysis of Floodplain-Management Alternatives

We now develop a model that provides a framework for evaluating the reduction in flood damages due to implementation of any combination of the floodplain-management measures listed in Table 10-16. This model can be represented by the coaxial graph shown in Figure 10-28. One version of this graph is construct-

**TABLE 10-17**
Significant Floods in the United States since 1889.[a]

| Year | Date | Location[b] | Type[c] | Deaths | Damages[d] |
|---|---|---|---|---|---|
| 1889 | May | Johnstown, PA | D | 3,000 | ? |
| 1900 | 8 Sep | Galveston, TX [H unnamed] | S | 6,000 | $30 |
| 1903 | May–Jun | lower Missouri & upper Mississippi rivers | R | 100 | $40 |
| 1903 | 14 Jun | Willow Creek, OR | F | 225 | ? |
| 1913 | Mar–Apr | Ohio River | R | 467 | $147 |
| 1919 | 14 Sep | Corpus Christi, TX [H unnamed] | S | 600–900 | $22 |
| 1921 | Jun | Arkansas River, CO | D, F | 120 | $25 |
| 1921 | Sep | Texas | R | 215 | $19 |
| 1927 | Apr–May | Lower Mississippi River | R | 313 | $230 |
| 1927 | Nov | Vermont, New Hampshire | R | 88 | $46 |
| 1928 | 12 Mar | St. Francis Dam, CA | D | 420 | $14 |
| 1928 | 13 Sep | Lake Okeechobee, FL [H unnamed] | R | 1,836 | $26 |
| 1935 | May–Jun | Republican & Kansas rivers, KS | R | 110 | $18 |
| 1936 | Mar | Northeast | R | 107 | $270 |
| 1937 | Jan–Feb | Ohio & lower Mississippi rivers | R | 137 | $418 |
| 1938 | Mar | Southern California | R | 79 | $25 |
| 1938 | 21 Sep | New England [H unnamed] | S, R | 494 | $306 |
| 1939 | Jul | Licking & Kentucky rivers, KY & OH | F | 78 | $2 |
| 1940 | Aug | Virginia, Carolinas, Tennessee | F | 40 | $12 |
| 1947 | May–Jun | Lower & middle Mississippi River | R | 29 | $235 |
| 1951 | Jun–Jul | Kansas & Missouri | R | 15 | $800 |
| 1955 | Aug | New England [H Carol, Diane] | R, S | 187 | $714 |
| 1955 | Dec | West Coast | R | 61 | $155 |
| 1957 | 27–30 Jun | Texas & Louisiana [H Audrey] | R,S | 390 | $150 |
| 1963 | Mar | Ohio River | R | 26 | $98 |
| 1964 | Jun | Montana | R | 31 | $54 |
| 1964 | Dec | California & Oregon | R | 47 | $430 |
| 1965 | Jun | Sanderson, TX | F | 26 | $3 |
| 1965 | Jun | S. Platte & Arkansas rivers, CO | R | 24 | $670 |
| 1965 | 10 Sep | Florida, Louisiana [H Betsy] | S | 75 | $1,420 |
| 1969 | Jan–Feb | California | R | 60 | $399 |
| 1969 | Aug | James River, VA | R | 154 | $116 |
| 1969 | Aug | Gulf Coast [H Camille] | S | 259 | $1,400 |
| 1970 | 30 Jul–5 Aug | Texas [H Celia] | S | 11 | $453 |
| 1971 | Aug | New Jersey | R | 3 | $139 |
| 1972 | 2 Feb | Buffalo Creek, WV | D | 125 | $60 |
| 1972 | 9–10 Jun | Rapid City, SD | F | 237 | $160 |
| 1972 | Jun | Middle Atlantic [H Agnes] | R | 117 | $3,200 |
| 1973 | Spring | Mississippi River basin | R | 33 | $1,155 |
| 1975 | Jun–Jul | Red River of the North, ND, MN, | R | <10 | $273 |
| 1975 | Sep | Northeast [H Eloise] | R,S | 50 | $470 |
| 1976 | 5 Jun | Teton Dam, ID | D | 11 | $400 |
| 1976 | 31 Jul | Big Thompson River, CO | F | 144 | $39 |
| 1977 | Apr | Southern Appalachians | F | 22 | $424 |
| 1977 | 19–20 Jul | Conemaugh River, PA | F | 78 | $300 |
| 1977 | Sep | Kansas City, MO & KS | R | 25 | $80 |
| 1977 | 8 Nov | Toccoa Creek, GA | D | 39 | $3 |
| 1979 | Apr | Mississippi & Alabama | R | 10 | $500 |
| 1979 | Sep | Gulf Coast [H Frederic] | S | 13 | $2,000 |
| 1980 | 18 May | Toute & Cowley rivers, WA [Mt. St. Helens] | V | 60 | ? |
| 1983 | Apr–Jun | Great Salt Lake, UT | R | ? | $621 |
| 1983 | May | Mississippi | R | 1 | $500 |
| 1983 | Sep | Arizona | R | 13 | $416 |
| 1985 | Nov | Virginia, West Virginia | R | 69 | $1,250 |
| 1986 | Winter | California | R | 17 | $270 |
| 1990 | Apr | Texas, Arkansas, Oklahoma | R | 17 | $1,000 |

**TABLE 10-17**
Significant Floods in the United States since 1889. (*continued*)

| Year | Date | Location[b] | Type[c] | Deaths | Damages[d] |
|------|------|----------|-------|--------|----------|
| 1990 | Jun | eastern Ohio | F | 21 | ? |
| 1993 | Jan | Arizona | R | ? | $400 |
| 1993 | May–Sep | Mississippi River basin | R | 48 | $20,000 |
| 1995 | May | South Central | R | 32 | $5,500 |
| 1995 | Jan–Mar | California | R | 27 | $3,000 |
| 1996 | Feb | Pacific Northwest | R | 9 | $1,000 |
| 1996 | Dec | Pacific Northwest | R | 36 | $2,500 |
| 1997 | Mar | Ohio River basin | R | >50 | $500 |
| 1997 | Apr–May | Red River of the North, ND, MN | R | 8 | $2,000 |
| 1999 | Sep | Eastern North Carolina [H Floyd] | R | 42 | $6,000 |

[a]Data from National Oceanographic and Atmospheric Administration, *Climatological Data, National Summary* (1977) 28 (3); Keller (2000), and Perry (2000).

[b]"H *name*" indicates hurricane.

[c]R = regional flood; F = flash flood; D = dam-break flood; S = storm surge; V = flood induced by volcanic eruption.

[d]Millions (uninflated).

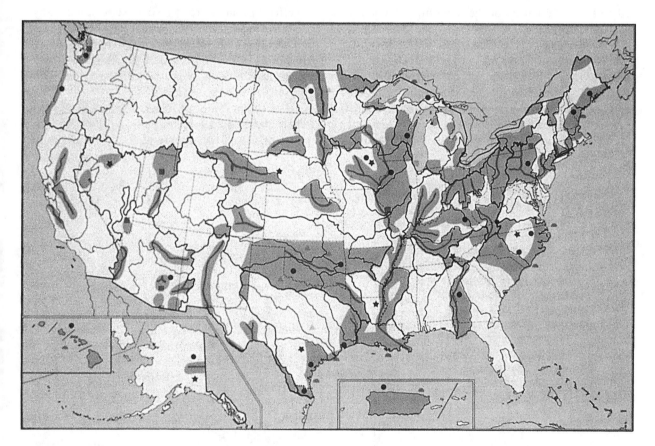

**FIGURE 10-26**
Shaded areas are those identified by the U.S. Water Resources Council (1978) as subject to major flood damages. Heavy lines are major streams with periodic damages due to overbank flooding. Symbols show locations of specific problems described in the report.

**FIGURE 10-27**
Flooding and flood susceptibility, downtown Lisbon, NH (population 1400). (a) The record 1927 flood on the Ammonoosuc River. (b) Levels of floods with indicated return periods. From U.S. Natural Resources Conservation Service (1983), courtesy of George Stevens and Gerry Lang.

ed to show the effects of the measures included in each plan at each of several selected future times.

Analysis of damage reduction begins with the relation between flood magnitude and exceedence probability in the upper right quadrant. Methods for constructing this relation when flood frequencies are not to be modified (e.g., by construction of dams) are based on statistical analyses of existing data as discussed in the subsequent section. Where

the effects of dams are to be evaluated, detailed hydrologic studies based on climatological analysis of precipitation (Section 4.4.4) and perhaps snowmelt (Section 5.5.2), runoff models (Section 9.5), reservoir-storage models, and streamflow-routing models (Section 9.3) are required.

The magnitude of flood damages depends largely on the depth of flooding, so the upper left quadrant of Figure 10-28 relates flood depth to

**FIGURE 10-28**
Coaxial graph showing how relations between flood discharge and exceedence probability (upper-right quadrant), discharge and depth (upper left), and depth and flood damages (lower left) are combined to develop the damage-exceedence probability relation (lower right). The method is identical to that described in Box 10-5 for constructing monotonic-increasing WRIDCs.

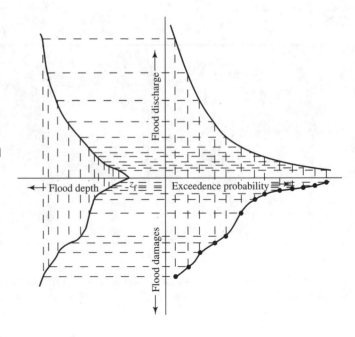

flood discharge. This relation is usually developed via detailed surveys of the stream and floodplain and application of hydraulic principles. [See, for example, Dingman (1984).] The relation between the depth of water in a stream, $Y$, and discharge, $Q$, can usually be expressed by an equation of the form

$$Y = c \cdot Q^f, \qquad \text{(10-15)}$$

where $c$ and $f$ are empirical constants that apply to a particular location. Then the elevation of the water level, $Z_w$, can be calculated as

$$Z_w = Z_b + Y, \qquad \text{(10-16)}$$

and the depth of flooding, $Z_f$, as

$$Z_f = \begin{cases} Z_w - Z_{bf} & \text{for} \quad Z_w > Z_{bf}; \\ 0 & \text{otherwise,} \end{cases} \qquad \text{(10-17)}$$

where $Z_b$ is the elevation of the bottom of the channel and $Z_{bf}$ is the elevation at which flooding begins (bankfull elevation for the channel). The curve in the upper left quadrant of Figure 10-28 is a typical relation between $Z_f$ and $Q$.

The lower left quadrant of Figure 10-28 relates flooding depth to damages for the particular floodplain. Information to define such relations is based on field or aerial-photographic surveys of the types and locations of existing buildings or on plans or projections for future developments. Thus the form of the depth–damage curve varies widely from one location to another.

The curve in the lower right quadrant of Figure 10-28 relates damages to exceedence frequency. This curve is established from the positions of the curves in the other three quadrants exactly as described for constructing water-resource-index duration curves in Box 10-5.

Once the damage–frequency relation has been established, we can calculate the average annual damages. Conceptually, this average value is proportional to the area between the damage–frequency curve and the frequency axis. If both scales of the damage–frequency curve are arithmetic, this area can be measured directly using a planimeter or digitizing device; alternatively, it can be estimated from points on the curve as

$$\mu_D = \sum_{i=1}^{n} [EP(D_i) - EP(D_{i-1})] \cdot \left( \frac{D_i + D_{i-1}}{2} \right), \qquad \text{(10-18)}$$

where $\mu_D$ is the average annual damages, $EP(D_i)$ is the exceedence probability for damage value $D_i$, and the subscript $i$ indicates individual points (increasing from 1 to $n$, from left to right).

We now show how this coaxial model can be used to estimate the damage reduction due to the various floodplain-management alternatives listed in Table 10-16.[6] The references in brackets in the following discussion indicate the quadrant of Figure 10-29 in which each effect occurs.

**Flood-Control Dams:** Flood-control dams reduce the peak-flood discharge associated with a given exceedence probability at locations downstream from the dam (upper right). Their effects decrease rapidly with downstream distance, and tend to be relatively less for larger floods.

**Dikes and Levees:** Dikes and levees maintain a zero level of flooding in the area behind them until they are overtopped or fail (upper left). As noted previously, the constriction caused by dikes and levees can increase downstream flood levels and velocities.

**Channelization:** Channelization involves straightening, deepening, or lining the stream channel so that it can accommodate larger discharges without overtopping its banks. Thus its effect is to lower the flood depth associated with a given discharge (upper left). Experience has shown that channelization is seldom warranted because its beneficial effects are temporary and its adverse environmental effects are often severe. Indeed, stream-restoration projects are underway in many regions to undo the environmental insults produced by channelization projects.

**Floodproofing:** Floodproofing is the construction of buildings such that they can accommodate flooding with minimal damage. Such construction can involve "retrofitting" existing buildings, or designing future buildings to withstand floods. Its effect is to reduce the damages associated with a given level of flooding (lower left).

**Removal of Structures:** Damages associated with a given flood level can also be reduced by removing structures from the floodplain (lower left).

---

[6] Exercises 10-9 and 10-10 give you an opportunity to apply this methodology (in tabular rather than graphical form) using data from an actual situation.

**FIGURE 10-29**
Use of a coaxial graph to estimate the reduction in damages expected from construction of flood-control dams, construction of dikes or levees, channelization, floodproofing, removal of structures, flood warning, and flood-plain zoning. See text.

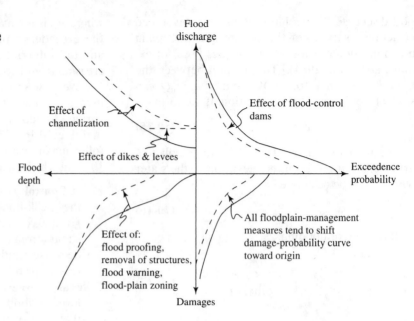

Flood discharge

Effect of channelization

Effect of flood-control dams

Effect of dikes & levees

Flood depth

Exceedence probability

Effect of:
flood proofing,
removal of structures,
flood warning,
flood-plain zoning

All floodplain-management measures tend to shift damage-probability curve toward origin

Damages

**Flood Warning:** Timely warnings of imminent floods can reduce damages by giving the opportunity to remove or provide temporary protection for damageable property (lower left). However, warnings are probably most valuable in reducing loss of lives. Effective warning systems can be readily developed for large rivers that respond relatively slowly and predictably to rain and snowmelt. Warnings for smaller streams, which tend to be "flashy" (i.e., respond quickly to rainfall), present a challenge to meteorologists and hydrologists.

**Floodplain Zoning:** Floodplain zoning is used here as a general term for any form of land-use control that limits future damageable developments on floodplains. Thus its effect is to reduce the damages associated with a given flood level (lower left).

Each of these measures moves one of the curves in Figure 10-29 such that, singly or in combination, they shift the damage–frequency curve (lower right) toward the origin, reducing the area underneath that curve and hence reducing the average annual damages. As noted earlier, decisions as to which, if any, of the measures should be implemented are based on a comparison of this damage reduction to the economic, environmental, and social costs of implementation.

### 10.4.4 Flood-Frequency Analysis

Physical-hydrologic analysis of floods is based on understanding the meteorology and climatology of precipitation (Section 4.4.4) and snowmelt (Section 5.4), and the mechanisms by which water travels to streams to induce response hydrographs (Section 9.2). Such analyses are particularly important in the development of models for predicting the flood response to a specified hypothetical input event and for forecasting the flood peak that will result from an actual event. Here we focus on the statistical analysis of floods (prediction of the first type as defined in Section 2.9.2) to provide information for the design of floodplain-management strategies. The basic objective is to estimate the magnitude–frequency relation that forms the basis for evaluating floodplain-management alternatives in Figures 10-28 and 10-29. Approaches to flood characterization differ depending on whether a record of measured streamflows long enough for statistical analysis is available for the stream reach of interest.

#### Flood Analysis at Gaging Stations

In the United States, the U.S. Geological Survey (Department of the Interior) operates a network of some 9,000 stream-gaging stations. Where a sufficiently long (10 yr would be an absolute minimum)

record of measured streamflows exists, analysis of floods is based on the statistical characterization of time series of instantaneous peak flows.

Most commonly, such time series are constructed as **annual series** by selecting the highest instantaneous peak flow in each year of record ($M(Q)$ in Figure 2-5, with $\Delta t = 1$ yr). Alternatively, a **partial-duration series** consisting of all flood peaks over a specified threshold value (typically, the value at which overbank flow begins) is used. The relation between the two types of series is discussed by Dunne and Leopold (1978); we confine our discussion to annual series. In any case, peak flows can be obtained for analysis from the U.S. Geological Survey via the Internet, as described for average daily flows in Box 2-2. Usually the standard U.S. Geological Survey water year (1 October–30 September) is used. (See Section 2.5.2.)

Magnitude–exceedence probability relations for annual flood peaks are estimated as described in Box 10-8, and Exercise 10-8 gives you a chance to apply the method to annual flood data. It should be noted that these methods are based on the assumption that annual floods are statistically independent

---

## BOX 10-8

· · · · · · · ·

### Estimation of Exceedence Probability at Gaging Stations

We assume we have a time series of annual flood peaks or annual minimum $d$-day flows, $q_j$, for each of $j = 1, 2, ...., N$ years. There are two basic approaches to estimating exceedence probability from such a sample: (1) non-parametric and (2) parametric.

**Non-Parametric Approach**

This approach is essentially identical to that described in Box C-1. It does not assume a probability distribution for flood peaks or minimum $d$-day flows and consists of three steps:

1. Rank the flows from lowest ($i = 1$) to highest ($i = N$). The $i$th-ranked flow is designated $q_{(i)}$.
2. Estimate the quantile value for each flow value in the sample from a plotting-position formula (Equation CB1-2 or CB1-3). For low flows, we are interested in the non-exceedence probability, $F_Q(q_{(i)})$. For flood peaks, we are interested in the exceedence probability, $EP_Q(q_{(i)})$ or return period, $TR_Q(q_{(i)})$. The relations among these quantities are given in Equations (C-31) and (C-32).
3. Plot the $q_{(i)}$, $EP_Q(q_{(i)})$ (floods) or $q_{(i)}$, $F_Q(q_{(i)})$ (low flows) values on a graph and fit a smooth curve through the points to allow interpolation.

In Step 3, it is best to use a special type of graph paper called **probability paper,** which is designed such that, if $Q$ follows a normal distribution (see Section C.7), a plot of $q_{(i)}$ vs. $EP_Q(q_{(i)})$ or $F_Q(q_{(i)})$ will define a straight line. Even if the values are not normally distributed, such paper is useful because the probability scale is expanded at the low and high ends, allowing for easier reading and more accurate interpolation of values in the ranges that are of most interest. However, estimation of exceedence or non-exceedence probabilities, like quantiles, is very uncertain in the highest and lowest ranges [Equation (C-47)], and extrapolation to values greater than $q_{(N)}$ or less than $q_{(1)}$ is not advisable.

**Parametric Approach**

This approach involves fitting an appropriate probability distribution to the data. Basic approaches to this are described in Section C.5. A regional approach to determining the appropriate distribution, rather than analysis of individual station data, is strongly recommended; see Hosking and Wallis (1997) and Stedinger et al. (1992). The advantage of the parametric approach is that one can extrapolate beyond the actual data to estimate the high (floods) or low (low flows) exceedence-probability values. However, such extrapolations are still subject to considerable uncertainty because one cannot be sure that the assumed distribution is the correct one.

As discussed in Section C.9, it is always wise to determine whether the data have significant persistence (Box C-6) before carrying out the above procedures. If persistence is present, exceedence-probability estimates are not reliable.

(i.e., that the magnitude of one year's peak is not related to the magnitude of the preceding year's peak) and that there are no trends or periodicities in the data induced by climate or land-use changes. Hydrologists are increasingly questioning these assumptions as evidence of hydrologic impacts of large-scale climate fluctuations such as the El-Nino-Southern Oscillation and human-induced global warming accumulates (Sections 3.1.4 and 3.2.9; Box 3-3). However, Douglas et al. (2000) found no trends in flood magnitudes in the United States except in the Midwest.[7]

### Flood Analysis at Ungaged Reaches

It is often necessary to develop flood magnitude–exceedence probability relations for stream reaches where no streamflow measurements have been made. Development of such relations begins with the magnitude–exceedence probability relations previously established for a number of gaging stations in the region using the methods of Box 10-8. Equations relating the magnitudes of floods with specified exceedence probabilities at those gaging stations to measurable characteristics of their drainage basins (e.g., area, slope, location, elevation, geology) or channel size (e.g., bankfull width) are then usually developed by regression analysis [discussed in most statistics texts; e.g., Haan (1977)]. The resulting equations are then applied to ungaged locations.

Box 10-9 gives an example of regression relations derived for New Hampshire and Vermont; relations for all 50 states are summarized in the U.S. Geological Survey's *Water-Resources Investigations Report* 94-4002 (1994).

Other approaches to developing magnitude–probability relations for floods involve the use of rainfall–runoff models such as the Rational Method or the U.S. Soil Conservation Service (SCS) method described in Section 9.6.2.

---

[7] One test for independence, which should be applied before doing the probability analysis described in Box 10-8, is to compute the autocorrelation coefficient and test whether it is significantly different from 0, as described in Box C-6. However, it is also important to account for spatial correlation (Douglas et al. 2000).

## 10.5 HYDROLOGIC ANALYSIS: LOW STREAMFLOWS AND DROUGHTS

Characterization of the magnitude, frequency, and duration of low streamflows and droughts is vital for assessing the reliability of flows for all instream and withdrawal uses and for defining resource shortages and drought shortages (Table 10-5).

### 10.5.1 Definitions and Basic Concepts

#### Low Streamflows

The objective of low-flow analysis is to estimate the frequency or probability with which streamflow in a given reach will be less than various levels. Thus the flow–duration curve (Figure 10-11) is an important tool of low-flow analysis; from it one can readily determine the flow associated with any exceedence or non-exceedence probability. As noted, the flow exceeded 95% of the time, $q_{.95}$, is a useful index of water availability that is often used for design purposes.

For purposes of statistical analysis, low flows are defined as annual minimum flows averaged over consecutive-day periods of varying length. The most commonly used averaging period is $d = 7$ days, but analyses are often carried out for $d = 1, 3, 15, 30, 60, 90,$ and 180 days as well. Low-flow quantile values are cited as "$dQp$," where $p$ is now the annual *non*-exceedence probability (in percent) for the flow averaged over $d$ days. The 7-day average flow that has an annual non-exceedence probability of 0.10 (a recurrence interval of 10 yr), called "*7Q10*," is commonly used as a low-flow design value. The *7Q10* value is interpreted as follows:

*In any year, there is a 10% probability that the lowest 7-consecutive-day average flow will be less than the 7Q10 value.*

The other dQp values have analogous interpretations.

Methods for determining *dQp* values are described in Section 10.5.3.

#### Droughts

Droughts are extended severe dry periods. To qualify as a drought, a dry period must have a duration of at least a few months and be a significant departure

---

## BOX 10-9

· · · · · · · ·

### Estimation of Flood Magnitude–Frequency Relations at Ungaged Locations

Dingman and Palaia (1999) compared three approaches to estimating flood quantiles ($QT$, where $T$ is return period) at ungaged locations in New Hampshire and Vermont:

Model 1:   $QT = f(\text{bankfull channel width})$;
Model 2:   $QT = f(\text{drainage area, mean elevation})$;

and

Model 3:   $QT = f(\text{drainage area})$.

Each model was developed via systematic regression analysis. Other predictors, such as slope and 2-yr, 24-hr rainfall were tested, but did not prove useful.

    The prediction equations for the three models are given in the following table in which discharges are in ft$^3$ s$^{-1}$, $W$ is bankfull width in ft, $A$ is drainage area in mi$^2$, and $Z$ is mean basin elevation in ft.

    Figure 10-30 compares the uncertainties of the estimates obtained with each model and the uncertainties of estimates based on parametric analysis of gaging-station data (Box 10-8). Clearly, estimates based on actual data are most precise, and precision increases with record length. However, where gage data are lacking, estimates based on field measurement of bankfull width are superior to those based on drainage area and elevation. (This has been found by other studies, too.)

The U.S. Geological Survey, in cooperation with state water agencies and/or highway departments has developed similar equations for most states; references to these are listed in Ponce (1989).

| Return Period, $T$ (yr) | Model: Equation | $r^2$ |
|---|---|---|
| | 1: $Q2.33 = 0.874 \cdot W^{1.82}$ | 0.923 |
| 2.33 | 2: $Q2.33 = 38.7 \cdot A^{0.792} \cdot 10^{0.000254 \cdot Z}$ | 0.929 |
| | 3: $Q2.33 = 85.1 \cdot A^{0.832}$ | 0.854 |
| | 1: $Q10 = 1.74 \cdot W^{1.78}$ | 0.921 |
| 10 | 2: $Q10 = 74.8 \cdot A^{0.750} \cdot 10^{0.000269 \cdot Z}$ | 0.894 |
| | 3: $Q10 = 173.4 \cdot A^{0.792}$ | 0.806 |
| | 1: $Q20 = 2.33 \cdot W^{1.76}$ | 0.912 |
| 20 | 2: $Q20 = 97.6 \cdot A^{0.792} \cdot 10^{0.000276 \cdot Z}$ | 0.877 |
| | 3: $Q20 = 231.2 \cdot A^{0.774}$ | 0.782 |
| | 1: $Q50 = 3.32 \cdot W^{1.73}$ | 0.895 |
| 50 | 2: $Q50 = 134 \cdot A^{0.708} \cdot 10^{0.000284 \cdot Z}$ | 0.852 |
| | 3: $Q50 = 325.8 \cdot A^{0.752}$ | 0.748 |
| | 1: $Q100 = 4.27 \cdot W^{1.71}$ | 0.878 |
| 100 | 2: $Q100 = 168 \cdot A^{0.691} \cdot 10^{0.000289 \cdot Z}$ | 0.832 |
| | 3: $Q100 = 415.9 \cdot A^{0.736}$ | 0.721 |

---

from normal. Droughts must be expected as part of the natural variability of climate, even in the absence of any long-term climate change. However, "permanent" droughts due to natural climate shifts do occur, and appear to have been responsible for large-scale migrations and declines of civilizations through human history. The possibility of regional droughts associated with climatic shifts due to human-induced global warming cannot be excluded.

    As shown in Figure 10-31, droughts begin with a deficit in precipitation that is unusually extreme and prolonged relative to the usual climatic conditions (**meteorological drought**). This is often, but not always, accompanied by unusually high temperatures, high winds, low humidity, and high solar radiation that result in increased evapotranspiration.

These conditions commonly produce extended periods of unusually low soil moisture, which affect agriculture and natural plant growth and the moisture of the forest floor (**agricultural drought**). As the precipitation deficit continues, stream discharge, lake, wetland, and reservoir levels, and water-table elevations decline to unusually low levels (**hydrological drought**). When precipitation returns to more normal values, drought recovery follows the same sequence: meteorological, agricultural, hydrological.

    Meteorological drought is usually characterized as a precipitation deficit. For example, Figure 10-32 shows the difference between actual precipitation and normal (i.e., climatic average) precipitation during the severe drought of the 1960s in the

**FIGURE 10-30**
Comparison of uncertainty (percent of true value) in estimating flood quantiles at stream gages with 88-, 47.5-, and 22-yr records and at ungaged sites using the three statistical models discussed in Box 10-9. From Dingman and Palaia (1999), used with permission.

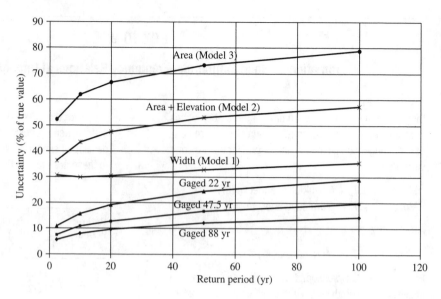

northeastern United States. In the most severely affected area, the deficit amounted to about one-year's precipitation out of a four-year period. (Compare to Figure 4-38.)

A commonly used measure of agricultural drought, and drought generally, is the **Palmer Drought Severity Index** (PDSI) and its variants. The PDSI is computed from a conceptual model similar to the Thornthwaite model (Box 7-3) that uses actual precipitation and temperature data to compute the water balance for a climatic region, usually on a monthly basis. The values are cumulated and compared with climatic normals to compute a number between −6 (extreme drought) and +6 (extreme wetness). Although these computed values have a variable and often uncertain relation to other measures and perceptions of drought[8] (Guttman et al. 1992), the PDSI is widely used because it can be computed from readily available data and used to develop long time series for climatic analysis (Figure 10-33a).

Hydrological droughts usually have the broadest and most severe impacts, and are our principal focus here. Hydrologic droughts are described and quantified in terms of their severity, duration, and intensity, as described in more detail in Section 10.5.4. In contrast with floods, droughts usually

have subtle beginnings and endings that may be difficult to detect except in retrospect, and they usually affect larger areas than do floods.

### 10.5.2 Overview of Major Low-Flow and Drought Issues

As shown in Figures 3-32 to 3-34, streamflow variations tend to be synchronous over wide regions. This is further illustrated in Figures 10-33 and 10-34. Figure 10-33a is a time series of the percentage of the area of the conterminous United States with PDSI values < −4, denoting "severe" or "extreme" drought. The 1930s were the driest period in the last century, with almost two-thirds of the country in extreme or severe drought in 1934 (Figure 10-34a). Drought conditions were also widespread in the mid-1950s and in several other periods.

Figure 10-33b shows the number of states in the 48 conterminous United States for which the U.S. Geological Survey reported "major and other memorable" droughts for each year during the period 1920–1988. There have been some six multi-year periods since 1920 when half or more of the United States have experienced droughts; thus the frequency of such events is about one per decade. The minimum proportion of drought years reported was 26.5% (Arizona and Rhode Island), 19 states reported droughts in more than half the years, and one state (New Mexico) reported droughts in over 72% of the years!

---

[8] In part, this is because they do not distinguish between precipitation in the form of rain and snow (Section 4.1.4).

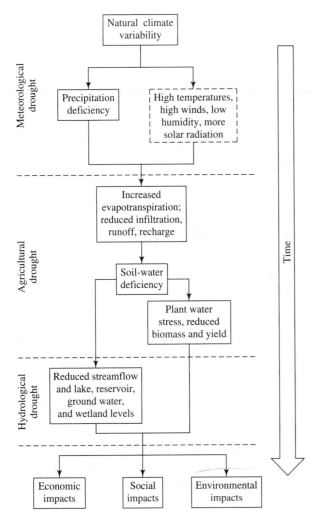

**FIGURE 10-31**
Sequence of drought impacts. Modified after National Drought Mitigation Center Web page (http://enso.unl.edu/ndmc, 2000).

Figure 10-34 shows the PDSI for the conterminous United States for three years of significant drought: 1934, when the drought of the 1930s was at its most intense and widespread and the Dust Bowl was causing widespread social disruption; 1965, when the water supplies of the urban areas of the Northeast were severely impacted; and 1988, when flows on the Mississippi River were at record lows and river navigation had to be suspended.

Clearly drought—and the perception of drought—is a common phenomenon in both time and space. As discussed in Section 10.6, growing population and demand for water will mean that the socio-economic impacts of droughts will steadi-

ly increase in the coming years whether or not the frequency and severity of droughts increases.

### 10.5.3   Low-Flow Frequency Analysis

As noted earlier, the objective of low-flow frequency analysis is to estimate quantiles of annual $d$-day-average minimum flows. As with floods, such estimates are usually required for reaches without long-term streamflow records. These estimates are developed by first analyzing low flows at gaging stations.

#### *Low-Flow Analysis at Gaging Stations*

For a gaged reach, low-flow analysis involves development of a time series of annual $d$-day low flows, where $d$ is the averaging period. As shown in Table 10-18, the analysis begins with a time series of average daily flows for each year.[9] Then, overlapping $d$-day averages are computed for the $d$ values of interest. For each value of $d$, this creates $365 - (d - 1)$ values of consecutive $d$-day averages for each year.[10] The smallest of these values is then selected to produce an annual time series of minimum $d$-day flows. It is this time series that is then subjected to frequency analysis to estimate the quantiles of the annual $d$-day flows, as described in Box 10-8.

#### *Low-Flow Analysis at Ungaged Reaches*

As with floods, estimates of low-flow quantiles are usually required for stream reaches where there are no long-term gaging-station records. There are two basic approaches to developing such estimates:

1. Relate $dQp$ values to drainage-basin characteristics via regression analysis, as described for floods in Box 10-9. For example, in New Hampshire and Vermont, Dingman and Lawlor (1995) found that $7Q10$ (ft$^3$ s$^{-1}$) could be satisfactorily estimated from drainage area ($A$, mi$^2$), mean drainage-basin elevation ($Z$, ft a.s.l.), and the fraction of the drainage basin covered by sand-and-gravel deposits ($D$):

---

[9] In order to minimize the likelihood of a year ending during an annual minimum, it is standard practice in the United States to base the analysis on the "low-flow year" of 1 April–30 March rather than the usual 1 October–30 September water year.
[10] $366 - (d - 1)$ values for leap years.

**FIGURE 10-32**
Precipitation deficiency (normal precipitation minus actual precipitation) in the Northeast from October 1961 through December 1965 (inches). Climatic normal precipitation for the region is 40 to 45 in. yr$^{-1}$ (Figure 4-38).

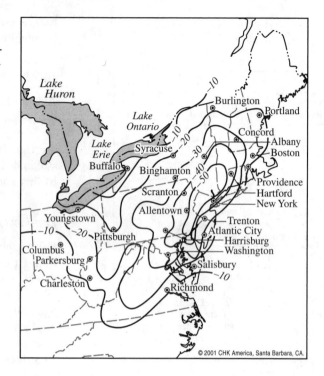

$$7Q10 = 0.109 \cdot A^{1.25} \cdot 10^{0.000174 \cdot Z + 0.647 \cdot D}.$$

**(10-19)**

Studies in other regions have related various low-flow quantiles to these and/or other drainage-basin characteristics.

2. During the low-flow season, make a number of spot measurements of discharge at the ungaged stream reach where the $dQp$ estimate is needed. Then relate those flows to concurrent flows at a nearby gaging station using

$$q_u = a + b \cdot q_g,$$

**(10-20)**

where $q_u$ is the flow at the ungaged site, $q_g$ is the concurrent flow at the gaged site, and $a$ and $b$ are estimated via regression analysis (Figure 10-35). Then estimate the $dQp$ at the ungaged site, $dQp_u$, as

$$dQp_u = a + b \cdot dQp_g,$$

**(10-21)**

where $dQp_g$ is the $dQp$ value established by frequency analysis at the gaged site.[11] In order to minimize errors when using this

procedure, each pair of flows used to establish Equation (10-20) should be from a separate hydrograph recession, the $r^2$ value for the relation of Equation (10-20) should be at least 0.70, and the two basins should be similar in size, geology, topography, and climate.

### 10.5.4 Drought Analysis[12]

The objective of drought analysis is to characterize the magnitude, duration, and severity of meteorological, agricultural, or hydrological drought in a region of interest. The analysis process can be structured in terms of five questions:

1. What type of drought is of interest?
2. What averaging period will be used?
3. How will "drought" be quantitatively defined?
4. What are the magnitude–frequency relations of drought characteristics?
5. How are regional aspects of drought addressed?

---

[11] In many cases, using the logarithms of the flows in Equations (10-20) and (10-21) gives better results than using the flows themselves.

[12] This discussion closely follows the paper by Dracup et al. (1980).

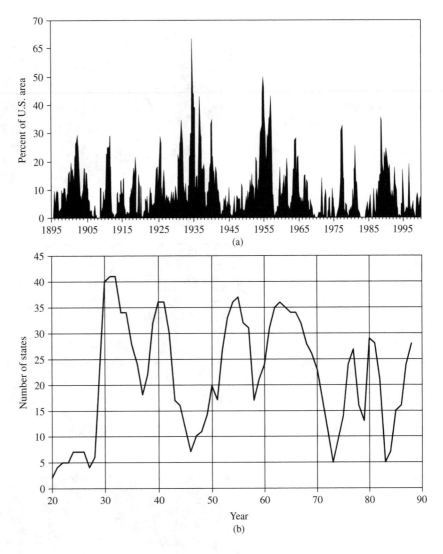

**FIGURE 10-33**
(a) Percent area of the conterminous United States in severe and extreme drought (Palmer Drought Severity Index < −4), 1895–1999. From National Drought Mitigation Center Web page (http://enso.unl.edu/ndmc, 2000). (b) Number of states experiencing "significant" drought, 1920–1988. Data from U.S. Geological Survey *Water-Supply Paper* 2375.

## Drought Type

As noted, one may be interested in one or more of the basic types of drought, each reflected in time series of particular types of data: meteorological (precipitation); agricultural (soil moisture or PDSI); or hydrological (streamflow, reservoir levels, or ground-water levels).

## Averaging Period

As with time-series analysis generally (Section 2.7.1), drought analysis requires selection of an averaging period ($\Delta t$). Since droughts by definition have significant duration, one would usually select $\Delta t = 1$ month, 3 months (a season), or 1 yr, with the choice depending on the available data and the purposes of the analysis. For a given record length the

selection of $\Delta t$ involves a trade-off in the uncertainty of the analysis (Sections C.8 and C.9):

$$\text{longer } \Delta t \nearrow \begin{array}{l} \text{smaller sample size} \rightarrow \text{greater} \\ \text{uncertainty} \end{array}$$
$$\searrow \begin{array}{l} \text{lower autocorrelation} \rightarrow \text{smaller} \\ \text{uncertainty} \end{array}$$

These effects are illustrated in Table 10-19.

## Drought Definition

Figure 10-36 shows a time series of a selected quantity, $X$ (e.g., precipitation, PDSI, streamflow, ground-water level), averaged over an appropriate $\Delta t$. The quantitative definition of drought is determined by the **truncation level,** $X_0$, selected by the analyst: Values of $X < X_0$ are defined as droughts. Typical values for $X_0$ might be

**FIGURE 10-34**
Palmer Drought Severity Index values for three years of significant drought in the conterminous United States. (a) Nationwide 1934; (b) Northeast 1965; (c) Upper Midwest 1988. From NOAA Global Data Center Web site (http//www.ngdc.noaa.gov, 2000).

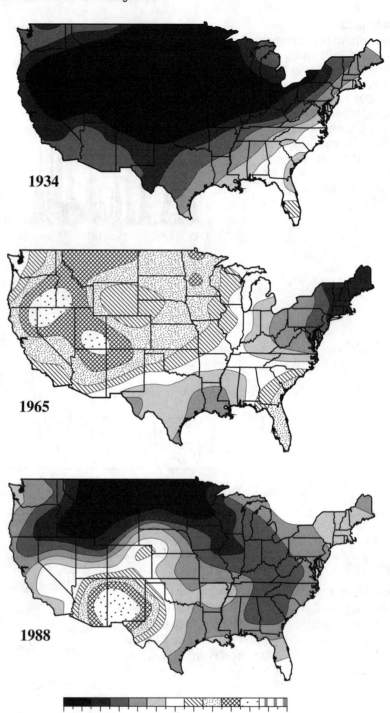

1934

1965

1988

-6    -4    -2    0    2    4    6

© 2001 CHK America, Santa Barbara, CA.

**TABLE 10-18**

Example of Computation of *d*-Consecutive-Day Averages for Low-Flow Analysis. Values in **bold** are minimums for the period shown.

| Day | 1-Day Average | 3-Day Average | 7-Day Average | 15-Day Average |
|-----|--------------|---------------|---------------|----------------|
| 1 | 203 | | | |
| 2 | 196 | 197 | | |
| 3 | 193 | 192 | | |
| 4 | 188 | 189 | 189 | |
| 5 | 185 | 185 | 185 | |
| 6 | 182 | 181 | 187 | |
| 7 | 176 | 178 | 195 | |
| 8 | 176 | 187 | 201 | 198 |
| 9 | 208 | 211 | 204 | 197 |
| 10 | 249 | 230 | 206 | 199 |
| 11 | 234 | 229 | 207 | 202 |
| 12 | 204 | 210 | 209 | 204 |
| 13 | 192 | 195 | 203 | 204 |
| 14 | 188 | 190 | 205 | 204 |
| 15 | 191 | 191 | 206 | 203 |
| 16 | 194 | 202 | 206 | 203 |
| 17 | 220 | 221 | 204 | 200 |
| 18 | 248 | 227 | 201 | 196 |
| 19 | 212 | 246 | 198 | **194** |
| 20 | 188 | 192 | 190 | 195 |
| 21 | 175 | 178 | **181** | 195 |
| 22 | 171 | 173 | **181** | 215 |
| 23 | 172 | **170** | 184 | 240 |
| 24 | **167** | 173 | 187 | |
| 25 | 181 | 187 | 231 | |
| 26 | 213 | 202 | 288 | |
| 27 | 211 | 207 | | |
| 28 | 197 | 296 | | |
| 29 | 479 | 414 | | |
| 30 | 567 | | | |

**FIGURE 10-35**

Concurrent flows for 16 different base-flow periods at two reaches in Tennessee. Shoal Creek has a long-term gaging record ($q_g$); Buffalo River does not ($q_u$). The linear relation fit to the points determines the values of *a* and *b* in Equation (10-20). These values can then be used to estimate $dQp$ values for Buffalo River via Equation (10-21). Data from Riggs (1972).

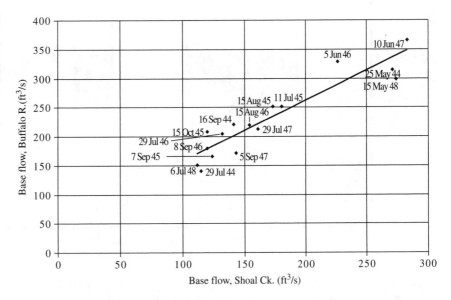

**TABLE 10-19**

Example of Effect of $\Delta t$ on Drought Sample Size ($N$) and Autocorrelation Coefficient ($r_{1x}$) for Three California Streams. Data from Dracup et al. (1980).

| Stream | $\Delta t = 1$ month | | $\Delta t = 3$ month | | $\Delta t = 1$ yr | |
|---|---|---|---|---|---|---|
| | $N$ | $r_{1x}$ | $N$ | $r_{1x}$ | $N$ | $r_{1x}$ |
| Kings R. | 63 | 0.71 | 25 | 0.04 | 12 | −0.07 |
| Sacramento R. | 57 | 0.53 | 19 | 0.17 | 8 | −0.02 |
| Woods Ck. | 72 | 0.48 | 23 | 0.04 | 12 | −0.07 |

$X_0 = \mu_X$ (mean value of $X$),
$X_0 = X_{.50}$ (median value of $X$),

or

$X_0 = \mu_X - \sigma_X$ (mean minus one standard deviation).

Dracup et al. (1980) suggested choosing $X_0 = \mu_X$ because it standardizes the analysis and gives more significance to extreme events, which are usually of most interest. Note, however, that defining $X_0$ as the mean or median usually results in about 50% of the time periods being "droughts" (exactly 50% if $X_0 = x_{.50}$). In these cases, we are actually analyzing *dry periods* rather than droughts, and the term "drought" should be restricted to the more extended and severe dry periods.

Once $X_0$ is determined, each period for which $X < X_0$ constitutes a "drought," and each

"drought" is characterized by the following measures (Figure 10-36):

**Duration,** $D \equiv$ length of period for which $X < X_0$;
**Severity,** $S \equiv$ cumulative deviation from $X_0$;
**Intensity** (or **magnitude**), $I \equiv \dfrac{S}{D}$.

Note that if $X$ is streamflow $[L^3\,T^{-1}]$, then the dimensions of $S$ are $[L^3\,T^{-1}] \times [T] = [L^3]$ and the dimensions of $I$ are $[L^3\,T^{-1}]$.

### Magnitude–Frequency Relations

Once the severities, durations, and intensities of "droughts" have been determined for a given time series, the magnitude–frequency characteristics of each of those quantities can be analyzed as indicated in Example 10-3.

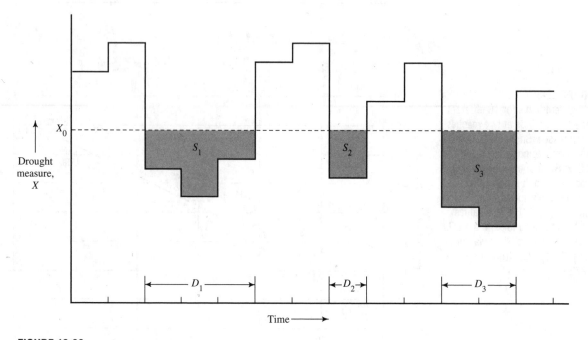

**FIGURE 10-36**
Quantitative definition of droughts. $X$ is a drought measure, $X_0$ is the truncation level. $D_1, D_2, D_3$ are durations of droughts 1, 2, and 3. The areas $S_1, S_2, S_3$ are the severities of droughts 1, 2, and 3.

**FIGURE 10-37**

Annual average flows for the Pemigewasset River at Plymouth, NH, 1904–1996. Horizontal line is average flow for the period, used as the truncation level in Example 10-3.

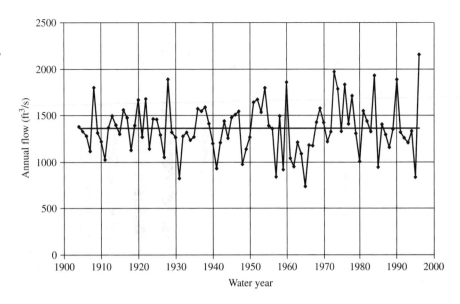

## EXAMPLE 10-3

The Pemigewasset River at Plymouth, NH, has been gaged since 1904; its annual flows ($\Delta t = 1$ yr) are plotted in Figure 10-37. Using a truncation level equal to the mean flow ($X_0 = \mu_X$), we find that 46 of the 94 years had flows less than the mean and that a total of 21 "droughts" occurred. Table 10-20 lists and Figure 10-38 shows the distributions of the durations, severities, and intensities of these annual "droughts." Durations and severities are approximately exponentially distributed; the

droughts of 1929–1935 and 1961–1967 had durations of 7 yr, and the 1960s drought was by far the most severe. The distribution of intensities is irregular, with the most intense droughts being the single-year events of 1995, 1957, and 1959.

Although droughts are sometimes discussed in terms of exceedence probabilities or return periods (e.g., the "100-yr drought"), computation of those values is not straightforward due to the variability

**TABLE 10-20**

Drought Severities ($S$), Durations ($D$), and Intensities ($I$) Determined from Analysis of 1904–1996 Annual Flows for the Pemigewasset River at Plymouth, NH. Truncation level equal to mean flow (Figure 10-37). Values in parentheses are ranks (1 = largest).

| Drought Years | Severity, $S$ (ft$^3$ s$^{-1}$ yr) | Duration, $D$ (yr) | Intensity, $I$ (ft$^3$ s$^{-1}$) |
|---|---|---|---|
| 1906–1907 | 261 (12) | 2 (7) | 130 (13) |
| 1909–1911 | 428 (7) | 3 (3) | 143 (12) |
| 1915 | 29 (20) | 1 (12) | 29 (20) |
| 1918 | 201 (14) | 1 (12) | 201 (8) |
| 1921 | 61 (19) | 1 (12) | 61 (18) |
| 1923 | 187 (16) | 1 (12) | 187 (9) |
| 1926–1927 | 315 (11) | 2 (7) | 158 (11) |
| 1929–1935 | 793 (2) | 7 (1) | 113 (14) |
| 1940–1942 | 648 (3) | 3 (3) | 216 (6) |
| 1944 | 73 (18) | 1 (12) | 73 (16) |
| 1948–1950 | 607 (4) | 3 (3) | 202 (7) |
| 1957 | 490 (6) | 1 (12) | 490 (2) |
| 1959 | 412 (8) | 1 (12) | 412 (3) |
| 1961–1967 | 1917 (1) | 7 (1) | 274 (5) |
| 1971–1972 | 113 (17) | 2 (7) | 56 (19) |
| 1979–1980 | 345 (10) | 2 (7) | 173 (10) |
| 1983 | 1 (21) | 1 (12) | 1 (21) |
| 1985 | 385 (9) | 1 (12) | 385 (4) |
| 1987–1988 | 204 (13) | 2 (7) | 102 (15) |
| 1991–1993 | 198 (15) | 3 (3) | 66 (17) |
| 1995 | 493 (5) | 1 (12) | 493 (1) |

**FIGURE 10-38**
Distributions of annual drought
characteristics, Pemigewasset
River at Plymouth, NH, 1904–1996
(Table 10-20). (a) Severity; (b)
Duration; (c) Intensity.

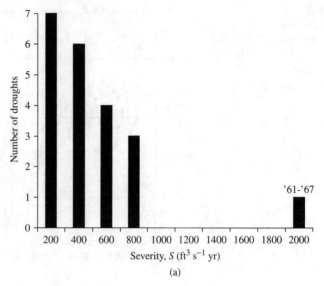

(a)

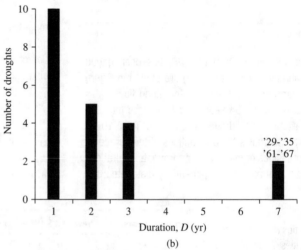

(b)

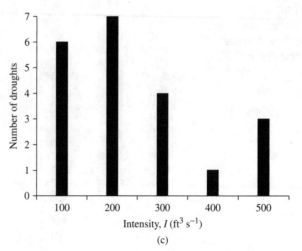

(c)

of the three measures of drought magnitude. One can apply probability analysis via the "theory of runs" to gain further insight into droughts; this theory is described in some statistics texts (e.g., Swan and Sandilands 1995).

### *Regional Aspects*

Dracup et al. (1980) recommended that drought analyses be carried out on a regional basis. This means that the analysis methods described earlier should be applied to several precipitation, PDSI, streamflow, and/or ground-water records within a region that is defined on the basis of climate, physiographic province, or political subdivision.

### 10.5.5   Concluding Comment

Figures 3-33, 3-34, and 10-33 do not suggest that hydrologic droughts are becoming more frequent or severe, and in fact the data cited in Box 3-3 indicate that streamflow is increasing in many parts of the world. Instead, the phenomenon that is occurring in many regions, even in the absence of climate change, is that resource shortages are being created by population growth and the concomitant increasing demand for water, particularly for irrigation. This creates the situation illustrated in Figure 10-39, where the intersections of natural streamflow variability with rising demand produce more frequent shortages. The next section explores current and projected water use in the United States and the world.

## 10.6   CURRENT AND PROJECTED WATER USE

### 10.6.1   Basic Concepts

As noted earlier, projections of population growth and economic activity form the basis for water-use projections. Withdrawals for domestic water use at some future time, $t$, $W_D(t)$, are estimated as

$$W_D(t) = u_D(t) \cdot N(t), \qquad (10\text{-}22)$$

where $N(t)$ is the projected population and $u_D(t)$ is the projected per-capita rate of water use (e.g., for the United States, $u_D(1995) = 179$ gal person$^{-1}$ day$^{-1}$ = 247 m$^3$ person$^{-1}$ yr$^{-1}$ from public water supplies). Future withdrawals for a particular industry, $W_I(t)$, are similarly estimated as

$$W_I(t) = u_I(t) \cdot I(t), \qquad (10\text{-}23)$$

where $I(t)$ is the projected rate of production for that industry (number of units of product produced per unit time) and $u_I(t)$ is the projected amount of water needed to produce each unit of product. Table 10-21 gives $u_I$ values for some selected products; information for other products can be found in Glieck (1993).

The sum of $W_D(t)$ plus the $W_I(t)$ values for all relevant industries are usually taken as the projected future withdrawal use at a given use site. However, as noted earlier, withdrawal uses within a drainage basin are not additive because water that is withdrawn, but not consumed, returns to the hydrologic system and can be withdrawn at down-

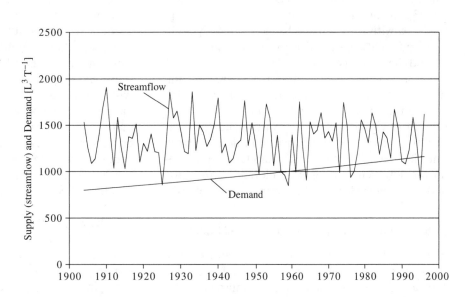

**FIGURE 10-39**
The intersection of growing demand with natural variability. Demand curve is increasing exponentially at 4% per year; streamflow is represented by a random series of normally-distributed values with the same mean and standard deviation as the Pemigewasset River at Plymouth, NH.

**TABLE 10-21**
Per-Unit Water Withdrawal Requirements for Various Products. Data from Todd (1970) and *Water News* (March 1987).

| Product | Requirement | |
| --- | --- | --- |
| | SI Units | English Units |
| **Industrial** | | |
| Beer | 15 L L$^{-1}$ | 15 gal gal$^{-1}$ |
| Pulp and paper | 236,000 L T$^{-1}$ | 56,500 gal ton$^{-1}$ |
| Gasoline | 8 L L$^{-1}$ | 8 gal gal$^{-1}$ |
| Cotton | 1 L m$^{-1}$ | 1 gal yd$^{-1}$ |
| Wool | 400,000 L T$^{-1}$ | 95,900 gal ton$^{-1}$ |
| Automobiles | 38,000 L unit$^{-1}$ | 10,000 gal unit$^{-1}$ |
| Steel | 86,600 L T$^{-1}$ | 21,000 gal ton$^{-1}$ |
| Cement | 900 L T$^{-1}$ | 210 gal ton$^{-1}$ |
| Electricity[a] | 2.5 L kW · hr$^{-1}$ | 0.7 gal kW · hr$^{-1}$ |
| Electricity[b] | 125 L kW · hr$^{-1}$ | 33 gal kW · hr$^{-1}$ |
| **Agricultural** | | |
| Grapes | 14,640 L kg$^{-1}$ | 1,758 gal lb$^{-1}$ |
| Cherries | 2,980 L kg$^{-1}$ | 358 gal lb$^{-1}$ |
| Corn syrup | 1,120 L kg$^{-1}$ | 135 gal lb$^{-1}$ |
| Milk | 1,080 L kg$^{-1}$ | 130 gal lb$^{-1}$ |
| Oats | 983 L kg$^{-1}$ | 118 gal lb$^{-1}$ |
| Wheat | 883 L kg$^{-1}$ | 106 gal lb$^{-1}$ |
| Apricots | 790 L kg$^{-1}$ | 95 gal lb$^{-1}$ |
| Corn | 740 L kg$^{-1}$ | 89 gal lb$^{-1}$ |
| Apples | 410 L kg$^{-1}$ | 49 gal lb$^{-1}$ |
| Oranges | 390 L kg$^{-1}$ | 47 gal lb$^{-1}$ |
| Grapefruit | 220 L kg$^{-1}$ | 26 gal lb$^{-1}$ |
| Potatoes | 190 L kg$^{-1}$ | 23 gal lb$^{-1}$ |
| Strawberries | 140 L kg$^{-1}$ | 17 gal lb$^{-1}$ |

[a]Thermoelectric generation with recycled cooling water.

[b]Thermoelectric generation without recycled cooling water.

stream use sites. Thus the total withdrawal usage in a drainage basin can exceed the available supply.

Consumptive uses are projected in the same way as withdrawal uses, using appropriate consumptive-use multipliers in Equations (10-22) and (10-23). In contrast with withdrawal uses, consumptive uses within a drainage basin are strictly additive, and total consumptive use in a basin cannot exceed the available supply.

Instream uses, like withdrawal uses, are not additive: A given quantity of water in a stream can serially or simultaneously generate hydroelectric power, support navigation, transport wastes, and (if water quality is adequate) provide fish and wildlife habitat. Thus instream flow needs are projected by forecasting the flows required for each relevant instream use and selecting the largest of these individual values.

### 10.6.2   Current and Projected Use: United States

Figure 10-40 summarizes the sources and uses of fresh-water withdrawals in the United States in 1995 and Tables 10-22 and 10-23 summarize the re-

gional distribution of water withdrawals. The total freshwater withdrawal, 341 billion gal/day (= 471 km$^3$ yr$^{-1}$), amounts to about 6.5% of average precipitation and 21.5% of average streamflow. Withdrawals for the cooling of thermoelectric (nuclear and fossil-fuel) power plants and for irrigation each account for about 40% of the total, with thermoelectric cooling dominating in the East and agriculture in the West. Domestic, commercial, and industrial usage together account for about 8% of withdrawal. Figure 10-40 shows the percentage of each type of withdrawal that was consumed; agricultural uses account for about 85% of the nation's consumptive use. Annual reviews of water-resource problems on a state-by-state basis are published by the National Water Summary Program of the U.S. Geological Survey (U.S. Geological Survey 1984; 1985; 1986; 1988).

Brown (2000) analyzed past trends of water use in the United States and made projections to the year 2040. As shown in Figure 10-41, total withdrawals for the largest uses have declined in recent years, in spite of significant increases in population

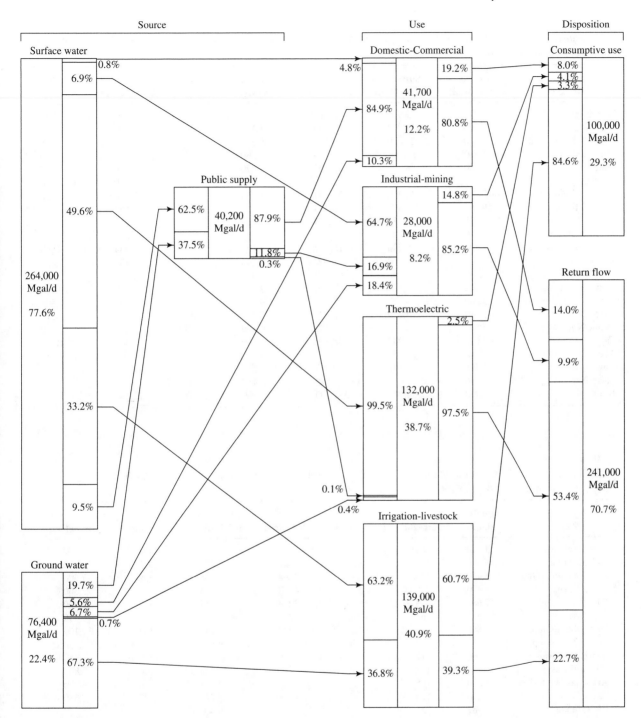

**FIGURE 10-40**

Source, use, and disposition of freshwater withdrawals in the United States, 1995. The lines and arrows indicate the distribution of water from source to disposition. For example, surface water was 77.6% of total freshwater withdrawn; 9.5% of this water was used in public supply, amounting to 62.5% of public supplies. 87.9% of public supply was used in the domestic-commercial sector, amounting to 84.9% of water use in that sector. 19.2% of domestic-commercial use was consumptive (8.0% of all consumptive use), and 80.8% was return flow (14.0% of all return flow). From Solley et al. (1998).

**TABLE 10-22**

Total Offstream Use of Freshwater by Source and Disposition for U.S. Water-Resources Regions (Figure 10-1), 1995. Mgd = $10^6$ gal/day; gd = gal/day. Data from Solley et al. (1998).

| Region | Ground Water (Mgd) | Surface Water (Mgd) | Total (Mgd) | Consumptive Use (Mgd) | Recycled (Mgd) | Conveyance Losses (Mgd) | Per Capita Use (gd) |
|---|---|---|---|---|---|---|---|
| New England | 725 | 2,980 | 3,710 | 388 | 0 | 0 | 289 |
| Mid-Atlantic | 2,690 | 18,900 | 21,600 | 1,170 | 72 | 1.9 | 509 |
| South Atlantic-Gulf | 7,110 | 25,000 | 32,100 | 5,570 | 237 | 33 | 848 |
| Great Lakes | 1,510 | 31,100 | 32,700 | 1,580 | 0 | 0.1 | 1,500 |
| Ohio | 1,980 | 28,100 | 30,100 | 1,870 | 1.1 | 0.7 | 1,330 |
| Tennessee | 258 | 8,730 | 8,980 | 289 | 0.3 | 0 | 2,140 |
| Upper Mississippi | 2,570 | 20,700 | 23,300 | 1,660 | 11 | 0 | 1,050 |
| Lower Mississippi | 9,180 | 10,800 | 20,000 | 7,740 | 0.7 | 553 | 2,720 |
| Souris-Red-Rainy | 115 | 138 | 253 | 122 | 0 | 1.8 | 364 |
| Missouri Basin | 9,320 | 26,700 | 36,000 | 14,200 | 22 | 7,840 | 3,380 |
| Arkansas-White-Red | 7,490 | 8,590 | 16,100 | 8,190 | 37 | 944 | 1,800 |
| Texas-Gulf | 5,960 | 11,700 | 17,700 | 7,340 | 71 | 390 | 1,050 |
| Rio Grande | 1,930 | 4,740 | 6,670 | 2,960 | 7.2 | 1,360 | 2,600 |
| Upper Colorado | 116 | 7,310 | 7,420 | 2,520 | 1.7 | 1,940 | 10,400 |
| Lower Colorado | 3,000 | 4,970 | 7,960 | 4,520 | 187 | 1,090 | 1,500 |
| Great Basin | 1,610 | 4,420 | 6,030 | 3,260 | 33 | 1,140 | 2,510 |
| Pacific Northwest | 5,500 | 26,500 | 32,000 | 10,600 | 0.1 | 8,050 | 3,220 |
| California | 14,600 | 21,900 | 36,500 | 25,300 | 330 | 1,860 | 1,140 |
| Alaska | 58 | 154 | 211 | 25 | 0 | 0.1 | 350 |
| Hawaii | 515 | 497 | 1010 | 542 | 6.2 | 98 | 853 |
| Caribbean | 156 | 433 | 588 | 189 | 0 | 15 | 152 |
| Total | 76,400 | 264,000 | 341,000 | 100,000 | 1,020 | 25,300 | 1,280 |

**TABLE 10-23**

Total Offstream Use of Freshwater by Use Category for U.S. Water-Resources Regions (Figure 10-1), 1995. All values in $10^6$ gal day$^{-1}$. Data from Solley et al. (1998).

| Region | Public Supply | Domestic | Commercial | Industrial | Irrigation and Livestock | Mining | Thermoelectric |
|---|---|---|---|---|---|---|---|
| New England | 1,440 | 169 | 90 | 153 | 165 | 24 | 1,670 |
| Mid-Atlantic | 6,000 | 486 | 283 | 1,430 | 427 | 321 | 12,600 |
| South Atlantic-Gulf | 5,470 | 719 | 130 | 2,790 | 5,005 | 339 | 17,600 |
| Great Lakes | 4,420 | 355 | 152 | 4,170 | 385 | 390 | 22,800 |
| Ohio | 2,680 | 328 | 170 | 3,690 | 245 | 327 | 22,600 |
| Tennessee | 574 | 64 | 22 | 1,070 | 253 | 11 | 6,990 |
| Upper Mississippi | 1,880 | 311 | 208 | 988 | 739 | 134 | 19,100 |
| Lower Mississippi | 1,070 | 73 | 36 | 2,890 | 9,140 | 5.3 | 6,730 |
| Souris-Red-Rainy | 66 | 17 | 0.3 | 22 | 108 | 1.4 | 38 |
| Missouri Basin | 1,570 | 138 | 34 | 152 | 25,000 | 306 | 8,800 |
| Arkansas-White-Red | 1,550 | 105 | 115 | 438 | 9,650 | 56 | 4,170 |
| Texas-Gulf | 2,840 | 115 | 42 | 1,060 | 5,740 | 197 | 7,680 |
| Rio Grande | 487 | 25 | 19 | 10 | 6,060 | 55 | 18 |
| Upper Colorado | 141 | 12 | 6.2 | 6.4 | 7,080 | 23 | 146 |
| Lower Colorado | 1,170 | 45 | 30 | 47 | 6,500 | 152 | 63 |
| Great Basin | 605 | 14 | 25 | 91 | 5,200 | 74 | 24 |
| Pacific Northwest | 1,910 | 260 | 1,070 | 1,080 | 27,200 | 35 | 385 |
| California | 5,610 | 124 | 396 | 541 | 29,600 | 78 | 205 |
| Alaska | 81 | 8.7 | 11 | 55 | 1.1 | 24 | 31 |
| Hawaii | 214 | 3.7 | 46 | 19 | 662 | 0.5 | 67 |
| Caribbean | 437 | 13 | 3.4 | 14 | 113 | 4.5 | 2.2 |
| Total | 40,200 | 3,390 | 2,890 | 20,700 | 139,000 | 2,560 | 132,000 |

**FIGURE 10-41**
Freshwater withdrawals by use category from 1950 to 1995. Data from Solley et al. (1998).

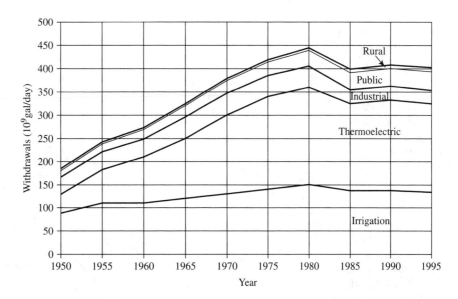

and industrial and agricultural outputs. These decreases are due to increased efficiencies of water use at thermoelectric plants, in many industries, and in irrigation. Analyzing individual water-use sectors and projecting continued increases in water-use efficiencies, Brown (2000) forecast a 2.5% increase in total withdrawals by 2020 and a 7% increase by 2040. These are relatively modest increases and, as shown in Table 10-24, are considerably lower than previous projections. Note that most of the earlier forecasts, all of which were based on elaborate analyses, were gross overestimates—this underscores the uncertainties attendant on economic and population projections. (See also Osborn et al. 1986.)

### 10.6.3   Current and Projected Use: Global

Falkenmark and Lindh (1993) analyzed global water use using the metric of per-capita water availability for a region or country at time $t$, $A(t)$:

$$A(t) = \frac{P - ET + Q_{in}}{N(t)}. \quad \textbf{(10-24)}$$

Here, $P$ is average precipitation, $ET$ is average evapotranspiration, and $Q_{in}$ is the average rate of inflow of river and ground water from surrounding regions.

Figure 10-42a compares current water demand (use) with water availability for several countries and two regions of the United States; the slanting lines show equal percentages of usage of the available water, or "mobilization levels." The vertical lines in Figure 10-42 separate levels of "water competition" (circled numbers), which are depicted in Figure 10-43. In Figure 10-42b, the Roman numerals indicate the basic water-management situation for countries or regions with various levels of water use and mobilization; these are described in Table 10-25.

Vörösmarty et al. (2000) combined a global runoff model, a global climate model, and population projections to compare the relative effects on

**TABLE 10-24**
Forecasts of United States Water Withdrawals by Brown (2000) Compared with Those of Previous Analyses. Values in $10^9$ gal/day. From Brown (2000); see that paper for original references.

| Source | Year Forecast Made | Year of Forecast 2000 | 2020 | 2040 |
|---|---|---|---|---|
| Senate Select Committee on Water Resources | 1961 | 888 | | |
| U.S. Water Resources Council | 1968 | 804 | 1368 | |
| Wollman and Bonem | 1971 | 563 | 897 | |
| U.S. National Water Commission | 1973 | 1,000 | | |
| U.S. Water Resources Council | 1978 | 306 | | |
| Guldin | 1989 | 385 | 461 | 527 |
| Brown | 2000 | 341* | 350 | 364 |

*Actual value [Solley et al. (1998)].

**FIGURE 10-42**

Global per capita water demand and availability (Note logarithmic scale.) Slanting lines show equal percentages of usage of available water ("mobilization level"). Vertical lines separate levels of "water competition" (circled numbers; see Figure 10-43). (a) Points show demand and availability for selected countries (data from mid-1970s). (b) Roman numerals indicate basic water-management situations described in Table 10-25. From Falkenmark and Lindh (1993). *Water in Crisis: A Guide to the World's Fresh Water Resources*, edited by Peter H. Gleick, © 1993 by Oxford University Press, Inc. Used by permission of Oxford University Press, Inc.

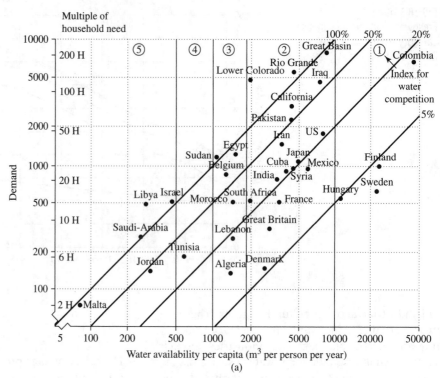

(a)

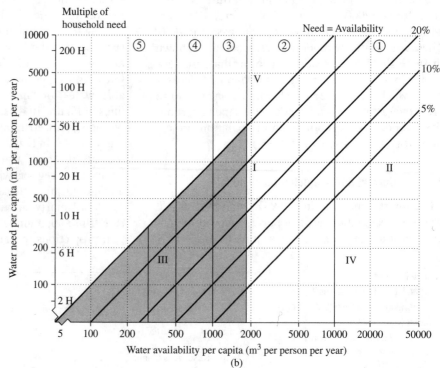

(b)

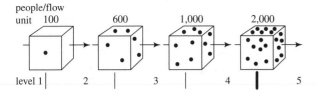

people/flow
unit    100        600      1,000      2,000

level 1 |      2    |    3    |    4    |    5

Quality and dry          Water stress   Absolute water scarcity
season problems

**FIGURE 10-43**
Visualization of levels of water competition (circled numbers on Figure 10-42). Each cube represents $10^6$ $m^3$ $yr^{-1}$ of water availability; each dot represents 100 people. From Falkenmark and Lindh (1993). *Water in Crisis: A Guide to the World's Fresh Water Resources,* edited by Peter H. Gleick, © 1993 by Oxford University Press, Inc. Used by permission of Oxford University Press, Inc.

water availability of projected climate change due to global warming (see Section 3.2.9) and population growth in the year 2025. Their metric was the dimensionless water stress, $W$, given by

$$W = \frac{U}{P - ET + Q_{in}}, \qquad (10\text{-}25)$$

where $U$ is water use. (Note that $W$ is essentially the water mobilization as used in Figure 10-42 and Table 10-25.) Table 10-26 shows the numbers of people currently (1995) living with various degrees of water stress.

The models used by Vörösmarty et al. (2000) forecast an overall reduction of global runoff of 6%, resulting in a 4% increase in water stress due to climate change alone. However, the forecast increase in stress due to population growth and economic development was far larger: 50%. Hence they concluded that, over the next 25 yr, population and economic growth will produce much greater stress on water resources than will climate change. In evaluating the geographic distribution of the forecast stress (Figure 10-44), they further concluded that

> Many parts of the developing world will experience large increases in relative water demand. In water-rich areas such as the wet tropics, the challenge will not be in providing adequate quantities of water, but in providing clean supplies that minimize public health problems. Arid and semi-arid regions face the additional challenge of absolute water scarcity. Pro-

**TABLE 10-25**
Generalized Water-Management Situation for Regions with Various Levels of Water Availability, Demand, and Mobilization Shown in Figure 10-42b. From Falkenmark and Lindh (1993).

| Region of Figure 10-42b | Water-Management Situation | Options |
|---|---|---|
| I | Moderate availability<br>Moderate demand<br>High mobilization | Reduce demand by increasing<br>    water-use efficiency<br>Control pollution<br>    Integrate land-use and water-resources management |
| II | High availability<br>Moderate demand<br>Low mobilization<br>Possible difficulty<br>    in meeting future demand | Solve regional imbalances<br>    Control pollution |
| III | Low availability<br>Low-to-moderate demand<br>Critical availability problems<br>    in face of population growth | Increase storage via artificial<br>    recharge of ground water and<br>    reservoirs<br>Integrate land use and water-resources management<br>Increase water usability by major<br>    pollution-reduction programs |
| IV | High availability<br>Low demand<br>Low mobilization<br>Significant flexibility to meet demands of<br>    growing population | Assure equitable distribution of<br>    water<br>Protect against storage loss due to<br>    reservoir siltation |
| V | Moderate availability<br>High demand<br>High mobilization<br>Water wastage<br>Depletion of ground water | Reduce demand by increasing<br>    water-use efficiency<br>Control pollution |

**TABLE 10-26**

Current (2000) World Population Living under Various Levels of Water Stress (Mobilization) as Estimated by Vörösmarty et al. (2000).

| Water Stress [Equation (10-25)] | Stress Level | Population (billions) |
|---|---|---|
| < 0.1 | low | 1.72 |
| 0.1 to 0.2 | moderate | 2.08 |
| 0.2 to 0.4 | medium-high | 1.44 |
| > 0.4 | high | 0.46 |

jected increases in scarcity will be focused on rapidly expanding cities. (Vörösmarty et al., 2000; p. 287).

# EXERCISES

Exercises marked with ** have been programmed in EXCEL on the CD that accompanies this text. Exercises marked with * can be advantageously executed on a spreadsheet, but you will have to construct your own worksheets to do this.

**10-1.** Obtain a water-resources planning document for your state, region, or municipality. (a) Are the goals, objectives, and boundaries of the plans clearly stated? (b) How do the goals, objectives, and boundaries of the plans relate to those discussed in Section 10.1.1? (c) Can you suggest modifications of the goals, objectives, and boundaries of the plans that seem more appropriate? (d) Does the plan appear to be developed using the approach outlined in Figure 10-2? (e) What types of hydrologic analysis are used?

**10-2.** Obtain a water-resources planning document that focuses on water supply and demand for your state, region, or municipality. (a) What definitions of supply ("safe yield") and demand are used? (b) How are projections of water use made? (c) Are connections between ground and surface water realistically included in the plan? (d) Are connections between water supply and water quality realistically included in the plan?

**10-3.** Review U.S. Geological Survey *Water Supply Papers* 2325, 2350, 2375, 2400, and 2425 for a state of interest and write a paper (length and format as specified by your instructor) describing the major hydrological features and water-resources issues for that state.

***10-4.** Following the procedure in Box 2-2, download at least 10 yr of daily streamflow values for a stream gage of interest. Use the data and the methods of Box 10-2 to construct (a) a period-of-record flow-duration curve (FDC) and (b) a median-annual FDC. Plot both curves on probability paper if available. How do the $q_{.95}$ values differ for the two FDCs?

***10-5.** (a) Construct a reservoir storage–yield curve for a potential reservoir location in your region using Equation (10-13). Determine the mean, variance, and coefficient of variation of annual flows by either (i) downloading annual streamflow values for a stream gage in the region (Box 2-2) and computing the values using the methods described in Box C-2 or (ii) using the appropriate regression equation from Table 10-7. (b) Compute the $M$ value (Equation 10-12) for various storage sizes. At what size would the reservoir become dominated by carry-over storage?

****10-6.** Retrieve the spreadsheet file called DOSag.xls on the text CD, which contains the Streeter–Phelps Model described in Box 10-6. Compute and plot a WRIDC for dissolved oxygen (DO)(minimum DO concentration, $c_{min}$) and determine the fraction of time $c_{min} > 4$ mg $L^{-1}$ where the FDC for the reach is given by the following values:

| $EP_Q(q)$ | $q$ (m³ s⁻¹) |
|---|---|
| 0.98 | 0.25 |
| 0.95 | 0.40 |
| 0.90 | 0.60 |
| 0.80 | 1.00 |
| 0.70 | 1.40 |
| 0.60 | 2.00 |
| 0.50 | 2.60 |
| 0.40 | 3.70 |
| 0.30 | 4.80 |
| 0.20 | 7.20 |
| 0.10 | 14.0 |
| 0.05 | 30.0 |
| 0.01 | 70.0 |

(a) Assume that the BOD concentration in the sewage-treatment-plant effluent equals 100 mg $L^{-1}$, the DO concentration in the effluent equals 0 mg $L^{-1}$, and the DO concentration in the river upstream of the effluent discharge is at saturation. (b) Repeat the exercise assuming that effluent BOD is reduced to 50 mg $L^{-1}$ by wastewater treatment.

****10-7.** Retrieve the spreadsheet file called PolPot.xls on the text CD, which contains the lake-pollution model described in Box 10-7 (Example 10-2). Values for Squam Lake, New Hampshire, are entered, and a time scale from 0 to 15 years is in cell range C35:C50. The model assumes

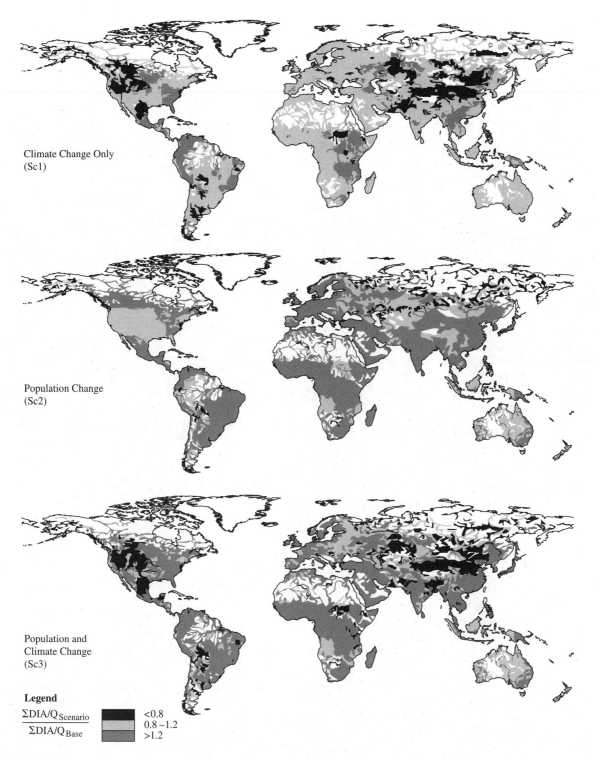

Climate Change Only
(Sc1)

Population Change
(Sc2)

Population and
Climate Change
(Sc3)

**Legend**

$\dfrac{\Sigma DIA/Q_{Scenario}}{\Sigma DIA/Q_{Base}}$

    <0.8
    0.8–1.2
    >1.2

**FIGURE 10-44**

Global distribution of forecast change in water-use index (ratio of total use to available water) by 2025 due to (a) climate change only; (b) population change only; (c) climate change and population change. Dark areas will have improved conditions (2025 index less than 80% of 1985 index); light-gray areas will have significantly worsened conditions (2025 index more than 20% greater than 1985 index). Reprinted with permission from Vörösmarty et al., *Science* 289: 284–288. © 2000 American Association for the Advancement of Science.

that the values in Table 10-15 apply to phosphorus export from forested (4,100 g km$^{-2}$ yr$^{-1}$) and developed (15,000 g km$^{-2}$ yr$^{-1}$) land. Alter the fraction of watershed in urban development, (cells D24:D29) from 0 to 0.4 to 0.8, and print the worksheet for each value. (a) How does the equilibrium concentration, $C_{eq}$, change? (b) Graph the lake concentration, $CL$ (cells D35:D50), as a function of time for each level of development. (c) How many years does it take to reach the critical concentration for eutrophication, 0.01 g m$^{-3}$, for each value?

*10-8. Following the procedure in Box 2-2, download annual peak flood values for a stream gage of interest. (a) Use the non-parametric method of Box 10-8 to construct and plot a flood–frequency graph for this station. Use probability paper if available. (b) Using the curve fit to your plotted points, estimate the discharges associated with the following exceedence probabilities: 0.5, 0.2, 0.1, 0.04, 0.02, and 0.01. (c) Note which of the values are extrapolated beyond the data and discuss the uncertainty in the estimates.

**10-9. Retrieve the spreadsheet file called Lisbon$$.xls on the text CD. Cell range A11:H26 contains a tabular version of the coaxial flood-damage model described in Figures 10-28 and 10-29. The data in the table are for Lisbon, NH (see Figure 10-27) (U.S. Soil Conservation Service, 1983; 1986) for 1983 conditions, with no damage-reduction measures in place. Compare the reduction in average annual damages to existing structures due to (a) a flood-control dam that reduces peaks for all exceedence probabilities by 10%, (b) a levee system that raises the elevation at which flooding begins by 4.0 ft, (c) channelization that lowers the channel bottom by 1.5 ft, (d) floodproofing that halves flood damages below an elevation of 579 ft, (e) removal of structures to eliminate flood damages at exceedence probabilities of 0.1 and greater, and (f) a flood-warning system that reduces damages at all levels by 20%. (g) Qualitatively evaluate the costs and environmental effects of these measures and decide which, if any, you would recommend.

**10-10. This exercise continues the analysis of the flood-damage data in Exercise 10-9. Model a projected 50% growth in the next 20 years by multiplying the damages at all levels by 1.5. This increases average annual damages to $82,800 if no flood-damage-reduction measures are implemented. If this growth were prevented by floodplain zoning, the benefits of zoning would amount to $82,800 − $55,200 = $27,600. (a) Compare this benefit with the measures in Exercise 10-9(a)–(f) and this new level of damages. (b) Qualitatively evaluate the costs and environmental effects of these measures, including floodplain zoning, and decide which, if any, you would recommend.

*10-11. Following the procedure in Box 2-2, download annual average flow values for a stream gage of interest. Use these data to conduct an analysis of annual hydrologic drought as in Example 10-3.

# A

# Hydrologic Quantities

Hydrology is a quantitative earth science—a branch of geophysics—and both practical and theoretical problems require manipulation of numerical quantities. Correct treatment of such quantities requires an understanding of the *qualitative* aspects of numbers—the concepts of dimensions, units, and numerical precision. Hydrologists also encounter quantities measured in various unit systems and must become adept at converting measurements and equations made in one system to other systems.

This appendix summarizes rules for the correct treatment of dimensions and units, relative and absolute measurement precision, and unit and equation conversion.

## A.1 DIMENSIONS AND UNITS

### A.1.1 Dimensions

The fundamental dimensional character of quantities encountered in physical hydrology can be expressed as

$$[M^a L^b T^c \Theta^d]$$

or

$$[F^e L^f T^g \Theta^h],$$

where [M] indicates the dimension of mass, [L] the dimension of length, [T] the dimension of time, [Θ]

the dimension of temperature, and [F] the dimension of force; and the exponents $a, b, ..., h$ are rational numbers[1] or zero.

The choice of whether to use force or mass is a matter of convenience. Dimensions expressed in one system are converted to the other system via Newton's Second Law of Motion:

$$[F] = [M L T^{-2}]; \quad \text{(A-1a)}$$
$$[M] = [F L^{-1} T^2]. \quad \text{(A-1b)}$$

The dimensions of energy are [F L] or [M L² T⁻²].[2] The dimensional character of some relations in this text will be clearer if we use [E] to designate the dimensions of energy; thus,

$$[E] \equiv [M L^2 T^{-2}] = [F L]. \quad \text{(A-2)}$$

Quantities obtained either by counting or as the ratio of measurable quantities with identical dimensions are **dimensionless;** their dimensional character is denoted as [1]. Quantities obtained as logarithmic, exponential, and trigonometric functions are also dimensionless.[3]

A table on the inside front cover gives the dimensional character of quantities commonly encountered in hydrology. Those with dimensions involving only length are classed as "geometric"

---

[1] Rational numbers are the positive and negative integers, ratios of integers, and zero.
[2] One can readily remember this latter formula by recalling Einstein's famous formula $E = m \cdot c^2$, where $E$ is the amount of energy in a quantity of mass, $m$, and $c$ is the velocity of light.
[3] The arguments of exponential and trigonometric functions should be dimensionless, though one sometimes encounters empirical relations in which this rule is violated.

(angle is included here also); those involving length and time, or only time, are "kinematic"; those involving mass or force are "dynamic"; and those involving temperature are "thermal" (latent heat is included here also). The properties of water listed in the table are more fully described in Section B.2.

### A.1.2    Units

Units are the arbitrary standards in which the magnitudes of quantities are expressed. When we give the units of a quantity, we are expressing the ratio of its magnitude to the magnitude of an arbitrary standard with the same fundamental dimension.

A table on the front endpaper gives the units of the fundamental dimensions in the three systems of units that have been in common use in science and engineering. The **Système International** (SI) is the international standard for all branches of science. Hydrologists may also encounter the centimeter-gram-second (cgs) system, which was an earlier standard and which is especially convenient for measurements involving water, and the English system, which is still widely used in the United States.

It would be sufficient for purposes of physical hydrology to have only four units, one for each of the fundamental dimensions. It has proven convenient, however, to define many other units. The table on the inside front cover gives the units in which the quantities most commonly encountered in hydrology are usually measured.

## A.2    PRECISION AND SIGNIFICANT FIGURES

**Precision** is the "fineness", or degree, to which a quantity is measured. The precision of any measured value can be expressed in both absolute and relative terms.

**Absolute precision** is expressed in terms like "to the nearest $x$," where $x$ is some measurement unit.

**Relative precision** is expressed as the number of **significant figures** in the numerical expression of a measured quantity; this number is equal to the number of digits, counted as beginning with the leftmost non-zero digit and proceeding to the right to include all digits warranted by the precision of the measurement.

All measured quantities have finite precision, which must be appropriately expressed and considered in calculations, as described in the following sections.

### A.2.1    Absolute Precision

If we were to measure a distance to the nearest centimeter, we would have to report it as, say, 21 cm. If we were to report the measurement as 21.0 cm or 21.00 cm, we would be implying that it had been made to the nearest 0.1 cm or 0.01 cm, respectively. If a measurement is given as, say 200 m, the precision is not clear, because we don't know whether the measurement was made to the nearest meter, 10 meters, or 100 meters. One way of avoiding this ambiguity is to use scientific notation and express the quantity as $2 \times 10^2$ m, $2.0 \times 10^2$ m, or $2.00 \times 10^2$ m, as appropriate. Otherwise, additional information, usually in the form of other analogous measurements, is required to clarify the situation.

In adding or subtracting measured values, we must be concerned with absolute precision, so we observe the following rule:

**Rule 1** The absolute precision of a sum or difference equals the absolute precision of the *least* absolutely precise number involved in the calculation.

Hydrologists often deal with streamflow data collected by the U.S. Geological Survey (USGS), and it is the policy of the USGS to report such measurements with the absolute precision shown in Table A-1. The example below shows how this precision should be treated in adding flows.

**TABLE A-1**

Absolute Precision of Streamflow Data Reported By The U.S. Geological Survey

| Discharge Range (ft$^3$ s$^{-1}$) | Precision (ft$^3$ s$^{-1}$) |
|---|---|
| < 1 | 0.01 |
| 1 to 9.9 | 0.1 |
| 10 to 999 | 1 |
| > 1000 | 3 significant figures |

## EXAMPLE A-1

(a) Suppose the average flow for two consecutive days is reported as 102 ft$^3$ s$^{-1}$ and 3.2 ft$^3$ s$^{-1}$. What is the total for the two days?

*Solution* Adding the reported values gives 105.2 ft$^3$ s$^{-1}$, but the larger flow was measured only to the nearest 1 ft$^3$ s$^{-1}$, so we must report the total as 105 ft$^3$ s$^{-1}$.

(b) Suppose the flows for the two days were 1020 ft$^3$ s$^{-1}$ and 3.2 ft$^3$ s$^{-1}$. What is the total?

*Solution* Here the sum must be reported as 1020 ft$^3$ s$^{-1}$, because the larger flow was measured to the nearest 10 ft$^3$ s$^{-1}$.

(c) Given the following daily flows (in ft$^3$ s$^{-1}$), what is the total flow for the period? 27, 104, 12, 2310, 6.4, 0.11, 256.

*Solution* Adding all these values gives 2715.51 ft$^3$ s$^{-1}$, but the largest value was measured only to the nearest 10 ft$^3$ s$^{-1}$, so we must report the sum as 2720 ft$^3$ s$^{-1}$.

### A.2.2   Relative Precision

In the reporting of a measured value, any digits farther to the right than is warranted by the measurement precision are **nonsignificant figures.** Only the significant figures should be included in stating a measured value; thus 21, 21.0, and 21.00 cm represent two, three, and four significant figures, respectively.

In multiplication and division, we must be concerned with relative precision; thus we observe the following rule:

**Rule 2** The number of significant figures of a product or quotient equals the number of significant figures of the *least* relatively precise number involved in the calculation.

## EXAMPLE A-2

Suppose the water-surface width of a stream is measured as 20.4 m, the average depth as 1.2 m, and the average velocity as 1.7 m s$^{-1}$. What is the discharge?

*Solution* Discharge, $Q$, equals the product of width, average depth, and average velocity, so

$$Q = 20.4 \text{ m} \times 1.2 \text{ m} \times 1.7 \text{ m s}^{-1} = 41.616 \text{ m}^3 \text{ s}^{-1}.$$

However, because of Rule 2, we report the discharge to two significant figures as $Q = 42$ m$^3$ s$^{-1}$.

## A.3   UNIT CONVERSION

Because of the common use of three systems of units and the proliferation of units within each system, hydrologists must become expert in the skill of converting from one set of units to another.

A table on the front endpaper gives factors for converting among many of the units used in this text. These factors are used as either numerators or denominators in fractions whose physical value is 1.000 . . . , but whose numerical value is some other number. For example, in terms of actual lengths,

$$\frac{1 \text{ ft}}{0.3048 \ldots \text{ m}} = 1.000 \ldots ;$$

$$\frac{0.3048 \ldots \text{ m}}{1 \text{ ft}} = 1.000 \ldots .$$

Rule 2 must be followed in all unit conversions. However, because all conversion factors have infinite precision, it is only the precision of the measured quantities—not that of the conversion factors—that determines the number of significant figures in the converted value. Thus the following rule must be observed in doing unit conversions:

**Rule 3** In unit conversions, the number of digits retained in the conversion factors must be greater than the number of significant digits in any of the measured quantities involved.

Except for commonly used temperature units (discussed below), a zero value in one unit system is a zero value in the other systems. Conversion in these cases is simply a matter of multiplying by the appropriate conversion factor, and the decision of whether to put the factor in the numerator or denominator is determined by the direction of the conversion. Rules 2 and 3 must be followed in all unit conversions, as indicated in Examples A-3–A-5.

## EXAMPLE A-3

Suppose a distance is measured as 9.6 mi. How is that same distance expressed in meters?

**Solution**  The table of conversion factors on the front endpaper indicates that we multiply 9.6 mi times 1609 m mi$^{-1}$:

$$9.6 \text{ mi} \times \frac{1609 \dots \text{m}}{1.000 \dots \text{mi}} = 15{,}446.4 \text{ m} \rightarrow 15{,}000 \text{ m}.$$

Note that the conversion factor has four digits, so Rule 3 is observed. Following Rule 2, we round the converted value to two significant figures.

Clearly, it would be misleading to express the result as 15,446.4 m, because this would imply that we know the distance to a precision of 0.1 m, whereas the original measurement was known only to 0.1 mi or about 161 m. However, in following Rule 2 we have in fact lost some absolute precision: stating the distance as 15,000 m implies an absolute precision of 1000 m, which is considerably less precise than the original precision of 161 m. Still, this is the accepted procedure—if we had instead stated the converted distance as 15,400 m we would be exaggerating the true precision of the originally measured value.

Generally, we accept the loss in absolute precision that results from applying Rule 2. An alternative that more accurately conveys the precision of the original measurement is to state the converted value with an explicit absolute precision—in the given example, as 15,400 ± 200 m. This is seldom done, however.

## EXAMPLE A-4

Express the measured distance of 855.26 m in (a) kilometers, and (b) miles.

**Solution**  Observing Rules 2 and 3 yields the following calculations:

(a)

$$855.26 \text{ m} \times \frac{1.000 \dots \text{km}}{1000 \dots \text{m}} = 0.85526 \text{ km}.$$

(b)

$$855.26 \text{ m} \times \frac{1.000 \dots \text{mi}}{1609.34 \dots \text{m}} = 0.53144 \text{ mi}.$$

Note that in (b) there is again a loss of precision: The original measurement was to the nearest 0.01 m, but 0.00001 mi ≈ 0.016 m.

Example A-5 applies Rules 2 and 3 to a case in which two unit conversions are required.

## EXAMPLE A-5

Convert 19 mi hr$^{-1}$ to m s$^{-1}$.

**Solution**

$$19 \text{ mi hr}^{-1} \times \left( \frac{1609 \dots \text{m}}{1.000 \dots \text{mi}} \right) \times \left( \frac{1.000 \dots \text{hr}}{3600 \dots \text{s}} \right)$$
$$= 8.4919.. \text{ m s}^{-1} \rightarrow 8.5 \text{ m s}^{-1}$$

Conversion of actual temperatures from one system to another involves addition or subtraction, because the zero points differ. Example A-6 illustrates the procedure; note that actual Celsius and Fahrenheit temperatures are written here with the degree sign before the letter symbol (read "degree Celsius" or "degree Fahrenheit"), while temperature *differences*—distances on the temperature scale—for each system are written with the symbol after the letter (read "Celsius degree" or "Fahrenheit degree"). The zero point for the Kelvin scale is absolute zero, so the degree sign is not used in that system.

## EXAMPLE A-6

(a) Convert −37 °F to °C.

**Solution**

$$(-37° \text{ F} - 32.000 \dots °\text{F}) \times \frac{1.000 \dots \text{C}°}{1.80 \dots \text{F}°}$$
$$= -38.33 \dots °\text{C} \rightarrow -38 °\text{C}.$$

(b) Convert −37 °C to °F.

**Solution**

$$(-37 °\text{C}) \times \frac{1.800 \dots \text{F}°}{1.000 \dots \text{C}°} + 32.000 \dots °\text{F}$$
$$= -34.6 \dots °\text{F} \rightarrow -35 °\text{F}.$$

(c) Convert −37 °C to K.

**Solution**

$$(-37 °\text{C}) \times \frac{1.000 \dots \text{K}}{1.000 \text{C}°} + 273.16 \dots \text{K}$$
$$= 236.16 \text{ K} \rightarrow 236 \text{ K}.$$

(d) Convert 295 K to °C.

*Solution*

$$(295 \text{ K}) \times \frac{1 \text{ C}°}{1 \text{ K}} - 273.2 \text{ K} = 21.8 \text{ °C} \rightarrow 22 \text{ °C}.$$

Note that Rule 1 was observed.

Conversion of temperature *differences* does not involve addition or subtraction, because we are dealing only with distances on the temperature scales. Thus conversion of temperature differences follows the same procedures illustrated in Example A-3, as shown in the following example:

---

## EXAMPLE A-7

(a) Convert a temperature difference of 3.4 F° to C°.

*Solution*

$$3.4 \text{ F}° \times \frac{1.000\ldots \text{ C}°}{1.800\ldots \text{ F}°} = 1.888\ldots \text{ C}° \rightarrow 1.9 \text{ C}°.$$

(b) Convert a temperature difference of 3.4 C° to F°.

*Solution*

$$3.4 \text{ C}° \times \frac{1.800\ldots \text{ F}°}{1.000\ldots \text{ C}°} = 6.12\ldots \text{ F}° \rightarrow 6.1 \text{ F}°.$$

(c) Convert a temperature difference of 3.4 C° to K.

*Solution*

$$3.4 \text{ C}° \times \frac{1.000\ldots \text{ K}}{1.000\ldots \text{ C}°} = 3.4 \text{ K}.$$

---

Hydrologists should also observe the following three rules concerning significant figures:

**Rule 4** Unless it is clear that greater precision is warranted, assume no more than three-significant-figure precision in measured hydrologic quantities.

As is noted in Table A-1, there are many cases where only two-significant-figure precision is warranted.

As was noted by Harte (1985, p. 4),

Nonsignificant figures have a habit of accumulating in the course of a calculation, like mud on a boot, and you must wipe them off at the end. It is still good policy to keep one or two nonsignificant figures during a calculation, however, so that the rounding off at the end will yield a better estimate.

Therefore, one should always observe the following rule:

**Rule 5** In unit conversions, statistical computations, and other computations involving several steps, do not round off to the appropriate number of significant figures until you get to the final answer.

The numbers on computer printouts and calculator displays almost always have more digits than is warranted by the precision of measured hydrologic quantities. Thus you are seldom justified in simply reporting the numbers directly as given by those devices without appropriate rounding-off. We summarize this situation as follows:

**Rule 6** Computers and calculators don't know anything about significant figures.

---

## A.4 EQUATIONS: DIMENSIONAL PROPERTIES AND CONVERSION

### A.4.1 Dimensional Properties

**Rule 7** An equation that completely and correctly describes a physical relation has the same dimensions on both sides of the equals sign.

Equations that conform to Rule 7 are **dimensionally homogeneous.** A corollary of this rule is that only quantities with identical dimensional quality can be added or subtracted.

While there are no exceptions to Rule 7, there are some important qualifications:

**Qualification 7a** A dimensionally homogeneous equation may not correctly and completely describe a physical relation.

This situation can arise when terms of negligible magnitude are not included in the formulation, or when a dimensionally homogeneous equation is simply incorrect due to error.

**Qualification 7b** Equations that are not dimensionally homogeneous can be very useful approximations of physical relationships.

The magnitudes of hydrologic quantities are commonly determined by the complex interaction of many factors, and it is often virtually impossible to formulate the physically correct equation or to measure all the relevant independent variables. Thus hydrologists are often forced to develop and rely on relatively simple empirical equations that are dimensionally inhomogeneous.

An example of Qualification 7b is the following equation relating the velocity, $U$ [L T$^{-1}$], of a stream to its average depth, $Y$ [L], and water-surface slope, $S$, expressed as the tangent of the slope angle [1]:

$$U = \frac{Y^{2/3} \cdot S^{1/2}}{n}. \qquad \text{(A-3)}$$

In this equation, $n$ is a factor reflecting the frictional resistance to flow offered by the channel bed and banks, and it is treated as a dimensionless number. This inhomogeneous empirical relation is commonly taken as the equation of motion for open-channel flows [Equation (9-19) and Table 9-6. [The nature of Equation (A-3) is discussed more fully in Example A-8.]

**Qualification 7c** Equations can be dimensionally homogeneous but not unitarily homogeneous. (However, all unitarily homogeneous equations are of course dimensionally homogeneous.)

This situation can arise because each system of units includes "superfluous" units, such as miles (= 5280 ft), kilometers (= 1000 m), acres (= 43,560 ft$^2$), hectares (= $10^4$ m$^2$), and liters (= $10^{-3}$ m$^3$). Thus the equation

$$Q = 1000 \cdot U \cdot A, \qquad \text{(A-4)}$$

where $Q$ is streamflow rate in L s$^{-1}$, $U$ is stream velocity in m s$^{-1}$, and $A$ is stream cross-sectional area in m$^2$, is dimensionally homogeneous but not unitarily homogeneous. Clearly, the multiplier 1000 in Equation (A-4) is a unit-conversion factor (L$^3$ m$^{-3}$) required to make the equation correct for the specified units.

As was noted, dimensionally and/or unitarily inhomogeneous empirical equations are frequently encountered in hydrology. It is extremely important that the practicing hydrologist cultivate the habit of checking every equation for dimensional and unitary homogeneity, because of the following requirement:

**Rule 8a** If an inhomogeneous equation is given, the units of each variable in it MUST be specified.

This rule is one of the main reasons you should train yourself to examine each equation you encounter for homogeneity: If you use an inhomogeneous equation with units other than those for which it was given, you will get the wrong answer. Surprisingly, it is not uncommon to encounter in the earth-sciences and the engineering literature inhomogeneous equations for which units are not specified—so *caveat calculator*!

Rule 8a has an equally important corollary:

**Rule 8b** At least one of the coefficients or additive numbers in a unitarily inhomogeneous equation must change when the equation is to be used with different systems of units.

In practice, for example in writing a computer program to make a series of calculations, we often want to use an inhomogeneous equation with quantities measured in units different from those used in developing the equation. Similarly, we might want to compare inhomogeneous empirical equations that were developed for differing units. The steps for determining the new numerical values when an inhomogeneous equation is to be used with different units are detailed in the following section.

## A.4.2   Equation Conversion

The guiding principle in equation conversion is the following:

**Rule 9** In equations, the dimensions and units of quantities are subjected to the same mathematical operations as the numerical magnitudes.

Careful execution of the following steps will assure that equation conversion is done correctly.

1. Write out the equation with the NEW units next to each term.
2. Next to each new unit, write the factor for converting the NEW units to the OLD units. (This might seem backwards, but it isn't.)
3. Perform the algebraic manipulations necessary to consolidate and simplify back to the original form of the equation.

In executing Steps 2 and 3, note that exponents are not changed in equation conversion and that the conversion factors are subject to the same exponentiation as the variables they accompany.

## EXAMPLE A-8a

Convert the inhomogeneous Equation (A-3), which is written for $U$ in m s$^{-1}$ and $Y$ in m, for use with $U$ in ft s$^{-1}$ and $Y$ in ft.

***Solution***  Following the preceding steps, we find:

1.

$$(U \text{ ft s}^{-1}) = \frac{(Y \text{ ft})^{2/3} \cdot S^{1/2}}{n} \; ;$$

2.

$$\left(U \text{ ft s}^{-1} \cdot \frac{0.3048\ldots \text{m}}{1.000\ldots \text{ft}}\right) = \frac{\left(Y \text{ ft} \cdot \dfrac{0.3048\ldots \text{m}}{1.000\ldots \text{ft}}\right)^{2/3} \cdot S^{1/2}}{n} \; ;$$

3.

$$0.3048\ldots \cdot U = \frac{Y^{2/3} \cdot 0.4529\ldots \cdot S^{1/2}}{n} \; ;$$

$$U = \frac{1.49 \cdot Y^{2/3} \cdot S^{1/2}}{n} .$$

Thus the implicit coefficient 1.000... in Equation (A-3) is changed to 1.49 for use with the new units. Note that although this coefficient has infinite precision, it is usually expressed to three significant figures, in conformance with Rule 4.

One should always check to make sure that a conversion was done correctly. To do this, follow these steps:

1. Pick an arbitrary set of values in the original units for the variables on the right-hand side of the equation, enter them in the original equation, and calculate the value of the dependent variable in the original units.
2. Convert the values of the independent variables to the new units. (Dimensionless quantities do not change value.)

3. Enter the converted independent variable values from Step 2 into the converted equation and calculate the value of the dependent variable in the new units.
4. Convert the value of the dependent variable calculated in Step 3 back to the old units and check to see that it is identical to that calculated in Step 1.

Example A-8b shows these steps for the equation converted in Example A-8a:

## EXAMPLE A-8b

1. Enter the arbitrary values $Y = 2.40$ m, $S = 0.00500$, and $n = 0.040$ into the original equation and calculate $U$ in m s$^{-1}$:

$$U = \frac{2.40^{2/3} \times 0.00500^{1/2}}{0.040} = 3.17 \text{ m s}^{-1}.$$

2. Convert

$$Y = 2.40 \text{ m} \times \frac{1.000\ldots \text{ft}}{0.3048\ldots \text{m}} = 7.87 \text{ ft}.$$

The $S$ and $n$ values do not change, because they are dimensionless.

3. Substitute the converted values into the new equation:

$$U = \frac{1.49 \times 7.87^{2/3} \times 0.00500^{1/2}}{0.040} = 10.42 \text{ ft s}^{-1}.$$

4. Convert this value of $U$ back to the old units and compare with the value in Step 1:

$$10.42 \text{ ft s}^{-1} \times \frac{0.3048\ldots \text{m}}{1.000\ldots \text{ft}} = 3.18 \text{ m s}^{-1} \quad \text{OK.}[4]$$

---

[4] The difference between this value and the original value is due only to roundoff error.

# B

# Water as a Substance

Forces acting on water cause it to move through the hydrologic cycle, and the physical properties of water determine the qualitative and quantitative relations between those forces and the resulting motion. The physical properties of water, as well as its interactions with the environment, are in turn determined by its atomic and molecular structures. Thus, although the detailed study of these structures and properties is outside the traditional scope of hydrology, it is important for the student of hydrology to have some understanding of them.

Water is a very unusual substance with anomalous properties. This strangeness is the reason it is so common at the earth's surface—a topic that is entertainingly elaborated on by van Hylckama (1979). The abundance of water, and its existence in all three phases, makes our planet unique (see Figure B-1) and makes the science of hydrology vital to understanding and managing the environment and our relation to it.

## B.1   STRUCTURE OF WATER

### B.1.1   Molecular and Inter-Molecular Structure

The water molecule is formed by the combination of two hydrogen atoms (Group Ia, with one electron in the outer shell) with one oxygen atom (Group VIa, with six electrons in the outer shell); hence, it has the chemical formula $H_2O$. As is shown in Figure B-2a, the outer shell of oxygen can accommodate eight electrons, so it has two vacancies. The outer (and only) shell of hydrogen can hold two electrons, so it has one vacancy. The electron vacancies of two hydrogen atoms and one oxygen atom can be mutually filled by the sharing of outer-shell electrons, shown schematically in Figure B-2b. This sharing is known as a **covalent bond.**

The two most important features of the water molecule are (1) that its covalent bonds are very strong (i.e., much energy is needed to break them) and (2) that the molecular structure is asymmetric, with the hydrogen atoms attached on one "side" of the oxygen atom and an angle of about 105° between the two hydrogens (Figure B-3).

The asymmetry of the water molecule causes it to have a positively charged end (the "side" where the hydrogens are attached) and a negatively charged end (the "side" opposite the hydrogens), much like the poles of a magnet. Most of the unusual properties of water are ultimately the result of its being made up of these **polar molecules.** The polarity produces an attractive force between the positively charged end of one molecule and the negatively charged end of another, as is shown in Figure B-4. This force, called a **hydrogen bond,** is absent from most other liquids.

Although the hydrogen bond is only about one-twentieth the strength of the covalent bond (Stillinger 1980), it is far stronger than the intermolecular bonds that are present in liquids with symmetrical, nonpolar molecules. We get an idea of this strength when we compare the melting/freezing temperature and the boiling/condensation temperature of the hydrides of all the Group VIa elements: oxygen (O), sulfur (S), selenium (Se), and tellurium

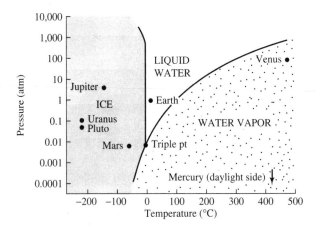

**FIGURE B-1**

Surface temperatures and pressures of the planets plotted on the phase diagram for water. From *Opportunities in the Hydrologic Sciences* © 1991 by the National Academy of Sciences. Reprinted with permission of the National Academy Press.

(Te). These elements are all characterized by an outer electron shell that can hold eight electrons but has two vacancies. Thus they all form covalent bonds with two hydrogens but, except for water, the resulting molecules are nearly symmetrical and therefore nonpolar.

In the absence of strong intermolecular forces that result from polar molecules, the melting/freezing and boiling/condensation temperatures of these compounds would be expected to rise as their atomic weights increased. As is shown in Figure B-5, these expectations are fulfilled, except—strikingly—in the case of $H_2O$. The reason for this departure from expectations is the hydrogen bonds, which attract one molecule to another and which can be loosened (as in melting) or broken (as in evaporation) only when the vibratory energy of the molecules is large—that is, when the temperature is high. Because of its high melting and boiling temperatures, water is one of the very few substances that exists in three physical states—solid, liquid, and gas—at earth-surface temperatures.

## B.1.2   Freezing and Melting

At temperatures below 0 °C, the vibratory energy of water molecules is sufficiently low that the hydrogen bonds can lock the molecules into the regular three-dimensional crystal lattice of ice (Figure B-6). In this lattice, each molecule is hydrogen-bonded to

four adjacent molecules. The angle between the hydrogen atoms in each molecule remains at 105°, but each molecule is oriented so that a puckered honeycomb of perfect hexagons is visible when the lattice is viewed from one direction. Thus ice is a hexagonal crystal, and snowflakes show infinite variation on a theme of six-fold symmetry.

When ice is warmed to 0 °C, further additions of heat cause melting, in which about 15% of the hydrogen bonds break (Stillinger 1980). Because of the rupturing of some of the hydrogen bonds, the rigid ice lattice partially collapses, and a given number of molecules takes up less space in the liquid phase than in the solid. As a result, the density of ice is less than that of water (91.7% of the density of liquid water at 0 °C). Very few substances have a lower density in the solid state than in the liquid, and this property is of immense importance: If rivers and lake froze from the bottom up instead of from the top down, biological and hydrological conditions in higher latitudes would be markedly altered.

Although melting always occurs when ice at earth-surface pressure is warmed to 0 °C, freezing may not always take place when liquid water is cooled to 0 °C. If the liquid contains no impurities and is not in contact with preexisting ice, it is possible to supercool it to temperatures as low as −41 °C. This resistance to freezing occurs because the water molecules form various types of nonhexagonal hydrogen-bonded polyhedra (Figure B-4), which prevent the formation of the ice lattice. However, if ice particles (or common impurities like clay minerals, which have a crystal structure like that of ice) are present, they provide templates that act as growth nuclei to trigger the formation of ice at 0 °C. Significant supercooling is quite common in clouds, where effective growth nuclei may be lacking, and it is an important factor in the development of cloud particles into raindrops and snowflakes (Section D.5). In rivers, supercooling of 0.01 to 0.1 °C often occurs during the formation of ice in fast-flowing reaches (Michel 1971).

### B.1.3   Evaporation and Condensation

At temperatures less than 100 °C, some molecules at the liquid–air interface with greater-than-average energy sever all hydrogen bonds with their neighbors and fly off from the bulk liquid. Thus evaporation from ice and liquid water can take place at all temperatures below the boiling point, and the loss of

**FIGURE B-2**
(a) Schematic diagram of a hydrogen atom (left) and an oxygen atom (right). (b) Schematic diagram of a water molecule showing covalent bonding. ⊖ symbols represent electrons. From *Fluvial Hydrology* by S. Lawrence Dingman. © 1984 by W.H. Freeman and Company. Reprinted with permission.

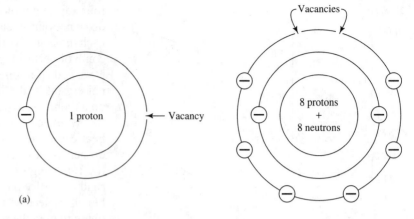

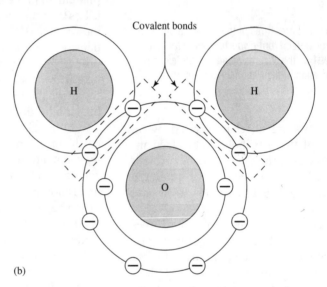

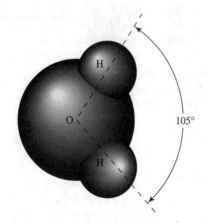

**FIGURE B-3**
Diagram of a water molecule, showing the angle between the hydrogen atoms. After Davis and Day (1961).

these high-energy molecules results in a lowering of the average energy, and hence the temperature, of the remaining solid or liquid. When liquid water is heated to 100 °C, further additions of energy cause the eventual breaking of all the remaining hydrogen bonds, and the liquid is transformed entirely into a gas consisting of relatively widely spaced, mostly non-bonded individual $H_2O$ molecules.

### B.1.4 Dissociation

An **ion** is an elemental or molecular species with a net positive or negative electrical charge. At any given instant, a fraction of the molecules of liquid water are **dissociated** into positively charged hydrogen ions, designated $H^{+1}$, and negatively charged hydroxide ions, designated $OH^{-1}$. In spite of their

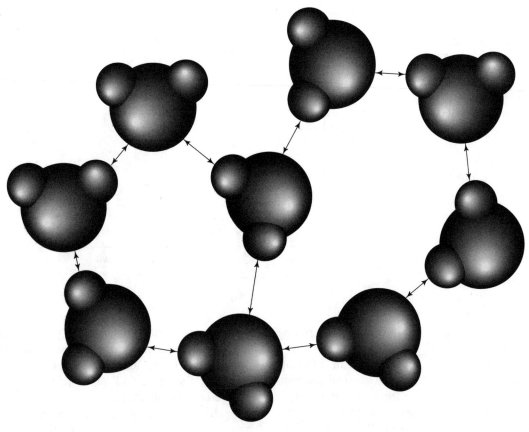

**FIGURE B-4**
Schematic diagram of water molecules in the liquid state. Arrows indicate hydrogen bonds between the opposite-
ly charged ends of adjacent molecules. From *Fluvial Hydrology* by S. Lawrence Dingman. © 1984 by W.H.
Freeman and Company. Reprinted with permission.

generally very low concentrations, these ions participate in many important chemical reactions.

Hydrogen ions are responsible for the acidity
of water, and acidity is usually measured in terms of
the quantity called **pH,** which is defined as

$$pH \equiv -\log_{10}([H^{+1}]),   \textbf{(B-1)}$$

where $[H^{+1}]$ designates the concentration of hydrogen ions in mg $L^{-1}$. The concentration of hydrogen ions in pure water at 25 °C is $10^{-7.00}$ mg $L^{-1}$
(pH = 7.00). As $[H^{+1}]$ increases above this value
(pH decreases below 7.00), water becomes more
acidic; as $[H^{+1}]$ decreases (pH > 7.00), it becomes
more basic.

Certain chemical reactions change the concentration of hydrogen ions, causing the water to become more or less acidic. The degree of acidity, in
turn, determines the propensity of the water to dissolve many elements. The pH of cloud water in
equilibrium with the carbon dioxide in the atmosphere is about 5.7; additional reactions make the
natural pH of rainwater fall in the range from 4.5 to
5.6, depending on location (Turk 1983).

### B.1.5   Isotopes

**Isotopes** of an element have the same number of
protons and electrons, but differing numbers of
neutrons; thus they have similar chemical behavior,
but differ in atomic weight. Some isotopes are **radioactive,** and decay naturally to other atomic forms
at a characteristic rate, while others are **stable.**

Table B-1 gives the properties and abundances
of the isotopes of hydrogen and oxygen, from which

**FIGURE B-5**
Melting/freezing and boiling/condensation temperatures of Group VIa hydrides. In the absence of hydrogen bonds, water would have a melting/freezing point of −100 °C and a boiling/condensation point of −91 °C. After Davis and Day (1961).

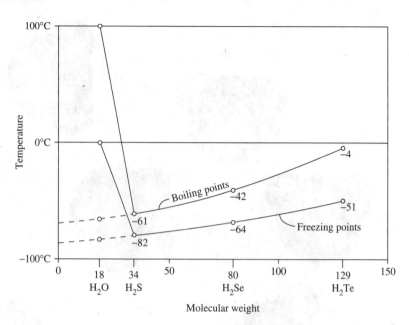

it can be calculated that 99.73 % of all water consists of "normal" $^1H_2^{16}O$.[1]

The various isotopes are involved in differing proportions in phase changes and chemical and biological reactions, so they are **fractionated** as water moves through the hydrologic cycle. (See Fritz and Fontes 1980; Drever 1982). Thus the relative concentrations of these isotopes can be used in some hydrologic situations to identify the sources of water in glaciers, aquifers, or streams. (See, e.g., Perry and Montgomery 1982; also see Box 9-1.)

The isotope $^3H$, called **tritium,** is radioactive; it decays with a half life of 12.5 yr to $^3He$. It is produced in very small concentrations by natural processes and in larger concentrations by nuclear reactions. It has potential for use in dating relatively recent water in aquifers and glaciers. (See, e.g., Davis and Murphy 1987; also see Section 8.5.1.) Figure B-7 shows the annual rates of production of bomb tritium.

## B.2   PROPERTIES OF WATER

In this section, we briefly describe the bulk properties of liquid water that influence its movement through the hydrologic cycle and its interactions

with the terrestrial environment. More detailed discussions of these properties can be found in Dorsey (1940) and Davis and Day (1961). Properties of water in the solid and vapor form are discussed where relevant in the text.

The variation of water's properties with temperature is important in many hydrologic contexts. Thus, in the following discussion, the values of each property at 0 °C are given in the three unit systems, and their relative variations with temperature are

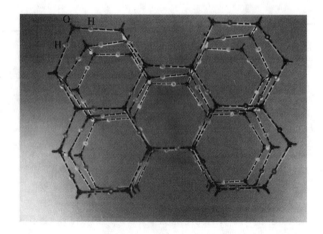

**FIGURE B-6**
A model of the crystal lattice of ice, showing its hexagonal structure. White circles are hydrogen atoms, dark circles are oxygen atoms, dashed lines are hydrogen bonds. Photo by author.

---

[1] By convention, the atomic weight is written to the upper left of the element symbol.

## TABLE B-1
Characteristics of Isotopes of Hydrogen and Oxygen

| Isotope | Natural Abundance (%) | Stability |
|---------|----------------------|-----------|
| $^1H$ | 99.985 | stable |
| $^2H$ (Deuterium) | 0.015 | stable |
| $^3H$ (Tritium) | trace | radioactive |
| $^{16}O$ | 99.76 | stable |
| $^{17}O$ | 0.04 | stable |
| $^{18}O$ | 0.20 | stable |

shown in Table B-2. Empirical equations for computing the values of the properties as functions of temperature are also given.

### B.2.1   Density

**Mass density,** $\rho$, is the mass per unit volume $[M\,L^{-3}]$ of a substance; **weight density,** $\gamma$, is the weight per unit volume $[F\,L^{-3}]$. These are related by Newton's Second Law (i.e., force equals mass time acceleration), so that

$$\gamma = \rho \cdot g, \qquad \text{(B-2)}$$

where $g$ is the acceleration due to gravity $[L\,T^{-2}]$ ($g = 9.81$ m s$^{-2}$ = 32.2 ft s$^{-2}$).

In the SI system of units, the kilogram is defined as the mass of 1 m$^3$ of pure water at its temperature of maximum density, 3.98 °C. For water at 0 °C,

$$\rho = 999.87\,\text{kg m}^{-3} = 0.99987\,\text{g cm}^{-3} = 1.9397\,\text{slug ft}^{-3}$$

and

$$\gamma = 9799\,\text{N m}^{-3} = 979.9\,\text{dyn cm}^{-3} = 62.46\,\text{lb ft}^{-3}.$$

The **specific gravity** of a substance is the ratio of its weight density to the weight density of pure water at 3.98 °C; thus it is dimensionless.

Because gravitational force and momentum are proportional to mass, and pressure depends on weight, either $\rho$ or $\gamma$ appears in most equations describing the motion of fluids.

The change in density of water with temperature is unusual (Table B-2) and environmentally significant. As was noted, liquid water at 0 °C is denser than ice. As liquid water is warmed from 0 °C its density initially *increases*, whereas most other substances become less dense as they warm. This anomalous increase continues until density reaches a maximum value of 1000 kg m$^{-3}$ at 3.98 °C; beyond this, the density decreases with temperature, as with most other substances. These density variations can be approximated as

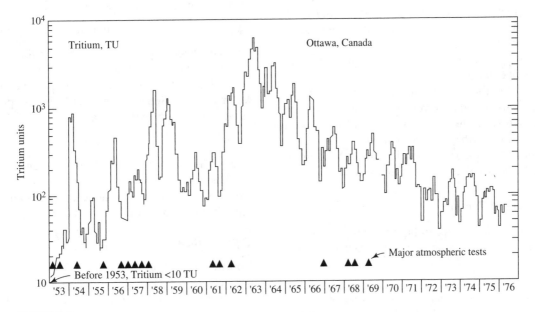

### FIGURE B-7
Tritium concentrations in precipitation at Ottawa, Canada. This is the longest continuous record; it is representative of trends in the northern hemisphere. From R. Allan Freeze and John A. Cherry, *Groundwater*, © 1979, p. 137. Reprinted with permission of Prentice Hall, Englewood Cliffs, NJ.

**TABLE B-2**

Relative Values of Properties of Pure Liquid Water as Functions of Temperature

| Temperature (°C) | $\rho, \gamma$ | $\mu$ | $\nu$ | $\sigma$ | $c_p$ | $\lambda_v$ |
|---|---|---|---|---|---|---|
| 0 | 1.00000 | 1.0000 | 1.0000 | 1.0000 | 1.0000 | 1.0000 |
| 3.98 | 1.00013 | | | | | |
| 5 | 1.00012 | 0.8500 | 0.8500 | 0.9907 | 0.9963 | 0.9953 |
| 10 | 0.99986 | 0.7314 | 0.7315 | 0.9815 | 0.9940 | 0.9904 |
| 15 | 0.99926 | 0.6374 | 0.6379 | 0.9722 | 0.9924 | 0.9857 |
| 20 | 0.99836 | 0.5607 | 0.5616 | 0.9630 | 0.9915 | 0.9810 |
| 25 | 0.99720 | 0.4983 | 0.4997 | 0.9524 | 0.9910 | 0.9763 |
| 30 | 0.99580 | 0.4463 | 0.4482 | 0.9418 | 0.9907 | 0.9715 |

$\rho \equiv$ mass density,    $\gamma \equiv$ weight density,    $\mu \equiv$ dynamic viscosity,    $\nu \equiv$ kinematic viscosity, $\sigma \equiv$ surface tension, $\lambda_v \equiv$ latent heat of vaporization, $c_p \equiv$ specific heat.

$$\rho = 1000 - 0.019549 \cdot |T - 3.98|^{1.68}, \quad \textbf{(B-3)}$$

where $T$ is temperature in °C and $\rho$ is in kg m$^{-3}$ (Heggen 1983). The variation of $\gamma$ with temperature can be approximated via Equations (B-2) and (B-3).

In lakes where temperatures reach 3.98 °C, the density maximum controls the vertical distribution of temperature and causes an annual or semi-annual overturn of water that has a major influence on biological and physical processes. However, except for lakes, the variation of density with temperature is small enough that it can usually be neglected in hydrological calculations.[2]

The addition of dissolved or suspended solids to water increases its density in proportion to the density of the solids and their concentration. Again, the density effects of dissolved materials can be important in lakes, but they are not usually significant in other environments.[3] However, high concentrations of suspended matter can significantly increase the effective density of water in rivers.

Note that the kilogram and gram are commonly used as units of force as well as of mass: 1 kilogram (gram) of force is the weight of a mass of 1 kilogram (gram) at the earth's surface, where $g = 9.81$ m s$^{-2}$ (981 cm s$^{-2}$). Thus, from Equation (B-2), 1 kilogram of force = 9.810 N; 1 gram of force = 981.0 dyn.

### B.2.2   Surface Tension

Molecules in the surface of liquid water are subjected to a net inward force due to hydrogen bonding with the molecules below the surface (Figure B-8). **Surface tension,** $\sigma$, is equal to the magnitude of that force divided by the distance over which it acts; thus its dimensions are [F L$^{-1}$]. It can also be viewed as the work required to increase the surface area of a liquid by a unit amount ([F L]/[L$^2$] = [F L$^{-1}$]).

Surface tension and the closely related phenomenon of **capillarity** significantly influence fluid motion where a water surface is present and where the flow scale is less than a few millimeters—for example, in porous media that are partially saturated or ones in which there is an interface between water and an immiscible liquid (e.g., hydrocarbons).

As might be expected from the strong intermolecular forces, water has a surface tension higher than that of most other liquids; its value at 0 °C is

$$\sigma = 0.0756 \text{ N m}^{-1} = 75.6 \text{ dyn cm}^{-1} = 0.00518 \text{ lb ft}^{-1}.$$

Surface tension decreases rapidly as temperature increases (Table B-2), and this effect can be important when one is considering the movement of water in soils. (See Section 6.5.4.) The temperature effect can be approximated as

$$\sigma = 0.001 \cdot (20987 - 92.613 \cdot T)^{0.4348}, \quad \textbf{(B-4)}$$

where $T$ is in °C and $\sigma$ is in N m$^{-1}$ (Heggen 1983).

Dissolved substances can also increase or decrease surface tension, and certain organic compounds have a major effect on its value.

#### Capillary Rise

Consider the small (diameter a few millimeters or less) cylindrical tube immersed in a body of water with a free surface[4] shown in Figure B-9. If the ma-

---

[2] An exception to this generalization can occur when one is considering deep regional ground-water flow systems on geologic time scales. In these systems, density variations related to geothermal heat flow can have significant effects on the flow.
[3] Density gradients due to the dissolution of minerals like salt can significantly affect deep ground-water flows on geological time scales.

---

[4] A "free surface" is a surface of liquid water at atmospheric pressure. In diagrams, such a surface is designated by the inverted triangular *hydrat* symbol, $\nabla$.

## FIGURE B-8

Intermolecular forces acting on typical surface (S) and nonsurface (B) molecules. From *Fluvial Hydrology* by S. Lawrence Dingman. © 1984 by W.H. Freeman and Company. Reprinted with permission.

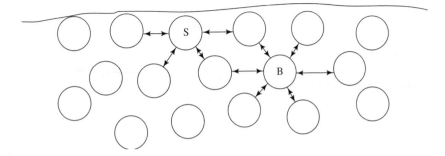

terial of the tube is such that the hydrogen bonds of the water are attracted to it, the molecules in contact with the tube are drawn upward. The degree of attraction between the water and the tube is reflected in the **contact angle** between the water surface, or **meniscus,** and the tube: the stronger the attraction, the smaller the angle. Because of the intermolecular hydrogen bonds, the entire mass of water within the tube will be also drawn upward until the adhesive force between the molecules of the tube and those of the water is balanced by the downward force due to the weight of the water suspended within the tube.

The height to which the water will rise in the tube can thus be calculated by equating the upward and downward forces. The upward force, $F_u$, equals the vertical component of the surface tension times the distance over which that force acts:

$$F_u = \sigma \cdot \cos(\theta_c) \cdot 2 \cdot \pi \cdot r, \qquad \textbf{(B-5)}$$

where $\theta_c$ is the contact angle between the water surface and the tube and $r$ is the radius of the tube. The downward force due to the weight of the column of water, $F_d$, is

$$F_d = \gamma \cdot \pi \cdot r^2 \cdot h_{cr}, \qquad \textbf{(B-6)}$$

where $\gamma$ is the weight density of water and $h_{cr}$ is the height of the column.

Equating $F_u$ and $F_d$ and solving for $h_{cr}$ yields

$$h_{cr} = \frac{2 \cdot \sigma \cdot \cos(\theta_c)}{\gamma \cdot r}. \qquad \textbf{(B-7)}$$

Thus the height of capillary rise is inversely proportional to the radius of the tube and directly proportional to the surface tension and the cosine of the contact angle.

Table B-3 gives the contact angle for water in contact with air and selected solids. Note that the contact angle for most soil materials is close to 0°.

Under natural conditions, the interconnected pores between particles of granular geologic materials act as capillary tubes and cause a **tension-satu-**

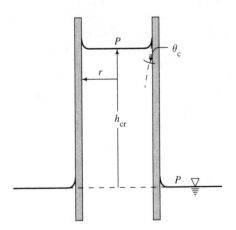

## FIGURE B-9

Sketch of phenomenon of capillary rise in a circular tube of radius $r$, $\theta_c$ is the contact angle between the meniscus and the wall, and $P$ is atmospheric pressure. After Dingman (1984).

**TABLE B-3**

Surface-Tension Contact Angles for Water–Air Interfaces Against Various Solids

| Solid | Contact Angle, $\theta_c$ (°) | cos ($\theta_c$) |
|---|---|---|
| Glass | 0 | 1.0000 |
| Most silicate minerals | 0 | 1.0000 |
| Ice | 20 | 0.9397 |
| Platinum | 63 | 0.4540 |
| Gold | 68 | 0.3746 |
| Talc | 86 | 0.0698 |
| Paraffin | 105 to 110 | −0.2588 to −0.3420 |
| Shellac | 107 | −0.2924 |
| Carnauba wax | 107 | −0.2924 |

Data from Dorsey (1940) and Jellinek (1972).

**rated zone** or **capillary fringe** above the free surface (Section 6.4.2). The height of this fringe can be calculated approximately via Equation (B-7), with $r$ equal to the radius of the granular particles; it ranges from a few centimeters in gravel and sand to well over a meter in silt. For clay soils, however, additional electrical forces become significant and Equation (B-7) does not hold.

### Pressure Relations

In unsaturated porous media of silt size and larger ($\geq 0.002$ mm; see Section 6.1.1), pressure is determined by the radius of curvature of menisci, and the relation between the two can be developed from further consideration of the capillary-rise phenomenon.

When capillary rise has ceased in Figure B-9, the column of water is suspended from the meniscus, which is in turn attached to the walls by hydrogen bonds. Thus the water is under **tension,** which is defined as negative (i.e., less than atmospheric) pressure. The weight of the water suspended beneath the plane that is tangent to the lowest point of the meniscus is equal to $F_d$ [Equation (B-6)]. The area of that plane is $\pi \cdot r^2$, so the pressure in the suspended water at that plane, $P_m$, is

$$P_m = -\frac{\gamma \cdot \pi \cdot r^2 \cdot h_{cr}}{\pi \cdot r^2} = -\gamma \cdot h_{cr}. \quad \textbf{(B-8)}$$

Substituting Equation (B-7) into Equation (B-8) gives

$$P_m = -\frac{2 \cdot \sigma \cdot \cos(\theta_c)}{r}. \quad \textbf{(B-9)}$$

Thus the pressure difference across a meniscus, like the height of capillary rise, is inversely proportional to the radius of a capillary opening. In unsaturated soils of silt size or larger, this radius is determined by the soil texture and by the amount of water present. (See Section 6.3.2.) The moisture-characteristics curve (Figure 6-7) is the empirical manifestation of the pressure–radius relation of Equation (B-9) in the range $0 > P_m > -0.9$ atmospheres ($= -930$ cm $= -91.2$ kPa). Tensions many times stronger than 1 atmosphere can exist in soils, but they are due to attractive forces between the water and the mineral grains and do not represent capillary pressure (Gray and Hassanizadeh 1991).

### B.2.3   Boundary-Layer Flow, Viscosity, and Turbulence

#### Boundary-Layer and Potential Flows

When water flows over a solid boundary, hydrogen bonds cause the fluid molecules adjacent to the boundary to "stick" so that the velocity at the boundary is zero. This phenomenon is called the **no-slip condition.** This condition produces a frictional drag that is transmitted through the fluid for considerable distances normal to the boundary.

In virtually all open-channel flows of interest to hydrologists, the retarding effects of the boundary are present throughout the flow, and such flows are therefore called **boundary-layer flows.**

Flows within the pores of porous media are also boundary-layer flows, but mathematically they are not treated as such. As described in Section 8.1.1, Darcy's Law describes bulk flow through a representative "chunk" of the medium rather than through individual pores. Thus the effects of intra-pore viscous resistance are represented by the bulk hydraulic conductivity, and the boundary of the overall flow (i.e., the aquifer boundary) is not a source of flow resistance. Thus ground-water flow is mathematically represented as a **potential flow.**

#### Viscosity and Laminar Flow

For flows that are very close to a boundary and are very slow, the water moves in parallel layers (Figure B-10a), and the flow is called **laminar.** The **dynamic viscosity,** $\mu$, is the friction between the layers that transmits the boundary friction through the fluid and resists the forces tending to cause flow. The layers move successively faster as one moves away from the boundary, so that a velocity gradient exists, and $\mu$ is the proportionality between this gradient and the frictional force per unit area, called the **shear stress,** $\tau(y)$. Thus we have

$$\tau(y) = \mu \cdot \frac{d\,u(y)}{d\,y}, \quad \textbf{(B-10)}$$

where $u(y)$ is the velocity at a distance $y$ from the boundary.

Equation (B-10) shows that, for a given boundary shear stress, a higher viscosity causes a smaller velocity gradient—that is, the velocity increases more slowly as one moves away from the boundary.

**FIGURE B-10**
Paths of individual water particles in (a) laminar flow, (b) turbulent flow (highly schematic). After Dingman (1984).

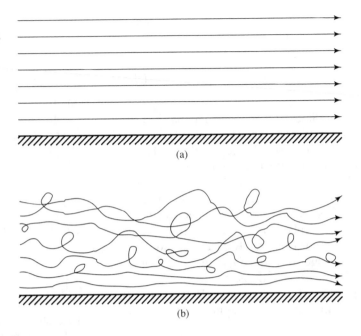

(a)

(b)

Note that, because $\tau(y)$ has the dimensions $[F\ L^{-2}]$, $\mu$ has the dimensions $[F\ T\ L^{-2}] = [M\ L^{-1}\ T^{-1}]$.

The dynamic viscosity is due to inter-molecular attractions, and so it is also called the **molecular viscosity.** In spite of the strength of the hydrogen bonds, water's viscosity is relatively low because of the rapidity with which the hydrogen bonds spontaneously break and reform (about once every $10^{-12}$ s). Viscosity at 0 °C is

$$\mu = 0.001787\ \text{N s m}^{-2} = 0.01787\ \text{dyn s cm}^{-2}$$
$$= 3.735\ 10^{-5}\ \text{lb s ft}^{-2}.$$

In many contexts, the ratio $\mu/\rho$ arises; thus it is convenient to define the **kinematic viscosity,** $\nu\ [L^2\ T^{-1}]$:

$$\nu \equiv \frac{\mu}{\rho}. \qquad \textbf{(B-11)}$$

Values of $\nu$ at 0 °C are

$$\nu = 1.787 \times 10^{-6}\ \text{m}^2\ \text{s}^{-1} = 1.787 \times 10^{-2}\ \text{cm}^2\ \text{s}^{-1}$$
$$= 1.926 \times 10^{-5}\ \text{ft}^2\ \text{s}^{-1}.$$

As is shown in Table B-2, viscosity decreases rapidly as temperature increases. The temperature effect can be approximated as

$$\mu = 2.0319 \times 10^{-4} + 1.5883 \times 10^{-3} \cdot \exp\left[-\left(\frac{T^{0.9}}{22}\right)\right], \qquad \textbf{(B-12)}$$

where $T$ is in °C and $\mu$ is in N s m$^{-2}$ (Heggen 1983).

Some dissolved constituents increase viscosity, others decrease it, but these effects are usually negligible at the concentrations found in nature.[5] However, moderate-to-high concentrations of suspended material can significantly increase the effective viscosity of the fluid.

### Turbulence

As distance from a boundary and velocity increase, the inertia of the moving water increasingly overcomes viscous friction and the flow paths of individual water "particles" increasingly deviate from the parallel layers of laminar flow. At relatively modest distances and velocities, all semblance of parallel flow disappears and the water moves in highly irregular eddies. This is the phenomenon called **turbulence** (Figure B-10b).

Because the water in turbulent eddies moves in directions other than the main flow direction, turbulence consumes some of the energy that would otherwise drive the main flow. This energy loss can be thought of as an addition to the internal friction of the fluid and therefore as an addition to the ef-

---

[5] As with density, thermal and solute effects on viscosity can have a significant influence on deep ground-water flow over geologic time scales.

fective viscosity. This effect is called the **eddy viscosity,** $\varepsilon$, which can be defined as

$$\varepsilon \equiv k^2 \cdot \rho \cdot y^2 \cdot \frac{d\,u(y)}{d\,y}, \qquad \textbf{(B-13)}$$

where $k$ is a dimensionless constant that has been determined by experiment to equal 0.4. Note that the dimensions of $\varepsilon$ are identical to those of $\mu$.

Physically, the effect of viscosity is always present, and it is the ultimate mechanism by which the retarding effect of a boundary is transmitted into the fluid. Thus the flow resistances due to eddy viscosity and molecular viscosity are additive. However, the magnitude of $\varepsilon$ greatly exceeds that of $\mu$ for even small values of $y$ and $du(y)/dy$, so $\mu$ is usually neglected for turbulent flows.

Replacing $\mu$ in Equation (B-10) with the analogous quantity $\varepsilon$ from Equation (B-13) gives

$$\tau(y) = k^2 \cdot \rho \cdot y^2 \cdot \left[\frac{d\,u(y)}{d\,y}\right]^2. \qquad \textbf{(B-14)}$$

Thus when turbulence is present and viscosity can be ignored, the relation between velocity gradient and friction becomes a function of the density, the location within the flow, and the velocity gradient.

### Criteria for Laminar or Turbulent Flow State

Comparison of Equations (B-10) and (B-14) shows that the nature and magnitude of internal flow resistance differ greatly between laminar and turbulent flows. For open-channel flows, the equations of motion for the two states differ (as will be discussed more fully later), and the presence in turbulent flows of eddies with vertical velocity components makes possible the suspension of sediment from stream beds. For ground-water flows, all quantitative descriptions are based on Darcy's Law, which applies only to laminar flow. Thus it is essential to be able to determine the state of flows in the hydrologic cycle.

The criterion for distinguishing laminar from turbulent flow can be developed by examining the ratio of eddy viscosity, $\varepsilon$, to dynamic viscosity, $\mu$:

$$\frac{\varepsilon}{\mu} = \frac{k^2 \cdot \rho \cdot y^2}{\mu} \cdot \frac{du(y)}{dy}. \qquad \textbf{(B-15)}$$

The larger this ratio is, the more important are the inertial forces and the greater is the degree of turbulence.

However, it is clear from Equation (B-15) that the ratio $\varepsilon/\mu$ will take on different values at different positions within a given flow. To convert that equation into a form that will characterize an entire flow, we can (1) replace $y$ with a characteristic length scale of the flow, $L$, and (2) replace $du(y)/dy$ with a characteristic velocity-gradient scale given by $U/L$, where $U$ is the average flow velocity. After making these generalizations and dropping the numerical constant in Equation (B-15), we define the dimensionless ratio $\Re$ as

$$\Re \equiv \frac{\rho \cdot U \cdot L}{\mu} = \frac{U \cdot L}{\nu}, \qquad \textbf{(B-16)}$$

where $\Re$ is called the **Reynolds number,** after the English hydraulician Osborne Reynolds, who studied the transition from laminar to turbulent flow. Reynolds numbers appear in many hydraulic contexts, with the precise velocity and length defined by convention for each situation. We now examine the forms of the Reynolds numbers for porous-media and open-channel flows.

For flows in porous media, the Reynolds number $\Re_{pm}$ is defined as

$$\Re_{pm} \equiv \frac{\rho \cdot V \cdot d}{\mu} = \frac{V \cdot d}{\nu}, \qquad \textbf{(B-17)}$$

where $V$ is the "Darcy velocity" (volumetric flow rate per unit cross-sectional area) and $d$ is the average diameter of the grains making up the medium. These flows are laminar when $\Re_{pm} < 1$, turbulent when $\Re_{pm} > 10$, and transitional in between.

The Reynolds number applicable to open-channel flows, $\Re_{oc}$, is defined as

$$\Re_{oc} \equiv \frac{\rho \cdot U \cdot Y}{\mu} = \frac{U \cdot Y}{\nu}, \qquad \textbf{(B-18)}$$

where $U$ is the average flow velocity and $Y$ is the average flow depth. Observation has shown that such flows are laminar when $\Re_{oc}$ is less than about 500, and full turbulence is present when $\Re_{oc} > 2000$. Flows with $500 < \Re_{oc} < 2000$ are characterized as "transitional." Substitution of some typical values of $U$ and $Y$ will show that virtually all streams of any size have turbulent flow.

## B.2.4 Thermal Capacity

**Thermal capacity** (or **heat capacity**), $c_p$, is the property that relates a temperature change of a substance to a change in its heat-energy content. It is defined as the amount of heat energy, $\Delta H$, absorbed (released) by a mass $M$ of a substance when its temperature is raised (lowered) by an amount $\Delta T$:

$$c_p \equiv \frac{\Delta H}{M \cdot \Delta T}. \qquad \text{(B-19)}$$

Thus its dimensions are $[\text{L}^2\text{T}^{-2}\,\Theta^{-1}] = [\text{E M}^{-1}\Theta^{-1}]$.

The thermal capacity of water at 0 °C is

$$c_p = 4216 \text{ J kg}^{-1}\text{ K}^{-1} = 1.007 \text{ cal g}^{-1}\text{ C}^{\circ}$$
$$= 32.43 \text{ Btu slug}^{-1}\text{ F}^{\circ -1}$$
$$(= 1.007 \text{ Btu lb}^{-1}\text{ F}^{\circ -1})$$

and, as is shown in Table B-2, its value decreases slowly with temperature.[6] In the range of temperatures usually encountered in hydrology, the temperature dependence of $c_p$ can be approximated as

$$c_p = 4175 + 1.666 \cdot$$
$$\left[ \exp\left(\frac{34.5 - T}{10.6}\right) + \exp\left(-\frac{34.5 - T}{10.6}\right) \right], \qquad \text{(B-20)}$$

where $T$ is in °C and $c_p$ is in J kg$^{-1}$ K$^{-1}$ (Heggen 1983).

The temperature of a substance reflects the vibratory energy of its molecules. The thermal capacity of water is very high relative to that of most other substances because, when heat energy is added to it, much of the energy is used to break hydrogen bonds rather than to increase the rate of molecular vibrations. This high specific heat has a profound influence on organisms and the global environment: It makes it possible for warm-blooded organisms to regulate their temperatures, and it makes the oceans and other bodies of water moderators of the rates and magnitudes of ambient temperature changes.

## B.2.5 Latent Heats

Latent heat is energy that is released or absorbed when a given mass of substance undergoes a change of phase. Thus its dimensions are energy per mass,

$[\text{E M}^{-1}]$ or $[\text{L}^2\text{ T}^{-2}]$. The term "latent" is used because no temperature change is associated with the gain or loss of heat. The large amounts of energy that are required to break hydrogen bonds during melting and vaporization (and conversely are released by the formation of bonds during freezing and condensation) make water's latent heats very large relative to those of other substances.

The **latent heat of fusion,** $\lambda_f$, is the quantity of heat energy that is added (released) when a unit mass of substance melts (freezes). For water, this is a significant quantity:

$$\lambda_f = 3.34 \times 10^5 \text{ J kg}^{-1} = 79.7 \text{ cal g}^{-1}$$
$$= 4620 \text{ Btu slug}^{-1}$$
$$(= 144 \text{ Btu lb}^{-1}).$$

Latent heat of fusion plays an important role in the dynamics of freezing and thawing of water bodies and of water in the soil: Once the temperature is raised (lowered) to 0 °C, this heat must be conducted to (from) the melting (freezing) site in order to sustain the melting (freezing) process.

The **latent heat of vaporization,** $\lambda_v$, is the quantity of heat energy that is added (released) when a unit mass of substance vaporizes (condenses). Vaporization involves the complete breakage of hydrogen bonds, and water has one of the largest latent heats of vaporization of any substance:

$$\lambda_v = 2.495 \text{ MJ kg}^{-1} = 595.9 \text{ cal g}^{-1}$$
$$= 3.457 \times 10^4 \text{ Btu slug}^{-1}$$
$$(= 1074 \text{ Btu lb}^{-1}).$$

The latent heat of vaporization decreases with temperature, approximately as

$$\lambda_v = 2.495 - (2.36 \times 10^{-3}) \cdot T, \qquad \text{(B-21)}$$

where $T$ is in °C and $\lambda_v$ is in MJ kg$^{-1}$.

At 100 °C, the latent heat of vaporization is 2.261 MJ kg$^{-1}$, more than six times the latent heat of fusion and more than five times the amount of energy it takes to warm the water from the melting point to the boiling point.

Water's enormous latent heat of vaporization plays important roles: as a factor in global heat transport (Section 3.1.2); as a source of energy that drives the precipitation-forming process (Section 4.1 and D.5), and as a mechanism for transferring large amounts of heat from the earth's surface into the atmosphere (Section 3.1.1).

---

[6] Thermal capacity actually reaches a minimum value of 4178.42 J kg$^{-1}$ K$^{-1}$ at 37 °C and slowly increases at higher temperatures.

### B.2.6 Solvent Power

Because of the unique polar structure of the water molecules and the existence of hydrogen bonds, almost every substance is soluble in water to some degree. Salts, such as sodium chloride (NaCl), readily form ions that are maintained in solution because the positive and negative ends of the water molecules attach to the oppositely charged ions. Each ion is thus surrounded by a cloud of water molecules that prevents the $Na^{+1}$ and $Cl^{-1}$ ions from recombining. Other substances, particularly organic compounds such as sugars, alcohols, and amino acids, are soluble because the molecules form hydrogen bonds with the water molecules.

The importance of the solvent power of water to biogeochemical processes cannot be overstated. The first steps in the process of erosion involve the dissolution and aqueous alteration of minerals, and a significant portion of all the material transported by rivers from land to oceans is carried in solution (Section 3.2.7). Virtually all life processes take place in water and depend on the delivery of nutrients and the removal of wastes in solution. In plants, the carbon dioxide necessary for photosynthesis enters in dissolved form; in animals, the transport and exchange of oxygen and carbon dioxide essential for metabolism take place in solution.

## B.3 FLOW EQUATIONS

Hydrologic flows are mathematically described by (1) an equation of energy or momentum that relates the rate of flow (velocity) to the driving forces and the internal and external resistance to flow and (2) an equation of conservation of mass (continuity).

### B.3.1 Ground-Water Flows

Because ground water flows through very small pores at low velocities, its flows are almost always laminar. The energy equation for laminar ground-water flows is Darcy's Law [Equation (8-1) for saturated flows; Equation (6-8) for unsaturated flows].

The continuity equation for saturated ground-water flows is derived in Box 8-1; this is combined with Darcy's Law to give the general equation of ground-water flow, Equation (8-14). The general

equation may be simplified to apply to specific situations [Equations (8-15) – (8-18)].

The Richards Equation (Box 6-1) is a combination of Darcy's Law and the continuity equation for vertical unsaturated flow.

### B.3.2 Open-Channel Flows

To develop relations analogous to Darcy's Law, we take a macroscopic view of open-channel flows. We will be concerned with average-flow characteristics, where the average is taken over the entire cross-section at right angles to the general flow direction. Because virtually all open-channel flows are turbulent, we consider only the turbulent energy equation.

#### *Energy Equation*

The energy of flowing water consists of potential energy and kinetic energy. These energies are conveniently expressed as heads (energy per weight of water, [F L]/[F] = [L]). As in porous-media flows, total potential head is the sum of (1) gravitational potential head and (2) pressure potential head. The gravitational potential head at any point is given by $Z$, the elevation of the bottom above an arbitrary datum. The pressure potential head is given by $Y$, the average flow depth at that point (we assume hydrostatic pressure conditions). Kinetic energy is expressed by the **velocity head,** equal to $U^2/(2 \cdot g)$, where $U$ is average velocity and $g$ is acceleration due to gravity.

From the second law of thermodynamics, total energy $H$ must decrease downstream due to dissipation by friction. This loss, $dH$, is the sum of decreases in the three energy components, so

$$
\begin{aligned}
dH &= -\left[ d\left(\frac{U^2}{2 \cdot g}\right) + dY + dZ \right] \\
&= -\left[ \frac{U}{g} \cdot dU + dY + dZ \right].
\end{aligned}
\quad \textbf{(B-22)}
$$

Dividing Equation (B-22) by $dx$ then gives the downstream rate of change of head:

$$
\frac{dH}{dx} = -\left( \frac{U}{g} \cdot \frac{dU}{dx} + \frac{dY}{dx} + \frac{dZ}{dx} \right).
\quad \textbf{(B-23)}
$$

For unsteady flows, there is an additional energy loss due to changes in average velocity with time, $t$, so the total energy loss is

$$\frac{dH}{dx} = -\left(\frac{U}{g} \cdot \frac{\partial U}{\partial x} + \frac{\partial Y}{\partial x} + \frac{\partial Z}{\partial x} + \frac{1}{g} \cdot \frac{\partial U}{\partial t}\right). \tag{B-24}$$

If we note that $\partial Z/\partial x$ equals the negative of the tangent of the channel slope, $S_c \equiv \tan(\beta_c)$, we can also write Equation (B-24) as

$$\frac{dH}{dx} = S_c - \frac{\partial Y}{\partial x} - \frac{U}{g} \cdot \frac{\partial U}{\partial x} - \frac{1}{g} \cdot \frac{\partial U}{\partial t}. \tag{B-25}$$

The downstream gradient of total head, $dH/dx$, is called the **energy slope**, and Equation (B-25) is known as the **energy equation** for open-channel flow.

All the terms in the energy equation represent forces tending to accelerate the flow: (1) the channel slope gives the force due to gravitational acceleration; (2) the $\partial Y/\partial x$ term represents the force due to a pressure gradient in the $x$-direction; (3) $\partial U/\partial x$ represents a force due to **convective acceleration;** and (4) $\partial U/\partial t$ represents a force due to **local acceleration.**

For turbulent flows, the flow rate is proportional to the square root, rather than to the first power, of the energy gradient:

$$\begin{pmatrix}\text{flow}\\\text{velocity}\end{pmatrix} = \begin{pmatrix}\text{channel}\\\text{conductivity}\end{pmatrix} \times \begin{pmatrix}\text{energy}\\\text{gradient}\end{pmatrix}^{1/2}. \tag{B-26a}$$

The channel conductivity for turbulent open-channel flows depends on the channel roughness, irregularity, and tortuosity as well as on the depth, and Equation (B-26a) is usually given as either the Chézy Equation [Equation (9-18)],

$$U = u_C \cdot C \cdot Y^{1/2} \cdot \left(\frac{dH}{dx}\right)^{1/2}, \tag{B-26b}$$

or the Manning Equation [Equation (9-19)],

$$U = \left(\frac{u_M \cdot Y^{2/3}}{n}\right) \cdot \left(\frac{dH}{dx}\right)^{1/2}, \tag{B-26c}$$

where $u_C$ and $u_M$ are unit-conversion factors, $C$ is Chézy's resistance factor, $n$ is Manning's resistance factor, and $n$ and $C$ are related via Equation (9-20). Combining Equations (B-25) and (B-26c) yields

$$U = \left(\frac{u_M \cdot Y^{2/3}}{n}\right) \cdot \left[S_c - \frac{\partial Y}{\partial x} - \frac{U}{g} \cdot \frac{\partial U}{\partial x} - \frac{1}{g} \cdot \frac{\partial U}{\partial t}\right]^{1/2}. \tag{B-27}$$

At a given cross-section, where $n$ and $S_c$ are constant, Equation (B-27) indicates that velocity is largely a function of depth and slope, modified by spatial and temporal gradients of depth and velocity. If these gradients are small enough to neglect, Equation (B-27) reduces to

$$U_u = \left(\frac{u_M \cdot Y^{2/3}}{n}\right) \cdot S_c^{1/2}, \tag{B-28}$$

and velocity becomes a single-valued function of $Y$. A steady flow with no downstream change in depth (or velocity) is a **uniform flow,** so the velocity given by Equation (B-28), $U_u$, is called the **uniform-flow velocity.**

### Continuity Equation

Figure B-11 is an element of a stream channel receiving lateral inflows (which can be surface or subsurface flows). The downstream ($x$-direction) extent of the element is $dx$, and the conservation of mass for the element during time increment $dt$ is

$$M_{in} - M_{out} = \Delta M, \tag{B-29}$$

where $M_{in}$ and $M_{out}$ are the mass of water entering and leaving, respectively, the volume element in time increment $dt$ and $\Delta M$ is the change in the amount of mass stored in the element over $dt$.

$M_{in}$ is the sum of the discharge from upstream and the lateral inflow; that is,

$$M_{in} = \rho \cdot U \cdot A \cdot dt + \rho \cdot q \cdot dx \cdot dt, \tag{B-30}$$

where $\rho$ is the mass density of water (assumed constant), $U$ is the average velocity and $A$ the cross-sectional area at the upstream boundary, and $q$ is the rate of lateral inflow per unit channel length, $[L^2 T^{-1}]$.

Both velocity and area can change in the $x$-direction, so

$$M_{out} = \rho \cdot \left(U + \frac{\partial U}{\partial x} \cdot dx\right) \cdot \left(A + \frac{\partial A}{\partial x} \cdot dx\right) \cdot dt. \tag{B-31}$$

Neglecting terms involving $(dx)^2$, $\Delta M$ is

$$\Delta M = \rho \cdot \frac{\partial A}{\partial y} \cdot dt \cdot dx. \tag{B-32}$$

Substituting Equations (B-30)–(B-32) into (B-29), simplifying, and again neglecting terms involving $(dx)^2$ yields

$$q - U \cdot \frac{\partial A}{\partial x} - A \cdot \frac{\partial U}{\partial x} = \frac{\partial A}{\partial t} \tag{B-33}$$

**FIGURE B-11**
Definitions of terms for deriving the continuity equation for open-channel flow.

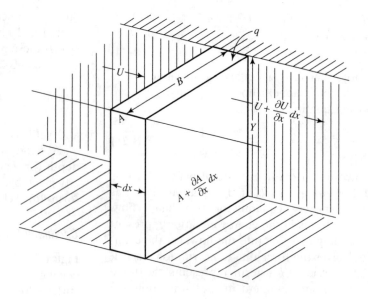

as one form of the continuity equation. However, this can be put into a more useful form by noting that discharge, $Q$, equals $U \cdot A$; so,

$$U \cdot \frac{\partial A}{\partial x} + A \cdot \frac{\partial U}{\partial x} = \frac{\partial Q}{\partial x}. \qquad \textbf{(B-34)}$$

Thus Equation (B-34) is equivalent to

$$q - \frac{\partial Q}{\partial x} = \frac{\partial A}{\partial t}. \qquad \textbf{(B-35)}$$

Another version of the continuity equation can be given for the case when the channel width, $B$, is assumed to remain constant. Here, $A = B \cdot Y$, where $Y$ is average depth, so we have

$$q - \frac{\partial Q}{\partial x} = B \cdot \frac{\partial Y}{\partial t}. \qquad \textbf{(B-36)}$$

### Complete Equations

The complete one-dimensional equations of open-channel flow are given by the most general forms of the energy and continuity equations [Equations (B-27) and (B-34)]. These are often called the **St. Venant Equations.** Approaches to simultaneous numerical solution of these equations were discussed by Strelkoff (1970). However, situations commonly arise where one or more of the terms on the right-hand side of Equation (B-27) can be neglected, and in these cases the energy and continuity equations can be combined into a form that can be solved analytically or by simpler numerical techniques. These simplifications are described in the following sections.

### Convection–Diffusion Equations

If the acceleration terms (those involving $\partial U / \partial x$ and $\partial U / \partial t$ ) in Equation (B-27) are neglected, the energy equation simplifies to

$$\frac{dH}{dx} = S_c - \frac{\partial Y}{\partial x}. \qquad \textbf{(B-37)}$$

This relation can be combined with Equation (B-34) into a single expression:

$$\frac{q}{B} - \frac{3 \cdot U_u}{2} \cdot \frac{\partial Y}{\partial x} + \frac{U_u \cdot Y}{2 \cdot S_c} \cdot \frac{\partial^2 Y}{\partial x^2} = \frac{\partial Y}{\partial t}. \qquad \textbf{(B-38)}$$

[See Dingman (1984) for the detailed derivation of Equation (B-38).]

The third term of Equation (B-38) represents the rate at which the crest of the flood wave declines with time (and with distance as the wave travels downstream) in the absence of lateral inflow. This term is often called the **diffusive term** by analogy to convection–diffusion equations that arise in many physical phenomena, and a **flood diffusivity,** $D_f$, can be defined as

$$D_f \equiv \frac{U_u \cdot Y}{2 \cdot S_c} = \frac{u_M \cdot Y^{5/3}}{2 \cdot n \cdot S_c^{1/2}}. \qquad \textbf{(B-39)}$$

Equation (B-39) indicates that the tendency for flood peaks to flatten as they move downstream, due to pressure forces, increases with depth and decreases with channel roughness and slope.

The form of Equation (B-38) is identical to that of a convective–diffusion equation that arises in many physical contexts and for which analytical solutions are available for many boundary and initial conditions.

### Kinematic-Wave Equations

If the pressure-force term $\partial Y/\partial x$ as well as the acceleration terms are negligibly small, the equation of motion reduces to the uniform-flow equation [Equation (B-28)]. When this simplification is combined with the continuity relation, the flow equation becomes

$$\frac{q}{B} - \frac{5 \cdot U_u}{3} \cdot \frac{\partial Y}{\partial x} = \frac{\partial Y}{\partial t}. \quad \textbf{(B-40)}$$

Thus when the terms in the energy equation other than channel slope are negligible, there is no diffusive term in the flow equation. This implies that the flood wave moves downstream as a **kinematic wave**[7] with a velocity $U_F$, where

$$U_F = \frac{5 \cdot U_u}{3}. \quad \textbf{(B-41)}$$

Equation (B-40) is called the **kinematic-wave equation.** Equation (B-41) shows that, when Equation (B-40) applies, a flood wave moves downstream at a rate 1.67 times as fast as that of the water itself. [See Equation (9-24).]

Where Equation (B-40) applies, none of the points on a flood wave grows or dissipates except as a result of lateral inflows; in the absence of such inflows a kinematic wave moves through a channel without changing shape. When there is lateral inflow, the sign and magnitude of $\partial Y/\partial t$ at the crest will be given by the rate of lateral inflow per unit width: when lateral inflows are positive (from surface or subsurface inputs) the peak increases as it moves downstream; when they are negative (due to seepage into the bottom for losing streams, bank storage for gaining streams, or overbank flows for flooding streams), the peak decreases downstream.

Solutions to Equation (B-40) can be found by using the method of characteristics or by finite-difference approaches, as described by Eagleson (1970) and Stephenson and Meadows (1986). Woolhiser and Liggett (1967) and Woolhiser (1981) found that Equation (B-40) was a good approximation to the full Saint-Venant Equations when

$$\frac{g \cdot L \cdot S_c}{U^2} = \frac{g \cdot L \cdot n}{u_M^2 \cdot Y^{4/3}} > 10 \quad \textbf{(B-42)}$$

and

$$\frac{L \cdot S_c}{Y} > 5, \quad \textbf{(B-43)}$$

where $L$ is the length of flow path being modeled.

---

[7] The term *kinematic* refers to motion exclusive of the influences of mass and force; i.e., without accelerations due to the characteristics of the flow itself, such as changes in depth and velocity.

# C

# Statistical Concepts Useful in Hydrology

Hydrologic observations are measurements of a particular quantity through time at a fixed location (e.g., streamflow at a gaging station) or of a quantity distributed in space at a particular instant or over a particular time period (e.g., rainfall at a number of rain gages in a region). In either case, the values of the quantity at other times or locations are subject to chance fluctuations and cannot be known exactly. Thus virtually all hydrologic quantities are **random variables,** and hydrologic observations must be regarded as **samples** taken from some **probability distribution,** which is a theoretical relation between magnitude and probability.

Estimates of the spatial or temporal frequency with which values of hydrologic variables occur are the basis for estimates of risk and economic impact that are essential for sound water-resource-management decisions. Thus statistical analysis—particularly estimation of the probability distributions that quantify magnitude–frequency relations—constitutes the principal means by which hydrologic observations are transformed into terms that are useful for water-resource management.

## C.1  PROBABILITY AND RANDOM VARIABLES

If outcome $A$ of an observation (experiment or measurement) is defined as the occurrence of a particular value or range of values of a random variable $X$, the **probability** of $A$, $\Pr\{A\}$, is defined as

$$\Pr\{A\} \equiv \lim_{N \to \infty} \frac{N(A)}{N}, \qquad \textbf{(C-1)}$$

where $N(A)$ is the number of times outcome $A$ occurs in $N$ outcomes. Thus probability is a number between 0 (impossible) **and** 1 (certain) that is proportional to the relative frequency with which the stated value or range of values would occur in a very large number of outcomes.

If outcomes $A, B, C, \ldots$ of a particular experiment or measurement (a **trial**) are mutually exclusive, then

$$\begin{aligned}\Pr\{A \text{ or } B \text{ or } C \text{ or } \ldots\} = \\ \Pr\{A\} + \Pr\{B\} + \Pr\{C\} + \ldots, \end{aligned} \qquad \textbf{(C-2)}$$

and the sum of the probabilities of all possible outcomes is unity. If the outcomes of successive or separate trials are **independent** (i.e., the results of one trial do not affect the probabilities of outcomes in other trials), then

$$\begin{aligned}\Pr\{A \text{ and } B \text{ and } C \text{ and } \ldots\} = \\ \Pr\{A\} \cdot \Pr\{B\} \cdot \Pr\{C\} \cdot \ldots \end{aligned} \qquad \textbf{(C-3)}$$

If a random variable can take on only specific exact numerical values, it is called a **discrete random variable.** Discrete random variables are those that are determined by counting; for example, the number of days with rainfall greater than 25 mm in a year. A sequence of such values observed at successive time intervals is a **discrete time series,** such as was discussed in Section 2.7.1.

If a particular random variable can take on any numerical value within some interval on the real-number line, it is called a **continuous random vari-**

**able.** For example, streamflow is a continuous random variable, because it can potentially assume any non-negative value. (The fact that measurement precision dictates that streamflow can be expressed only to two or three significant figures does not affect this reasoning.)

Statistical methods are applicable to quantities that are treated conceptually as discrete individual outcomes, even though the actual quantity is continuous in time or space. This gives rise to the need to convert continuous time traces into discrete time series by averaging or by selecting extreme values, as was explained in Section 2.7.1 (Figure 2-5). Thus the discrete values of streamflow listed in Table 2-2 constitute a time series of continuous random variables.

## C.2 PROBABILITY DISTRIBUTIONS

The magnitudes of spatially and temporally variable hydrologic quantities are random variables, so their magnitude–frequency relations are quantitatively described in terms of probability distributions. The exact definition of probability distributions differs between discrete and continuous random variables.

### C.2.1 Discrete Random Variables

For a discrete random variable $X$, the probability distribution $p_X(x_i)$ is an equation or table that gives the probability that $X$ takes on a particular value $x_i$:

$$p_X(x_i) \equiv \Pr\{X = x_i\}. \qquad \text{(C-4)}$$

$p_X(x_i)$ is called the **probability function** of $X$. The sum of the probabilities associated with all possible values of $X$ must equal 1:

$$\sum_{all\ X} p_X(x_i) = 1. \qquad \text{(C-5)}$$

It is common to express a discrete probability distribution in terms of a **cumulative probability function,** $P_X(x_i)$, which is the probability that $X$ takes a value less than or equal to $x_i$:

$$P_X(x_i) \equiv \Pr\{X \le x_i\} = \sum_{all\ X \le x_1} p_x(x_i). \qquad \text{(C-6)}$$

The probability of occurrence of values of $X$ in the interval between $x_j$ and $x_k$ is

$$\Pr\{x_j \le X \le x_k\} = \sum_{x_i = x_j}^{x_k} p_X(x_i). \qquad \text{(C-7)}$$

### C.2.2 Continuous Random Variables

Because continuous random variables can theoretically take on any value within an interval on the real number line, an infinite number of possible values exists, and the probability of occurrence of any particular *exact* value is zero. Thus we cannot define a probability function like Equation (C-4) for a continuous variable; probabilities can be defined only over intervals on the real number line. To express these probabilities, we define the **cumulative distribution function** (cdf) of a continuous variable $X$, $F_X(x)$:

$$F_X(x) \equiv \Pr\{X \le x\}. \qquad \text{(C-8)}$$

The relative probability that a continuous random variable $X$ takes on particular values $x$ is expressed by the **probability density function** (pdf), $f_x(x)$, defined as

$$f_X(x) = \frac{d\,F_X(x)}{d\,x}; \qquad \text{(C-9)}$$

thus

$$F_X(x) = \int_{-\infty}^{x} f_X(x) \cdot dx. \qquad \text{(C-10)}$$

The probability that $X$ takes on a value between $x = a$ and $x = b$, $a < b$, is given by

$$\Pr\{a < X \le b\} = \int_{a}^{b} f_X(x) \cdot dx$$
$$= F_X(b) - F_X(a). \qquad \text{(C-11)}$$

Clearly,

$$\int_{-\infty}^{+\infty} f_X(x) \cdot dx = 1. \qquad \text{(C-12)}$$

The cdf is a complete specification of the statistical properties of a random variable, and it is usually expressed as a mathematical function, $\phi$, of $x$ and of one or more **parameters,** $\theta_i$:

$$F_X(x) = \phi(x, \theta_i), \quad i = 1, 2, \ldots, k. \qquad \textbf{(C-13)}$$

Table C-1 summarizes the formulas for several of the distributions commonly used for hydrologic variables; most of these are two- or three-parameter distributions (i.e., $k = 2$ or 3). Haan (1977), Stedinger et al. (1992), and Hosking and Wallace (1997) give information on other distributions.

Probability distributions can be characterized by quantiles, product moments, and L-moments, as will be described in subsequent sections and is indicated in Table C-1. The most common applications of statistics in hydrology involve continuous variables like streamflow, so we will restrict the subsequent discussion to continuous distributions.

### C.2.3  Expectation

Functions of a random variable are also random variables. Denote an arbitrary function of $X$ as $\psi(X)$; then the **expected value** of $\psi(X)$, $E[\psi(X)]$, is defined as

$$E[\psi(X)] \equiv \int_{-\infty}^{+\infty} \psi(x) \cdot f_X(x) \cdot dx. \qquad \textbf{(C-14)}$$

As will be shown, the expected value of $\psi(X)$ is the mean, or average, value of $\psi(X)$.

### C.2.4  Quantiles

One of the simplest ways of describing the distribution of a random variable is to give the value of

**TABLE C-1**
Some Probability Distributions Commonly Used in Hydrology[a]

| Distribution cdf and Inverse | Range | Product Moments | L-Moments |
|---|---|---|---|
| **Uniform (U):** | | | |
| $F_X(x) = \dfrac{x - \alpha}{\beta - \alpha}$ | $\alpha \leq x \leq \beta$ | $\mu_X = \dfrac{\alpha + \beta}{2}$ | $\lambda_1 = \dfrac{\alpha + \beta}{2}$ |
| $x = \alpha + (\beta - \alpha) \cdot F$ | | $\sigma_X^2 = \dfrac{(\beta - \alpha)^2}{12}$ | $\lambda_2 = \dfrac{\beta - \alpha}{6}$ |
| | | | $\tau_3 = 0; \ \tau_4 = 0$ |
| **Normal (N):** | | | |
| $F_Z(z) = \displaystyle\int_{-\infty}^{z} f_Z(\xi) \cdot d\xi$ (Table C-4) | | $\mu_X$ | $\lambda_1 = \mu_X$ |
| $z(F) \approx 5.0633 \cdot [F^{0.135} - (1 - F)^{0.135}]$ | $-\infty < x < \infty$ | | $\lambda_2 = \dfrac{\sigma_X}{\pi^{1/2}}$ |
| $Z \equiv \dfrac{X - \mu_X}{\sigma_X}$ | | $\sigma_X$ | $\tau_3 = 0; \ \tau_4 = 0.1226$ |
| **Lognormal (LN):** | | | |
| $F_{Z'}(z') = \displaystyle\int_{-\infty}^{z'} f_{Z'}(\xi) \cdot d\xi$ (Table C-4) | | $\mu_X = \exp\!\left(\mu_Y + \dfrac{\sigma_Y^2}{2}\right)$ | $\lambda_1 = \mu_Y$ |
| $z'(F) \approx 5.0633 \cdot [F^{0.135} - (1 - F)^{0.135}]$ | $x > 0$ | $\sigma_X^2 = \mu_X^2 \cdot [\exp(\sigma_Y^2) - 1]$ | $\lambda_2 = \exp\!\left(\mu_Y + \dfrac{\sigma_Y^2}{2}\right) \cdot \mathrm{erf}\!\left(\dfrac{\sigma_Y}{2}\right)$ |
| $Z' \equiv \dfrac{Y - \mu_Y}{\sigma_Y}$ | | $\gamma_X = 3 \cdot CV_X + CV_X^2$ | $\tau_3, \tau_4$: see Hosking and Wallis (1997) |
| $Y \equiv \ln(X)$ | | | |
| **Exponential (E):** | | | |
| $F_X(x) = 1 - \exp[-\eta \cdot (x - \xi)]$ | $\eta > 0$ | $\mu_X = \xi + \dfrac{1}{\eta}$ | $\lambda_1 = \xi + \dfrac{1}{\eta}$ |
| $x(F) = \xi - \dfrac{1}{\eta} \cdot \ln(1 - F)$ | $x > \xi$ | $\sigma_X = \dfrac{1}{\eta^2}$ | $\lambda_2 = \dfrac{1}{2 \cdot \eta}$ |
| | | $\gamma_X = 2$ | $\tau_3 = 0.3333; \ \tau_4 = 0.1667$ |

several **quantiles** of the distribution. The $q$th quantile of the variable $X$ is the value $x_q$ that is larger than $100 \cdot q$ percent of all values; thus, for a continuous variable,

$$F_X(x_q) = q. \qquad \textbf{(C-15)}$$

The **quantile function** is the inverse of the probability distribution [i.e., Equation (C-13) solved for $x$]:

$$x = x(F_X) = F_X^{-1}(x) = \phi^{-1}[F_X(x), \theta_i]. \qquad \textbf{(C-16)}$$

The most commonly reported quantiles are the **median**, $x_{.50}$, the **lower quartile**, $x_{.25}$, and the **upper**

quartile, $x_{.75}$. The median is one measure of **central tendency**; $X$ has a 0.5 probability of being less than $x_{.50}$. The interval $[x_{.25}, x_{.75}]$ is the **interquartile range**; $X$ has a 0.5 probability of being in this interval. Thus the difference $x_{.75} - x_{.25}$ is one measure of the spread, or **dispersion**, of the data around the central value, and the ratio of the interquartile range to the median is a relative measure of dispersion.

In general, we do not know the distribution from which a sample was obtained, and so we do not know the true ("population") values of its quantiles. Thus we are forced to estimate these values from the sample, as is described in Box C-1 and illustrated in Example C-1.

**TABLE C-1**
(Continued)

| Distribution cdf and Inverse | Range | Product Moments | L-Moments |
|---|---|---|---|
| **Gumbel (G):** $F_X(x) = \exp\left[-\exp\left(\dfrac{x - \xi}{\alpha}\right)\right]$ $x = \xi - \alpha \cdot \ln[-\ln(F)]$ | $-\infty < x < \infty$ | $\mu_X = \xi + 0.5772 \cdot \alpha$ $\sigma_X^2 = 1.645 \cdot \alpha^2$ $\gamma_X = 1.1396$ | $\lambda_1 = \xi + 0.5772 \cdot \alpha$ $\lambda_2 = 0.6931 \cdot \alpha$ $\tau_3 = 0.1699$ $\tau_4 = 0.1504$ |
| **Generalized Extreme Value (GEV):** $F_X(x) = \exp\left\{-\left[1 - \dfrac{\kappa \cdot (x - \xi)}{\alpha}\right]^{1/\kappa}\right\}$ $x = \xi + \dfrac{\alpha}{\kappa} \cdot \{1 - [-\ln(F)]^{\kappa}\}$ | $\kappa > 0:$ $x < \left(\xi + \dfrac{\alpha}{\kappa}\right)$ $\kappa < 0:$ $x > \left(\xi + \dfrac{\alpha}{\kappa}\right)$ | $\mu_X = \xi + \dfrac{\alpha}{\kappa} \cdot [1 - \Gamma(1 + \kappa)]$ $\sigma_X^2 = \left(\dfrac{\alpha}{\kappa}\right)^2 \cdot \{\Gamma(1 + 2 \cdot \kappa) - [\Gamma(1 + \kappa)]^2\}$ | $\lambda_1 = \xi + \dfrac{\alpha}{\kappa} \cdot [1 - \Gamma(1 + \kappa)]$ $\lambda_2 = \dfrac{\alpha}{\kappa} \cdot (1 - 2^{-\kappa}) \cdot \Gamma(1 + \kappa)$ $\tau_3, \tau_4$ see Stedinger et al. (1992) |
| **Generalized Pareto (GPA):** $F_X(x) = 1 - \left[1 - \kappa \cdot \dfrac{(x - \xi)}{\alpha}\right]^{1/\kappa}$ $x = \xi + \dfrac{\alpha}{\kappa} \cdot [1 - (1 - F)^{\kappa}]$ | $\kappa < 0:$ $\xi \le x$ $\kappa > 0:$ $\xi \le x \le$ $\xi + \dfrac{\alpha}{\kappa}$ | $\mu_X = \xi + \dfrac{\alpha}{1 + \kappa}$ $\sigma_X^2 = \dfrac{\alpha^2}{(1 + \kappa^2) \cdot (1 + 2 \cdot \kappa)}$ | $\lambda_1 = \xi + \dfrac{\alpha}{1 + \kappa}$ $\lambda_2 = \dfrac{\alpha}{(1 + \kappa) \cdot (2 + \kappa)}$ $\tau_3 = \dfrac{1 - \kappa}{3 + \kappa}$ $\tau_4 = \dfrac{(1 - \kappa) \cdot (2 - \kappa)}{(3 + \kappa) \cdot (4 + \kappa)}$ |

[a]From Haan (1977), Stedinger et al. (1992), and Hosking and Wallis (1997).

## BOX C-1

### Estimation of Quantiles from a Sample

Consider a sample of $N$ measured values of a random variable $X$. To estimate the quantiles of the probability distribution from which the sample was taken, first rank the values as $x_{(1)} \leq x_{(2)} \leq \ldots \leq x_{(N)}$.

**Estimating the Median**

The median of the distribution is estimated as the median of the sample, $\hat{x}_{.50}$. If $N$ is an even number, calculate $k = N/2$ and find $\hat{x}_{.50}$ as

$$\hat{x}_{.50} = 0.5 \cdot (x_{(k)} + x_{(k+1)}). \qquad \textbf{(CB1-1a)}$$

If $N$ is odd, calculate $k = (N - 1)/2$ and find $\hat{x}_{.50}$ as

$$\hat{x}_{.50} = x_{(k+1)}. \qquad \textbf{(CB1-1b)}$$

**Estimating Other Quantiles**

For each of the values in the sample, estimate $x_q$ as $x_{(i)}$, where $i$ indicates rank and

$$q = \frac{i - 0.4}{N + 0.2}. \qquad \textbf{(CB1-2)}$$

Then interpolate between observations to obtain $\hat{x}_p$ for the desired $p$.

Equation (CB1-2) is known as the **Cunnane plotting-position formula;** it gives approximately unbiased quantile estimates (Stedinger et al. 1992). Another commonly used plotting-position formula is the **Weibull plotting-position formula,** which gives unbiased estimates of exceedence frequencies:

$$q = \frac{i}{N + 1}. \qquad \textbf{(CB1-3)}$$

Plotting positions provide estimates only for values between $i = 1$ and $i = N$ and are not reliable near either limiting value. Unfortunately, there is no superior method for quantile estimation that does not require an assumption about the probability distribution from which the sample was drawn.

## EXAMPLE C-1

Estimate (a) the median and (b) the interquartile range for the time series of average annual streamflow (Series a) in Table 2-2.

**Solution** The data are ranked from lowest ($i = 1$) to highest ($i = 25$) in column 2 of Table C-2. (This is readily done by using the "sort" command in a spreadsheet.) Column 3 gives the values of $q$ calculated via Equation CB1-2.

(a) $N = 25$, an odd number, so $k = 12$, and we find $\hat{x}_{.50}$ as the value at rank 13:

$$\hat{x}_{.50} = x_{(13)} = 18.8 \text{ ft}^3 \text{ s}^{-1}.$$

(b) Linearly interpolating between $x_{(6)} = 15.6 \text{ ft}^3 \text{ s}^{-1}$ ($q = 0.222$) and $x_{(7)} = 16.8 \text{ ft}^3 \text{ s}^{-1}$ ($q = 0.262$) yields $16.4 \text{ ft}^3 \text{ s}^{-1}$ as the estimate of $x_{.25}$. Interpolating between $x_{(19)} = 22.5 \text{ ft}^3 \text{ s}^{-1}$ ($q = 0.738$) and $x_{(20)} = 23.2 \text{ ft}^3 \text{ s}^{-1}$ ($q = 0.778$) yields $22.7 \text{ ft}^3 \text{ s}^{-1}$

**TABLE C-2**

Ranked Average Annual Streamflow Quantiles, $x_q$ (ft$^3$ s$^{-1}$) from Column 2 of Table 2–2, with Estimated $q = i/(N + 1)$

| Rank, $i$ | Average Annual Flow, $x_q$ | Estimated $q$ | Rank, $i$ | Average Annual Flow, $x_q$ | Estimated $q$ | Rank, $i$ | Average Annual Flow, $x_q$ | Estimated $q$ |
|---|---|---|---|---|---|---|---|---|
| 1 | 9.1 | 0.038 | 10 | 17.5 | 0.385 | 18 | 22.5 | 0.692 |
| 2 | 10.2 | 0.077 | 11 | 17.8 | 0.423 | 19 | 22.5 | 0.731 |
| 3 | 10.3 | 0.115 | 12 | 18.0 | 0.462 | 20 | 23.2 | 0.769 |
| 4 | 14.1 | 0.154 | 13 | 18.8 | 0.500 | 21 | 23.3 | 0.808 |
| 5 | 15.3 | 0.192 | 14 | 18.9 | 0.538 | 22 | 24.9 | 0.846 |
| 6 | 15.6 | 0.231 | 15 | 20.5 | 0.577 | 23 | 25.0 | 0.885 |
| 7 | 16.8 | 0.269 | 16 | 21.2 | 0.615 | 24 | 29.7 | 0.923 |
| 8 | 17.3 | 0.308 | 17 | 22.0 | 0.654 | 25 | 31.5 | 0.962 |
| 9 | 17.5 | 0.346 | | | | | | |

as the estimate of $x_{.75}$. The inter-quartile range is thus $22.7 - 16.4 = 6.3 \text{ ft}^3 \text{ s}^{-1}$, and the relative range is $6.3/18.8 = 0.335$.

Because of the importance of floods and droughts, hydrologists often need to estimate quantiles in the "tails" of the distribution of a precipitation or streamflow time series, e.g., $x_{.05}$ or $x_{.99}$. As will be noted in Section C.8, such estimates are subject to considerable uncertainty and in general should not be made by the methods of Box C-1. Rather they are best made by fitting a probability distribution to the data, as is described in Section C.5.

## C.2.5 Product Moments

A probability distribution can also be characterized in terms of its **product moments.** The general definition of the $r$th product moment about the value $X = x^*$, $_{x^*}M_r(X)$, is

$$_{x^*}M_r(X) \equiv \mathrm{E}\left[(X - x^*)^r\right]$$
$$= \int_{-\infty}^{+\infty} (x - x^*)^r \cdot f_X(x) \cdot dx. \quad \textbf{(C-17)}$$

The first moment about $X = 0$ of a distribution is called the **mean,** or **expected value,** of $X, \mathrm{E}(X)$. It is denoted $\mu_X$ and defined as

$$\mu_X \equiv \int_{-\infty}^{+\infty} x \cdot f_X(x) \cdot dx. \quad \textbf{(C-18)}$$

The mean, like the median, is a measure of central tendency, and it determines the location of the central part of the distribution along the real-number line (Figure C-1a).

Our discussions of product moments for $r > 1$ will always be for $x^* = \mu_X$, so we henceforth shorten the notation to $M_r(X)$.

The second moment about the mean, $M_2(X)$, is called the **variance;** it is denoted $\sigma_X^2$ and defined as

$$\sigma_X^2 \equiv M_2(X) = \mathrm{E}[(X - \mu_X)^2]$$
$$= \int_{-\infty}^{+\infty} [x_i - \mu_X]^2 \cdot f_X(x_i) \cdot dx. \quad \textbf{(C-19)}$$

The **standard deviation,** $\sigma_X$, is the square root of the variance. Like the inter-quartile range, it is a measure of the spread of the distribution around the central value (Figure C-1b). A relative measure of spread derived from product moments is the dimensionless **coefficient of variation,** $CV_X$:

$$CV_X \equiv \frac{\sigma_X}{\mu_X}. \quad \textbf{(C-20)}$$

The third moment about the mean, $M_3(X)$, is a measure of the symmetry of a distribution. To facilitate comparison among distributions of variables of different magnitudes, symmetry is usually characterized by the dimensionless **skewness,** $\gamma_X$:

$$\gamma_X \equiv \frac{M_3(X)}{\sigma_X^3}. \quad \textbf{(C-21)}$$

Figure C-1c shows the relation between the sign of the skewness and the shape of a distribution.

The fourth moment about the mean provides information about the "peakedness" of the central portion of the distribution. It is usually characterized by the dimensionless ratio $\kappa_X$, called the **kurtosis:**

$$\kappa_X \equiv \frac{M_4(X)}{\sigma_X^4}. \quad \textbf{(C-22)}$$

As with quantiles, we can never know the true values of the moments of a distribution; we must estimate them from the sample. Box C-2 shows how this is done.

---

### EXAMPLE C-2

Calculate the mean, standard deviation, coefficient of variation, skewness, and kurtosis of the average annual streamflows in Series a of Table 2-2.

**Solution** Application of the equations in Box C-2 yields the following results:

Equation (CB2-1): $m_X = 19.3 \text{ ft}^3 \text{ s}^{-1}$
Equation (CB2-2): $\hat{M}_2(X) = 29.22000 \ (\text{ft}^3 \text{ s}^{-1})^2$
Equation (CB2-3): $\hat{M}_3(X) = 23.63578 \ (\text{ft}^3 \text{ s}^{-1})^3$
Equation (CB2-4): $\hat{M}_4(X) = 2482.248 \ (\text{ft}^3 \text{ s}^{-1})^4$
Equation (CB2-5): $s_X = 5.52 \text{ ft}^3 \text{ s}^{-1}$
Equation (CB2-6): $\hat{C}V_X = 0.285$
Equation (CB2-7): $g_X = 0.159$
Equation (CB2-8): $k_X = 3.45$

Commands for computing the sample mean, standard deviation, skewness, and kurtosis are available on most spreadsheet programs. In some cases, the kurtosis values are computed as $k_X - 3$, because the kurtosis of the normal distribution is 3 (Section C.7).

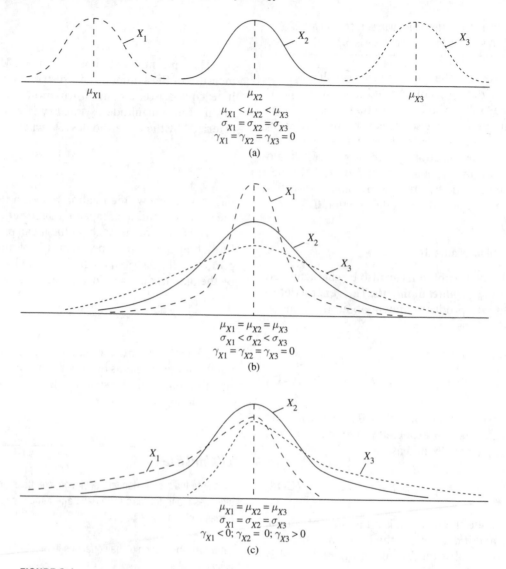

**FIGURE C-1**
Effects of increasing magnitudes of the (a) mean, $\mu_x$, (b) standard deviation, $\sigma_x$, and (c) skewness, $\gamma_x$ on the location and shape of probability distributions

Note that Rules 1, 2, and 5 of Appendix A regarding significant figures are followed in reporting these values. (Note also that most spreadsheet programs do not follow those rules.)

### C.2.6 Probability-Weighted Moments and L-Moments

Because of the generally small sample sizes available for characterizing hydrologic time series, estimates of the third and higher product moments are usually very uncertain. This has led to the use of an alternative system for characterizing probability distributions, called **L-moments.** (See Stedinger et al. 1992; Hosking and Wallis 1997.)

L-moments are based on **probability-weighted moments** (PWM). The $r$th PWM, $\beta_r$, is defined as

$$\beta_r \equiv \mathrm{E}\{X \cdot [F_X(X)]^r\}$$
$$= \int_{-\infty}^{+\infty} x \cdot [F_X(x)]^r \cdot f_X(x) \cdot dx. \quad \textbf{(C-23)}$$

L-moments, $\lambda_r$, are linear combinations of PWMs.

**BOX C-2**

. . . . . . .

**Estimation of Product Moments from a Sample**

Consider a sample of $N$ measured values of a random variable $X$ ($x_1, x_2, ..., x_N$). To estimate the product moments of the probability distribution from which the sample was taken, compute the sample mean,

$$m_X = \frac{1}{N} \cdot \sum_{i=1}^{N} x_i; \qquad \text{(CB2-1)}$$

the sample 2nd moment,

$$\hat{M}_2(X) = \frac{1}{N} \cdot \sum_{i=1}^{N} (x_i - m_X)^2; \qquad \text{(CB2-2)}$$

the sample 3rd moment,

$$\hat{M}_3(X) = \frac{1}{N} \cdot \sum_{i=1}^{N} (x_i - m_X)^3; \qquad \text{(CB2-3)}$$

and the sample 4th moment,

$$\hat{M}_4(X) = \frac{1}{N} \cdot \sum_{i=1}^{N} (x_i - m_X)^4; \qquad \text{(CB2-4)}$$

where the "hat" (^) indicates an estimate.

The sample standard deviation, $s_X$, the coefficient of variation, $\hat{C}V_X$, the skewness, $g_X$, and the kurtosis, $k_X$, are then found as

$$s_X = \hat{\sigma}_X = \left[ \frac{N}{N-1} \cdot \hat{M}_2(X) \right]^{1/2}, \qquad \text{(CB2-5)}$$

$$\hat{C}V_X = \frac{s_X}{m_X}, \qquad \text{(CB2-6)}$$

$$g_X = \hat{\gamma}_X = \frac{N^2 \cdot \hat{M}_3(X)}{(N-1) \cdot (N-2) \cdot s_X^3}, \qquad \text{(CB2-7)}$$

and

$$k_X = \hat{\kappa}_X = \frac{N^3 \cdot \hat{M}_4(X)}{(N-1) \cdot (N-2) \cdot (N-3) \cdot s_X^4} \qquad \text{(CB2-8)}$$

(Haan 1977).

Kirby (1974) pointed out that the sample estimates of $CV(X)$ and $\gamma_X$ cannot exceed certain values that depend on sample size, no matter what the true values of those statistics are:

$$\hat{C}V_X < (N-1)^{1/2}; \qquad \text{(CB2-9)}$$

$$|g_X| < \frac{N-2}{(N-1)^{1/2}}. \qquad \text{(CB2-10)}$$

Thus, with $N = 20$, $\hat{C}V_X < 4.36$ and $|g_X| < 4.13$ even if $CV_X$ and $|\gamma_X|$ are much larger.

---

The first four L-moments are computed as

$$\lambda_1 = \beta_0, \qquad \text{(C-24)}$$
$$\lambda_2 = 2 \cdot \beta_1 - \beta_0, \qquad \text{(C-25)}$$
$$\lambda_3 = 6 \cdot \beta_2 - 6 \cdot \beta_1 + \beta_0, \qquad \text{(C-26)}$$

and

$$\lambda_4 = 20 \cdot \beta_3 - 30 \cdot \beta_2 + 12 \cdot \beta_1 - \beta_0. \qquad \text{(C-27)}$$

From Equations (C-18), (C-23), and (C-24), we see that $\lambda_1 = \mu_X$. The higher L-moments also reflect probability-distribution properties analogous to higher product-moment properties: $\lambda_2$ is a measure of the dispersion, or spread, of the distribution; $\lambda_3$ is a measure of asymmetry; and $\lambda_4$ is a measure of peakedness. (See Hosking and Wallis 1997.) We can also define dimensionless ratios of L-moments that are analogous to the coefficient of variation, skewness, and kurtosis:

L−coefficient of variation (L−CV):  $\tau_2 \equiv \frac{\lambda_2}{\lambda_1}$;

$$\text{(C-28)}$$

L−skewness (L−SK):  $\tau_3 \equiv \frac{\lambda_3}{\lambda_2}$;  (C-29)

L−kurtosis (L−KU):  $\tau_4 \equiv \frac{\lambda_4}{\lambda_2}$.  (C-30)

Note that PWMs involve raising values of $F_X(X)$, rather than of $X$, to powers [Equation (C-23)]. Because $F_X(X) \leq 1$, estimates of sample PWMs, L-moments, and L-moment ratios are much less

---

## BOX C-3

· · · · · · ·

### Sample L-Moments

Consider a sample of $N$ measured values of a random variable $X$. To estimate the L-moments of the probability distribution from which the sample was taken, first rank the values as $x_{(1)} \leq x_{(2)} \leq \dots \leq x_{(N)}$ as in Box C-1. Estimate the probability-weighted moments as

$$b_0 = m_X, \qquad \textbf{(CB3-1)}$$

$$b_1 = \frac{1}{N \cdot (N-1)} \cdot \sum_{i=2}^{N} (i-1) \cdot x_{(i)}, \qquad \textbf{(CB3-2)}$$

$$b_2 = \frac{1}{N \cdot (N-1) \cdot (N-2)}$$
$$\cdot \sum_{i=3}^{N} (i-1) \cdot (i-2) \cdot x_{(i)}, \qquad \textbf{(CB3-3)}$$

and

$$b_3 = \frac{1}{N \cdot (N-1) \cdot (N-2) \cdot (N-3)}$$
$$\cdot \sum_{i=4}^{N} (i-1) \cdot (i-2) \cdot (i-3) \cdot x_{(i)} \qquad \textbf{(CB3-4)}$$

(Hosking and Wallis 1997).

The sample estimates of the first four L-moments, denoted $L_1$–$L_4$, are then calculated by substituting $b_0$–$b_3$ for $\beta_0$–$\beta_s$ in Equations (C-24)–(C-27). The sample estimates of L-coefficient of variation, L-skewness, and L-kurtosis are computed by using Equations (C-28)–(C-30).

---

susceptible to the influences of a few large or small values in the sample. Hence they are generally preferable to product moments for characterizing probability distributions of hydrologic variables.

Box C-3 explains how L-moments are computed from a sample.

Subsequent examples will show how these L-moment values are used to identify probability distributions that characterize data and estimate the parameters of distributions.

### *EXAMPLE C-3*

Calculate the first four L-moments of the average annual streamflows in Series a of Table 2-2.

***Solution:*** The formulas for computing L-moment statistics can be readily programmed for spreadsheets. Application of the equations in Box C-3 yields the following results:

$$b_0 = 19.3 \text{ ft}^3 \text{ s}^{-1};$$
$$b_1 = 11.2 \text{ ft}^3 \text{ s}^{-1};$$
$$b_2 = 8.03 \text{ ft}^3 \text{ s}^{-1};$$
$$b_3 = 6.29 \text{ ft}^3 \text{ s}^{-1};$$
$$L_1 = 19.3 \text{ ft}^3 \text{ s}^{-1};$$
$$L_2 = 3.13 \text{ ft}^3 \text{ s}^{-1};$$
$$L_3 = 0.075 \text{ ft}^3 \text{ s}^{-1};$$
$$L_4 = 0.554 \text{ ft}^3 \text{ s}^{-1};$$
$$t_2 = 0.162;$$
$$t_3 = 0.024;$$
$$t_4 = 0.177.$$

## C.3  EXCEEDENCE PROBABILITY AND RETURN PERIOD

Hydrologists commonly express probabilities in terms of **exceedence probability,** $EP_X(x)$, defined as

$$EP_X(x) \equiv \text{Pr}\{X > x\} = 1 - F_X(x). \qquad \textbf{(C-31)}$$

When applied to time series, $EP_X(x)$ is the probability that $X > x$ within any time interval of a specified length, $\Delta t$.

When $EP_X(x)$ remains constant for all time intervals, it can be shown that the average number of intervals between occurrences of the event $X > x$ is equal to $1/EP_X(x)$. Thus exceedence probability is frequently expressed as **return period** (or **recurrence interval**), $TR_X(x)$, where

$$TR_X(x) \equiv \frac{1}{EP_X(x)} = \frac{1}{1 - F_X(x)}. \qquad \textbf{(C-32)}$$

When $\Delta t = 1$ yr (the most common choice in hydrologic analysis), $TR_X(x)$ is the *average* number of years between years in which $X > x$. In spite of the use of the words "return" or "recurrence", there is no regularity or periodicity in occurrences of exceedences involved in the definition of $TR_X(x)$.

Box C-4 applies the rules of probabilities for independent events [Equation (C-3)] to show how various exceedence probabilities are calculated for specified time periods $n \cdot \Delta t$ (e.g., over a period of $n$ yr).

$$1 - [1 - 0.01]^{40} = 0.331.$$

Values of $EP_X(x)$ are usually estimated from estimates of quantiles.

## C.4 COVARIANCE AND CORRELATION

The tendency for two variables $X$ and $Y$ to fluctuate in parallel is measured by the **covariance,** $COV_{X,Y}$, defined as

$$COV_{X,Y} \equiv \mathrm{E}\{[X - \mathrm{E}(X)] \cdot [Y - \mathrm{E}(Y)]\}$$
$$= \mathrm{E}(X \cdot Y) - \mathrm{E}(X) \cdot \mathrm{E}(Y). \quad \textbf{(C-33)}$$

The magnitude of the covariance depends on the scale of the variables and their units of measurement, so a more meaningful measure of this tendency is the dimensionless **correlation coefficient,** $\rho_{X,Y}$:

$$\rho_{X,Y} \equiv \frac{COV_{X,Y}}{\sigma_X \cdot \sigma_Y}. \quad \textbf{(C-34)}$$

Note that $\rho_{X,Y} = \rho_{Y,X}$.

The range of $\rho_{X,Y}$ is $-1 \le \rho_{X,Y} \le 1$. As is shown in Figure C-2, $\rho_{X,Y} > 0$ indicates that relatively high

### EXAMPLE C-4

By definition, the "100-yr flood" is the annual peak discharge with an exceedence probability of 0.01 [Equation (C-32)]. (a) What is the probability that the next exceedence of the 100-yr flood will occur in the 10th year from now? (b) What is the probability that the 100-yr flood will be exceeded at least once in the next 40 yr?

### Solution

(a) By using Equation (CB4-5), we can calculate

$$[1 - 0.01]^9 \times 0.01 = 0.00916.$$

(b) By using Equation (CB4-6), we can calculate

---

### BOX C-4
· · · · · · ·
### Exceedence Probabilities for a Specified Number of Time Intervals

In the following calculations, we assume that $X$ is a continuous random variable that is distributed in time and that $EP_X(x)$ is constant for all time periods, which are of length $\Delta t$. Thus, by definition [Equation (C-31)],

$$\mathrm{Pr}\{X > x \text{ in the next time interval}\} = EP_X(x) \quad \textbf{(CB4-1)}$$

and

$$\mathrm{Pr}\{X \le x \text{ in the next time interval}\} = 1 - EP_X(x) = F_X(x). \quad \textbf{(CB4-2)}$$

Applying the multiplicative rule for successive independent trials [Equation (C-3)] gives the following:

$$\mathrm{Pr}\{X > x \text{ in all of the next } n \text{ time intervals}\} = [EP_X(x)]^n; \quad \textbf{(CB4-3)}$$

$$\mathrm{Pr}\{X \le x \text{ in all of the next } n \text{ time intervals}\} = [1 - EP_X(x)]^n; \quad \textbf{(CB4-4)}$$

$$\mathrm{Pr}\{\text{ next interval in which } X > x \text{ is } n \text{ intervals from now }\} = [1 - EP_X(x)]^{n-1} \cdot EP_X(x); \quad \textbf{(CB4-5)}$$

and

$$\mathrm{Pr}\{X > x \text{ in at least one interval of next } n \text{ intervals}\} = 1 - [1 - EP_X(x)]^n. \quad \textbf{(CB4-6)}$$

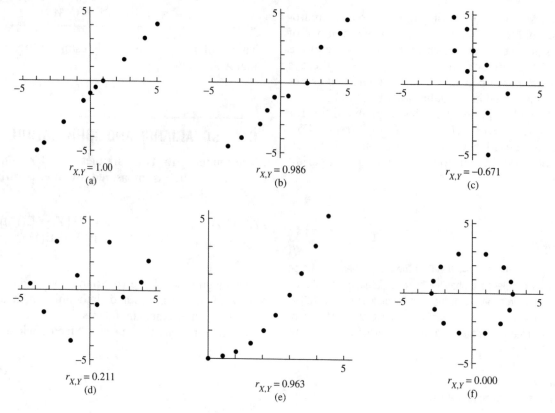

**FIGURE C-2**
Values of $r_{X,Y}$ for various degrees of linear relation; after Haan (1977)

(low) values of $X$ tend to be associated with high (low) values of $Y$; $\rho_{X,Y} < 0$ indicates that high (low) values of $X$ are associated with low (high) values of $Y$. When $|\rho_{X,Y}| = 1$, there is a perfect linear relation between $X$ and $Y$ (Figure C-2a). When $\rho_{X,Y} = 0$, there is no degree of linear relation; however, $X$ and $Y$ could be related in other ways (Figure C-2f).

Estimates of $\rho_{X,Y}$, designated $r_{X,Y}$, are made from samples of pairs of $(X,Y)$ values, as described in Box C-5. Most spreadsheets include built-in functions that calculate $r_{X,Y}$.

---

### EXAMPLE C-5

(a) Compute the degree of linear correlation between the annual values of average flows ($Q$) and flood peaks ($P$) for the Oys-

ter River near Durham, NH. (b) Determine whether the correlation is statistically significant. The data are in Table 2-2 and are plotted in Figure C-3.

**Solution**

(a) Using Equation (CB5-1), we calculate $r_{Q,P} = 0.443$.
(b) Using Equation (CB5-2), we calculate

$$r^* = \frac{1.96}{(22)^{1/2}} = 0.418.$$

Because $r_{Q,P} > r^*$, we conclude that there is a statistically significant degree of linear relation between annual flood peaks and annual average flows at this location. More precisely, there is less than a 5% chance that we would compute a value of $r_{Q,P} > 0.418$ if the true correlation $\rho_{Q,P} = 0$. Visually, Figure C-3 confirms these conclusions—it suggests that there is a definite tendency for high peak flows to occur in years with high average flows, but there is considerable scatter.

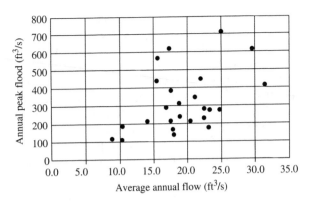

**FIGURE C-3**
Scatter plot of annual flood peaks ($P$) vs. annual average streamflow ($Q$) for the Oyster River near Durham, NH; data in Table 2-2

## BOX C-5

· · · · · · · ·

### Sample Correlation Coefficients

If $X$ and $Y$ are two variables associated in space or time (for example, any two of the time series in Table 2-2), and if we have samples $(x_1, x_2, ..., x_N)$ and $(y_1, y_2, ..., y_N)$, where the $(x_i, y_i)$ are pairs of associated values (e.g., they occur at the same time), then their correlation coefficient, $\rho_{X,Y}$, can be estimated as

$$r_{X,Y} = \hat{\rho}_{X,Y} = \frac{\sum_{i=1}^{N}(x_i - m_X) \cdot (y_i - m_Y)}{(N-1) \cdot s_X \cdot s_Y}. \quad \textbf{(CB5-1)}$$

If $N \geq 25$ and $|r_{X,Y}| > r^*$, where

$$r^* = \frac{1.96}{(N-3)^{1/2}}, \quad \textbf{(CB5-2)}$$

then we say that $r_{X,Y}$ is "significantly different from zero" or, more precisely, that there is less than a 5% chance that the true correlation coefficient, $\rho_{X,Y}$, for the two populations equals zero (Haan 1977). Thus, when $|r_{X,Y}| > r^*$, we conclude that there is a degree of linear relation between $X$ and $Y$. [The significance test for $\rho_{X,Y}$ is done differently when $N < 25$; see Haan (1977)].

## C.5 DATA ANALYSIS: IDENTIFYING AN APPROPRIATE PROBABILITY DISTRIBUTION

Various probability distributions have been assumed to characterize various random variables in hydrology (Table C-1); the most commonly used distributions were conveniently summarized by National Environment Research Council (1975), Stedinger et al. (1992), and Hosking and Wallis (1997).

There may be theoretical reasons for believing that a given hydrologic variable follows a particular distribution, but we can never know for certain whether such a belief is correct. To see whether a given variable appears to be from a particular distribution, which represents the hypothetical population from which the sample was taken, the best one

can do is to examine the sample of measured values of the variable and conduct statistical tests based on sample quantiles, moments, or L-moments.

A comprehensive description of these tests and an extensive review of probability distributions are beyond the scope of this text. We will, however, describe two approaches that have wide application in hydrology for identifying distributions that characterize a sample of data. More detailed discussion of these and other approaches can be found in Haan (1977), Stedinger et al. (1992), and Helsel and Hirsch (1992).

### C.5.1 Sample Quantiles

The first step in both procedures is to compute the quantiles for the sample. Consider a sample of size $N$, $(x_1, x_2, ..., x_N)$, of a hydrologic variable $X$. The sample values are ranked (sorted) in ascending order, such that $x_{(1)} \leq x_{(2)} \leq ... \leq x_{(N)}$. The quantile $F_X(x_{(i)})$ associated with each sample value is then estimated by using a plotting-position formula such as Equation (CB1-2) (as in Table C-2).

### C.5.2 The Probability-Plot Correlation Coefficient Approach

The probability-plot correlation coefficient (PPCC) approach involves calculating the correlation coefficient (Box C-5) between the sample values and the values that would exist at corresponding exceedence probabilities if the data were from a proposed distribution. If the correlation coefficient is sufficiently close to 1, the hypothesis that the data are from the proposed distribution is not rejected.

The values associated with a specified exceedence probability for the proposed distribution, $z_i$, are calculated from the distribution's quantile function (Stedinger et al. 1992; Hosking and Wallis 1997):

$$z_i = F^{-1}(x_i). \qquad \textbf{(C-35)}$$

The value of the correlation coefficient between $x_{(i)}$ and $z_i$, $r_{x,z}$, is then compared with values tabulated for the particular distribution (Table C-3). These values are a function of the sample size, $N$, and the **significance level**, $\alpha$. The significance level is the probability that a sample of size $N$ drawn from the proposed distribution would have a value of $r_{x,z}$ less than the tabulated value. Thus, if the calculated value of $r_{x,z}$ is less than the appropriate tabulated value, one can assume that there is less than a $100 \cdot \alpha\%$ chance that the data came from the proposed distribution, and the hypothesis that the data came from that distribution is rejected. Otherwise, the hypothesis is not rejected. Standard statistical procedure requires selecting the $\alpha$ value before conducting the analysis; values of $\alpha = 0.05$ or 0.10 are typically chosen.

### EXAMPLE C-6

Use the PPCC approach to judge whether the mean annual streamflow data in Table C-2 can be represented by the normal distribution.

**Solution:** We select $\alpha = 0.10$. There is no analytical relation for the inverse (quantile function) of the normal distribution, but it can be approximated as (Stedinger et al. 1992)

$$z_i = 5.0633 \cdot \{ F_X(x_i)^{0.135} - [1 - F_X(x_i)]^{0.135} \}.$$

Using this formula, we obtain the following results:

| Rank, $i$ | Average Annual Flow, $x_i$ (ft$^3$ s$^{-1}$) | Estimated Quantile $F_X(x_i)$ | $z_i$ |
|---|---|---|---|
| 1 | 9.1 | 0.024 | −1.9899 |
| 2 | 10.2 | 0.063 | −1.5289 |
| 3 | 10.3 | 0.103 | −1.2632 |
| 4 | 14.1 | 0.143 | −1.0655 |
| 5 | 15.3 | 0.183 | −0.9028 |
| · | · | · | · |
| · | · | · | · |
| · | · | · | · |
| 25 | 31.5 | 0.976 | 1.9899 |

The correlation coefficient $r_{x,z} = 0.9868$. This value exceeds the value in Table C-3 for $N = 25$, $\alpha = 0.10$, so we do not reject the hypothesis that the data came from a normal distribution.

**TABLE C-3**

Critical Values of the Correlation Coefficient, $r_{x,z}$, for the PPCC Test for the Normal Distribution as a Function of Sample Size, $N$[a]

| | Significance Level, $\alpha$ | | | | Significance Level, $\alpha$ | | |
|---|---|---|---|---|---|---|---|
| $N$ | 1% | 5% | 10% | $N$ | 1% | 5% | 10% |
| 10 | 0.8760 | 0.9170 | 0.9340 | 100 | 0.9812 | 0.9870 | 0.9893 |
| 20 | 0.9250 | 0.9500 | 0.9600 | 200 | 0.9902 | 0.9930 | 0.9942 |
| 30 | 0.9470 | 0.9640 | 0.9700 | 300 | 0.9935 | 0.9953 | 0.9960 |
| 40 | 0.9580 | 0.9720 | 0.9770 | 500 | 0.9958 | 0.9970 | 0.9975 |
| 50 | 0.9650 | 0.9770 | 0.9810 | 1000 | 0.9976 | 0.9982 | 0.9985 |
| 60 | 0.9700 | 0.9800 | 0.9830 | 2000 | 0.9989 | 0.9992 | 0.9993 |
| 70 | 0.9740 | 0.9825 | 0.9854 | 3000 | 0.9992 | 0.9995 | 0.9995 |
| 80 | 0.9770 | 0.9843 | 0.9870 | 5000 | 0.9996 | 0.9997 | 0.9997 |
| 90 | 0.9793 | 0.9858 | 0.9882 | 10000 | 0.9998 | 0.9998 | 0.9999 |

[a]Values from Vogel (1986). Interpolate using

$$r_{X,z} = 1 - 0.001 \cdot \exp(A) \cdot N^B,$$

where $A$ is constant and $B$ is the intercept of the regression of ln $[1000 \cdot (1 - r_{X,z})]$ vs. ln($N$) in the range of $N$ values of interest.

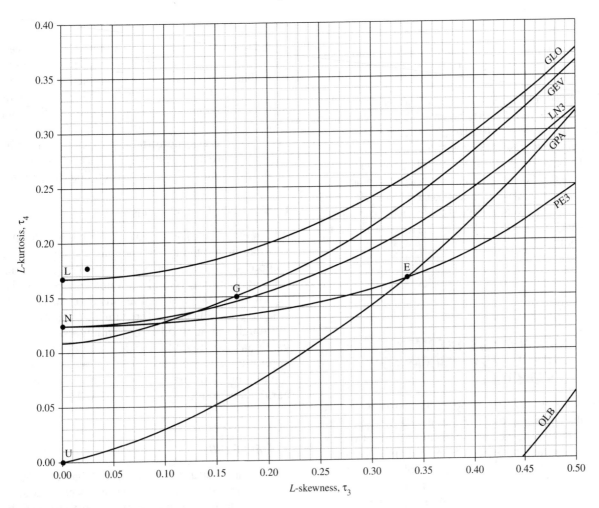

**FIGURE C-4**
The L-moment diagram is a plot of $\tau_4$ vs. $\tau_3$ [Equations (C-29) and (C-30)] used to help decide what distribution characterizes data. Two-parameter distributions plot as points, three-parameter distributions as lines. (See Table C-1.) Sample values of $t_3$, $t_4$ are plotted and compared with the locations of points and lines for various distributions; the dot shows where the data of Example C-7 plot. "OLB" indicates limit of possible values. From *Regional Frequency Analysis*, by Hosking and Wallis © 1997 Cambridge University Press, used with permission.

### C.5.3   L-Moment Approach

The L-moment approach to identifying the probability distribution that appears to characterize a given set of data is based on an **L-moment diagram**, which is a plot of theoretical L-kurtosis vs. L-skewness values for candidate distributions (Figure C-4). Two-parameter distributions plot as a single point and three-parameter distributions as a line on an L-moment diagram.

To identify candidate distributions for a given sample, the sample L-kurtosis and L-skewness values $t_3$ and $t_4$ are calculated as in Box C-3 and plotted on the diagram. Candidate distributions for a given data set are those that are "close" to the sample point. The quantitative definition of "close" is a somewhat involved matter (Hosking and Wallis 1997), beyond the scope of this discussion. Qualitatively, the best procedure is to plot a number of samples for the quantity of interest (e.g., annual average flows) from the region of interest, so that one can get a sense of the sampling variability and any trends that may suggest a particular distribution. Once a candidate distribution is selected, it can be quantitatively evaluated via the PPCC approach described above.

## EXAMPLE C-7

Use the L-moment approach to identify candidate distributions for the mean annual flow data of Table C-2.

**Solution** From Example C-3, we have L-skewness = 0.024 and L-kurtosis = 0.177. As is shown in Table C-1, the normal distribution has values of 0 and 0.123, respectively. The sample point thus plots reasonably close to the point for the normal distribution (Figure C-4), and thus appears consistent with the results of the PPCC analysis. In the absence of other information, it appears reasonable to conclude that these data can be represented by a normal distribution. To confirm this, annual flow data for several other stations in the region should be analyzed similarly.

## C.6   DATA ANALYSIS: ESTIMATING PARAMETERS OF PROBABILITY DISTRIBUTIONS

As was noted earlier, probability distributions are usually expressed as equations giving the pdf or cdf as a function of the variable value, $x$, and of one or more parameters, $\theta_i$ [Equation (C-13)]. In order to use the equations to compute probabilities for a given variable, values of the parameters must be estimated from sample values of the variable.

There are three general approaches to estimating parameter values from samples: (1) the method of moments; (2) the method of maximum likelihood; and (3) the method of L-moments. A brief description of each is given here; more detailed discussions of their statistical properties can be found in Haan (1977), in Yevjevich (1972), and in general statistics texts.

### C.6.1   Method of Moments

As is indicated in Table C-1, moment values can be related mathematically to parameters. In the method of moments, the sample moments are calculated as in Box C-2 and substituted in those relations. The relations are then solved to give the parameter values. Generally, the number of moments needed equals the number of parameters in the distribution.

### C.6.2   Method of Maximum Likelihood

Maximum-likelihood (ML) estimators are derived by finding the parameter values that maximize the probability of obtaining the sample at hand. In some cases, ML estimators are identical to moment estimators. When they differ, the ML estimators generally have better statistical properties. However, in some cases the formulas for computing ML estimators are complex; consult statistical texts (e.g. Stedinger et al. 1992) for more detailed discussion.

### C.6.3   Method of L-Moments

The method of L-moments is analogous to the method of moments: the sample L-moments are computed (Box C-3) and substituted into equations relating L-moments and parameters (Table C-1). Again, the number of L-moments that must be calculated generally equals the number of parameters. Although this method is not widely used in conventional statistics, "In a wide range of hydrologic applications, L moments provide simple and reasonably efficient estimators . . . of a distribution's parameters" (Stedinger et al. 1992, p. 18.6).

## EXAMPLE C-8

Assuming the average annual flow data of Table C-2 follow a normal distribution, estimate the parameters of the distribution by using first the method of moments and then the method of L-moments.

**Solution**

(a) The normal distribution has two parameters, the mean, $\mu_X$, and the standard deviation, $\sigma_X$. In the method of moments, these are simply equated to the sample mean, $m_X$, and the sample standard deviation, $s_X$, respectively. Both of these were calculated in Example C-2:

$$\hat{\mu}_X = m_X = 19.3 \text{ ft}^3 \text{ s}^{-1};$$
$$\hat{\sigma}_X = s_X = 5.52 \text{ ft}^3 \text{ s}^{-1}.$$

For the normal distribution, the method-of-moments estimators are identical to the ML estimators.

(b) Table C-1 gives the relations between L-moments and parameters for the normal distribution:

$$\lambda_1 = \mu_X;$$
$$\lambda_2 = \frac{\sigma_X}{\pi^{1/2}} = 0.5642 \cdot \sigma_X.$$

Thus we find that

$$\hat{\mu}_X = m_X = 19.3 \text{ ft}^3 \text{ s}^{-1},$$

which is, of course, identical to the method-of-moments estimator; also,

$$\hat{\sigma}_X = \pi^{1/2} \cdot L_2 = 1.77 \times 3.134 = 5.55 \text{ ft}^3 \text{ s}^{-1},$$

which is very close to the method-of-moments estimator.

## C.7    THE NORMAL DISTRIBUTION

The normal distribution plays a large role in statistical analysis. One important application is in the characterization of measurement error and uncertainty, as was discussed in Section 2.5.2. In addition, many hydrologic variables (e.g., the average annual flows in Table C-2) can be characterized by the normal or log-normal distribution.

### C.7.1    Normal pdf and cdf

As was noted, the normal distribution has two parameters, the mean, $\mu_X$, and the standard deviation, $\sigma_X$. Its pdf is

$$f_X(x) = \frac{1}{(2 \cdot \pi)^{1/2} \cdot \sigma_X} \cdot \exp\left[ -\frac{1}{2} \cdot \left( \frac{x - \mu_X}{\sigma_X} \right)^2 \right].$$
(C-36)

Thus the value of the pdf can be calculated if $\mu_X$ and $\sigma_X$ are specified. The **standard normal variate,** $Z$, has a normal distribution with $\mu_Z = 0$ and $\sigma_Z = 1$. Hence, its pdf is

$$f_Z(z) = \frac{1}{(2 \cdot \pi)^{1/2}} \cdot \exp\left( -\frac{z^2}{2} \right).$$
(C-37)

By making the transformation

$$Z = \frac{X - \mu_X}{\sigma_X},$$
(C-38)

we can state that

$$f_X(x) = f_Z(z)$$
(C-39a)

and

$$F_X(x) = F_Z(z).$$
(C-39b)

Note that the normal distribution is symmetrical about the value $x = \mu_X$ or $z = 0$; thus it has skewness $\gamma_X = 0$ and L-skewness $\tau_3 = 0$. The kurtosis, $\kappa_X$, of the normal distribution = 3, and the L-kurtosis $\tau_4 = 0.123$.

The normal cdf is the integral of Equation (C-36) or (C-37) [Equation (C-10)]; however, this integral can be evaluated only as a series expansion, so tables are provided giving $F_Z(z)$ as a function of $z$ (Table C-4). With Equations (C-11), (C-38), and (C-39) and the symmetry property, Table C-4 can be used to calculate the cdf for any normally-distributed variable and the probability that a value will fall into any interval, as is demonstrated in Example C-9. Table C-5 gives the probability of $Z$ between $-z^*$ and $+z^*$ for specified values of $z^*$.

### EXAMPLE C-9

Suppose the annual streamflow data in Series $a$ of Table 2-2 are from a normally distributed population with mean and standard deviation equal to the sample values of those statistics calculated in Example C-8. (a) If a year with an average flow less than the 10-percentile value is considered a drought year, what average-annual flow is the upper limit for drought years? (b) Suppose a drought year has occurred, and a flow of 30 ft³ s⁻¹ or more in the following year will provide enough water to satisfy annual demand and re-fill the water-supply reservoir. What is the probability of such an event? (c) Suppose a planning study concludes that, with projected population growth, a year with flows less than 15 ft³ s⁻¹ will cause significant stress on the water-supply system. What is the probability of this event? (d) Suppose a study concludes that years with flows less than 8 ft³ s⁻¹ will have insufficient dilution of sewage effluent discharged into the river. What is the probability of this event?

### Solution:

(a) From Table C-4, we see that $F_Z(z) \approx 0.9$ when $z = 1.28$. Then, from symmetry, we know $F_Z(z) \approx 0.1$ when $z = -1.28$. From Equation (C-38), we find

$$-1.28 = \frac{x_1 - 19.3}{5.55},$$

and

$$x_1 = 12.2 \text{ ft}^3 \text{ s}^{-1}.$$

(b) From Equation (C-38), we find

$$z = \frac{30 - 19.3}{5.55} = 1.93.$$

From Table C-4, $F_Z(1.93) = 0.9732$. Thus the probability of $Z$ being greater than 1.93 = $(1 - 0.9732) = 0.0268$, so the probability of a year with flow greater than 30 ft$^3$ s$^{-1}$ is about 0.027.

(c) From Equation (C-38), we find

$$z = \frac{15 - 19.3}{5.55} = -0.775.$$

From Table C-4, $F_Z(0.775) \approx 0.781$. Thus $F_Z(-0.775) = (1 - 0.781) = 0.219$, so the probability of a year with flow less than 15 ft$^3$ s$^{-1}$ is about 0.22.

(d) From Equation (C-38), we find

$$z = \frac{8 - 19.3}{5.55} = -2.04.$$

**TABLE C-4**
Values of the Cumulative Distribution Function of the Normal Distribution for the Standard Normal Variate, $z$, for $z > 0$

| $z$ | $F_Z(z)$ | $z$ | $F_Z(z)$ | $z$ | $F_Z(z)$ | $z$ | $F_Z(z)$ |
|------|---------|------|---------|------|---------|------|---------|
| 0.00 | 0.5000 | 0.41 | 0.6591 | 0.82 | 0.7939 | 1.23 | 0.8907 |
| 0.01 | 0.5040 | 0.42 | 0.6628 | 0.83 | 0.7967 | 1.24 | 0.8925 |
| 0.02 | 0.5080 | 0.43 | 0.6664 | 0.84 | 0.7995 | 1.25 | 0.8944 |
| 0.03 | 0.5120 | 0.44 | 0.6700 | 0.85 | 0.8023 | 1.26 | 0.8962 |
| 0.04 | 0.5160 | 0.45 | 0.6736 | 0.86 | 0.8051 | 1.27 | 0.8980 |
| 0.05 | 0.5199 | 0.46 | 0.6772 | 0.87 | 0.8078 | 1.28 | 0.8997 |
| 0.06 | 0.5239 | 0.47 | 0.6808 | 0.88 | 0.8106 | 1.29 | 0.9015 |
| 0.07 | 0.5279 | 0.48 | 0.6844 | 0.89 | 0.8133 | 1.30 | 0.9032 |
| 0.08 | 0.5319 | 0.49 | 0.6879 | 0.90 | 0.8159 | 1.31 | 0.9049 |
| 0.09 | 0.5359 | 0.50 | 0.6915 | 0.91 | 0.8186 | 1.32 | 0.9066 |
| 0.10 | 0.5398 | 0.51 | 0.6950 | 0.92 | 0.8212 | 1.33 | 0.9082 |
| 0.11 | 0.5438 | 0.52 | 0.6985 | 0.93 | 0.8238 | 1.34 | 0.9099 |
| 0.12 | 0.5478 | 0.53 | 0.7019 | 0.94 | 0.8264 | 1.35 | 0.9115 |
| 0.13 | 0.5517 | 0.54 | 0.7054 | 0.95 | 0.8289 | 1.36 | 0.9131 |
| 0.14 | 0.5557 | 0.55 | 0.7088 | 0.96 | 0.8315 | 1.37 | 0.9147 |
| 0.15 | 0.5596 | 0.56 | 0.7123 | 0.97 | 0.8340 | 1.38 | 0.9162 |
| 0.16 | 0.5636 | 0.57 | 0.7157 | 0.98 | 0.8365 | 1.39 | 0.9177 |
| 0.17 | 0.5675 | 0.58 | 0.7190 | 0.99 | 0.8389 | 1.40 | 0.9192 |
| 0.18 | 0.5714 | 0.59 | 0.7224 | 1.00 | 0.8413 | 1.41 | 0.9207 |
| 0.19 | 0.5753 | 0.60 | 0.7257 | 1.01 | 0.8438 | 1.42 | 0.9222 |
| 0.20 | 0.5793 | 0.61 | 0.7291 | 1.02 | 0.8461 | 1.43 | 0.9236 |
| 0.21 | 0.5832 | 0.62 | 0.7324 | 1.03 | 0.8485 | 1.44 | 0.9251 |
| 0.22 | 0.5871 | 0.63 | 0.7357 | 1.04 | 0.8508 | 1.45 | 0.9265 |
| 0.23 | 0.5910 | 0.64 | 0.7389 | 1.05 | 0.8531 | 1.46 | 0.9279 |
| 0.24 | 0.5948 | 0.65 | 0.7422 | 1.06 | 0.8554 | 1.47 | 0.9292 |
| 0.25 | 0.5987 | 0.66 | 0.7454 | 1.07 | 0.8577 | 1.48 | 0.9306 |
| 0.26 | 0.6026 | 0.67 | 0.7486 | 1.08 | 0.8599 | 1.49 | 0.9319 |
| 0.27 | 0.6064 | 0.68 | 0.7517 | 1.09 | 0.8621 | 1.50 | 0.9332 |
| 0.28 | 0.6103 | 0.69 | 0.7549 | 1.10 | 0.8643 | 1.51 | 0.9345 |
| 0.29 | 0.6141 | 0.70 | 0.7580 | 1.11 | 0.8665 | 1.52 | 0.9357 |
| 0.30 | 0.6179 | 0.71 | 0.7611 | 1.12 | 0.8686 | 1.53 | 0.9370 |
| 0.31 | 0.6217 | 0.72 | 0.7642 | 1.13 | 0.8708 | 1.54 | 0.9382 |
| 0.32 | 0.6255 | 0.73 | 0.7673 | 1.14 | 0.8729 | 1.55 | 0.9394 |
| 0.33 | 0.6293 | 0.74 | 0.7704 | 1.15 | 0.8749 | 1.56 | 0.9406 |
| 0.34 | 0.6331 | 0.75 | 0.7734 | 1.16 | 0.8770 | 1.57 | 0.9418 |
| 0.35 | 0.6368 | 0.76 | 0.7764 | 1.17 | 0.8790 | 1.58 | 0.9429 |
| 0.36 | 0.6406 | 0.77 | 0.7794 | 1.18 | 0.8810 | 1.59 | 0.9441 |
| 0.37 | 0.6443 | 0.78 | 0.7823 | 1.19 | 0.8830 | 1.60 | 0.9452 |
| 0.38 | 0.6480 | 0.79 | 0.7852 | 1.20 | 0.8849 | 1.61 | 0.9463 |
| 0.39 | 0.6517 | 0.80 | 0.7881 | 1.21 | 0.8869 | 1.62 | 0.9474 |
| 0.40 | 0.6554 | 0.81 | 0.7910 | 1.22 | 0.8888 | 1.63 | 0.9484 |

From Table C-4, $F_Z(2.04) = 0.9793$. Thus $F_Z(-2.04) = (1 - 0.9793) = 0.0207$, so the probability of a year with flow less than 8 ft$^3$ s$^{-1}$ is about 0.021.

The computations in Example C-9 assume that the annual flows are normally distributed and that the mean and standard deviation are known. Both the PPCC test and the L-moment calculations suggest that assuming a normal distribution is reasonable, and many studies have indicated that annual flows in humid regions are normally distributed. However, we should be aware that our estimates of the mean and standard error are made from the particular sample of 25 years that we happen to have, and thus are subject to uncertainty, called

**TABLE C-4**
(Continued)

| $z$ | $F_Z(z)$ | $z$ | $F_Z(z)$ | $z$ | $F_Z(z)$ | $z$ | $F_Z(z)$ |
|------|----------|------|----------|------|----------|------|----------|
| 1.64 | 0.9495 | 2.05 | 0.9798 | 2.46 | 0.9931 | 2.87 | 0.9979 |
| 1.65 | 0.9505 | 2.06 | 0.9803 | 2.47 | 0.9932 | 2.88 | 0.9980 |
| 1.66 | 0.9515 | 2.07 | 0.9808 | 2.48 | 0.9934 | 2.89 | 0.9981 |
| 1.67 | 0.9525 | 2.08 | 0.9812 | 2.49 | 0.9936 | 2.90 | 0.9981 |
| 1.68 | 0.9535 | 2.09 | 0.9817 | 2.50 | 0.9938 | 2.91 | 0.9982 |
| 1.69 | 0.9545 | 2.10 | 0.9821 | 2.51 | 0.9940 | 2.92 | 0.9982 |
| 1.70 | 0.9554 | 2.11 | 0.9826 | 2.52 | 0.9941 | 2.93 | 0.9983 |
| 1.71 | 0.9564 | 2.12 | 0.9830 | 2.53 | 0.9943 | 2.94 | 0.9984 |
| 1.72 | 0.9573 | 2.13 | 0.9834 | 2.54 | 0.9945 | 2.95 | 0.9984 |
| 1.73 | 0.9582 | 2.14 | 0.9838 | 2.55 | 0.9946 | 2.96 | 0.9985 |
| 1.74 | 0.9591 | 2.15 | 0.9842 | 2.56 | 0.9948 | 2.97 | 0.9985 |
| 1.75 | 0.9599 | 2.16 | 0.9846 | 2.57 | 0.9949 | 2.98 | 0.9986 |
| 1.76 | 0.9608 | 2.17 | 0.9850 | 2.58 | 0.9951 | 2.99 | 0.9986 |
| 1.77 | 0.9616 | 2.18 | 0.9854 | 2.59 | 0.9952 | 3.00 | 0.9987 |
| 1.78 | 0.9625 | 2.19 | 0.9857 | 2.60 | 0.9953 | 3.20 | 0.9993 |
| 1.79 | 0.9633 | 2.20 | 0.9861 | 2.61 | 0.9955 | 3.40 | 0.9997 |
| 1.80 | 0.9641 | 2.21 | 0.9864 | 2.62 | 0.9956 | 3.60 | 0.9998 |
| 1.81 | 0.9649 | 2.22 | 0.9868 | 2.63 | 0.9957 | 3.80 | 0.9999 |
| 1.82 | 0.9656 | 2.23 | 0.9871 | 2.64 | 0.9959 | 4.00 | 1.0000 |
| 1.83 | 0.9664 | 2.24 | 0.9875 | 2.65 | 0.9960 | | |
| 1.84 | 0.9671 | 2.25 | 0.9878 | 2.66 | 0.9961 | | |
| 1.85 | 0.9678 | 2.26 | 0.9881 | 2.67 | 0.9962 | | |
| 1.86 | 0.9686 | 2.27 | 0.9884 | 2.68 | 0.9963 | | |
| 1.87 | 0.9693 | 2.28 | 0.9887 | 2.69 | 0.9964 | | |
| 1.88 | 0.9699 | 2.29 | 0.9890 | 2.70 | 0.9965 | | |
| 1.89 | 0.9706 | 2.30 | 0.9893 | 2.71 | 0.9966 | | |
| 1.90 | 0.9713 | 2.31 | 0.9896 | 2.72 | 0.9967 | | |
| 1.91 | 0.9719 | 2.32 | 0.9898 | 2.73 | 0.9968 | | |
| 1.92 | 0.9726 | 2.33 | 0.9901 | 2.74 | 0.9969 | | |
| 1.93 | 0.9732 | 2.34 | 0.9904 | 2.75 | 0.9970 | | |
| 1.94 | 0.9738 | 2.35 | 0.9906 | 2.76 | 0.9971 | | |
| 1.95 | 0.9744 | 2.36 | 0.9909 | 2.77 | 0.9972 | | |
| 1.96 | 0.9750 | 2.37 | 0.9911 | 2.78 | 0.9973 | | |
| 1.97 | 0.9756 | 2.38 | 0.9913 | 2.79 | 0.9974 | | |
| 1.98 | 0.9761 | 2.39 | 0.9916 | 2.80 | 0.9974 | | |
| 1.99 | 0.9767 | 2.40 | 0.9918 | 2.81 | 0.9975 | | |
| 2.00 | 0.9772 | 2.41 | 0.9920 | 2.82 | 0.9976 | | |
| 2.01 | 0.9778 | 2.42 | 0.9922 | 2.83 | 0.9977 | | |
| 2.02 | 0.9783 | 2.43 | 0.9925 | 2.84 | 0.9977 | | |
| 2.03 | 0.9788 | 2.44 | 0.9927 | 2.85 | 0.9978 | | |
| 2.04 | 0.9793 | 2.45 | 0.9929 | 2.86 | 0.9979 | | |

**TABLE C-5**

Relation between $Z$ and $\Pr\{-z^* \leq z \leq z\}$ for the Standard Normal Variate

| $z^*$ | $\Pr\{-z^* \leq z \leq z^*\}$ |
|-------|------------------------------|
| 0.67 | 0.50 |
| 1.00 | 0.683 |
| 1.15 | 0.75 |
| 1.28 | 0.80 |
| 1.44 | 0.85 |
| 1.64 | 0.90 |
| 1.96 | 0.95 |
| 2.00 | 0.954 |
| 2.33 | 0.98 |
| 2.50 | 0.988 |
| 2.57 | 0.99 |
| 3.00 | 0.997 |
| 3.50 | 0.9996 |
| $\infty$ | 1.0000 |

"sampling error". Section C.8 explores how this uncertainty is assessed.

### C.7.2 Log-Normal Distribution

Most hydrologic variables cannot be negative numbers, and many have pdfs that are strongly positively skewed, like $X_3$ in Figure C-1c. In many cases, such variables can be well represented by the **log-normal distribution,** in which the logarithms of the variable values are normally distributed, rather than the values themselves. Thus, with the definition

$$LX \equiv \ln(X), \qquad \textbf{(C-40)}$$

the pdf for the log-normal distribution is

$$f_X(x) = \frac{1}{(2 \cdot \pi)^{1/2} \cdot \sigma_{LX}} \cdot \exp\left[-\frac{1}{2} \cdot \left(\frac{Lx - \mu_{LX}}{\sigma_{LX}}\right)^2\right]. \qquad \textbf{(C-41)}$$

It can be shown (Haan 1977) that

$$\mu_{LX} = \left(\frac{1}{2}\right) \cdot \ln\left(\frac{\mu_X^2}{CV_X^2 + 1}\right) \qquad \textbf{(C-42)}$$

and

$$\sigma_{LX}^2 = \ln(CV_X^2 + 1). \qquad \textbf{(C-43)}$$

If $X$ is log-normally distributed and $\mu_X$ and $\sigma_X$ are specified, the probability that $x$ is in any range can be readily calculated by using Equation (C-

40)– (C-43) and the tables of the normal distribution, as is shown in Example C-10. The PPCC test and L-moment ratios can be used to test the hypothesis that a given sample came from a log-normal distribution.

---

### EXAMPLE C-10

Suppose the annual peak streamflow data in Series $b$ of Table 2-2 are from a log-normally distributed population with the mean and the standard deviation of the annual floods equal to the sample values of those statistics. (a) What is the probability that a particular future year has a peak streamflow exceeding 500 ft$^3$ s$^{-1}$? (b) What is the return period of a flood of 500 ft$^3$ s$^{-1}$?

**Solution** Calculate $m_X = 319.6$ ft$^3$ s$^{-1}$ and $s_X = 166.3$ ft$^3$ s$^{-1}$ from the data in Table 2-2. Then, from Equation (C-20),

$$CV_X = \frac{166.3}{319.6} = 0.520.$$

From Equations (C-42) and (C-43),

$$m_{LX} = \left(\frac{1}{2}\right) \cdot \ln\left(\frac{319.6^2}{0.520^2 + 1}\right) = 5.6474;$$
$$s_{LX}^2 = \ln(0.520^2 + 1) = 0.2396;$$
$$s_{LX} = 0.4895.$$

(a) For $x = 500$ ft$^3$ s$^{-1}$, $Lx = 6.2146$, so, from Equation (C-38),

$$z = \frac{6.2146 - 5.6474}{0.4895} = 1.1588.$$

From Table C-4, the probability of $z \leq 1.1588$ is 0.877, so the probability that $z > 1.1588 = (1 - 0.877) = 0.023$. This is also the probability that $Lx > 6.2146$ and that $x > 500$ ft$^3$ s$^{-1}$.

(b) The exceedence probability of a flood of 500 ft$^3$ s$^{-1}$ is 0.023 (assuming a log-normal distribution), so we see from Equation (C-32) that

$$TR_X = \frac{1}{0.023} = 43.5 \text{ yr.}$$

---

### C.8 SAMPLING ERROR

In general, the true values of the quantiles and moments of, and correlation coefficients between, populations of hydrologic quantities are unknowable.

However, we can estimate the true values by taking sample(s) from the population(s) and calculating the sample quantiles, moment statistics, or correlation coefficients, using the formulas in Boxes C-1, C-2, or C-3. **Sampling error** is the uncertainty inherent in these sample estimates of population quantities. Here, we consider approaches to quantifying that uncertainty.

Note that, in contrast to the discussion of *measurement* error in Section 2.5.2, *sampling* error refers to the uncertainty in estimates of population values based on samples of finite size. Measurement errors in the sample values are not accounted for.

Imagine that we could take an infinite number of samples of size $N$ from a population and calculate the statistics of interest for each sample.[1] The values of each statistic so calculated are functions of random variables (the measured values in the sample), and thus can also be considered populations of random variables. Each such population has a probability distribution, with its own quantiles and moment statistics. Because they are derived by sampling, these distributions are known as **sampling distributions.** The underlying population from which the sample values were taken is called the **parent distribution.**

The standard deviations of sampling distributions are called **standard errors** of the sample estimates. Standard errors can be calculated, at least approximately, for all the statistics discussed here; they provide relative measures of the uncertainty in the estimates. If the form of the sampling distribution is known, this information, along with the standard error, can be used to compute absolute measures of the uncertainty in the form of **confidence intervals.**

### C.8.1   Standard Errors

If the parent distribution is normal, formulas for computing standard errors can be developed from statistical theory. In all cases, the theoretical standard errors are functions of the population moments and are inversely related to the sample size.

[1]It is not possible to do this with actual hydrologic data, but repeated sampling from a known distribution can readily be simulated by using computer-generated data available in most spreadsheets. This method of empirical statistical analysis is called **Monte-Carlo simulation.**

Sample estimates of standard errors are thus estimated from the sample size and from various sample statistics.

Even though sampling from time series seems quite different from sampling from objects distributed in space, the same concepts apply. In general, however, we do not have the option of reducing uncertainty (i.e., the standard error) by increasing the sample size: We must use the measurements that have been made to date. Special considerations that apply to estimating sampling error for time series that display a property called "persistence" are discussed in Sections C.9.3 and C.9.4.

### Mean

As is shown in most statistics books, the population of the sample mean, $m_X$, has standard deviation, $\sigma_{mX}$,

$$\sigma_{mX} = \frac{\sigma_X}{N^{1/2}}; \qquad \text{(C-44)}$$

thus $\sigma_{mX}$ is the **standard error of the mean.** It is estimated by substituting $s_X$ for $\sigma_X$ in Equation (C-44).

### Standard Deviation

The general formula for the **standard error of the sample standard deviation,** $\sigma_{sX}$, involves estimation of the fourth moment of the $X$ distribution, which usually cannot be reliably estimated for the sample sizes encountered by hydrologists. If $X$ is approximately normally distributed, $\sigma_{sX}$ is given by

$$\sigma_{sX} = \frac{\sigma_X}{(2 \cdot N)^{1/2}} \qquad \text{(C-45)}$$

(Yevjevich 1972). We will assume that this relation is sufficiently accurate for our purposes. Again, the sample estimate of $\sigma_{sX}$, $s_{sX}$, is found by substituting $s_X$ for $\sigma_X$ in Equation (C-45).

### Skewness

As with the standard deviation, the general formula for the **standard error of the skewness,** $\sigma_{gX}$, involves higher moments and so is not of practical use in hydrology. However, if $X$ is normally distributed, then

$$\sigma_{gX} = \left(\frac{6}{N}\right)^{1/2} \qquad \text{(C-46)}$$

(Yevjevich 1972).

### Quantiles

The standard error of a quantile, $\sigma_{qX}$, is

$$\sigma_{qX} = \frac{1}{f_X(x_q)} \cdot \left[ \frac{q \cdot (1-q)}{N} \right]^{1/2} \qquad \textbf{(C-47)}$$

(Yevjevich 1972). If the form of the parent distribution is known or assumed, $f_X(x_q)$ can be calculated from the appropriate pdf.

### Correlation Coefficient

For $\rho_{X,Y}$ near zero and $N > 25$, $\sigma_{rXY}$ can be estimated as

$$\sigma_{rXY} = \frac{1 - r_{X,Y}}{N^{1/2}} \qquad \textbf{(C-48)}$$

(Yevjevich 1972).

    However, because $\rho_{X,Y}$ is bounded between $-1$ and $+1$, it is most general to use the standard error of the transformed variable $W_{X,Y}$, where

$$W_{X,Y} \equiv 0.5 \cdot \ln\left[ \frac{1 + \rho_{X,Y}}{1 - \rho_{X,Y}} \right]. \qquad \textbf{(C-49)}$$

Table C-6 gives $W_{X,Y}$ as a function of $r_{X,Y}$. The standard error of $\hat{W}_{X,Y}$, $\sigma_{WXY}$, is

$$\sigma_{WXY} = \frac{1}{(N-3)^{1/2}} \qquad \textbf{(C-50)}$$

(Haan 1977).

### C.8.2 Sampling Distributions

The sampling distributions of moment and quantile statistics are "asymptotically normal" (Yevjevich 1972). This means that, as the sample size, $N$, gets larger, the sampling distribution approaches the normal distribution. However, at small and medium values of $N$, the sampling distribution can differ significantly from the normal. The value of $N$ for which the sampling distribution can usefully be approximated by the normal distribution is larger (1) the higher is the moment or the further the quantile is from the mean and (2) the higher is the skew of the parent distribution.

    We will examine two sampling distributions that have application to quantifying the uncertainty of sample estimates of the moment statistics $\mu_X$ and $\sigma_X$ for the sample sizes and distributions commonly found in hydrology. The details of these applications are discussed in Section C.8.3.

**TABLE C-6**

Relation between $W$ and $\rho_{X,Y}$ [Equation (C-49)]

| $\rho_{X,Y}$ | $W$ | $\rho_{X,Y}$ | $W$ | $\rho_{X,Y}$ | $W$ | $\rho_{X,Y}$ | $W$ |
|---|---|---|---|---|---|---|---|
| −1.00 | −∞ | −0.50 | −0.5493 | 0.00 | 0.0000 | 0.50 | 0.5493 |
| −0.98 | −2.2976 | −0.48 | −0.5230 | 0.02 | 0.0200 | 0.52 | 0.5763 |
| −0.96 | −1.9459 | −0.46 | −0.4973 | 0.04 | 0.0400 | 0.54 | 0.6042 |
| −0.94 | −1.7380 | −0.44 | −0.4722 | 0.06 | 0.0601 | 0.56 | 0.6328 |
| −0.92 | −1.5890 | −0.42 | −0.4477 | 0.08 | 0.0802 | 0.58 | 0.6625 |
| −0.90 | −1.4722 | −0.40 | −0.4236 | 0.10 | 0.1003 | 0.60 | 0.6931 |
| −0.88 | −1.3758 | −0.38 | −0.4001 | 0.12 | 0.1206 | 0.62 | 0.7250 |
| −0.86 | −1.2933 | −0.36 | −0.3769 | 0.14 | 0.1409 | 0.64 | 0.7582 |
| −0.84 | −1.2212 | −0.34 | −0.3541 | 0.16 | 0.1614 | 0.66 | 0.7928 |
| −0.82 | −1.1568 | −0.32 | −0.3316 | 0.18 | 0.1820 | 0.68 | 0.8291 |
| −0.80 | −1.0986 | −0.30 | −0.3095 | 0.20 | 0.2027 | 0.70 | 0.8673 |
| −0.78 | −1.0454 | −0.28 | −0.2877 | 0.22 | 0.2237 | 0.72 | 0.9076 |
| −0.76 | −0.9962 | −0.26 | −0.2661 | 0.24 | 0.2448 | 0.74 | 0.9505 |
| −0.74 | −0.9505 | −0.24 | −0.2448 | 0.26 | 0.2661 | 0.76 | 0.9962 |
| −0.72 | −0.9076 | −0.22 | −0.2237 | 0.28 | 0.2877 | 0.78 | 1.0454 |
| −0.70 | −0.8673 | −0.20 | −0.2027 | 0.30 | 0.3095 | 0.80 | 1.0986 |
| −0.68 | −0.8291 | −0.18 | −0.1820 | 0.32 | 0.3316 | 0.82 | 1.1568 |
| −0.66 | −0.7928 | −0.16 | −0.1614 | 0.34 | 0.3541 | 0.84 | 1.2212 |
| −0.64 | −0.7582 | −0.14 | −0.1409 | 0.36 | 0.3769 | 0.86 | 1.2933 |
| −0.62 | −0.7250 | −0.12 | −0.1206 | 0.38 | 0.4001 | 0.88 | 1.3758 |
| −0.60 | −0.6931 | −0.10 | −0.1003 | 0.40 | 0.4236 | 0.90 | 1.4722 |
| −0.58 | −0.6625 | −0.08 | −0.0802 | 0.42 | 0.4477 | 0.92 | 1.5890 |
| −0.56 | −0.6328 | −0.06 | −0.0601 | 0.44 | 0.4722 | 0.94 | 1.7380 |
| −0.54 | −0.6042 | −0.04 | −0.0400 | 0.46 | 0.4973 | 0.96 | 1.9459 |
| −0.52 | −0.5763 | −0.02 | −0.0200 | 0.48 | 0.5230 | 0.98 | 2.2976 |
|  |  |  |  |  |  | 1.00 | ∞ |

### The t Distribution

The mathematical definition of the $t$ distribution can be found in most statistics books (e.g., Haan 1977). For our purposes, it is important to note that the exact form of the distribution is determined by a quantity called the **degrees of freedom,** $DF$, which can be calculated directly from the sample size, $N$. In the situations encountered in this appendix,

$$DF = N - 1. \qquad \textbf{(C-51)}$$

The $t$ distribution is useful because it can be shown that, *if the parent distribution of X is normal*, the quantity

$$T = \frac{N^{1/2} \cdot [m_X - \mu_X]}{s_X} \qquad \textbf{(C-52)}$$

has a $t$ distribution with $DF = N - 1$. $T$ is thus analogous to the standard normal variate, $Z$ [Equation (C-38)].

The $t$ distribution is symmetrical about its mean value $\mu_T = 0$, and it has a variance $\sigma_T^2 = DF/(DF - 2)$. The quantiles of the distribution can readily be calculated if $DF$ (i.e., $N$) is given, and these quantiles are tabulated in statistics books. Table C-7 gives the quantiles $t_{0.025}$ and $t_{0.975}$ for selected values of $DF$. As $DF$ gets large, the $t$ distribution approaches the standard normal distribution; for $DF > 120$, $t_q = z_q$ to a close approximation.

Even though $T$ follows the $t$ distribution only when the parent distribution is normal, it is a good approximation even for skewed parent distributions if $N$ is 30 or more (Barrett and Goldsmith 1976).

### The Chi-Squared ($\chi^2$) Distribution

Like the $t$ distribution, the form of the $\chi^2$ distribution depends on the degrees of freedom, as given by Equation (C-51). The $\chi^2$ distribution is important because, *if the parent distribution of X is normal*, the quantity

$$\chi^2 = \frac{(N - 1) \cdot s_X^2}{\sigma_X^2} \qquad \textbf{(C-53)}$$

has a $\chi^2$ distribution with $DF = N - 1$.

The mean of the distribution $\mu_{\chi^2} = DF$, and its variance $\sigma_{\chi^2}^2 = 2 \cdot DF$. However, this distribution is asymmetrical; values of the quantiles $\chi^2_{0.025}$ and $\chi^2_{0.975}$ are tabulated in Table C-7. As $N$ gets large,

**TABLE C-7**
0.025 and 0.975 Quantiles of the $t$ Distribution and of the $\chi^2$ Distribution as a Function of Degrees of Freedom, $DF$

| | t-Distribution | | $\chi^2$-Distribution | |
|---|---|---|---|---|
| **DF** | $t_{DF,\,0.025}$ | $t_{DF,\,0.975}$ | $\chi^2_{DF,\,0.025}$ | $\chi^2_{DF,\,0.975}$ |
| 10 | −2.23 | 2.23 | 3.25 | 20.5 |
| 11 | −2.20 | 2.20 | 3.82 | 21.9 |
| 12 | −2.18 | 2.18 | 4.40 | 23.3 |
| 13 | −2.16 | 2.16 | 5.01 | 24.7 |
| 14 | −2.14 | 2.14 | 5.63 | 26.1 |
| 15 | −2.13 | 2.13 | 6.26 | 27.5 |
| 16 | −2.12 | 2.12 | 6.91 | 28.8 |
| 17 | −2.11 | 2.11 | 7.56 | 30.2 |
| 18 | −2.10 | 2.10 | 8.23 | 31.5 |
| 19 | −2.09 | 2.09 | 8.91 | 32.9 |
| 20 | −2.09 | 2.09 | 9.59 | 34.2 |
| 21 | −2.08 | 2.08 | 10.3 | 35.5 |
| 22 | −2.07 | 2.07 | 11.0 | 36.8 |
| 23 | −2.07 | 2.07 | 11.7 | 38.1 |
| 24 | −2.06 | 2.06 | 12.4 | 39.4 |
| 25 | −2.06 | 2.06 | 13.1 | 40.6 |
| 30 | −2.04 | 2.04 | 16.8 | 47.0 |
| 40 | −2.02 | 2.02 | 24.4 | 59.3 |
| 50 | −2.01 | 2.01 | 32.4 | 71.4 |
| 60 | −2.00 | 2.00 | 40.5 | 83.3 |
| 100 | −1.98 | 1.98 | 74.2 | 129.6 |

(From Haan 1977.)

the $\chi^2$ distribution approaches the normal distribution (Yevjevich 1972), but the approximation is not very close for the sample sizes usually available in hydrology.

### C.8.3  Confidence Intervals

Absolute uncertainty of statistics computed from sample values is expressed as

$$\Pr\{L_\alpha \le \Theta \le U_\alpha\} = 1 - \alpha, \qquad \textbf{(C-54)}$$

where $\Theta$ is the true value of the measured quantity or of the population statistic of interest, $L_\alpha$ and $U_\alpha$ are respectively the lower and upper **confidence limits** that define the **confidence interval,** $1 - \alpha$ is the **confidence level,** and $0 \le \alpha \le 1$. In words, Equation (C-54) can be stated as "I am $100 \cdot (1 - \alpha)\%$ sure that the interval $L_\alpha$ to $U_\alpha$ contains $\Theta$." If the distribution of the estimator of $\Theta$ depends on $\Theta$ and is known, $L_\alpha$ and $U_\alpha$ can be determined for any selected value of $\alpha$ as functions of the sample values.

## Mean

Because $T$ in Equation (C-52) follows the $t$ distribution, we can write

$$\Pr\left\{t_{N-1,\alpha/2} \leq \frac{N^{1/2} \cdot [m_X - \mu_X]}{s_X} \leq t_{N-1,1-\alpha/2}\right\}$$
$$= 1 - \alpha. \qquad \text{(C-55)}$$

Because of the symmetry of the $t$ distribution, $t_{N-1,\alpha/2} = -t_{N-1,1-\alpha/2}$ and Equation (C-55) can be written as

$$\Pr\left\{m_X - \frac{t_{N-1,1-\alpha/2} \cdot s_X}{N^{1/2}} \leq \mu_X \leq m_X \right.$$
$$\left. + \frac{t_{N-1,1-\alpha/2} \cdot s_X}{N^{1/2}}\right\} = 1 - \alpha. \qquad \text{(C-56)}$$

Example C-11 shows how Equation (C-56) is applied to compute the confidence interval for the mean.

---

### EXAMPLE C-11

Compute the 95% confidence intervals for the mean of the population of average annual streamflow (Series $a$) in Table 2-2.

**Solution** From Example C-2, $N = 25$, $m_X = 19.3$ ft$^3$ s$^{-1}$, and $s_X = 5.52$ ft$^3$ s$^{-1}$. From Table C-7, $t_{24,0.975} = 2.06$. Thus,

$$\frac{t_{N-1,1-\alpha/2} \cdot s_X}{N^{1/2}} = \frac{2.06 \times 552}{25^{1/2}} = 2.27 \text{ ft}^3 \text{ s}^{-1}.$$

Referring to Equations (C-54) and (C-56), we calculate

$$L_{.05} = 19.3 - 2.27 = 17.0 \text{ ft}^3 \text{ s}^{-1}$$

and

$$U_{.05} = 19.3 + 2.27 = 21.6 \text{ ft}^3 \text{ s}^{-1},$$

and we state that we are 95% confident that the true mean flow is between 17.0 and 21.6 ft$^3$ s$^{-1}$.

These limits are approximate, because (1) we have assumed that the parent distribution is normal and (2) we used sample estimates of the statistics.

---

## Standard Deviation

Because $\chi^2$ in Equation (C-53) has a $\chi^2$ distribution, we can write

$$\Pr\left\{\chi^2_{N-1,\alpha/2} \leq \frac{(N-1) \cdot s_X^2}{\sigma_X^2} \leq \chi^2_{N-1,1-\alpha/2}\right\}$$

$$= 1 - \alpha. \qquad \text{(C-57)}$$

Manipulating the quantities inside the braces, we get

$$\Pr\left\{\frac{(N-1) \cdot s_X^2}{\chi^2_{N-1,1-\alpha/2}} \leq \sigma_X^2 \leq \frac{(N-1) \cdot s_X^2}{\chi^2_{N-1,\,\alpha/2}}\right\}$$
$$= 1 - \alpha. \qquad \text{(C-58)}$$

Example C-12 shows how Equation (C-58) is used to estimate the confidence intervals for the standard deviation.

---

### EXAMPLE C-12

Estimate the 95% confidence intervals for the standard deviation of the population of average annual streamflows (Series $a$) in Table 2-2.

**Solution** From Example C-2, $N = 25$ and $s_X^2 = 30.47$ (ft$^3$ s$^{-1}$)$^2$. From Table C-7, $\chi^2_{24,0.025} = 12.4$ and $\chi^2_{24,0.975} = 39.4$. Referring to Equations (C-54) and Equation (C-58), we calculate for the variance

$$L_{.05} = \frac{24 \times 30.47}{39.4} = 18.56 \text{ (ft}^3 \text{ s}^{-1})^2$$

and

$$U_{.05} = \frac{24 \times 30.47}{12.4} = 58.97 \text{ (ft}^3 \text{ s}^{-1})^2.$$

Taking the square roots gives the 95% confidence interval for the standard deviation,

$$L_{.05} = 18.56^{1/2} = 4.31 \text{ ft}^3 \text{ s}^{-1}$$

and

$$U_{.05} = 58.97^{1/2} = 7.68 \text{ ft}^3 \text{ s}^{-1},$$

and we say that we are 95% confident that the true standard deviation is between 4.31 and 7.68 ft$^3$ s$^{-1}$.

Note that these confidence intervals are not symmetrical about $s_X$. They are approximate for the reasons given in Example C-11 but, because they are for a statistic based on the second moment, the degree of approximation for a non-normal parent distribution is less good than in the case of the mean.

---

## Skewness and Kurtosis

The sampling distributions for the skewness, $\gamma_X$, and the kurtosis, $\kappa_X$, are not known for the sample sizes usually available in hydrology. Monte-Carlo experiments have shown that estimates of these quantities are highly unreliable even for very large $N$. Furthermore, it has been shown that sample estimates of skewness and kurtosis have upper bounds that depend on sample size and are independent of the actual population skewness [Equations (CB2-9) and (CB2-10)].

## Quantiles

As is explained in Box C-1, sample estimates of quantiles are based on the ranks of the sample values. Precise calculation of specified confidence intervals for these estimates is not possible. However, Loucks et al. (1981) show that the probability that the $q$th quantile of $X$, $x_q$, lies between the $j$th-ranked sample value, $x_{(j)}$, and the $k$th-ranked sample value, $x_{(k)}$, with $j < k$, is given by

$$\Pr\{x_{(j)} \leq x_q \leq x_{(k)}\}$$
$$= \sum_{i=j}^{k-1} \frac{N!}{i!(N-i)!} \cdot q^i \cdot (1-q)^{N-i}. \quad \textbf{(C-59)}$$

Table C-8 shows the at-least-90% confidence intervals for the median ($x_{50}$) as calculated via Equation (C-59). Note (1) that the confidence intervals are symmetric in terms of ranks, but not necessarily in terms of $x$ values and (2) that the spread is wide—for $N = 25$, we would estimate the median as the 13th-ranked value, but the 95.7% confidence interval is between the 8th-ranked and 18th-ranked values (inclusive). As $q$ gets further from the median, the confidence intervals get wider.

## Correlation Coefficient

Confidence intervals for correlation coefficients are computed by using the transformed variable $W_{X,Y}$ defined in Equation (C-49). If $N > 25$, $\hat{W}_{X,Y}$ has a normal distribution with mean

$$\mu_{WXY} = 0.5 \cdot \ln\left[\frac{1 + \rho_{X,Y}}{1 - \rho_{X,Y}}\right] \quad \textbf{(C-60)}$$

and with $\sigma_{WXY}$ given by Equation (C-50) (Haan 1977). Thus $Z_W$ is a standard normal variate, where

**TABLE C-8**
At-least-90% Confidence Intervals for the Median Calculated via Equation (C-59)[a]

| $N$ | Lower Confidence Limit, $x_{(j)}$ | Median Estimate [Eq. (CB1-1)] | Upper Confidence Limit, $x_{(k)}$ | $\alpha$ | $1 - \alpha$ |
|---|---|---|---|---|---|
| 7 | $x_{(1)}$ | $x_{(4)}$ | $x_{(7)}$ | 0.016 | 0.984 |
| 8 | $x_{(2)}$ | $(x_{(4)} + x_{(5)})/2$ | $x_{(6)}$ | 0.070 | 0.930 |
| 9 | $x_{(2)}$ | $x_{(5)}$ | $x_{(7)}$ | 0.039 | 0.961 |
| 10 | $x_{(2)}$ | $(x_{(5)} + x_{(6)})/2$ | $x_{(9)}$ | 0.021 | 0.979 |
| 11 | $x_{(3)}$ | $x_{(6)}$ | $x_{(9)}$ | 0.065 | 0.935 |
| 12 | $x_{(3)}$ | $(x_{(6)} + x_{(7)})/2$ | $x_{(10)}$ | 0.039 | 0.961 |
| 13 | $x_{(4)}$ | $x_{(7)}$ | $x_{(10)}$ | 0.092 | 0.908 |
| 14 | $x_{(4)}$ | $(x_{(7)} + x_{(8)})/2$ | $x_{(11)}$ | 0.057 | 0.943 |
| 15 | $x_{(4)}$ | $x_{(8)}$ | $x_{(12)}$ | 0.035 | 0.965 |
| 16 | $x_{(5)}$ | $(x_{(8)} + x_{(9)})/2$ | $x_{(12)}$ | 0.077 | 0.923 |
| 17 | $x_{(5)}$ | $x_{(9)}$ | $x_{(13)}$ | 0.049 | 0.951 |
| 18 | $x_{(6)}$ | $(x_{(9)} + x_{(10)})/2$ | $x_{(13)}$ | 0.096 | 0.904 |
| 19 | $x_{(6)}$ | $x_{(10)}$ | $x_{(14)}$ | 0.064 | 0.936 |
| 20 | $x_{(6)}$ | $(x_{(10)} + x_{(11)})/2$ | $x_{(15)}$ | 0.041 | 0.959 |
| 21 | $x_{(7)}$ | $x_{(11)}$ | $x_{(15)}$ | 0.078 | 0.922 |
| 22 | $x_{(7)}$ | $(x_{(11)} + x_{(12)})/2$ | $x_{(16)}$ | 0.017 | 0.983 |
| 23 | $x_{(8)}$ | $x_{(12)}$ | $x_{(16)}$ | 0.093 | 0.907 |
| 24 | $x_{(8)}$ | $(x_{(12)} + x_{(13)})/2$ | $x_{(17)}$ | 0.064 | 0.936 |
| 25 | $x_{(8)}$ | $x_{(13)}$ | $x_{(18)}$ | 0.043 | 0.957 |
| 30 | $x_{(11)}$ | $(x_{(15)} + x_{(16)})/2$ | $x_{(20)}$ | 0.099 | 0.901 |
| 35 | $x_{(13)}$ | $x_{(18)}$ | $x_{(23)}$ | 0.090 | 0.910 |
| 40 | $x_{(15)}$ | $(x_{(20)} + x_{(21)})/2$ | $x_{(26)}$ | 0.082 | 0.918 |
| 45 | $x_{(17)}$ | $x_{(23)}$ | $x_{(29)}$ | 0.074 | 0.926 |
| 50 | $x_{(19)}$ | $(x_{(25)} + x_{(26)})/2$ | $x_{(32)}$ | 0.066 | 0.934 |
| 60 | $x_{(24)}$ | $(x_{(30)} + x_{(31)})/2$ | $x_{(37)}$ | 0.093 | 0.907 |

[a]That is, $\Pr\{x_{(j)} \leq x_5 \leq x_{(k)}\} \leq 0.90$.

After Loucks et al. (1981).

$$Z_W \equiv \frac{W_{X,Y} - \mu_{WXY}}{\sigma_{WXY}} \qquad \text{(C-61)}$$

and

$$\Pr\{z_{\alpha/2} \le Z_W \le z_{1-\alpha/2}\} = 1 - \alpha. \qquad \text{(C-62)}$$

Substituting Equation (C-61) into (C-62), rearranging, and using sample estimates then gives

$$\Pr\{\hat{W}_{X,Y} - z_{1-\alpha/2} \cdot s_{WXY} \le \mu_{WXY}$$
$$\le \hat{W}_{X,Y} + z_{1-\alpha/2} \cdot s_{WXY}\} = 1 - \alpha. \quad \text{(C-63)}$$

Example C-13 shows how Equation (C-63) is applied to compute confidence intervals for $\rho_{X,Y}$.

---

### EXAMPLE C-13

The sample correlation coefficient between the average annual flows and the annual flood peaks in Table 2-2 is $r_{X,Y} = 0.442$. What are the 95% confidence intervals for $\rho_{X,Y}$?

**Solution:** First, we calculate $\hat{W}$ via Equation (C-49):

$$\hat{W} = 0.5 \cdot \ln\left[\frac{1 + 0.442}{1 - 0.442}\right] = 0.447.$$

From Equation (C-50),

$$s_{WXY} = \frac{1}{(25 - 3)^{1/2}} = 0.213.$$

From Table C-4, $z_{1-\alpha/2} = 1.96$. Referring to Equations (C-54) and (C-63), we have, for $W_{X,Y}$,

$$L_{.05} = 0.477 - 1.96 \times 0.213 = 0.060$$

and

$$U_{.05} = 0.477 + 1.96 \times 0.213 = 0.894.$$

From the relation between $W_{X,Y}$ and $\rho_{X,Y}$ (Table C-6), this transforms to the following confidence limits for $\rho_{X,Y}$:

$$L_{.05} = 0.06$$

and

$$U_{.05} = 0.71.$$

Thus, we can state that we are 95% confident that the true $\rho_{X,Y}$ is between 0.06 and 0.71. Because the confidence interval does not include zero, we can conclude that there is less than a 5% chance that $\rho_{X,Y} = 0$.

---

## C.9 PERSISTENCE AND AUTOCORRELATION

### C.9.1 Definition and Estimation

Recall that a time series is a list of successive values of a variable. **Persistence** is the tendency for high values of the variable to follow high values, and low values to follow low values.[2] Thus the presence of persistence indicates that successive values of the time series are not independent, but are instead related in some way to one or more preceding values. If persistence is present, it must be accounted for in a statistical analysis of the time series, as discussed below.

The presence of the most common type of persistence is reflected in the **autocorrelation coefficient** (or **serial correlation coefficient**), $\rho_{kX}$, defined as

$$\rho_{kX} \equiv \frac{E[(x_{[i]} - \mu_X) \cdot (x_{[i+k]} - \mu_X)]}{\sigma_X^2}, \qquad \text{(C-64)}$$

where the subscripts in brackets denote successive values in the time series. By comparison with Equation (C-34), we see that $\rho_{kX}$ is simply the correlation coefficient between time-series values at a spacing of $k$ time intervals. Thus, like $\rho_{X,Y}$, $\rho_{kX}$ is a dimensionless number that can take on values between $-1$ and $+1$.

As with other statistics, $\rho_{kX}$ cannot be known, but must be estimated by calculating $r_{kX}$ as described in Box C-6. Since $r_{kX}$ could be greater than 0 in a sample from a population for which $\rho_{kX} = 0$, it is essential that $r_{kX}$ be tested for "significance" against a critical value that decreases with increasing length of record, $T$ (Box C-6).

### C.9.2 Causes and Significance

Variables with seasonal variation tend to have significant autocorrelation at a lag equal to the period of the seasonality. For example, monthly average streamflow values in many regions have significant correlation at lag $k = 12$. Such obvious seasonal signals can conceal more subtle persistence induced by longer-term climatic influences (e.g., the ENSO teleconnections described in Section 3.1.4). These longer-term effects can be explored by normalizing

---

[2]"High" and "low" values are relative to the average value.

---

## BOX C-6

· · · · · · ·

### Sample Autocorrelation Coefficients

If we designate the successive values of a time series $X$ as $(x_{[1]}, x_{[2]}, x_{[3]}, ..., x_{[T]})$, the **autocorrelation coefficient at lag $k$,** $\rho_{kX}$, can be estimated by first calculating

$$r'_{kX} = \frac{\sum_{i=1}^{T-k}[(x_{[i]} - m_X) \cdot (x_{[i+k]} - m_X)]}{(T-1) \cdot s_X^2}. \quad \textbf{(CB6-1)}$$

However, estimates of $\rho_{kX}$ computed by Equation (CB6-1) are biased low (Stedinger et al. 1992) and, for estimating $\rho_{1X}$, an unbiased estimate should be computed as

$$r_{1X} = \hat{\rho}_{1X} = \frac{r'_{1X} \cdot T + 1}{T - 4} \quad \textbf{(CB6-2)}$$

(Wallis and O'Connell 1972). Equation (CB6-2) shows that bias in estimating $\rho_{1X}$ is considerable even for long record lengths.

If $|r_{1X}| > r_1^*$, where

$$r_1^* = \frac{2}{T^{1/2}}, \quad \textbf{(CB6-3)}$$

we say that $r_{1X}$ is "significantly different from zero" or, more precisely, that there is less than a 5% chance that the actual autocorrelation for the population equals zero. Thus, if $|r_{1X}| > r_1^*$ and $r_{1X} > 0$, we accept the hypothesis that persistence is present (Chatfield 1984).

---

the original monthly data to create a new "de-seasonalized" time series, $z_i$, as

$$z_i \equiv \frac{Q_{m,y} - m_{Qm}}{s_{Qm}}, \quad \textbf{(C-65)}$$

where $Q_{m,y}$ is the flow for month $m$ in year $y$, $m_{Qm}$ is the mean flow for month $m$ (e.g., mean flow for all Januarys), and $s_{Qm}$ is the standard deviation for month $m$.

Persistence is reflected by a positive value for $\rho_{1X}$; usually it is produced by storage with a residence time at least as great as the time interval of the time series—for example, where a hydrologic variable represents outflow from a substantial reservoir (watershed, lake, aquifer), as was discussed in Section 2.8.2. A zero value of $\rho_{1X}$ indicates that successive values of $X$ are not linearly related. A negative value of $\rho_{1X}$ indicates that high values of $X$ tend to be followed by low values and vice versa; this reflects the opposite of persistence and is rare in hydrologic time series.

For variables with no seasonal variation, $\rho_{kX}$ values typically decline quasi-exponentially as $k$ increases (i.e., values are less correlated as they are farther apart in time). Thus attention is usually focused on the lag-1 autocorrelation coefficient, $\rho_{1X}$.

### EXAMPLE C-14

Figure C-5 is a plot of the time series of annual average flows, $Q$, of the Squam River, the outlet of Squam Lake in central New Hampshire, for water years 1940–1995 ($T = 56$ yr). Does the storage in this lake (residence time $\approx 5$ yr) induce significant autocorrelation in the annual flows?

**Solution:** Annual average flows contain no seasonality, so no adjustment is necessary. Using Equation (CB6-1), we find that

$$r'_{1Q} = 0.288.$$

The unbiased estimate $r_{1Q}$ is computed via Equation (CB6-2):

$$r_{1Q} = \frac{0.288 \times 56 + 1}{56 - 4} = 0.330.$$

To test for significant persistence, we compute $r_1^*$ via Equation (CB6-3):

$$r_1^* = \frac{2}{56^{1/2}} = 0.267.$$

Because $r_{1Q} > r_1^*$, we conclude that there is less than a 5% chance that there is no persistence (i.e., that $\rho_{1Q} = 0$) and assume that persistence is present.

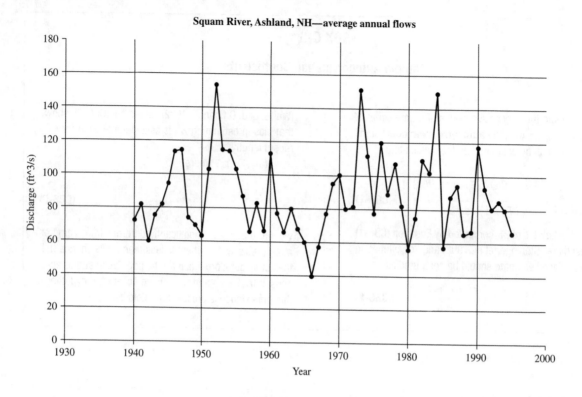

**FIGURE C-5**
Time series of annual streamflows of the Squam River near Ashland, NH, 1940–1995 (Example C-14)

### C.9.3   Effects of Persistence on Uncertainty of Time-Series Statistics

The presence of persistence makes it more likely that a time-series sample was taken from a period when values tended to be higher or lower than average, and when the variability was not representative of the population. Thus persistence reduces the confidence with which one can estimate the quantile and moment statistics of the time series.

The effect of persistence on quantile or L-moment estimates has not been well documented. However, it is intuitively clear that such estimates made by the methods described in Box C-1 could be highly inaccurate when $\rho_{1X}$ is significant, unless $T$ is quite large.

The effects of autocorrelation on uncertainty in estimates of the mean and standard deviation can approximately be evaluated quantitatively (Yevjevich 1972). For a given statistic $\theta$, this is done by calculating the **effective record length,** $T_{E\theta}$, for a time

series with persistence. $T_{E\theta}$, which is always smaller than $T$, is then used to calculate the standard error of the statistic via Equation (C-44) or (C-45), but with $T_{E\theta}$ substituted for $N$.

For the simplest type of persistent time series, where the value of $\rho_{1X}$ completely characterizes the persistence, the effective record length for estimating the standard error of the mean, $T_{Em}$, is

$$T_{Em} = T \cdot \frac{1 - \rho_{1X}}{1 + \rho_{1X}}. \tag{C-66}$$

The effective record length for estimating the standard error of the standard deviation, $T_{Es}$, under the same conditions is

$$T_{Es} = T \cdot \frac{1 - \rho_{1X}^2}{1 + \rho_{1X}^2}. \tag{C-67}$$

In practice, the estimate $r_{1X}'$ is used in Equations (C-66) and (C-67) when $r_{1X}' > r_1^*$ as given in Equation (CB6-3).

**FIGURE C-6**
Relative effects of autocorrelation on effective sample lengths for estimating standard errors of (a) the mean [Equation (C-66)] and (b) the standard deviation [Equation (C-67)].

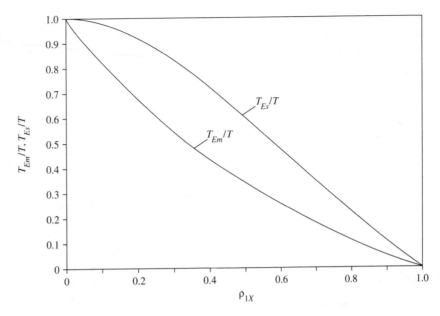

Figure C-6 shows how these effective record lengths vary with $r_{1X}'$, and Example C-15 shows how these relations are applied in the estimation of uncertainty.

---

## EXAMPLE C-15

Estimate the 95% confidence intervals for estimates of the mean and standard deviation for the average annual flows of the Squam River, accounting for autocorrelation.

**Solution:** For the annual flows of the Squam River, $m_Q = 88.3$ ft$^3$ s$^{-1}$ and $s_Q = 24.4$ ft$^3$ s$^{-1}$. From Equation (C-66),

$$T_{Em} = 56 \text{ yr} \times \frac{1 - 0.330}{1 + 0.330} = 28.2 \text{ yr.}$$

We now compute the confidence intervals for the mean as in Example C-11. Table C-7 gives $t_{27,0.975} = 2.05$, so

$$\frac{t_{27,0.975} \cdot s_Q}{T_{Em}} = \frac{2.05 \times 24.2}{28.2^{1/2}} = 9.3 \text{ ft}^3 \text{ s}^{-1}.$$

Referring to Equations (C-54) and (C-56), we can calculate

$$L_{.05} = 88.3 \text{ ft}^3 \text{ s}^{-1} - 9.3 \text{ ft}^3 \text{ s}^{-1} = 79.0 \text{ ft}^3 \text{ s}^{-1}$$

and

$$U_{.05} = 88.3 \text{ ft}^3 \text{ s}^{-1} + 9.3 \text{ ft}^3 \text{ s}^{-1} = 97.6 \text{ ft}^3 \text{ s}^{-1}.$$

Thus we are 95% confident that $79.0 \text{ ft}^3 \text{ s}^{-1} \leq \mu_Q \leq 97.6 \text{ ft}^3 \text{ s}^{-1}$.
For the standard deviation, we use Equation (C-67) to find

$$T_{Es} = 56 \text{ yr} \times \frac{1 - 0.330^2}{1 + 0.330^2} = 45.0 \text{ yr.}$$

From Table C-7, we see that that $\chi^2_{45,0.975} = 65.4$ and $\chi^2_{45,0.025} = 28.3$. Then, following the steps of Example C-12, we have

$$L_{.05} = \frac{T_{Es} \cdot s_Q^2}{\chi^2_{45,0.975}} = \frac{45.0 \times 595.47}{65.4} = 409.72 \text{ (ft}^3 \text{ s}^{-1})^2$$

and

$$U_{.05} = \frac{T_{Es} \cdot s_Q^2}{\chi^2_{45,0.025}} = \frac{45.0 \times 595.47}{28.3} = 946.86 \text{ (ft}^3 \text{ s}^{-1})^2.$$

Thus,

$$L_{.05} = 409.72^{1/2} = 20.2 \text{ ft}^3 \text{ s}^{-1}$$

and

$$U_{.05} = 946.86^{1/2} = 30.8 \text{ ft}^3 \text{ s}^{-1}.$$

Thus we are 95% confident that $20.2 \text{ ft}^3 \text{ s}^{-1} \leq \sigma_Q \leq 30.8 \text{ ft}^3 \text{ s}^{-1}$.

---

Adjustment for the skewness is given by Stedinger et al. (1992).

### C.9.4 Effects of Persistence on Uncertainty of Correlation Estimates

If two time-series variables $X$ and $Y$ have persistence (i.e., $\rho_{1X} > 0$ and $\rho_{1Y} > 0$), they will tend to appear to be more highly linearly correlated than they actually are. This "inflation" in the value of $r_{X,Y}$

can be accounted for by computing an effective sample size, $T_{Er}$, for the significance test of $r_{X,Y}$ [Equation (CB5-2)]. From Yevjevich (1972), $T_{Er}$ is approximately

$$T_{Er} = \frac{T}{1 + 2 \cdot r_{1X} \cdot r_{1Y}}, \qquad \textbf{(C-68)}$$

and we define $r^{*\prime}$ by analogy with Equation (CB5-2) as

$$r^{*\prime} = \frac{1.96}{(T_{Er} - 3)^{1/2}}. \qquad \textbf{(C-69)}$$

### EXAMPLE C-16

Suppose one has records of the average annual level of water in an observation well ($G$) and the average annual precipitation over the region ($P$) for 30 yr. The sample correlation coefficient between the two time series is $r_{G,P} = 0.433$. The ground-water levels have a sample autocorrelation $r_{1G}{}' = 0.545$, and the precipitation has a sample autocorrelation $r_{1P}{}' = 0.402$. Are the two time series significantly linearly correlated?

**Solution:** First, we test whether the two sample autocorrelation coefficients are significantly greater than zero by using Equation (CB6-3):

$$r_1^* = \frac{2}{30^{1/2}} = 0.365.$$

Because $r_{1G}{}' > r_1^*$ and $r_{1P}{}' > r_1^*$, we conclude that both time series are autocorrelated. Thus we use Equation (C-68) to calculate $T_{Er}$:

$$T_{Er} = \frac{30}{1 + 2 \times 0.402 \times 0.545} = 20.9.$$

Next, we compute $r^{*\prime}$ via Equation (C-69):

$$r^{*\prime} = \frac{1.96}{(20.9 - 3)^{1/2}} = 0.463.$$

Because $r_{G,P} < r^{*\prime}$, we conclude that there is no significant correlation between $G$ and $P$. Note that our conclusion would have been the opposite had we not accounted for the autocorrelations in the two time series.

## C.10 STATISTICAL CRITERIA FOR MODEL "CALIBRATION" AND "VALIDATION"

Here we introduce three criteria that have been used in "calibrating" and "validating" hydrologic models (see Section 2.9.4) and for comparing performance among models. These criteria are particularly appropriate for models that simulate continuous time series of daily (or other time-period) streamflows rather than peak flows or drought flows. Application and further discussion of these and other criteria can be found in Nash and Sutcliffe (1970), Garrick et al. (1978), World Meteorological Organization (1986), and Martinec and Rango (1989).

In the following, we consider that we have a time series of $i = 1, 2, ..., N$ measured values $Q_i$ and simulated values $\hat{Q}_i$.

### C.10.1 Nash–Sutcliffe Coefficient

Nash and Sutcliffe (1970) proposed the criterion $R_{NS}^2$, where

$$R_{NS}^2 \equiv 1 - \frac{\sum_{i=1}^{N}(Q_i - \hat{Q}_i)^2}{\sum_{i=1}^{N}(Q_i - m_Q)^2} \qquad \textbf{(C-70)}$$

and $m_Q$ is the average value of $Q$ for the period being simulated. If several years are being used, Martinec and Rango (1989) recommended that $m_Q$ be the average for those years and that values of $R_{NS}^2$ be calculated separately for each year and averaged to give an overall measure.

### C.10.2 Coefficient of Gain from Daily Means

This statistic, designated $R_{DG}^2$, is

$$R_{DG}^2 \equiv 1 - \frac{\sum_{i=1}^{N}(Q_i - \hat{Q}_i)^2}{\sum_{i=1}^{N}(Q_i - m_{Q_i})^2}, \qquad \textbf{(C-71)}$$

where $m_{Qi}$ is the average value of $Q$ *for time period i*. The World Meteorological Office (1986) recommended that, for both calibration and validation, $m_{Qi}$ be the values computed for the calibration period.

## C.10.3 Evaluation Measures Used in BROOK90

To improve evaluation of the fit and to decide how parameters should be adjusted in calibration, the BROOK90 model (Federer 1995) compares measured and simulated flows separately for three "seasons" (February–May; June–September; October–January) and, within each season, for "high" (> 5 mm), "medium" (0.5 to 5 mm), and "low" (< 0.5 mm) flows. Three-day running-mean values are used for all comparisons to account for the one-day time step, so, in the following, the $Q_i$ and $\hat{Q}_i$ values are averages over days $i - 2, i - 1$, and $i$.

For each season and flow range, $\hat{Q}_i - Q_i$ is separated into a systematic or **bias error,** $e_{bi}$, and a scatter or **dispersion error,** $e_{di}$; that is,

$$\hat{Q}_i - Q_i = e_{bi} + e_{di}, \qquad \text{(C-72)}$$

where

$$e_{bi} \equiv Q_i \cdot \frac{m(\hat{Q})}{m(Q)} - Q_i \qquad \text{(C-73)}$$

and

$$e_{di} \equiv \hat{Q}_i - Q_i \cdot \frac{m(\hat{Q})}{m(Q)}, \qquad \text{(C-74)}$$

in which we now indicate the mean of a quantity as $m(.)$.

Then the **mean bias error,** $MBE$, and **mean dispersion error,** $MDE$, are found as

$$MBE = m(e_{bi}) = m(\hat{Q}) - m(Q) \qquad \text{(C-75)}$$

and

$$MDE = m(|e_{di}|). \qquad \text{(C-76)}$$

The model also calculates the **normalized bias error,** $NBE$, and **normalized dispersion error,** $NDE$, as

$$NBE(\%) \equiv 100 \cdot \frac{MBE}{m(Q)} \qquad \text{(C-77)}$$

and

$$NDE(\%) \equiv 100 \cdot \frac{MDE}{m(Q)}. \qquad \text{(C-78)}$$

# D
# Water and Energy in the Atmosphere

## D.1 PHYSICS OF RADIANT ENERGY

All matter at a temperature above absolute zero radiates energy in the form of electromagnetic waves that travel at the speed of light. The rate at which this energy is emitted is given by the **Stefan-Boltzmann Law,**

$$Q_r = \varepsilon \cdot \sigma \cdot T^4, \qquad \text{(D-1)}$$

where $Q_r$ is the rate of energy emission per unit surface area per unit time $[\text{E L}^{-2} \text{T}^{-1}]$[1], $T$ is the *absolute* temperature of the surface $[\Theta]$, $\sigma$ is a universal constant called the **Stefan-Boltzmann constant** $[\text{E L}^{-2} \text{T}^{-1} \Theta^{-4}]$ ($\sigma = 4.90 \times 10^{-9}$ MJ m$^{-2}$ day$^{-1}$ K$^{-4}$ = $5.67 \times 10^{-8}$ W m$^{-2}$ K$^{-4}$ = $1.38 \times 10^{-12}$ cal cm$^{-2}$ s$^{-1}$ K$^{-4}$), and $\varepsilon$ is a dimensionless quantity called the **emissivity** of the surface.

The value of $\varepsilon$ ranges from 0 to 1, depending on the material and texture of the surface. A surface with $\varepsilon = 1$ is called a **blackbody;** most earth materials have emissivities near 1 (Table D-1).

The wavelength, $\lambda$ [L], and frequency, $f$ [T$^{-1}$], of electromagnetic radiation are inversely related as

$$\lambda \cdot f = c, \qquad \text{(D-2)}$$

where $c$ is the speed of light [L T$^{-1}$] ($c = 2.998 \times 10^8$ m s$^{-1}$). The spectrum of electromagnetic radiation extends over 21 orders of magnitude, as shown in Figure D-1. However, only radiation in the **near-ul-** traviolet (wavelengths from 0.2 to 0.4 $\mu$m), the **visible** (0.4 to 0.7 $\mu$m), and the **infrared** (0.7 to 80 $\mu$m) ranges plays a role in the earth's energy balance and climate.

The wavelength of the energy emitted by a radiating surface decreases as its temperature increases according to Planck's Law; the wavelength at which the maximum energy radiation occurs, $\lambda_{\max}$, is related to the absolute temperature via **Wien's Displacement Law,**

$$\lambda_{\max} \cdot T = 2897, \qquad \text{(D-3)}$$

where $\lambda_{\max}$ is in $\mu$m and $T$ is in K.

Electromagnetic energy is transmitted through a vacuum undiminished. However, when it strikes matter, portions of it may be reflected and/or absorbed. The following terms are used to describe the interactions of matter and radiant energy at a given wavelength:

**Absorptance,** $\alpha(\lambda)$, is the fraction of the incident energy at wavelength $\lambda$ that is absorbed by a surface; this energy raises the temperature of the matter and/or causes a phase change (melting or evaporation).

**Reflectance,** $\rho(\lambda)$, is the fraction of the incident energy at wavelength $\lambda$ that is reflected by the surface; this energy does not affect the matter and continues traveling undiminished in a new direction.

**Transmittance,** $\tau(\lambda)$, is the fraction of the incident energy at wavelength $\lambda$ that is transmitted through the matter; this energy does not affect

---

[1]The symbol [E] is used for the dimensions of energy; $[\text{E}] \equiv [\text{M L}^2 \text{T}^2] = [\text{F L}]$.

**TABLE D-1**
Emissivities of Various Forms of Water and Various Earth Materials.

| Surface | Conditions | Emissivity, $\varepsilon$ |
|---|---|---|
| Clouds | Dense | 0.99 |
| Liquid water | | 0.95 |
| Ice | At 0°C | 0.97 |
| Sand | Dry | 0.95 |
| | Wet | 0.98 |
| Peat | Dry | 0.97 |
| | Wet | 0.98 |
| Rock | Light sandstone | 0.98 |
| | Limestone, gravel | 0.92 |
| Grass | Typical fields | 0.95 |
| | Lawn | 0.97 |
| Crops | Corn, beans | 0.94 |
| | Cotton, tobacco | 0.96 |
| | Sugar cane | 0.99 |
| Cactus | | 0.98 |
| Trees | Deciduous forest | 0.95 |
| | Coniferous forest | 0.97 |

Data from Lee (1980).

the matter and continues traveling undiminished in the original direction.

The values of $\alpha(\lambda)$, $\rho(\lambda)$, and $\tau(\lambda)$ must each be between zero and one and, for a given surface and wavelength, must always sum to one:

$$\alpha(\lambda) + \rho(\lambda) + \tau(\lambda) = 1. \qquad \textbf{(D-4)}$$

The values of $\alpha(\lambda)$, $\rho(\lambda)$, and $\tau(\lambda)$ are determined by the nature of the matter and vary with wavelength for a given material. The reflectance integrated over the visible wavelengths (0.4 to 0.7 $\mu$m) is called the **albedo** (i.e., "whiteness"). Table D-2 gives the albedos of various forms of water and some earth materials.

## D.2   COMPOSITION AND VERTICAL STRUCTURE OF THE ATMOSPHERE

### D.2.1   Composition

The atmosphere is a mixture of gases in which liquid and solid particles are suspended. The propor-

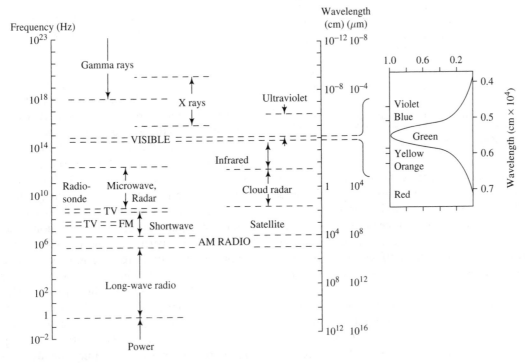

**FIGURE D-1**
The electromagnetic spectrum. After Miller et al. (1983).

**TABLE D-2**

Visible-Range Reflectance (albedo) of Various Forms of Water and Various Earth Materials.

| Surface | Conditions | | Albedo, a |
|---|---|---|---|
| Clouds | Low overcast: | 100 m thick | 0.40 |
| | | 200 m thick | 0.50 |
| | | 500 m thick | 0.70 |
| Liquid water | Smooth; solar angle 60° | | 0.05 |
| | | 30° | 0.10 |
| | | 20° | 0.15 |
| | | 10° | 0.35 |
| | | 5° | 0.60 |
| | Wavy; solar angle 60° | | 0.10 |
| Solid water | Fresh snow; | low density | 0.85 |
| | | high density | 0.65 |
| | Old snow; | clean | 0.55 |
| | | dirty | 0.45 |
| | Glacier ice; | clean | 0.35 |
| | | dirty | 0.25 |
| Sand | Dry, light; | high sun | 0.35 |
| | | low sun | 0.60 |
| | Gray; | wet | 0.10 |
| | | dry | 0.20 |
| | White; | wet | 0.25 |
| | | dry | 0.35 |
| Soil | Organic; dark | | 0.10 |
| | Clay | | 0.20 |
| | Sandy; light | | 0.30 |
| Grass | Typical fields | | 0.20 |
| | Dead; wet | | 0.20 |
| | dry | | 0.30 |
| Tundra, heather | | | 0.15 |
| Crops | Cereals, tobacco | | 0.25 |
| | Cotton, potato, tomato | | 0.20 |
| | Sugar cane | | 0.15 |
| Trees | Rain forest | | 0.15 |
| | Eucalyptus | | 0.20 |
| | Red pine forest | | 0.10 |
| | Mixed hardwoods in leaf | | 0.18 |

Data from Lee (1980).

tions of the major constituents, and many of the minor ones, are effectively constant in time and space (Table D-3). The variable constituents include those that are most significant hydrologically (especially liquid water, ice, water vapor, dusts, and carbon dioxide), because they affect the energy balance of the atmosphere and the formation of clouds and precipitation.

As indicated in Table D-3, the concentrations of many of these variable constituents are increasing due to human activity and may thereby be affecting global climate (Ramanathan 1988; Section 3.2.9). An additional concern is the depletion of the stratospheric ozone layer, which blocks a biologi-

cally damaging portion of the sun's radiant energy, due to reactions involving industrial chemicals (Cicerone 1987).

### D.2.2   Vertical Structure

The earth's atmosphere extends from the surface to a height of about 700 km. This gaseous envelope is characterized by the typical vertical distributions of pressure and temperature shown in Figure D-2. The pattern of vertical temperature gradients delineates three distinct "spheres" between which there is little mixing. Only the troposphere and stratosphere are involved in processes that directly influence climate, weather, and hydrologic activity, so in this text "atmosphere" refers to the troposphere and stratosphere.

Vertical temperature and pressure gradients vary with latitude, season, and local weather patterns, but generally maintain the structure shown in Figure D-2. For example, the boundary between the troposphere and the stratosphere is higher (15–16 km) near the equator and lower (5–6 km) over the poles, and is higher in summer than in winter.

### D.2.3   Pressure–Temperature Relations

Temperature, pressure, and density in the atmosphere are related via the Ideal Gas Law,

$$\frac{P}{T_a \cdot \rho_a} = R_a, \qquad \textbf{(D-5)}$$

where $P$ is atmospheric pressure in kPa, $T_a$ is air temperature in K, $\rho_a$ is the mass density of air in kg m$^{-3}$, and $R_a$ is the **gas constant** for air, which for the units given has the value $R_a = 0.288$. As a consequence of this law, an increase (decrease) in pressure is always accompanied by an increase (decrease) in temperature and density. Thus when a "parcel" of air moves horizontally from a region of high atmospheric pressure to a region of low pressure, or moves vertically upward to lower pressure, it expands (density decreases) and cools. The opposite changes take place when the air moves horizontally or vertically from low to high pressure. These temperature changes do not involve removal or input of heat, and are called **adiabatic** cooling or warming.

Horizontal gradients of atmospheric pressure are important in wind dynamics, but are too small to induce significant adiabatic temperature changes. Vertical gradients are far larger and, as discussed in

**TABLE D-3**

Composition of the Atmosphere.

| Permanent Constituents | Volume Percent | Variable Constituents | Volume Percent |
|---|---|---|---|
| Nitrogen ($N_2$) | 78.084 | Water vapor ($H_2O$) | <4 |
| Oxygen ($O_2$) | 20.946 | Water (liquid and solid) | <1 |
| Argon (Ar) | 0.934 | Carbon dioxide ($CO_2$) | 0.0345[a] |
| Neon (Ne) | 0.001818 | Methane ($CH_4$) | 0.00017[b] |
| Helium (He) | 0.000524 | Sulfur dioxide ($SO_2$) | <0.0001 |
| Krypton (Kr) | 0.000114 | Nitrous oxide ($N_2O$) | 0.0000304[c] |
| Hydrogen ($H_2$) | 0.00005 | Carbon monoxide (CO) | 0.00002 |
| Xenon (Xe) | 0.0000087 | Dusts (soot, soil, salts) | <0.00001 |
| Radon (Rn) | $6 \times 10^{-18}$ | Ozone ($O_3$) | <0.000007[d] |
| | | Nitrogen dioxide ($NO_2$) | <0.000002[e] |
| | | Ammonia ($NH_4$) | Trace |

Data from Miller et al. (1983) excepts as noted.

[a]Global average is 0.0345%: concentration is increasing about 0.001 volume % per year (Ramanathan 1988) due to burning of fossil fuels, cement manufacture, and deforestation (Rosenberg 1987).

[b]Concentration is increasing about 0.00002 volume % per year (Ramanathan 1988) due to increased rice-paddy cultivation, domestic animals, and other causes (Rosenberg 1987).

[c]Concentration is increasing about 0.000001 volume % per year (Ramanathan 1988) due to increased ferilizer use and burning of fossil fuels (Rosenberg 1987).

[d]Concentration in the stratospheric ozone layer appears to be decreasing (Callis and Natarajan 1986) due to chemical reactions involving chlorinated fluorocarbon compounds used as refrigerants and propellents. Concentrations in the troposphere are increasing (Ramanathan 1988) due to industrial activity and automobile use.

[e]Concentrations in the stratosphere are increasing about 0.0000003 volume % per year (Callis and Natarajan 1986).

**FIGURE D-2**

Atmospheric "spheres" and the average vertical distribution of temperature and pressure.

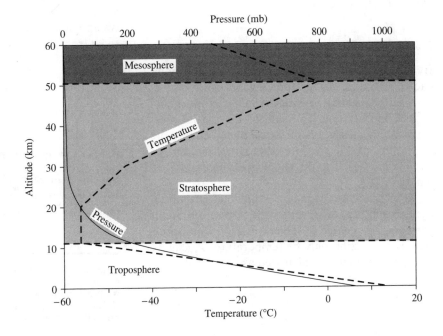

Section D.5.1, play a major role in precipitation formation. The vertical temperature gradient in the troposphere, called the **lapse rate,** has an average value of $-6.5$ °C km$^{-1}$ ($-3.5$ °F/1000 ft$^{-1}$). Under some conditions, the temperature gradient at or near the surface may temporarily reverse direction; this phenomenon is called an **inversion.**

From Equation (D-5), the density of air at normal sea-level pressure ($P = 101.3$ kPa) is

$$\rho_a = \frac{352}{T_a + 273.2},\qquad \textbf{(D-6)}$$

where $\rho_a$ is in kg m$^{-3}$ and $T_a$ is in °C (Table D-4).

## D.3  WATER VAPOR

Water in vapor form consists of separate $H_2O$ molecules, mixed among the molecules of the other gases of the air. All of these molecules are in constant motion, independent of the bulk motion of the air, with velocities on the order of 500 m s$^{-1}$ and an average distance between intermolecular collisions of $5 \times 10^{-8}$ m (Shuttleworth 1979).

Evaporative processes affect and are affected by the amount of water vapor in the air near the evaporating surface, and there are several ways of expressing that amount.

### D.3.1  Vapor Pressure

By virtue of their molecular motion and collisions, each constituent of a mixture of gases exerts a pressure, called a **partial pressure,** which is proportional to its concentration. The sum of the partial pressures of the gases in the atmosphere equals the total atmospheric pressure. The partial pressure of water vapor is called the **vapor pressure** and is designated $e$ [F L$^{-2}$].

The maximum vapor pressure that is thermodynamically stable is called the **saturation vapor pressure,** designated $e*$. The saturation vapor pressure is a function only of temperature, $T$, (Figure D-3); its value can be calculated as

$$e* = 0.611 \cdot \exp\left(\frac{17.3 \cdot T}{T + 237.3}\right),\qquad \textbf{(D-7)}$$

where $e*$ is in kPa and $T$ is in °C.[2]

Under most natural conditions, $e*$ represents the maximum amount of water vapor that the atmosphere can hold at temperature $T$, and the addition of more water vapor or the lowering of the temperature results in **condensation** via the formation of liquid droplets or ice crystals in the air. The process of condensation is considered in Section D.5.2.

### D.3.2  Absolute Humidity

The **absolute humidity,** also called the **vapor density,** is the mass concentration of water vapor in a volume of air. The Ideal Gas Law provides the relation between vapor pressure and absolute humidity, namely,

$$\frac{e}{T_a \cdot \rho_v} = R_v,\qquad \textbf{(D8-a)}$$

where $e$ is vapor pressure (kPa), $T_a$ is air temperature (K), $\rho_v$ is the absolute humidity (kg m$^{-3}$), and $R_v$ is the gas constant for water vapor ($R_v = 0.463$ for the units given). Since the molecular weight of water vapor is 0.622 times the molecular weight of air ($R_a/R_v = 0.622$), the relation between $e$ and $\rho_v$ can also be expressed as

$$\frac{e}{\rho_v} = \frac{P}{0.622 \cdot \rho_a},\qquad \textbf{(D-8b)}$$

where $e$ and $P$ are in the same units, and $\rho_v$ and $\rho_a$ are in the same units. A third useful relation between $e$ and $\rho_v$ at atmospheric pressure can be derived from the Ideal Gas Law:

$$\frac{e}{\rho_v} = \frac{T_a}{2.17},\qquad \textbf{(D-8c)}$$

where $e$ is in kPa, $\rho_v$ is in kg m$^{-3}$, and $T_a$ is in K.

### D.3.3  Specific Humidity

**Specific humidity,** $q$, is the concentration of water vapor expressed as the mass of water vapor per unit mass of air. Thus we have

---

[2]Equation (D-6) is an empirical relation that is sufficiently accurate for hydrologic computations. The 237.3 in the denominator is correct; it is not a misprint for 273.2, the adjustment from °C to K. The true relation between $e*$ and $T$ is a very lengthy formula called the Goff–Gratch Equation.

See note Bottom pg 273

**TABLE D-4**

Density of Air, $\rho_a$ (kg m$^{-3}$), at Sea Level as a Function of Temperature, $T_a$ (°C)
[Equation (D-6); $P = 101.3$ kPa].

| $T$ | $-20$ | $-15$ | $-10$ | $-5$ | 0 | 5 | 10 | 15 | 20 | 25 | 30 |
|---|---|---|---|---|---|---|---|---|---|---|---|
| $\rho_a$ | 1.39 | 1.36 | 1.34 | 1.31 | 1.29 | 1.26 | 1.24 | 1.22 | 1.20 | 1.18 | 1.16 |

$$q = \frac{\rho_v}{\rho_a} = \frac{0.622 \cdot e}{P}. \qquad \textbf{(D-9)}$$

### D.3.4   Relative Humidity

If the temperature and vapor pressure of a "parcel" of air lie below the curve representing saturation (Point A in Figure D-3), the air is unsaturated, and its **relative humidity,** $W_a$, is the ratio (usually expressed as a percent) of its actual vapor pressure, $e_a$, to its saturation vapor pressure:

$e_a = rel\ hum \cdot e_a^*$

$$W_a \equiv \frac{e_a}{e_a^*}. \qquad \textbf{(D-10)}$$

If this parcel of air is now cooled, its state moves along the heavy dashed line in Figure D-3. In this process, its vapor pressure does not change, but its relative humidity increases (Point B). At Point C, the parcel has reached its saturation vapor pressure and its relative humidity is therefore 100%.

### D.3.5   Dew Point

The temperature to which a parcel with a given vapor pressure has to be cooled in order to reach saturation is called the **dew point,** $T_d$; it can be calculated for unsaturated conditions as

$$T_d = \frac{\ln(e) + 0.4926}{0.0708 - 0.00421 \cdot \ln(e)}, \qquad \textbf{(D-11)}$$

where $T_d$ is in °C and $e$ is in kPa.[3]

As noted earlier, under most natural conditions the cooling of a parcel of air to its dew point induces condensation. However, it is possible under certain conditions to cool a parcel of air below its dew point without causing condensation. Under these conditions the air is **supersaturated,** and its state is represented by a point above the curve in Figure D-3 (e.g., Point D).

## D.4   PHYSICS OF EVAPORATION

### D.4.1   Mass (Water) Transfer

In Figure D-4, dry air with a temperature of $T_a$ lies above a horizontal water surface with a temperature of $T_s$. The molecules at the surface are attracted to those in the body of the liquid by hydrogen bonds, but some of the surface molecules have sufficient energy to sever the bonds and enter the air. The number of molecules with this "escape energy" increases as $T_s$ increases.

The water molecules entering the air move in random motion, and as these molecules accumulate in the layer of air immediately above the surface, some will re-enter the liquid. The rate at which they re-enter is proportional to the concentration of molecules in this layer. At equilibrium, the rates of escape and re-entry are equal, and the vapor pressure in the molecular layer immediately above the surface is the saturation vapor pressure at the temperature of the surface $e_s^*$ [Equation (7-3)].

The rate of evaporation is the rate at which molecules move from the saturated surface layer into the air above, and that rate is proportional to the difference between the vapor pressure of the surface layer and the vapor pressure of the overlying air, $e_a$; that is,

$$E \propto e_s^* - e_a, \qquad \textbf{(D-12)}$$

where $E$ is evaporation rate, $e_a$ is measured at some representative height, and the proportionality depends on that height and on the factors controlling the diffusion of water vapor in the air (Section D.6.6). Equation (D-12) is known as **Dalton's Law**[4]; it is the basis for the mass-transfer approach for estimating evapotranspiration [Equation (7-17)].

Although Equation (D-12) is straightforward, we should note two important implications:

> **1.** Depending on the temperature of the surface and the temperature and humidity of

---

[3]Equation (D-11) is an empirical relation derived by solving Equation (D-7) for temperature.

[4]Discovered by John Dalton, the English chemist, in 1802.

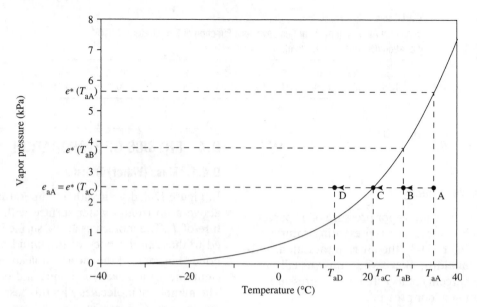

**FIGURE D-3**
The curve is the relation between saturation vapor pressure, $e_a^*(T_a)$, and air temperature, $T_a$ [Equation (D-7)]. The heavy dashed line traces the state of a parcel of air that initially has vapor pressure $e_{aA}$, temperature $T_{aA}$, and relative humidity $W_{aA} = e_{aA}/e_a^*(T_{aA})$ (point A). When the air is cooled to $T_{aB}$, its relative humidity has increased to $W_{aB} = e_{aA}/e_a^*(T_{aB})$; when it is cooled to $T_{aC}$, its relative humidity has increased to $W_{aC} = e_{aA}/e_a^*(T_{aC}) = 1$. Thus, $T_{aC}$ is the dew point for this parcel of air. At point D, the parcel is supersaturated ($W_{aD} > 1$).

the air, the difference between the two vapor pressures can be positive, zero, or negative. If $e_s^* > e_a$, evaporation is occurring; if $e_s^* < e_a$, there is water condensing on the surface; and if $e_s^* = e_a$, neither condensation or evaporation is occurring.

2. The value of $e_a$ can be less than or equal to the saturation vapor pressure at the air temperature, $e_a^*$ (i.e., relative humidity can be less than or equal to 100%). Evaporation will occur even if the relative humidity equals 100%, as long as $e_s^* > e_a^*$. However,

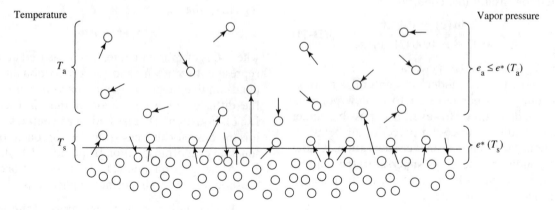

**FIGURE D-4**
Schematic diagram of flux of water molecules over a water surface. The vapor pressure at the surface is $e_s^*(T_s)$; the vapor pressure of the overlying air is less than or equal to $e_a^*(T_a)$. The rate of evaporation is proportional to $(e_s^* - e_a)$ [Equation (D-12)].

under these conditions the evaporating water will normally condense in the overlying air to form a fog or mist.

Note that evaporation occurs in exactly the same way whether the underlying surface is liquid water or ice. The only difference between the two situations is that the saturation vapor pressure for an ice surface at a given temperature is slightly lower than that given by Equation (D-7) (Figure D-5).

### D.4.2  Latent-Heat Transfer

The **latent heat of vaporization,** $\lambda_v$, is the quantity of heat energy that must be absorbed to break the hydrogen bonds when evaporation takes place; this same quantity is released when the bonds are reformed upon condensation (Section B.2.5). Thus evaporation is always accompanied by a transfer of heat out of the water body, and condensation on the surface by an addition of heat to the water body. This process is called **latent-heat transfer.**

Because of this coupling, the rates of latent-heat and mass (water) transfer are directly proportional. When the underlying surface is liquid water,

$$LE = \lambda_v \cdot E, \qquad \textbf{(D-13a)}$$

where $LE$ is the rate of latent-heat transfer $[E\,T^{-1}]$, $E$ is the rate of evaporation or condensation $[M\,T^{-1}]$, and $\lambda_v$ is the latent heat of vaporization $[E\,M^{-1}]$. If condensation takes place in the air, latent heat is liberated to warm the air at the rate $LE$.

If the underlying surface is snow or ice, energy is required to disrupt the molecular structure of ice as well as to sever hydrogen bonds with neighboring molecules. Thus in this case latent-heat transfer involves the **latent heat of fusion,** $\lambda_f$, as well as the latent heat of vaporization:

$$LE = (\lambda_v + \lambda_f) \cdot E. \qquad \textbf{(D-13b)}$$

Note that if $E$ is expressed in dimensions of $[L\,T^{-1}]$, Equation (D-13) becomes

$$LE = \rho_w \cdot \lambda_v \cdot E \qquad \textbf{(D-14a)}$$

and

$$LE = \rho_w \cdot (\lambda_v + \lambda_f) \cdot E, \qquad \textbf{(D-14b)}$$

where $LE$ is in $[E\,L^{-2}\,T^{-1}]$.

### D.5  PHYSICS OF PRECIPITATION

In order to produce hydrologically significant rates of precipitation, a sequence of four processes must occur: (1) cooling to the dew point, (2) condensation, (3) droplet growth, and (4) importation of water vapor (Gilman 1964). An examination of

**FIGURE D-5**
Difference between vapor pressure over liquid water and ice at temperatures below 0 °C.

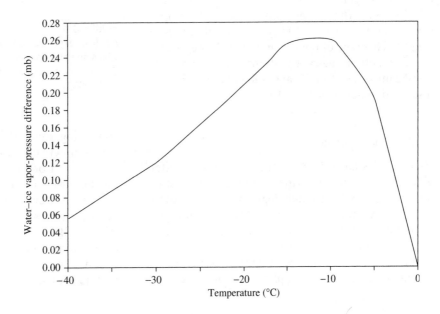

these processes will help the hydrologist understand regional and seasonal patterns of inputs to the hydrologic cycle, the potential effects of local and global climate change on those patterns, the ways in which pollutants may become incorporated in precipitation (acid rain), and attempts to increase precipitation by cloud seeding as a water-resource management technique.

### D.5.1 Cooling

To begin the formation of precipitation, air containing water vapor must be cooled to its dew point. A parcel of air can lose heat by (1) radiation to cooler surroundings, (2) mixing with a cooler body of air, (3) conduction to a cool surface (e.g., cool ocean water or snow), (4) adiabatic cooling by horizontal movement to a region of lower pressure, and (5) adiabatic cooling by vertical uplift. Cooling by the first four of these processes may produce fog or drizzle, but only vertical uplift can cause rates of cooling high enough to produce significant precipitation.

Figure D-6a shows how the temperature and volume (density) of a parcel of air change adiabatically as it is vertically displaced: If no condensation occurs, its temperature will decrease at the **dry adiabatic lapse rate** of 1 °C/100 m (5.5 °F/1000 ft). If condensation occurs (Figure D-6b), its temperature will decrease at a lower rate due to the liberation of latent heat. This **moist adiabatic lapse rate** varies with temperature, initial vapor pressure, and elevation, but is typically about one-half the dry rate: 0.5 °C/100 m. The average lapse rate in the troposphere, about 0.65 °C/100 m (Figure D-2), is a weighted average of the dry and moist lapse rates.

The meteorological situations that can produce significant rates of uplift and adiabatic cooling are discussed in Sections 4.1.1–4.1.3.

### D.5.2 Condensation

Experiments have shown that air containing water vapor, but with no impurities in the form of dust or ions, can be cooled to high degrees of supersaturation (relative humidities of up to 800%) without the formation of droplets. This is because air at its dew point contains fewer than 4% water molecules (Table D-3), so the chance of enough of them collecting to form a stable droplet ($10^{-4}$ mm diameter,

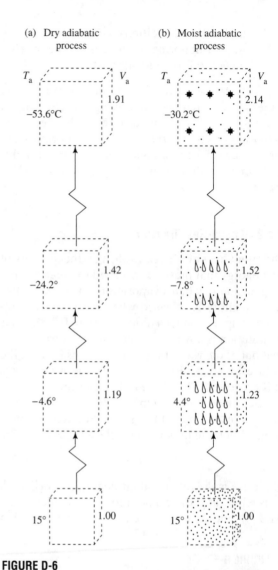

**FIGURE D-6**

Adiabatic expansion and cooling during uplift of (a) air in which no condensation occurs and (b) air saturated at 15 °C. $T_a$ is temperature and $V_a$ is relative volume ($1/\rho_a$). After Miller et al. (1983).

containing some $10^8$ molecules) by random collision is effectively zero (Miller et al. 1983).

Thus, in order for condensation to occur near the dew point, foreign particles larger than $10^{-4}$ mm to which water molecules are attracted via hydrogen bonds must be present to provide a substrate. Air usually contains thousands of such particles, called **cloud condensation nuclei** (CCN), per cubic centimeter (Figure D-7), so that surfaces of the size required for stability are usually present. When the

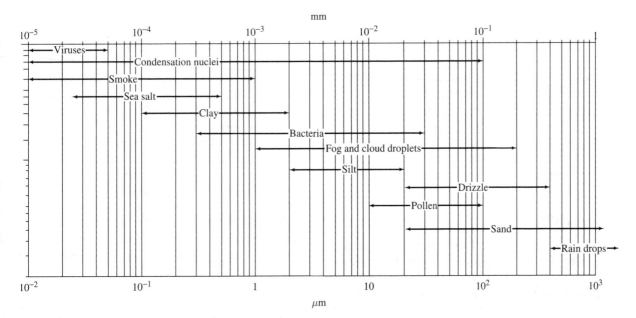

**FIGURE D-7**
Size ranges of nongaseous atmospheric constituents (logarithmic scale). Data from Miller et al. (1983).

dew point is reached and the water vapor is thermodynamically ready to condense, these surfaces provide the sites on which condensation occurs.[5]

The major natural sources of CCN are meteoric dust, windblown clay and silt, volcanic material, smoke from forest fires, and sea salt. It is estimated that the natural concentrations of CCN are about 100 per cubic centimeter (cm$^{-3}$) over the oceans, and about 300 cm$^{-3}$ over the continents (Radke and Hobbs 1976). Human activities produce CCN (chiefly combustion products, including the sulfur and nitrogen compounds that produce acid precipitation) in concentrations of 3500 cm$^{-3}$ or more. Radke and Hobbs (1976) estimated that the global anthropogenic production rate of CCN may be comparable to the natural production rate and that in some industrial areas, where concentrations can reach 30,000 cm$^{-3}$ (Miller et al. 1983), nuclei are dominated by those produced anthropogenically.

Although the concentration of CCN is usually not a limiting factor in the precipitation-forming process, the practice of "cloud seeding" (Section 4.4.5) involves the artificial introduction of nuclei to induce droplet formation and ultimately precipitation in cases where insufficient numbers of nuclei are present.

### D.5.3   Droplet Growth

The preceding discussion shows that condensation to form clouds occurs when (1) water vapor is present, (2) sufficient CCN are present, and (3) there is a sufficient degree of uplift to bring about cooling to the dew point. Cloud droplets have diameters in the range 0.001 to 0.2 mm (Figure D-7); droplets of this size range have fall velocities between 0.01 and 70 cm s$^{-1}$.

In order for precipitation to fall from clouds to earth, some of the droplets must grow to a size such that their fall velocity exceeds the rate of uplift and such that they can survive evaporation as they fall. Thus we must consider the processes by which cloud droplets can grow several orders of magnitude to form raindrops of 0.4 to 4 mm diameter or snowflakes of even larger size. There are two processes by which this growth occurs: (1) drop collision and (2) ice-crystal growth.

---

[5]Water molecules are especially attracted to certain types of particles, which are called **hygroscopic** nuclei. Condensation may occur on these nuclei even at relative humidities as low as 76%. Sea salt is the most common of the hygroscopic nuclei, but it is usually present only in low concentrations (Miller et al. 1983).

### Droplet Collision

Condensation at temperatures above 0 °C produces cloud-water droplets of varying sizes. The larger droplets have larger fall velocities than the smaller ones and, because of this velocity differential, collisions occur between droplets of different sizes. Many of these collisions result in the coalescence of the two droplets and hence the gradual growth of the larger ones. Ultimately these become large enough to fall out of the cloud.

### Ice-Crystal Growth

If the air is saturated and the temperature is less than −40 °C, the vibrational energy of the $H_2O$ molecules is low enough such that clusters of them can spontaneously form ice crystals. At temperatures between −40 °C and 0 °C, however, a "template" with a molecular structure similar to that of ice is required to nucleate ice-crystal growth. Certain types of CCN, particularly clay minerals, can provide this template and therefore induce the formation of ice crystals at temperatures near the freezing point. Water substance condensing on other types of nuclei remains in the thermodynamically unstable liquid form, so that clouds below the freezing point often consist of a mixture of ice crystals and supercooled water droplets.

As shown in Figure D-5, the saturation vapor pressure of an ice surface at a given temperature is somewhat less than the saturation vapor pressure of a liquid-water surface at that temperature. This differential causes $H_2O$ molecules to evaporate from the liquid particles and condense on the ice particles. The ice crystals thus tend to grow at the expense of the liquid droplets and eventually acquire a size and fall velocity that allows them to fall through the cloud as snowflakes. Further growth by collision may also occur as the flakes fall. If the temperature below the cloud is above 0 °C, the snowflakes will melt and reach the earth as raindrops; otherwise, they will fall as snow.

Because CCN that are effective seeds for ice-crystal formation are less common than those suitable for liquid-droplet condensation, one of the most common cloud-seeding practices has been the introduction of silver–iodide particles into supercooled clouds. Silver iodide has a crystal structure similar to that of ice, and triggers droplet growth by providing nuclei for ice-crystal formation.

## D.5.4 Importation of Water Vapor

The concentration of liquid water and/or ice in most clouds is in the range of 0.1 to 1 g m$^{-3}$ (Gilman 1964). The following computation will show that even if all the water in a very thick cloud were to fall as rain, the total depth of precipitation would be relatively small.

Consider a 10,000-m thick cumulonimbus (i.e., thunderhead) cloud, so that the total cloud volume above each 1 m$^2$ of ground surface is 10,000 m$^3$. If the concentration of water substance in the cloud is 0.5 g m$^{-3}$, the total volume of cloud water above each 1 m$^2$ is 5000 cm$^3$. If all this water fell as precipitation, its depth would be 0.5 cm. This value represents a near maximum for the amount of precipitation produced by all the water initially present in a cloud.

Thus, since most rain-producing clouds are less than 10,000 m in thickness and water concentrations are less than 0.5 g m$^{-3}$, a final requirement for the occurrence of significant amounts of precipitation is that a continual supply of water vapor be imported into the cloud to replace that which falls out. This inflow of moisture is provided by winds that converge on the precipitation-producing clouds (Figure D-8).

One of the reasons that cloud-seeding may fail to produce significant precipitation is that, although

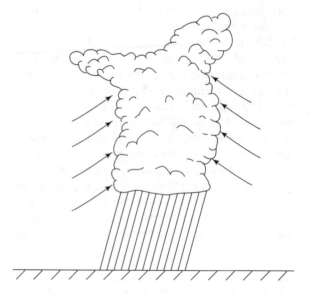

**FIGURE D-8**

In order to produce significant amounts of precipitation, vapor-bearing winds must provide a continual supply of water to the precipitation-producing clouds.

it may induce droplet formation and growth, it cannot induce the importation of water vapor necessary to sustain the process of precipitation.

## D.6   PHYSICS OF TURBULENT TRANSFER NEAR THE GROUND

### D.6.1   Planetary Boundary Layer

The **planetary boundary layer** is the lowest layer of the atmosphere in which the winds, which are induced by horizontal pressure gradients, are affected by the frictional resistance of the surface. The thickness of this layer varies in space and time from a few tens of meters to one or two kilometers, depending on the topography and roughness of the surface, the wind velocity, and the rate of heating or cooling of the surface (Peixoto and Oort 1992). The frictional resistance produces turbulent eddies,

which are irregular and chaotic motions with vertical components (Figure D-9). These vertical components are the means by which momentum, sensible heat, and water vapor and its accompanying latent heat are exchanged between the atmosphere and the land surface.

### D.6.2   Turbulent Velocity Fluctuations

Because of turbulent eddies, the instantaneous horizontal wind velocity at any level, $v_a$, fluctuates in time, and we can separate the instantaneous velocity into a time-averaged component, $\overline{v_a}$, and a deviation from that average, $v_a'$, caused by the eddies:

$$v_a = \overline{v_a} + v_a'. \qquad \textbf{(D-15)}$$

Note that $v_a'$ may be positive or negative and the overbar denotes time-averaging. By definition, the time-averaged value of the fluctuations equals zero, so

$$\overline{v_a'} = 0 \qquad \textbf{(D-16)}$$

**FIGURE D-9**
Conceptual diagram of the process of momentum transfer by turbulent diffusion. Friction caused by surface roughness reduces average velocities (straight arrows) near the surface and produces turbulent eddies (circular arrows), resulting in a net downward transfer of momentum. The vertical component of the eddies moves heat and water vapor upward or downward, depending on the directions of temperature and vapor-pressure gradients.

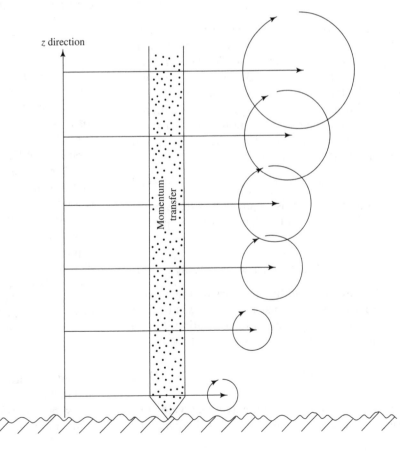

and

$$\overline{\overline{v_a} + v_a'} = \overline{v_a} + \overline{v_a'} = \overline{v_a}. \qquad \textbf{(D-17)}$$

The wind speed discussed in this text is the value averaged over a period appropriate for defining the "prevailing conditions" for the analysis (usually one hour to one day). To simplify the notation, we use the symbol $v_a$ rather than $\overline{v_a}$ to denote this time average.

Under prevailing conditions producing a given average horizontal wind speed at a given level, turbulent eddies are also reflected in a time-varying vertical air velocity, designated $w_a$, that can be similarly separated into a time-averaged component, $\overline{w_a}$, and an instantaneous deviation from that average, $w_a'$:

$$w_a = \overline{w_a} + w_a'. \qquad \textbf{(D-18)}$$

However, there is no net vertical air movement, so the time average of both components is zero and we have

$$\overline{w_a} = 0 \qquad \textbf{(D-19)}$$

and

$$\overline{w_a'} = 0. \qquad \textbf{(D-20)}$$

Because friction due to the surface causes the average wind speed to increase upward in the planetary boundary layer (Section D.6.3), a positive (negative) vertical velocity deviation, $w_a'$, reduces (increases) the horizontal velocity at the new level and produces a negative (positive) fluctuation in the horizontal velocity, $v_a'$. The intensity of turbulence can be characterized by a quantity called the **friction velocity,** $u_*$, defined as

$$u_* \equiv (-\overline{v_a' \cdot w_a'})^{1/2}, \qquad \textbf{(D-21)}$$

where the minus sign is required because simultaneous values of $w_a'$ and $v_a'$ have opposite signs.

### D.6.3    Vertical Distribution of Wind Velocity

Measurements in the planetary boundary layer show that the vertical distribution of wind can be well represented by a logarithmic relation that has been developed from theoretical considerations:

$$v_a = \frac{1}{k} \cdot u_* \cdot \ln\left(\frac{z - z_d}{z_0}\right), \; z > z_d + z_0. \; \textbf{(D-22)}$$

In this equation, $v_a$ is the time-averaged velocity at height $z$ above the ground surface, $[\text{L T}^{-1}]$; $u_*$ is the

friction velocity $[\text{L T}^{-1}]$; the height $z_d$ is called the **zero-plane displacement** [L]; the height $z_o$ is called the **roughness height,** [L] (Figure D-10); and $k$ is a dimensionless constant. Note that $v_a = 0$ at $z = z_d + z_0$, which is the effective or nominal surface level.

Equation (D-22) is known as the **Prandtl–von Karman Universal Velocity-Distribution** for turbulent flows. Experimental data have shown that $k = 0.4$ and that the values of $z_d$ and $z_0$ are approximately proportional to the average height of the vegetation or other roughness elements covering the ground surface [Equations (7-50) and (7-51)].

### D.6.4    Diffusion

**Diffusion** is the process by which constituents of a fluid, such as its momentum, heat content, or a dissolved or suspended constituent, are transferred from one position to another within the fluid. Such transfers occur whenever there are differences in concentrations of the constituent in different parts of the fluid. The rate of transfer of a constituent $X$ in the direction $z$ is directly proportional to the gradient of concentration of $X$ in the $z$ direction; that is,

$$F_z(X) = -D_X \cdot \frac{dC(X)}{dz}, \qquad \textbf{(D-23)}$$

where $F_z(X)$ is the rate of transfer of $X$ in direction $z$ per unit area per unit time (called the **flux** of $X$), $C(X)$ is the concentration of $X$, and $D_X$ is called the **diffusivity** of $X$ in the fluid (Figure D-11). Equation (D-23) is the mathematical expression of Fick's First Law of Diffusion (Table 2-1).

The minus sign in Equation (D-23) indicates that $X$ always moves from regions where its concentration is higher to regions where its concentration is lower. Diffusivity always has dimensions $[\text{L}^2 \text{T}^{-1}]$, while the dimensions of the other quantities in Equation (D-23) depend on the nature of the property $X$. As will be developed later, the value of the diffusivity depends on (1) the nature of the property $X$ and (2) the physical mechanism by which the property is transferred.

We have already seen that water vapor is transferred between the surface and the air whenever there is a difference in the vapor pressure between the surface and the overlying air [Equation (D-12)] and that a transfer of latent heat always accompanies the vapor transfer [Equations (D-13) and (D-14)]. A second mode of non-radiant heat transfer occurs

**FIGURE D-10**

Vertical distribution of wind velocity over a vegetative surface of height $z_{veg}$. The profile follows the logarithmic relation of Equation (D-22). The zero-plane displacement, $z_d$, is about $0.7 \cdot z_{veg}$, and the roughness height, $z_0$, is about $0.1 \cdot z_{veg}$. Note that Equation (D-22) gives $v_a = 0$ when $z = z_d + z_0$.

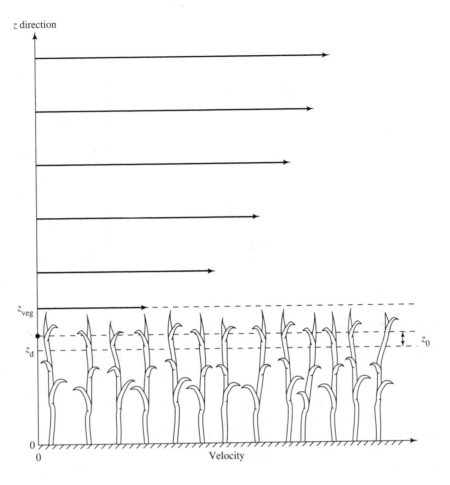

in the form of **sensible heat,** that is, heat energy that can be directly sensed via measurement of the temperature and application of Equation (B-19). Thus, sensible-heat transfer occurs whenever there is a temperature difference between the surface and the air; the relation that is analogous to Equation (D-12) is

$$H \propto T_s - T_a, \qquad \textbf{(D-24)}$$

where $H$ is the rate of sensible-heat transfer from surface to air per unit area of surface [E L$^{-2}$ T$^{-1}$], and $T_s$ and $T_a$ are the surface and air temperatures, respectively. When $T_s < T_a$, $H$ is negative and sensible heat moves from the air to the surface.

Equation (D-23) is the basis for a quantitative understanding of the processes that transfer water vapor and latent and sensible heat between the surface and the overlying atmosphere. However, several steps are required to combine that equation with Equations (D-12), (D-13), and (D-24) to produce a useful quantitative formulation of these processes.

We will be concerned here only with diffusion in the vertical direction, and $z$ will henceforth represent the height above the ground surface [L].

Since the absolute humidity $\rho_v$ is the concentration of water vapor in the air, the diffusion equation for water vapor, $(V)$, is

$$F_z(V) = -D_V \cdot \frac{d\rho_v}{dz}, \qquad \textbf{(D-25)}$$

where $D_V$ is the diffusivity of water vapor. Since latent heat and water vapor are directly coupled [Equation (D-13)], the diffusion equation for latent heat is

$$F_z(LE) = -D_V \cdot \lambda_v \cdot \frac{d\rho_v}{dz}. \qquad \textbf{(D-26)}$$

From Equation (B-19), the concentration of sensible heat, $h$ [E L$^{-3}$], in air can be expressed as

$$h = \rho_a \cdot c_a \cdot (T_a - T_b), \qquad \textbf{(D-27)}$$

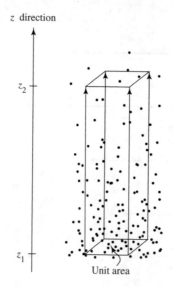

z direction

$z_2$

$z_1$

Unit area

**FIGURE D-11**
Conceptual diagram of the general diffusion process. Concentration of dots indicate concentration of quantity $X$. More of the property $X$ moves from $z_1$ to $z_2$ in a given time when the diffusivity, $D_x$, is higher.

where $c_a$ is the heat capacity of air (at constant pressure), $T_a$ is the air temperature, and $T_b$ is an arbitrary base temperature, usually taken as $0\ °C$. Thus Equation (D-23) is written for the diffusion of sensible heat, $H$, as

$$F_z(H) = -D_H \cdot \frac{d(\rho_a \cdot c_a \cdot T_a)}{dz}, \qquad \textbf{(D-28)}$$

where $D_H$ is the diffusivity of sensible heat in turbulent air.

To evaluate $D_V$ and $D_H$, we must understand the mechanism of diffusion in the planetary boundary layer. As noted earlier, diffusion is effected by the bulk vertical movement of air in turbulent eddies, which carries latent heat (in the form of water vapor) and sensible heat vertically in the directions dictated by Equations (D-12) and (D-24) (Figure D-9). These eddies are also the mechanism by which the frictional drag of the surface is transmitted into the air flow to reduce the wind velocity; this process can also be viewed as a downward diffusion of momentum (mass times velocity).

Thus, in order to quantitatively express the effectiveness of turbulent eddies in energy and water-vapor transfer, we must examine the vertical transfer of momentum. Theory and field measurements indicate that when the actual lapse rate

equals the adiabatic lapse rate (called "neutral" conditions), the diffusivities of momentum, water vapor, and heat are equal. Thus we develop the basic heat- and vapor-transfer relations for neutral conditions, and then show how these can be modified to represent transfers in stable and unstable conditions.

### D.6.5  Momentum Transfer

Like latent and sensible heat, momentum transfer is also governed by Equation (D-23). Recall that momentum equals mass times velocity, so that the concentration of momentum (momentum per unit volume) at any level equals the mass density of the air times the velocity, $\rho_a \cdot v_a$. Thus Equation (D-23) becomes

$$F_z(M) = -D_M \cdot \frac{d(\rho_a \cdot v_a)}{dz}, \qquad \textbf{(D-29)}$$

where $D_M$ is the diffusivity of momentum in turbulent air.

In the lowest levels of the atmosphere, $\rho_a$ can be considered constant at the prevailing air temperature, so Equation (D-29) can be simplified to

$$F_z(M) = -D_M \cdot \rho_a \cdot \frac{dv_a}{dz}. \qquad \textbf{(D-30)}$$

Velocity always increases with height because frictional drag slows air movement near the ground [Equation (D-22)]. Therefore, $dv_a/dz$ is always positive. Thus $F_z(M)$ is always negative, reflecting the fact that momentum is being transferred downward via turbulent eddies from where velocities are higher to where they are lower.

$F_z(M)$ has the dimensions of a force per unit area $[F\ L^{-2}]$ and physically represents the horizontal shear stress due to differences of wind velocity at adjacent levels. Because momentum is conserved, this stress does not vary with height.

It can be shown that shear stress is directly proportional to the square of the friction velocity:

$$F_z(M) = -\rho_a \cdot u_*^2. \qquad \textbf{(D-31)}$$

Thus we can combine Equations (D-30) and (D-31) and solve for the diffusivity of momentum, $D_M$:

$$D_M = \frac{u_*^2}{dv_a/dz}. \qquad \textbf{(D-32)}$$

To evaluate the denominator of Equation (D-32), we see from Equation (D-22) that

$$\frac{dv_a}{dz} = \frac{u_*}{k \cdot (z - z_d)}. \quad \textbf{(D-33)}$$

Substitution of Equation (D-33) into Equation (D-32) then gives

$$D_M = k \cdot u_* \cdot (z - z_d). \quad \textbf{(D-34)}$$

Thus the diffusivity of momentum increases in proportion to distance above the zero-plane displacement height and in proportion to the velocity of the eddies, which is characterized by $u_*$.

Without elaborate instrumentation to record the rapid fluctuations of horizontal and vertical velocity, $u_*$ and hence $D_M$ cannot be directly measured. However, we can evaluate $D_M$ from measurements of wind velocity, $v_m$, at some height, $z_m$, as follows. From Equation (D-22),

$$u_* = \frac{k \cdot v_m}{\ln\left(\dfrac{z_m - z_d}{z_0}\right)}. \quad \textbf{(D-35)}$$

This value can then be substituted into Equation (D-34) to give

$$D_M = \frac{k^2 \cdot v_m \cdot (z_m - z_d)}{\ln\left(\dfrac{z_m - z_d}{z_0}\right)}. \quad \textbf{(D-36)}$$

We can also use Equations (D-33) and (D-34) to show that

$$D_M = k^2 \cdot (z - z_d)^2 \cdot \frac{dv_a}{dz}. \quad \textbf{(D-37)}$$

This relation will be useful in developing relations for calculating latent- and sensible-heat transfer.

### D.6.6   Latent-Heat Transfer

The upward flux of water vapor equals the evaporation rate, $E$, and this equivalence along with Equation (D-8b) can be used to rewrite Equation (D-26) as

$$\rho_w \cdot E = -D_V \cdot \frac{0.622 \cdot \rho_a}{P} \cdot \frac{de}{dz}. \quad \textbf{(D-38)}$$

The upward latent-heat transfer rate, $LE$, can then be found directly as

$$LE = \lambda_v \cdot \rho_w \cdot E = -D_V \cdot \lambda_v \cdot \frac{0.622 \cdot \rho_a}{P} \cdot \frac{de}{dz}. \quad \textbf{(D-39)}$$

[See Equation (D-26).]

As noted earlier, vapor and latent heat are transported vertically by the same turbulent eddies that are involved in momentum transfer. Thus, in order to develop a useful relation for calculating the rate of latent-heat transfer, we need to combine the relations that apply to the vertical velocity profile and the transfer of momentum with the diffusion equation for latent heat [Equation (D-39)].

As noted, $D_V = D_M$ under neutral lapse-rate conditions. Thus we can replace $D_V$ in Equation (D-39) with Equation (D-37) to give

$$LE = -k^2 \cdot (z - z_d)^2 \cdot \lambda_v \cdot \frac{0.622 \cdot \rho_a}{P} \cdot \frac{de}{dz} \cdot \frac{dv_a}{dz}. \quad \textbf{(D-40)}$$

Assuming a logarithmic profile for wind velocity and vapor pressure and integrating between two observational heights $z_1$ and $z_2$, for $z_2 > z_1 \geq (z_d + z_0)$, then yields

$$LE = -\lambda_v \cdot \frac{0.622 \cdot \rho_a}{P} \cdot \frac{k^2}{\left[\ln\left(\dfrac{z_2 - z_d}{z_1 - z_d}\right)\right]^2}$$
$$\cdot (v_2 - v_1) \cdot (e_2 - e_1), \quad \textbf{(D-41)}$$

where $v_i \equiv v_a(z_i)$ and $e_i \equiv e(z_i)$.

Equation (D-41) can be used for determining latent-heat transfer rate using observations of wind speed and vapor pressure at two heights, $z_1$ and $z_2$. If we take the lower height as that of the nominal surface (i.e., $z_1 = z_d + z_0$, $v_1 = 0$, and $e_1 = e_s$) and use the subscript $a$ in place of 2, Equation (D-41) becomes

$$LE = -\lambda_v \cdot \frac{0.622 \cdot \rho_a}{P} \cdot \frac{k^2}{\left[\ln\left(\dfrac{z_a - z_d}{z_0}\right)\right]^2}$$
$$\cdot v_a \cdot (e_a - e_s), \quad \textbf{(D-42)}$$

which requires wind-speed measurement at only one level.

As noted, $k = 0.4$. At typical near-surface conditions, we can assume that $\lambda_v = 2.47$ MJ kg$^{-1}$, $\rho_a = 1.24$ kg m$^{-3}$, and $P = 101.3$ kPa and define a **bulk latent-heat-transfer coefficient,** $K_{LE}$, as

$$K_{LE} \equiv \frac{3.01 \times 10^{-3}}{\left[\ln\left(\dfrac{z_a - z_d}{z_0}\right)\right]^2}, \quad \textbf{(D-43)}$$

where $K_{LE}$ has units of MJ m$^{-3}$ kPa$^{-1}$ and depends only on the measurement height and the nature of the surface. For a given experimental situation, $z_a$, $z_d$, and $z_0$ will be fixed. Then we can write Equation (D-41) as

$$LE = K_{LE} \cdot (v_2 - v_1) \cdot (e_1 - e_2). \quad \textbf{(D-44)}$$

and Equation (D-42) as

$$LE = K_{LE} \cdot v_a \cdot (e_s - e_a). \quad \textbf{(D-45)}$$

(Note that the minus signs have been eliminated by reversing the vapor pressures.)

Equation (D-45) shows that the upward rate of latent-heat transport depends on the product of the wind speed and the difference between the vapor pressure at the surface and the vapor pressure in the overlying air.

Note that Equations (D-44) and (D-45) give the evaporation rate, $E$, if their right-hand sides are divided by the latent heat, $\lambda_v$. In this form, these equations are used to estimate evaporation from snowpacks [Equations (5-45) and (5-46)] and open-water surfaces [Equation (7-17)], and evapotranspiration from vegetated surfaces [Equation (7-76)]. Thus Equations (D-44) and (D-45) are more precise versions of Dalton's Law [Equation (D-12)].

### D.6.7 Sensible-Heat Transfer

Again assuming that density and specific heat are essentially constant under prevailing conditions, we can simplify the diffusion equation for sensible heat [Equation (D-28)] to

$$F_z(H) = H = -D_H \cdot \rho_a \cdot c_a \cdot \frac{dT_a}{dz}. \quad \textbf{(D-46)}$$

If $D_H = D_M$, we can again substitute Equation (D-37) into Equation (D-46) to give

$$H = -k^2 \cdot \rho_a \cdot c_a \cdot (z - z_d)^2 \cdot \frac{dv_a}{dz} \cdot \frac{dT_a}{dz}. \quad \textbf{(D-47)}$$

Integrating between levels $z_1$ and $z_2$ yields

$$H = -\rho_a \cdot c_a \cdot \frac{k^2}{\left[ \ln \left( \dfrac{z_2 - z_d}{z_1 - z_d} \right) \right]^2} \cdot (v_2 - v_1) \cdot (T_2 - T_1). \quad \textbf{(D-48)}$$

Again, we can take level 1 as the nominal surface and write

$$H = -\rho_a \cdot c_a \cdot \frac{k^2}{\left[ \ln \left( \dfrac{z_a - z_d}{z_0} \right) \right]^2} \cdot v_a \cdot (T_a - T_s). \quad \textbf{(D-49)}$$

Equations (D-48) and (D-49) are exactly analogous to Equations (D-41) and (D-42) for latent-heat transfer. And, as before, we can assume that $\rho_a = 1.24$ kg m$^{-3}$ and $c_a = 1005$ J kg$^{-1}$ K$^{-1}$ and define a bulk sensible-heat transfer coefficient $K_H$ as

$$K_H \equiv \frac{1.99 \times 10^{-4}}{\left[ \ln \left( \dfrac{z_a - z_d}{z_0} \right) \right]^2}, \quad \textbf{(D-50)}$$

with units MJ m$^{-3}$ K$^{-1}$. As with $K_{LE}$, $K_H$ depends only on the measurement height and the nature of the surface, and for a given experimental situation, $z_a$, $z_d$, and $z_0$ will be fixed. Then we can write Equations (D-48) as

$$H = K_H \cdot (v_2 - v_1) \cdot (T_1 - T_2) \quad \textbf{(D-51)}$$

and Equation (D-49) as

$$H = K_H \cdot v_a \cdot (T_s - T_a). \quad \textbf{(D-52)}$$

(Again, the minus signs have been eliminated by reversing the surface and air temperatures.)

Equation (D-52) shows that the upward rate of sensible-heat transport depends on the product of the wind speed and the difference between the temperature at the surface and the temperature in the overlying air.

### D.6.8 Effects of Atmospheric Stability on Heat and Vapor Transfer

Figure D-12 shows unstable, neutral, and stable lapse rates near the ground. When a parcel of air is transported upward in a turbulent eddy, it cools adiabatically. Thus if the actual lapse rate is steeper than adiabatic (unstable), the air in the eddy is warmer and, hence, less dense than the surrounding air and will continue to rise due to buoyancy, enhancing vertical transport of heat. If the actual lapse rate is less than the adiabatic (stable), the air in the eddy will be cooler and denser than the surroundings and will sink toward the surface, reducing vertical transport.

Under neutral conditions, the diffusivities of water vapor and sensible heat are identical to the

**FIGURE D-12**
Unstable, neutral (adiabatic), and stable lapse rates near the surface. Air at surface is at surface temperature, $T_s$. When air moves upward in a turbulent eddy, it cools along the adiabatic line. If the actual gradient (lapse rate) is steeper than the adiabatic gradient (i.e., is in the unstable region), vertical turbulent motion is enhanced by buoyancy effects. If the actual gradient is less steep than the adiabatic gradient (i.e., is in the stable region), vertical turbulent motion is suppressed by buoyancy effects. Inversions are stable gradients in which the air temperature increases with height.

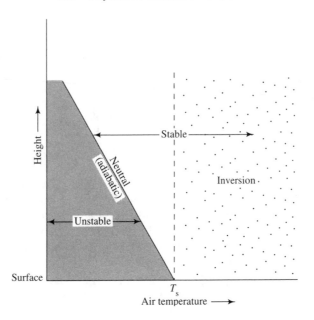

diffusivity of momentum (i.e., $D_V/D_M = 1$ and $D_H/D_M = 1$), because the same turbulent eddies are responsible for the transport of all three quantities. However, under unstable conditions, the vertical movement of eddies is enhanced beyond that due to the wind velocity, and there can be significant vertical transport of water vapor and/or sensible heat but little transport of momentum, so $D_V/D_M > 1$ and $D_H/D_M > 1$. These conditions typically occur when wind speed is low and the surface is strongly heated by the sun, inducing strong convection. (See Section 4.1.2.) Conversely, when the lapse rate near the ground is stable, turbulence is suppressed and $D_V/D_M < 1$ and $D_H/D_M < 1$. This situation is typical when warm air overlies a cold surface, such as a snowpack.

In estimating evaporation, the effect of instability can be accounted for by using a value of $b_o > 0$ in the empirical mass-transfer equation [Equation (7-17)] or by using a relation such as Equation (7-20). Here we present more general approaches to accounting for the effects of non-neutral lapse rates, following the discussion by Anderson (1976) and Cline (1997).

Stability-correction factors for momentum, water vapor, and sensible heat, designated $\Phi_M$, $\Phi_V$, and $\Phi_H$, respectively, can be incorporated into the general equations for latent- and sensible-heat transfer as follows:

*Latent heat and evaporation* [Equation (D-41)]:

$$LE = -\lambda_v \cdot \frac{0.622 \cdot \rho_a}{P \cdot \Phi_M \cdot \Phi_V} \cdot \frac{k^2}{\left[\ln\left(\dfrac{z_2 - z_d}{z_1 - z_d}\right)\right]^2}$$

$$\cdot (v_2 - v_1) \cdot (e_2 - e_1). \qquad \textbf{(D-53)}$$

*Sensible heat* [Equation (D-48)]:

$$H = -\frac{\rho_a \cdot c_a}{\Phi_M \cdot \Phi_H} \cdot \frac{k^2}{\left[\ln\left(\dfrac{z_2 - z_d}{z_1 - z_d}\right)\right]^2}$$

$$\cdot (v_2 - v_1) \cdot (T_2 - T_1). \qquad \textbf{(D-54)}$$

The $\Phi$ factors are related to the stability condition of the atmosphere, which is characterized by the dimensionless **Richardson number,** $Ri$, given by

$$Ri = \frac{2 \cdot g \cdot (z_2 - z_1) \cdot (T_2 - T_1)}{(T_2 + T_1 + 2 \cdot 273.2) \cdot (v_2 - v_1)^2}, \qquad \textbf{(D-55)}$$

where the subscripts again refer to measurements at two heights, $z_2 > z_1 \geq z_d + z_0$. Neutral conditions exist when $Ri = 0$, stable conditions when $Ri < 0$, and unstable conditions when $Ri > 0$. To determine the $\Phi$-factor values, first calculate $Ri$, then use the relations given in Table D-5.

### D.6.9 Eddy Correlation

We have seen that vertical transfer of momentum, energy, and water vapor in the planetary boundary layer is effected by turbulent eddies. Using notation

**TABLE D-5**

Formulas for Stability Factors ($\Phi_M$, $\Phi_V$, and $\Phi_H$) for Computing Latent- and Sensible-Heat Transfer as Functions of Richardson number ($Ri$). See Equations (D-53)–(D-55). From Cline (1997).

| Stability Factor | $Ri < -0.03$ | $-0.03 \leq Ri \leq 0$ | $0 < Ri < 0.19$ |
|---|---|---|---|
| $\Phi_M$ | $(1 - 18 \cdot Ri)^{-1/4}$ | $(1 - 18 \cdot Ri)^{-1/4}$ | $(1 - 5.2 \cdot Ri)^{-1}$ |
| $\Phi_V, \Phi_H$ | $1.3 \cdot (1 - 18 \cdot Ri)^{-1/4}$ | $(1 - 18 \cdot Ri)^{-1/4}$ | $(1 - 5.2 \cdot Ri)^{-1}$ |

analogous to that of Equation (D-15) for velocity, we can also express the instantaneous values of specific humidity, $q$, and temperature, $T_a$, at any level as sums of a time-average value (denoted by the overbar) and an instantaneous fluctuation from the average (denoted by the prime). We then have

$$q = \bar{q} + q' \qquad \textbf{(D-56)}$$

and

$$T_a = \bar{T}_a + T_a', \qquad \textbf{(D-57)}$$

where $\bar{q'} = \bar{T_a'} = 0$.

If positive (negative) fluctuations of the vertical velocity, $u_a'$, are accompanied by positive (negative) values of specific humidity, $q'$, then $u_a'$ and $q'$ are positively correlated, water vapor is being transferred upward, and evaporation is occurring. Under these conditions, the time averaged product $\overline{u_a' \cdot q'}$ is greater than 0.

Conversely, if positive (negative) velocity fluctuations are accompanied by negative (positive) humidity fluctuations, then $u_a'$ and $q'$ are negatively correlated, water vapor is being transferred downward, and condensation is occurring. Under these conditions, $\overline{u_a' \cdot q'}$ is less than 0.

Thus, if the simultaneous fluctuations $u_a'$ and $q'$ can be measured, the evaporation rate $E$ [L T$^{-1}$] can be computed as

$$E = \frac{\rho_a}{\rho_w} \cdot \overline{u_a' \cdot q'}. \qquad \textbf{(D-58)}$$

Analogous reasoning can be used for temperature, so that upward sensible-heat transfer rate, $H$, can be computed from simultaneous measurement of $T_a'$ and $u_a'$ as

$$H = \rho_a \cdot c_a \cdot \overline{u_a' \cdot T_a'}. \qquad \textbf{(D-59)}$$

Equations (D-58) and (D-59) are the basis for the eddy-correlation method of measuring sensible- and latent-heat transfer and evaporation (Sections 7.3.3 and 7.8.3).

# E

# Estimation of Daily Clear-Sky Solar Radiation on Sloping Surfaces

Solar radiation is usually an important contributor to the energy balance at the earth's surface and hence is a parameter in models of snowmelt [Equations (5-26)–(5-29)] and evapotranspiration [Equations (7-26)–(7-27)]. Because solar radiation is routinely measured at only widely spaced locations (see Figures E-1 and E-2) and at certain research stations, it is useful to have a means of estimating its magnitude from more readily available quantities.

This appendix develops a model for estimating the clear-sky radiation incident on a slope in the absence of vegetation. We first develop the relations giving the radiation incident on a horizontal plane at an arbitrary point on the earth's surface. (Quantities applying to a horizontal plane are written with the "prime" throughout.) These relations are then modified to account for the effects of slope inclination and aspect. This model has been programmed in EXCEL format on the CD accompanying this text as file SolarRad.xls. Empirical adjustments for the effects of cloud cover and vegetation can be made as indicated in Equations (5-28)–(5-33).

## E.1  RADIATION INCIDENT ON A HORIZONTAL PLANE

### E.1.1  Extraterrestrial Radiation

We first develop the astronomical relations that give the daily radiation incident at the top of the atmosphere directly above an arbitrary point; this is called the **extraterrestrial radiation.** Most of these relations are taken from Iqbal (1983).

### Solar Constant

The average radiation flux on a plane perpendicular to the solar beam at the upper surface of the atmosphere is called the **solar constant,** $I_{sc}$:

$$I_{sc} = 1367 \text{ W m}^{-2} = 118.1 \text{ MJ m}^{-2} \text{ day}^{-1}$$
$$= 2821 \text{ cal cm}^{-2} \text{ day}^{-1}.$$

### Day Angle

The extraterrestrial radiation flux incident on a plane tangent to the earth's surface is a function of (1) the radiation flux on a plane perpendicular to the solar beam and (2) the angle of the tangent plane relative to the beam. Because of the eccentricity of the earth's orbit and the 23.5° angle of its rotational axis relative to the orbital plane, both of these factors vary regularly through the year as a function of the earth's location in its orbit (Figure E-3). The position of the earth in its orbit is given by the **day angle,** $\Gamma$, where

$$\Gamma = \frac{2 \cdot \pi \cdot (J - 1)}{365} \tag{E-1}$$

and $J$ is the **day number** of the year ($J = 1$ on 1 January and 365 on 31 December).

### Eccentricity

The relative earth–sun distance changes regularly during the year. Because radiative flux follows the

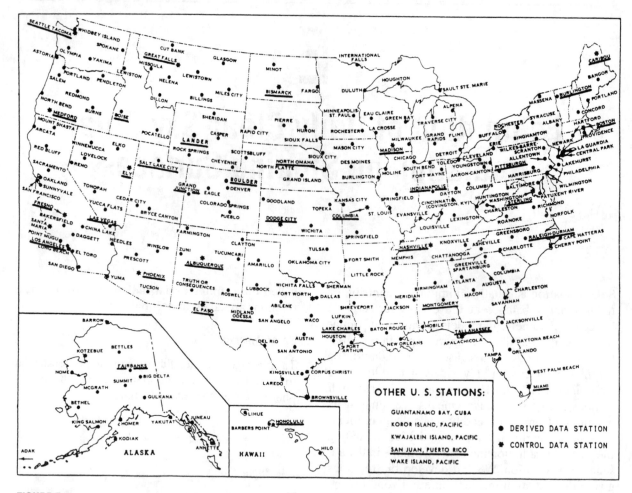

**FIGURE E-1**
Stations at which solar radiation data are routinely collected by the U.S. National Weather Service. Underlined locations are stations where global and direct radiation are individually measured. From Iqbal (1983).

inverse-square law, this distance is most usefully expressed as the **eccentricity correction,** $E_0$, which is the square of the ratio of the average distance, $r_0$, to the distance at any time, $r$. $E_0$ can be calculated for any day of the year as

$$E_0 = (r_0/r)^2 = 1.000110 + 0.034221 \cdot \cos(\Gamma)$$
$$+ 0.001280 \cdot \sin(\Gamma) + 0.000719 \cdot \cos(2 \cdot \Gamma)$$
$$+ 0.000077 \cdot \sin(2 \cdot \Gamma).$$

$$(\text{E-2})$$

### *Declination*

The angle between a horizontal (tangent) plane and the solar beam is determined by the latitude, $\Lambda$, of the plane and the **declination** of the sun, $\delta$. The sun's declination is the latitude at which the sun is direct-

ly overhead at noon; due to the 23.5° tilt of the earth's rotational axis, this latitude changes regularly between +23.5° and −23.5° as the earth revolves around the sun. The declination is given by

$$\delta = (180/\pi) \cdot [0.006918 - 0.399912 \cdot \cos(\Gamma)$$
$$+ 0.070257 \cdot \sin(\Gamma) - 0.006758 \cdot \cos(2 \cdot \Gamma)$$
$$+ 0.000907 \cdot \sin(2 \cdot \Gamma) - 0.002697 \cdot \cos(3 \cdot \Gamma)$$
$$+ 0.00148 \cdot \sin(3 \cdot \Gamma)],$$

$$(\text{E-3})$$

where $\delta$ is in degrees.

### *Solar Noon, Sunrise, and Sunset*

The angle between a line from an observer on the earth to the sun and a vertical line extending from the observer is called the **zenith angle,** $\theta$. Solar noon

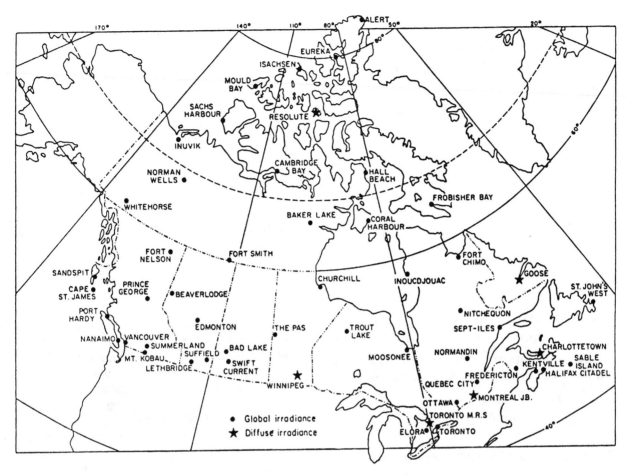

**FIGURE E-2**

Stations at which solar radiation data are routinely collected by Environment Canada. Starred locations are stations where global and direct radiation are indivdually measured. From Iqbal (1983).

occurs when this angle is at its minimum value for the day. Between sunrise and sunset,

$$\theta = \cos^{-1}[\sin(\Lambda) \cdot \sin(\delta) + \cos(\Lambda) \cdot \cos(\delta) \cdot \cos(\omega \cdot t)], \quad \textbf{(E-4)}$$

where $\Lambda$ is latitude, $t$ is the number of hours before ($-$) or after ($+$) true solar noon and $\omega$ is the angular velocity of the earth's rotation ($15°$ hr$^{-1}$ = $0.2618$ radian hr$^{-1}$).

Thus solar noon occurs when $\omega \cdot t = 0$, and the times of sunrise, $T_{hr}$, and sunset, $T_{hs}$, occur at equal times before and after solar noon. These times can determined from Equation (E-4) as

$$T_{hr} = -\frac{\cos^{-1}[-\tan(\delta) \cdot \tan(\Lambda)]}{\omega} \quad \textbf{(E-5a)}$$

and

$$T_{hs} = +\frac{\cos^{-1}[-\tan(\delta) \cdot \tan(\Lambda)]}{\omega}. \quad \textbf{(E-5b)}$$

At latitudes greater than $66.5°$, the absolute value of $[-\tan(\delta) \cdot \tan(\Lambda)]$ can exceed one, and Equations (E-5) cannot be used. When this happens, there is either no sunrise (winter) or no sunset (summer), and $T_{hr}$ and $T_{hs}$ must be assigned the value 0 or 12 hr, respectively.

### Radiation on a Horizontal Plane

The instantaneous extraterrestrial radiation flux on a horizontal plane, $k'_{ET}$, can be calculated as

**FIGURE E-3**

Motion of the earth around the sun. AU stands for **astronomical unit,** the average earth–sun distance; $\delta$ is the declination. From Iqbal (1983).

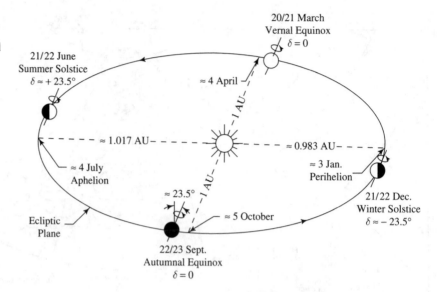

$$k'_{ET} = I_{sc} \cdot E_o \cdot [\cos(\delta) \cdot \cos(\Lambda) \cdot \cos(\omega \cdot t) + \sin(\delta) \cdot \sin(\Lambda)]. \quad \textbf{(E-6)}$$

Integrating Equation (E-6) between $T_{hr}$ and $T_{hs}$ yields the daily solar radiation flux, $K'_{ET}$, on such a plane:

$$K'_{ET} = 2 \cdot I_{sc} \cdot E_0 \cdot [\cos(\delta) \cdot \cos(\Lambda) \cdot \sin(\omega \cdot T_{hr})/\omega + \sin(\delta) \cdot \sin(\Lambda) \cdot T_{hr}]. \quad \textbf{(E-7)}$$

### E.1.2   Direct (Beam) Radiation at the Surface

As it passes through the atmosphere, the energy in the solar beam is reduced due to absorption and reflection (scattering) by the gaseous and solid particles in the atmosphere. Here we calculate the degree of reduction as a function of the path length through the atmosphere, specifically accounting for the effects of water vapor and dust.[1]

The energy flux in the solar beam (**direct solar radiation**) reaching a horizontal plane at the surface, $K'_{dir}$, is

$$K'_{dir} = \tau \cdot K'_{ET}, \quad \textbf{(E-8)}$$

where $\tau$ is the total atmospheric transmissivity [see Equation (D-4)]. Following Bolsenga (1964), $\tau$ is estimated as

$$\tau = \tau_{sa} - \gamma_{dust}, \quad \textbf{(E-9)}$$

where $\tau_{sa}$ is the transmissivity due to scattering and absorption by water vapor and the constant atmospheric gases, and $\gamma_{dust}$ is the attenuation due to dust.

$\tau_{sa}$ depends on the path length of the solar beam and the water-vapor content of the atmosphere. Relative path length is expressed as the dimensionless **optical air mass,** $M_{opt}$, which can be found as a function of latitude and declination from Figure E-4. Water-vapor content is expressed as **precipitable-water content,** $W_p$, which can be estimated from an empirical relation such as

$$W_p = 1.12 \cdot \exp(0.0614 \cdot T_d), \quad \textbf{(E-10)}$$

where $T_d$ is the surface dew point in °C [see Equation (D-11)] and $W_p$ is in cm (Bolsenga 1964).

Empirical relations are then used to estimate $\tau_{sa}$:

$$\tau_{sa} = \exp(a_{sa} + b_{sa} \cdot M_{opt}), \quad \textbf{(E-11)}$$

where

$$a_{sa} = -0.124 - 0.0207 \cdot W_p \quad \textbf{(E-12)}$$

and

$$b_{sa} = -0.0682 - 0.0248 \cdot W_p. \quad \textbf{(E-13)}$$

It is difficult to generalize about the appropriate values of $\gamma_{dust}$. Appropriate values can be estimated from information provided by Bolsenga (1964), who cited values in the range 0.0 to 0.05 for remote stations, 0.03 to 0.10 for moderate-sized cities, and up to 0.13 for large metropolitan areas.

---

[1]Recall that the effects of clouds are considered in Sections 5.4.2 and 7.3.4. Additional reductions can also occur from abnormal concentrations of gases and particles, as from volcanoes or pollution.

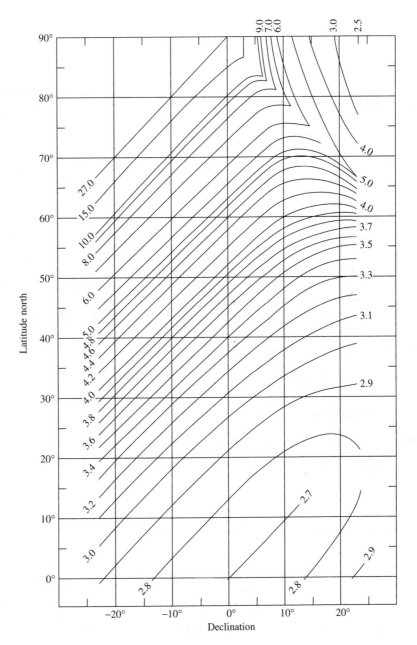

His tables give clear-sky radiation values for $0 < \gamma_{dust} < 0.20$.

### E.1.3 Diffuse Radiation

About one-half the energy scattered from the solar beam reaches the surface as **diffuse radiation,** $K'_{dif}$. Thus this quantity can be calculated as

$$K'_{dif} = 0.5 \cdot \gamma_s \cdot K'_{ET}, \qquad \textbf{(E-14)}$$

where $\gamma_s$ is the attenuation of the solar beam due to scattering by water vapor and permanent atmospheric constituents. Adding the effects of dust yields

$$\gamma_s = 1 - \tau_s + \gamma_{dust}. \qquad \textbf{(E-15)}$$

The value of $\tau_s$, like that of $\tau_{sa}$, can be estimated from the optical air mass (Figure E-4) and the precipitable water content. The resulting equation is

$$\tau_s = \exp(a_s + b_s \cdot M_{opt}), \qquad \textbf{(E-16)}$$

where

$$a_s = -0.0363 - 0.0084 \cdot W_p \qquad \textbf{(E-17)}$$

and

$$b_s = -0.0572 - 0.0173 \cdot W_p. \qquad \textbf{(E-18)}$$

### E.1.4   Global Radiation

The **global radiation,** $K'_g$, is the sum of direct plus diffuse radiation:

$$K'_g \equiv K'_{dir} + K'_{dif}. \qquad \textbf{(E-19)}$$

### E.1.5   Backscattered Radiation

Of the solar radiation striking the surface, a portion given by the albedo, $a$, (Table D-2) is reflected back to the atmosphere. Of this, about one-half is reflected again from the atmosphere to the surface to increase the total radiation flux. This **backscattered radiation,** $K'_{bs}$, can be estimated as

$$K'_{bs} = 0.5 \cdot \gamma_s \cdot a \cdot K'_g. \qquad \textbf{(E-20)}$$

### E.1.6.   Total Incident Radiation

The total daily clear-sky radiation incident on a horizontal plane at the surface, $K'_{cs}$, can be estimated by combining Equations (E-8), (E-14), (E-19), and (E-20):

$$K'_{cs} = (\tau + 0.5 \cdot \gamma_s \cdot a \cdot \tau + 0.5 \cdot \gamma_s \\ + 0.25 \cdot \gamma_s^2 \cdot a) \cdot K'_{ET}. \qquad \textbf{(E-21)}$$

## E.2   RADIATION ON A SLOPING PLANE

### E.2.1   Equivalent Slope

Extraterrestrial radiation on a sloping plane is computed using the concept of the **equivalent slope** (Lee 1964):

> The angle of incidence of the solar beam on a sloping plane at latitude $\Lambda$ and longitude $\Omega$ is the same as the angle of incidence on a horizontal plane at a point on a great circle passing through the slope at right angles to it and as many degrees removed from it as the angle of the slope.

The difference in longitude between the location of the original slope and the equivalent plane, $\Delta\Omega$, is given by

$$\Delta\Omega = \\ \tan^{-1}\left[\frac{\sin(\beta) \cdot \sin(\alpha)}{\cos(\beta) \cdot \cos(\Lambda) - \sin(\beta) \cdot \sin(\Lambda) \cdot \cos(\alpha)}\right], \\ \textbf{(E-22)}$$

where $\beta$ is the angle of inclination of the slope (positive downward) and $\alpha$ is the azimuth (orientation) of the slope (clockwise from north). The **equivalent latitude,** $\Lambda_{eq}$, of the horizontal plane is

$$\Lambda_{eq} = \sin^{-1}[\sin(\beta) \cdot \cos(\alpha) \cdot \cos(\Lambda) \\ + \cos(\beta) \cdot \sin(\Lambda)]. \qquad \textbf{(E-23)}$$

### E.2.2   Solar Noon, Sunrise, and Sunset

The concept of equivalent slope permits computation of the times of solar noon, sunrise, and sunset on the slope of interest using relations analogous to those for a horizontal plane. Thus solar noon occurs at $-\Delta\Omega/\omega$, and the times of sunrise, $T_{sr}$, and sunset, $T_{ss}$, are

$$T_{sr} = -\frac{\cos^{-1}[-\tan(\Lambda_{eq}) \cdot \tan(\delta)] - \Delta\Omega}{\omega} \\ \textbf{(E-24a)}$$

and

$$T_{ss} = +\frac{\cos^{-1}[-\tan(\Lambda_{eq}) \cdot \tan(\delta)] - \Delta\Omega}{\omega}. \\ \textbf{(E-24b)}$$

Note that neither sunrise nor sunset on a slope can have an absolute value greater than that for a horizontal surface at the same latitude. There are situations for slopes at high latitudes, and where $\beta + \Lambda_{eq} > 90°$ for north-facing slopes, when there are two sunrises and two sunsets in a single calendar day. In these cases the times of first sunrise and last sunset are the same as those for a horizontal surface at the same latitude.

### E.2.3   Extraterrestrial Radiation

By direct analogy with Equation (E-7), the daily extraterrestrial radiation on a sloping plane is given by

$$K_{ET} = I_{sc} \cdot E_0 \cdot \{\cos(\Lambda_{eq}) \cdot \cos(\delta)$$

$$\cdot \frac{[\sin(\omega \cdot T_{ss} + \Delta\Omega) - \sin(\omega \cdot T_{sr} + \Delta\Omega)]}{\omega}$$

$$+ \sin(\Lambda_{eq}) \cdot \sin(\delta) \cdot (T_{ss} - T_{sr})\} \qquad \textbf{(E-25)}$$

### E.2.4   Total Incident Radiation at the Surface

Since only the direct (beam) radiation is dependent on slope and aspect, the total clear-sky solar radiation incident on a sloping plane, $K_{cs}$, is

$$K_{cs} = \tau \cdot K_{ET} + K'_{\text{dif}} + K'_{bs}, \qquad \textbf{(E-26)}$$

where $\tau$ is found via Equations (E-9)–(E-13), $K_{ET}$ is from Equation (E-25), $K'_{dif}$ is from Equation (E-14), and $K'_{bs}$ is from Equation (E-20).

The relative radiation received on a slope can be expressed as a **slope factor,** $f_{sl}$, where

$$f_{sl} \equiv \frac{K_{cs}}{K'_{cs}}. \qquad \textbf{(E-27)}$$

# F

# Stream-Gaging Methods for Short-Term Studies

Discharge, $Q$, is the volume rate of flow $[L^3 T^{-1}]$ through a stream cross-section at right angles to the flow direction:

$$Q = U \cdot A = U \cdot B \cdot Y. \qquad \text{(F-1)}$$

Here, $U$ is average velocity through the cross-section $[L T^{-1}]$, $A$ is the area of the cross-section $[L^2]$, $B$ is the water-surface width at the cross-section $[L]$, and $Y$ is the average depth at the cross-section $[L]$. The process of measuring discharge is called **stream gaging.**

Methods for determining the discharge occurring at the time of observation can be classified as shown in Table F-1. Discharge can be measured directly by several methods or determined indirectly by (1) observing the **stage,** $Z_s$, defined as

$$Z_s \equiv Z_w - Z_0, \qquad \text{(F-2)}$$

where $Z_w$ is the elevation of the water surface and $Z_0$ is the elevation of an arbitrary datum, and (2) using a previously established relation between stage and discharge. The stage–discharge relation is called a **rating curve** (or **rating table**); its form is determined by the configuration of the stream channel in the measurement reach. This configuration may be that of the natural channel (**natural control**) or it may be that of an artificial structure such as a weir or flume (**artificial control**).

This appendix describes stream-gaging methods that are suitable for short-term and special-purpose studies: direct measurement via the velocity-area method (Section F.2) and dilution gaging (Section F.3), indirect measurement via weirs (Section F.4)

and flumes (Section F.5), and the measurement of stage (Section F.6). Section F.7 describes methods for estimating the discharge of a peak flow that occurred prior to the time of measurement.

The details of constructing permanent stream-gaging installations for long-term measurement stations were described by Reinhart and Pierce (1964), Buchanan and Somers (1969), Gregory and Walling (1973), Herschy (1999), and Shaw (1988). Herschy (1999) provided a complete discussion of measurement errors using various techniques.

## F.1 SELECTION OF MEASUREMENT LOCATION

The overall purposes of a study determine the general location of a stream-gaging location. For example, when water-balance information is sought, the gage might be sited near the outlet of a drainage basin or at the mouths of major streams entering a lake. When information on flows is to be used in the design of structures or land-use plans, the gage should be located near the site of the structure or plan. Gages used as part of a general-purpose hydrologic network may be variously placed to provide information on a range of drainage-basin types and sizes.

The specific location of a stream gage depends on (1) accessibility; (2) minimization of flow bypassing the section as ground water beneath the channel or in flood channels at high flows; (3) absence of

**TABLE F-1**

Classification of Stream-Gaging Methods.

I. Direct measurement
  A. Volumetric gaging
  B. Velocity–area gaging
  C. Dilution gaging
II. Indirect measurement via stage–discharge relation
  A. Empirical rating curve (natural control)
  B. Theoretical rating curve (artificial control)
    1. Weirs
    2. Flumes

backwater conditions due to high water levels in a stream or lake to which the gaged stream is tributary, or to tidal fluctuations; and (4) the presence of conditions suitable for the particular measurement method to be used.

## F.2 VELOCITY-AREA METHOD

The **velocity-area method** involves direct measurement of the components of discharge [Equation (F-1)] at successive locations (called **verticals**) along a stream cross-section and numerical integration of the measured values to give the total discharge.

Figure F-1 defines the quantities involved. The more verticals used in a measurement ($N$), the more accurate it will be (other things equal);

$N \geq 25$ is usually recommended, with the spacing of verticals adjusted such that $< 5\%$ of the total discharge occurs in each subsection. Thus, in general, spacings between verticals will vary, being closer together where the flow is deeper and faster and farther apart where the flow is slower and shallower.

Velocity at each vertical ($U_i$) is measured by a current meter suspended from a rod or cable, depth ($Y_i$) is measured by rod or weighted cable, and cross-section location ($X_i$) by measurement tape or range finder.

### F.2.1 Selection of Measurement Section

The quality of a velocity-area measurement is strongly influenced by the nature of the measurement cross-section. Accuracy and precision are enhanced in sections with the following characteristics:

1. converging flow (i.e., cross-sectional area decreasing downstream) without areas of near-zero velocity or eddies,
2. absence of backwater conditions (due, for example, to high water levels in a stream or lake to which the gaged stream is tributary),
3. smooth cross-section with minimal flow obstructions upstream or downstream, and
4. velocities and depths not exceeding the range for which the velocity- and depth-measuring devices give accurate results and for which one can safely negotiate the section.

**FIGURE F-1**

Definitions of terms for measurement of discharge by the velocity–area method [Equations (F-3) and (F-4)]. Dashed lines delineate individual subsections, numbered consecutively $i = 1, 2, ..., N$. Left and right edges of water (LEW and REW) are defined for an observer facing *downstream*; sections can be numbered starting on left or right bank.

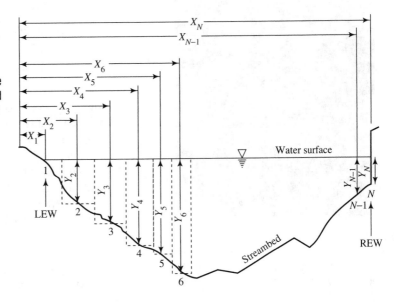

The accuracy of measurements at natural sections can be improved by removing obstructions in and above and below the section, since this will not affect discharge. If this is done, however, the measured velocities and depths will not be representative of the natural values.

Cross-sections may be negotiated by wading, by boat, or from bridges or specially constructed cableways (Buchanan and Somers 1969; Herschy 1999). In wading measurements, the current meter is fixed to a hand-held vertical rod (called a **wading rod**) that is also used to gage the depth. In measurements from bridges, cableways, and boats, the current meter is suspended on a weighted cable and depth is measured by sounding the bottom with the weight.

In wading measurements (Figure F-2), the location of the cross-section is usually marked by a measuring tape, and the location of a vertical is measured against the tape. For small streams, the accuracy measurements can be improved by making them from a specially constructed temporary wooden bridge on which cross-stream distances are permanently marked (Figure F-3).

Cross-sections should be perpendicular to the dominant velocity vector, and all velocities used in the computation should be perpendicular to the

**FIGURE F-3**
Temporary structure used in velocity–area gaging via wading rod. Note the cross-section distances permanently fixed to the structure. Photo by author.

section. Where the vector is not perpendicular to the section at a particular vertical, the perpendicular velocity should be measured or the velocity used in the computations must be corrected by multiplying by the cosine of the angle between the perpendicular and the actual velocity vector.

### F.2.2 Method of Integration

Discharge through the cross-section is given by

$$Q = \sum_{i=1}^{N} U_i \cdot A_i, \qquad \textbf{(F-3)}$$

where the area of each subsection, $A_i$, is computed as

$$A_i = Y_i \cdot \frac{|X_{i+1} - X_{i-1}|}{2}, \qquad \textbf{(F-4)}$$

the $X_i$ are cross-stream distances to successive verticals measured from an arbitrary horizontal datum, the $U_i$ are the vertically averaged velocities at each vertical, and the $Y_i$ are the depths at each vertical. $X_1$ and $X_N$ are located at the ends of the section (left and right edges of water[1]), and $X_{N-1}$ and $X_{N+1}$ are taken as zero. The velocities at the ends of the section ($U_1$ and $U_N$) will always equal zero, and $Y_1$

**FIGURE F-2**
Velocity–area stream gaging using a wading rod. The tape marking the cross section is visible just in front of the observer. Photo by J.V.Z. Dingman.

---

[1]"Left" and "right" edges of water are defined relative to an observer facing *downstream*.

and $Y_N$ will also equal zero (unless the corresponding bank is vertical).

Equation (F-4) is called the **mid-section method,** and it is the method used by the U.S. Geological Survey, which is the federal agency responsible for collecting streamflow data in the United States. It has been shown (Hipolito and Loureiro 1988) that integration using Equation (F-4) gives the most precise measurements.

### F.2.3   Measurement of Velocity

#### Vertical Velocity Profile

The vertically averaged velocity is usually estimated by assuming that the velocity is logarithmically related to distance above the bottom, as for wind flow over the ground surface [Equation (D-22)]:

$$u(y_i) = \frac{1}{k} \cdot u_{*i} \cdot \ln\left(\frac{y_i}{y_{0i}}\right). \qquad \textbf{(F-5)}$$

Here $u(y_i)$ is velocity at a distance $y_i$ above the bottom at vertical $i$, $u_{*i}$ is the friction velocity at vertical $i$, and $y_{0i}$ is the roughness height at vertical $i$. (The zero-plane displacement height is taken as zero in open-channel flows.) The derivation of Equation (F-5) was discussed by Dingman (1984).

The local friction velocity for water flow can be directly calculated as

$$u_{*i} = (g \cdot Y_i \cdot S_c)^{1/2}, \qquad \textbf{(F-6)}$$

where $g$ is gravitational acceleration, $Y_i$ is local flow depth, and $S_c$ is channel slope.

The value of $y_{0i}$ is determined by the height of the roughness elements on the channel bed, which can be represented by the median diameter of the bed particles, $d_{50}$. If

$$d_{50} < \frac{4 \cdot \mu}{\rho \cdot u_{*1}}, \qquad \textbf{(F-7)}$$

where $\mu$ is viscosity and $\rho$ is mass density, then the flow is **hydraulically smooth,** and

$$y_{0i} = \frac{0.11 \cdot \mu}{\rho \cdot u_{*i}} \qquad \textbf{(F-8)}$$

Where the criterion of Equation (F-7) is not fulfilled, the flow is **hydraulically rough,** and

$$y_{0i} \approx 0.033 \cdot d_{50}. \qquad \textbf{(F-9)}$$

#### Estimating Average Velocity in a Vertical

**Six-Tenths-Depth Method**   If Equation (F-5) applies and $y_{0i}$ is very small relative to the flow depth, it can be shown that the average velocity occurs at a distance of $0.368 \cdot Y_i$ above the bottom (Dingman 1984). Based on this, the **Six-Tenths-Depth Method** assumes that the velocity measured at a distance of $0.6 \cdot Y_i$ below the surface ($0.4 \cdot Y_i$ above the bottom) is the average velocity at that point in the cross-section.

Standard U.S. Geological Survey practice is to use the Six-Tenths-Depth Method where $Y_i < 2.5$ ft (0.75 m).

**Two-Tenths-and-Eight-Tenths-Depth Method**   If the velocity is given by Equation (F-5), it can be shown that

$$u(0.4 \cdot Y_i) = \frac{u(0.2 \cdot Y_i) + u(0.8 \cdot Y_i)}{2}. \qquad \textbf{(F-10)}$$

Thus average vertical velocity can be estimated as the average of the velocities at $0.2 \cdot Y_i$ and $0.8 \cdot Y_i$.

It has been found that the Two-Tenths-and-Eight-Tenths-Depth Method gives more accurate estimates of average velocity than does the Six-Tenths-Depth Method (Carter and Anderson 1963), and standard U.S. Geological Survey practice is to use the Two-Tenths-and-Eight-Tenths-Depth Method where $Y_i > 2.5$ ft (0.75 m).

**General Two-Point Method**   If velocity is measured at two points, each an arbitrary fixed distance above the bottom, the relative depths of those sensors will change as the discharge changes. Again assuming a logarithmic vertical velocity distribution, Walker (1988) derived the following expression for calculating the average vertical velocity from two sensors fixed at arbitrary distances above the bottom, $y_{i1}$ and $y_{i2}$, where $y_{i2} > y_{i1}$:

$$U_i = \frac{[1 + \ln(y_{i2})] \cdot u(y_{i1}) + [1 + \ln(y_{i1})] \cdot u(y_{i2})}{\ln(y_{i2}/y_{i1})}. \qquad \textbf{(F-11)}$$

Walker (1988) also calculated the error in estimating $U_i$ for sensors located at various combinations of relative depths.

**Multi-Point Method**   As noted, the standard formulas for calculating average vertical velocity assume a logarithmic vertical velocity distribution with $y_{0i} \ll Y_i$. These assumptions may not be appropriate for channels with roughness elements (boulders, aquatic vegetation) whose heights are a significant fraction of depth and which have significant obstructions upstream and downstream of the measurement section. In these cases, Buchanan and Somers (1969) recommended estimating $U_i$ as

$$U_i = 0.5 \cdot u_i(0.4 \cdot Y_i) + 0.25 \cdot [u_i(0.2 \cdot Y_i) + u_i(0.8 \cdot Y_i)]. \qquad \textbf{(F-12)}$$

However, the highest accuracy in these situations is assured by measuring velocity several heights at each vertical, with averages found by numerical integration over each vertical or over the entire cross section. Alternatively, a statistical sampling approach may be appropriate (Dingman 1989).

**Surface-Velocity (Float) Method**  If it is not possible to accurately measure velocities at various depths, the average velocity can be estimated by observing the time it takes floats inserted at representative locations across the stream to travel a given distance. The measurement distance should be about 10 times the stream width.

The average velocity for each path can then be estimated as

$$U_i = f\left(\frac{d_{50}}{Y_i}\right) \cdot u(Y_i), \qquad \textbf{(F-13)}$$

where $u(Y_i)$ is the velocity of the $i$th float and $f(d_{50}/Y_i)$ is a proportion that depends on the ratio of the average height of channel roughness elements, $d_{50}$, to the flow depth. Figure F-4 shows this relation, again assuming a logarithmic velocity profile. A reasonable general value is $f(d_{50}/Y_i) = 0.85$.

### Current Meters

Several types of current meters are available, including horizontal-axis (propeller or screw type), vertical-axis (Price-type; this is standard for the U.S. Geological Survey), and electromagnetic instruments. Accurate measurements require carefully calibrated and maintained current meters (Smoot and Novak 1968) and, at each measurement point, averaging the velocity over time (usually 30 to 60 s) to eliminate fluctuations due to turbulence.

### F.2.4  Accuracy

Carter and Anderson (1963) evaluated the accuracy of velocity-area measurements using standard U.S. Geological Survey techniques. They determined discharge from measurements of velocity and depth at over 100 verticals in 127 cross-sections in different streams, then recomputed the discharge using smaller numbers of observations at each site. Care-

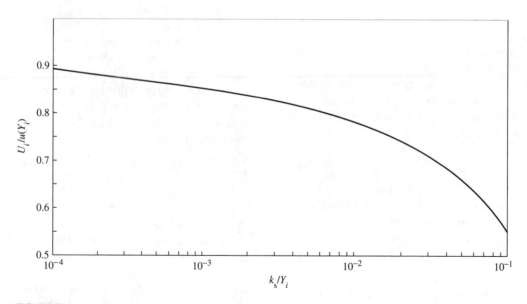

**FIGURE F-4**
Theoretical ratio of average velocity, $U_i$, to surface velocity, $u(Y_i)$, as a function of the ratio of roughness height, $d_{50}$, to depth, $Y_i$, for a logarithmic velocity profile.

**FIGURE F-5**
Standard deviation of percentage errors as a function of number of verticals used in velocity–area measurements using the Six-Tenths-Depth Method and the Two-Tenths-and-Eight-Tenths-Depth Method. From Carter and Anderson (1963).

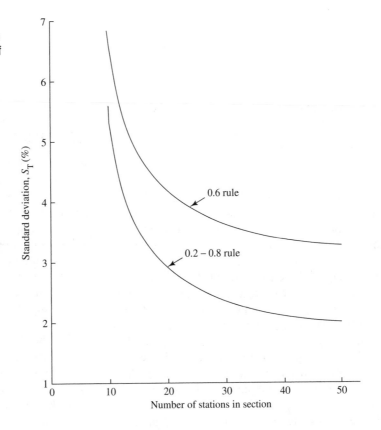

fully calibrated current meters were used in these observations.

The results are summarized in Figure F-5: They show the standard deviation of the error as a function of number of verticals ($N$) for the Six-Tenths and Two-Tenths-and-Eight-Tenths methods. About two-thirds of discharge measurements should have percentage errors less than the values given by the curves, and about 95% of measurements should have errors less than twice the values given by the curves. Although Carter and Anderson (1963) gave no information about the nature of the streams they measured, it is likely that errors for highly irregular channels would be larger than indicated by Figure F-5.

## F.3   DILUTION GAGING

Dilution gaging involves introducing a tracer into the flow at an upstream location and measuring the rate of arrival of the tracer at a downstream loca-

tion. It is usually a more accurate method than velocity-area gaging in small, highly turbulent streams with rough, irregular channels. As with velocity-area measurements, the method is most accurate when there is no change in discharge during the measurement. Kilpatrick and Cobb (1985) and Herschy (1999) gave a complete discussion of the method.

The distance between the two measurement locations must be long enough to allow complete mixing of the tracer with the flow, but short enough so that the change in discharge is insignificant. It is best to determine the distance required for complete mixing empirically, by observing the mixing of a visible dye such as fluorescein in the reach of interest. Alternatively, this length, $L_{\mathrm{mix}}$, may be estimated as

$$L_{\mathrm{mix}} = K_{\mathrm{mix}} \cdot \frac{C \cdot B^2}{g^{1/2} \cdot Y}, \qquad \textbf{(F-14)}$$

where $C$ is Chézy's $C$ [estimated from Table 9-6 and Equation (9-20)], $B$ is average reach width, $Y$ is average reach depth, $g$ is gravitational acceleration, and $K_{\mathrm{mix}}$ is found from Table F-2 (Kilpatrick and Cobb 1985).

**TABLE F-2**

Mixing Coefficients for Dilution Gaging.

| Number and Location of Injection Points | $K_{mix}$ |
|---|---|
| 1 point at center of flow | 0.500 |
| 2 points, 1 at center of each half of flow | 0.125 |
| 3 points, 1 at center of each third of flow | 0.055 |
| 1 point at edge of flow | 2.00 |

From Kilpatrick and Wilson (1989)

Gregory and Walling (1973) summarized the requirements for a tracer. The substance should (1) be readily soluble, (2) have zero or very low natural concentration in the stream, (3) not be chemically reactive with or physically absorbed by substances in the stream, (4) be easily detectable at low concentrations, (5) be harmless to the observer and to stream life, and (6) be of reasonable cost. Sodium chloride (NaCl) is probably the best choice in most situations; its concentration can usually be readily detected by developing a calibration curve between electrical conductivity and concentration in the stream water and subsequently measuring the conductivity.

There are two techniques for dilution gaging: (1) constant-rate injection and (2) slug (or gulp) injection (Figure F-6). Constant-rate injection requires a somewhat more elaborate installation than slug injection. The tracer solution is injected at a constant rate, $Q_T$, for a period of time sufficient for the downstream concentration to reach a steady equilibrium value, $C_{eq}$. Then the discharge, $Q$, is calculated as

$$Q = Q_T \cdot \frac{C_T - C_{eq}}{C_{eq} - C_b}, \qquad \textbf{(F-15)}$$

where $C_T$ is the concentration of the tracer solution and $C_b$ is the natural background concentration of the tracer in the stream.

Slug injection involves dumping a volume, $V_T$, of tracer solution with concentration $C_T$ into the stream at the upstream site. Concentration at the downstream site, $C_d(t)$, is then measured as a function of time until it recedes to its background value $C_b$. Stream discharge is then given by

$$Q = \frac{(C_T - C_b) \cdot V_T}{\displaystyle\int_0^\infty [C_d(t) - C_b] \cdot dt}, \qquad \textbf{(F-16)}$$

where the integral is evaluated by graphically measuring the area under the $C_d(t)$ vs. $t$ curve.

## F.4  SHARP-CRESTED V-NOTCH WEIRS

Accurate discharge measurements in small streams can be made using portable V-notch weirs in temporary installations (Figure F-7). The weir plate can be constructed of plywood or metal. The notch should be sharp-edged so that the water "springs free" even at low discharges. Thus, if plywood is used, the notch itself should be formed of metal strips.

The weir should be installed in the stream such that all the flow is diverted through it, and such that a virtually horizontal pool extends some distance upstream. The weir plate must be carefully leveled so that the V-notch is symmetric about a vertical line.

Discharge through a weir is determined by measuring the elevation of the water surface, $Z_w$, above the point of the V-notch, $Z_v$, and relating that elevation to discharge. Thus precise measurement of $Z_w$ is essential.

$Z_w$ should be measured where the water surface is virtually horizontal in the pool formed by the weir; this should ideally be at an upstream distance at least twice the vertical dimension of the weir opening. However, acceptably approximate $Z_w$ measurements may be obtained by observing the water level on a scale fixed to the upstream face of the weir plate far enough from the notch to avoid the effects of drawdown. This latter arrangement eliminates the need for precise leveling to establish the relation between elevations observed upstream and the elevation of the point of the V-notch.

The relation between $Z_w$ and discharge, $Q$, is given by

$$Q = C_w \cdot g^{1/2} \cdot \tan\left(\frac{\theta_v}{2}\right) \cdot (Z_w - Z_v), \qquad \textbf{(F-17)}$$

where $g$ is gravitational acceleration, $\theta_v$ is the angle of the V-notch, $Z_v$ is the elevation of the base of the V-notch, and $C_w$ is a weir coefficient. When $(Z_w - Z_v) > 0.3 \cdot (Z_v - Z_b)$, where $Z_b$ is the elevation of the streambed at the upstream face of the weir, $C_w = 0.43$. Values of $C_w$ for smaller values of

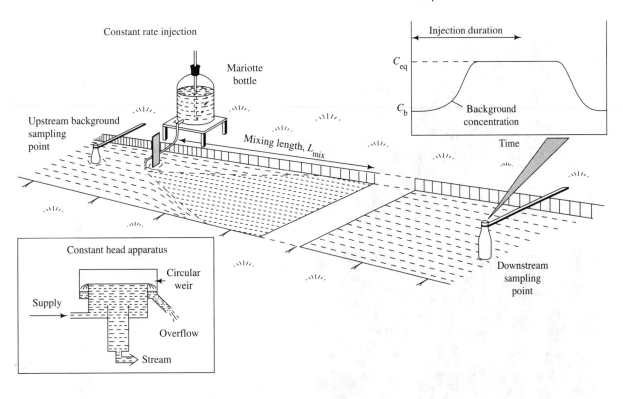

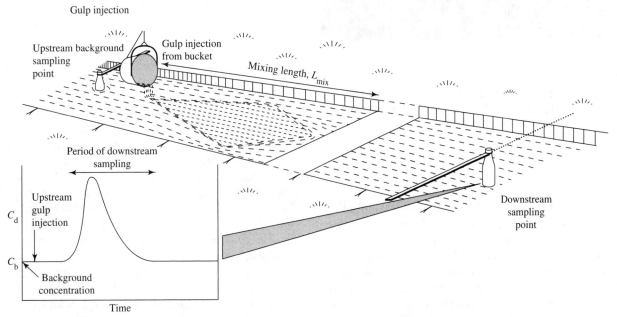

**FIGURE F-6**
Dilution-gaging techniques. From Gregory and Walling (1973).

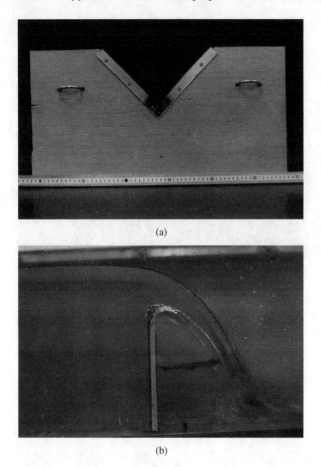

(a)

(b)

**FIGURE F-7**
(a) Construction of a V-notch weir in a plywood weir plate. (b) The notch should have a sharp edge so that the water springs free even at low discharges. Photos by author.

**TABLE F-3**
Ranges of Flows Measurable by V-Notch Weirs.[a]

|  | Maximum $Q$ | | |
|---|---|---|---|
|  | $(m^3\ s^{-1})$ | $(ft^3\ s^{-1})$ | $(L\ s^{-1})$ |
| **90° Notch** | | | |
| Maximum $H_w$ (m) | | | |
| 0.15 | 0.0117 | | 11.7 |
| 0.25 | 0.0421 | | 42.1 |
| 0.30 | 0.0664 | | 66.4 |
| 0.50 | 0.238 | | 238 |
| Maximum $H_w$ (ft) | | | |
| 0.50 | | 0.431 | |
| 0.80 | | 1.40 | |
| 1.00 | | 2.44 | |
| 1.50 | | 6.72 | |
| **60° Notch** | | | |
| Maximum $H_w$ (m) | | | |
| 0.15 | 0.00677 | | 6.77 |
| 0.25 | 0.0243 | | 24.3 |
| 0.30 | 0.0383 | | 38.3 |
| 0.50 | 0.137 | | 137 |
| Maximum $H_w$ (ft) | | | |
| 0.50 | | 0.249 | |
| 0.80 | | 0.806 | |
| 1.00 | | 1.41 | |
| 1.50 | | 3.88 | |

[a]$H_w \equiv Z_w - Z_v$ in Equation (F-17).

$(Z_w - Z_v)$ should ideally be obtained by calibration. [See Dingman (1984); Herschy (1999).]

The maximum vertical opening and angle of a V-notch weir are dictated by the anticipated range of flows and the required measurement precision. Narrower angles give greater precision at low flows, but reduce the overall range of measurement for a given size of opening. Table F-3 gives the capacity of 90° and 60° V-notch weirs with various maximum dimensions. Weirs with compound angles can be constructed to give more optimal combinations of low-flow sensitivity and range; these must be calibrated to obtain the stage–discharge relation.

## F.5 FLUMES

Flumes are devices that conduct the streamflow through a short reach with a constricted cross-section that accelerates the flow and provides a fixed stage–discharge relation. Large flumes are used in permanent gaging stations, and portable flumes can be used for short-term measurements in small streams. Both flumes and weirs give stable rating curves, but flumes are advantageous where one wishes to avoid inducing sediment deposition or inundating upstream areas.

There are many flume designs, each with its own rating curve (Herschy 1999). Some types may be commercially purchased. Figure F-8 and Tables F-4 and F-5 show the design and ranges of one of the most commonly used portable designs, the **Parshall flume;** and Figure F-9 gives the design of the **modified Parshall flume** used by the U.S. Geological Survey for temporary gaging of small streams.

As with weirs, flumes must be properly leveled, and are usually installed in temporary dams

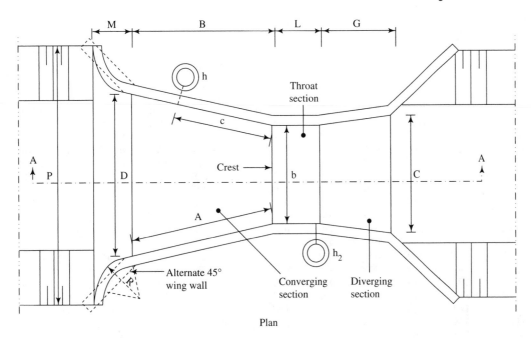

Plan

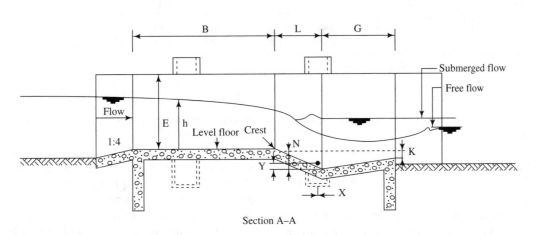

Section A–A

**FIGURE F-8**
Configuration of the Parshall flume. Actual dimensions corresponding to letters are given in Table F-4 and discharge ranges in Table F-5. From Herschy (1985).

that assure that all the flow passes through the measuring device.

# F.6 STAGE MEASUREMENT

Direct measurement of discharge is difficult and time-consuming, and recording it is impossible; however, observing and recording water-surface el-

evation, or stage [Equation (F-2)], is relatively simple. Thus the stage–discharge relation (rating curve) is an essential component of discharge measurement for which repeated or continuous records are required.

## F.6.1. Methods of Measurement

Stage is most simply determined by observing the position of the water surface on a ruler-like **staff gage** (Figure F-10). The zero level on the staff gage

**TABLE F-4**
Dimensions of Standard Parshall Flumes[a]

| | | | | | | | | | | | $h_2$ | |
|---|---|---|---|---|---|---|---|---|---|---|---|---|
| b | D | C | B | L | G | E | N | K | A | c | X | Y |
| 0.025 | 0.167 | 0.093 | 0.357 | 0.076 | 0.204 | 0.153–0.229 | 0.029 | 0.019 | 0.363 | 0.241 | 0.008 | 0.013 |
| 0.051 | 0.213 | 0.135 | 0.405 | 0.114 | 0.253 | 0.153–0.253 | 0.043 | 0.022 | 0.415 | 0.277 | 0.016 | 0.025 |
| 0.076 | 0.259 | 0.178 | 0.457 | 0.152 | 0.30 | 0.305–0.610 | 0.057 | 0.025 | 0.466 | 0.311 | 0.025 | 0.038 |
| 0.152 | 0.396 | 0.393 | 0.610 | 0.30 | 0.61 | 0.61 | 0.114 | 0.076 | 0.719 | 0.415 | 0.051 | 0.076 |
| 0.229 | 0.573 | 0.381 | 0.862 | 0.30 | 0.46 | 0.76 | 0.114 | 0.076 | 0.878 | 0.588 | 0.051 | 0.076 |
| 0.305 | 0.844 | 0.610 | 1.34 | 0.61 | 0.91 | 0.91 | 0.228 | 0.076 | 1.37 | 0.914 | 0.051 | 0.076 |
| 0.457 | 1.02 | 0.762 | 1.42 | 0.61 | 0.91 | 0.91 | 0.228 | 0.076 | 1.45 | 0.966 | 0.051 | 0.076 |
| 0.610 | 1.21 | 0.914 | 1.50 | 0.61 | 0.91 | 0.91 | 0.228 | 0.076 | 1.52 | 1.01 | 0.051 | 0.076 |
| 0.914 | 1.57 | 1.22 | 1.64 | 0.61 | 0.91 | 0.91 | 0.228 | 0.076 | 1.68 | 1.12 | 0.051 | 0.076 |
| 1.22 | 1.93 | 1.52 | 1.79 | 0.61 | 0.91 | 0.91 | 0.228 | 0.076 | 1.83 | 1.22 | 0.051 | 0.076 |
| 1.52 | 2.30 | 1.83 | 1.94 | 0.61 | 0.91 | 0.91 | 0.228 | 0.076 | 1.98 | 1.32 | 0.051 | 0.076 |
| 1.83 | 2.67 | 2.13 | 2.09 | 0.61 | 0.91 | 0.91 | 0.228 | 0.076 | 2.13 | 1.42 | 0.051 | 0.076 |
| 2.13 | 3.03 | 2.44 | 2.24 | 0.61 | 0.91 | 0.91 | 0.228 | 0.076 | 2.29 | 1.52 | 0.051 | 0.076 |
| 2.44 | 3.40 | 2.74 | 2.39 | 0.61 | 0.91 | 0.91 | 0.228 | 0.076 | 2.44 | 1.62 | 0.051 | 0.076 |
| 3.05 | 4.75 | 3.66 | 4.27 | 0.91 | 1.83 | 1.22 | 0.34 | 0.152 | 2.74 | 1.83 | | |
| 3.66 | 5.61 | 4.47 | 4.88 | 0.91 | 2.44 | 1.52 | 0.34 | 0.152 | 3.05 | 2.03 | | |
| 4.57 | 7.62 | 5.59 | 7.62 | 1.22 | 3.05 | 1.83 | 0.46 | 0.229 | 3.50 | 2.34 | | |
| 6.10 | 9.14 | 7.31 | 7.62 | 1.83 | 3.66 | 2.13 | 0.68 | 0.31 | 4.27 | 2.84 | | |
| 7.62 | 10.67 | 8.94 | 7.62 | 1.83 | 3.96 | 2.13 | 0.68 | 0.31 | 5.03 | 3.35 | | |
| 9.14 | 12.31 | 10.57 | 7.92 | 1.83 | 4.27 | 2.13 | 0.68 | 0.31 | 5.79 | 3.86 | | |
| 12.19 | 15.48 | 13.82 | 8.23 | 1.83 | 4.88 | 2.13 | 0.68 | 0.31 | 7.31 | 4.88 | | |
| 15.24 | 18.53 | 17.27 | 8.23 | 1.83 | 6.10 | 2.13 | 0.68 | 0.31 | 8.84 | 5.89 | | |

[a]Letters refer to dimensions in Figure F-8. All values in meters.

From Herschy (1985).

may be taken as the datum if it is below the level of zero discharge.

Stage can also be determined via a float–counterweight system (Figure F-11) and with instruments that directly measure water pressure (pressure transducers).

Whatever system is used, one should always establish by survey the elevation of the gage relative to some point whose elevation will not change, and periodically check that elevation to avoid errors due to disturbance of the staff gage.

Continuous records of discharge are obtained by recording stage by means of a float or pressure sensor attached to an analog or digital recorder and using the rating curve to convert to discharge. Buchanan and Somers (1968) and Herschy (1999) described many approaches to measuring and recording stage.

### F.6.2 Measurement Location

To establish a useful stage–discharge relation, stage must be measured within a few stream widths of the discharge-measurement cross-section. Stage should be observed where it is sensitive to discharge variations (i.e., where $dZ_s/dQ$ is relatively large) and where it can be accurately measured (usually to within 0.01 ft or 0.003 m). These conditions are usually met in an area of quiet water not far upstream of a reach of accelerating flow and near the bank, where wave action is minimal.

If necessary, wave action can be virtually eliminated by measuring stage in a **stilling well.** Such a device can be constructed by surrounding the staff gage with a section of metal or plastic barrel into which holes have been punched, or simply with piled rocks. Construction of more elaborate stilling wells dug into stream banks and communicating to the stream via pipes is discussed by Buchanan and Somers (1968) and Herschy (1999).

### F.6.3 Stage–Discharge Relations at Natural Controls

When artificial controls are used the rating curve has a theoretical form [e.g., Equation (F-17)], although calibration of the relation by direct measurements may be required over at least some flow ranges. For natural controls, the rating curve is es-

**TABLE F-5**
Discharge Characteristics of Parshall Flumes

**TABLE F-5**
Discharge Characteristics of Parshall Flumes

| Throat width $b$ (m) | Discharge Range $(m^3\ s^{-1} \times 10^{-3})$ | | Equation $Q = Kh^u$ $(m^3\ s^{-1})$ |
|---|---|---|---|
| | Minimum | Maximum | |
| 0.025 | 0.09 | 5.4 | $0.0604h^{1.55}$ |
| 0.051 | 0.18 | 13.2 | $0.1207h^{1.55}$ |
| 0.076 | 0.77 | 32.1 | $0.1771h^{1.55}$ |
| 0.152 | 1.50 | 111 | $0.3812h^{1.58}$ |
| 0.229 | 2.50 | 251 | $0.5354h^{1.53}$ |
| 0.305 | 3.32 | 457 | $0.6909h^{1.522}$ |
| 0.457 | 4.80 | 695 | $1.056h^{1.538}$ |
| 0.610 | 12.1 | 937 | $1.428h^{1.550}$ |
| 0.914 | 17.6 | 1427 | $2.184h^{1.566}$ |
| 1.219 | 35.8 | 1923 | $2.953h^{1.578}$ |
| 1.524 | 44.1 | 2424 | $3.732h^{1.587}$ |
| 1.829 | 74.1 | 2929 | $4.519h^{1.595}$ |
| 2.134 | 85.8 | 3438 | $5.312h^{1.601}$ |
| 2.438 | 97.2 | 3949 | $6.112h^{1.607}$ |
| 3.048 | 0.16[a] | 8.28[a] | $7.463h^{1.60}$ |
| 3.658 | 0.19[a] | 14.68[a] | $8.859h^{1.60}$ |
| 4.572 | 0.23[a] | 25.04[a] | $10.96h^{1.60}$ |
| 6.096 | 0.31[a] | 37.97[a] | $14.45h^{1.60}$ |
| 7.620 | 0.38[a] | 47.14[a] | $17.94h^{1.60}$ |
| 9.144 | 0.46[a] | 56.33[a] | $21.44h^{1.60}$ |
| 12.192 | 0.60[a] | 74.70[a] | $28.43h^{1.60}$ |
| 15.240 | 0.75[a] | 93.04[a] | $35.41h^{1.60}$ |

[a]Values in $m^3\ s^{-1}$.

From Herschy (1985)

tablished empirically by concurrent direct measurement of discharge and observation of stage over a range of discharges.

At any cross-section, there is usually a relation between discharge, $Q$, and stage, $Z_s$, of the form

$$Q = a \cdot Z_s^b, \qquad \text{(F-18)}$$

where $a$ and $b$ are determined by the configuration of the stream reach, the exact location at which stage is measured, and the datum for $Z_s$. The datum for stage measurement, $Z_0$, [Equation (F-2)] is simply a convenient elevation below the elevation corresponding to zero discharge. The values of $a$ and $b$ in Equation (F-18) typically vary for different ranges of $Z_s$ in natural streams, and the rating curve must be empirically determined by measurement of $Z_s$ and $Q$ over the range of interest.

Ideally, stage should be measured where (1) hysteresis in the stage–discharge relation is minimal and (2) the relation will not change with time due to erosion or deposition in the measurement reach. However, these conditions may be difficult to find

in many streams. Figure F-12 shows a rating curve where shifts have occurred; this demonstrates that one must continually check the rating curve by frequent discharge measurements.

## F.7   SLOPE-AREA MEASUREMENTS

It is often possible to estimate the discharge of a recent flood peak from observations of high-water marks in a reach. The procedure involves surveying the slope of the water surface as revealed by the high-water marks and the cross-sectional area beneath the marks, and hence is known as the **slope-area method.**

### F.7.1   Standard Method

In the standard slope-area method, an estimated channel-resistance value is used along with the surveyed cross-sectional dimensions in the equation for uniform flow [Equation (9-18) or (9-19)] to compute the discharge. Dalrymple and Benson (1967) described the field measurements needed to apply the method and noted that the selection of a suitable reach (distance between upstream and downstream cross-sections) is the most important element of a slope-area measurement. The presence of unequivocal high-water marks is essential, and the stream reach should be relatively straight and uniform without pronounced constrictions, protrusions, or free overfalls. The accuracy of the method generally increases with reach length, and the reach should meet one or more of the following criteria: (1) reach length at least 75 times flow depth, (2) vertical water-surface fall through the reach greater than 0.15 m, and (3) water-surface fall greater than computed velocity heads.

In the following steps, we assume that we have information on two cross-sections; the subscript $u$ designates the upstream cross-section, and $d$ the downstream cross-section. These steps follow the approach described by Ponce (1989). Dalrymple and Benson (1967) also gave equations for applying the method using more than two cross-sections.

1. Survey the upstream and downstream cross-sections below the high-water marks and determine the cross-sectional areas ($A_u$, $A_d$) and hydraulic radii ($R_u$, $R_d$) for each,

**FIGURE F-9**
Configuration of a 3-inch modified Parshall flume. Its maximum discharge capacity is about 0.5 ft³ s⁻¹ (0.014 m³ s⁻¹ = 14 L s⁻¹). From Buchanan and Somers (1968).

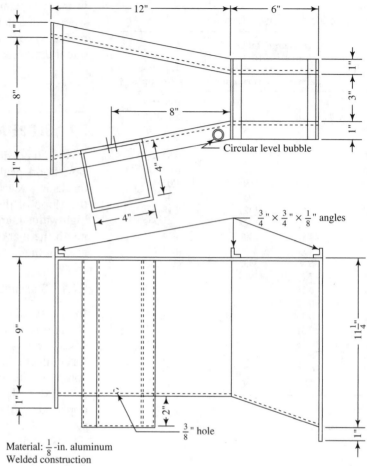

Material: $\frac{1}{8}$-in. aluminum
Welded construction
Note: This stilling well can accommodate
 a 3-in. float and be used with a recorder
 if continuous measurement is desired for a
 period.

**FIGURE F-10**
Staff gage installed on a small stream. Photo by author.

the distance between the sections ($L$), and the elevation difference between the high-water marks at the two sections ($\Delta z$).

2. Estimate the Manning resistance factor ($n$) for the reach from Table 9-6 or by comparison with photographs (Barnes 1967; Hicks and Mason 1991).

3. Compute the section **conveyances,** $K_u$ and $K_d$, as

$$K_u \equiv \frac{u_m \cdot R_u^{2/3} \cdot A_u}{n}, \qquad \textbf{(F-19a)}$$

and

$$K_d \equiv \frac{u_m \cdot R_d^{2/3} \cdot A_d}{n}, \qquad \textbf{(F-19b)}$$

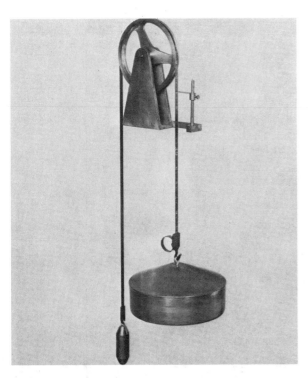

**FIGURE F-11**
Simple float–counterweight system for measuring water-surface elevation. The pulley can be attached to a recorder to obtain a continuous record. From Buchanan and Somers (1968).

where $u_m$ is the unit-conversion factor for the Manning equation (Table 9-6).

   **4.** Compute the average reach conveyance, $K$, as the geometric mean of the section conveyances:

$$K = (K_u \cdot K_d)^{1/2}. \qquad \text{(F-20)}$$

An iterative procedure is now used to compute $Q$. Successive iterations are denoted in the following steps by subscripts $i = 1, 2, ...$ Usually no more than five iterations are required for convergence to an effectively constant value.

   **5.** Compute the first estimate of the **friction slope,** $S_1$, as

$$S_1 = \frac{\Delta z}{L}. \qquad \text{(F-21)}$$

   **6.** Compute the first estimate of the discharge, $Q_1$, as

$$Q_1 = K \cdot S_1^{1/2}. \qquad \text{(F-22)}$$

   **7.** Estimate the average velocities ($U_{ui}, U_{di}$) at the two sections as

$$U_{ui} = \frac{Q_1}{A_u} \qquad \text{(F-23a)}$$

and

$$U_{di} = \frac{Q_1}{A_d}. \qquad \text{(F-23b)}$$

   **8.** Estimate the **velocity heads** ($h_{ui}, h_{di}$) at the two sections as

$$h_{ui} = \frac{U_{ui}^2}{2 \cdot g} \qquad \text{(F-24a)}$$

and

$$h_{di} = \frac{U_{di}^2}{2 \cdot g}. \qquad \text{(F-24b)}$$

where $g$ is gravitational acceleration.

   **9.** Compute a new estimate of the friction slope, $S_{i+1}$, as

$$S_{i+1} = \frac{\Delta z + k \cdot (h_{ui} - h_{di})}{L}, \qquad \text{(F-25)}$$

where $k$ is an **eddy coefficient,** empirically found to be given by

$$k = 0.5 \text{ if } A_d > A_u \qquad \text{(F-26a)}$$

and

$$k = 1 \text{ if } A_d \leq A_u. \qquad \text{(F-26b)}$$

   **10.** Compute a new estimate of the discharge, $Q_{i+1}$, as

$$Q_{i+1} = K \cdot S_{i+1}^{1/2}. \qquad \text{(F-27)}$$

   **11.** Compare $Q_{i+1}$ with $Q_i$ and repeat steps 7–11 until successive discharge estimates do not change.

Herschy (1985) indicated that the usual range of error with the slope-area method is 10 to 20%, and Kirby (1987) gave a complete analysis of the errors involved in slope-area estimates. Accuracy is improved by careful selection of the measurement reach and by using more cross-sections, but is inher-

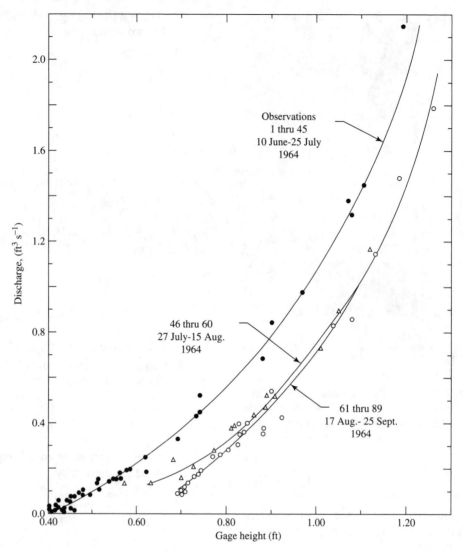

**FIGURE F-12**
Rating curve for a natural-cross section in a small stream in Alaska. Note that the relation has shifted with time due to siltation of the channel. From Dingman (1970).

ently limited by the need to estimate the channel resistance ($n$) and the fact that the flow is seldom truly uniform.

### F.7.2   Simplified Method

Based on statistical analysis of over 600 measured flows in natural channels, Dingman and Sharma (1997) developed the following simplified equation that eliminates the need to estimate channel resistance in slope-area calculations:

$$Q = 1.564 \cdot A^{1.173} \cdot Y^{0.400} \cdot S^{-0.056 \cdot \log(S)}. \quad \textbf{(F-28)}$$

In this equation, $Q$ is in m³ s⁻¹, $A$ is the average cross-sectional area for the reach in m², $Y$ is average depth in m, and $S$ is the water-surface slope tangent.

Equation (F-28) gave good estimates of discharge for in-bank flows over a wide range of channel sizes ($0.41 \text{ m}^2 \leq A \leq 8520 \text{ m}^2$) and slopes ($0.00001 \leq S \leq 0.0418$), but tended to over-estimate for flows in which both $Q < 3 \text{ m}^3 \text{ s}^{-1}$ and $Fr < 0.2$, where $Fr$ is the **Froude number,** defined as

$$Fr \equiv \frac{U}{(g \cdot Y)^{1/2}}. \quad \textbf{(F-29)}$$

## EXAMPLE F-1

The flood of 22 March 1948 on Esopus Creek at Coldbrook, NY, left high-water marks. Cross-sections were surveyed at two sections separated by a distance of $L = 78.7$ m. The difference in elevation of the high-water marks between the two stations was $\Delta Z = 0.35$ m; other survey data are summarized in the table. The estimated Manning's $n$ for the reach is $n = 0.043$. Determine the peak discharge of the flood.

| Section | Area (m²) | Width (m) | Hydraulic Radius (m) |
|---------|-----------|-----------|----------------------|
| $u$     | 135.7     | 55.2      | 2.43                 |
| $d$     | 136.6     | 53.4      | 2.52                 |

**Solution (Standard Method)** Information for Steps 1 and 2 is given. Subsequent steps are as follows:

3. Compute section conveyances [Equation (F-19)]:

$$K_u = \frac{1.00 \times (2.43 \text{ m})^{2/3} \times (135.7 \text{ m}^2)}{0.043} = 5704.2 \text{ m}^3 \text{ s}^{-1}$$

and

$$K_d = \frac{1.00 \times (2.52 \text{ m})^{2/3} \times (136.6 \text{ m}^2)}{0.043} = 5891.2 \text{ m}^3 \text{ s}^{-1}.$$

4. Compute average conveyance [Equation (F-20)]:

$$K = [(5704.2 \text{ m}^3 \text{s}^{-1}) \times (5891.2 \text{ m}^3 \text{s}^{-1})]^{1/2} = 5797.0 \text{ m}^3 \text{s}^{-1}.$$

5. Compute $S_1$ [Equation (F-21)]:

$$S_1 = \frac{0.35 \text{ m}}{78.7 \text{ m}} = 0.00446.$$

6. Compute $Q_1$ [Equation (F-22)]:

$$Q_1 = (5797.0 \text{ m}^3 \text{ s}^{-1}) \times (0.00446)^{1/2} = 387.0 \text{ m}^3 \text{ s}^{-1}.$$

7. Compute $U_{u1}$ and $U_{u2}$ [Equation (F-23)]:

$$U_{u1} = \frac{387.0 \text{ m}^3 \text{ s}^{-1}}{135.7 \text{ m}^2} = 2.85 \text{ m s}^{-1}$$

and

$$U_{d1} = \frac{387.0 \text{ m}^3 \text{ s}^{-1}}{136.6 \text{ m}^2} = 2.83 \text{ m s}^{-1}$$

8. Compute $h_{u1}$ and $h_{d1}$ [Equation (F-24)]:

$$h_{u1} = \frac{(2.85 \text{ m s}^{-1})^2}{2 \times 9.81 \text{ m s}^{-2}} = 0.415 \text{ m}.$$

and

$$h_{u1} = \frac{(2.83 \text{ m s}^{-1})^2}{2 \times 9.81 \text{ m s}^{-2}} = 0.409 \text{ m}.$$

9. Compute $S_2$ [Equation (F-25)]:
Since $A_d > A_u$, $k = 0.5$ [Equation (F-24)] and

$$S_2 = \frac{(0.35 \text{ m}) + 0.5 \times (0.415 \text{ m} - 0.409 \text{ m})}{78.7 \text{ m}}$$
$$= 0.00449.$$

10. Compute $Q_2$ [Equation (F-27)]:

$$Q_2 = (5797.0 \text{ m}^3 \text{ s}^{-1}) \times (0.00449)^{1/2} = 394.8 \text{ m}^3 \text{ s}^{-1}.$$

11. $Q_2$ is larger than $Q_1$. Using $Q_2$, we compute new estimates of the velocities, velocity heads, and friction slope and find $Q_3 = 394.9$ m³ s⁻¹. The difference between $Q_2$ and $Q_3$ is well within the uncertainty of the method (recall that $n$ is an estimated value), so we can end the computation and conclude that the peak flood discharge was 395 m³ s⁻¹.

**Solution (Simplified Method):** The average cross-sectional area for the reach is $A = (135.7 \text{ m}^2 + 136.6 \text{ m}^2)/2 = 136.2$ m²; average hydraulic radius, $R$, is 2.48 m; and the water-surface slope $S = 0.00446$ as found in Step 5. Entering these values into Equation (F-28) yields

$$Q = 1.564 \times 136.2^{1.173} \times 2.48^{0.400} \times 0.00446^{-0.056 \cdot \log(0.00446)}$$
$$= 360 \text{ m}^3 \text{ s}^{-1}.$$

For this example the standard and simplified estimates differ by less than 10%, within the usual uncertainty range for slope-area estimates.

# References

Abdul, A.S., and R.W. Gillham (1984). Laboratory studies of the effects of the capillary fringe on streamflow generation. *Water Resources Research* 20:691–698.

Abdul, A.S., and R.W. Gillham (1989). Field studies of the effects of the capillary fringe on streamflow generation. *Journal of Hydrology* 112:1–18.

Abrahams, A.D., A.J. Parsons, and S.-H. Luk (1986). Resistance to overland flow on desert hillslopes. *Journal of Hydrology* 88:343–363.

Adams, F.D. (1938). *The Birth and Development of the Geological Sciences.* Baltimore, MD: Williams and Wilkins.

Adams, W.P. (1976). Areal differentiation of snow cover in east central Ontario. *Water Resources Research* 12:1226–1234.

Allen, R.G., M.E. Jensen, J.L. Wright, and R.D. Burman (1989). Operational estimates of reference evapotranspiration. *Agronomy Journal* 81:650–662.

Alley, W.M. (1984). On the treatment of evapotranspiration, soil moisture accounting, and aquifer recharge in monthly water balance models. *Water Resources Research* 20:1137–1149.

Allison, G.B., and C.J. Barnes. (1983). Estimation of evaporation from non-vegetated surfaces using natural deuterium. *Nature* 301:143–145.

Allison, G.B., W.J. Stone, and M.W. Hughes. (1984). Recharge in karst and dune elements of a semi-arid landscape as indicated by natural isotopes and chloride. *Journal of Hydrology* 76:1–25.

Amarasekera, K.N., R.F. Lee, E.R. Williams, and E.A.B. Eltahir (1997). ENSO and the natural variability in the flow of tropical rivers. *Journal of Hydrology* 200:24–39.

Ambroise, B., K. Beven, and J. Freer (1996). Toward a generalization of the TOPMODEL concepts: Topographic indices of hydrological similarity. *Water Resources Research* 32:2135–2145.

American Meteorological Society (1992). Policy statement: Planned and inadvertent weather modification. *Bulletin of the American Meteorological Society* 73:331–337.

Amorocho, J., and W.E. Hart (1965). The use of laboratory catchments in the study of hydrologic systems. *Journal of Hydrology* 3:106–123.

Anderson, E.A. (1968) Development and testing of snow pack energy balance equations. *Water Resources Research* 4:19–37.

Anderson, E.A. (1973). National Weather Service River Forecast System - Snow Accumulation and Ablation Model. Silver Spring, MD: U.S. National Oceanic and Atmospheric Administration *NWS Technical Memorandum* HYDRO-17.

Anderson, E.A. (1976). A point energy and mass balance model of a snow cover. Silver Spring, MD: U.S. National Oceanic and Atmospheric Administration *NOAA Technical Report* NWS-19.

Anderson, E.A., H.J. Greenan, R.Z. Whipkey, and C.T. Machell. (1977). *NOAA - ARS Cooperative Snow Research Project Watershed Hydro-Climatology and Data for Water Years 1960–1974.* : U.S. National Oceanic and Atmospheric Administration and U.S. Agricultural Research Service.

Anderson, M.G., and T.P. Burt (1977). A laboratory model to investigate the soil moisture conditions on a draining slope. *Journal of Hydrology* 33:383–390.

Anderson, M.G., and T.P. Burt (1978). Toward more detailed field monitoring of variable source areas. *Water Resources Research* 14:1123–1131.

Anderson, M.G., and T.P. Burt (1980). Interpretation of recession flow. *Journal of Hydrology* 46:89–101.

Anderson, M.P., and X. Cheng (1993). Long- and short-term transience in a groundwater/lake system in Wisconsin, USA. *Journal of Hydrology* 145:1–18.

Anderson, M.P., D.S. Ward, E.G. Lappala, and T.A. Prickett (1992). Computer models for subsurface water. Chapter 22 in Maidment (1992).

Arcement, Jr., G.J., and V.R. Schneider (1989). Guide for selecting Manning's roughness coefficients in natural channels and flood plains. U.S. Geological Survey *Water-Supply Paper* 2339, 38 p.

Arkin, P., and P. Ardanuy (1989). Estimating climatic-scale precipitation from space: A review. *Journal of Climate* 2:1229–1238.

Arnold, J.G., and P.M. Allen (1999). Automated methods for estimating baseflow and ground water recharge from streamflow records. *Journal of the American Water Resources Association* 35:411–424.

Askew, A.J. (1970). Derivation of formulae for variable lag time. *Journal of Hydrology* 10:225–242.

Assouline, S., and Y. Mualem (1997). Modeling the dynamics of seal formation and its effect on infiltration as related to soil and rainfall characteristics. *Water Resources Research* 33:1527–1536.

Atkinson, T.C. (1978). Techniques for measuring subsurface flow on hillslopes. In Kirkby (1978):73–120.

Auer, A.H., Jr. (1974). The rain versus snow threshold temperature. *Weatherwise* 27:67.

Ball, J.E. (1994). The influence of storm temporal patterns on catchment response. *Journal of Hydrology* 158:285–303.

Barnes, H.B. (1967). Roughness characteristics of natural channels. U.S. Geological Survey *Water-Supply Paper* 1849, 213 p.

Barnett, T.P., L. Dumenii, U. Schlese, and E. Roeckner (1988). The effect of Eurasian snow cover on global climate. *Science* 239:504–507.

Barney, G.O., ed. (1980) *The Global 2000 Report to the President of the U.S.* New York, NY: Pergamon Press.

Barrett, J.P. and L. Goldsmith (1976). When is n large enough? *The American Statistician* 30:67–71.

Barron, E.J. (1989). Severe storms during earth history. *Geological Society of America Bulletin* 101:601–612.

Barros, A.P., and D.P. Lettenmaier (1994). Dynamic modeling of orographically induced precipitation. *Reviews of Geophysics* 32:265–284.

Barry, R.G., and R.J. Chorley (1982). *Atmosphere, Weather, and Climate.* New York, NY: Methuen & Co.

Barton, M. (1974). New concepts in snow surveying to meet expanding needs. *Advanced Concepts Technical Study on Snow and Ice Resources.* Washington, DC: National Academy of Sciences.

Bates, R.E., and M.A. Bilello (1966). Defining the cold regions of the Northern Hemisphere. Hanover, NH: U.S. Army Cold Regions Research and Engineering Laboratory *Technical Report* 178.

Baumgartner, A., and E. Reichel (1975). *The World Water Balance.* New York, NY: Elsevier Scientific Publishing Co.

Baver, L.D., W.H. Gardner, and W.R. Gardner (1972). *Soil Physics* (4th ed.) New York, John Wiley & Sons. 498 pp.

Bazemore, D.E., K.N. Eshleman, and K.J. Hollenbeck (1994). The role of soil water in stormflow generation in a forested headwater catchment: synthesis of natural tracer and hydrometric evidence. *Journal of Hydrology* 162:47–75.

Bear, J., and A. Verruijt (1987). *Modeling Groundwater Flow and Pollution.* Dordrecht, Netherlands: D. Reidel.

Bell, J.P., T.J. Dean, and M.G. Hodnett (1987). Soil moisture measurement by an improved capacitance technique, Part II. Field techniques, evaluation, and calibration. *Journal of Hydrology* 93:79–90.

Belt, C.B., Jr. (1975). The 1973 flood and man's constriction of the Mississippi River. *Science* 189:681–684.

Bengtsson, L. (1982). Percolation of meltwater through a snowpack. *Cold Regions Science and Technology* 6:73–81.

Bengtsson, L. (1988). Movement of meltwater in small basins. *Nordic Hydrology* 19:237–244.

Benton, G.S., R.T. Blackburn, and V.O. Snead (1950). The role of the atmosphere in the hydrologic cycle. *American Geophysical Union Transactions* 31:61–73.

Beran, M.A., and J.V. Sutcliffe (1972). An index of flood-producing rainfall based on rainfall and soil-moisture deficit. *Journal of Hydrology* 17:229–236.

Bergstrom, S., and L.P. Graham (1998). On the scale problem in hydrologic modeling. *Journal of Hydrology* 211:253–265.

Bergstrom, S., and M. Brandt (1985). Measurement of areal water equivalent of snow by natural gamma radiation - experiences from northern Sweden. *Hydrological Sciences Journal* 30:465.

Berkowicz, R., and L.P. Prahm (1982). Sensible heat flux estimated from routine meteorological data by the resistance method. *Journal of Applied Meteorology* 21:1845–1864.

Berndt, H.W., and W.B. Fowler (1969). Rime and hoarfrost in upper-slope forests of eastern Washington. *Journal of Forestry* 67: 92–95.

Berndtsson, R. (1987). Application of infiltration equations to a catchment with large spatial variability in infiltration. *Hydrological Sciences Journal* 32:399–413.

Berndtsson, R., and M. Larson (1987). Spatial variability of infiltration in a semi-arid environment. *Journal of Hydrology* 90:117–133.

Berner, E.K., and R. A. Berner (1987). *The Global Water Cycle*. Englewood Cliffs, NJ: Prentice-Hall, Inc.

Berry, M.O. (1981). Snow and climate. Chap. 2 in Gray and Male (1981).

Bethlahmy, N. (1976). The two-axis method: A new method to calculate average precipitation over a basin. *Hydrological Sciences Bulletin* 21:379–385.

Betson, R.P. (1964). What is watershed runoff? *Journal of Geophysical Research* 69:1541–1552.

Beven, K. (1981). Kinematic subsurface stormflow. *Water Resources Research* 17:1419–1424.

Beven, K. (1982a). On subsurface stormflow: Predictions with simple kinematic theory for saturated and unsaturated flows. *Water Resources Research* 18:1627–1633.

Beven, K. (1982b). On subsurface stormflow: An analysis of response times. *Hydrological Sciences Journal* 4:505–521.

Beven, K. (1983). Surface water hydrology - Runoff generation and basin structure. *Reviews of Geophysics* 21:721–730.

Beven, K. (1984). Infiltration into a class of vertically non-uniform soils. *Hydrological Sciences Journal* 29:425–434.

Beven, K. (1989). Changing ideas in hydrology - the case of physically-based models. *Journal of Hydrology* 105:157–172.

Beven, K., and R.T. Clarke (1986). On the variation of infiltration into a homogeneous soil matrix containing a population of macropores. *Water Resources Research* 22:383–388.

Beven, K., and P. Germann (1982). Macropores and water flow in soils. *Water Resources Research* 18:1311–1325.

Beven, K., and M.J. Kirkby (1979). A physically-based variable contributing-area model of catchment hydrology. *Hydrological Science Bulletin* 24(1):43–69.

Beven, K., and E.F. Wood (1993). Flow routing and the hydrological response of channel networks. Chapter 5 in K. Beven and M.J. Kirkby, eds. *Channel Network Hydrology*. New York: John Wiley and Sons.

Bilello, M.A. (1984). Regional and seasonal variations in snow-cover density in the U.S.S.R. Hanover, NH: U.S. Army Cold Regions Research and Engineering Laboratory *Special Report* 84–22.

Bissell, V.C, and E.L. Peck (1973). Monitoring snow water equivalent by using natural soil radioactivity. *Water Resources Research* 9:885–890.

Biswas, A.K. (1970). *History of Hydrology*. Amsterdam: North-Holland Publishing Co.

Biswas, T.D., D.R. Nielsen, and J.W. Biggar (1966). Redistribution of soil water after infiltration. *Water Resources Research* 2:513–524.

Blöschl, G., D. Gutknecht, and R. Kirnbauer (1991b). Distributed snowmelt simulations in an Alpine catchment: 2. Parameter study and model predictions. *Water Resources Research* 27:3181–3188.

Blöschl, G., R. Kirnbauer, and D. Gutknecht (1991a). Distributed snowmelt simulations in an Alpine catchment: 1. Model evaluation on the basis of snow cover patterns. *Water Resources Research* 27:3171–3179.

Black, P.E. (1975). Runoff from watershed models. *Water Resources Research* 6:465–477.

Black, P.E., and J.W. Cronn, Jr. (1975). Hydrograph responses to watershed model size and similitude relations. *Journal of Hydrology* 26:255–266.

Black, T.A., W.R. Gardner, and C.B. Tanner (1970). Water storage and drainage under a row crop on a sandy soil. *Agronomy Journal* 62:46–51.

Bland, W.L., P.A. Helmke, and J.M. Baker. (1997). High-resolution snow-water equivalent measurement by gamma-ray spectroscopy. *Agricultural and Forest Meteorology* 83:27–36.

Bolsenga, S.J. (1964). Daily sums of global radiation for cloudless skies. Hanover, NH: U.S. Army Cold Regions Research and Engineering Laboratory *Research Report* 160.

Bonell, M. (1993). Progress in the understanding of runoff generation dynamics in forests. *Journal of Hydrology* 150:217–275.

Bonnor, G.M. (1975). The error of area estimates from dot grids. *Canadian Journal of Forestry Research* 5:10–17.

Boonstra, J., and M.N. Bhutta (1996). Groundwater recharge in irrigated agriculture: the theory and practice of inverse modeling. *Journal of Hydrology* 174:357–374.

Bosch, J.M., and J.D. Hewlett (1982). A review of catchment experiments to determine the effect of vegetation changes on water yield and evapotranspiration. *Journal of Hydrology* 55:3–23.

Bouchet, R.B. (1963). Evapotranspiration réelle et potentielle, signification climatique. Gentbrugge, Belgium: International Association of Scientific Hydrology *Publication* 62:134–142.

Boughton, W.C. (1989). A review of the USDA SCS curve number method. *Australian Journal of Soil Research* 27:511–523.

Boughton, W.C. (1994). An error in the SCS curve number method. Unpublished Technical Note, Griffith University, Brisbane, Australia.

Boussinesq, J. (1877). Essai sur la théorie des eaux courantes. *Memoires Acadèmie des Sciences Institute de France* 23:252–260.

Bouwer, H. (1986). Intake rate: Cylinder infiltrometer. In A. Klute, ed., *Methods of Soil Analysis; Part 1: Physical and Mineralogical Methods*. Madison, WI: Soil Science Society of America.

Bowen, I.S. (1926). The ratio of heat losses by conduction and by evaporation from any water surface. *Physical Review* 27:777–787.

Boyd, M.J. (1978). A storage-routing model relating drainage basin hydrology and geomorphology. *Water Resources Research* 14:921–928.

Bradley, R.S., H.F. Diaz, J.K. Eischeid, P.D. Jones, P.M. Kelly, and C.M. Goodess (1987). Precipitation fluctuations over northern hemisphere land areas since the mid-19th century. *Science* 237:171–176.

Brakensiek, D.L. (1977). Estimating the effective capillary pressure in the Green-and-Ampt infiltration equation. *Water Resources Research* 13:680–682.

Brakensiek, D.L., H.B. Osborn, and W.J. Rawls, eds. (1979). *Field Manual for Research in Agricultural Hydrology*. Science and Education Administration, U.S. Department of Agriculture.

Branson, F.A., G.F. Gifford, K.G. Renard, and R.F. Hadley (1981). *Rangeland Hydrology*. Dubuque, IA: Kendall/Hunt Publishing Co.

Bras, R.L. (1990). *Hydrology*. Reading, MA: Addison-Wesley Publishing Co.

Bras, R.L., and I. Rodriguez-Iturbe (1976a). Network design for the estimation of areal mean of rainfall events. *Water Resources Research* 12:1185–1195.

Bras, R.L., and I. Rodriguez-Iturbe (1976b). Rainfall network design for runoff prediction. *Water Resources Research* 12:1197–1208.

Bras, R.L., and I. Rodriguez-Iturbe (1984). *Random Functions in Hydrology*. Reading, MA: Addison-Wesley Publishing Co.

Bredehoeft, J.D., S.S. Papadopulos, and H.H. Cooper, Jr. (1982). Groundwater: The water-budget myth. In *The Scientific Basis of Water-Resource Management*. Washington, DC: National Academy Press.

Bristow, K. L. (1992). Prediction of daily mean vapor density from daily minimum air temperature. *Agricultural and Forest Meteorology* 59:309–317.

Bromley, J., W.M. Edmunds, E. Fellman, J. Brouwer, S.R. Gaze, J. Sudlow, and J.-D. Taupin (1997). Estimation of rainfall inputs and direct recharge to the deep unsaturated zone of southern Niger using the chloride profile method. *Journal of Hydrology* 188/189:139–154.

Brooks, C.F. (1940). Hurricanes into New England: Meteorology of the storm of September 21, 1938. Washington. DC: Smithsonian Institution *Report for 1939*:241–251.

Brooks, R.H., and A.T. Corey (1964). Hydraulic properties of porous media. Fort Collins, CO, *Colorado State University Hydrology Paper* 3.

Brown, J., S.L. Dingman, and R.J. Lewellen (1968). Hydrology of a drainage basin on the Alaskan Coastal Plain. U.S. Army Cold Regions Research and Engineering Lab., Hanover, NH, *Research Report* 240.

Brown, R.J.E, and T. L. Pewe (1973). Distribution of permafrost in North America and its relationship to the environment: A review, 1963–1973. In *Permafrost: North American Contribution to the Second International Conference*. Washington, DC: National Academy of Sciences.

Brown, T.C. (2000) Projecting U.S. freshwater withdrawals. *Water Resources Research* 36:769–780.

Brubaker, K.L., A. Rango, and W. Kustas (1996). Incorporating radiation inputs into the Snowmelt Runoff Model. *Hydrological Processes* 10:1329–1343.

Brubaker, K.L., D. Entekhabi, and P.S. Eagleson (1993). Estimation of continental precipitation recycling. *Journal of Climate* 6:1077–1089.

Bruce, J.P., and R.H. Clark (1966). *Introduction to Hydrometeorology*. Long Island City, NY: Pergamon Press.

Brutsaert, W. (1975). On a derivable formula for long-wave radiation from clear skies. *Water Resources Research* 11:742–744.

Brutsaert, W. (1982). *Evaporation into the Atmosphere*. Dordrecht, Holland: D. Reidel Publishing Company.

Brutsaert, W. (1988). The parameterization of regional evaporation - some directions and strategies. *Journal of Hydrology* 102:407–426.

Brutsaert, W. (1994). The unit response of groundwater outflow from a hillslope. *Water Resources Research* 30:2759–2763.

Brutsaert, W., and J.P. Lopez (1998). Basin-scale geohydrologic drought flow features of riparian aquifers in the southern Great Plains. *Water Resources Research* 34:233–240.

Brutsaert, W., and J.L. Nieber (1977). Regionalized drought flow hydrographs from a mature glaciated plateau. *Water Resources Research* 13:637–643.

Brutsaert, W., and H. Stricker (1979). An advection-aridity approach to estimate actual regional evapotranspiration. *Water Resources Research* 15:443–450.

Buchanan, T.J., and W.P. Somers (1968). Stage measurement at gaging stations. U.S. Geological Survey *Techniques of Water-Resources Investigations* Book 3 Chapter A7.

Buchanan, T.J., and W.P. Somers (1969). Discharge measurement at gaging stations. U.S. Geological Survey *Techniques of Water-Resources Investigations* Book 3 Chapter A8.

Budyko, M.I. (1974). Climate and Life. *International Geophysics Series*, v. 18, San Diego, CA: Academic Press.

Burgy, R.H., and J.N. Luthin (1956). A test of the single- and double-ring types of infiltrometers. *American Geophysical Union Transactions* 37:189–191.

Busby, M.W. (1963). Yearly variations in runoff for the conterminous United States, 1931–1960. Washington DC: U.S. Geological Survey *Water-Supply Paper 1669-S*.

Busenberg, E., and L.N. Plummer (1992). Use of chlorofluorocarbons ($CCl_3F$ and $CCl_2F_2$) as hydrologic tracers and age-dating tools: The alluvium and terrace system of central Oklahoma. *Water Resources Research* 28:2257–2283.

Buttle, J.M., and D.A. House (1997). Spatial variability of saturated hydraulic conductivity in shallow macroporous soils in a forested basin. *Journal of Hydrology* 203:127–142.

Buttle, J.M., and J.J. McDonnell (1987). Modeling the areal depletion of snowcover in a forested catchment. *Journal of Hydrology* 90:43–60.

Caine, N. (1975). An elevational control of peak snowpack variability. *Water Resources Research* 11:613–621.

Caissie, D., T.L. Pollock, and R.A. Cunjak (1996). Variation in stream water chemistry and hydrograph separation in a small drainage basin. *Journal of Hydrology* 178:137–157.

Calder, I.R. (1977). A model of transpiration and interception loss from a spruce forest in Plynlimon, central Wales. *Journal of Hydrology* 33:247–265.

Calder, I.R. (1978). Transpiration observations from a spruce forest and comparisons with predictions from an evaporation model. *Journal of Hydrology* 38:33–47.

Calder, I. R. (1992). Hydrologic effects of land-use change. Chapter 13 in D.R. Maidment, ed., *Handbook of Hydrology*, New York, McGraw-Hill.

Calder, I.R., and P.T.W. Rosier (1976). The design of large plastic-sheet net-rainfall gauges. *Journal of Hydrology* 30:403–405.

Callis, L.B., and M. Natarajan (1986). Ozone and nitrogen dioxide changes in the stratosphere during 1979–84. *Nature* 323:772–775.

Calvo, J.C. (1986). An evaluation of Thornthwaite's water balance technique in predicting stream runoff in Costa Rica. *Hydrological Sciences Journal* 31:51–60.

Campbell, G.S. (1974) A simple method for determining unsaturated conductivity from moisture retention data. *Soil Science* 117:311–314.

Carlson, R.E., J.W. Enz, and D.G. Baker (1994). Quality and variability of long term climate data relative to agriculture. *Agricultural and Forest Meteorology* 69:61–74.

Carlson, T. (1991). Modeling stomatal resistance:an overview of the 1989 workshop at the Pennsylvania State University. *Agricultural and Forest Meteorology* 54:103–106.

Carroll, S.S, and T.R. Carroll (1989). Effect of uneven snow cover on airborne snow water equivalent estimates obtained by measuring terrestrial gamma radiation. *Water Resources Research* 25:1505–1510.

Carroll, S.S., and N. Cressie (1996). A comparison of geostatistical methodologies used to estimate snow water equivalent. *Water Resources Bulletin* 32: 267–278.

Carroll, T., and G.D. Voss (1984). Airborne snow-water equivalent over a forested environment using terrestrial gamma radiation. 41st Annual Meeting of the Eastern Snow Conference *Proceedings*.

Carter, R.W., and I.E. Anderson (1963). Accuracy of current meter measurements. American Society of Civil Engineers *Proceedings, Journal of the Hydraulics Division* 89:105–115.

Cedergren, H. (1989). *Seepage, Drainage, and Flow Nets*. New York, NY: John Wiley & Sons, Inc.

Cess, R.D. et al. (1995). Absorption of solar radiation by clouds. *Science* 267:496–499.

Cey, E.E., D.L. Rudolph, G.W. Parkin, and R. Aravena (1998). Quantifying groundwater discharge to a small perennial stream in southern Ontario, Canada. *Journal of Hydrology* 210:21–37.

Chang, M. (1981). A survey of the U.S. national precipitation network. *Water Resources Bulletin* 17:241–243.

Chang, M., and R. Lee (1974). Objective double-mass analysis. *Water Resources Research* 10:1123–1126.

Changnon, S.A., ed. (1981). METROMEX: A review and summary. *Meteorological Monographs* 18:(40).

Changnon, S.A. (1985). Research agenda for floods to solve a policy failure. American Society of Civil Engineers *Proceedings, Journal of Water Resources Planning and Management* 111:54–64.

Chapman, P.J., B. Reynolds, and H.S. Wheater (1993). Hydrochemical changes along stormflow pathways in a small moorland headwater catchment in Mid-Wales, UK. *Journal of Hydrology* 151:241–265.

Chapman, T. 1999. A comparison of algorithms for stream flow recession and baseflow separation. *Hydrological Processes* 13:701–714.

Chapman, W.L., and J.E. Walsh (1993). Recent variations of sea ice and air temperatures in high latitudes. *Bulletin of the American Meteorological Society* 74:33–47.

Chatfield, C. (1984). *Time Series Analysis: An Introduction* (3rd ed.) New York, NY: Chapman and Hall.

Chen, C.-L. (1982). Infiltration formulas by curve number procedure. American Society of Civil Engineers *Proceedings, Journal of the Hydraulics Division* 108:823–829.

Cheng, X., and M.P. Anderson (1994). Simulating the influence of lake position on groundwater fluxes. *Water Resources Research* 30:2041–2049.

Cherkauer, D.S., and D.C. Nader (1989). Distribution of groundwater seepage to large surface-water bodies: The effect of hydraulic heterogeneities. *Journal of Hydrology* 109:151–165.

Cherkauer, D.S., and J.P. Zager (1989). Groundwater interaction with a kettle-hole lake: Relation of observations to digital simulations. *Journal of Hydrology* 109:167–184.

Chery, D.L., Jr. (1967). A review of rainfall-runoff physical models as developed by dimensional analysis and other methods. *Water Resources Research* 3:881–889.

Chery, D.L., Jr. (1968). Output response to pulse inputs of a scaled laboratory watershed system. U.S. Department of Agriculture Southeast Watershed Research Center, Tucson, AZ, 45 p.

Chiew, F.H.S., and T.A. McMahon (1990). Estimating groundwater recharge using a surface watershed modeling approach. *Journal of Hydrology* 114:285–304.

Chiew, F.H.S., T.A. McMahon, and I.C. O'Neill. (1992). Estimating groundwater recharge using an integrated surface and groundwater modeling approach. *Journal of Hydrology* 131:151–186.

Chorley, R.J. (1978). The hillslope hydrologic cycle. In Kirkby (1978):1–42.

Chow, T.L. (1992). Performance of an ultrasonic level sensing system for automated monitoring of snowcover depth. *Agricultural and Forest Meteorology* 62:75–85.

Chow, T.L. (1994). Design and performance of a fully automated evaporation pan. *Agricultural and Forest Meteorology* 68:187–200.

Chow, V.T. (1959). *Open Channel Hydraulics*. New York, NY: McGraw-Hill.

Chow, V.T., ed. (1964). *Handbook of Applied Hydrology*. New York, NY: McGraw-Hill.

Chu, S.T. (1978). Infiltration during an unsteady rain. *Water Resources Research* 14:461–466.

Cicerone, R.J. (1987). Changes in stratospheric ozone. *Science* 237:35–42.

Claasen, H.C., and D.R. Halm (1996). Estimates of evapotranspiration or effective moisture in Rocky Mountain watersheds from chloride ion concentrations in stream baseflow. *Water Resources Research* 32:363–372.

Claassen, H.C., M.M. Reddy, and D.R. Halm (1986). Use of the chloride ion in determining hydrologic-basin water budgets - A 3-year case study in the San Juan Mountains, Colorado, U.S.A. *Journal of Hydrology* 85:49–71.

Clapp, R.B., and G.M. Hornberger (1978). Empirical equations for some soil hydraulic properties. *Water Resources Research* 14:601–604.

Climate Analysis Center (1989). *Weekly Climate Bulletin No. 89/28* (15 July 1989). National Weather Service.

Cline, D.W. (1997). Snow surface energy exchanges and snowmelt at a continental midlatitude Alpine site. *Water Resources Research* 33:689–701.

Cohen, P., O.L. Franke, and B.L. Foxworthy (1968). An atlas of Long Island's water resources. Albany, NY: New York Water Resources Commission *Bulletin 62*.

Colbeck, S.C. (1971). One-dimensional water flow through snow. Hanover, NH: U.S. Army Cold Regions Research and Engineering Laboratory *Research Report 296*.

Colbeck, S.C. (1978). The physical aspects of water flow through snow. *Advances in Hydroscience* 11:165–206.

Colonell, J.M., and G.R. Higgins (1973). Hydrologic response of Massachusetts watersheds. *Water Resources Bulletin* 9:793–800.

Cong, C., and E.A.B. Eltahir (1996). Sources of moisture for rainfall in West Africa. *Water Resources Research* 32:3115–3121.

Conway, H., and R. Benedict (1994). Infiltration of water into snow. *Water Resources Research* 30:641–649.

Cook, P.G., et al. (1998). Water balance of a tropical woodland system, Northern Australia: A combination of micro-meteorological, soil physical, and groundwater chemical approaches. *Journal of Hydrology* 210:161–177.

Cooper, J.D. (1980). Measurement of moisture fluxes in unsaturated soil in Thetford Forest. Wallingford, Oxon., U.K.: Natural Environment Research Council Institute of Hydrology *Report 66*.

Cosby, B.J., G.M. Hornberger, R.B. Clapp, and T.R. Ginn (1984). A statistical exploration of the relationships of soil moisture characteristics to the physical properties of soils. *Water Resources Research* 20:682–690.

Court, A., and M.T. Bare (1984). Basin precipitation estimates by Bethlahmy's two-axis method. *Journal of Hydrology* 68:149–158.

Cowan, W.I. (1956). Estimating hydraulic roughness coefficients. *Agricultural Engineering* 37:473–475.

Cowling, E.B. (1983). International aspects of acid deposition. in R. Herrmann and A.I. Johnson, eds., *Acid Rain - A Water Resources Issue for the 80's*. Bethesda, MD: American Water Resources Association.

Cox, L.M. (1971). Field performance of the universal surface precipitation gage. 39th Annual Meeting of the Western Snow Conference *Proceedings*.

Cox, L.M., et al. (1978). The care and feeding of snow pillows. 46th Annual Meeting of the Western Snow Conference *Proceedings*.

Crayosky, T.W., D.R. DeWalle, T.A. Seybert, and T.E. Johnson (1999). Channel precipitation dynamics in a forested Pennsylvania headwater catchment (USA). *Hydrological Processes* 13:1303–1314.

Creutin, J.D., and C. Obled (1982). Objective analyses and mapping techniques for rainfall fields: An objective comparison. *Water Resources Research* 18:413–431.

Croley, T.E., II (1989). Verifiable evaporation modeling on the Laurentian Great Lakes. *Water Resources Research* 25:781–792.

Croley, T.E., II (1992). Long-term heat storage in the Great Lakes. *Water Resources Research* 28:69–81.

Croley, T.E., II, and H.C. Hartmann (1985). Resolving Thiessen polygons. *Journal of Hydrology* 76:363–379.

Crowe, A.S., and F.W. Scwartz (1985). Application of a lake-watershed model for the determination of water balance. *Journal of Hydrology* 81:1–26.

Currie, D.J., and V. Paquin (1987). Large-scale biogeographical patterns of species richness of trees. *Nature* 329:326–327.

Custer, S.G., P. Farnes, J.P. Wilson, and R.D. Snyder (1996). A comparison of hand- and spline-drawn precipitation maps for mountainous Montana. *Water Resources Bulletin* 32:393–405.

Dai, A., A.D. Del Genio, and I.T. Fung (1997). Clouds, precipitation, and temperature range. *Nature* 386:665–666.

Dalrymple, T., and M.A. Benson (1967). Measurement of peak discharge by the slope-area method. U.S. Geological Survey *Techniques of Water-Resources Investigations* Book 3 Chapter A2.

Daluz Vieira, J.H. (1983). Conditions governing the use of approximations for the St. Venant equations for shallow surface water flow. *Journal of Hydrology* 60:43–58.

Daniel, J.F. (1976). Estimating groundwater evapotranspiration from streamflow records. *Water Resources Research* 12:360–364.

Darling, W.G., and A.H. Bath (1988). A stable isotope study of recharge processes in the English Chalk. *Journal of Hydrology* 101:31–46.

Das Gupta, A., and G.N. Paudyal (1988). Estimating aquifer recharge and parameters from water level observations. *Journal of Hydrology* 99:103–116.

Davies, J.A., and C.D. Allen (1973). Equilibrium, potential, and actual evaporation from cropped surfaces in southern Ontario. *Journal of Applied Meteorology* 12:647–657.

Davis, K.S., and J.A. Day (1961). *Water: The Mirror of Science*. New York: Doubleday-Anchor.

Davis, R.T. (1973). Operational snow sensors. 30th Annual Meeting of the Eastern Snow Conference *Proceedings*.

Davis, S.N., and E. Murphy, eds. (1987). Dating ground water and the evaluation of repositories for radioactive waste. U.S. Nuclear Regulatory Commission NUREG/CR-4912.

Day, J.A., and G.L. Sternes (1970). *Climate and Weather*. Reading, MA: Addison-Wesley.

Dean, T.J., J.P. Bell, and A.J.B. Baty (1987). Soil moisture measurement by an improved capacitance technique, Part I. Sensor design and performance. *Journal of Hydrology* 93:67–78.

Dedkov, A.P., and V.I. Mozzherin (1984). *Eroziya i Stok Nanosov na Zemle*. Kazan', USSR: Izdatelstvo Kazanskogo Universiteta.

Deevey, E.E., Jr. (1970). Mineral cycles. In *The Biosphere*. San Francisco, CA: W.H. Freeman and Co.

Derecki, J.A. (1976). Multiple estimates of Lake Erie evaporation. *Journal of Great Lakes Research* 2:124–149.

Derecki, J.A. (1981). Operational estimates of Lake Superior evaporation based on IFYGL findings. *Water Resources Research* 17:1453–1462.

Devillez , F., and H. Laudelot (1986). Application d'une modèle hydrologique à une bassin versant forestière de Wallonie. *Annales Science Forestière* 43:475–504.

DeVries, J.J., and T.V. Hromadka (1992). Computer models for surface water. Chapter 21 in Maidment (1992).

Dey, B. and O.S.R.U. Bhanu Kumar (1983). Himalayan winter snow cover area and summer monsoon rainfall over India. *Journal of Geophysical Research* 88:5471–5474.

Dickinson, W.T., M.E. Holland, and G.L. Smith (1967). An experimental rainfall-runoff facility. Colorado State University *Hydrology Paper 25*, 81 p.

Dickson, R.R. (1984). Eurasian snow cover versus Indian monsoon rainfall - An extension of the Hahn-Shukla results. *Journal of Climatology and Applied Meteorology* 23:171–173.

Dingman, S.L. (1970). Hydrology of the Glenn Creek watershed, Tanana River drainage, central Alaska. Hanover, NH: U.S. Army Cold Regions Research and Engineering Laboratory *Research Report 297*.

Dingman, S.L. (1973). Effects of permafrost on streamflow characteristics in the discontinuous permafrost zone of central Alaska. In *Permafrost: North American Contribution to the Second International Conference*. Washington, DC: National Academy of Sciences.

Dingman, S.L. (1975). Hydrologic effects of frozen ground. Hanover, NH: U.S. Army Cold Regions Research and Engineering Laboratory *Special Report 218*.

Dingman, S.L. (1978a). Drainage density and streamflow: A closer look. *Water Resources Research* 14:1183–1187.

Dingman, S.L. (1978b). Synthesis of flow-duration curves for unregulated streams in

New Hampshire. *Water Resources Bulletin* 14:1481–1502.

Dingman, S.L. (1981). Elevation: A major influence on the hydrology of New Hampshire and Vermont, USA. *Hydrological Sciences Bulletin* 26:399–413.

Dingman, S.L. (1984). *Fluvial Hydrology*. New York, NY: W.H. Freeman and Co.

Dingman, S.L., and A. H. Johnson (1971). Pollution potential of some New Hampshire lakes. *Water Resources Research* 7:1208–1215.

Dingman, S.L., and S.C. Lawlor (1995). Estimation of low-flow quantiles in New Hampshire and Vermont. *Water Resources Bulletin* 31:243–256.

Dingman, S.L., and K.J. Palaia (1999). Comparison of models for estimating flood quantiles in New Hampshire and Vermont. *Journal of the American Water Resources Association* 35:1233–1243.

Dingman, S.L., D. Seely-Reynolds, and R.C. Reynolds III (1988). Application of kriging to estimating mean annual precipitation in a region of orographic influence. *Water Resources Bulletin* 24:329–339.

Dingman, S.L., and K.P. Sharma (1997). Statistical development and validation of discharge equations for natural channels. *Journal of Hydrology* 199:13–35.

Dingman, S.L., et al. (1971). Hydrologic reconnaissance of the Delta River and its drainage basin, Alaska. Hanover, NH: U.S. Army Cold Regions Research and Engineering Laboratory *Research Report* 262.

Dingman, S.L., et al. (1980). Climate, snow cover, microclimate, and hydrology. In J. Brown, P.C. Miller, L.L. Tieszen, and F.L. Bunnell, eds., *An Arctic Ecosystem: The Coastal Tundra at Barrow, Alaska*. Stroudsburg, PA: Dowden, Hutchinson, and Ross.

Diskin, M.H. (1970). On the computer evaluation of Thiessen weights. *Journal of Hydrology* 11:69–78.

Dolman, A.J., J.B. Stewart, and J.D. Cooper (1988). Predicting forest transpiration from climatological data. *Agricultural and Forest Meteorology* 42:337–353.

Donahue, R.L., R.W. Miller, and J.C. Shickluna (1983). *Soils: An Introduction to Soils and Plant Growth* (5th ed.). Englewood Cliffs, NJ: Prentice-Hall, Inc.

Donald, J.R., E.D. Soulis, N. Kouwen, and A. Pietrino (1995). A land cover-based snow cover representation for distributed hydrologic models. *Water Resources Research* 31:995–1009.

Dooge, J. (1975). The water balance of bogs and fens. In *Hydrology of Marsh-Ridden Areas; Proceedings of Minsk Symposium 1972*. Paris, France: UNESCO Press.

Dooge, J. (1986). Looking for hydrologic laws. *Water Resources Research* 22:46S-58S.

Dorsey, N.E. (1940). *Properties of Ordinary Water Substance in All Its Phases: Water, Water-Vapor and All the Ices*. New York, NY: Reinhold.

Doss, P.K. (1993). The nature of a dynamic water table in a system of non-tidal, freshwater coastal wetlands. *Journal of Hydrology* 141:107–126.

Dovers, S.R., and J.W. Handmer (1995). Ignorance, the precautionary principle, and sustainability. *Ambio* 24(2):92–97.

Dowdeswell, J.A., et al. (1998). The mass balance of circum-Arctic glaciers and recent climate change. *Quaternary Research* 48:1–14.

Dracup, J.A., and E. Kahya (1994). The relationships between U.S. streamflow and La Niña events. *Water Resources Research* 30:2133–2141.

Dracup, J.A., K.S. Lee, and E.G. Paulson, Jr. (1980). On the definition of droughts. *Water Resources Research* 16:297–302.

Draper, N.R., and H. Smith (1981). *Applied Regression Analysis* (2nd ed.). New York, NY: Wiley, 709 pp.

Dreibelbis, F.R. (1962). Some aspects of watershed hydrology as determined from soil moisture data. *Journal of Geophysical Research* 67:3425–3435.

Drever, J.I. (1982). *The Geochemistry of Natural Waters*. Englewood Cliffs, NJ: Prentice-Hall, Inc.

Dunkle, S.A., et al. (1993). Chlorofluorocarbons ($CCl_3F$ and $CCl_2F_2$) as dating tools and hydrologic tracers in shallow groundwater of the Delmarva Peninsula, Atlantic Coastal Plain, United States. *Water Resources Research* 29:3837–3860.

Dunne, T. (1978). Field studies of hillslope flow processes. In Kirkby (1978):227–293.

Dunne, T., and R.D. Black (1970). Partial area contributions to storm runoff in a small New England watershed. *Water Resources Research* 6:1296–1311.

Dunne, T., and L.B. Leopold (1978). *Water in Environmental Planning*. San Francisco, CA: W.H. Freeman and Co.

Dunne, T., T.R. Moore, and C.H. Taylor (1975). Recognition and prediction of runoff-producing zones in humid regions. *Hydrological Sciences Bulletin* 20:305–327.

Dunne, T., A.G. Price, and S.C. Colbeck (1976). The generation of runoff from subarctic snowpacks. *Water Resources Research* 12:675–694.

Dynesius, M., and C. Nilsson (1994). Fragmentation and flow regulation of river systems in the northern third of the world. *Science* 266:753–762.

Eagleson, P.S. (1970). *Dynamic Hydrology*. New York, NY: McGraw-Hill Book Co.

Eagleson, P.S. (1972). Dynamics of flood frequency. *Water Resources Research* 8:878–898.

Eagleson, P.S. (1978). Climate, soil, and vegetation: 3. A simplified model of soil moisture movement in the liquid phase. *Water Resources Research* 14:722–730.

Eagleson, P.S. (1982). Ecological optimality in water-limited soil-vegetation systems: 1. Theory and hypothesis. *Water Resources Research* 18:323–340.

Eagleson, P.S. (1986). The emergence of global-scale hydrology. *Water Resources Research* 22:6S-14S.

Eagleson, P.S., et al. (1991). *Opportunities in the Hydrologic Sciences*. Washington, DC: National Academy Press.

Eichinger, W.E., M.B. Parlange, and H. Stricker (1996). On the concept of equilibrium evaporation and the value of the Priestley-Taylor coefficient. *Water Resources Research* 32:161–164.

Eller, H., and A. Denoth (1996). A capacitive soil moisture sensor. *Journal of Hydrology* 185:137–146.

Ellins, K.K., A. Roman-Mas, and R. Lee (1990). Using $^{222}$Rn to examine groundwater/surface discharge interaction in the Rio Grande de Manati, Puerto Rico. *Journal of Hydrology* 115:319–341.

Elrick, D.E., W.D. Reynolds, H.R. Geering, and K.-A. Tan (1990). Estimating steady infiltration rate times for infiltrometers and permeameters. *Water Resources Research* 26:759–769.

Elsenbeer, H., D. Lorieri, and N. Bonell (1995). Mixing model approaches to estimate storm flow sources in an overland flow-dominated tropical rain forest catchment. *Water Resources Research* 3:2267–2278.

Eltahir, E.A.B. (1996). El Niño and the natural variability in the flow of the Nile River. *Water Resources Research* 32:131–137.

Eltahir, E.A.B., and R.L. Bras (1994). Precipitation recycling in the Amazon basin. *Quarterly Journal of the Royal Meteorological Society* 120:861–880.

Eltahir, E.A.B., and R.L. Bras (1996). Precipitation recycling. *Reviews of Geophysics* 34:367–378.

Emmett, W.W. (1978). Overland flow. In Kirkby (1978):145–176.

Enfield, D.B. (1989). El Nino, past and present. *Reviews of Geophysics* 27:159–187.

Engman, E.T. (1966). Rainfall-runoff studies in the Sleepers River Research Watershed. Paper presented at Northeast Water Resources Symposium, Burlington, VT: University of Vermont.

Engman, E.T., and N. Chauhan (1995). Status of microwave soil moisture measurements with remote sensing. *Remote Sensing and Environment* 51:189–198.

Engman, E.T., and D.M. Hershfield (1969). Precipitation climatology of the Sleepers River Watershed near Danville, Vermont. Washington, DC. U.S. Agricultural Research Service *Paper* ARS 41–148.

Erskine, A.D., and A. Papaioannou (1997). The use of aquifer response rate in the assessment of groundwater resources. *Journal of Hydrology* 202:373–391.

Eshleman, K.N., J.S. Pollard, and A.K. O'Brien (1993). Determination of contributing areas for saturation overland flow from chemical hydrograph separation. *Water Resources Research* 19:3577–3587.

Evans, C., T.D. Davies, and P.S. Murdoch (1998). Component flow processes at four streams in the Catskill Mountains, New York, analyzed using episodic concentration/discharge relationships. *Hydrological Processes* 13:563–575.

Falkenmark, M., and G. Lindh (1993). Water and economic development. Chapter 7 in Gleick (1993).

Farnsworth, R.K., and E.S. Thompson (1982). Mean monthly, seasonal, and annual pan evaporation for the United States. U.S. National Atmospheric and Oceanic Administration *Technical Report* NWS 34.

Farnsworth, R.K., E.S. Thompson, and E.L. Peck (1982). Evaporation atlas for the contiguous 48 United States. U.S. National Atmospheric and Oceanic Administration *Technical Report* NWS 33.

Federer, C.A. (1979). A soil-plant-atmosphere model for transpiration and availability of soil water. *Water Resources Research* 15:555–562.

**628**

Federer, C.A. (1982). Transpirational supply and demand: Plant, soil, and atmospheric effects evaluated by simulation. *Water Resources Research* 18:355–362.

Federer, C.A. (1995). *BROOK90: A simulation model for evaporation, soil water, and streamflow, Version 3.1. Computer freeware and documentation.* Durham, NH: U.S. Forest Service.

Federer, C.A., and D. Lash (1978a). BROOK: A hydrologic simulation model for eastern forests. Durham, NH: University of New Hampshire Water Resources Research Center *Research Report* No. 19.

Federer, C.A., and D. Lash (1978b). Simulated streamflow response to possible differences in transpiration among species of hardwood trees. *Water Resources Research* 14:1089–1097.

Federer, C.A., C. Vörösmarty, and B. Fekete (1996). Intercomparison of methods for calculating evaporation in regional and global water balance models. *Water Resources Research* 32:2315–2321.

Federer, C.A., et al. (1990). Thirty years of hydrometeorologic data at the Hubbard Brook Experimental Forest, New Hampshire. Radnor, PA: U.S. Forest Service, Northeastern Forest Experiment Station *General Technical Report* NE-141.

Fennessey, N.M., and R.M. Vogel (1990). Regional flow-duration curves for ungauged sites in Massachusetts. *Journal of Water Resources Planning and Management* 116:530–549.

Fetter, C.W. (1994). *Applied Hydrogeology* (3rd ed.). Upper Saddle River, NJ: Prentice-Hall, 691 p.

Ficke, J.F. (1972). Comparison of evaporation computation methods, Pretty Lake, Lagrange County, northeastern Indiana. U.S. Geological Survey *Professional Paper* 686-A.

Finch, J.W. (1998). Estimating direct groundwater recharge using a simple water balance model - sensitivity to land surface parameters. *Journal of Hydrology* 211:112–125.

Fontaine, T.A., and D.E. Todd, Jr. (1993). Measuring evaporation with ceramic Bellani plate atmometers. *Water Resources Bulletin* 29:785–795.

Forster, F., and H.M. Keller (1988). Hydrologic simulation of forested catchments using the BROOK model. In *Recent Advances in the Modelling of Hydrologic Systems*, NATO Advanced Study Institute.

Foster, J.L., D.K. Hall, and A.T.C. Chang (1987). Remote sensing of snow. *Eos* 11 Aug 1987:681–684.

Frederick, R.H., V.A. Myers, and E.P Auciello (1977). Five to 60-minute precipitation frequency for eastern and central United States. Silver Spring, MD: U.S. National Oceanic and Atmospheric Administration *Technical Memo NWS HYDRO-35*.

Freeze, R.A. (1968). Quantitative interpretation of regional groundwater flow patterns as an aid to water balance studies. Gentbrugge, Belgium: International Association of Scientific Hydrology *Publication* 77.

Freeze, R.A. (1969). The mechanism of natural groundwater recharge and discharge: 1. One-dimensional, vertical, unsteady, un-saturated flow above a recharging or discharging groundwater flow system. *Water Resources Research* 5:153–171.

Freeze, R.A. (1971). Three-dimensional, transient, saturated-unsaturated flow in a groundwater basin. *Water Resources Research* 7: 929–941.

Freeze, R.A. (1972a). Role of subsurface flow in generating surface runoff: 1. Baseflow contributions to channel flow. *Water Resources Research* 8:609–623.

Freeze, R.A. (1972b). Role of subsurface flow in generating surface runoff: 2. Upstream source areas. *Water Resources Research* 8:1272–1283.

Freeze, R.A. (1974). Streamflow generation. *Reviews of Geophysics and Space Physics* 12:627–647.

Freeze, R.A., and J. Banner (1970). The mechanism of natural groundwater recharge and discharge: 2. Laboratory column experiments and field measurements. *Water Resources Research* 6:138–155.

Freeze, R.A., and J.A. Cherry (1979). *Groundwater.* Englewood Cliffs, NJ: Prentice-Hall, Inc.

Freeze, R.A., and P. A. Witherspoon (1967). Theoretical analysis of regional groundwater flow: 2. Effect of water-table configuration and subsurface permeability variation. *Water Resources Research* 3: 623–634.

Freeze, R.A., and P.A. Witherspoon (1968). Theoretical analysis of regional groundwater flow: 3. Quantitative interpretations. *Water Resources Research* 4: 581–590.

Fretwell, J.D., J.S. Williams, and P.J. Redman (1996). National water summary on wetland resources. U.S. Geological Survey *Water-Supply Paper* 2425.

Freyberg, D.L., J.W. Reeder, J.B. Franzini, and I. Remson (1980). Application of the Green-Ampt model to infiltration under time-dependent surface water depths. *Water Resources Research* 16:517–528.

Fritz, P., and J.C. Fontes, eds. (1980). *Handbook of Environmental Isotope Geochemistry.* New York, NY: Elsevier.

Galloway, J.N. (1989). Atmospheric acidification: Projections for the future. *Ambio* 18:161–166.

Garrick, M., C. Cunnane, and J.E. Nash (1978). A criterion of efficiency for rainfall-runoff models. *Journal of Hydrology* 36:375–381.

Garstka, W.U. (1964). Snow and snow survey. Section 10 in Chow (1964).

Gary, H.L. (1972). Rime contributes to water balance in high-elevation aspen forests. *Journal of Forestry* 70: 93–97.

Gash, J.H.C. (1979). An analytical model of rainfall interception in forests. *Quarterly Journal of the Royal Meteorological Society* 105:43–55.

Gash, J.H.C., and A.J. Morton (1978). An application of the Rutter model to the estimation of the interception loss from Thetford forest. *Journal of Hydrology* 38:47–58.

Gash, J.H.C., I.R. Wright, and C.R. Lloyd (1980). Comparative estimates of interception loss from three coniferous forests in Great Britain. *Journal of Hydrology* 48:87–105.

Gburek, W.J. (1990). Initial contributing area of a small watershed. *Journal of Hydrology* 118:387–403.

Gentilli, J. (1955). Estimating dew point from minimum temperature. *Bulletin of the American Meteorological Society* 36:128–131.

Georgievsky, V. Yu., S.A. Zhuravin, and A.V. Ezhov (1995). *Proceedings* American Geophysical Union 15th Annual Hydrology Days. Atherton, CA: Hydrology Days Publications:47–58.

Germann, P.F. (1989). Macropores and hydrologic hillslope processes. In M.G. Anderson and T.P. Burt, eds., *Surface and Subsurface Processes in Hydrology.* New York, NY: John Wiley and Sons.

Giambelluca, T.W., D.L. McKenna, and P.C. Ekern (1992). An automated recording atmometer: 1. Calibration and testing. *Agricultural and Forest Meteorology* 62:109–125.

Gibbs, R.J. (1970). Mechanisms controlling world water chemistry. *Science* 170:1088–1090.

Gillham, R.W. (1984). The capillary fringe and its effect on water-table response. *Journal of Hydrology* 67:307–324.

Gilman, C.S. (1964). Rainfall. Section 9 in Chow (1959).

Giorgi, F., and R. Avissar (1997) Representation of heterogeneity effects in earth system modeling: Experience from land surface modeling. *Reviews of Geophysics* 35:413–437.

Giusti, L., and C. Neal (1993). Hydrological pathways and solute chemistry of storm runoff at Dargall Lane, southwest Scotland. *Journal of Hydrology* 142:1–27.

Glass, N.R., et al. 1982. Effects of acid precipitation. *Environmental Science and Technology* 16:162A-169A.

Gleick, P.H., ed. (1992). *Water in Crisis.* New York, Oxford University Press.

Glover, R.E. (1964). The pattern of freshwater flow in a coastal aquifer. U.S. Geological Survey *Water-Supply Paper* 1613-C.

Golding, D.L., and R.H. Swanson (1986). Snow distribution patterns in clearings and adjacent forest. *Water Resources Research* 22:1931–1940.

Goodison, B.E. (1981). Compatibility of Canadian snowfall and snowcover data. *Water Resources Research* 17:893–900.

Goodison, B.E., H.L. Ferguson, and G.A. McKay (1981). Measurement and data analysis. In Gray and Male (1981).

Goodman, A.S. (1984). *Principles of Water Resources Planning.* Englewood Cliffs, NJ: Prentice-Hall.

Grasty, R.L. (1979). One flight snow-water equivalent measurement by airborne gamma-ray spectrometry. Paper presented at World Meteorological Organization Workshop on Remote Sensing of Snow and Soil Moisture by Nuclear Techniques, Voss, Norway, April 1979.

Gray, D.M. and D.H. Male, eds. (1981) *Handbook of Snow.* Elmsford, NY: Pergamon Press.

Gray, W.G., and S.M. Hassanizadeh (1991). Paradoxes and realities in unsaturated flow theory. *Water Resources Research* 27:1847–1854.

Green, W.H., and G.A. Ampt (1911). Studies on soil physics, 1: The flow of air and water through soils. *Journal of Agricultural Science* 4:1–24.

Gregory, K.J., and D.E. Walling (1973). *Drainage Basin Form and Process*. New York, NY: John Wiley and Sons.

Grimmond, C.S.B., S.A. Isard, and M.J. Belding (1992). Development and evaluation of continuously weighing mini-lysimeters. *Agricultural and Forest Meteorology* 62:205–218.

Groisman, P. Ya., and D.R. Easterling (1994). Variability and trends of total precipitation and snowfall over the United States and Canada. *Journal of Climate* 7: 184–205.

Groisman, P. Ya., T.R. Karl, and R.W. Knight (1994). Observed impact of snow cover on the heat balance and the rise of continental spring temperatures. *Science* 263:198–200.

Groisman, P. Ya., and D.R. Legates (1994). The accuracy of United States precipitation data. *Bulletin of the American Meteorological Society* 75:215–227.

Grove, M., J. Harbor, and B. Engel (1998). Composite vs. distributed curve numbers: Effects on estimates of storm runoff depths. *Journal of the American Water Resources Association* 34:1015–1023.

Groves, J.R. (1989). A practical soil-moisture profile model. *Water Resources Bulletin* 25:875–880.

Grundy, W.D. (1988). *Using the USGS STATPAC Programs: A Hands-On Course*. Denver, CO: The Computer-Oriented Geological Society.

Gupta, V.K., E. Waymire, and C.T. Wang (1980). Representation of an instantaneous unit hydrograph from geomorphology. *Water Resources Research* 16:855–862.

Gupta, V.L., S.M. Afaq, J.W. Fordham, and J.M. Federici (1979). Unsteady streamflow modeling guidelines. *Journal of Hydrology* 43:79–97.

Guttman, N.B., J.R. Wallis, and J.R.M. Hosking (1992). Spatial comparability of the Palmer drought severity index. *Water Resources Bulletin* 28:1111–1119.

Haan, C.T. (1977). *Statistical Methods in Hydrology*. Ames, IA: Iowa State University Press.

Haitjema, H.M. (1995). On the residence time distribution in idealized groundwatersheds. *Journal of Hydrology* 172:127–146.

Hakenkamp. C.C., H.M. Valett, and A.J. Boulton (1993). Perspectives on the hyporheic zone: integrating hydrology and biology - concluding remarks. *Journal of the North American Benthological Society* 12:94–99.

Hall, F.R. (1968). Base-flow recessions - A review. *Water Resources Research* 4:973–983.

Halpert, M.S., and C.F. Ropelewski (1992). Surface temperature patterns associated with the Southern Oscillation. *Journal of Climate* 5:577–593.

Hammer, M.J. and K.A. MacKichan (1981). *Hydrology and Quality of Water Resources*. New York, NY: J. Wiley & Sons.

Hamon, R.W. (1963). Computation of direct runoff amounts from storm rainfall. Wallingford, Oxon., U.K.: International Association of Scientific Hydrology *Publication* 63.

Hamon, R.W., L.L. Weiss, and W.T. Wilson (1954). Insolation as an empirical function of daily sunshine duration. *Monthly Weather Review* 82:141–146.

Hansen, E.M. (1987). Probable maximum precipitation for design floods in the United States. *Journal of Hydrology* 96:267–278.

Hansen, E.M., L.C. Schreiner, and J.F. Miller (1982). Application of probable maximum precipitation estimates - United States east of the 105th meridian. U.S. National Oceanographic and Atmospheric Administration *Hydrometeorological Report* 52.

Harbeck, G.E. (1972). A practical field technique for measuring reservoir evaporation utilizing mass-transfer theory. U.S. Geological Survey *Professional Paper* 272-E.

Harbeck, G.E., et al. (1954). Water-loss investigations: Lake Hefner studies, technical report. U.S. Geological Survey *Professional Paper* 269.

Harlin, J.M. (1984). Watershed morphometry and time to hydrograph peak. *Journal of Hydrology* 67:141–154.

Harr, R.D. (1982). Fog drip in the Bull Run Municipal Watershed, Oregon. *Water Resources Bulletin* 18:785–789.

Harris, R.N., and D.S. Chapman (1997). Borehole temperatures and a baseline for 20th-century global warming estimaes. *Science* 275:1618–1621.

Harrison, D.E., and N.K. Larkin (1998). El Niño-Southern Oscillation sea surface temperature and wind anomalies, 1946–1993. *Reviews of Geophysics* 36:353–399.

Harte, J. (1985). *Consider A Spherical Cow*. Los Altos, CA: William Kaufmann, Inc.

Hartley, S. (1990). An investigation into the validity of the assumption that long-term average change in storage is zero in the application of the water-balance equation to estimates of long-term evapotranspiration. Unpublished manuscript, Earth Sciences Department, University of New Hampshire, Durham, NH.

Hays, J.D., J. Imbrie, and N.J. Shackleton (1976). Variations in the Earth's orbit: Pacemaker of the ice ages. *Science* 194:1121–1132.

Heath, R.C. (1982). Basic ground-water hydrology. U.S. Geological Survey *Water-Supply Paper* 2220.

Heggen, R.J. (1983). Thermal dependent physical properties of water. *Journal of Hydraulic Engineering* 109:298–302.

Helsel, D.R., and R.M. Hirsch (1992). *Statistical Methods in Water Resources*. Amsterdam, Elsevier.

Helvey, J.D. (1971). A summary of rainfall interception by certain conifers of North America. In *Biological Effects in the Hydrological Cycle*, Proceedings of Third International Seminar for Hydrology Professors. West Lafayette, IN: Purdue University Agricultural Experiment Station: 103–113.

Helvey, J.D., and J.H. Patric (1965a). Design criteria for interception studies. Wallingford, Oxon., U.K.: International Association of Hydrologic Sciences *Publication* 67:131–137.

Helvey, J.D., and J.H. Patric (1965b). Canopy and litter interception of rainfall by hardwoods of eastern United States. *Water Resources Research* 1:193–206.

Helvey, J.D., and J.H. Patric (1983). Sampling accuracy of pit vs. standard rain gages on the Fernow Experimental Forest. *Water Resources Bulletin* 19:87–89.

Hem, J.D. (1985). Study and interpretation of the chemical characteristics of natural water. U.S. Geological Survey *Water-Supply Paper* 2254.

Henderson-Sellers, A. (1992). Continental cloudiness changes this century. *GeoJournal* 27:255–262.

Hendrick, R.L., and R.J. DeAngelis (1976). Seasonal snow accumulation, melt and water input - a New England model. *Journal of Applied Meteorology* 15:715–727.

Hendrick, R.L., R.J. DeAngelis, and S.L. Dingman (1978). The role of elevation in determining spatial distributions of precipitation, snow, and water input at Mt. Mansfield, Vermont. Hanover, NH: U.S. Army Cold Regions Research and Engineering Laboratory *Proceedings of a Workshop on Modelling of Snow-Cover Runoff*.

Herschy, R.W. (1985). *Streamflow Measurement*. New York, NY: Elsevier Applied Science Publishers Ltd.

Hershfield, D.M. (1961). Rainfall frequency atlas of the United States for durations from 30 minutes to 24 hours and return periods from 1 to 100 years. U.S. Weather Bureau *Technical Paper* 40.

Hess, K.M. (1988). Sewage plume in a sand and gravel aquifer, Cape Cod, Massachusetts. In U.S. Geological Survey *Water-Supply Paper* 2325:87–92.

Hewlett, J.D. (1982). *Principles of Forest Hydrology* (2nd ed.). Athens, GA: University of Georgia Press.

Hewlett, J.D., and A.R. Hibbert (1961). Increases in water yield after several types of forest cutting. *International Association of Scientific Hydrology Bulletin* 6:5–17.

Hewlett, J.D., and A.R. Hibbert (1963). Moisture and energy conditions within a sloping soil mass during drainage. *Journal of Geophysical Research* 68:1081–1087.

Hicks, D.M., and P.J. Mason (1991). *Roughness Characteristics of New Zealand Rivers*. Wellington, N.Z.: New Zealand Water Resources Survey, 329 pp.

Hillel, D. (1980a). *Fundamentals of Soil Physics*. New York, NY: Academic Press.

Hillel, D. (1980b). *Applications of Soil Physics*. New York, Academic Press.

Hillel, D., and C.H.M. van Bavel (1976). Dependence of profile water storage on soil hydraulic properties: a simulation model. *Soil Science Society of America Journal* 40:807–815.

Hino, M., K. Fujita, and H. Shutto (1987). A laboratory experiment on the role of grass for infiltration and runoff processes. *Journal of Hydrology* 90:303–325.

Hipolito, J.N., and J.M. Loureiro (1988). Analysis of some velocity-area methods for calculating open-channel flow. *Hydrological Sciences Journal* 33:311–320.

Holmes, B.H. (1972). A History of Federal Water Resources Programs, 1800–1960. U.S. Department of Agriculture *Miscellaneous Publication* 1233.

Holmes, B.H. (1979). History of Federal Water Resources Programs, 1961–1970. U.S. Department of Agriculture *Miscellaneous Publication* 1379.

Holtan, H.N., and D.E. Overton (1963). Analyses and application of simple hydrographs. *Journal of Hydrology* 1:250–264.

Holtan, H.N., and D.E. Overton (1964). Storage-flow hysteresis in hydrograph synthesis. *Journal of Hydrology* 2:309–323.

Holzman, B. (1937). Sources of moisture for precipitation in the United States. U.S. Department of Agriculture *Technical Bulletin* 589.

Hooper, R.P, N. Christopherson, and N.E. Peters (1990). Modeling streamwater chemistry as a mixture of soil-water end-members — An application to the Panola Mountain catchment, Georgia, U.S.A. *Journal of Hydrology* 116:321–343.

Hoover, M.D. (1971). Snow interception and redistribution in the forest. In *Biological Effects in the Hydrological Cycle*, Proceedings of Third International Seminar for Hydrology Professors. West Lafayette, IN: Purdue University Agricultural Experiment Station: 114–122.

Hopkins, C.D., Jr. (1973). Estimates of precipitation, flood stages, and frequency. In *What If Agnes Had Hit the Connecticut River Basin?* Boston, MA: New England River Basins Commission.

Hornbeck, J. W. (1986). Modeling the accumulation and effects of chemicals in snow. In Morris (1986).

Hornbeck, J.W., C.A. Federer, and R.S. Pierce (1987). Effects of whole-tree clearcutting on streamflow can be adequately estimated by simulation. International Association of Scientific Hydrology *Publication* 1676: 565–573.

Hornbeck, J.W., et al. (1986). Clearcutting northern hardwoods: Effects on hydrologic and nutrient ion budgets. *Forest Science* 32:667–686.

Hornberger, G.M., K.J. Beven, B.J. Cosby, and D.E. Sappington (1985). Shenandoah Watershed Study: Calibration of a topography-based, variable contributing area hydrological model to a small forested catchment. *Water Resources Research* 21:1841–1850.

Horne, F.E., and M.L. Kavvas (1997). Physics of the spatially averaged snowmelt process. *Journal of Hydrology* 191:179–207.

Horton, R.E. (1931). The field, scope, and status of the science of hydrology. American Geophysical Union *Transactions*, Reports and Papers, Hydrology:189–202.

Horton, R.E. (1933). The role of infiltration in the hydrologic cycle. *American Geophysical Union Transactions* 14:446–460.

Horton, R.E. (1945). Erosional development of streams and their drainage basins: Hydrophysical approach to quantitative morphology. *Geological Society of America Bulletin* 56:275–370.

Hosking, J.R.M., and J.R. Wallis (1997). *Regional Frequency Analysis*. New York, NY: Cambridge University Press.

Hostetler, S.W., and P.J. Bartlein (1990). Simulation of lake evaporation with application to modeling lake level variations of Harney-Malheur Lake, Oregon. *Water Resources Research* 26:2603–2612.

Howell, D.G., and R.W. Murray (1986). A budget for continental growth and denudation. *Science* 233:446–449.

Hubbert, M.K. (1940). The theory of groundwater motion. *Journal of Geology* 48:785–944.

Huber, W.C. (1992). Contaminant transport in surface water. Chapter 14 in Maidment (1992).

Hudson, J.A. (1988). The contribution of soil-moisture storage to the water balances of upland forested and grassland catchments. *Hydrological Sciences Journal* 33:287–309.

Huff, F.A. (1955). Comparison between standard and small orifice rain gages. *American Geophysical Union Transactions* 36:689–694.

Huff, F.A. (1977). Effects of the urban environment on heavy rainfall distribution. *Water Resources Bulletin* 13:807–816.

Huff, F.A. (1995). Characteristics and contributing causes of an abnormal frequency of flood-producing rainstorms at Chicago. *Water Resources Bulletin* 31: 703–714.

Huff, F.A., and S.A. Chagnon (1973). Precipitation modification by major urban areas. *Bulletin of the American Meteorological Society* 54:1220–1232.

Hunt, B. (1990). An approximation of the bank storage effect. *Water Resources Research* 26:2769–2775.

Hunt, L.A., L. Kuchar, and C.J. Swanton (1998). Estimation of solar radiation for use in crop modeling. *Agricultural and Forest Meteorology* 91:293–300.

Hunt, R.J., D.P. Krabbenhoft, and M.P. Anderson (1996). Groundwater inflow measurements in wetland systems. *Water Resources Research* 32:495–507.

Iqbal, M. (1983). *An Introduction to Solar Radiation*. New York, NY: Academic Press.

Jackson, C.R. (1992). Hillslope infiltration and lateral downslope unsaturated flow. *Water Resources Research* 28:2533–2539.

Jackson, T.J. (1988). Research toward an operational passive microwave remote sensing system for soil moisture. *Journal of Hydrology* 102:95–112.

Jacobsen, O.H., and P. SchjΔnning (1993). Field evaluation of time domain reflectometry for soil water measurements. *Journal of Hydrology* 152:159–172.

James, I.C., III, B.T. Bower, and N.C. Matalas (1969). Relative importance of variables in water resources planning. *Water Resources Research* 5:1165–1173.

James, W.P., J. Warinner, and M. Reedy (1992) Application of the Green-Ampt infiltration equation to watershed modeling. *Water Resources Bulletin* 28:623–635.

Jarvis, P.G. (1976). The interpretations of the variations in leaf water potential and stomatal conductance found in canopies in the field. Royal Society of London *Philosophical Transactions Series B* 273:593–610.

Jayatilaka, C.J., R.W. Gilham, D.W. Blowes, and R.J. Nathan (1996). A deterministic-empirical model of the effect of the capillary fringe on near-stream area runoff. 2. Testing and application. *Journal of Hydrology* 184:317–336.

Jenkins, C.T. (1968). Techniques for computing rate and volume of stream depletion by wells. *Groundwater* 6(2):37–46.

Jensen, M.E., R.D. Burman, and R.G. Allen, eds. (1990). Evapotranspiration and irrigation water requirements. New York, NY: American Society of Civil Engineering *Manuals and Reports of Engineering Practice* No. 70.

Jobson, H.E. (1972). Effect of using averaged data on the computed evaporation. *Water Resources Research* 8:513–518.

Johansson, P.-O. (1986). Diurnal groundwater level fluctuations in sandy till - A model analysis. *Journal of Hydrology* 87:125–134.

Johansson, P.-O. (1987). Estimation of groundwater recharge in sandy till with two different methods using groundwater level fluctuations. *Journal of Hydrology* 90:183–198.

Johnson, A.I. (1963). A field method for measurement of infiltration. U.S. Geological Survey *Water-Supply Paper* 1544-F.

Johnston, R.H. (1986). Factors affecting ground-water quality. In U.S. Geological Survey *Water-Supply Paper* 2325:71–86.

Jones, P.D., and K.R. Briffa (1995). Growing season temperatures over the former Soviet Union. *International Journal of Climatology* 15:943–960.

Jordan, J.P. (1994). Spatial and temporal variability of stormflow generation processes on a Swiss catchment. *Journal of Hydrology* 153:357–382.

Kabat, P., R.W.A. Hutjes, and R.A. Feddes (1997). The scaling characteristics of soil parameters: From plot scale heterogeneity to subgrid parameterization. *Journal of Hydrology* 190:363–396.

Karl, T.R., et al. (1993). Asymmetric trends of daily maximum and minimum temperature. *Bulletin of the American Meteorological Society* 74:1007–1023.

Karl, T.R., P. Ya. Groisman, R.W. Knight, and R.R. Heim, Jr. (1993). Recent variations in snow cover and snowfall in North America and their relation to precipitation and temperature variations. *Journal of Climate* 6:1327–1344.

Karl, T.R., R.W. Knight, and N. Plummer. 1995. Trends in high-frequency climate variability in the twentieth century. *Nature* 377: 217–220.

Karl, T.R., and R.G. Quayle (1988). Climate change in fact and theory: Are we collecting the facts? *Climate Change* 13:5–17.

Karl, T.R., and W.E. Riebsame (1989). The impact of decadal fluctuations in mean precipitation and temperature on runoff: A sensitivity study over the United States. *Climatic Change* 15:423–447.

Kaufmann, R.K., and D.I. Stern (1997). Evidence for human influence on climate from hemispheric temperature relations. *Nature* 388:39–44.

Kazmann, R.G. (1988). *Modern Hydrology* (3rd ed.). Dublin, OH: The National Water Well Association.

Keller, E.A. (2000). *Environmental Geology* (8th ed.) Upper Saddle River, NJ; Prentice-Hall.

Kelliher, F.M., R. Leuning, M.R. Raupach, and E.-D. Schulze (1995). Maximum conductances for evaporation from global vegetation types. *Agricultural and Forest Meteorology* 73:1–16.

Khosla, B.K. (1980). Comparison of calculated and in situ measured unsaturated hydraulic conductivity. *Journal of Hydrology* 47:325–332.

Kilpatrick, F.A., and E.D. Cobb (1985). Measurement of discharge using tracers. U.S. Geological Survey *Techniques of Water-Resources Investigations* Book 3 Chapter A16.

Kilpatrick, F.A., and J.F. Wilson, Jr. (1989). Measurement of time of travel in streams by dye tracing. U.S. Geological Survey *Techniques of Water-Resources Investigations* Book 3 Chapter A9.

King, H.W., C.O. Whisler, and G. Woodburn (1960). Hydraulic Similitude and Dimensional Analysis. Pp. 318–330 in *Hydraulics*, 5th edition,. New York, John Wiley & Sons.

Kirby, W. (1974). Algebraic boundedness of sample statistics. *Water Resources Research* 10:220–222.

Kirby, W.H. (1987). Linear error analysis of slope-area discharge determinations. *Journal of Hydrology* 96:125–138.

Kirkby, M.J. (1976). Tests of the random network model and its application to basin hydrology. *Earth Surface Processes* 1:197–212.

Kirkby, M.J., ed. (1978). *Hillslope Hydrology*. New York, NY: John Wiley and Sons.

Kirkby. M.J. (1988). Hillslope runoff processes and models. *Journal of Hydrology* 100:315–339.

Kirkby, M.J. (1993). Network hydrology and geomorphology. Chapter 1 in K. Beven and M.J. Kirkby, eds. *Channel Network Hydrology*. New York, NY: John Wiley and Sons.

Kitanidis, P.K. (1992). Geostatistics. Chapter 20 in Maidment (1992).

Klemeš, V. (1986a). Dilettantism in hydrology: Transition or destiny? *Water Resources Research* 22:177S-188S.

Klemeš, V. (1986b). Operational testing of hydrological simulation models. *Hydrological Sciences Journal* 31:13–24.

Klock, G.O. (1972). Snowmelt temperature influence on infiltration and soil water retention. *Journal of Soil and Water Conservation* 27:12–14.

Knighton, D. (1984). *Fluvial Forms and Processes*. London, U.K.: Edward Arnold, Ltd.

Kohler, M.A., and L.H. Parmele (1967). Generalized estimates of free-water evaporation. *Water Resources Research* 3:997–1005.

Kohler, M.A., T.J. Nordenson, and W.E. Fox (1955). Evaporation from pans and lakes. U.S. Weather Bureau *Research Paper* 38.

Kumar, S., and S.C. Jain (1982). Application of SCS infiltration model. *Water Resources Bulletin* 18:503–507.

Kundzewicz, Z.W., L. Gottschalk, and B. Webb, eds. (1987). Hydrology 2000. International Association of Hydrologic Sciences *Publication* No. 171.

Kustas, W.P., A. Rango, and R. Uijlenhoet (1994). A simple energy budget algorithm for the snowmelt runoff model. *Water Resources Research* 30:1515–1527.

Kuusisto, E. (1986). The energy balance of a melting snow cover in different environments. In Morris (1986).

Kuznetsova, L.P. (1990). Use of data on atmospheric moisture transport over continents and large river basins for the estimation of water balances and other purposes. Paris:

UNESCO International Hydrological Programme (IHP-III Project 1.1).

LaBaugh, J.W. (1984). Uncertainty in phosphorus retention, Williams Fork Reservoir, Colorado. *Water Resources Research* 21:1684–1692.

LaBaugh, J.W. (1986). Wetland ecosystem studies from a hydrologic perspective. *Water Resources Bulletin* 22:1–10.

LaBaugh, J.W., et al. (1997). Hydrological and chemical estimates of the water balance of a closed-basin lake in north central Minnesota. *Water Resources Research* 33:2799–2812.

Lamb, H.H. (1982). *Climate, History and the Modern World*. New York, NY: Methuen, Inc.

Larkin, R.G., and J. M. Sharp, Jr. (1992). On the relationship between river-basin geomorphology, aquifer hydraulics, and ground-water flow direction in alluvial aquifers. *Geological Society of America Bulletin* 104:1608–1620.

Larson, G.J., M.R. Delcore, and S. Offer (1987). Application of the tritium interface method for determining recharge rates to unconfined drift aquifers, I. Homogeneous case. *Journal of Hydrology* 91:59–72.

Larson, L.L., and E. L. Peck (1974). Accuracy of precipitation measurements for hydrologic modeling. *Water Resources Research* 10:857–863.

Lawrence, G.B., M.B. David, and W.C. Shortle (1995). A new mechanism for calcium loss in forest-floor soils. *Nature* 378:162–165.

Lean, J., and D.A. Warrilow (1989). Simulation of the regional climatic impact of Amazon deforestation. *Nature* 342:411–413.

Leaney, F.W., K.R.J. Smettem, and D.J. Chittleborough (1993). Estimating the contribution of preferential flow to subsurface runoff from a hillslope using deuterium and chloride. *Journal of Hydrology* 147:83–103.

Leathers, D.J., and D.A. Robinson (1993). The association between extremes in North American snow cover extent and United States temperatures. *Journal of Climate* 6:1345–1355.

Lebel, T., G. Bastin, C. Obled, and J.D. Creutin (1987). On the accuracy of areal rainfall estimation: A case study. *Water Resources Research* 23:2123–2134.

LeBoutillier, D.W., and P.R. Waylen (1993). A stochastic model of flow-duration curves. *Water Resources Research* 29:3434–3541.

Lee, D.R. (1977). A device for measuring seepage flux in lakes and estuaries. *Limnology and Oceanography* 22:140–147.

Lee, D.R., and H.B.N. Hynes (1978). Identification of groundwater discharge zones in a reach of Hillman Creek in southern Ontario. *Water Pollution Research Canada* 13:121–133.

Lee, R. (1964). Potential insolation as a topoclimatic characteristic of drainage basins. *International Association of Scientific Hydrology Bulletin* 9:27–41.

Legates, D.R., and T.L. DeLiberty (1993). Precipitation measurement biases in the United States. *Water Resources Bulletin* 29:855–861.

Lemeur, R., and L. Zhang (1990). Evaluation of three evapotranspiration models in

terms of their applicability for an arid region. *Journal of Hydrology* 114:395–411.

Lemon, E.R. (ed.) (1983). *$CO_2$ and Plants: The Response of Plants to Rising Levels of Carbon Dioxide*. Boulder, CO: Westview. American Association for the Advancement of Science Symposium No. 84.

Leonard, R.E., and A.R. Eschner (1968). Albedo of intercepted snow. *Water Resources Research* 4:931–935.

Leopold, L.B., and T. Maddock (1954). *The Flood Control Controversy*. New York, NY: The Ronald Press Company.

Leopold, L.B., M.G. Wolman, and J.P. Miller (1964). *Fluvial Processes in Geomorphology*. San Francisco, CA: W.H. Freeman and Co.

Lesack, L.F.W. (1993). Water balance and hydrologic characteristics of a rain forest catchment in the central Amazon basin. *Water Resources Research* 29:759–773.

Lettenmaier, D.P., and E.F. Wood (1992). Hydrologic forecasting. Chapter 26 in Maidment (1992).

Lettenmaier, D.P., E.F. Wood, and J.R. Wallis (1994). Hydro-climatological trends in the continental United States, 1948–88. *Journal of Climate* 7:586–607.

Levine, J.B., and G.D. Salvucci (1999). Equilibrium analysis of groundwater-vadose zone interactions and the resulting spatial distribution of hydrologic fluxes across a Canadian prairie. *Water Resources Research* 35:1369–1383.

Lewis, W.M., Jr. et al. (1995). *Wetlands: Characteristics and Boundaries*. Washington, DC: National Academy Press.

Likens, G.E., C.T. Driscoll, and D.C. Buso (1996). Long-term effects of acid rain: Response and recovery of a forested ecosystem. *Science* 252:244–246.

Linacre, E.T. (1993). Data-sparse estimation of lake evaporation, using a simplified Penman equation. *Agricultural and Forest Meteorology* 64:237–256.

Lindberg, S.E., and C.T. Garten (1988). Sources of sulphur in forest canopy throughfall. *Nature* 336:148–151.

Lindroth, A. (1985). Canopy conductance of coniferous forests related to climate. *Water Resources Research* 21:297–304.

Lindsey, S.D., and R. K. Farnsworth (1997). Sources of solar radiation estimates and their effect on daily potential evaporation for use in streamflow modeling. *Journal of Hydrology* 201:348–366.

Lins, H.F., and P.J. Michaels (1994). Increasing U.S. streamflow linked to greenhouse forcing. *EOS* 75:281,284,285.

Linsley, R.K., and M.A. Kohler (1951). Predicting the runoff from storm rainfall. U.S. Weather Bureau *Research Paper* 34.

Linsley, R.K., M.A. Kohler, and J.L.H. Paulhus (1982). *Hydrology for Engineers* (3rd ed.). New York, NY: McGraw-Hill Book Co.

Lloyd, C.R., J.H.C. Gash, W.J. Shuttleworth, and A. de O. Marques F. (1988). The measurement and modeling of rainfall interception by Amazonian rain forest. *Agricultural and Forest Meteorology* 43:277–294.

Loague, K. (1990). Simple design for simultaneous steady-state infiltration experiments with ring infiltrometers. *Water Resources Bulletin* 26:935–938.

Loague, K., and G.A. Gander (1990). R-5 revisited:1. Spatial variability of infiltration on a small rangeland catchment. *Water Resources Research* 28:957–971.

Loague, K.M., and R.A. Freeze (1985). A comparison of rainfall-runoff modeling techniques on small upland catchments. *Water Resources Research* 21:229–248.

Loaiciga, H.A., et al. (1996). Global warming and the hydrologic cycle. *Journal of Hydrology* 174:83–127.

Lohman, S.W. (1979). Ground-water hydraulics. U.S. Geological Survey *Professional Paper* 708, 70 p.

Loijens, H.S., and R.L. Grasty (1973). Airborne measurement of snow-water equivalent using natural gamma radiation over southern Ontario, 1972–1973. Ottawa, Ontario: Inland Waters Directorate, Environment Canada, *Science Series No. 34.*

Lottes, A.L., and A.M. Ziegler (1994). World peat occurrence and the seasonality of climate and vegetation. *Paleogeography, Paleoclimatology, Paleoecology* 106:23–37.

Loucks, D.P., J.R. Stedinger, and D.A. Haith (1981). *Water Resource Systems Planning and Analysis.* Englewood Cliffs, NJ: Prentice-Hall.

Loukas, A., and M.C. Quick (1996). Physically-based estimation of lag time for forested mountainous watersheds. *Hydrological Sciences Journal* 41:1–19.

Lovett, G.M., W.A. Reiners, and R.K. Olson (1982). Cloud droplet deposition in subalpine balsam fir forests: Hydrological and chemical inputs. *Science* 218:1303–1304.

Lundberg, A., and S. Halldin (1994). Evaporation of intercepted snow: Analysis of governng factors. *Water Resources Research* 30:2587–2598.

Lundberg, A., I. Calder, and R. Harding (1998). Evaporation of intercepted snow: measurement and modeling. *Journal of Hydrology* 206:151–163.

L'vovich, M.I. (1974). *World Water Resources and Their Future.* Translated by R.L. Nace. Washington DC: American Geophysical Union.

Lyford, F.P., and H.K. Qashu (1969). Infiltration rates as affected by desert vegetation. *Water Resources Research* 5:1373–1376.

Maidment, D.R. ed. (1992) *Handbook of Hydrology*, New York, McGraw-Hill.

Male, D.H., and D.M. Gray (1981). Snowcover ablation and runoff. In Gray and Male (1981).

Maller, R.A., and M.L. Sharma (1981). An analysis of areal infiltration considering spatial variability. *Journal of Hydrology* 52:25–37.

Malmstrom, V.H. (1969). A new approach to the classification of climate. *Journal of Geography* 68:351–357.

Manley, R.E. (1977). The soil moisture component of mathematical catchment simulation models. *Journal of Hydrology* 35:341–356.

Marchand, M., and F.H. Toornstra (1986). *Ecological Guidelines for River Basin Development.* Leiden, Netherlands; Centrum voor Milienkunde, Dept 28, Rijksuniversiteit.

Margaritz, M. et al. (1990). A new method to determine regional evapotranspiration. *Water Resources Research* 26:1757–1762.

Markham, C.G. (1970). Seasonality of precipitation in the United States. *Annals of the Association of American Geographers* 60:593–597.

Marsh, P., and M.K. Woo (1985). Meltwater movement in natural heterogeneous snow covers. *Water Resources Research* 21:1710–1716.

Martinec, J., and A. Rango (1989). Merits of statistical criteria for the performance of hydrological models. *Water Resources Bulletin* 25:421–432.

Martz, L.W. and J. Garbrecht (1992). Numerical definition of drainage network and subcatchment areas from digital elevation models. *Computers in Geoscience* 18:747–761.

Matalas, N.C., and T. Maddock III (1976). Hydrologic semantics. *Water Resources Research* 12:123.

Mattikali, N.M., E.T. Engman, T.J. Jackson, and L.R. Ahuja (1998). Microwave remote sensing of temporal variations of surface soil moisture during Washita '92 and its application to the estimation of soil physical properties. *Water Resources Research* 34:2289–2299.

McBride, M.S., and H.O. Pfannkuch (1975). The distribution of seepage within lakes. U.S. Geological Survey *Journal of Research* 3:505–512.

McCaffrey, S.C. (1993). Water, politics, and international law. Chapter 8 in Gleick (1993).

McCuen, R.H. (1998). *Hydrologic Analysis and Design* (2nd ed.). Upper Saddle River, NJ: Prentice-Hall, 814 p.

McCulloch, J.S.G., and M. Robinson (1993). History of forest hydrology. *Journal of Hydrology* 150:189–216.

McDonnell, J. J. (1990). A rationale for old water discharge through macropores in a steep, humid catchment. *Water Resources Research* 26:2821–2832.

McGuinness, C.L. (1963). The role of ground water in the national water situation. U.S. Geological Survey *Water-Supply Paper* 1800.

McGurk, B.J., and D.L. Azuma (1992). Correlation and prediction of snow water equivalent from snow sensors. USDA Forest Service Pacific Southwest Research Station *Report* PSW-RP-211, 13 pp.

McGurk, B.J., T.J. Edens, and D.L. Azuma (1993). Predicting wilderness snow water equivalent with nonwilderness snow sensors. *Water Resources Bulletin* 29:85–94.

McKay, D.S., and L.E. Band (1998). Extraction and representation of nested catchment areas from digital elevation models in lake-dominated topography. *Water Resources Research* 34:897–901.

McKay, G.A. (1970). Precipitation. In D.M. Gray, ed. *Handbook on the Principles of Hydrology.* Port Washington, NY: Water Information Center, Inc.

McKay, G.A., and D.M. Gray (1981). The distribution of snowcover. In Gray and Male (1981).

McMahon, T.A. (1992). Hydrologic design for water use. Chapter 27 in Maidment (1992).

McMillan, W.D., and R.H. Burgy (1960). Interception loss from grass. *Journal of Geophysical Research* 65:2387–2394.

McNamara, J.P., D.L. Kane, and L.D. Hinzman (1997). Hydrograph separations in an Arctic watershed using mixing model and graphical techniques. *Water Resources Research* 33:1701–1719.

McQueen, I.S. (1963). Development of a hand-portable rainfall-simulator infiltrometer. U.S. Geological Survey *Circular* 482.

Meiman, J.R. (1968). Snow accumulation related to elevation, aspect and forest canopy. In Fredericton, New Brunswick, Canada: University of New Brunswick *Workshop Seminar on Snow Hydrology Proceedings.*

Mein, R.G., and C.L. Larson (1973). Modeling infiltration during a steady rain. *Water Resources Research* 9:384–394.

Meinzer, O.E. (1932). Outline of methods for estimating ground-water supplies. U.S. Geological Survey *Water-Supply Paper* 638-C.

Mellor, M. (1964). Snow and Ice at the Earth's Surface. Hanover, NH: U.S. Army Cold Regions Research and Engineering Laboratory Cold Regions *Science and Engineering Monograph* II-C1.

Merriam, R.A. (1961). Surface water storage on annual ryegrass. *Journal of Geophysical Research* 66:1833–1838.

Meyboom, P. (1966). Unsteady groundwater flow near a willow ring in hummocky terrain. *Journal of Hydrology* 4:38–62.

Michel, B. (1971). Winter regime of rivers and lakes. Hanover, NH: U.S. Army Cold Regions Research and Engineering Laboratory *Cold Regions Science and Engineering Monograph* III-B1a.

Milhous, R.T., J.M. Bartholow, M.A. Updike, and A.R. Moos (1990). Reference manual for the generation and analysis of habitat time series - Version II. U.S. Fish and Wildlife Service *Instream Flow Information Paper* No. 27.

Miller, A., J.C. Thompson, R.E. Peterson, and D.R. Haragan (1983). *Elements of Meteorology.* Columbus, OH: C.E. Merrill Publishing Co.

Miller, J.F., R.H. Frederick, and R.J. Tracey (1973). Precipitation frequency atlas of the conterminous western United States (by states). Silver Spring, MD: U.S. National Weather Service *NOAA Atlas* 2 (11 volumes).

Miller, N., and R. Cronshey (1989). Runoff curve numbers - the next step. In B.C. Yen, ed., *Channel Flow and Catchment Runoff*, Department of Civil Engineering, University of Virginia, p. 910–916.

Milliman, J.D., and J.P.M. Syvitski (1992). Geomorphic/tectonic control of sediment discharge to the ocean: The importance of small mountain rivers. *Journal of Geology* 100:525–544.

Milly, P.C.D. (1988). Advances in modeling of water in the unsaturated zone. *Transport in Porous Media* 3:491–514.

Mimikou, M. and S. Kamaki (1985). Regionalization of flow duration curve characteristics. *Journal of Hydrology* 82:77–91.

Mishra, S., J.C. Parker, and N. Singhal (1989). Estimation of soil hydraulic properties and their uncertainty from particle size distribution data. *Journal of Hydrology* 108:1–18.

Mitchell, J.F.B. (1989). The "greenhouse" effect and climatic change. *Reviews of Geophysics* 27:115–139.

Mitchell, J.F.B., T.C. Johns, J.M. Gregory, and S.F.B. Tett (1995). Climate response to increasing levels of greenhouse gases and sulphate aerosols. *Nature* 376:501–504.

Mitsch, W.J., and J.G. Gosselink (1993). *Wetlands* (2nd ed.). New York, NY: Van Nostrand Reinhold.

Moench, A.F., V.B. Sauer, and E.M. Jennings (1974). Modification of routed streamflow by channel loss and base flow. *Water Resources Research* 10:963–968.

Monteith, J.L. (1965). Evaporation and environment. In New York, NY:Cambridge University Press, *Proceedings of the 19th Symposium of the Society for Experimental Biology* p. 205–233.

Morris, D.A., and A.I. Johnson (1967). Summary of hydrologic and physical properties of rock and soil materials, as analyzed by the Hydrologic Laboratory of the U.S. Geological Survey 1948–1960. U.S. Geological Survey *Water-Supply Paper* 1839-D.

Morris, E.M., ed. (1986). Modeling snowmelt-induced processes. Wallingford, Oxon, U.K.: International Association of Hydrological Sciences *IAHS Publication* 155.

Morrissey, D.J. (1987). Estimation of the recharge area contributing water to a pumped well in a glacial-drift, river-valley aquifer. U.S. Geological Survey *Open-File Report* 86–543.

Morrissey, M.L., J.A. Maliekal, J.S. Greene, and J. Wang (1995). The uncertainty of simple spatial averages using rain gauge networks. *Water Resources Research* 31:2011–2017.

Mosley, M.P. (1979). Streamflow generation in a forested watershed, New Zealand. *Water Resources Research* 15:795–806.

Mosley, M.P. (1982). Subsurface flow velocities through selected forest soils, South Island, New Zealand. *Journal of Hydrology* 55:65–92.

Mostaghimi, S., and J.K. Mitchell (1982). Peak runoff model comparison on central Illinois watersheds. *Water Resources Bulletin* 18:9–18.

Motoyama, H. (1990). Simulation of seasonal snowcover based on air temperature and precipitation. *Bulletin of the American Meteorological Society* 29:1104–1110.

Mualem, Y. (1976). A new model for predicting the hydraulic conductivity of unsaturated porous media. *Water Resources Research* 12:513–522.

Mukammal, E.I., and H.H. Neumann (1977). Application of the Priestly-Taylor evaporation model to assess the influence of soil moisture on the evaporation from a large weighing lysimeter and Class A pan. *Boundary-Layer Meteorology* 12:243–256.

Mulholland, P.J. (1993). Hydrometric and stream chemistry evidence of three storm flowpaths in Walker Branch Watershed. *Journal of Hydrology* 151:291–316.

Munley, W.G., Jr., and L.E. Hipps (1991). Estimation of regional evaporation for a tall-grass prairie from measurements of properties of the atmospheric boundary layer. *Water Resources Research* 27:225–230.

Musiake, K., Y. Oka, and M. Koike (1988). Unsaturated zone soil moisture behavior under temperate humid conditions - ten-

siometric observations and numerical simulations. *Journal of Hydrology* 102:179–200.

Myneni, R.B., et al. (1997). Increased plant growth in the northern high latitudes from 1981 to 1991. *Nature* 386:698–702.

Nace, R.L. (1974). General evolution of the concept of the hydrological cycle. In Paris: UNESCO-World Meteorological Organization-International Association of Hydrological Sciences, *Three Centuries of Scientific Hydrology*, p. 40–48.

Nash, J.E., and J.V. Sutcliffe (1970). River flow forecasting through conceptual models, Part I: A discussion of principles. *Journal of Hydrology* 10:282–290.

Nassif, S.H., and E.M. Wilson (1975). The influence of slope and rain intensity on runoff and infiltration. *Hydrological Sciences Bulletin* 20:539–553.

National Academy of Sciences (1983). *Safety of Existing Dams: Evaluation and Improvement*. Washington, DC: National Academy Press.

National Environment Research Council (1975). *Flood Studies Report*. London, U.K.: National Environment Research Council.

Neuman, S. (1976). Wetting front pressure head in the infiltration model of Green and Ampt. *Water Resources Research* 12:564–566.

Newbury, R.W., J.A. Cherry, and R.A. Cox (1969). Groundwater-streamflow systems in Wilson Creek Experimental Watershed, Manitoba. *Canadian Journal of Earth Sciences* 6:613–623.

Newson, M. (1992). *Land, Water, and Development: River Basin Systems and Their Sustainable Development*. London, Routledge, 351 p.

Nichols, W.D. (1993). Estimating discharge of shallow groundwater by transpiration from greasewood in the northern Great Basin. *Water Resources Research* 29:2771–2778.

Nichols, W.D. (1994). Groundwater discharge by phreatophyte shrubs in the Great Basin as related to depth to groundwater. *Water Resources Research* 30:3265–3274.

Nielsen, D.R., D. Kirkham, and W.R. Van Wijk (1961). Diffusion equation calculations of field soil water infiltration profiles. *Soil Science Society of America Proceedings* 25:165–168.

Nielsen, D.R., M.T. van Genuchten, and J.W. Biggar (1986). Water flow and solute transport processes in the unsaturated zone. *Water Resources Research* 22:89S–108S.

Novakowski, K.S., and R.W. Gillham (1988). Field investigations of the nature of water-table response to precipitation in shallow water-table environments. *Journal of Hydrology* 97:23–32.

Nutter, W.L. (1975). Moisture and energy conditions in a draining soil mass. Atlanta, GA: Georgia Institute of Technology Environmental Resources Center *ERC* 0875.

Nwaogazie, F.I.L. and A.K. Tyagi (1984). Unified streamflow routing by finite elements. *American Society of Civil Engineers Proceedings, Journal of the Hydraulics Division* 110:1595–1611.

Olin, M.H.E., and C. Svensson (1992). Evaluation of geological and recharge parameters for an aquifer in southern Sweden. *Nordic Hydrology* 23:305–314.

Oliver, J.E., and J.J. Hidore (1984). *Climatology: An Introduction*. Columbus, OH: C.E. Merrill.

O'Loughlin, E.M. (1981). Saturation regions in catchments and their relations to soil and topographic properties. *Journal of Hydrology* 53:229–246.

Omernik, J.M. (1976). The influence of land use on stream nutrient levels. U.S. Environmental Protection Agency. *Ecological Research Series* EPA-600/3–76–014.

Omolayo, A.S. (1993). On the transposition of areal reduction factors for rainfall frequency estimation. *Journal of Hydrology* 145:191–205.

Ophori, D., and J. Toth (1990). Relationships in regional groundwater discharge to streams: An analysis by numerical simulation. *Journal of Hydrology* 119:215–244.

Ormsbee, L.E., and A.Q. Khan (1989). A parametric model for steeply sloping forested watersheds. *Water Resources Research* 25:2053–2065.

Orville, H.D. (1995). Report on the Sixth WMO Scientific Conference on Weather Modification. *Bulletin of the American Meteorological Society* 76: 372–373.

Osborn, C.T., J.E. Schefter, and L. Shabman (1986). The accuracy of water use forecasts: Evaluation and implications. *Water Resources Bulletin* 22:101–109.

Osterkamp, W.R., L.J. Lane, and C.S. Savard (1994). Recharge estimates using a geomorphic/distributed-parameter simulation approach, Amargosa River Basin. *Water Resources Bulletin* 30:493–507.

Paetzold, R.F., G.A. Matzkanin, and A. De Los Santos (1985). Surface soil water content measurement using pulsed nuclear magnetic resonance techniques. *Soil Science Society of America Journal* 49:537–540.

Parlange, M.B., and G.G. Katul (1992a). Estimation of diurnal variation of potential evaporation from a wet bare soil surface. *Journal of Hydrology* 132:71–89.

Parlange, M.B., and G.G. Katul (1992b). An advection-aridity evaporation model. *Water Resources Research* 28:127–132.

Paulhus, J.L.H. (1965). Indian Ocean and Taiwan rainfalls set new records. *Monthly Weather Review* 93:331–335.

Peck, A.J., and R.M. Rabbidge (1966). Soil water potential: Direct measurement by a new technique. *Science* 151:1385–1386.

Peck, E.L. (1997). Quality of hydrometeorological data in cold regions. *Journal of the American Water Resources Association* 33:125–134.

Peltier, W.R., and A.M. Tushingham (1991). Influence of glacial isostatic adjustment on tide gauge measurement of secular sea level change. *Journal of Geophysical Research* 96:6779–6796.

Penman, H.L. (1948). Natural evaporation from open water, bare soil, and grass. *Royal Society of London Proceedings*, Series A, 193:120–145.

Penman, H.L. (1956). Evaporation: An introductory survey. *Netherlands Journal of Agricultural Science* 4:7–29.

**634**

Perrens, S.J., and K.K. Watson (1977). Numerical analysis of two-dimensional infiltration and redistribution. *Water Resources Research* 13:781–790.

Perry, C.A. (2000). Significant floods in the United States during the 20th Century. U.S. Geological Survey *Fact Sheet* 024–00.

Perry, E.C., Jr., and C.W. Montgomery, eds. (1982). *Isotope Studies of Hydrologic Processes*. DeKalb, IL: Northern Illinois University Press.

Person, M.A., J.Z. Taylor, and S.L. Dingman (1998). Sharp interface models of salt water intrusion and wellhead delineation on Nantucket Island, Massachusetts. *Groundwater* 36:731–742.

Peters, D.L., J.M. Buttle, C.H. Taylor, and B.D. Lazerte (1995). Runoff production in a forested, shallow soil, Canadian Shield basin. *Water Resources Research* 31:1291–1304.

Petersen, M.S. (1984). *Water Resource Planning and Development*. Englewood Cliffs, NJ: Prentice-Hall.

Peterson, T.C., V.S. Golubev, and P. Ya. Groisman (1995). Evaporation losing its strength. *Nature* 377:687–688.

Philip, J.R. (1957). The theory of infiltration: 4. Sorptivity and algebraic infiltration equations. *Soil Science* 84:257–264.

Philip, J.R. (1969). Theory of infiltration. *Advances in Hydroscience* 5:215–290.

Piechota, T.C., J.A. Dracup, and R.G. Fovell (1997). Western US streamflow and atmospheric circulation patterns during El Niño-Southern Oscillation. *Journal of Hydrology* 201:249–271.

Piexoto, J.P. and Oort, A.H. (1992). *Physics of Climate*. New York, NY: American Institute of Physics.

Pilgrim, D.H. (1976). Travel times and nonlinearity of flood runoff from tracer measurements on small watersheds. *Water Resources Research* 12:487–496.

Pilgrim, D.H., and I. Cordery (1992). Flood runoff. Chapter 9 in Maidment (1992).

Pitman, J.I. (1989). Rainfall interception by bracken litter - relationship between biomass, storage, and drainage rate. *Journal of Hydrology* 111:281–291.

Poff, N.L., et al. (1997). The natural flow regime: A paradigm for river conservation and restoration. *BioScience* 47:769–784.

Ponce, V.M. (1989). *Engineering Hydrology*. Englewood Cliffs, NJ: Prentice-Hall.

Postel, S.L., G.C. Daily, and P.R. Ehrlich (1996). Human appropriation of renewable fresh water. *Science* 271:785–788.

Priestley, C.H.B., and R.J. Taylor (1972). On the assessment of surface heat flux and evaporation using large-scale parameters. *Monthly Weather Review* 100:81–92.

Prince, K.R., T.E. Reilly, and O.L. Franke (1989). Analysis of the shallow groundwater flow system near Connetquot Brook, Long Island, New York. *Journal of Hydrology* 107:223–250.

Probst, J.L., and Y. Tardy (1987). Long range streamflow and world continental runoff fluctuations since the beginning of this century. *Journal of Hydrology* 94:289–311.

Putuhena, W.M., and I. Cordery (1996). Estimation of interception capacity of the forest floor. *Journal of Hydrology* 180:283–299.

Quimpo, R.G., A.A. Alejandrino, and T.A. McNally (1983). Regionalized flow-duration curves for Philippines. *Journal of Water Resources Planning and Management* 109:320–330.

Radke, L.F., and P.V. Hobbs (1976). Cloud condensation nuclei on the Atlantic seaboard of the United States. *Science* 193:999–1002.

Ragab, R., J. Finch, and R. Harding (1997). Estimation of groundwater recharge to chalk and sandstone aquifers using simple soil models. *Journal of Hydrology* 190:19–41.

Ragan, R.M. (1966). An experimental investigation of partial area contributions. International Association of Scientific Hydrology *Publication* 76:241–249.

Rainey, R.H. (1967). Natural displacement of pollution from the Great Lakes. *Science* 155:1241–1243.

Ramanathan, V. (1988). The greenhouse theory of climatic change: A test by an inadvertent global experiment. *Science* 240:293–299.

Randall, A.D., and A.I. Johnson, eds. (1988). The Northeast glacial aquifers. American Water Resources Association *Monograph Series* No. 11.

Rango, A., and J. Martinec (1994). Model accuracy in snowmelt-runoff forecasts. *Water Resources Bulletin* 30:463–470.

Rasmussen, A.H., M. Hondzo, and H.G. Stefan (1995). A test of several evaporation equations for water temperature simulations in lakes. *Journal of the American Water Resources Association* 31:1023–1028.

Rasmussen, E. M. (1985). El Nino and variations in climate. *American Scientist* 73:168–177.

Rasmussen, W.C., and G.E. Andreasen (1959). Hydrologic budget of the Beaverdam Creek basin, Maryland. U.S. Geological Survey *Water-Supply Paper* 1472.

Raven, P.H., R.F. Evert, and H. Curtis (1976). *The Biology of Plants* (2nd ed.). New York, NY: Worth Publishers, Inc.

Rawls, W.J., L.R. Ahuja, D.L. Brakensiek, and A Shirmohammadi (1992). Infiltration and soil water movement. Chapter 5 in Maidment (1992).

Reid, G.C. (1987). Influence of solar variability on global sea surface temperatures. *Nature* 329:142–143.

Reidel, J.T., and L.C. Schreiner (1980). Comparison of generalized estimates of probable maximum precipitation with greatest observed rainfalls. Silver Spring, MD: U.S. National Weather Service *Technical Report NWS* 25.

Reinhart, K.G., and R.S. Pierce (1964). Stream-gaging stations for research on small watersheds. U.S. Department of Agriculture *Agriculture Handbook* 268.

Reisner, M. (1986). *Cadillac Desert*. New York, NY: Viking Penguin.

Rice, K., and G.M. Hornberger (1998). Comparison of hydrochemical tracers to estimate source contributions to peak flow in a small, forested, headwater catchment. *Water Resources Research* 34:1755–1766.

Richards, A.L. (1931). Capillary conduction of liquids through porous media. *Physics* 1:316–333.

Richey, J.E., C. Nobre, and C. Deser (1989). Amazon River discharge and climate variability: 1903 to 1985. *Science* 246:101–103.

Riehl, H., and J. Meitin (1979). Discharge of the Nile River: A barometer of short-period climatic fluctuation. *Science* 206:1178–1179.

Riggs, H.C. (1972). Low-flow investigations. Chapter B1, Book 4 of *Techniques of Water-Resources Investigations of the United States Geological Survey*.

Riggs, H.C., and K.D. Harvey (1990) Temporal and spatial variability of streamflow. In M.G. Wolman and H.C. Riggs, eds. *Surface Water Hydrology* (Vol. O-1, *Geology of North America*). Boulder, CO: The Geological Society of America.

Rikhter, G.D. (1954). *Snowcover, Its Formation and Properties*. Moskva, USSR: Izdatel'stvo Akademia Nauk SSSR. Translation No. 6, US Army Snow, Ice, and Permafrost Research Establishment.

Roberge, J., and A. P. Plamondon (1987). Snowmelt runoff pathways in a boreal forest hillslope, The role of pipe throughflow. *Journal of Hydrology* 95:39–54.

Robinson, D.A., K.F. Dewey, and R.R. Heim, Jr. (1993). Global snow cover monitoring: an update. *Bulletin of the American Meteorological Society* 74:1689–1696.

Rodda, J.C. (1985). Precipitation research. *Eos* (*American Geophysical Union Transactions*) 8 Jan 1985:10.

Rodriguez-Iturbe, I. (1993). The geomorphological unit hydrograph. Chapter 3 in K. Beven and M.J. Kirkby, eds. *Channel Network Hydrology*, New York, John Wiley and Sons.

Rodriguez-Iturbe, I., G. Devoto, and J.B. Valdes (1979). Discharge response analysis and hydrologic similarity: The interrelation between the geomorphologic IUH and the storm characteristics. *Water Resources Research* 15:1435–1444.

Rodriguez-Iturbe, I., and J.M. Mejia (1974). Design of rainfall networks in time and space. *Water Resources Research* 10:713–728.

Rodriguez-Iturbe, I. and J.B. Valdes (1979). The geomorphologic structure of hydrologic response. *Water Resources Research* 15:1409–1420.

Rorabaugh, M.I. (1964). Estimating changes in bank storage and ground-water contribution to streamflow. *International Association of Scientific Hydrology Publication* 63:432–441.

Rosenberg, N.J., M.S. McKenney, and P. Martin (1989). Evapotranspiration in a greenhouse-warmed world: A review and a simulation. *Agricultural and Forest Meteorology* 47:303–320.

Rosenberry, D.O. and T.C. Winter (1997). Dynamics of water-table fluctuations in an upland between two prairie-pothole wetlands in North Dakota. *Journal of Hydrology* 191:266–289.

Rosenberry, D.O., A.M. Sturrock, and T.C. Winter (1993). Evaluation of the energy budget method of determining evaporation at Williams Lake, Minnesota, using alternative instrumentation and study approaches. *Water Resources Research* 29:2473–2483.

Rosenthal, W., and J. Dozier (1996). Automated mapping of montane snow cover at subpixel resolution from the Landsat Thematic Mapper. *Water Resources Research* 32:115–130.

Roth, K., R. Schulin, H. Fluhler, and W. Attinger (1990). Calibration of time domain reflectometry for water content measurement using a composite dielectric approach. *Water Resources Research* 26:2267–2273.

Rouse, W.R., and R.G. Wilson (1972). A test of the potential accuracy of the water-budget approach to estimating evapotranspiration. *Agricultural Meteorology* 9:421–446.

Rubin, J. (1967). Numerical method for analyzing hysteresis-affected post-infiltration redistribution of soil moisture. *Soil Science Society of America Proceedings* 31:13–20.

Rushton, K.R., and C. Ward (1979). The estimation of groundwater recharge. *Journal of Hydrology* 41:345–361.

Rutter, A.J., K.A. Kershaw, P.C. Robins, and A.J. Morton (1971). A predictive model of rainfall interception in forests; 1. Derivation of the model from observations in a plantation of Corsican pine. *Agricultural Meteorology* 9:367–384.

Sabol, G.V., and T.J. Ward (1983). Discussion of Cheng (1982). *American Society of Civil Engineers Proceedings, Journal of the Hydraulics Division* 109:344–346.

Sacks, L.A., J.S. Herman, L.F. Konikow, and A.L. Vela (1992). Seasonal dynamics of groundwater-lake interactions at Doñana National Park, Spain. *Journal of Hydrology* 130: 123–134.

Sahagian, D.L., F.W. Schwartz, and D.K. Jacobs (1994). Direct anthropogenic contributions to sea level rise in the twentieth century. *Nature* 367:54–57.

Salama, R.B., I. Tapley, T. Ishii, and G. Hawkes (1994). Identification of areas of recharge and discharge using Landsat-TM satellite imagery and aerial photography mapping techniques. *Journal of Hydrology* 162:119–141.

Salas, J.D. (1992). Analysis and modeling of hydrologic time series. Chapter 19 in Maidment (1992).

Salvucci, G.D. (1996). Series solution for Richards equation under concentration boundary conditions and uniform initial conditions. *Water Resources Research* 32:2401–2407.

Salvucci, G.D. (1997). Soil and moisture independent estimation of stage-two evaporation from potential evaporation and albedo or surface temperature. *Water Resources Research* 33:111–122.

Salvucci, G.D., and D. Entekhabi (1994a). Explicit expressions for Green-Ampt (delta function diffusivity) infiltration rate and cumulative storage. *Water Resources Research* 30:2661–2661.

Salvucci, G.D., and D. Entekhabi (1994b). Equivalent steady soil moisture profile and the time compression approximation in water balance modeling. *Water Resources Research* 30:2737–2749.

Salvucci, G.D., and D. Entekhabi (1995). Ponded infiltration into soils bounded by a water table. *Water Resources Research* 31:2751–2759.

Santer, B., et al. (1996). A search for human influences on the thermal structure of the atmosphere. *Nature* 382:39–46.

Savenije, H.H.G. (1995). New definitions for moisture recycling and the relationship with land-use changes in the Sahel. *Journal of Hydrology* 167:57–78.

Schaefer, G.L., and J.G. Werner (1996). SNO-TEL into the year 2000. Paper presented at American Meteorological Society 12th Conference on Biometeorology and Aerobiology, Atlanta, GA, 28 Jan-2 Feb 1996.

Schimmrich, S.H. (1997). Exploring geology on the World-Wide Web - hydrology and hydrogeology. *Journal of Geoscience Education* 45:173–176.

Schindler, D.W. (1988). Effects of acid rain on freshwater ecosystems. *Science* 239:149–157.

Schmidt, R.A., and D.R. Gluns (1991). Snowfall interception on branches of three conifer species. *Canadian Journal of Forest Research* 21: 1262–1269.

Schmidt, R.A., and C.A. Troendle (1989). Snowfall into a forest clearing. *Journal of Hydrology* 110:335–348.

Schneider, W.J., and J.E. Goddard (1974). Extent and development of urban flood plains. U.S. Geological Survey *Circular* 601-J.

Schumm, S.A., and G.C. Lusby (1963). Seasonal variation of infiltration capacity and runoff on hillslopes in western Colorado. *Journal of Geophysical Research* 68: 3655–3666.

Scotter, D.R., B.E. Clothier, and E.R. Harper (1982). Measuring saturated hydraulic conductivity and sorptivity using twin rings. *Australian Journal of Soil Research* 20:295–304.

Searcy, J.K. (1959). Flow-duration curves. U.S. Geological Survey *Water-Supply Paper* 1542-A.

Searcy, J.K., and C.H. Hardison (1960). Double-mass curves. U.S. Geological Survey *Water-Supply Paper* 1541-B.

Sellers, W.D. (1965). *Physical Climatology.* Chicago, IL: University of Chicago Press.

Sellmann, P.V., and S.L. Dingman (1970). Prediction of stream frequency from maps. *Journal of Terramechanics* 7:101–115.

Sene, K.J., J.H.C. Gash, and D.D. McNeil (1991). Evaporation from a tropical lake: Comparison of theory with direct measurements. *Journal of Hydrology* 127:193–217.

Shafer, B.A., and L.E. Dezman (1982). Development of a surface water supply index (SWSI) to assess the severity of drought conditions in snowpack runoff areas. In *50th Annual Western Snow Conference* (Reno, NV) *Proceedings.*

Sharma, M.L., G.A. Gander, and G.C. Hunt (1980). Spatial variability of infiltration in a watershed. *Journal of Hydrology* 45:101–122.

Shaw, E.M, and P.P. Lynn (1972). Areal rainfall evaluation using two surface fitting techniques. *Hydrological Sciences Bulletin* 17:419–433.

Shaw, E.M. (1988). *Hydrology in Practice* (2nd ed.). London, U.K.: Van Nostrand-Reinhold.

Shaw, R.E., and E.E. Prepas (1990). Groundwater-lake interactions: I. Accuracy of seepage meter estimates of lake seepage. *Journal of Hydrology* 119:105–120.

Shiklomanov, I.A., and A.A. Sokolov (1983). Methodological basis of world water balance investigation and computation. In *New Approaches in Water Balance Computations.* International Association for Hydrological Sciences Publ. No. 148. (Proceedings of the Hamburg Symposium).

Shukla, J., and Y. Mintz (1982). Influence of land-surface evapotranspiration on the earth's climate. *Science* 215:1498–1500.

Shuttleworth, W.J. (1979). Evaporation. Wallingford, Oxon., U.K.: Institute of Hydrology *Report* 56.

Shuttleworth, W.J. (1988). Macrohydrology - the new challenge for process hydrology. *Journal of Hydrology* 100:31–56.

Shuttleworth, W.J. (1992). Evaporation. Chapter 4 in Maidment (1992).

Shuttleworth, W.J. and J.S. Wallace (1985). Evaporation from sparse crops—an energy combination theory. *Quarterly Journal of the Royal Meteorological Society* 111:838–855.

Silliman, S.E., J. Ramirez, and R.L. McCabe (1995). Quantifying downflow through creek sediments using temperature time series: one-dimensional solution incorporting measured surface temperature. *Journal of Hydrology* 167:99–119.

Simpson, J.J., J.R. Stitt, and M. Sienko (1998). Improved estimates of the areal extent of snow cover from AVHRR data. *Journal of Hydrology* 204:1–23.

Simpson, R.H., and H. Riehl (1981). *The Hurricane and Its Impact.* Baton Rouge, LA: Louisiana State University Press.

Singh, K.P. (1971). Model flow duration and streamflow variability. *Water Resources Research* 7:1031–1036.

Singh, P., G. Spitzbart, H. Hübl, and H.W. Weinmeister (1997). Hydrological response of snowpack under rain-on-snow events: A field study. *Journal of Hydrology* 202:1–20.

Singh, V.P., and P.K. Chowdhury (1986). Comparing some methods of estimating mean areal rainfall. *Water Resources Bulletin* 22:275–282.

Singh, V.P., L. Bengtsson, and G. Westerstrom (1997). Kinematic wave modeling of saturated basal flow in a snowpack. *Hydrological Processes* 11:177–187.

Sklash, M.G., and R.N. Farvolden (1979). The role of groundwater in storm runoff. *Journal of Hydrology* 43:45–65.

Slatyer, R.O., and I.C. McIlroy (1961). *Practical Microclimatology.* Melbourne, Australia: CSIRO.

Slaughter, C.W. (1966). *Evaporation from snow and evaporation retardation by monomolecular films.* Colorado Springs, CO: Watershed Management Unit, Colorado State University.

Sloan, P.G., and I.D. Moore (1984). Modeling subsurface stormflow on steeply sloping forested watersheds. *Water Resources Research* 20:1815–1822.

Smith, L., and S.W. Wheatcraft (1992). Groundwater flow. Chapter 6 in Maidment (1992).

Smith, P.J., and R.S. Wikramaratna (1981). A method for estimating recharge and boundary flux from groundwater level observations. *Hydrological Sciences Bulletin* 26:113–136.

Smith, R.A., R.B. Alexander, and K.J. Lanfear (1993). Stream water quality in the conterminous United States – Status and trends of selected indicators during the 1980s. In. U.S. Geological Survey *Water-Supply Paper* 2400:67–92.

Smith, R.A., R.B. Alexander, and M.G. Wolman (1987a). Analysis and interpretation of water-quality trends in major U.S. rivers. U. S. Geological Survey *Water-Supply Paper* 2307.

Smith, R.A., R.B. Alexander, and M.G. Wolman (1987b). Water-quality trends in the nation's rivers. *Science* 235:1607–1615.

Smoot, G.F., and C.E. Novak (1968). Calibration and maintenance of vertical-axis type current meters. U.S. Geological Survey *Techniques of Water-Resources Investigations* Book 8 Chapter B2.

Solley, W.B., R.R. Pierce, and H.A. Perlman (1998). Estimated use of water in the United States in 1995. U.S. Geological Survey *Circular* 1200, 71 p.

Sommerfield, R.A., and J.E. Rocchio (1993). Permeability measurements on new and equitemperature snow. *Water Resources Research* 29:2485–2490.

Sophocleous, M. (1985). The role of specific yield in ground-water recharge estimations. *Ground Water* 23:52–58.

Sophocleous, M., and C.A. Perry (1984). Experimental studies in natural groundwater-recharge dynamics: Assessment of recent advances in instrumentation. *Journal of Hydrology* 70:369–382.

Sophocleous, M., and C.A. Perry (1985). Experimental studies in natural groundwater-recharge dynamics: The analysis of observed recharge events. *Journal of Hydrology* 81:297–332.

Souch, C., C.P. Wolfe, and C.S.B. Grimmond (1996). Wetland evaporation and energy partitioning: Indiana Dunes National Lakeshore. *Journal of Hydrology* 184:189–208.

Space, M.L., N.L. Ingraham, and J.W. Hess (1991). The use of stable isotopes in quantifying groundwater discharge to a partially diverted creek. *Journal of Hydrology* 129:175–193.

Spittlehouse, D.L., and T.A. Black (1981). A growing season water balance model applied to two Douglas fir stands. *Water Resources Research* 17:1651–1656.

Springer, E.P., and G.F. Gifford (1980). Spatial variability of rangeland infiltration rates. *Water Resources Bulletin* 16:550–552.

Stannard, D.I., and D.O. Rosenberry (1991). A comparison of short-term measurements of lake evaporation using eddy correlation and energy budget methods. *Journal of Hydrology* 122:15–22.

Stedinger, J. R., R.M. Vogel, and E. Foufoula-Georgiu (1992). Frequency analysis of extreme events. Chapter 18 in Maidment (1992).

Steenhuis, T.S., and W.H. Van Der Molen (1986). The Thornthwaite-Mather procedure as a simple engineering method to predict recharge. *Journal of Hydrology* 84:221–229.

Steenhuis, T.S., C.D. Jackson, S.K.J. Kung, and W. Brutsaert (1985). Measurement of groundwater recharge on eastern Long Island, New York, U.S.A. *Journal of Hydrology* 79:145–169.

Stein, J., H.G. Jones, J. Roberge, and W. Sochanska (1986). The prediction of both runoff quality and quantity by the use of an integrated snowmelt model. In Morris (1986).

Stephens, D.B. (1995). *Vadose Zone Hydrology*. Boca Raton, FL, CRC Press, Inc. 339 pp.

Stephens, D.B., and R. Knowlton, Jr. (1986). Soil water movement and recharge through sand at a semiarid site in New Mexico. *Water Resources Research* 22:881–889.

Stephenson, D., and M.E. Meadows (1986). *Kinematic Hydrology and Modeling*. Developments in Water Science 26. New York, NY: Elsevier Publishing Co, Inc.

Stewart, J.B. (1977). Evaporation from the wet canopy of a pine forest. *Water Resources Research* 13:915–921.

Stewart, J.B. (1988). Modeling surface conductance of pine forest. *Agricultural and Forest Meteorology* 43:17–35.

Stewart, J.B. (1989). On the use of the Penman-Monteith equation for determining areal evapotranspiration. In T.A. Black et al., eds. Estimation of Areal Evapotranspiration. Wallingford, Oxon., U.K.: International Association of Hydrological Sciences *Publication* 177:3–12.

Stewart, J.B., and L.W. Gay (1989). Preliminary modelling of transpiration from the FIFE site in Kansas. *Agricultural and Forest Meteorology* 48:305–315.

Stillinger, F.H. (1980). Water revisited. *Science* 209:451–455.

Stoddard, J.L. et al. (1999). Regional trends in aquatic recovery from acidification in North America and Europe. *Nature* 401:575–578.

Strahler, A.N. (1952). Hypsometric (area-altitude) analysis of erosional topography. *Geological Society of America Bulletin* 63:1117–1142.

Streeter, H.W., and E.B. Phelps (1925). A study of the pollution and natural purification of the Ohio River, III. Factors concerned in the phenomena of oxidation and reaeration. U.S. Public Health Service *Bulletin* 146.

Strelkoff, T. (1970). Numerical solution of Saint Venant equations. *American Society of Civil Engineers Proceedings, Journal of the Hydraulics Division* 96:223–252.

Stresky, S.J. (1991). Morphology and Flow Characteristics of Pipes in a Forested New England Hillslope. Durham, NH: Thesis, M.Sc. in Hydrology, University of New Hampshire.

Swan, A.R.H., and M. Sandilands (1995). *Introduction to Geological Data Analysis*. Oxford, UK: Blackwell Scientific.

Swartzendruber, D. (1997). Exact mathematical derivation of a two-term infiltration equation. *Water Resources Research* 33:491–496.

Swartzendruber, D., and T.C. Olson (1961a). Sand-model study of buffer effects in the double-ring infiltrometer. *Soil Science Society of America Proceedings* 25:5–8.

Swartzendruber, D., and T.C. Olson (1961b). Model study of the double-ring infiltrometer as affected by depth of wetting and particle size. *Soil Science* 92:219–225.

Swistock, B.R., D.R. DeWalle, and W.E. Sharpe (1989). Sources of acidic storm flow in an Appalacian headwater stream. *Water Resources Research* 25:2139–2147.

Tabios, G.Q, and J.D. Salas (1985). A comparative analysis of techniques for spatial interpolation of precipitation. *Water Resources Bulletin* 21:365–380.

Tallaksen, L.M. (1995). A review of baseflow recession analysis. *Journal of Hydrology* 165:349–370.

Tani, M. (1997). Runoff generation processes estimated from hydrological observations on a steep forested hillslope with a thin soil layer. *Journal of Hydrology* 200:84–109.

Taniguchi, M., and M.L. Sharma (1990). Solute and heat transport experiments for estimating recharge rate. *Journal of Hydrology* 119:57–69.

Tarboton, D.G. (1997). A new method for the determination of flow directions and upslope areas in grid digital elevation models. *Water Resources Research* 33:309–319.

Taylor, G.H., C. Daly, and W.P. Gibson (1993). Development of a new Oregon annual precipitation map using the PRISM model. *The State Climatologist* 17(2):1–4.

Tchobanoglous, G., and E.D. Schroeder (1985). *Water Quality*. Reading, MA: Addison-Wesley.

Tett, F.B., J.F.B. Mitchell, D.E. Parker, and M.R. Allen (1996). Human influence on the atmospheric vertical structure: Detection and observations. *Science* 274:1170–1173.

Theis, C.V. (1935). The relation between the lowering of the piezometric surface and the rate and duration of discharge of a well using groundwater storage. *American Geophysical Union Transactions* 2:519–524.

Thiery, D. (1988). Forecast of changes in piezometric levels by a lumped hydrological model. *Journal of Hydrology* 97:129–148.

Thiessen, A.H. (1911). Precipitation for large areas. *Monthly Weather Review* 39:1082–1084.

Thornthwaite, C.W. (1948). An approach toward a rational classification of climate. *Geographical Review* 38:55–94.

Thornthwaite, C.W., and J.R. Mather (1955). The water balance. Philadelphia, PA: Drexel Institute of Technology, Climatological Laboratory *Publication* 8.

Todd, D.K. (1953). Sea water intrusion in coastal aquifers. *American Geophysical Union Transactions* 34:749–754.

Todd, D.K., ed. (1970). *The Water Encyclopedia*. Port Washington, NY: Water Information Center, Inc.

Topp, G.C., J.L. Davis, and A.P. Annan (1980). Electromagnetic determination of soil water content: Measurements in coaxial transmission lines. *Water Resources Research* 16:574–582.

Torres, R., et al. (1998). Unsaturated zone processes and the hydrologic response of a steep, unchanneled catchment. *Water Resources Research* 34:1865–1879.

Tóth, J. 1962. A theory of groundwater motion in small drainage basins in central Alberta, Canada. *Journal of Geophysical Research* 67:4375–4387.

Tóth, J. 1963. A theoretical analysis of ground-water flow in small drainage basins. *Journal of Geophysical Research* 68:4795–4812.

Trenberth, K.E., G.W. Branstator, and P.A. Arkin (1988). Origins of the 1988 North American drought. *Science* 242:1640–1645.

Tricker, A.S. (1978). The infiltration cylinder: Some comments on its use. *Journal of Hydrology* 36:383–391.

Tricker, A.S. (1979). The design of a portable rainfall simulator infiltrometer. *Journal of Hydrology* 41:143–147.

Tricker, A.S. (1981). Spatial and temporal patterns of infiltration. *Journal of Hydrology* 49:261–277.

Trimble, S.W. (1975). Denudation studies: Can we assume stream steady state? *Science* 188:1207–1208.

Troch, P.A., F.P. De Troch, and W. Brutsaert (1993). Effective water table depth to describe initial conditions prior to storm rainfall in humid regions. *Water Resources Research* 29:427–434.

Troxell, H.C., et al. (1942). Characteristic spatial variation of orographic precipitation. U.S. Geological Survey *Water-Supply Paper* 844.

Tseng, P.-H., T.H. Illangasekare, and M.F. Meier (1994) Modeling of snow melting and uniform wetting front migration in a layered subfreezing snowpack. *Water Resources Research* 30:2363–2376.

Tsonis, A.A. (1996). Widespread increases in low-frequency variability of precipitation over the past century. *Nature* 382:700–702.

U.S. Geological Survey (1954). Water-loss investigations: Lake Hefner studies, base data report. U.S. Geological Survey *Professional Paper* 270.

U.S. Geological Survey (1984). National water summary 1983 - Hydrologic events and issues. U.S. Geological Survey *Water-Supply Paper* 2250.

U.S. Geological Survey (1985). National water summary 1984 - Hydrologic events, selected water-quality trends, and ground-water resources. U.S. Geological Survey *Water-Supply Paper* 2275.

U.S. Geological Survey (1986). National water summary 1985 - Hydrologic events and surface-water resources. U.S. Geological Survey *Water-Supply Paper* 2300.

U.S. Geological Survey (1988). National water summary 1986 - Hydrologic events and ground-water quality. U.S. Geological Survey *Water-Supply Paper* 2325.

U.S. Soil Conservation Service (1972). Hydrology. Section 4, *SCS National Engineering Handbook*. U.S. Soil Conservation Service.

U.S. Soil Conservation Service (1983). *Flood Plain Management Plan Study, Ammonoosuc River, Lisbon NH*. Durham, NH: U.S. Soil Conservation Service and NH Office of State Planning.

U.S. Soil Conservation Service (1986). Lisbon flood prevention RC&D measure plan and environmental assessment. Woodsville, NH: Town of Lisbon and Grafton County Conservation District, NH *RC&D Measure Number* 33–6001–009–340.

U.S. Water Resources Council (1973). Water and related land resources: Establishment of principles and standards for planning. *U.S. Federal Register* 38(174-III) (10 September 1973).

U.S. Water Resources Council (1978). *The Nation's Water Resources*. U.S. Water Resources Council.

U.S. Water Resources Council (1982). Economic and environmental principles and guidelines for water and related land resources implementation studies. *U.S. Federal Register* 47(55) (22 March 1982).

Urbonas, B.R., and L.A. Roesner (1992). Hydrologic design for urban drainage and flood control. Chapter 28 in Maidment (1992).

Vörösmarty, C.V., et al. (1997a). The storing and aging of continental runoff in large reservoir systems of the world. *Ambio* 26:210–219.

Vörösmarty, C.V., C.A. Federer, and A.L. Schloss (1998). Potential evaporation functions compared on US watersheds: Possible implications for global-scale water balance and terrestrial ecosystem modeling. *Journal of Hydrology* 207:147–169.

Vörösmarty, C.V., M. Meybeck, B. Fekete, and K. Sharma (1997b). The potential impact of neo-Castorization on sediment transport by the global network of rivers. In *Human Impact on Erosion and Sedimentation*, International Association of Hydrologic Sciences *Publication* 245.

Vörösmarty, C.V., P. Green, J. Salisbury, and R.B. Lammers (2000). The vulnerability of global water resources: Major impacts from climate change or human development?. *Science* 289:284–288.

Valdes, J.B., M. Diaz-Granados, and R.L. Bras (1990). A derived PDF for the initial soil moisture in a catchment. *Journal of Hydrology* 113:163–176.

Valdes, J.B., Y. Fiallo, and I. Rodriguez-Iturbe (1979). A rainfall-runoff analysis of the geomorphologic IUH. *Water Resources Research* 15:1421–1434.

Valente, F., J.S. David, and J.H.C. Gash (1997). Modelling interception loss for two sparse eucalypt and pine forests in central Portugal using reformulated Ruttter and Gash analytical models. *Journal of Hydrology* 190:141–162.

Van Bavel, C.H.M. (1966). Potential evaporation: The combination concept and its experimental verification. *Water Resources Research* 2:455–467.

Van de Greind, A.A., and E.T. Engman (1985). Partial area hydrology and remote sensing. *Journal of Hydrology* 81:211–251.

Van Genuchten, M. T. (1980). A closed-form equation for predicting the hydraulic conductivity of unsaturated soils. *Soil Science Society of America Journal* 44:892–898.

van Hylckama, T.E.A. (1979). Water, something peculiar. *Hydrological Sciences Bulletin* 24:499–507.

Van Tonder, G.J., and J. Kirchner (1990). Estimation of natural groundwater recharge in the Karroo aquifers of South Africa. *Journal of Hydrology* 121:395–419.

Viessman, W., Jr., G.L. Lewis, and J.W. Knapp (1989). *Introduction to Hydrology* (3rd ed.). New York, NY: Harper & Row, Publishers.

Villholth, K.G., K.H. Jensen, and J. Fredericia (1998). Flow and transport processes in a macroporous subsurface-drained glacial till soil. I. Field investigations. *Journal of Hydrology* 207:98–120.

Viswanathan, M.N. (1984). Recharge characteristics of an unconfined aquifer from the rainfall-water table relationship. *Journal of Hydrology* 70:233–250.

Vogel, R.M., and N.M. Fennessey (1994). Flow-duration curves, I: New interpretation and confidence intervals. *Journal of Water Resources Planning and Management* 120:485–504.

Vogel, R.M., and N.M. Fennessey (1995). Flow-duration curves, II: A review of applications in water resources planning. *Water Resources Bulletin* 31:1029–1039.

Vogel, R.M., M. Lane, R.S. Ravindiran, and P. Kirshen (1999a). Storage reservoir behavior in the United States. *Journal of Water Resources Planning and Management* 125:245–254.

Vogel, R.M. and T.R. McMahon (1996). Approximate reliability and resilience indices of overyear reservoirs fed by AR(1) gamma and normal inflows. *Hydrological Sciences Journal* 41:75–96.

Vogel, R.M., I. Wilson, and C. Daly (1999b). Regional regression models of annual streamflow for the United States. *Journal of Irrigation and Drainage Engineering* 125:148–157.

Waddington, J.M., N.T. Roulet, and A.R. Hill (1993). Runoff mechanisms in a forested groundwater discharge wetland. *Journal of Hydrology* 147:37–60.

Walker, C.D., and J.-P. Brunel (1990). Examining evapotranspiration in a semi-arid region using stable isotopes of hydrogen and oxygen. *Journal of Hydrology* 118:55–75.

Walker, G.R., I.D. Jolly, and P.G. Cook (1991). A new chloride leaching approach to the estimation of diffuse recharge following a change in land use. *Journal of Hydrology* 128:49–67.

Walker, J.F. (1988). General two-point method for determining velocity in open channel. *American Society of Civil Engineers Proceedings, Journal of Hydraulic Engineering* 114:801–805.

Walkotten, W.J. (1972). A recording soil moisture tensiometer. Portland, OR: U.S. Forest Service Pacific Northwest Forest and Range Experiment Station *Research Note* PNW-180.

Walling, D.E., and B.W. Webb (1987). Material transport by the world's rivers: Evolving perspectives. In *Water for the Future: Hydrology in Perspective*. International Association for Hydrological Sciences Publ. No. 164. (Proceedings of the Rome Symposium).

Wallis, J.R., and K.L. Bowden (1962). A rapid method for getting area-elevation information. Berkeley, CA: U.S. Forest Service Pacific Southwest Forest and Range Experiment Station *Research Note* 208.

Wallis, J.R., and P.E. O'Connell (1972). Small sample estimation of $\rho_1$. *Water Resources Research* 8:707–712.

Walsh, J.E. (1984). Snow cover and atmospheric variability. *American Scientist* 72:50–57.

Wang, H.F, and M.P. Anderson (1982). *Introduction to Groundwater Flow Modeling.* San Francisco, CA: W.H. Freeman and Co.

Wang, Q.J. and J.C. Dooge (1994). Limiting cases of water flux at the land surface. *Journal of Hydrology* 155:429–440.

Ward, R.C. (1984). On the response to precipitation of headwater streams in humid areas. *Journal of Hydrology* 74:171–189.

Waring, E.A., and J.A.A. Jones (1980). A snowmelt and water equivalent gauge for British conditions. *Hydrological Sciences Bulletin* 25:129–134.

Warrick, A.W., D.O. Lomen, and A. Islas (1990). An analytical solution to Richards' equation for a draining soil profile. *Water Resources Research* 26:253–258.

Wellings, S.R. (1984). Recharge of a chalk aquifer at a site in Hampshire, England - 1. Water Balance and Unsaturated Flow. *Journal of Hydrology* 69:259–273.

Wenzel, H.G., Jr. (1982). Rainfall for urban storm water design. in D.F. Kibler, ed., Urban stormwater hydrology. American Geophysical Union *Water Resources Monograph* 7.

Weyman, D.R. (1970). Throughflow on hillslopes and its relation to the stream hydrograph. *International Association of Scientific Hydrology Bulletin* 15(2):25–32.

Whipkey, R.Z., and Kirkby, M.J. (1978). Flow within the soil. In Kirkby (1978):121–144.

Whisler, F.D., and H. Bouwer (1970). Comparison of methods for calculating vertical drainage and infiltration in soils. *Journal of Hydrology* 10:1–19.

White, G.W. (1975). Flood hazard in the United States: A research assessment. Boulder, CO: University of Colorado Institute of Behavioral Science *Monograph* NSF-RA-E-75-006.

White, W.N. (1932). A method of estimating groundwater supplies based on discharge by plants and evaporation from soil. U.S. Geological Survey *Water-Supply Paper* 659-A.

Whiting, P.J., and M. Pomeranets (1997). A numerical study of bank storage and its contribution to streamflow. *Journal of Hydrology* 202:121–136.

Whittaker, R.H. (1975). *Communities and Ecosystems* (2nd Ed.). New York, NY: Macmillan.

Wigley, T.M.L., and P.D. Jones (1985). Influence of precipitation changes and direct $CO_2$ effects on streamflow. *Nature* 314:149–151.

Wigmosta, M.S. and S.J. Burges (1997). An adaptive modeling and monitoring approach to describe the hydrologic behavior of small catchments. *Journal of Hydrology* 202:48–77.

Wikramaratna, R.S., and C.E. Reeve (1984). A modeling approach to estimating aquifer recharge on a regional scale. *Hydrological Sciences Journal* 29:327–337.

Wilcock, D.N., and C.I. Essery (1984). Infiltration measurements in a small lowland catchment. *Journal of Hydrology* 74:191–204.

Wilf, P., S.L. Wing, D.R. Greenwood, and C.L. Greenwood (1998). Using fossil leaves as paleoprecipitation indicators: An Eocene example. *Geology* 26:203–206.

Williams, T.H.L. (1978). An automatic scanning and recording tensiometer system. *Journal of Hydrology* 39:175–183.

Williams, T.H.L. (1980). An automatic electrical resistance soil-moisture measuring system. *Journal of Hydrology* 46:385–390.

Wilmott, C.J., and D.R. Legates (1991). Rising estimates of terrestrial and global precipitation. *Climate Research* 1:179–186.

Wilson, B.N., and J.W. Brown (1992). Development and evaluation of a dimensionless unit hydrograph. *Water Resources Bulletin* 28:397–408.

Wilson, C. (1969). Climatology of the cold regions - Northern Hemisphere II. Hanover, NH: U.S. Army Cold Regions Research and Engineering Laboratory *Cold Regions Science and Engineering Monograph* I-A3b.

Winter, T.C. (1976). Numerical simulation analysis of the interaction of lakes and ground water. U.S. Geological Survey *Professional Paper* 1001.

Winter, T.C. (1981). Uncertainties in estimating the water balance of lakes. *Water Resources Bulletin* 17:82–115.

Winter, T.C. (1983). The interaction of lakes with variably saturated porous media. *Water Resources Research* 19:1203–1218.

Winter, T.C. (1984). Geohydrologic setting of Mirror Lake, West Thornton, New Hampshire. U.S. Geological Survey *Water-Resources Investigations Report* 84–4266.

Winter, T.C., J.S. Eaton, and G.E. Likens (1989). Evaluation of inflow to Mirror Lake, New Hampshire. *Water Resources Bulletin* 25: 991–1008.

Winter, T.C., J.W. LaBaugh, and D.O. Rosenberry (1988). The design and use of a hydraulic potentiomanometer for direct measurement of difference in hydraulic head between groundwater and surface water. *Limnology and Oceanography* 33:1209–1214.

Winter, T.C., D.O. Rosenberry, and A.M. Sturrock (1995). Evaluation of 11 equations for determining evaporation for a small lake in the north central United States. *Water Resources Research* 31:983–993.

Woo, M.-K., and P. Steer (1986). Monte Carlo simulation of snow depth in a forest. *Water Resources Research* 22:864–868.

Wood, E.F., M. Sivapalan, and K. Beven (1990). Similarity and scale in catchment storm response. *Reviews of Geophysics* 28:1–18.

Wood, M.K., and W.H. Blackburn (1984). An evaluation of the hydrologic soil groups as used in the SCS runoff method on rangelands. *Water Resources Bulletin* 20:379–389.

Woodruff, J.F., and J.D. Hewlett (1970). Predicting and mapping the average hydrologic response for the eastern United States. *Water Resources Research* 6:1312–1326.

Woods, R., and M. Sivapalan (1999). A synthesis of space-time variability in storm response: Rainfall, runoff generation, and routing. *Water Resources Research* 35:2469–2485.

Woodward, F.I. (1987). *Climate and Plant Distribution.* Cambridge, U.K.: Cambridge University Press.

Woolhiser, D.A. (1981). Physically based models of watershed runoff. In V.P. Singh, ed., *Rainfall Runoff Relationships.* Fort Collins, CO: Water Resources Publications.

Woolhiser, D.A. and J.A. Liggett (1967). Unsteady one-dimensional flow over a plane - The rising hydrograph. *Water Resources Research* 3:753–771.

Work, R.A, H.J. Stockwell, T.G. Freeman, and R.T. Beaumont (1965). Accuracy of field snow surveys. Hanover, NH: U.S. Army Cold Regions Research and Engineering Laboratory *Technical Report 163.*

Workman, S.R., and S.E. Serrano (1999). Recharge to alluvial valley aquifers from overbank flow and excess infiltration. *Journal of the American Water Resources Association* 35:425–432.

World Commission on Environment and Development (1989). *Our Common Future.* Oxford, U.K.: Oxford University Press.

World Meteorological Organization (1973). Manual for estimation of probable maximum precipitation. Geneva, Switzerland: World Meteorological Organization *Operational Hydrology Report 1 (WMO 332).*

World Meteorological Organization (1981). *Guide to Hydrological Practices, vol. 1, Data Acquisition and Processing.* Geneva, Switzerland: World Meteorological Organization.

World Meteorological Organization (1986). Intercomparison of models of snowmelt runoff. Geneva, Switzerland: World Meteorological Organization *Operational Hydrology Report 23.*

World Meteorological Organization (1992). Simulated real time intercomparison of hydrological models. Geneva, Switzerland: World Meteorological Organization *Operational Hydrology Report 38.*

Wu, J., R. Zhang, and J. Yang (1996). Analysis of rainfall-recharge relationships. *Journal of Hydrology* 177:143–160.

Yang, D., B.E. Goodison, S. Ishida, and C.S. Benson (1998). Adjustment of daily precipitation data at 10 climate stations in Alaska: Application of World Meteorological Organization intercomparison results. *Water Resources Research* 34:241–256.

Yevjevich, V. (1972). *Probability and Statistics in Hydrology.* Fort Collins, CO: Water Resources Publications.

Youngs, E.G. (1988). Soil physics and hydrology. *Journal of Hydrology* 100:411–431.

Youngs, E.G. and A. Poulovassilis (1976). The different forms of moisture profile development during the redistribution of soil water after infiltration. *Water Resources Research* 12:1007–1012.

Yu, C., et al. (1997). Two- and three-parameter calibrations of time domain reflectometry for soil moisture measurement. *Water Resources Research* 33:2417–2421.

Zegelin, S.J., I. White, and D.R. Jenkins (1989). Improved field probes for soil water content and electrical conductivity measurement using time domain reflectometry. *Water Resources Research* 25:2367–2376.

# Index